Elements of Ecology

Elements of Ecology

THIRD EDITION

Robert Leo Smith

West Virginia University

Sponsoring Editor: Glyn Davies
Project Editor: Shuli Traub
Editorial Assistant: Madelyn Elliott
Design Supervisor: Dorothy Bungert/Mary Archondes
Cover Design: Mary Archondes
Cover Photo: Okeefenokee Swamp, Waterlillies, Marsh and Cypress, by Wendell Metzen, Bruce Coleman, Inc.
Photo Researcher: Carol Parden
Production Manager/Assistant: Kewal Sharma/Jeffrey Taub
Compositor: Ruttle, Shaw & Wetherill
Printer and Binder: R. R. Donnelley & Sons Company
Cover Printer: The Lehigh Press

Elements of Ecology, Third Edition

Library of Congress Cataloging-in-Publication Data

Smith, Robert Leo.
Elements of ecology/Robert L. Smith.—3rd ed.
p. cm.
Includes bibliographical references and index.
ISBN 0–06–046328–7
1. Ecology. I. Title.
QH541.S624 1991 91–29919
574.5—dc20 CIP

95 9 8 7

For Alice

Contents

PART III CONDITIONS FOR LIFE 55

PART IV POPULATION ECOLOGY *155*

Introduction: Looking at Populations *156*

Preface

Readers will discover a major shift in emphasis in this third edition. I have reoriented the text for nonmajors with an emphasis on applied ecology and environmental problems for two reasons. First, I wanted to shift *Elements* away from *Ecology and Field Biology.* Second, a need exists for an ecology text oriented toward nonmajors, who have little interest in the theoretical topics in major's texts, but are alert to major ecological issues such as drilling for oil in the Arctic National Wildlife Refuge, building dams, and clearcutting forests. Although environmental topics are covered in environmental science courses, these courses and texts typically place their strongest emphasis on social, political, and economic aspects and are weak in ecology and a scientific approach. Further, there is a growing trend to provide and even require an ecology course for students in engineering, law, business, journalism, liberal arts, and other fields, a course that will meet science requirements and at the same time increase the student's understanding of environmental issues.

What do these students need to know? For a start they need to know what ecology is—that it is not recycling and pollution control, and it is not environmental science. They should have some idea of how organisms respond to changing environmental conditions; how and why populations increase, decrease, and go extinct; how organisms compete for resources; how parasites and diseases spread; and why wolves prey on deer. They ought to understand what happens when we drain a swamp, straighten a stream, dam a river, or cut a forest, why the landscape changes over time, why and how toxic elements get into a food chain. My approach, then, has been twofold: Introduce enough pertinent theoretical ecology to provide a strong basic foundation and then apply these ecological principles and concepts to the students' own experiences and environmental issues.

NEW TOPICS

Users of the second edition will find the overall format of the text much the same, so they will be on familiar ground. But they will discover some old chapters and topics missing and new ones added. Chapter 2, "Natural Selection and Evolution," now includes a discussion of the genetics of small populations. The chapter on fire has been dropped and the material incorporated into Chapter 21, "Natural Disturbance and Human Impact." Four new chapters appear. Chapter 17 is on parasites and mutualism. Chapter 18 covers human control of natural populations, including weed and pest control, integrated pest management, management (and mismanagement) of exploited populations, and the restoration of endangered species. Chapter 21, on natural and human-caused disturbances to natural communities, discusses the effects of logging, cultivation, mining, urbanization, and human-caused fires, with an example of the Yellowstone fire of 1988. Chapter 25, "Human Intrusions upon Ecological Cycles," covers such topics as acid deposition and its possible relationship to forest decline, ozone depletion, and water pollution.

The last section of the book is on ecosystem diversity, a hallmark of both of my texts. I consider it an important and essential part of a nonmajor's text. For this reason I have added a discussion of the effects of human impacts on these systems. Many of the ecological and environmental problems and controversies focus on the fate of natural ecosystems: the destruction of tropical and old-growth temperate forests, encroachment on deserts by recreation and urban development, overgrazing of rangeland, oil drilling on the tundra, drainage of wetlands, and estuarine pollution. Such issues cannot be discussed intelligently without some knowledge of the structure and function of those ecosystems and how human intrusions affect them.

Given the current diversity of introductory ecology courses and lack of consensus on what they should cover, it is difficult to develop an ecology text that will meet each instructor's needs. To be flexible, this text presents enough material in sufficient detail to allow instructors to mold their own courses from it.

ORGANIZATION

I have divided this text into seven interrelated parts with numerous cross-references. Part I, "Introduction," explains what ecology is and provides a brief discussion of the history and methods of ecology. Part II deals with natural selection, evolution, genetics of small populations (of particular relevance in this day of diminishing, fragmented populations of many species), and speciation. The placement of this topic is unusual. You could also teach these chapters at the beginning of Part IV, "Population Ecology." I prefer these chapters early for two reasons. First, all aspects of ecology reside within the context of evolution. Individual organisms are the focal point of natural selection, which involves the differential ability of individual organisms to respond and adapt to the environment discussed in Part III. Second, it makes sense to introduce the concept of species and speciation as soon as possible. It is integral to ecology and crucial to the management of endangered species today.

Part III, "Conditions for Life," covers the relationship of organisms to their physical environment. Chapter 4 introduces some broad aspects of world climate and how it affects the pattern of life on Earth. Chapter 5 examines how organisms deal with temperature and its extremes. Chapter 6 looks at the way organisms confront moisture problems, including responses of plants to flooding. The material on light is divided into two chapters. Chapter 7 considers photosynthesis, its ecological implications, and shade tolerance. Chapter 8 covers the influence of light on daily and seasonal periodicities in plants and animals. Chapter 9 presents some basics of soils, an essential topic often ignored. This chapter discusses the new eleventh member of the Great Soil Groups.

Part IV widens the scope from individuals to populations. The introduction defines populations. Chapter 10 covers density, distribution, and age structure. Age structure leads into Chapter 11 on mortality, natality, and survivorship. These attributes lead to population growth, treated in Chapter 12. What regulates population growth is discussed in Chapter 13. It emphasizes competition among individuals of a species and the forms that competition takes. These relationships are reflected in the various life-history patterns, including mating and reproductive strategies, covered in Chapter 14. Much of this chapter falls into the category of behavioral ecology. Competition among individuals of different species is the subject of Chapter 15. This chapter introduces the concept of the niche and resource partitioning. Some species use others as food or as both habitat and food. These relationships fall under the topics of predation, parasites, and mutualism. Chapter 16 deals with predation in both forms, herbivory and carnivory. Chapter 17 discusses parasitism and disease, especially as it relates to human, wildlife, and parasite interactions, and social parasitism. The section on mutualism emphasizes its importance in interpopulation relationships and species well-being. Chapter 18 details the impact of humans on natural populations, especially as it relates to population increase and decrease, and restoration, conservation, and pest control. This chapter contains material essential to any nonmajor's course.

Part V broadens the focus to the community level. This section deals largely with community structure. A thin line separates population ecology from community ecology. I find that a good deal of community ecology in other texts fits more naturally under population ecology and even ecosystem ecology. Chapter 19 considers the spatial structure of communities with emphasis on edge, patchy environment, and habitat fragmentation, another topic essential in a nonmajor's course. Chapter 20 covers succession, community development, and change over time. Chapter 21 discusses the role of disturbance in the maintenance of natural communities and the effects of human disturbance on those communities.

Part VI explores ecosystem dynamics. The introduction presents the concept of the ecosystem, followed by Chapter 22 on primary and secondary production, the outcome of energy fixation. How energy flows through the ecosystem is the topic of Chapter 23. It discusses some of the latest concepts of food webs, a topic often considered under community ecology. Chapter 24 explores the major ecological cycles, including the water cycle. This chapter provides the foundation for Chapter 25, which examines how humans have intruded on natural cycles.

Part VII is concerned with a diversity of ecosystems. The introduction covers the classification of terrestrial ecosystems and the chapters that follow discuss terrestrial, freshwater, and marine ecosystems. I broke this material, which appeared in several concentrated chapters in the second edition, into a number of smaller chapters easier to grasp. Each chapter considers the structure and function of a set of ecosystems and how they are impacted by human activity. These chapters should form an integral part of a nonmajor's course because they deal directly with some of the most pressing and controversial ecological problems we face in a diminishing world.

I have arranged these chapters into related groups. Terrestrial ecosystems cover five chapters. Chapter 26 deals with grasslands and savannas; Chapter 27 shrublands and deserts; Chapter 28 the tundra and its northern neighbor the boreal forest; Chapter 29 temperate coniferous and deciduous forests; and Chapter 30 tropical forests. Freshwater ecosystems involve three chapters. Chapter 31 explores lakes and ponds and the effect of pollution on them. Chapter 32 discusses wetlands, their value and demise. Chapter 33 emphasizes the unique characteristics of flowing-water ecosystems and the impacts of dams and channelization on them. The last three chapters explore the marine environment and the impact of pollution on it. Chapter 34 presents the major physical and ecological characteristics of the marine environment. Chapter 35 covers the rocky and sandy shores and the effects of human activity on them. The final chapter looks at estuaries and associated salt marshes.

SPECIAL FEATURES

This edition of *Elements of Ecology* retains the features that increase its pedagogical usefulness:

Chapter Outlines

Outlines provide a quick survey and page-finding guide to topics within the chapter.

Objectives

The list of objectives alerts readers to what they should gain from the chapter.

Summary

Chapter summaries provide a quick review of the main points of the chapters. They may also serve as an abstract of the material contained in the chapter.

Review and Study Questions

Each chapter ends with a set of questions intended to reinforce the objectives for that chapter. The questions are of two types. Review questions provide a guide to the study of the material. Special study questions, marked with an asterisk (*), are designed to stimulate the student to investigate, think, or relate the material covered in the chapter to real situations—to apply ecological principles and concepts to issues and problems.

Selected References

Each chapter ends with a list of references selected from the voluminous literature. I based the selections of books on what I consider the basic, readily accessible sources. This list is supplemented by a number of journal papers and articles. I have annotated most of the selections; for those not annotated, the titles sufficiently describe the contents. Some of the government publications, those of the U.S. Fish and Wildlife Service and U.S. Forest Service, may be less accessible, for many college and university libraries will not have them in their collections. These titles, however, are available from the Superintendent of Documents, Washington, DC. They belong in a good ecological library. They are also easily available on interlibrary loan from university (usually land grant) and public libraries designated as government depositories.

I omitted literature citations in the text from this edition. Although I consider such citations essential in a major's text, they can be an unnecessary distraction in a nonmajors's text. However, sources of specific examples are cited in the list of selected references. The sources of tables and illustrations are fully cited in the source acknowledgments at the back of the text.

Boxed Material

Boxed material, found at certain places in the text, presents information whose inclusion would interrupt the main text, yet expands the text discussion. Most of the boxed material is quantitative, useful for analyzing results in a related field project.

Illustrations

This edition retains a number of the illustrations from earlier editions, but many new spot drawings, line diagrams, and photographs augment or replace previous illustrations.

A common criticism of American ecology texts is their heavy North American orientation. This provincialism is most evident in the illustrations, particularly the photographs. I have made a special effort to give this text an international flavor, not only in the written material, but also in the illustrations. Photographs are from Australia, the Near East, Southeast Asia, Central Europe, South America, and Africa, as well as North America.

Glossary

The glossary contains over 400 key terms used in the text. This glossary can serve as an abridged dictionary of ecology.

ACKNOWLEDGMENTS

To revise a textbook the author must depend heavily on the input of users and reviewers. This is especially true for this edition, which represents a departure in emphasis from the second edition. Had I tried to incorporate all the ideas, the book would have grown considerably. The idea was to reduce and not expand the text, so I did what an author has to do: weigh all suggestions and pick some of the best.

The following reviewers provided detailed critiques and suggestions: Edmund E. Bedicarrax, City College of San Francisco; Robert Cashner, University of New Orleans; Lee Christianson, University of the Pacific; Courtney T. Hackney, University of North Carolina at Wilmington; Ronald Hofstetter, University of Miami; Frank McCormick, University of Tennessee; Joseph Moore, California State University at Northridge; Fred E. Smeins, Texas A & M University; Margaret Stewart, State University of New York at Albany; O. W. Van Auken, The University of Texas at San Antonio; and Robert A. Wright, West Texas State University.

My son Robert Leo Smith, Jr., did the new illustrations under some pressure between his other assignments and work in medical illustration at the University of Virginia. My second son, Dr. Thomas M. Smith, an ecologist and ecological modeler, also at the University of Virginia, provided insights into successional theory and global climate change, as well as supplying some international photographs.

I am indebted to four of my former graduate students—Mohd. Nawayai Yasak, Mohd. Tajaddin Abdullah, Zabba Zabidin, and Burhanuddin Mohd. Nor of the Department of Wildlife and National Parks—and to Mohd. Khan bin Momin Khan, Director of the Department of Wildlife and National Parks, for inviting me to travel in peninsular Malaysia on two occasions and introducing me to the tropical forest and its problems in Southeast Asia. I am also indebted to another graduate student, Alexine Keuroghlian, of São Paulo, Brazil, for insights into tropical deforestation in Brazil.

The book could not have arrived at its present stage without the help and encouragement of the staff at HarperCollins. Glyn Davies, Biology Acquisitions Editor at HarperCollins, kept pushing this revision along with deadlines I thought impossible to make. Again I am indebted to Kristin Zimet, freelance copy editor, who had exceptional insights into ecology. She picked up inconsistencies and errors, eliminated redundancies, and suggested some organizational changes and points of emphasis that greatly improved the text.

The difficult task of putting the book together falls to the project editor, whose task it is to pull together the manuscript, art, captions, galleys, and pages, and to prod author, designers, illustrators, and compositors, as well as to pick up and correct all the nagging little problems that rear up along the way. Shuli Traub handled the task in an exceptional manner. It was a pleasure to work with her.

As usual my wife Alice, who seems to have spent most of her life as book widow, took care of all the problems of living over the years. To her go my thanks and deep appreciation for putting up with it all.

Robert L. Smith

PART I
Introduction

Outline

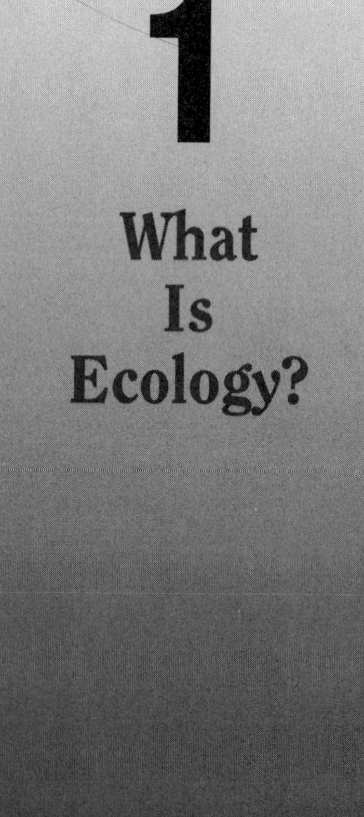

CHAPTER 1

What Is Ecology?

Objectives

On completion of this chapter, you should be able to:

1. Trace the development of various areas of ecology.
2. Discuss the relationship of ecology to environmental science and resource management.
3. Describe ecology as a science.

Ecology. For years the term was known only by specialists, a sort of sideline to biology, overshadowed, especially in the 1950s and 1960s, by molecular biology. Ecology was scarcely recognized by the academic world. Then the environmental movement began in the late 1960s and early 1970s. Because it is concerned with environmental relationships, ecology became popular. The term appeared everywhere—in newspapers, magazines, and books. Ecology became a household word, oversimplified, misused, and abused.

Ecology still is mistakenly equated with the environment. Ecology relates to the environment, but it is not the same as environmental studies or environmental science. The latter combine ecology, geology, economics, sociology, and political science. Ecology itself is a science, concerned with doing experimental studies in field and laboratory, making mathematical and statistical analyses of data, testing hypotheses, and drawing conclusions.

Ecology, by the usual definition, is the study of the relationship between organisms and their environment. Even that definition can be faulted unless you consider the words *relationship* and *environment* in their fullest meaning. Environment includes not only the physical but also the biological conditions under which an organism lives; and relationships involve interactions with the physical world as well as with members of other species and the same species.

The word *ecology* was coined by the German zoologist Ernst Haeckel in 1869. He called the study of the relationship of animals to their environment "Oekologie." Specifically, he was trying to relate animal body structure to Darwin's theory of evolution. The word did not come into general use, however, until it appeared in Warming's book *Plantesamfund: Grundtrak af den okologiske Plantegeografi* in 1895. Its modern meaning was largely fixed by that publication.

The term *ecology* comes from the Greek words *oikos*, "the family household," and *logy*, "the study of." Literally, ecology is the study of the household. It has the same root word as economics, or "management of the household." In effect, ecology could be considered as the study of the economics of nature. In fact, some economic concepts—such as resource allocation, cost-benefit ratios, and optimization theory—have crept into ecology.

THE ROOTS OF ECOLOGY

Ecology is a hybrid, more difficult to trace to definitive roots (Figure 1.1) than mathematics, chemistry, microbiology, and other sciences. One can argue that ecology goes back to the ancient Greek scholar Theophrastus, a friend of Aristotle who wrote about the interrelations between organisms and the environment. On the other hand, ecology as we know it today has its strongest roots in plant geography and natural history, including the study of plants, birds, mammals, fish, and insects.

Although the study of natural history provided much of the knowledge upon which ecology was based, the real impetus came from the explorations of plant geographers. They discovered that although plants differed in various regions of the world, the vegetation assumed certain similarities and differences that demanded explanation. One of the early influential plant geographers was Carl Ludwig Willdenow (1765–1812). He pointed out that similar climates supported similar vegetation. Willdenow was a major influence on a wealthy young Prussian, Friedrich Heinrich Alexander von Humboldt (1769–1859), for whom Humboldt Current was named. Sponsored by King Carlos IV of Spain—who provided the equipment needed to measure altitude, latitude, elevation, humidity, and temperature—young Humboldt spent five years exploring Latin America. He traveled in Mexico, Cuba, Venezuela, and Peru and explored the Orinoco and Amazon rivers. He described his travels in a 30-volume work, *Voyage to the Equatorial Regions*. Of these volumes, 14 were devoted to plants. Humboldt described vegetation in terms of outward appearance, correlated vegetation types with environmental characteristics, and coined the term *plant association*.

Among a second generation of plant geographers was Johannes Warming (1841–1924) at the University of Copenhagen. Like Humboldt, Warming traveled to South America to study the tropical vegetation of Brazil. The outcome was a book on Brazilian vegetation notable for its modern approach to plant ecology. Warming went on to write the first text on plant ecology, *Plantesamfund*, eventually translated into four languages, including English. In it Warming drew plant morphology, physiology, taxonomy, and biogeography

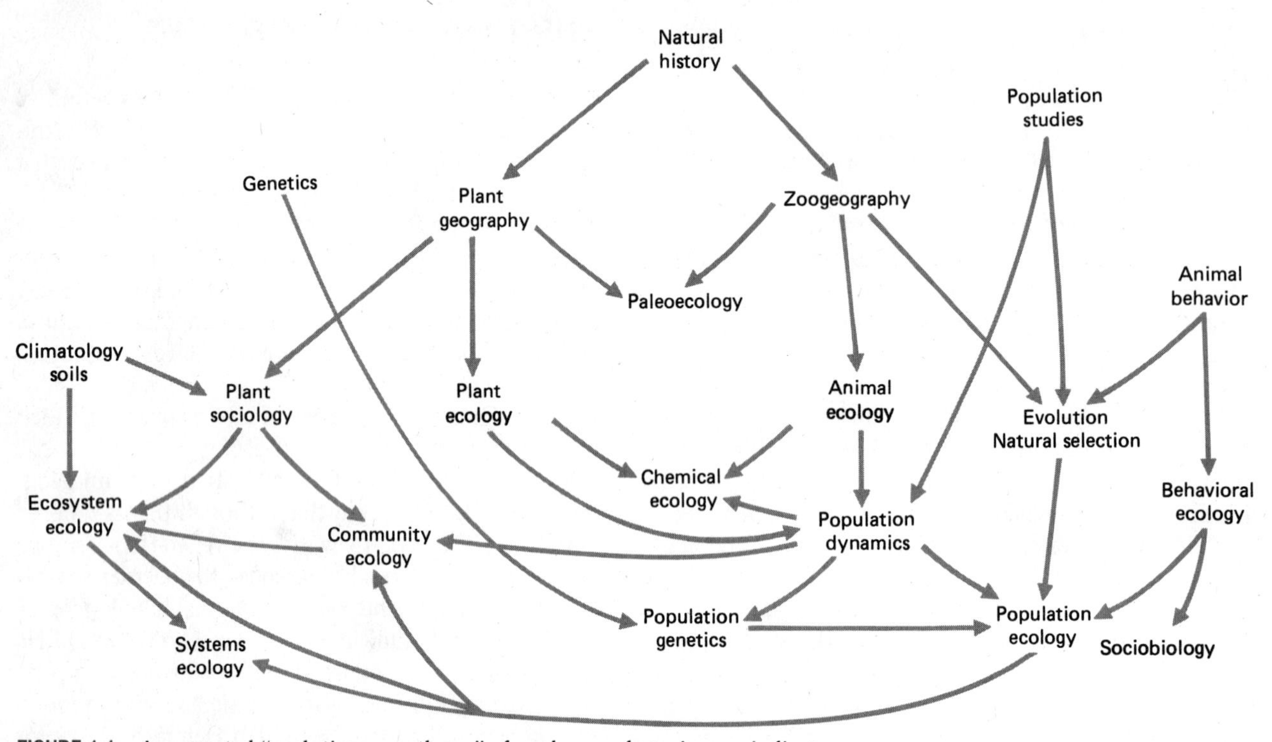

FIGURE 1.1 A suggested "evolutionary pathway" of modern ecology. Arrows indicate the sequence of development. The wide range of topics makes it difficult, if not impossible, for ecology to become a definitive science like physics or chemistry.

into a coherent whole. The book had a tremendous influence on the development of ecology.

Plant geographers and botanists built a solid foundation upon which modern ecology would arise. In Europe they influenced the development of plant sociology, the study of plant associations; and in the United States they stimulated the flowering of plant ecology, dominated by F. E. Clements, who emphasized vegetational dynamics and plant succession.

Early plant ecologists were concerned mostly with land plants. Another group of European biologists was interested in the relationship between aquatic organisms, plant and animal, and their environment. Prominent among them were A. Thienemann and F. A. Forel. Thienemann introduced the ideas of organic nutrient cycling and trophic feeding levels, using the terms *producers* and *consumers*. Forel was more interested in the physical characteristics of freshwater habitats, particularly lakes. He introduced the term *limnology* for the study of freshwater life.

Studies of energy accumulation and flow through various feeding groups in lakes influenced the research of a young limnologist at the University of Minnesota, R. A. Lindeman. He investigated the food relations of organisms in Cedar Creek Bogs with the idea of integrating food cycle dynamics with the principles of community succession. He traced "energy-available" relationships within the community. His 1942 paper, "The Trophic-Dynamic Aspect of Ecology," marked the beginning of ecosystem ecology, the study of whole living systems.

Lindeman's theory stimulated considerable research on energy flow and nutrient budgets both in North America and Europe. The increasing ability to measure energy flows and nutrient cycling by means of radioactive tracers and to analyze large amounts of data with computers permitted the development of *systems ecology*, which is the application of general systems theory and methods to ecology.

As plant ecology was evolving out of plant geography, activities in other areas of natural history were assuming an important role. One was the voyage of Charles Darwin on the *Beagle*, during which he collected numerous biological specimens, made detailed notes, and mentally framed his view of life on Earth. Influenced by the works of the geologist Charles Lyell, who proposed that Earth changed through time, Darwin noted how life, too, apparently changed through time. Working for years on his notes and collections, Darwin observed the relationships between organisms and environments, the similarities and dissimilarities of organisms within and between continents. He attributed these differences to geological barriers separating inhabitants. He noted how successive groups of plants and animals, distinct yet obviously related, replaced one another.

Developing his theory of evolution and the origin of species, Darwin was influenced by the writings of Thomas Malthus (1766–1834). An economist, Malthus advanced the principle that populations grew in geometric fashion, doubling at regular intervals. Experiencing such rapid growth, a population would outstrip its food supply. Ultimately, the population would be restrained by a "strong, constantly operating force—among plants and animals the waste of seed, sickness, and premature death. Among mankind misery and vice." From this concept Darwin developed the idea of "the survival of the fittest" as a mechanism of natural selection and evolution.

Meanwhile, unknown to Darwin, an Austrian monk, Gregor Mendel (1822–1884), was studying the transmission of inheritable characters from one generation of pea plants to another in his garden. The work of Mendel would have answered a number of Darwin's questions on the mechanism of inheritance and provided his theory of natural selection with the firm base it needed. Belatedly, Darwin's theory of evolution and Mendelian genetics were combined to form the study of evolution and adaptation, two central themes in ecology. Three major geneticists, Sewell Wright, R. A. Fisher, and J. H. Haldane, developed the field of *population genetics*.

Interest in Malthus stimulated the study of *population dynamics*, which developed in two directions. *Population ecology* is concerned with how populations grow (including birth and death rates), fluctuate, spread, and interact. *Population biology* is concerned with how natural selection and evolution affect populations.

The ideas of natural selection, evolution, and population dynamics, were applied first not to plants but to animals. The beginnings of animal ecology can be traced to R. Hesse's *Tiergeographie auf okologische Grundlage*, and Charles Elton's classic little book *Animal Ecology*, published in 1927 (and still in print). He defined animal ecology as the sociology and economics of animals.

In North America two pioneering animal ecologists were Charles Adams and Victor Shelford. Adams wrote the first text on animal ecology in 1913. In the same year Shelford published *Animal Communities in North America*, a landmark work that stressed the relationship between plants and animals and the idea of communities. The community concept was central to ecology until Tansley advanced the concept of the ecosystem.

Paralleling the growth of plant and animal ecology was *physiological ecology*. It is concerned with responses of individual organisms to temperature, moisture, light, and other such stresses. Physiological ecology dates to Justus Von Liebig (1803–1873), who studied the role of limited supplies of nutrients in the growth and development of plants. Half a century later the idea of limiting factors was extended to maximums—too much of a good thing—by F. F. Blackman. Then V. E. Shelford applied the concept of limiting factors to animals in a law of tolerance.

Natural history observations also spawned *behavioral ecology*. Early behavioral studies included those on ants by William Wheeler and on South American monkeys by Charles Carpenter. Konrad Lorenz and Niko Tinbergen gave a strong impetus to the field with their pioneering studies on the role of imprinting and instinct in the social life of animals, particularly birds. Behavioral ecology gave rise to a controversial offspring, *sociobiology*, which holds that genetics controls behavior.

Other observations led to investigations of chemical substances in the natural world. Scientists began to explore the use and nature of chemicals in animal recognition, trail making, and courtship, and in plant and animal defense. Such work has grown into the specialized field of *chemical ecology*.

Ecology has so many different roots that it probably

will always remain many-sided—as the ecological historian, Robert McIntosh, calls it, "a polymorphic discipline." We may never reduce it to a set of basic principles. Ecology ranges over many diverse areas—marine, freshwater aquatic, and terrestrial. It involves all taxonomic groups, from bacteria and protozoa to mammals and forest trees, and it deals with them at different levels: individuals, populations, and ecosystems. It studies these levels and groups from various points of view: behavioral, physiological, mathematical, and chemical. As a result, specialists may have little grasp of one another's work. On the other hand, insights from many directions are sure to enrich this still-growing field.

MODERN ECOLOGY

Modern ecology has developed along two major lines, ecosystem ecology and population ecology. *Ecosystem ecology* is holistic in its approach. Holists think ecosystems are too complex to study in isolated bits; they are best studied as functional units. *Population ecology* is reductionist in its approach. Reductionists argue that if you discover how each part of the system functions, you know how the whole system operates. However far apart these two approaches appear, they are really closely related. We must depend upon both to gain insight into the natural world.

Both ecosystem and population ecology are interested in three basic questions: what, how, and why. The answer to "what" is description. We must assemble facts, especially if the area being studied is new. Early papers in ecology provided descriptions of the structure of forests, grasslands, wetlands, and other environments, plant succession on sand dunes and old fields, the numbers of species in temperate and tropical regions, social structure, and dispersions of populations. *Descriptive ecology*, then, looks at the structure of populations, communities, and ecosystems.

Descriptive ecology may tell us what is there and what it is like, but how does it work? How are nutrients cycled through an ecosystem? How do populations respond to predation and to environmental changes? The study of how ecosystems, populations, and organism function makes up *functional ecology*.

The study of function involves experimentation. Experiments in the laboratory or in the field involve some form of manipulation to test a hypothesis. A hypothesis is a statement about a causative agent that can be tested experimentally, then accepted or rejected. For example, an ecologist might hypothesize that frass (droppings) deposited by the forest-defoliating insect, the gypsy moth, will stimulate understory vegetation and thus retain nitrogen and other nutrients in the forest ecosystem. Testing such a hypothesis would require both laboratory and field experiments.

Laboratory experiments are the most basic, because the researcher can reduce and control the variables, studying a few aspects in detail. In the case of our gypsy moth hypothesis, the ecologist can run laboratory tests on the frass to determine its nutrient content and set up experimental units to determine the rapidity with which the frass decomposes. The researcher can also measure the uptake of nitrogen and other nutrients by plants fertilized with gypsy moth frass. Rarely are the results of such experiments directly applicable to the field, but they can provide basic insights.

Most ecological experiments are carried out in the field. Field experiments involve the manipulation of some of the variables in some manner—adding or removing one of the species, erecting exclosures to prevent access to space or a resource, or adding or withholding nutrients and water. To test the gypsy moth hypothesis in the field, the ecologist would have to establish a pre-defoliation baseline of nutrient movement through the forest by measuring nutrients in forest streams. During defoliation the ecologist would measure output of nutrients from the forest in the same manner, record understory woody plant growth (complicated by the fact that plants would also respond to increased light to the forest floor), determine nutrient content in leaves, and note nitrogen accumulation in the forest soil. Then the researcher might be able to accept or reject the hypothesis. Either way, the ecologist would gain insight into the impact of gypsy moths on the forest ecosystem and undoubtedly discover new ways to study the problem.

Once you have some idea how ecosystems and populations function, you inevitably wonder why. Answering why falls within the realm of *evolutionary ecology*. Evolutionary answers are much harder to come by. Because evolution is still going on, the ecologist can study the effects of environmental variables such as drought and climate change on natural selection within populations and on survival rates of genotypes. In the case of the gypsy moth problem, the evolutionary ecologist might be interested in how oak trees respond to the selective pressures of defoliation, including mortality, survival, chemical defensive tactics of surviving trees, and competition among seedlings stimulated by the increased light and nutrients.

Based on their experimental studies, ecologists develop models. Models are abstractions and simplifications of natural phenomena developed to predict new phenomena or to provide insights into existing ones. The two basic types of models are analytical and simulation. *Analytical models*, common in population ecology, are based on straightforward assumptions. Mathematical formulas based on those assumptions allow predictive solutions, limited by the restrictions in the model. Examples are formulas for population growth and predation. *Simulation models* require the use of a computer. They allow the quantification of many more variables in much richer detail. Such models are used to examine interrelations of populations and habitat, energy flow and nutrient cycling, insect outbreaks, and production and management problems in forestry, wildlife, and fisheries.

At this point you probably realize the problems of ecological research rest in the complexity of living systems. Lawyers and politicians demand simple answers that ecologists cannot give. Because medical research can say with a high degree of certainty what causes AIDS, because engineering research can tell what effects stress and heat have on materials, politicians and the public expect definitive answers from ecologists. They want proof that acid rain is responsible for declines in forest and crop growth, that chlorofluorocarbons are causing the hole in the ozone layer, that destruction of old growth forests will drive a species to extinction. Ecologists cannot give definitive answers. Their research points out that burning of fossil fuels does produce acid rain and that acid rain does promote forest decline; their simulation models point out the close relationship between acid rain and dying forests; but acid rain does not work alone. Because ecologists cannot prove acid rain alone is responsible, politicians demand still more research.

We will never arrive at certainty. We must do our best with the insights ecology alone can give us, hoping to act wisely and in time on behalf of the natural world.

THE APPLICATION OF ECOLOGY

In the 1970s ecology, popularly described as a new science, became involved in social, political, and economic issues. People became aware of pollution, overpopulation, and degraded environments. Although the public treated these environmental issues as if they were new, ecologists had been involved with them for years. The problem was that people were not listening. For example, in 1864 George Perkin Marsh called attention to the effects of poor land use on the human environment in his dramatic book *Man and Nature*. F. E. Clements in the 1930s urged that the Great Plains be managed as grazing land and not be broken by the plow, which they ultimately were, resulting in a dust bowl. Paul B. Sears wrote *Deserts on the March* (1935) in response to the dust bowl of the 1930s, and William Vogt's *Road to Survival* (1948) and Fairfield Osburn's *Our Plundered Planet* (1948) both called attention to the growing population-resource problem. Aldo Leopold's *A Sand County Almanac* (1949), which called for an ecological land ethic, was read largely by those interested in wildlife management until the 1970s, when it became the bible of the ecology movement.

In 1932 H. L. Stoddard introduced the idea of the role of fire in the control of plant succession in his book *The Bobwhite Quail*. Aldo Leopold expounded the application of ecological principles to the management of wildlife in his classic *Game Management* (1933). In *Forest Soils* (1954), H. L. Lutz and R. F. Chandler discussed nutrient cycles and their role in the forest ecosystem, and J. Kittredge pointed out the impact of forests on the environment in *Forest Influences* (1948).

Rachel Carson probably did more than anyone else to bring environmental problems to the attention of the public. Since the publication of her book *Silent Spring* (1962), people have become more aware of the chemical poisons and other pollutants being recycled. Once castigated as more fiction than fact, Carson's predictions came only too true as carnivorous birds especially fell victim to pesticides. With a ban on the use of certain pesticides, at least in the United States, some species of hawks and fish-eating birds are gradually making a comeback.

Application of ecosystem and theoretical ecology to resource management has increased in the past decade or so, even though economics often takes precedence over ecology. Forestry, once concerned mostly with how to grow commercial species of trees, now emphasizes biomass accumulation, nutrient cycling, the effects of timber harvesting on nutrient budgets and on other species, and the role of fire in forest ecosystems. Range management is interested in the functioning of grassland ecosystems, the effects of grazing intensities on above-ground and below-ground production by plants, and species structure. Wildlife management, once emphasizing only game species, now considers the entire wildlife spectrum, including species hunted and not hunted. The range of interest covers both the population ecology of wildlife and the maintenance and management of plant communities as wildlife habitat. In the past few years, wildlife management has developed an interest in population genetics, with an emphasis on the effects of hunting on the genetics of game species.

The National Environmental Protection Act (NEPA), enacted in 1969, required the prediction of the environmental effects of all projects financed wholly or in part by federal funds. Later, state and local environmental laws extended such assessment to other situations. These laws required environmental impact statements (EISs) about many projects, ranging from industrial developments and power plants to dams and highway construction. EISs involve an interdisciplinary approach, stimulating a need for applied ecologists. *Applied ecology* is concerned with predicting ecological consequences of activities with goals other than biological ones and making recommendations to reduce, eliminate, or mitigate environmental damage. Generally, applied ecologists need baseline studies describing existing conditions and later studies documenting the effects of changes. Unfortunately, such studies often become corporate secrets. All of us must back up ecologists, insisting on the right to know such essential information.

Summary

Ecology, defined most simply and in its broadest sense, is the study of the interrelationship of organisms and the environment. Ecology is too often equated with environmental science, an amalgam of ecology, geology, economics, sociology, and political science. Eclectic in nature, ecology had its early beginnings in natural history, out of which stemmed both animal and plant ecology. Plant ecology developed earliest, growing out of plant geography, from which branched plant physiological ecology, plant sociology, and limnology. The latter was an impetus for the study of energy flow and nutrient cycling, the basis of ecosystem ecology. Darwin's theory of the origin of the species, the Malthusian theory of population growth, and genetics gave rise to population ecology and evolutionary ecology; the study of animal physiology provided the basis for environmental physiology of animals. The interrelationship of plants and animals, studied especially through herbivory, is the focus of community ecology. Ecology has developed from so many different sources that it will probably always remain a polymorphic discipline.

Modern ecology progresses along two lines, ecosystem ecology and population and evolutionary ecology. Each is concerned with the what, why, and how of ecological phenomena on its respective level. Ecosystem ecology takes a holistic approach, and population ecology a reductionist approach; both use descriptive, laboratory, and experimental approaches to develop analytical and simulation models. These models are abstract simplifications of natural phenomena, developed to gain insight into how the natural world functions. Because of the complexity of the natural world, ecologists have been unable to develop strongly predictive models to foretell the behavior of a chaotic world. Nevertheless, ecology does provide basic knowledge that may be applied in the management of the planet Earth, if its human inhabitants have the political courage and resolve to do so.

Review and Study Questions

1. In a book review that appeared in a well-known biological journal was the statement: "If the heat dissipated by a large expenditure of energy goes into, say, warming a river, it can indeed dislocate the ecology." What is wrong with this statement?
2. What are some major distinctions between ecology and environmental science? How does ecology relate to environmental problems?

*3. Read Aldo Leopold's *A Sand County Almanac* and Paul Sears's *Deserts on the March*. Do they have any application today? Have we learned any lessons from the past? Do we have a land ethic?

Selected References

Darwin, C. 1859. *The origin of species.* New York: Hill and Wang. Illustrated edition abridged by R. Leakey; excellent for the first reading.

Egerton, F. N., ed. 1977. *History of American ecology*. New York: Arno Press.

Eiseley, L. 1958. *Darwin's century: Evolution and the men who discovered it*. Garden City, NY: Doubleday.

Elton, C. 1927. *Animal ecology*. London: Sedwick & Jackson. A classic, easily read.

Gleason, H. A. 1975. Delving into the history of American ecology. *Bull. Ecol. Soc. Amer.* 56:5–10.

Kingsland, S. 1985. *Modeling nature.* Chicago: University of Chicago Press. A history of the development of population biology.

Kormondy, E. J., ed. 1974. *Readings in ecology.* Englewood Cliffs, NJ.: Prentice-Hall. Excerpts from a wide range of historically significant papers.

Lindeman, R. 1942. The trophic-dynamic aspect of ecology. *Ecology* 23:399–418.

McIntosh, R. P. 1985. *The background of ecology; concept and theory*. New York: Cambridge University Press. Excellent analysis of history of ecology in North America.

Sheall, J., ed. 1988. *Seventy-five years of ecology: The British Ecological Society*. Oxford: Blackwell Scientific Publishers. History of British ecology.

Tansley, A. G. 1935. The use and abuse of vegetational concepts and terms. *Ecology* 16:284–307. A historical paper.

Van Dyne, G. 1969. *Ecosystem concept in natural resource management.* New York: Academic Press.

Worster, D. 1977. *Nature's economy*. San Francisco: Sierra Club Books. An engaging history of ecology, a different view of history from that of McIntosh.

PART II

Natural Selection and Speciation

INTRODUCTION

Darwin and Wallace

In 1831—on the 27th of December, to be exact—young Charles Darwin, 22 years old, shipped aboard the HMS *Beagle*, a surveying ship of the British Navy, as a naturalist. The *Beagle* sailed from Plymouth, England, to the eastern coast of South America, through the Strait of Magellan, past Tierra del Fuego, up the west coast to Peru, and then across the water to the Galápagos Islands. From there the ship sailed to the South Sea Islands, New Zealand, and Australia, across the Indian Ocean, around the Cape of Good Hope, and on to England. Five years later Darwin returned, his notebooks filled with observations, his boxes filled with specimens of rocks, plants, and animals, and his head filled with ideas on evolution. On this trip he explored the jungles and pampas of South America and climbed the Andes. On the pampas he unearthed the fossil remains of prehistoric creatures—the glyptodon, the megatherium, and the guanaco—and he noted the similarities and differences between them and such species as the armadillo, the sloth, and the llama. He wondered that two such similar forms, one living and the other extinct, should exist in exactly the same part of the world. About it, Darwin later wrote (1887): "The wonderful relationship in the same continent between the dead and the living will, I do not doubt, hereafter, throw more light on the appearance of organic beings on our earth, and their disappearance from it, than any other class of facts."

During frequent stops along the South American coast, Darwin noticed that although individuals of the same species were identical or nearly so in one locality, they were slightly different in another. The nearer the locations were, the more closely did the two populations resemble each other. The more distant they were, the greater was their divergence. So often did Darwin observe this subtle difference within so many different species that he was convinced it was a general rule.

Darwin pondered these similarities and differences between fossil and living representatives and between populations of animals separated by space. He eventually came to the conclusion that species gradually became modified with the passage of time, an idea already advanced by the French biologist Lamarck. In a similar manner, he reasoned, when a species consisting of a homogeneous group increased its range into new habitats, the organisms would evolve or change in different ways, resulting in slightly different races in each region.

The questions "How did evolution occur?" and "What are its mechanisms?" still remained unexplained. Then in October 1838 Darwin happened to read "An Essay on Population" by Malthus, which considered the relationship between population size and food supply. Malthus (1798: 9–10) wrote:

> Through the animal and vegetable kingdoms, nature has scattered the seeds of life abroad with the most profuse and liberal hand. She has been comparatively sparing in the room and nourishment necessary to rear them. The germs of existence contained in this spot of earth, with ample food, and ample room to expand in, would fill millions of worlds in the course of a few thousand years. Necessity, that imperious all-pervading law of nature, restrains them within prescribed bounds. The race of plants and the race of animals shrink under this great restrictive law. And the race of man cannot, by any efforts of reason, escape from it. Among plants and animals its effects are waste of seed, sickness, and premature death. Among mankind, misery and vice.

From this idea Darwin developed his concept of "the struggle for existence," which forms an important part of his hypothesis regarding the mechanism of evolution.

Darwin reasoned from his observations on the differences among living things that some variations were more advantageous than others, that some variations enabled the organisms to occupy an area or to survive. Because only a few animals would survive, Darwin argued, those with more favorable variations would have better odds for survival. "Under these circumstances," wrote Darwin (1887:120), "favorable variations would tend to be preserved, and unfavorable ones to be destroyed. The result of this would be the formation of a new species. Here then I had at last got a theory by which to work. . . ."

This process that resulted in a greater survival of individuals possessing advantageous characteristics over those with less advantageous ones Darwin called *natural selection*. As a result of it, favorable variations will be retained in the population and will increase. Through time, selection will result in a population better adapted to its environment; if continued long enough, new and different species will evolve.

Although Darwin's theory of evolution is in essence accepted today, Darwin hesitated to publish it. He started to develop his hypotheses in 1837, and in 1842 he wrote a penciled 35-page abstract of his ideas, which he enlarged to 255 pages in 1844. He still withheld publication but discussed his ideas with two friends, the geologist Sir Charles Lyell and the botanist Dr. J. D. Hooker. In 1856 Lyell urged Darwin to publish his material, and by 1858 he had his book half completed.

Then in the summer of that year, Darwin received a letter from a fellow naturalist, Alfred Russel Wallace, who was exploring the Malay Archipelago. Its contents shocked Darwin, for with the letter, Wallace sent an essay entitled "On the Tendency of Varieties to Depart Indefinitely from the Original Type." This essay contained an excellent summary of Darwin's own theory of evolution by natural selection.

After much reflection by Darwin over the dilemma, and following the intervention of Lyell and Hooker, the two men, who had independently arrived at the same conclusion, presented their views jointly, through Lyell and Hooker, to the Linnaean Society of London on July 1, 1858, and subsequently published them in the society's journal.

Darwin pointed out that animals as well as plants produce more offspring than necessary to maintain the species. He concluded that the number of organisms must be held in check by their own struggle either with other individuals of the same species or different species or against "external nature."

Wrote Darwin (1884:119):

> Now, can it be doubted, from the struggle each individual has to obtain subsistence, that any minute variation in structure, habits, or instincts adapting that individual better to new conditions would tell upon its vigor and health? In the struggle it would have a better *chance* of surviving; and those of its offspring which inherited the variation, be it ever so slight, would also have a better *chance*. Yearly more are bred than can survive; the smallest grain in the balance, in the long run, must tell on which death shall fall, and which shall survive. Let this work of selection on one hand, and death on the other, go on for a thousand generations, who will pretend to affirm that it would produce no effect, when we remember what, in a few years, Blakewell effected in cattle, and Western in sheep, by this identical principle of selection?

Wallace stated much the same theory, except that he emphasized population control largely by food supply. Like Darwin, Wallace concluded that forms best adapted to their environment evolved through selection of the individual. Wallace (1858:274) argued that if any environmental change made existence difficult, then all of the individuals composing the species, those forming the least numerous and most feebly organized variety would suffer first, and, were the pressure severe, must soon become extinct. The same causes continuing in action, the parent species would next suffer, would gradually diminish in numbers, and with a recurrence of similar unfavorable conditions might also become extinct. The superior variety would then alone remain, and on a return to favorable circumstances would rapidly increase in numbers and occupy the place of the extinct species and variety.

Darwin further expounded his theory of natural selection and evolution in the *Origin of Species*, published in November 1859. Darwin received, as he should, the major credit for arriving at a theory of evolution, whereas Wallace is best known for his fundamental studies of the distribution of animals.

Darwin's theory of evolution can be summarized briefly: Variation exists among individuals of sexually reproducing species that affects their chances of survival and their reproductive rates. In addition, many species have such a high potential rate of increase that if unchecked they would exhaust both food and living space. Since food and space are limited, those with the most advantageous variations will have the better chance to survive. Constant selection of the better adapted and the elimination of the less fit result in the evolutionary change of populations. This outcome is "survival of the fittest."

Neither Wallace nor Darwin could explain adequately the nature and origin of variation or how these variations were transmitted from parent to offspring, although they realized that they were inherited. Had they known, much of the storm of controversy over the theory might never have developed. Ironically, the answers to Darwin's most pressing questions on variation and a basis for understanding the mechanisms of evolution were being discovered in a monastery garden by a contemporary—an Austrian monk, Gregor Mendel. His findings on the laws of inheritance, presented as a lecture to the Natural History Society of Brunn in 1866 and published in the *Transactions* of the society, were unappreciated by the scientific world at that time and remained lost in the obscure journal until 1900, when he was discovered independently by three biologists: deVries, Correns, and Tschermak.

CHAPTER 2

Natural Selection and Evolution

Outline

Objectives

On completion of this chapter, you should be able to:

1. Sketch the changes in Earth through geological time.
2. Explain the differences among adaptation, natural selection, and evolution.
3. Explain the importance of mutation and of recombination of genes as sources of genetic variation.
4. Distinguish among directional, disruptive, and stabilizing selection.
5. Discuss influences on evolution that lie outside natural selection.
6. Discuss the importance of genetic drift, inbreeding, and founder's effect on the conservation of species.
7. Explain what is meant by a viable population and effective population size.

THE EVOLUTION OF LIFE: AN OVERVIEW

To all of us Earth appears immutable. Waters of oceans wash the shores of continents that seemingly never change. Familiar mountain peaks have looked the same through centuries. Even local hills, rivers, and other topographic features have an air of permanence about them. But Earth was not always as it is today, nor will it be the same tomorrow, because changes are continually taking place, however slowly. Volcanoes erupt, new islands appear in oceans, beaches erode. Physical features of Earth—its oceans and land masses, its vegetation and animal life—have changed radically through time. These changes of the past have influenced life on Earth today.

Earth, geologists estimate, is some 4600 million years old. The first era, the Precambrian, which extended from 4600 million years ago to 570 million years ago, saw the production of an atmosphere and a hydrosphere, Earth's envelopes of air and water; the evolution of preliving compounds and, later, autotrophic forms of life that made their own food; the internal reorganization of Earth; and the development of ocean basins and continents.

In the early Paleozoic, some 540 million years ago, three separate land masses existed: Asia, North America and Europe, and Gondwanaland, which included modern-day Africa, South America, Australia, New Zealand, and Antarctica as well as scattered fragments of land. During the Paleozoic, 420 million years ago, South America and Africa lay close together around the South Pole, and the rest of Gondwanaland lay on the far side of the South Pole, pointing toward the equator. Slowly the land mass moved northward, so that by the Carboniferous age, 340 million years ago, the whole of Africa had moved across the South Pole and Antarctica lay in the region of the South Pole. Glaciers covered southern South America, South Africa, India, and Australia; and Europe and North America lay along the equator. There the climate was warm, humid, and seasonless, and a large part of the area was covered with swamps and tropical rain forests. During the Permian these three blocks joined, raising the Caledonian, Ural, and Appalachian mountain ranges and forming a single land mass, Pangaea.

As Pangaea moved northward, it began to break apart slowly, 5 to 10 cm a year. The first break in the single land mass apparently took place in the mid-Mesozoic age, approximately 180 million years ago, when North America and Africa parted to form the first narrow strip of the Atlantic. Africa and South America were still connected by the end of the Jurassic, but drift was taking place in the South Atlantic region. By the middle Cretaceous, about 100 million years ago, Africa and South America had split apart. South America and Antarctica still clung together until sometime in the early Cenozoic, finally becoming separated in the Eocene. Africa had already become separated from Antarctica between the middle Jurassic and the middle Cretaceous. Thus, by the end of the Cretaceous, Gondwanaland had broken up; the only intact land mass was North America/Eurasia, collectively known as Laurasia.

Until the lower Eocene, Laurasia remained intact. North America was connected to Europe by Greenland and Scandinavia. Then in the mid-Eocene, the North Atlantic joined the Arctic Ocean, separating Laurasia into North America and Europe.

The formation, breakup, and northward drift of continents resulted in broad climatic changes and the formation of geological barriers that affected evolving plant and animal life. Between 2700 million and 2000 million years ago, prokaryote (lacking chromosomes and nuclei) bacteria and photosynthetic blue-green algae developed. Eukaryote (with characteristic nuclei and chromosomes) organisms evolved 1800 million to 1000 million years ago. The Precambrian era came to a close with the sudden rise of Metazoa, a diverse and complex assemblage of animals in ancient seas that ushered in the Paleozoic era.

The Paleozoic saw the evolution of the earliest known fish, the first amphibians, the first reptiles, the first insects, and the first land plants as well as the rise of the great coal forests. The latter, dominant on the North American and European continents that at that time lay on the equator, consisted of such tall trees as *Lepidodendron, Sigillaria,* and *Cordaites*, forerunners of coniferous trees.

The end of the Paleozoic witnessed the expansion of primitive reptiles that ushered in the Mesozoic era, the age of reptiles. The Mesozoic, between 230 million and

70 million years ago, saw the rise and fall of dinosaurs and an explosive evolutionary radiation of the angiosperms, the seed-bearing plants. Throughout the Mesozoic no distinct floral or faunal regions existed. The land, even though partially separated, was one, with no effective barriers to dispersal of plants and animals. Mountain ranges lay only about the edges of Pangaea, and although shallow seas invaded the land, the invasions were short-lived, geologically. The climate, too, was warm and equitable, mostly tropical to subtropical, even to the coast of Alaska. These conditions allowed the great reptiles and early mammals to roam freely over continental land masses.

Then in the late Cretaceous, climatic conditions began to change as continental land masses drifted further apart. The warm, subtropical middle Cretaceous climate was replaced in the late Cretaceous by a warm, temperate climate. The cooling marked the end of the great reptiles. As the Cretaceous period moved into the Paleocene period of the Cenozoic, greater changes took place. Between the lower Cretaceous and the Eocene, the single, connected land mass inhabited by cone-bearing gymnosperms and reptiles was replaced by a divided land mass inhabited by flowering plants and mammals. Continental drift effectively separated plant and animal populations and hastened the development of different floras and faunas.

Although plants achieved a worldwide distribution before the continents broke apart, mammals did not. They never achieved any significance until the continents were well on their way to separation. One of the first groups of mammals, the marsupials, confronted competition from more advanced placental animals. Had the placentals achieved dominance before the breakup of continents, the marsupials might never have survived; but marsupials had spread to Antarctica and Australia before these land masses separated. Antarctica moved south, where the cold eliminated terrestrial mammalian life, while Australia, separate from all other land masses, became a final refuge for marsupial life. Free from competing placental forms, marsupials were able to radiate into a variety of forms and to fill niches similar to those occupied by placental mammals elsewhere. Meanwhile, marsupial mammals that existed in North America and South America were rapidly replaced by placental mammals, except for the opossum. South America, isolated from North America during the lower Cenozoic (Tertiary), supported diverse and unique forms of placental mammalian life. After a land bridge, the Isthmus of Panama, became exposed between the two continents, these mammals were unable to compete with the more advanced placentals that moved down from the north.

While competition resulted in the elimination of certain kinds of mammals, a major influence on the developing fauna was the rapidly changing climate of the northern land mass. In the lower Eocene, Europe, North America, and Asia were still connected in a manner that allowed animals to move from one to the other. At that time North America appeared to be the center of the early evolution of placental mammals. From North America they spread to Europe, but the cold climate inhibited the movement of mammals from North America to Asia. When North America became separated from Europe, new mammalian groups that evolved in Europe could not cross to North America, but they did have access to Asia. The mid-Eocene saw the return of a warm, moist climate; a semitropical rain forest extended to Alaska, and tropical conditions prevailed as far north as England. The climate of the Bering land bridge was benign enough to encourage the passage of mammals from Asia to North America. In the Oligocene the climate cooled again, restricting the migration of mammals to those tolerant of cold temperatures and eliminating tropical forms of plants from northern lands. From the Oligocene through the Pliocene, the cooling trend continued. The movement of animals from Eurasia to North America was restricted to such cold-tolerant species as mammoths and humans, and the flora of the continent was basically a modern one.

Meanwhile, the fauna of the Old World tropics remained free of the influences of great temperature changes, although the increasing aridity in Africa that began in the Oligocene and Miocene brought about the replacement of tropical forests by grasslands, on which evolved huge herds of ungulate fauna and eventually humans.

Cooling of the Pliocene continued until the Pleistocene ushered in the Ice Age. Uplift of continents, volcanism, and contrasts in climate placed severe stress on flora and fauna, but these factors were nothing compared to the great ice sheets. Northern Europe and northern North America were transformed periodically into arctic regions as glaciers formed. Temperate forests retreated southward, replaced first by boreal forests, then by tundra, and finally by ice. These ice ages occurred not once but at least four times. Ice sheets swept rock from one area and deposited it elsewhere, wore down mountains, filled in valleys, carved out lakes (including the Great Lakes), changed sea levels and water temperatures, changed the drainage patterns of rivers, and in the end left behind relict arctic climates in unglaciated regions that bordered the ice sheets. It was the age of woolly mammoth, woolly rhinoceros, mastodon, royal bison, camel, peccary, saber-toothed tiger, and dire wolf and of the invasion of the North American continent by humans. It was also the time when birds, fish, and insects isolated by the advance and retreat of glaciers tended to evolve into a number of new species.

HOW NATURAL SELECTION WORKS

Plants and animals have evolved throughout time. As environmental conditions slowly changed, flora and fauna responded—either adjusting to a new set of conditions or perishing. Earth is still a dynamic planet. Continents continue to drift, and environments are changing more rapidly than they did in the past because of massive human intervention. To survive as a species, organisms must adapt to the changing environment. The problem is that many conditions are changing faster than organisms can adapt to them.

ADAPTATION

Adaptation implies the ability of an organism to live in harmony or conformity with its environment. Adaptation comes about through the interaction of organisms with their environment. If an organism can tolerate a given set of conditions to such a degree that it can not only survive as an individual but also leave mature reproducing progeny in the population, it contributes its genetic traits to the population's gene pool. It is adapted to its environment. If an organism leaves few or no mature reproducing progeny, it contributes little or nothing to the gene pool. It is poorly adapted to its environment. Those individuals that contribute the most to the gene pool are said to be the most fit, and those that contribute little or nothing to the gene pool are said to be the least fit. The *fitness* of an individual is measured by its reproducing offspring, especially descendants, over several generations. That differential reproductive success of individual organisms is *natural selection*. It is the ability of individuals to leave the most reproducing offspring.

Adaptation is one of those terms in ecology that has been pushed nearly into a state of uselessness because it has been employed in different ways. The most common meaning of adaptation is any form of behavior that is assumed to be the result of natural selection. In other words, natural selection is responsible for most of what we observe in plants and animals. A second meaning is a change in the physical, physiological, or behavioral traits that results from some environmental pressure. The change supposedly improves the ability of one organism to survive and reproduce compared to another that did not change. A third meaning is any physical or physiological feature or form of behavior used to explain the ability of an organism to live where it does. Adaptations are observed because some ancestors left more descendants than others.

A current argument is that although some features of an organism are the result of natural selection, others have different origins. Some features and functions may have been borrowed for current use from a previous use. A structure or feature currently in use may have been derived from a structure that had no previous function. Not every feature of an organism is functional. If neutral in its effects, or if an accidental by-product of other selection changes, a feature may be retained because it is not subject to any selection pressures. It becomes a source of raw material for future evolution.

VARIATION

The community, aquatic or terrestrial, consists of many different, locally defined groups of individuals similar in structure and behavior. Individuals within these groups interbreed, oak tree with oak tree, white-footed mouse with white-footed mouse, largemouth bass with largemouth bass. Collectively, individuals within each group make up a genetic population, or *deme*.

Beyond one local population may be other similar demes. They may be separated by distance, or indistinctly separated—more or less adjacent and continuous over a wide area. Whatever the situation, hereditary material to a greater or lesser degree passes from one population to another. Some adjacent demes may interbreed so freely that they become essentially one. Others may not interbreed at all. If a local population of a plant or animal dies out, it will, if conditions are favorable, be replaced by individuals from surrounding populations. Individuals that die are replaced by their offspring, so the population tends to persist through the years, the inheritable features passing from one generation to the next.

Individuals that make up the deme are not identical. Just as wide individual variation exists inside human populations, so the same variation exists among individuals of sexually reproducing plants and animals. This variation is the raw material of natural selection.

That variation exists within a population can easily be demonstrated. All we need to do is to select some 100 or so specimens from a local population and observe and record the variations of a single character: the tail length of some species of mouse, the number of scales on the belly of a snake, the shapes and sizes of sepals and petals on flowers, the rows of kernels on ears of corn. These observations can be tabulated as frequency distributions (Figure 2.1). Many of the specimens will have characteristics with the same value. The value that is most common is called the *mode*. The average of all the values is the *mean*. Other values will vary above and below the mode. The frequencies of these values fall off away from the mode; that is, fewer and fewer individuals are in each class. The frequency distribution of these variable characters tends to follow a bell-shaped curve, the normal curve of probability. In some situations the distribution may deviate from the normal bell-shaped curve. Deviations from the center of the normal curve are measured as standard deviations. A standard deviation is obtained by squaring all the deviations, calculating their mean,

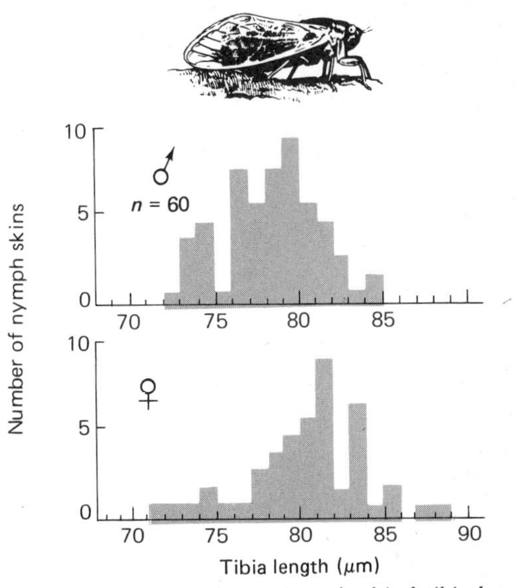

FIGURE 2.1 An example of variation: the hind tibia lengths of skins shed by nymphs of the periodical cicada, *Magicicada septendecim*.

and then finding the square root of the mean. These differences may point out some facts about variation within a population. The variations may be due to heredity, to environment, or, more often than not, to an interaction of both.

The characteristics of a species and variations in individuals are transmitted from parent to offspring. The sum of hereditary information carried by the individual is the *genotype*. The genotype directs the development of the individual and produces the characters that make up the individual's morphological, physiological, and behavioral characteristics. The external, observable expression of the genotype is the *phenotype*.

Some of the conspicuous individual variants in the phenotype—such as shortened tails, missing appendages, enlarged muscles, or other features that result from disease, injury, or constant use—are not inheritable. They are acquired characteristics, which the early evolutionist Lamarck hypothesized erroneously were passed down from one generation to another. However, some acquired characteristics that have been environmentally induced are inherited; or to say it

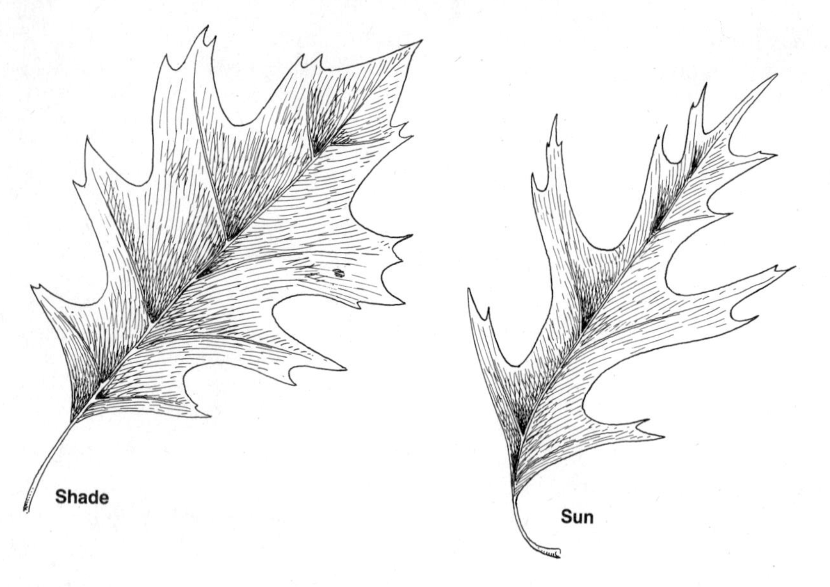

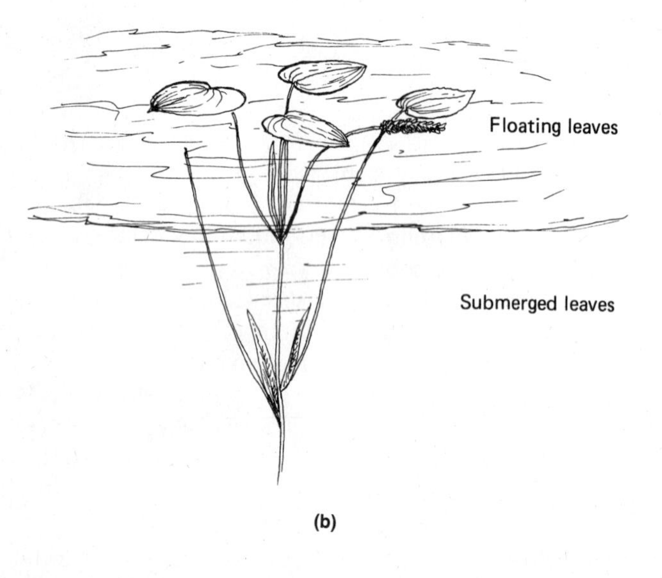

FIGURE 2.2 Many plants exhibit a plasticity of response to environmental conditions. (a) Leaves of red oak growing in the sun are more deeply lobed than leaves growing in the shade. Deeper lobing presents less surface area for the absorption of heat and more edge per surface area for dissipation of heat. (b) Some aquatic plants, such as the pondweed *Potamogeton,* have lance-shaped underwater leaves, more responsive to undersurface water movements, and broad, heart-shaped leaves that float on the surface.

more accurately, the ability of an organism to acquire such characteristics is inherited. The ability of a genotype to give rise to a range of phenotypic expressions under different environmental situations is known as *phenotypic plasticity*. Some genotypes have a narrow range of reaction to environmental conditions and therefore give rise to fairly constant phenotypic expression. Some of the best examples of phenotypic plasticity are found among plants. The size of plants, the ratio of reproductive tissue to vegetative tissue, and even the shape of the leaf may vary widely at different levels of nutrition, light, and moisture (Figure 2.2).

Sources of Variation

Of major importance to natural selection and adaptation are genetic, or inherited, variations within a population—variations that arise from mutations and especially from the shuffling of genes and chromosomes in sexual reproduction.

The genetic control mechanisms are in the chromosomes, found within the nucleus of plant and animal cells. Each chromosome consists mostly of long, coiled strands of deoxyribonucleic acid, DNA. DNA is the information template from which all cells in the organism are copied. It is a complex molecule in the shape of a double helix, resembling a twisted ladder in its construction. The long strands, comparable to the uprights of a ladder, are formed by an alternating sequence of deoxyribose sugar and phosphate groups. The connections between the strands, or the rungs, consist of pairs of nitrogen bases, adenine, guanine, cytosine, and thymine. In the formation of the rungs, adenine is always paired with thymine, and cytosine with guanine.

The DNA molecule is divided into smaller units, the *nucleotides*, which consist of three elements: phosphate, deoxyribose, and one of the nitrogen bases bonded to a strand at the deoxyribose sugar. The information of heredity is coded in the sequential pattern in which the base pairs occur. Each species is unique in that the base pairs are arranged in a different order and probably in different proportions from every other species.

Each chromosome carries units of heredity called *genes*, the informational units of the DNA molecule. Because chromosomes are paired in the body cells, genes are also paired. The position a gene occupies on a chromosome is known as its *locus*. Genes occupying the same locus on a pair of chromosomes are termed *alleles*. If alleles in a pair affect a given trait in the same manner, they are called *homozygous*. If alleles in a pair are different, the pair is called *heterozygous*.

When cells reproduce (a process of division called *mitosis*), each resulting cell nucleus receives the full complement of chromosomes, the double or *diploid* number. In organisms that reproduce sexually, the germ cells, or gametes (egg and sperm), result from the process of *meiosis*, in which the pairs of chromosomes are split so that the resulting cell nucleus receives only one-half of the full complement, or the *haploid* number. When egg and sperm unite to form a zygote, the diploid number is restored.

Recombination. When two gametes combine to form a zygote, the gene contents of the chromosomes of the parents are mixed in the offspring. Because the number of possible recombinations is infinitely large, recombination is the immediate and major source of variation. Recombination does not result in any change in genetic information, as mutations do, but it does provide different combinations of genes upon which selection can act. Because some combinations of interacting genes are more adaptive than others, selection determines the variations or new types that will survive in the population. The poorer combinations are eliminated by selection and the better ones retained.

The amount or degree of recombination influencing the amount of variability in a population is limited by a number of characteristics of a species. One limitation is the number of chromosomes and, therefore, the number of genes involved. Another is the frequency of crossing over, the exchanging of corresponding segments of homologous chromosomes during meiosis. Others include gene flow between populations, the length of generation time, and the type of breeding—for example, single versus multiple broods in a season in animals and self-pollination versus cross-pollination in plants.

Mutation. A *mutation* is an inheritable change of genetic material. It may be either a gene mutation (point mutation) or a chromosome mutation.

Gene, or point, mutation is an alteration in the sequence of one or more nucleotides. During meiosis the gene at a given locus is usually copied exactly and eventually becomes part of the egg or sperm. On occasion the precision of this duplication process breaks down, and the offspring DNA is not an exact replication of the parent DNA. The alteration may be a change in the order of nucleotide pairs, a substitution of one nucleotide pair for another, the deletion of a pair, or various kinds of transpositions.

Most gene mutations have very small or no apparent effects. Mutations of single genes that produce larger effects are usually deleterious. Point mutations are important because they restore and maintain variation in the gene pool, especially when selection reduces the frequency of unfavorable alleles or eliminates them, thus reducing the size of the gene pool. Such mutations, however, do not direct evolutionary change.

Chromosomal mutations, or macromutations, may result from a change in the number of chromosomes or a change in the structure of the chromosome. A change in chromosomal number can arise in two ways: (1) the complete or partial duplication of the diploid number rather than the transmission of the haploid number or (2) the deletion of some of the chromosomes.

Polyploidy is the duplication of entire sets of chromosomes. It can arise from an irregularity in meiosis or from the failure of the whole cell to divide at the end of the meiotic division of the nucleus. The individual body cell ordinarily is diploid ($2N$), or twice the haploid number. The union of a diploid gamete with a haploid gamete yields a triploid, usually sterile. Two diploid gametes ($2N$) may unite to form a tetraploid, also usually sterile. Few naturally occurring tetraploids have arisen within a single species (autotetraploid). One example is *Epilobium*, the common fireweed of open fields and thickets. The union of two diploid gametes from different species results in a hybrid tetraploid (allotetraploid). Such tetraploids are fertile because the chromosomes upon division behave as if they were diploids and produce balanced $2N$ chromosomes in the gametes (instead of the normal $1N$).

Allopolyploidy is common in plants. The condition is rare in animals because an increase in sex chromosomes would interfere with the mechanism of sex determination and the animal would be sterile. Polyploidy in plants would be rare, too, if plants depended on cross-fertilization. However, plants can be self-fertilized, and they reproduce vegetatively, which gives polyploidy a selective advantage among many species. Polyploid plants differ from the normal diploid individuals of the same species in appearance; they are generally larger, more vigorous, and more productive. Most of our agricultural plants are polyploids.

Another form of macromutation is duplication or deletion of a part of a normal complement of chromosomes. Such deletions or duplications result in abnormal phenotypic conditions. (One such condition is Down's syndrome in humans.)

Mutations may be neutral, beneficial, or disadvantageous, depending upon the environmental circumstances and the genetic background in which they arise. If the effect on the phenotype is deleterious, the mutant allele will be selected against. So will mutations that serve a function already filled by an established gene that appears in an established population adapted to its environment or in one that is highly heterozygous. However, if a mutation is advantageous or neutral, it may be retained, especially if it confers some selective advantage in a changing environment or if it appears in a highly inbred, homozygous population. Individual mutations, however, are rarely agents of recognizable change; that is usually brought about by an accumulation of mutations, each of which alters slightly the appearance or function of an organism. Most mutations just maintain variability in the gene pool of a population. Without mutations genetic variations over many generations would be so reduced that a population would not respond any further to natural selection.

Hardy-Weinberg Equilibrium

Variations found in a population are transmitted to the next generation through sexual reproduction when samples of the gene pool contained in gametes join to form zygotes. If genes occur in two forms, A and a, then any individual can fall into three possible diploid classes: AA, aa, and Aa. Individuals in which the alleles are the same—AA or aa—are homozygous. Haploid gametes produced by homozygous individuals are either all A or all a; those by heterozygous, half A and half a. They recombine in three possible ways in sexual reproduction: AA, Aa and aa (Figure 2.3). Thus the proportion of gametes carrying A and a is determined by the individual genotypes, the genes received from the parents.

Egg and sperm unite at random. If the mating pattern is also random—that is, if the chance that an individual will mate with another individual having a certain genotype is equal to the frequency of that genotype in the population—then we can predict the genotypes in the next generation.

Assume that a population homozygous for the dominant allele AA is mixed with an equal number from a population homozygous for the recessive aa. The allele frequency in the total population is 0.5 A and 0.5 a. The offspring of the matings, the F_1 generation, will all be heterozygous, Aa, and the frequency of the alleles will still be 0.5 A and 0.5 a. The offspring of these

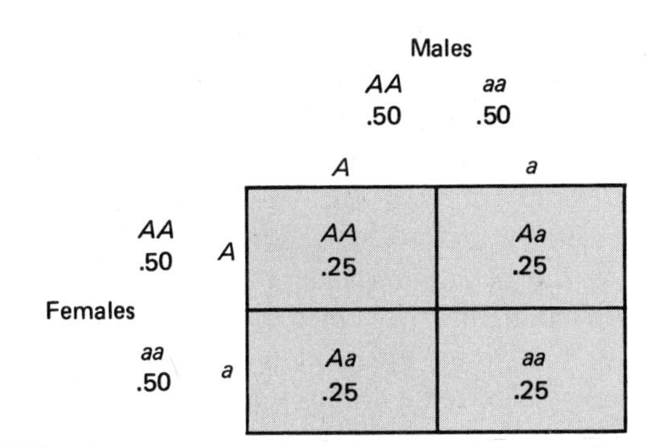

FIGURE 2.3 Mixing two homozygous populations, F_1 generation.

heterozygotes, the F_2 generation, will consist of 0.25 *AA*, 0.50 *Aa*, and 0.25 *aa*. These proportions represent the *genotypic frequencies*, but the allele frequency still remains at 0.5 for *A* and 0.5 for *a*.

These frequencies are predicted by one of the important statements in population genetics, the Hardy-Weinberg law. This law holds that the frequency of each member of a single pair of alleles will remain the same in successive generations if certain assumptions hold: (1) the organisms are diploid; (2) reproduction is sexual; (3) mating is random; (4) mutations do not occur or are in equilibrium ($A \rightarrow a$); (5) the population is large, so changes by chance in gene frequencies are insignificant; (6) natural selection does not occur; (7) migrations are negligible; and (8) generations do not overlap.

The frequency of the three genotypes can be expressed as the Hardy-Weinberg equilibrium. Essentially it is: Let p be the proportion of alleles that are *A*, and q the proportion of alleles that are *a*. Because all loci for that gene must be occupied either by *A* or *a*, $p + q = 1$. In our previous example, *A* (p) made up 50 percent of the alleles, and *a* (q) the other 50 percent, so $0.5 + 0.5 = 1.0$. In the F_1 generation, the genotypic frequency is .25 *AA* + .50 *Aa* + .25 *aa* (Figure 2.3). The frequency of the *AA* matings is $p \times p = p^2$; the frequency of *aa* matings is $q \times q = q^2$; and the frequency of the $A \times a$ matings is $(p \times q) + (p \times q)$. The frequency of the genotypes can be expressed as $p^2 + 2pq + q^2 = 1$. Thus $AA = p^2$, $Aa = 2pq$, and $aa = q^2$.

The Hardy Weinberg law, then, predicts that the

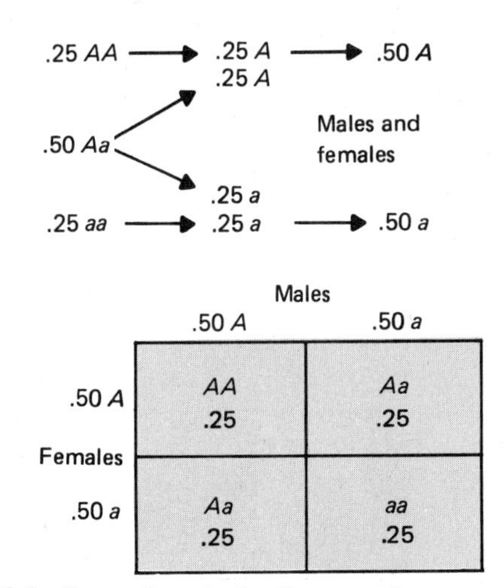

FIGURE 2.4 Proportions in the F_2 generation.

frequency of the alleles will not change even though the genotypic frequencies do. Consider the frequency of allele *A*, or p. In the homozygote it is p^2 and in the heterozygote it is $1/2(2pq)$. The frequency of *A* in the F_2 generation is $p^2 + 1/2(2pq) = p^2 + pq$. Because $q = (1 - p)$, the frequency equals:

$$p^2 + p(1 - p) = p^2 + p - p^2 = p$$

A similar situation exists for q, so as predicted by Hardy-Weinberg, the frequency of the alleles does not change (Figure 2.4).

To demonstrate we can use another example. This time let the parental generation have the genotypic frequency of .36*AA*, .48*Aa*, and .16*aa* (Figure 2.5). The allelic frequencies will be 0.6 for *A* and 0.4 for *a*; and the genotypic frequencies for the F_1 generation will still be .36*AA*, .48*Aa*, and .16*aa*. We can conclude that all succeeding generations will carry the same proportions of the three genotypes, provided our assumptions are met.

The assumptions are never fully met in any real population, so the Hardy-Weinberg law must be considered theoretical—a distribution against which actual observations can be compared. The beauty of the Hardy-Weinberg law is that it allows us to determine allele frequencies in a population when we know the frequency of one allele, q. Because q is the recessive

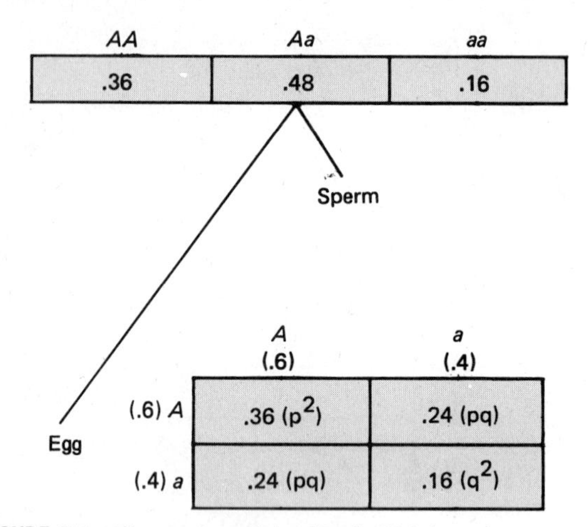

FIGURE 2.5 Illustration of the Hardy-Weinberg law.

or rarer allele, its expression in the homozygotic condition (q^2) provides all the information we need.

We find expressions of such alleles in blood types like Rh^+ and Rh^-. Let Rh^+ be indicated by *D*, the dominant allele, and Rh^- by *d*, the recessive allele. Genotypes *DD* and *Dd* are Rh^+ and genotype *dd* is Rh^-. Suppose we have a population in which 420 people are Rh^+ and 80 are Rh^-. The Hardy-Weinberg law allows us to calculate the allele frequency of *D*. Assuming random mating, which is usually the case for blood types, we can assume that the frequency of the homozygous recessive *d* is q^2. We can determine it by calculating the frequency of the Rh^- individuals in the population of 500: $80/500 = 0.16$. If q^2 is 0.16, we know that q is 0.40. Because $p = 1 - q$, $p = (1 - .40) = .60$. Thus the allele frequencies of *D* and *d* are .60 and .40, respectively. With this information we can determine the genotypic frequencies of *DD*, *Dd*, and *dd*. The frequency of *DD* is $p^2 = (.60)^2 = 0.36$; the frequency of the heterozygote, *Dd*, is $2(.60 \times .40) = .48$, and the frequency of *dd* is $q^2 = .16$. Because in this situation one gene is dominant, the observed genotypic frequencies match the Hardy-Weinberg frequencies.

That, of course, is not always the case. For example, blood types A, B. O, and AB are determined by several genes. The observable genotypic frequencies cannot be separated into homozygotes and heterozygotes. In these cases we can compare by statistical tests the goodness-of-fit of the observed frequencies with the Hardy-Weinberg frequencies, and thus discover whether or not the deviations result from chance alone.

The Hardy-Weinberg law is of fundamental importance in theoretical and applied population genetics. It allows us to state approximately what genotypic frequencies will be from a knowledge of gene frequencies alone. An important implication of this law is that in the absence of specific evolutionary or selective forces to change the frequencies, Mendelian inheritance alone is enough to maintain genetic variability in a population. Hardy-Weinberg equilibrium is attained in one generation of random mating in populations with nonoverlapping generations. With overlapping generations, Hardy-Weinberg equilibrium comes more slowly.

In natural populations the assumptions of the Hardy-Weinberg law are not met, and the equilibrium is violated. Mutations do occur, matings are rarely random, movements occur between populations, genetic drift (explained later) occurs, and natural selection does take place. All of these factors change genotypic and allelic frequencies from one generation to another, acting as evolutionary forces.

A DISTINCTION

Natural selection results when environmental forces, both abiotic and biotic, favor certain genotypes, as expressed by phenotypes, over others. It functions through nonrandom reproduction within a population. Not every individual in a population is able to contribute its share of genetic characteristics to the next generation or leave surviving offspring. Some fail to survive to reproduce. Others fail to mate or produce offspring. Some individuals leave more reproducing offspring than others. Their contribution to the gene pool or fitness increases at the expense of others.

An outcome of natural selection is a change in gene frequency in a population through time, or *evolution*. Evolution is a process of change in biological systems due mostly to forces in the environment: changes in moisture, temperature, food, habitat, interspecific and intraspecific competition, and predation. Although natural selection is a major force in evolution, natural selection is not evolution, nor is evolution natural selection.

TYPES OF SELECTION

Natural selection can proceed without any major changes in the phenotype. Random changes in gene composition take place in all populations. If the changes produce individuals that vary widely from the average characteristics of the population, they tend to be less fit and produce fewer offspring. The frequency of their genes is reduced in favor of more normal ones. Such selection is called *stabilizing* because it favors phenotypes near the population mean at the expense of the two extremes (Figure 2.6). Such selection is characteristic of relatively stable environments. However, that does not mean that some changes directed by natural selection are not taking place. Many plants and animals are faced with the problem of maintaining fitness in the face of changing environmental pressures. For them, selection may involve certain behavioral or physiological changes needed just to maintain fitness. Leigh Van Valen has made an analogy regarding such selection using the Red Queen in Alice's *Adventures through the Looking Glass:* "Now here, you see, it takes all kinds of running you can do to keep in the same place." Although there is variability in the genotype, the population mean does not vary.

In other situations selection may be disruptive or directional. Both result in a relatively rapid evolutionary change but can revert once selection pressures are relaxed.

In *disruptive selection* (Figure 2.6), both extremes are favored simultaneously, although not necessarily to the same degree. Such selection results when members of a species population are subject to different selection pressures operating in different environments, such as microhabitats. It usually results in *polymorphism*, in which a population contains two or more genotypes.

An example is the swallowtail butterfly *Papilio dardanus*, widely distributed across Africa (Figure 2.7). The females mimic an associated inedible species of butterfly that possesses a warning coloration. (A warning coloration tells a potential predator that the bearer is inedible. See Chapter 16.) The male is not a mimic. Instead, he retains a specific color pattern that the females recognize, essential for mating and reproduction. Within any given part of the species range, the females mimic an inedible species common to that region. At least three different female mimics exist. Intermediate female forms that do not bear any resemblance to the inedible models are selected against. In regions where inedible models are absent, so are mimicking females. In this example predation promotes disruptive selection in the population.

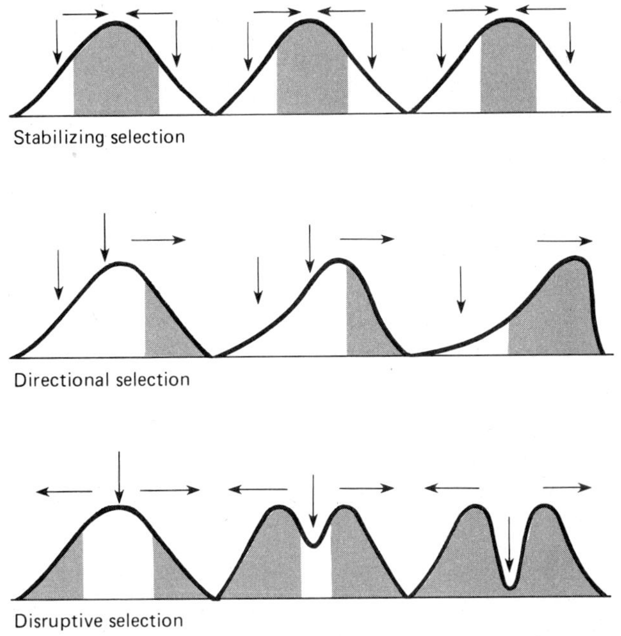

FIGURE 2.6 Three types of selection. Stabilizing selection favors organisms with values close to the population mean. Little or no change is produced. Directional selection moves the mean of the population toward one extreme. Disruptive selection increases the frequencies of both extremes. The shaded areas represent the phenotypes being selected for. Downward-pointing arrows represent selection pressures; horizontal arrows represent the direction of evolutionary change.

Selection is *directional* when one extreme phenotype is favored at the expense of others (Figure 2.6). In that case the mean phenotype is shifted toward one extreme, provided that heritable variations of an effective kind are present.

An example of directional selection is found in the survival of Darwin's medium ground finch (*Geospiza*

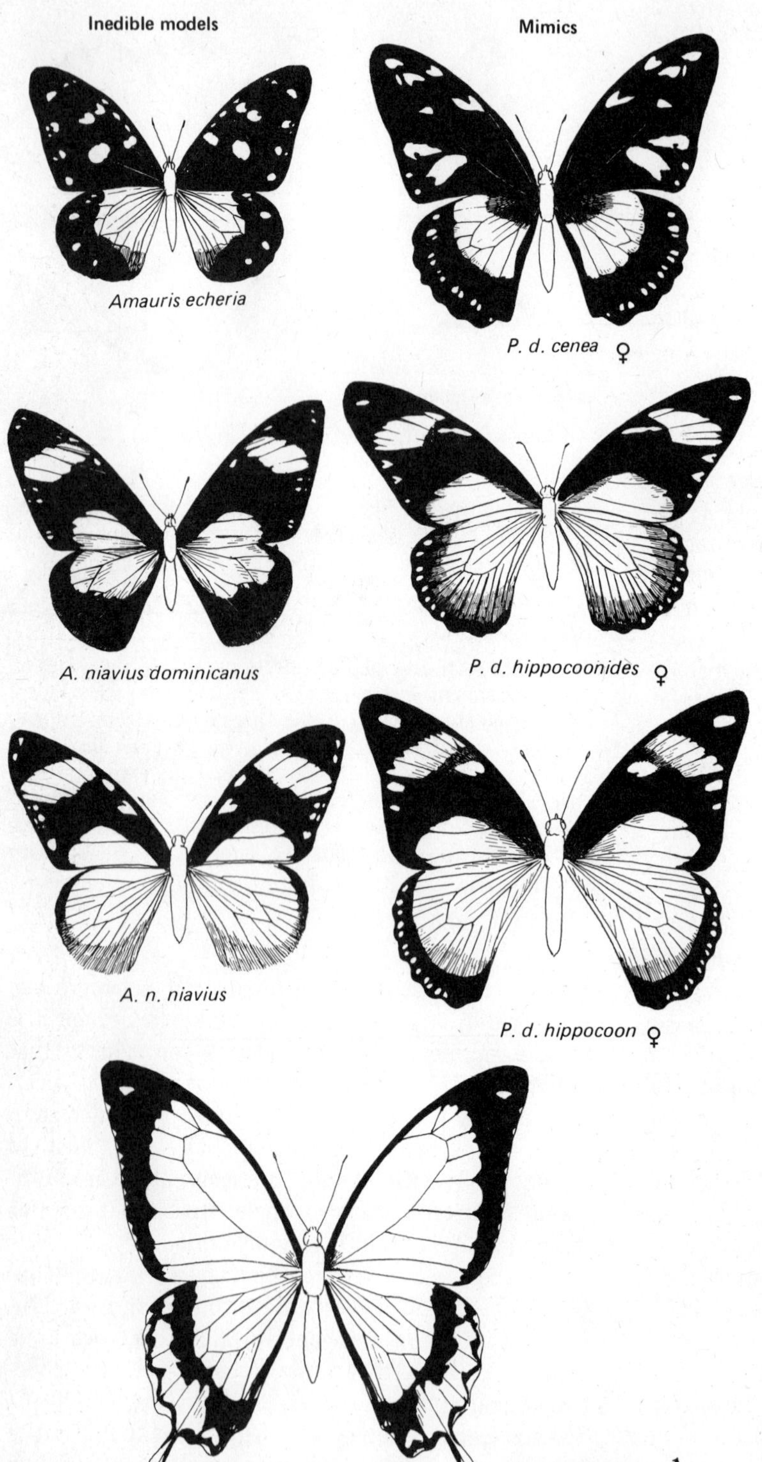

FIGURE 2.7 Disruptive selection in the African swallowtail butterfly *Papilio dardanus.* The African swallowtail is widely distributed across the continent as a number of races, including *dardanus* in west and southwest Africa, *cenea* in southern Africa, and *polytropus* and *tribullus* in east Africa. Males of all races are nonmimetic. Females of each race (right) have wing patterns and coloration that mimic inedible species in their own region.

fortis) on the 40-ha islet of Daphne Major, the Galápagos, from 1975 through 1978. During the 1970s, the island received regular rainfall (127–137 mm), resulting in an abundance of seeds and a large finch population (1500 banded birds). In 1977, however, only 24 mm of rain fell, and the island experienced drought. Plant production, and so seed production, which furnished the major food in the dry season, declined drastically. Small seeds in particular declined in abundance faster than large seeds, increasing the average size and hardness of available seed. The finches—which, in general, fed on smaller seeds—were forced to turn to the larger ones, ignored in normal years. Larger birds in the population ate the larger seeds, while smaller birds apparently had difficulty finding food. Large birds, especially males with large beaks, survived best because they were able to crack large, hard seeds. Females experienced heavy mortality. Overall, the population declined 85 percent from mortality and possibly emigration (Figure 2.8).

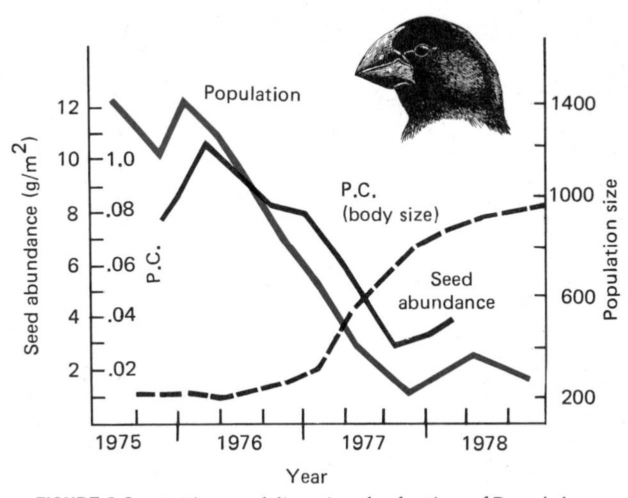

FIGURE 2.8 Evidence of directional selection of Darwin's medium ground finch *Geospiza fortis*. One graph presents the population estimate on the island of Daphne Major based on censuses of marked populations (standard deviation lines deleted), and a second graph estimates seed abundance, excluding two species of seeds never eaten by any Galápagos finches. Populations declined in the face of a seed scarcity brought about by a prolonged drought. The dashed line shows changes in principal component (P.C.) score 1, an index of body size. Note how the body size of birds, especially the males, increased during the drought period. That suggests that smaller-bodied birds were being selected against and large-bodied birds were being favored. Results suggest that the most intense selection in a species occurs under unfavorable environmental conditions.

This example suggests that natural selection exerts its greatest influence during periods of environmental stress or selection "bottlenecks" during a small portion of an organism's life history. It also suggests how a small, isolated, relatively sedentary, morphologically variable population of a species under selective pressure of a variable environment can experience a rapid evolution.

To cite another example, roadside populations of the common bitterweed *Helenium amarum* appear to be diverging from field populations of the same species. Bitterweed is common on disturbed sites throughout the southeastern coastal plain and piedmont of the United States. It is a winter annual whose seeds germinate in autumn immediately after seedfall. The plants overwinter as rosettes and in spring send up flower stalks after the daylength has exceeded 13 hours. Populations of roadside plants are under a different selection pressure than old field plants—roadside mowing. Mowing seems to favor rapidly growing plants that are shorter, branch at a lower level, and bear more seeds. These and other differences are retained by progeny grown in a greenhouse.

In any kind of selection, the characteristics selected are not necessarily the best of all possible traits but rather the most suitable of those available. Characteristics evolved under one set of circumstances may be inappropriate for subsequent environments. Present individuals are products of past selection and are stuck with options their ancestors have provided them.

CONSTRAINTS

Natural selection accounts for only a part of evolutionary change. Other forces help determine which organisms leave descendants and which do not. In doing so these forces influence evolutionary change, constraining or modifying the effects of natural selection.

One such force is the *founder effect*. It may come about when a fraction of a population is isolated from its parent population or when a patch of new or previously unoccupied habitat is colonized by a few individuals. These individuals carry only a small sample of the gene pool of the parent population. If chance is involved in the selection of founding individuals, the new populations will differ from the parent population and from each other. All genes in subsequent populations will stem from the few carried by the founders and from mutations and immigrants. The small size of the population enforces inbreeding among related individuals. Recessive genes, once shielded by the heterozygous condition, become exposed. Some may be deleterious, reducing survival. Other homozygous genes, at a selective disadvantage in the old environment, may enable the carrier to adapt to the new environment. Genes neutral in effect in the parent population also may advance the fitness of founder individuals. A small sample size and a limited diversity of genetic material may have a powerful influence on the development of a new species.

Also altering the genetic composition of a small population is *genetic drift*, a random change in gene frequency. The gene pool of each succeeding generation is a sample of the parental population that gave rise to it. As such, it is subject to sampling error. As in any sample, error is greatest in small populations. Eventually, that error in small populations results in a loss of genetic diversity and an increase in homozygosity. If drift lowers the fitness of the population, it will become extinct; but if the genes involved confer a selective advantage in the new environment, they will enhance the fitness of the population. Thus chance may play an important role in evolutionary change.

Another constraint is the nature of the original model and its embryological development. Evolution through natural selection has to act on already developed and more or less completely organized systems. Evolutionary processes can only tinker with the systems. What changes take place have to be within the limits imposed by the original. In general, response to selection will follow the line of least resistance, the easiest route, not necessarily the optimal one. Further, each population has its own limited genetic resources on which selection can act. The direction of evolution depends in part on those genes exposed to selection.

Organisms themselves can influence the direction of evolution. That is especially true among interacting pairs of species such as predator and prey, parasites and hosts, and competitors (see Chapters 16 and 17). In those cases one species acts as a selection force. In turn, the other reciprocates. For example, a host species over some period of time may evolve a way to resist attacks of a parasite. The parasite then evolves new ways of breaching the host's defenses.

Such an interaction, in which two genetically independent but ecologically related species influence evolutionary changes in each other, is *coevolution*. Coevolution may drive a member of a pair to become a specialist, which increases its probability of extinction because its narrow requirements may not be available in the future. Coevolution involves an interacting pair, which often can be difficult to prove in the field.

In other instances coevolutionary response is general. For example, plants may have evolved chemical defenses against herbivorous insects as far back as the Cretaceous period and transmitted those defenses to modern descendants. These chemical defenses now act to protect plants against modern insects. This evolution of a particular trait in a number of plants was guided by insect herbivory in the past, and the evolution of insects was guided by some prior evolution of plant defenses. The process of traits evolving in a number of species in response to traits in several other species is known as *diffuse coevolution*. The genetic changes in the interacting species are not highly coupled.

GENETICS OF SMALL POPULATIONS

Populations, however widespread, consist of rather tight-knit local groups or demes. The white-tailed deer, common as it is throughout its range, consists of fairly independent groups. This lack of interchange probably results from the matriarchal society of deer, the tendency of females to remain close to their birthplace, and short breeding movements even by males. Although such behavior suggests that inbreeding should be common, apparently enough movement of males among demes takes place that genetic variability is retained. In fact, the white-tailed deer exhibits high variability throughout its range. Collectively, demes across a region make up the population gene pool.

Now suppose that forest clearing, urban spread, industrial development, road construction, and the like fragment the population, eliminating some demes and isolating others. That is happening to many populations of plants and animals worldwide. Each isolated population represents only a sample of the total gene pool. The isolated population is now subject to inbreeding, genetic drift, or both.

INBREEDING

Inbreeding is mating between relatives, and in small populations individuals may be forced into this situation. With inbreeding, mates on the average are more closely related than they would be if they had been chosen at random from the population. Inbreeding occurs because of small populations, close proximity of potential mates, ecological preferences, and the like. Because inbreeding affects all loci, the outcome of inbreeding is an increase in homozygosity and decline in heterozygosity. The frequency of the genotypes becomes skewed toward the two homozygotes, but the frequency of the alleles does not change, as predicted by the Hardy-Weinberg law.

The extreme form of inbreeding is self-fertilization, which takes place among some plants. This extreme provides the basis for comparisons with less intense forms of inbreeding. With each generation of inbreeding, the homozygotes *AA* and *aa* breed true, and offspring from heterozygotes *Aa* will be one-half heterozygous *Aa*, one-quarter homozygous *AA*, and one-quarter homozygous *aa*. The same situation will occur the next generation. Add these new homozygotes to those already in the population, and you will arrive at the correct conclusion that eventually self-fertilized populations will become exclusively homozygous, *AA* and *aa*.

The more usual situation is close inbreeding, usually between brother and sister, parent and offspring, and first cousins, individuals who share a number of like genes. However, the possession of two like alleles does not imply inbreeding. Two alleles in a homozygote may be alike because of independent mutations. Inbred individuals possess alleles identical by descent. They can be traced back to a common ancestor. Each receives one-half of their genes from each parent, but each would receive a somewhat different sample (unless they are identical twins).

In normally outcrossing populations, close inbreeding is detrimental. Rare, recessive, deleterious genes become expressed. They can result in death, decreased fertility, smaller body size, loss of fertility, loss of vigor, reduced fitness, reduced pollen and seed fertility in plants, and the like. These consequences are termed *inbreeding depression*.

Of course, not all inbreeding is bad. Occasional inbreeding will fix certain rare alleles that otherwise might be lost. Inbreeding is used extensively in the breeding of domestic plants and animals to fix certain desirable genes that will breed true. These inbred lines are then outcrossed to produce hybrid vigor.

In nature close inbreeding is rare, less than 2 percent for natural populations for which there are data. Certain safeguards exist in nature to reduce inbreeding. They include spatial separation, or differences between sexes in the dispersal of young. One sex stays behind; the other leaves. Monogamous mating habits and the frequent loss of a mate during mating season or between seasons reduces inbreeding in birds. Female mammals tend to stay near their birthplace and young males leave or are driven away. Among prairie dogs, young males leave the family group before breeding and adult males may move to new breeding groups if adult daughters are in the home place. (Although dispersal does reduce inbreeding, sex-biased dispersal may have evolved for other reasons, such as enhanced reproductive success.) Kin recognition may also reduce close inbreeding. Because of their close association in early life, siblings recognize one another over time. Females mate with unrelated males, leave the group if a related male returns, or fail to come into estrus if their father is in the group. All these safeguards break down if the population is small and highly isolated.

GENETIC DRIFT

Small populations are more likely to experience genetic drift than inbreeding. As noted before, a large population is made up of subpopulations or demes. Each subpopulation has its own sample of the gene pool of the larger population. If the subpopulation is isolated or reduced, it will experience genetic drift. Over time each mating will include only a sample of the already small sample of the population's genes. If the population remains small, this sampling error will continue

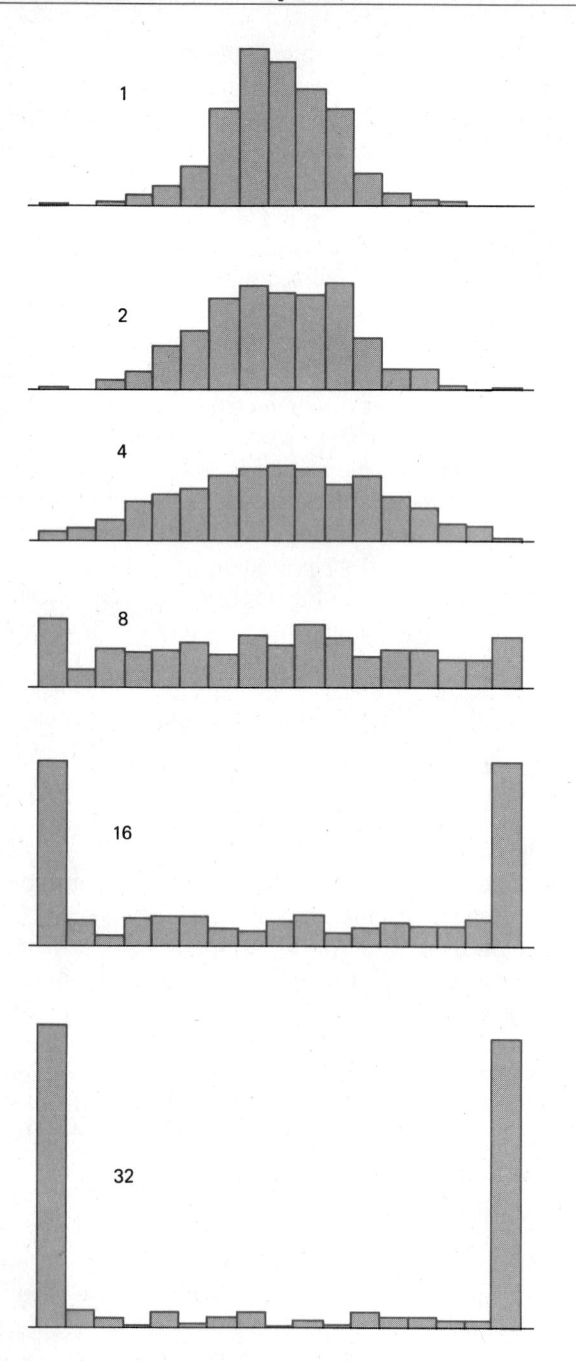

from one generation to another, because only a part of the population is involved in breeding. Through time some genes will become fixed or homozygous for one allele, other alleles will become lost, and some will continue to segregate. Allele frequencies spread out progressively as the proportion of fixed genes steadily increases (Figure 2.9), ultimately resulting in a homozygous population. The smaller the population, the more rapid genetic drift becomes.

Genetic drift is akin to inbreeding. The major difference is that inbreeding involves nonrandom mating, whereas genetic drift involves random mating. Of course, small populations experiencing genetic drift may also experience some inbreeding simply by chance. Both inbreeding and genetic drift result in an increase of homozygosity and a decrease of heterozygosity.

Measurements of genetic drift are based on an ideal population, characterized by a constant population size, equal sex ratios, equal probability of mating among all individuals, and a constant dispersal rate. Most populations, of course, are not ideal. They experience age-related differences in reproduction, and particularly in polygamous populations, the ratio of breeding males to females is unequal. In such populations the number of males is more important than the number of females in determining the amount of random drift. For this reason the actual size of a small population or a subpopulation is of little meaning. Of greatest importance is the genetically *effective population*

FIGURE 2.9 A computer simulation (Monte Carlo) of the dispersal of gene frequencies among 400 hypothetical populations over 32 generations. The genetic model had the following conditions: (1) Each population of eight diploid individuals—four randomly formed pairs—is constant from generation to generation; (2) each individual mates once and the number of offspring produced by each mating varies—a Poisson distribution with a mean of two; (3) selection and mutation are absent; (4) and each population started as 2 *AA*, 4 *Aa*, and 2 *aa*, so the initial gene frequency was 0.5 for each allele at a single autosomal locus. The populations were classed according to gene frequencies in generations 1, 2, 4, 8, 16, and 32. Because of chance variations—random genetic drift—in such small populations, gene frequencies show an increasing spread toward fixation of one allele or the other in most of the populations. This simulation demonstrates why small populations exhibit fixed genes and a lack of genetic diversity.

size, N_e. N_e is not the same as the actual number of breeding individuals, N. Unless the sexes are equal, N_e is less than N. The effective population size is defined as the size of an ideal deme (a randomly breeding one with a 1:1 sex ratio and with the number of progeny per family randomly distributed) that would undergo the same amount of random genetic drift as the actual population.

In a monogamous population in which one male mates with one female, all offspring are less related than in a polygamous population with, say, a ratio of one breeding male to four females. In the latter situation the offspring would be one-half or full sibs. The chance of an allele being lost or fixed and thus the amount of genetic drift is much greater in a polygamous population.

Population geneticists define the effective population size by the following formula:

$$N_e = \frac{4N_m N_f}{N_m + N_f}$$

where N_m and N_f are the numbers of breeding males and females, respectively. As the disparity in the ratio of males to females widens, the effective population size diminishes.

Consider a population of white-tailed deer consisting of 100 adult does and 40 adult bucks. The actual size of the population is 140, but because of the unequal sex ratio, the effective population size is $4(40 \times 100)/140 = 114$. In other words, an ideal population of 114 deer will experience the same amount of genetic drift or sampling error as the actual population of 140 deer. Genetically the actual population is smaller than the actual numbers imply.

Actual populations are dynamic, fluctuating in numbers over time. Under adverse environmental conditions or a sudden loss of habitat, the population may decline sharply or "crash." The survivors of the crash, the progenitors of future populations, possess only a sample of the original gene pool. This sharp reduction in numbers tends to reduce the average effective size over time and create a population *bottleneck* (Figure 2.10). This bottleneck can severely reduce genetic diversity in the remaining population and future generations. An example is the northern elephant seal (*Mirounga angustirostris*). Fur hunters reduced the

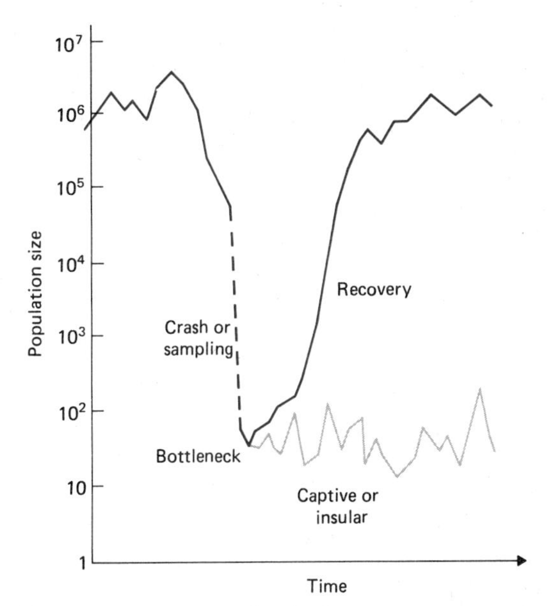

FIGURE 2.10 A population faced with an environmental catastrophe enters a "bottleneck," in which the surviving population consists of only a sample of the total gene pool. If the population makes a complete recovery, it may lack the genetic diversity found in the original population. If the population remains small, as in a captive or insular (fragmented, isolated) situation, it is subject to random genetic drift.

elephant seal population to about 20 animals in 1890. Once protected, the population has grown to over 30,000 individuals. Biologists examined the genetic variation among the seals by electrophoresis. They found no variation in a sample of 24 electrophoretic loci. In contrast, the population of southern elephant seal, never severely reduced, still retains high genetic variability. The lack of genetic variation in the northern elephant seal increases its vulnerability to environmental change.

DISPERSAL

Unless separated by a wide expanse of inhospitable habitat, some interchange takes place among subpopulations of organisms. If these immigrants enter the breeding population, they introduce a different genetic sample that tends to reduce or slow genetic drift and helps maintain genetic diversity. The influence of dispersal depends upon effective population size, amount of immigration, and the degree of monogamy and polygamy in the population; but little genetic interchange is needed to reduce genetic drift. Population geneticists believe that one successful immigrant per generation is the minimum needed to slow genetic drift and five immigrants the maximum, depending upon whether the population is monogamous or polygamous. Just as inbreeding and drift take place more swiftly in a polygamous population, so also will the spread of new genes occur more rapidly in such a population, especially when the harem size is large. New genes spread much more slowly in a monogamous population.

VIABLE POPULATIONS

The concepts of effective population size, inbreeding, genetic drift, and dispersal are of paramount importance in the conservation of species and maintenance of biological diversity. Rapidly increasing populations, fragmentation and elimination of habitats, and poaching are reducing populations of increasing numbers of species. Many are precariously low, including the black rhinoceros, Sumatran rhinoceros, tiger, chimpanzee, tamarin, red-cockaded woodpecker, spotted owl—the list could go on and on. Already the gene pool has been depleted, and the question is what size of populations is needed to save the species.

A certain threshold number of individuals is necessary to ensure the persistence of a subpopulation in a viable state for a period of several hundred years. This threshold number has been termed the *minimum viable population*. This population must be large enough to cope with chance variations in individual births and deaths, random series of environmental changes, genetic drift, and catastrophes.

Conservation geneticists consider a minimum viable population to be around 500 individuals with an absolute minimum of 50. Any smaller populations are subject to serious genetic drift and the loss of genetic variability needed to track environmental changes (Figure 2.11). This rule of thumb carries its own dangers. Many populations are already below the minimum viable population size. Do we give up on these species? The minimum viable population is no longer a useful rule of thumb, nor does effective population size lead to a magic number. The whooping crane was well below a minimum viable number, as was the Australian orange-bellied parrot before restoration began. Both are recovering. Nevertheless, estimates of viable populations suggest that for some species the number of individuals is very large, necessitating large reserves and intensive management.

Intensive management probably will be required to save many species, scattered as they are in small populations in islands of remaining habitat. Such manage-

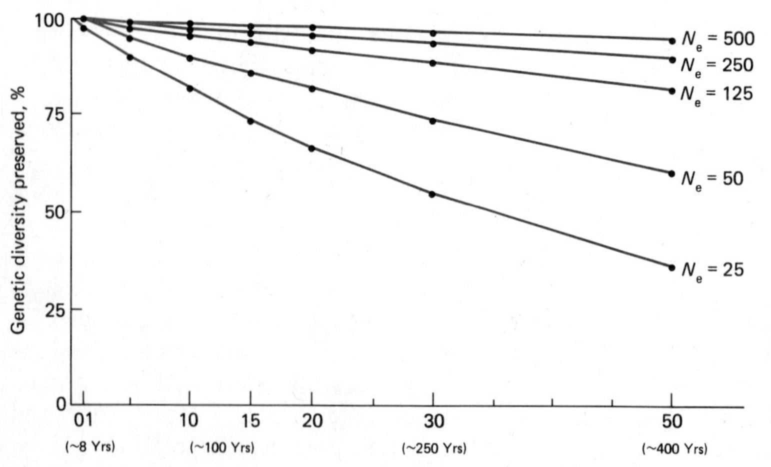

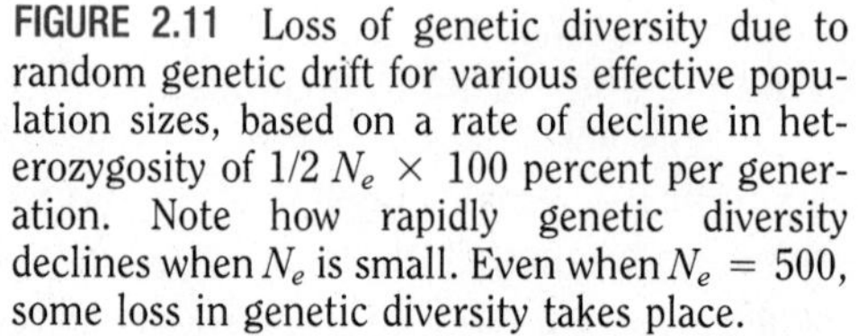

FIGURE 2.11 Loss of genetic diversity due to random genetic drift for various effective population sizes, based on a rate of decline in heterozygosity of 1/2 N_e × 100 percent per generation. Note how rapidly genetic diversity declines when N_e is small. Even when N_e = 500, some loss in genetic diversity takes place.

ment practices may include artificially dispersing breeding individuals among small populations to reduce inbreeding depression, maintaining a number of smaller populations to ensure some genetic variability, monitoring genetic variability, attaining viable population sizes as soon as possible, equalizing genetic contributions of founders, and undertaking rigorous efforts to preserve and expand habitats. It is much less costly in the long run to maintain viable habitats and ecosystems for species than to undertake artificial last-ditch efforts to save them.

Summary

Since its inception, life on Earth has changed through geological time. Land masses moved apart, seas drowned or retreated from continental land masses, mountains rose, climates changed. In such a changing environmental background, life evolved from simple forms to highly complex plants and animals. Certain groups flourished and disappeared, replaced by other forms more successful in new environments. These changes through time were the outcome of evolutionary processes.

The basis of evolution is genetic variation found in organisms as reflected in their phenotypes. Theoretically, in biparental populations variations in gene frequencies and gene ratios remain in equilibrium under certain conditions: random reproduction, equilibrium in mutation, no gene flow from one population to another, and large population size. In nature such conditions do not occur. Populations are often local units or demes that vary in size and are subjected to their own set of selection pressures. Mutations do occur, although they change gene composition slowly. In sexual reproduction, genes are scrambled and recombined in an immense number of variations. Mating and survival are nonrandom. All add up to a marked departure from genetic equilibrium, a change in the frequency of the gene pool. That change over time is evolution.

The direction evolution takes depends upon a number of conditions and events. One is natural selection—nonrandom reproduction. The successful are not those that survive but those that leave behind reproducing offspring. Another is population size. Small populations offer limited genetic variation on which selection can work. The nature of the organism, its pattern of development, and its structure place limitations on evolutionary change.

Natural selection tends to eliminate the less fit alleles and favor the most fit. If it favors those close to the population mean, selection is stabilizing and produces little change in the population. It is characteristic of more or less stable environments. If selection favors the extremes, it is disruptive or diversifying. Such selection may occur when segments of a population are subjected to different selection pressures. Selection also may favor one extreme and move the population mean in that direction. Characteristic of a changing environment, directional selection produces the most rapid phenotypic and genotypic changes.

Because of habitat fragmentation and human exploitation of the landscape, populations of many species of plants and animals are being reduced to isolated or semi-isolated small populations. These small populations carry only a sample of the genetic variability of the total population. Two potential outcomes are inbreeding and genetic drift.

Inbreeding, mating between relatives, brings out hidden genetic variation and increases homozygous rare alleles that often result in reduced fecundity, viability, and even death. The alleles involved at a particular locus are identical by descent.

Small populations are subject to random genetic drift, chance fluctuations in allele frequency as a result of random sampling among gametes. Drift results in the accumulation of fixed populations lacking in genetic variability. Genetic drift mimics inbreeding homozygosity, but it is the result of random mating, and alleles involved are not necessarily identical by descent. The effects of random drift are measured in terms of the heterozygosity of individuals.

Influencing the effects of genetic drift is the effective population size, N_e, the size of an ideal population with the same rate of increase in homozygosity as the population in question. In monogamous populations with an equal number of breeding males and females, the actual population size and the effective population size are the same. In polygamous situations, a wide disparity in sex ratios can strongly reduce the effective population size and increase genetic drift.

Exchange of individuals among populations (emigration and immigration) can reduce genetic drift and inhibit genetic divergence among subpopulations. A few exchanges per generation are sufficient to reduce random genetic drift. In order for a population to maintain genetic diversity and thus persist over time, it must not fall below a threshold number termed the minimum viable population. Population genetics has become important for the preservation and management of wild species in the face of human development of planet Earth, and the adverse effects of population fragmentation.

Review and Study Questions

1. Distinguish among natural selection, adaptation, and evolution. How do they relate?
2. What is meant by fitness?
3. What is a deme? What is the significance of the deme in natural selection and evolution?
4. What is the mean? Mode? A frequency distribution? How do they relate to variation in the physical characteristics of an organism?
5. Explain briefly the structure of a gene. What is DNA? RNA?
6. Distinguish between a genotype and a phenotype. On what does selection work? What is phenotypic plasticity?
7. Define allele, locus, homozygous, heterozygous, diploid, haploid, meiosis, gene pool.
8. What are the major sources of variation in the gene pool?
9. What is a mutation? Distinguish between a point mutation and a chromosomal mutation.
10. Contrast directional, stabilizing, and disruptive selection.
11. What is genetic drift? Founder's effect? Inbreeding? Population bottleneck? What is their significance? How do they act as constraints on natural selection?
12. What is coevolution?
13. What is meant by effective population size? Viable population? Minimum viable population?

*14. Consider the problem of introducing a new male into a small isolated population for the purpose of reducing inbreeding depression. What are some of the problems involved for his successful introduction and incorporation into the population? You might want to refer to Chapter 13 for some ideas.

*15. What problems, other than ecological ones, are involved in maintaining or reestablishing populations of endangered species? Do a report on the spotted owl, the desert tortoise or the sea turtle, reintroduction of wolves to Yellowstone, maintaining elephants in African national parks, or other species. Show how conservation of endangered species involves more than biology and ecology.

Selected References

Brown, J. H., and A. C. Gibson. 1983. *Biogeography*. St. Louis: Mosby. Good, brief discussion of past physical geography of Earth.

Chepko-Sade, B. D., and Z. T. Halpin, eds. 1987. *Mammalian dispersal patterns; The effect of social structure on population genetics*. Chicago: University of Chicago Press.

Frankel, O. H., and M. E. Soulé. 1981. *Conservation and evolution*. Cambridge, England: Cambridge University Press. A good introduction to population genetics relating to small populations.

Futuyma, D. J. 1984. *Evolutionary biology*. Sunderland, MA: Sinauer Associates. A superior treatment of the subject.

Hartl, D. 1988. *A primer of population genetics*. Sunderland, MA: Sinauer Associates. An excellent introduction.

Schonewald-Cox, C. M., S. M. Chambers, B. Macbryde, and W. L. Thomas, eds. 1983. *Genetics and conservation*. Menlo Park, CA: Benjamin/Cummings. A reference manual for managing wild animal and plant populations.

Soulé, M., ed. 1986. *Conservation biology: The science of scarcity and diversity.* Sunderland, MA: Sinauer Associates. A source book on ecology of small populations and extinction.

Soulé, M., ed. 1986. *Viable populations for conservation*. Cambridge, England: Cambridge University Press. Discussion of minimum viable population concepts and management of small populations.

Soulé, M., and B. Wilcox, eds. 1980. *Conservation biology; an evolutionary-ecological perspective*. Sunderland, MA: Sinauer Associates. A basic introduction.

Western, D., and M. C. Pearl, eds. 1989. *Conservation for the twenty-first century*. New York: Oxford University Press. Excellent overview of the problems of and possible solutions to the conservation of species and their habitat. A major reference.

Wilson, E. O., and W. H. Bossert. 1971. *A primer of population biology*. Sunderland, MA: Sinauer Associates. A good, basic introduction to population biology.

CHAPTER

3

Species and Speciation

Outline

Objectives

On completion of this chapter, you should be able to:

1. Define a species from three viewpoints: morphological, biological, evolutionary.
2. Explain what is meant by a cline, an ecotype, and a geographic isolate.
3. Define an isolating mechanism and describe several types.
4. Define speciation and distinguish three types.
5. Explain how polyploidy can produce an instant species.
6. Explain adaptive radiation and convergence.
7. Compare gradualism and punctuationalism.
8. Discuss the rate of evolution.

The discussion up to this point has concerned changes within a single species resulting from various evolutionary forces—genetic variation, mutation, and environmental changes. Of greater ecological interest is the splitting of one species into two or more genetic lines. Each species so formed might diverge still further into more new species. Such diversification eventually leads to the formation of new taxa, such as genera and families.

WHAT IS A SPECIES?

Field guide in hand, we have little difficulty distinguishing a robin from a wood thrush or a white oak from a red oak. Each has certain morphological characteristics that set it apart from other organisms. Each is an entity, a discrete unit to which a name has been given. That is the way Carl von Linné, who gave us our system of classification, or taxonomy, saw a great number of plants and animals. He, like others of his day, regarded the many organisms as fixed and unchanging units, the products of special creation. Differences and similarities were based on color pattern, structure, proportion, and other characteristics. From these criteria species were described, separated, and arranged into groups. Each species was monotypic; that is, it contained only those individuals that fairly well approximated the norm or the type for the species, the specimens from which the species was described. Some variation was permissible, but these variants were considered accidental, although some slight changes within the species were admitted to be possible. This *morphological species* is a classical concept still alive, useful, and necessary for classifying the vast number of plants and animals (Box 3.1).

Later, the studies of Darwin on variation, of Wallace on geographical distribution, and of Mendel on genetics emphasized that variation within a species was the rule. Because of sexual dimorphism, male and female of the same species looked like separate species. Many closely related forms replaced each other geographically. They intergraded so smoothly it was difficult to separate them precisely. A better definition of a species was needed.

Ernst Mayr, an evolutionary biologist and taxonomist, advanced the idea of the *biological species*, a group of actually or potentially interbreeding populations that is reproductively isolated from other groups. Interbreeding individuals live together in a similar environment in a given region and under similar ecological relationships. The individuals recognize each other as potential mates. They are a genetic unit, in which each individual holds for a short period of time a portion of the contents of an intercommunicating gene pool.

This definition, also known as the isolation species concept, has its limitations. It applies only to bisexual organisms, excluding asexual ones, and it overemphasizes between-population reproductive isolation. In reality a species consists of numerous local populations upon which natural selection acts. A species is made up of an array of local populations, each somewhat different from the other, but more or less linked by gene flow. These local populations experience enough interbreeding to provide a continuity and intergradation in characteristics from one population to another.

From an evolutionary point of view a species consists of populations that are united by gene flow, facilitated by species recognition cues and other phenomenona that promote genetic relatedness, that are separated from other species by reproductive isolating mechanisms, and that share a common evolutionary fate. Other distinguishable populations within a region that do not interbreed with that species are considered separate species. This definition, too, has shortcomings. A foolproof definition does not appear to be possible.

Species may be sympatric, allopatric, or parapatric. *Sympatric species* occupy the same area at the same time and thus have the opportunity to interbreed. The fact that they do not marks them as a "good" species, reproductively isolated from others. *Allopatric species* occupy areas separated by time and space. Because they do not have the opportunity to meet similar species, there is no indication whether they are capable of interbreeding with other allopatric species. Only if the barriers separating them are broken, allowing them to come together, can you test reproductive isolation. Often two allopatric species are not reproductively isolated, and when the barriers are eliminated, the two species behave as one. Such is the case, for example, with the red-shafted flicker and the yellow-shafted flicker. The two species, once allopatric, interbred when they became sympatric and are now considered one species, *Colaptes auratus*. *Parapatric species* are species or incipient (beginning) species that are in contact over a narrow zone along the borders of their range, local or regional. Minimal genetic interchange takes place between the two populations, and hybrids formed are at strong selective disadvantage.

Some sympatric species are *sibling species*—ones that are similar in appearance or morphology but do not interbreed. They may differ in behavior, physiology, or chromosomal structure. To the human eye the species may be virtually indistinguishable, but the differences are apparent to the animal. Examples of such sibling species are the three species of 17-year periodical cicadas, *Magicicada septendecim, M. cassini,* and *M. septendeculas*, and the three species of 13-year periodical cicadas, *M. tredecim, M. tredcassini*, and *M. tredecula*. The members of each group emerge in synchrony with each other. They maintain their separateness by differences in type and timing of their choruses. The chorus of *M. septendecim* is most intense in the morning, and it is an even, monotonous, nonsynchronized buzzing. The chorus of *M. cassini* is most intense in the afternoon and is a shrill, sibilant, synchronized buzzing that rises and falls in intensity. The chorus of the small *M. septendecula* is most intense about midday and is a more or less continuous, nonsynchronized repeating of short, separate buzzes with no regular fluctuations in pitch or intensity. The behavior and choruses of the 13-year species are similar.

WHY DEFINE A SPECIES?

Why, you may ask at this point, be concerned with the definition of a species? In this age of exploding human population growth, habitat destruction, and declining species populations increasingly vulnerable to extinction, the survival of many species depends upon human intervention and management. The species, then, becomes the evolutionary significant unit for conservation management. Are separate and isolated populations of a species, often called subspecies, genetically the same? Or are they different enough to warrant their consideration as a management unit? Did we lose a "species" when the dusky seaside sparrow, a taxonomically recognized subspecies of the seaside sparrow, went extinct? Is it in the best interests of species conservation to introduce individuals from remote populations into the threatened local population? Are they genetically similar, or are they separated by certain chromosomal differences? For example, the orangutan populations threatened on Sumatra and Borneo are chromosomally distinct, even though they look the same. When individuals of one population are introduced into the population of another, the hybrids between the chromosomal types experience outbreeding depression. A similar situation exists with spider monkeys, dik-diks (a small African antelope), and others. Not knowing such information, we can do more harm than good in our conservation efforts. The definition of a species is also critical as a legal issue, as it relates to efforts to preserve habitat and modify human activities to save threatened and endangered species. The concept of the species has an applied as well as a theoretical interest.

GEOGRAPHIC VARIATION IN A SPECIES

Because of the widespread variation in many morphological, physiological, behavioral, and genetic characteristics in a widely distributed species, significant differences often exist among populations in different regions. One group may differ, more or less, from other local populations. The greater the distance between populations, the more pronounced the differences may become. Geographic variants reflect the selective environmental forces acting on various genotypes, adapting each population to the locality it inhabits. Geographic variation shows up as clines, ecotypes, and geographic isolates.

CLINES

A *cline* is some measurable gradual change over a geographic region in the average of some phenotypic character such as size and coloration, or in gene and genotypic frequency. Clines are usually associated with an ecological gradient such as temperature, moisture,

BOX 3.1

Naming and Classifying Organisms

Although humans had been classifying plants and animals for centuries, there was no uniformity about any classification scheme. Botanists and naturalists had been trying since Aristotle's day (384–322 B.C.) to come up with a useful method. No system devised was satisfactory until a young Swedish naturalist and medical doctor, Carl von Linné, who Latinized his name to Carolus Linnaeus (1707–1778), developed one that is still in use today.

Linnaeus, who taught medicine and botany at the University of Uppsala, had a passion for classifying things—plants, animals, diseases, minerals, and many other items that came to his attention. He introduced a binomial, or two-name, system of naming plants and animals—a system that finally brought order out of chaos and led to a standardized method of naming all living organisms in Latin. Latin was then the universal language of science. Because it was a dead language, it would never change in grammar, syntax, or words. That gave it a great degree of uniformity and stability.

Linnaeus explored in Lapland and northern Europe; he had collectors send him plants from other parts of the world. Out of this research came a book in 1735, *Species Plantarum*. In it he listed and named all plants then known to science, giving each plant two Latin names—a generic name and a species name. Later, in 1758, he published a tenth edition, *Systema Naturae,* in which he consistently applied the same idea to animals.

Before Linnaeus could assign names to each organism, he had to sort out his plants and animals into broad general groups. First, all living things, he reasoned, belonged to two great groups, or kingdoms: plant and animal. The animal kingdom he subdivided into six parts: mammals, birds, reptiles, fishes, insects, and a catchall group, vermes (or worms). Among the mammals, for example, he noted that some organisms were similar. It was not too difficult to identify a cat as such, or a dog, although there are many different kinds of each. All have certain identifying characteristics. Cats have retractile claws, rounded heads, and well-developed but relatively small, rounded ears. Some have long tails; other have short tails, ear tufts, and a ruff on the cheeks and throat. All the long-tailed cats, then, were placed in one group, or genus, to which Linnaeus gave the Latin name *Felis*, for cat; and to the short-tailed cats he gave the name *Lynx*. Doglike mammals, too, are readily distinguishable, and he could break them down into groups of animals similar in appearance: the sharp-nosed little foxes, the doglike wolves, and their descendants, the true dogs. To the wolves and dogs, Linnaeus gave the name *Canis*, Latin for dog. Each genus then contained a number of different animals, all with some broad structural characteristics in common, yet with differences in the finer details: color, size, thickness of hair, and the like. Thus, the genus could be broken down into still finer groups, the

light, or altitude. Continuous variation results from gene flow from one population to another along the gradient. Because environmental selection pressures vary along the gradient, any one population along the gradient will differ genetically to some degree from the other, the difference increasing with the distance between the populations. For this reason, populations at the two extremes along the gradient may behave as different species. The length of the cline depends upon the gene flow and dispersal distance among the populations.

Clinal differences exist in size, body proportions, coloration, and physiological adaptations among animals. For example the white-tailed deer in North America exhibit a clinal variation in body weight. Deer in Canada and the northern United States are heaviest, weighing more than 136 kilograms (300 pounds), further south in Kansas 93 kilograms (207 pounds), 60 kilograms (130 pounds) in Louisiana, 46 kilograms (100 pounds) in Panama. The smallest, the Key deer in Florida, weighs less than 23 kilograms (50 pounds). Many species of plants exhibit clinal gradation, some individual species. To name them, Linnaeus gave another, second name to the species. The wolf became *Canis lupus*, the mountain lion *Felis concolor*, and the African lion *Felis leo*.

All catlike animals, according to the Linnean system, are grouped together into a still larger classification, the family, which takes its name from one of the genera it embraces. In the case of cats, the genus *Felis* was used. By adding the standardized suffix *idae* onto the root word *Felis*, the cat family became Felidae; and by a similar process the dog family became Canidae. Dogs and cats, skunks and weasels (Mustelidae), raccoons (Procyonidae), bears (Ursidae), seals (Phocidae), and so on were classified into families. All have several general features in common: five toes with claws, shearing teeth, well-developed canine teeth or fangs, and other skeletal structures. These animals were then grouped together and placed in an even higher category, the order (in this case the order Carnivora). All orders of animals giving birth to living young, possessing body hair, and nourishing young with milk were placed in an even higher category, the class Mammalia. All animals having a backbone were grouped in a phylum (in this instance Chordata), belonging to the animal kingdom.

The final result was a hierarchical scheme of classification in which similar species were grouped together and placed in a genus, similar genera were grouped into families, similar families into orders. The complete hierarchical classification of a mountain lion, *Felis concolor*, looks like this:

Kingdom: Animal
  Phylum: Chordata
    Class: Mammalia
      Order: Carnivora
        Family: Felidae
          Genus: *Felis*
            Specific epithet: *concolor*

As time went on, the number of known animals swelled, and the need arose to refine this hierarchy even further. Phyla were divided into subphyla, classes into subclasses and infraclasses, orders into suborders, families into subfamilies (and even enlarged into superfamilies), and species into subspecies.

The principal categories used in plant classification are somewhat different:

Kingdom: Plant
  Division: Tracheophyta
    Subdivision: Pteropsida
      Class: Angiospermae
        Subclass: Dicotyledoneae
          Order: Fagales
            Genus: *Fagus*
              Specific epithet: *grandifolia*

The plant classified here is the American beech, *Fagus grandifolia*.

in size, and others in time of flowering, growth, or other physiological responses to the environment. A number of prairie grasses, among them blue grama (*Bouteloua gracilis*), side-oats grama (*B. curtipendula*), big bluestem (*Andropogon gerardi*), and switch grass (*Panicum virgatum*), flower earlier in the northern and western regions of their ranges and progressively later toward the south and east. The seaside goldenrod (*Solidago sempervirens*) flowers progressively later in the fall from north to south along the Atlantic coast.

ECOTYPES

Some variations show marked discontinuities. Called step clines, they reflect abrupt changes in selection pressures. Such variants are called *ecotypes*. An ecotype is a genetic strain of a population that is adapted to its unique local environmental conditions. For example, a population inhabiting a mountaintop may differ from a population of the same species at the bottom of the slope with few intermediate forms. Often the ecological variants of a species will be scattered like a mosaic. That frequently is the situation where

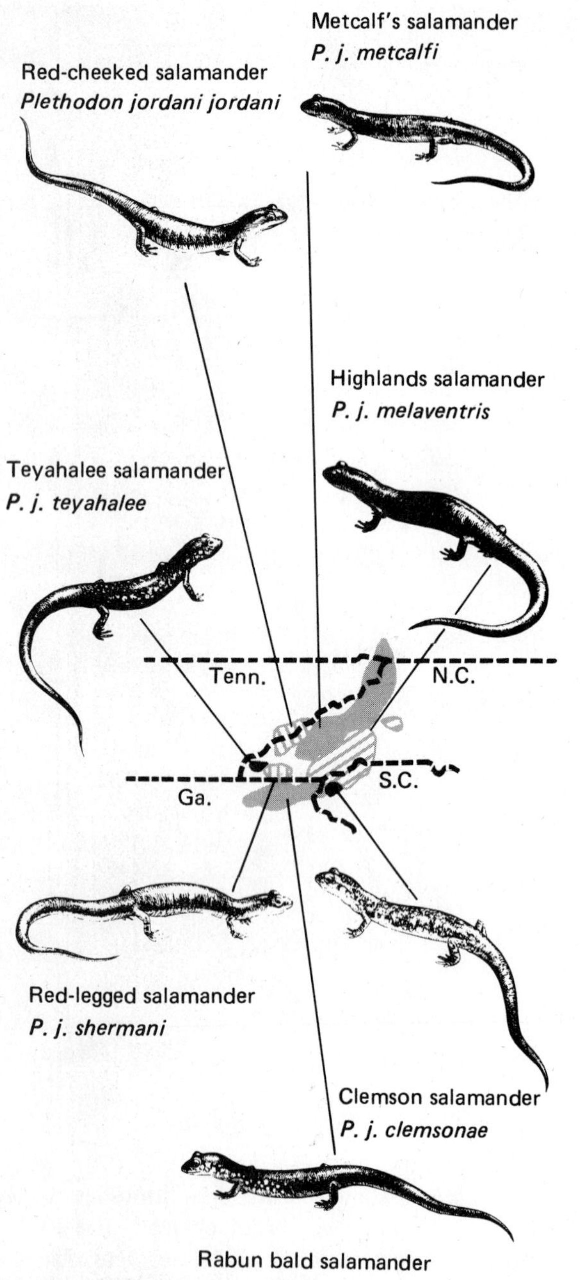

several habitats to which the species is adapted recur throughout the range of the species. Some of these ecotypes may have evolved independently from different local populations. (They are called polytypic ecotypes.)

The yarrow *Achillea* blankets the temperate and subarctic Northern Hemisphere with an exceptional number of ecological races. One species, *A. lanulosa*, occurs at all altitudes in the Sierra Nevada of California. It exhibits considerable variation, an adaptive response to different climates at various altitudes. Populations at lower altitudes are tall and have high seed production. Montane populations have a distinctive small size and low seed production.

GEOGRAPHIC ISOLATES

The southern Appalachian Mountains are noted for their diversity of salamanders, fostered in part by a rugged terrain, an array of environmental conditions, and the limited ability of salamanders to disperse through the mountains. Populations become isolated, preventing a free flow of genes among them. One species, the red-legged salamander, *Plethodon jordani*, is broken up into a number of semi-isolated populations, each characteristic of a particular part of the mountains (Figure 3.1). Groups of such populations make up *geographic isolates*, prevented by some extrinsic barrier—in the case of the salamanders, rivers and mountain ridges—from effecting a free flow of genes with other

FIGURE 3.1 Geographic isolates in *Plethodon jordani* salamanders of the Appalachian highlands. These salamanders originated when the population of the salamander *P. yonahlossee* became separated by the French Broad valley. The eastern population developed into Metcalf's salamander, which spread northeastward, the only direction in which any group member could find suitable ecological conditions. South, southwest, and northwest the mountains end abruptly, limiting the remaining *jordani*. Metcalf's salamander is the most specialized and ecologically divergent and the least competitive. Next the red-cheeked salamander was isolated from the red-legged and the rest of the group by the deepening of the Little Tennessee River. Remaining members are still somewhat connected.

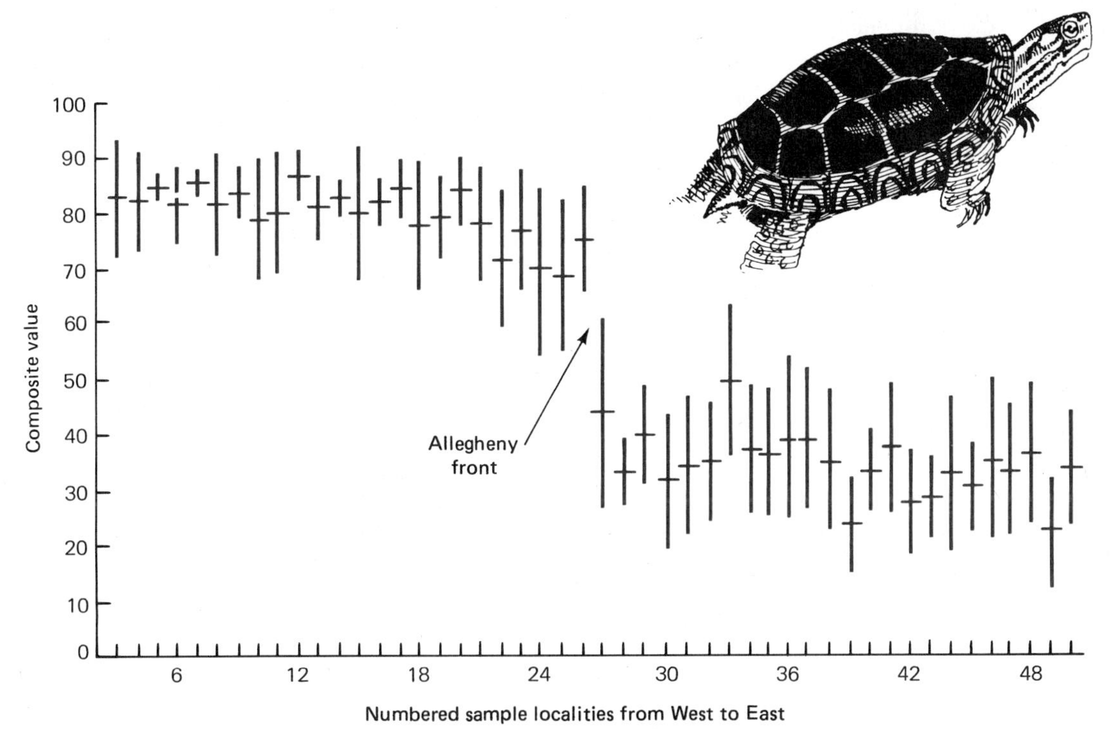

FIGURE 3.2 Effect of a mountain barrier on the subspeciation of the painted turtle (*C. p. picta*) and the midland painted turtle (*C. p. marginata*). North of the high Appalachians, the two subspecies intergrade. In West Virginia the high Appalachians separate them. This separation is illustrated in the plots of composite values of such measurable characteristics as seam alignment, width of plastron, plastron figure, and others. The horizontal bar is the mean value; the vertical bar is the standard deviation (average magnitude of deviations from the center of a normal curve).

subpopulations. The degree of isolation depends upon the barrier, but rarely is it complete (Figure 3.2).

These geographic isolates, and to some extent clinal variants, make up a subspecies, an aggregate of local populations differing taxonomically from other populations of the same species. Subspecies are just that—taxonomic entities. Evolutionary biologists have abandoned the subspecies as an evolutionary category. Strong isolates may have little gene flow between them and may be in the first stages of speciation. Other taxonomic geographical races of a continental species are, more often than not, connected by intermediate forms of intergrades. It is impossible to draw a line that will separate them all into subspecies.

POLYMORPHISM

The occurrence of several distinct forms of a species in the same habitat at the same time is *polymorphism*. It may involve differences in morphological characters and in physiology. The important feature about polymorphism is that the forms are distinct and the characteristic involved is discontinuous. There are no intermediates. Polymorphism may arise from disruptive or divergent selection and appears to be environmentally induced.

A classic example of genetic polymorphism as a result of changing environmental selection is industrial melanism in the peppered moth *Biston betularia* in

England. Before the middle of the nineteenth century, the moth, as far as is known, was always white with black speckling on the wings and body (Figure 3.3). In 1850, near the manufacturing center of Manchester, a black form of the species was caught for the first time. The black form, *carbonaria*, increased steadily through the years until it became extremely common, often reaching a frequency of 95 percent or more in Manchester and other industrial areas. From these places *carbonaria* spread into rural areas far from industrial cities. The form came about through the spread of dominant and semidominant mutant genes, none of which is recessive. The increased frequency and spread has been brought about through the selective pressure of predation. The original typical form of the peppered moth has a light color pattern that makes it inconspicuous when it rests on lichen-covered tree trunks. The dark form is very conspicuous on the lichen-covered trunks and is subject to heavy predation. After the rise of factories the grime and soot of the industrial areas killed or reduced the lichen on the trees and turned the bark a nearly uniform black. On such trees the black form had the selective advantage. In the polluted woods the light form bore the brunt of predation.

FIGURE 3.3 Normal and melanistic forms of the polymorphic moth *Biston betularia* at rest on a lichen-covered tree. The spread of the melanistic form, *carbonaria,* in industrial areas is associated with improved concealment of black individuals on soot-darkened, lichen-free tree trunks. Away from the industrial areas, the light color is most frequent because black individuals resting on lichen-covered trunks are subject to heavy predation by birds.

In the case of the peppered moth, environmental changes converted a disadvantageous allele or a mutant gene into an advantageous one, permitting its spread. Such a polymorphism will persist as long as environmental conditions favor the dark form. In other situations such polymorphisms will exist until the new advantageous form has completely replaced the original form or has so swamped it that the original can be maintained only by recurrent mutation. Such a situation is known as *transient polymorphism*.

Polymorphism may also result from environmental modification of gene action. That is possible only when the two environments—for example, two background colors—are present at the same time in the same place. Environmentally controlled polymorphism, favoring two or more forms, is the optimal expression of the characters concerned. All intermediates are at a disadvantage and are usually eliminated. The North American black swallowtail (*Papilio polyxenes*) and the European swallowtail (*P. machaon*) are good examples. Both swallowtails pupate either on green leaves and stems or on brown ones. Through selection both have acquired a genetic constitution that produces green-colored pupae in green environments and brown in brown environments. Green pupae would be quite conspicuous in winter, but butterflies emerge from the green in late summer. Those in brown pupae do not emerge until the following spring.

The swallowtails are examples of the most common type of polymorphism—*stable* or *balanced*. An apparent optimum proportion of two forms exists in the same habitat and is maintained by environmental heterogeneity. Any deviation in one direction or the other is a disadvantage.

There exists a good deal of circumstantial evidence that genetic polymorphism is related to a patchy environment. Not much experimental evidence exists, however, to support the hypothesis that a heterogeneous environment has a major role in maintaining or causing polymorphism.

HOW SPECIES STAY APART: ISOLATING MECHANISMS

Each spring there is a rush of courtship and mating activity in woods and fields, lakes and streams. Fish move into their spawning grounds, amphibians migrate to breeding pools, birds are singing. During this frenzy of activity each species remains distinct. Song sparrows mate with song sparrows, trout with trout, wood frogs with wood frogs, and few mistakes are made, even between species similar in appearance.

The means by which the many diverse species remain distinct are *isolating mechanisms*. They include morphological characters, behavioral traits, habitat selection, and genetic incompatibility. Isolating mechanisms may be premating or postmating. Premating mechanisms—those that prevent interspecific crosses—include ecological ones: habitat and seasonal isolation; behavior; and mechanical or structural incompatibility. Postmating mechanisms reduce the full success of interspecific crosses.

If two potential mates in breeding condition have little opportunity to meet, they are not likely to interbreed. Habitat selection, even on a local basis, reinforces this isolation among frogs and toads. Different calling and mating sites among concurrently breeding frogs and toads tend to keep the species separated. The upland chorus frog and the closely related southern chorus frog breed in the same pools, but ecological preferences tend to separate—partially, at least—the calling aggregations of the species. The southern chorus frog calls from concealed positions at the base of grass clumps or among vegetational debris, while the upland chorus frog calls from more open locations.

Temporal isolation (differences in the timing of the breeding and flowering seasons) separates some sympatric species. The American toad, for example, breeds early in the season, whereas the Fowler's toad breeds a few weeks later. Fluctuations in environmental stimuli can time mating seasons. Among the narrow-mouthed toads, *Microhyla olivacea* breeds only after rains, whereas *M. carolinensis* is little influenced by rain. Because temporal isolation is incomplete, call discrimination is also involved; nevertheless, some hybridization does occur.

Ethological barriers (differences in courtship and mating behavior) are the most important isolating mechanisms in animals. The males of animals have specific courtship displays, to which, in most instances, only females of the same species respond. These displays involve visual, auditory, and chemical stimuli. Some insects, such as certain species of butterflies and fruit flies, and some mammals possess species-specific scents. Birds, frogs and toads, some fish, and such "singing" insects as the crickets, grasshoppers, and cicadas have specific calls that attract the "correct" mates. Visual signals are highly developed in birds and some fish. Species-specific color patterns, structures, and displays, which give rise to a high degree of sexual dimorphism among such bird families as the hummingbirds and ducks, have apparently evolved under sexual selection. Among the insects, the flight paths and flash patterns of fireflies on a summer night are the most unusual visual stimuli. The light signals emitted by various species differ in timing, brightness, and color, which may range from white through blues, greens, yellows, orange, and red.

Mechanical isolating mechanisms make copulation or pollination between closely related species impossible. Although evidence for such mechanical isolation among animals is scarce, differences in floral structures and intricate mechanisms for cross-pollination commonly exist among plants. If hybrids should occur, they could possess such unharmonious combinations of floral structures that they would be unable to function together, either to attract insects to them or to permit insects to enter the flowers.

These three types of isolating mechanisms—ecological, behavioral, and mechanical—prevent the wastage of gametes, diminish the appearance of hybrids, and permit populations of incipient species to become wholly or partly sympatric.

A fourth type of isolating mechanism, the reduction of mating success, does not prevent the wastage of gametes, but it is highly effective in preventing crossbreeding. If hybrids do result, they may be sterile or at a selective disadvantage and thus be eliminated from the breeding population.

HOW SPECIES ARISE

The great diversity of plants and animals in the world causes us to wonder how all these species arose. Each is adapted to an ecological niche in the ecosystem to which it belongs, and each is genetically independent. The process by which each form becomes genetically isolated from the others, is *speciation*—the multiplication of species. Speciation is accomplished in most plants and animals by an interaction of heritable variation, natural selection, and barriers to gene flow.

ALLOPATRIC SPECIATION

The most usual type of speciation is *allopatric* or *geographic speciation*. The first step in geographic speciation is the splitting up of a single interbreeding population into two spatially isolated populations. Imagine a piece of land, warm and dry, occupied by species A. Then at some point in geologic time, mountains uplift, land sinks and becomes flooded with water, or some great vegetational catastrophe occurs, which splits the piece of land and separates a segment of species A from the rest of the population. The newly isolated segment will now become species A′. It now occupies an area of cool, moist climate in our imaginary land.

Because it represents only a random sample of the population of species A, A′ will possess a slightly different ratio of genetic combinations. The climatic conditions are different; the selective forces are different. Natural selection will favor any mutation or recombination of existing genes that will result in better adaptation to a cool, moist climate. Similar selection for a warm, dry climate will continue in population A on the original land mass. With different selective forces acting on them, the two populations will diverge. Accompanying genetic divergence will be changes in physiology, morphology, color, and behavior, resulting in ever-increasing external differences, until A′ becomes a geographic race. A′, however, is still capable of interbreeding with species A if given the opportunity.

If geographic barriers break down before isolating mechanisms are effective, then interbreeding takes place and the hybrid individuals produced are fully fertile and viable (Figure 3.4). If they are not at a selective disadvantage in competition with the parent populations, the genes of one race will be incorporated with the gene complex of the other. There will be a period of increased variability in the rejoined populations, and new adaptive forms will be established. Eventually, the variability will decrease to a normal amount, and once again there will be a single, freely interbreeding population.

If the barrier remains, however, further evolutionary diversification occurs. The two populations become increasingly different, and isolating mechanisms become more fully established. The two populations have now become *semispecies*, but genetic differences are not sufficient to prevent interbreeding if the two should come in contact. If the barrier falls at this stage, individuals of the two populations may interbreed and produce hybrid offspring. These offspring form a hybrid zone between the two semispecies populations.

Individuals in the hybrid zone may range from character combinations of species A to those of emerging species. Some hybrids may be indistinguishable from one parent stock or the other. Others may show a high degree of divergence. In some situations the hybrid zone may be relatively narrow. In others the hybrids may form a cline between the two parent populations. One or more characters may filter further than others into the range of the semispecies. Some genes acquired through hybridization are retained because they apparently confer some selective advantage, while disadvantageous genes are eliminated. This incorporation of genes from hybrids is *introgressive hybridization*. In other cases hybrids produced among semispecies are less fertile and viable than the parent stock because they contain discordant gene patterns. Their reproductive potential, if they are fertile at all, is low; they produce few offspring. Hybrids are at a selective disadvantage, while any color pattern, voice, behavior, or the like in parent stock that reinforces reproductive isolation will be favored. This selection against hybrids and reinforcement of isolating mechanisms continues until gene flow between the two populations has ceased. Then species A and new species A′ can invade each other's territory, occupy suitable niches—in our example a warm, dry environment and a cool, moist environment—and become wholly or partly sympatric.

A second, and probably more common, form of geographic speciation results from the establishment of a new colony by one founder (a gravid female) or a small number of founders (see the discussion of the founder principle, Chapter 2). Founder individuals may be "surplus" from a rapid population increase in which selection pressures have been relaxed. They may be small populations living on the periphery of the species range that have become isolated from the parent population, or they may be small populations that have found unex-

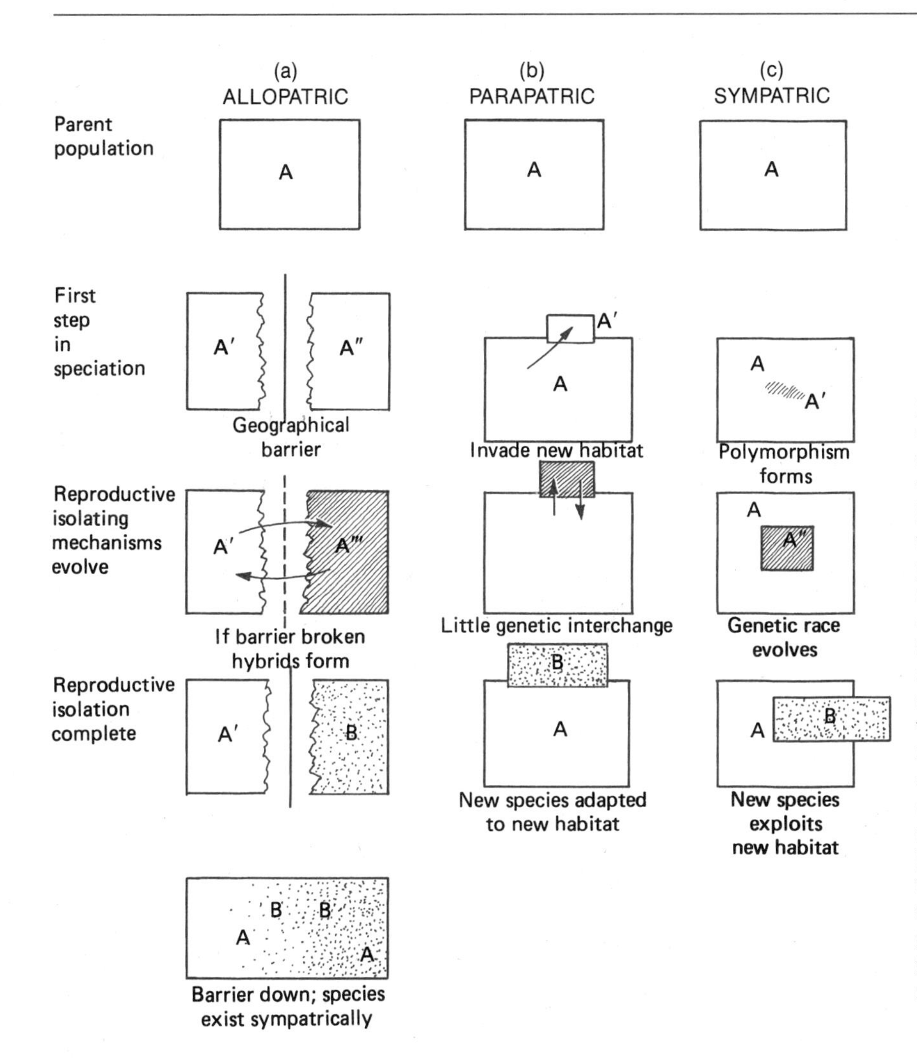

FIGURE 3.4 (a) Allopatric speciation. One species may become two as a result of isolation over a long period of time by a geographical barrier. If the barrier is moved before the species diverged from parental stock, they simply rejoin. If sufficient differences have evolved before the barrier breaks down, hybrids form. If the divergence is complete and the barriers break down, the species can exist sympatrically. (b) Parapatric speciation occurs when reproductive isolating mechanisms arise at the same time genetically unique individuals exploit or colonize a new environment. It generally occurs on the periphery between two habitats. (c) Sympatric speciation takes place within the center of a population in a patchy environment. Premating isolation mechanisms arise before the population shifts to a new niche.

ploited areas suitable for invasion. In all cases these small populations are away from genetically stable conditions in the center of the population, where variation in the gene pool is great and selection is for the heterozygotic condition. Founder populations have less genetic variation and experience greater selection for homozygosity. Because of small population size and inbreeding, any adaptive mutation carried by the population, especially by a dominant male, becomes fixed rapidly in the population. Such genetic changes may allow populations to exploit new habitats. Once a founder population is well established in a new habitat, reproductive isolation arises by chance. Postmating reproductive isolating mechanisms develop with the fixation of genetic changes. Premating reproductive isolation develops only if the new populations come in contact with the parent population. The numerous species of the fruit fly *Drosophila* in the Hawaiian Islands may have arisen in this fashion.

PARAPATRIC SPECIATION

There is a rapidly growing body of evidence that speciation can occur without geographic isolation. One such type is *parapatric speciation*, the evolution of a species as a contiguous (adjacent or nearby) population in a continuous cline (Figure 3.4b). The differentiated populations may meet in a narrow zone of overlap only a few hundred meters wide. It may take place among organisms that move very little, are capable of exploiting a somewhat different physical environment adjacent to the normal habitat, and become reproductively isolated from the rest of the population. Although suggestive of founder's effect, parapatric speciation differs; it requires no spatial isolation, and reproductive isolating mechanisms arise by selection at the same time genetically unique individuals exploit or colonize a new environment.

Parapatric speciation is characteristic of short-lived, rapidly reproducing species. It occurs in small to medium-sized populations living on the periphery of the main population. Chromosome macromutations frequently initiate speciation, with little or no genic differentiation even after speciation is complete. Such populations are characterized by high levels of homozygosity and the lack of long-range dispersal. Diverging populations are in close contact without the usual geographic isolation. Although gene flow does occur between populations, it rarely extends far because individuals experience strong directional selection in two different habitats or microhabitats.

Examples among animals are the flightless morabine grasshoppers of Australia, the sessile marine snail *Partula*, mole crickets (*Gryllotalpa*), mole rats (*Spalax*), and pocket gophers (*Thomomys*)—all diggers that remain in burrows for most of their lives. Other examples are cichlid fish in East Africa.

Parapatric speciation may be widespread among plants. Plants are rooted; even with seed dispersal by wind and animals, the level of gene flow between plant populations is so low that effective population size can be measured in meters. As a result, plant populations consist of isolated breeding units adapted to narrow local environmental conditions. It may be particularly prevalent among plants that colonize new habitats, such as soils heavily contaminated with nickel and cadmium from mine spoils.

SYMPATRIC SPECIATION

Sympatric speciation is the production of a new species within a population or within the dispersal range of a population (Figure 3.4c). It differs from parapatric speciation in three ways: (1) it occurs in the center of a population in a patchy environment rather than on the periphery; (2) premating reproductive isolating mechanisms arise before the population shifts to a new food source or habitat; and (3) it involves the evolution of a specialization to exploit an underused or novel resource and the invasion of an empty niche.

Sympatric speciation is most apt to occur among insects parasitic on plants and animals. About 70 percent of the some 900,000 known species of insects are parasitic, and many are host-specific. Such insects are small, have short life spans, possess high reproductive rates, have low competitive ability, readily adapt to new conditions, and experience sibmating and other forms of inbreeding.

The first condition for sympatric speciation is the formation of a stable polymorphism through disruptive selection. If individuals best adapted to a particular host, food source, or habitat tend to mate with one another, two reproductively isolated populations evolve. The genetic variation needed to establish a new host race or species has to be present in the population before the new host appears. A second condition is an underused or novel resource to which one polymorph may resort. Within the population some individuals must carry two types of genes—ones that enable them to select a new host and ones that enable them to counteract any chemical defense of the new host. Mutations at one or two loci would be enough to ensure survival on a host closely related to the original one.

Several other conditions facilitate sympatric speciation. Ideally, the new host plant or animal, or new food source, should be closely related to the original one. Because chemical recognition cues and chemical defenses would be similar, great genetic changes in the new host race would not be necessary. The new host should occur in abundance with the original host within the dispersal range of the new race. Then new genotypes could be tested on the hosts until the right combination was found. The new race should utilize the host as a site for courtship and mating, because mate selection would depend upon host selection. Because only homozygous individuals could locate and survive on the new host (heterozygous individuals would remain with the original host), a strong isolating mechanism would develop.

Sympatric speciation probably is a rare event, and it is difficult to separate from parapatric speciation; but the huge numbers of insect species suggest that it does happen. Possible examples of sympatric speciation may be found among treehoppers, milkweed longhorn beetles, codling moths—parasites on apples, walnuts, and plums—and the apple maggot.

POLYPLOIDY

Allopatric, parapatric, and sympatric speciation are more or less gradual processes. In sharp contrast is *abrupt speciation*, in which a new species arises spontaneously. The most common method by which abrupt speciation takes place is doubling the number of chromosomes, or *polyploidy* (see Chapter 2), most common in plants. An organism that has double the number of chromosomes in its gametes cannot produce fertile offspring with a diploid member of its ancestral population, but it can do so by mating with another polyploid. Thus, a polyploid has already achieved reproductive isolation. If the polyploid can spread into or exploit a new environment, a new species has been formed. Because plants favor asexual over sexual reproduction in unfavorable conditions, polyploid plants are not at a disadvantage, particularly among perennial herbs. In fact, polyploidy often enables plants to colonize and to tolerate more severe environments. Thus, the availability of new ecological niches favors the establishment of polyploid species.

Many of our common cultivated plants—potatoes, wheat, alfalfa, coffee, and grasses, to mention a few—are polyploids. Polyploidy is widespread among native plants, in which it produces a complex of species, such as blackberries (*Rubus*), willows, and birches. The common blue flag of northern North America is a polyploid that probably originated from two other species, *Iris virginica* and *I. setosa*, when the two, once wide-ranging, met during the retreat of the Wisconsin ice sheet. The sequoia is a relict polyploid, its diploid ancestors having become extinct.

ADAPTIVE RADIATION

All species are adapted to some particular environment; but because any environment is limited, the population can grow too large for available resources. That in itself is a selective force, for the time finally comes when those individuals able to use some unexploited environment and resource are at an advantage. Under reduced competition, those individuals have some opportunity to leave progeny behind. By eliminating disadvantageous genes, selection will strengthen the ability of the population to utilize the new niche it has occupied.

Not every organism can adapt to a new environment. Before a species can enter a new mode of life, it first must have physical access to the new environment. Having arrived there, the species must be capable of exploiting it. Here animals, particularly the vertebrates, have an advantage over plants. Most plants, although easily dispersed, are exacting in their habitat requirements. Animals, while finding it difficult to cross barriers, cope better with a new environment.

Once an organism has established a beachhead, it must possess sufficient genetic variation to establish itself under the selective pressures of climate and competition from other organisms. Adaptations that permit an organism to gain a foothold are only temporary; they must be altered, strengthened, and improved by selection. Finally, competition must be either absent or slight enough to allow the new invader to survive in its initial colonization. Such niches have been available to colonists of some remote islands, such as the Galápagos, the Hawaiian Islands, and the archipelagos of the South Pacific. The abundant empty niches in these diversified islands when the first colonists arrived encouraged the rapid evolution of species.

The Hawaiian honeycreepers (family Fringillidae, subfamily Drepanidinae) are examples of colonization and diversification in an unexploited environment (Figure 3.5). The ancestor of the honeycreepers was probably a nectar-feeding, insectivorous bird somewhat similar to the present-day *Himatione*. After colonizing one or two islands, stragglers undoubtedly invaded other surrounding islands. Because each group was under somewhat different selective pressures, the geographically isolated populations gradually diverged. By colonizing one island after another, and after reaching species level, recolonizing the islands from which they came (double invasion), the immigrants enriched the avifauna, especially on the larger island with more varied habitats. At the same time competition among

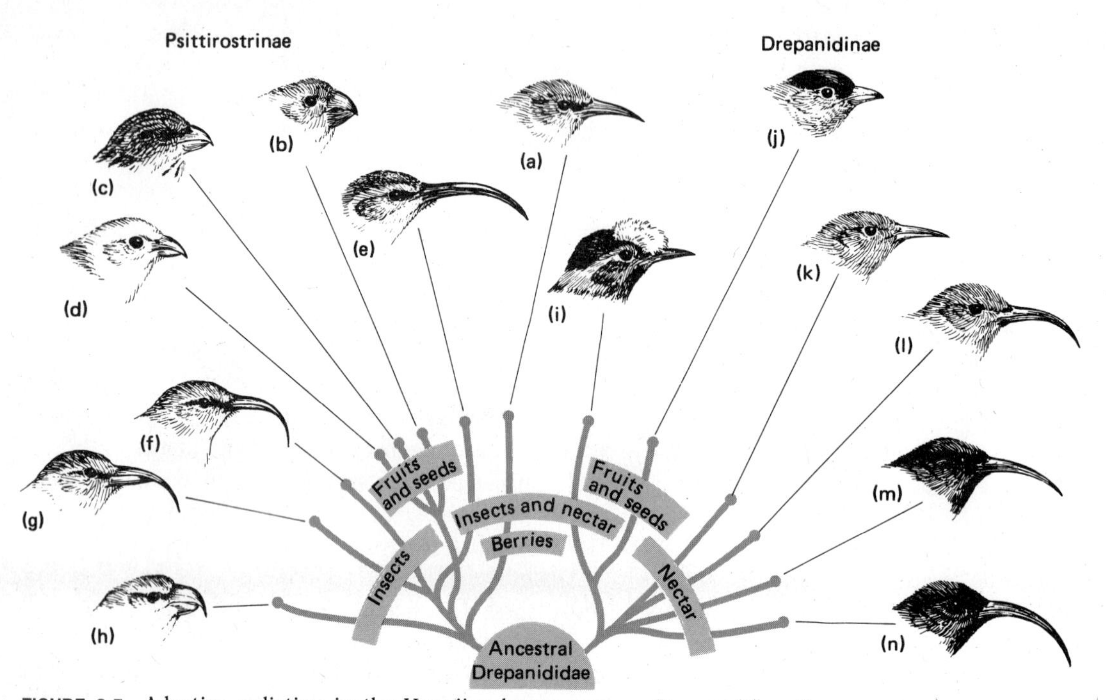

FIGURE 3.5 Adaptive radiation in the Hawaiian honeycreepers, Drepanididae. Selected representatives of two subfamilies, Drepanidinae and Psittirostrinae, illustrate how the family evolved through adaptive radiation from a common ancestral stock. Both subfamilies show a certain degree of parallel evolution based on diet. (a) *Loxops virens* probes for insects in the crevices of bark and folds of leaves and feeds on nectar and berries; (b) *Psittirostra kona,* a seed-eater, is extinct; (c) *Psittirostra cantans* feeds on a wide diet of seeds and fruit; (d) *Psittirostra psittacea,* a seedeater, feeds especially on the climbing screw pine (*Freycinetia arborea*); (e) *Hemignathus obscurus* is an insect and nectar-feeder; (f) *Hemignathus lucidus* is an insect-feeder; (g) *Hemignathus wilsoni* is an insect-feeder; (h) *Pseudonestor xanthophrys* feeds on larvae, pupae, and beetles of native Cerambycidae and grips branches with curved upper beak; (i) *Palmeria dolei,* feeds on insects and nectar of ohio (Metrosideros); (j) *Ciridops ana,* a fruit-eater and seed-eater, is extinct; (k) *Himatione sanguinea* feeds on the nectar of ohio; (l) *Vestiaria coccinae* feeds on the nectar from a variety of flowers; (m) *Drepanis funerea,* a nectar-feeder, is extinct; (n) *Drepanis pacifica,* a nectar-feeder, is extinct.

sympatric species placed a selective premium on divergence. The honeycreepers evolved into finchlike, creeperlike, and woodpeckerlike forms. Such divergence of one species into a number of different forms—each adapted to fit a different ecological niche, exploit a new environment, or tap a new source of food—is called *adaptive radiation*.

This principle is nicely illustrated by the genus *Hemignathus* (see Figure 3.5e, f, g), all members of which are primarily insectivorous, *Hemignathus obscurus*, whose lower mandible is about the same length as its upper mandible, uses its decurved bill like forceps to pick insects from crevices as it hops along the trunks and limbs of trees. The bill of *H. lucidus* is also decurved, but the lower mandible is much shorter and thickened. The bird uses the lower bill to chip and pry away loose bark as it seeks insects on the trunks of trees. In *H. wilsoni* the modification is carried even further. The lower mandible is straight and heavy. Holding its bill open to keep the slender upper mandible out of the way, the bird uses the lower mandible to pound, woodpeckerlike, into soft wood to expose insects. The bill of the genus *Hemignathus* was specialized at the start to feed on insects and nectar.

The honeycreepers also exhibit *parallel evolution*, or adaptive changes in different organisms with a common evolutionary heritage in response to similar environmental demands. The long, thin, decurved bill of *Hemignathus obscurus* is adapted for a diet of insects and nectar; so, too, are the bills of several members of the subfamily Drepanidinae (see Figure 3.5l, m, n).

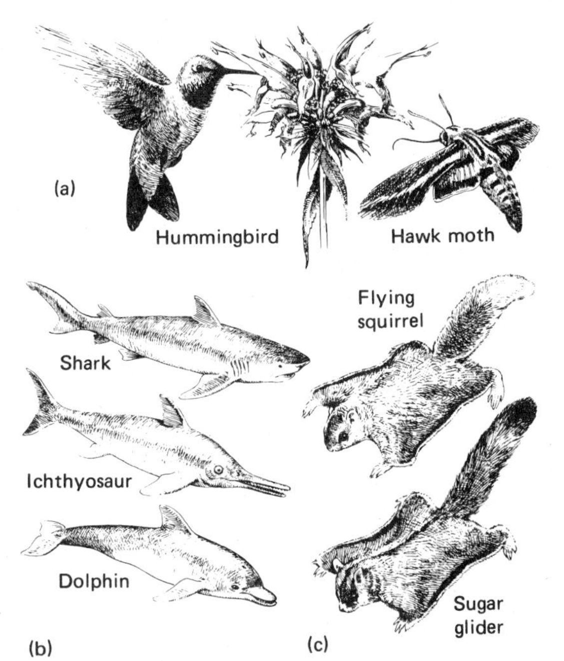

FIGURE 3.6 Convergent evolution in dissimilar organisms. (a) The hummingbird and the hawk moth both feed on nectar, the bird by day, and the moth by night. Both have adaptations of bill and mouth parts for probing flowers; both have a rapid, hovering mode of flight. (b) The prehistoric marine reptile, the ichthyosaur, the modern shark, and a marine mammal, the dolphin, all have the same streamlined shape for fast movement through the water. (c) A North American rodent, the flying squirrel, and an Australian marsupial, the sugar glider, both have a flat, bushy tail and an extension of skin between the foreleg and the hindleg that enable them to glide down from one tree limb to another.

CONVERGENCE

Instead of accentuating their differences in zones of overlap, some species tend to reduce their differences, and their characters converge (Figure 3.6). Instead of becoming less similar, they become more similar. This change apparently happens among species in which reproductive isolation is complete. Selective pressures may tend to favor an increasing resemblance among species as long as reproductive isolation is not upset and the similarities are advantageous. Such convergence can promote similar adaptations to the same environment or facilitate social relations among species.

There are a number of possible situations. Sympatric species may become more or less cryptic against background color. Duller coloration can conceal them, or more conspicuous color can facilitate flocking among individuals of the same and similar species such as the herons. Mimics sharing the same predators can reduce predation on each other. Social mimicry to facilitate mixed flocking is common among birds of the mountains of neotropical regions. For example, in parts of the northern Andes, most of the more common and conspicuous species in the mixed flocks of the humid temperate zone are predominantly brilliant blue or blue and yellow.

Similar convergence exists among plants. Within the chaparral vegetation of California, for example, the dominant plants belong to such diverse families as Ericaceae, Rhamnaceae, and Rosaceae; yet all are deep-rooted evergreen hard-leaved shrubs. In fact, throughout all areas with a Mediterranean type of climate, the vegetation has a similar appearance and character. Even though widely separated geographically and possessing different evolutionary histories, the vegetation has converged in both form and function. This convergence has been in response to similar selective forces, including fire, drought, high temperatures, and low rainfall.

GRADUALISM VERSUS PUNCTUATIONALISM

A central idea in Darwinian evolution is that populations are always evolving, a step-by-step process of selection from a pool of random variants. Descendants are linked to ancestors by a chain of intermediate genotypes (Figure 3.7). Some lines may be split from ancestral stock by geographic and other forms of isolation, gradually giving rise to new species. Other lines may slowly change into a new species, with the ancestral type becoming extinct. In any situation, speciation involves the repetition of favorable adaptations in several lines. Most morphological change takes place along with speciation. Refinement of phenotypic changes results from a gradual transformation brought about by natural selection. According to this *gradualistic model,* most species arise by a smooth transformation of an ancestral species. It changes continuously toward a descendant form. Such a continuous change results in an instability in the species. Thus evolutionary change takes place within fully established species.

A recent, alternative view is the *punctuational model*. It asserts that most change involves speciation in small populations of a species. It involves instantaneous speciation in small populations of a species. It involves instantaneous speciation followed by a long period of species stability or stasis. Adaptation is not necessary for evolutionary change; rather, new species arise by the total transformation of a lineage. The punctuational model is advanced by paleontologists, those who study the fossil record over geological time. Years of study have convinced some of them that species either remain the same for millions of years, changing little, such as the horseshoe crab; end their existence abruptly, geologically speaking, like the dinosaurs; or arise instantaneously. That interpretation appears to be the most reasonable explanation for the geologically short time gaps in the fossil record from one species to another. The fossil record for many lines (such as horses and elephants) is too complete and well-known for any intermediates (as postulated by gradualism) to exist.

A barrier to the acceptance of the punctuational model is the idea of instantaneous speciation. The term *instantaneous* is a relative one. An instant to a geologist or paleontologist is a long time to an evolutionary biologist. An instant in geological time is considered about 1 percent of the lifetime of a species. Most fossil species existed for 5 million to 11 million years, which allowed 50,000 to 100,000 years for speciation, a long time for species evolution. Species, according to the theory of punctuationalism, are rapidly established during periods of environmental instability. Old species become extinct, and new evolutionary lines appear. Once established, new species are relatively stable and resist essential change. Differential long-term evolutionary success, however, can still occasionally arise from natural selection acting on the phenotype.

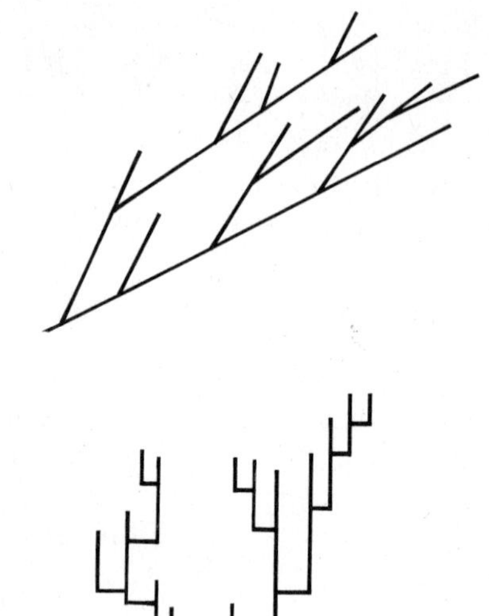

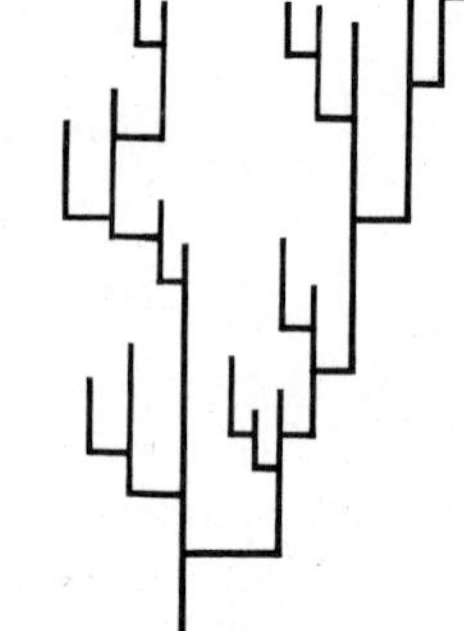

FIGURE 3.7 Two models of evolution. (a) The classical view, speciation through gradual change. (b) Abrupt speciation, or punctuationalism.

The two approaches may not be that far apart. Instantaneous speciation may require 10,000 to 100,000 years, a sufficient amount of time for natural selection to refine the products of speciation. Local populations would adjust to local environmental conditions and account for any fluctuations within a species during the period of stasis. Most species, once established, rarely evolve beyond differentiation into races or closely related species. All tend to be conservative and retain old habits and habitats. Species evolve new characteristics only when environmental changes force them to do so. Most species, when faced with that situation, become extinct.

HOW FAST DO SPECIES EVOLVE?

What are the rates of evolution? How fast do species arise? Those common questions have no definite answers. If we keep in mind that evolution is a change in gene frequency through time, and thus involves some change in phenotype, and that speciation is a multiplication of species, then the two related concepts can be kept apart.

Evolutionary changes involve some adaptations or changes in response to a changing environment. Such changes within a species can occur rapidly. (Consider how fast plant and animal breeders can change the phenotypes and growth efficiencies of domestic plants and animals through selective breeding, a form of goal-directed evolution brought about by human intervention.) In 50 years a single allele controlling industrial melanism in the peppered moth changed in frequency from 0 to 98 percent in populations exposed to industrial pollution. The house sparrow (*Passer domesticus*), introduced into the United States about 100 years ago, already has spread across North America and differentiated into a number of races. The continental population differs in size, color, and other morphological characteristics—differences that represent adaptations to different North American environments, from the eastern deciduous forest regions to the desert Southwest. Such changes involve only some features; other features may not change at all. Thus, each species is a mosaic of characteristics that have been passed on unchanged from ancestors and characteristics that have recently evolved.

Rates at which species evolve are difficult to determine. If most species arise by punctuationalism, or bursts of speciation during periods of environmental change, then determination of rates has little meaning. If some species result from isolation of small populations from parent populations, then rates vary. The numerous species of Hawaiian fruit flies (*Drosophila*) diverged in a few thousand years. Five endemic species of cichlid fish in the African Lake Nabugabo apparently developed in less than 4000 years, the length of time since the lake was isolated from Lake Victoria. Banana-feeding species of the Hawaiian moth genus *Hedylepta* diverged from a palm-feeding species in about 1000 years, the time since bananas were introduced to the islands by Polynesians. Other species change little. The American and Mediterranean species of the sycamore have evolved little since their separation at least 20 million years ago. When the two species were brought together in English gardens in the 1700s, they hybridized. The hybrid line, known as the London plane tree, is vigorous and fertile. Planted as ornamentals, the trees are escaping in places and becoming naturalized.

Summary

For the purposes of description and classification, taxonomists treat plants and animals as morphological species, discrete entities that exhibit little variation from the type of the species, the original specimen or specimens on which the description is based. However, organisms are highly variable over space and time. That fact has given rise to the concept of the biological species—a group of interbreeding individuals living together in a similar environment in a given region and in similar ecological relationships. This concept, limited as it is to bisexual organisms, is also filled with difficulties, such as the inability to distinguish, among some groups, just where one species begins and another ends. To get around that problem, evolutionary biologists suggest that the species be retained to identify kinds of organisms but that as a concept the species should be regarded as an evolutionary unit.

Species arise by the interaction of heritable variations, natural selection, and barriers to gene flow between populations, or isolating mechanisms. Isolating mechanisms include any morphological character, behavioral trait, habitat selection, or genetic incompatibility that is species-specific. If isolating mechanisms break down, hybridization results.

The most widely accepted mechanism of speciation is geographic isolation of one part of a population from another. Each part experiences different selection pressures. The two populations diverge and may ultimately become two distinct species. If geographic barriers break down, the two populations may hybridize, or they may evolve strong—usually premating—isolating mechanisms that reinforce their apartness. That process is known as allopatric speciation. Another type of speciation mechanism among populations of organisms that have low mobility and that inhabit contiguous but different and spatially separated habitats is parapatric speciation. It is probably most common among plants. In parapatric speciation both reproductive isolation and speciation take place at the same time. When reproductive isolation precedes differentiation and the process takes places within a population, the outcome is sympatric speciation.

An outcome of speciation is adaptive radiation, which results when populations of a species subdivide and diversify under the influence of selective pressures of new situations and occupy new ecological niches. When organisms with dissimilar ancestors evolve certain resemblances, they have experienced convergent evolution. Convergent evolution results from selection pressures in a similar environment.

Two views of speciation and evolutionary change are gradualism and punctualism. Gradualism says that species arise and differentiate by slow, continuous change within a species line. One species may slowly transform into another. Another species may arise by budding off from a parent poulation, forming an isolated population, and undergoing a series of changes largely brought about by natural selection. Punctuationalism argues that species arise "instantaneously" as distinct entities in a geological time span of 10,000 to 100,000 years during periods of great environmental instability.

How rapidly evolution and speciation take place is a topic of debate. Changes within a population at the level of a single allele can take place rapidly over a period of a number of generations. Truly instantaneous speciation, especially among plants, results from polyploidy, an alteration of the number of chromosomes. Some species apparently have never changed for millions of years. Fossil records suggest that new species arise in bursts of speciation, which would make a generalized rate of evolution meaningless. One condition seems certain: Evolution among some populations is always taking place somewhere on Earth.

Review and Study Questions

1. What is a species? A morphological species? A biological species?
2. Why is it difficult to define a species? Why is it important to try?
3. Distinguish between an allopatric species and a sympatric species; between those and a parapatric species.
4. What are sibling species? Can you come up with examples other than those given in the text?
5. What is a cline? How does it relate to the species problem?
6. What is an ecotype? A geographical isolate?
7. Define polymorphism. What is its ecological and evolutionary significance?
8. What is an isolating mechanism? Name and describe four classes of isolating mechanisms.
9. What happens if isolating mechanisms break down?
10. What is speciation? Distinguish among allopatric speciation, parapatric speciation, and sympatric speciation.
11. What is polyploidy? What is its evolutionary significance?
12. What property of plants makes them more likely than animals to form hybrid polyploid species?
13. What are adaptive radiation and convergence? Can you give some examples?
14. Contrast the concept of gradualism in speciation with that of punctuationalism. Why should punctuationalism be controversial?
15. Why is it difficult, if not impossible, to establish any definitive rate of evolution or speciation?

*16. Consider two conservation problems. The white rhinoceros of Africa is divided into two subspecies, northern and southern. There are only 18 free-ranging individuals of the northern subspecies left in the wild, but the southern form at present is in no real danger of extinction. Genetic and morphological studies suggest little difference between the two and throws the subspecies status into doubt. Should we divert the considerable amount of money necessary to save the 18 individuals or should we concentrate on the larger southern population?

Similarly, there appears to be little genetic difference between the Asiatic lion (*Panthera leo persica*) and the African lion (*P. l. leo*). The Asiatic subspecies is highly endangered, restricted to the Gir Forest Sanctuary in India. The African subspecies is widespread. Should we diminish our efforts to save the Asiatic subspecies? In each case discuss why and why not.

Selected References

Eisenberg, J. F. 1981. *The mammalian radiations*. Chicago: University of Chicago Press. A somewhat advanced compendium on mammalian evolution.

Futuyma, D. J. 1984. *Evolutionary biology*, 2nd ed. Sunderland, MA: Sinauer Associates. The best introduction to the subject.

Grant, P. 1986. *Ecology and evolution of Darwin's finches*. Princeton, NJ: Princeton University Press. Updates the story of Darwin's finches. A modern field study in evolution. Should be read with Lack 1974.

Grant, V. 1971. *Plant speciation*. New York: Columbia University Press. All aspects of speciation in plants. A basic reference.

Lack, D. 1974. *Darwin's finches: An essay on the general biological theory of evolution*. London: Cambridge University Press. A classic of biology.

Mayr, E. 1963. *Animal species and evolution*. Cambridge, MA: Harvard University Press. A classic major work on the subject.

Mayr, E. 1970. *Population, species, and evolution*. Cambridge, MA: Harvard University Press. A shorter, more accessible version of the above.

Otte, D., and J. Endler, eds. 1989. *Speciation and its consequences*. Sunderland, MA: Sinauer Associates. A series of papers that provide a major reference source on the subject.

Stanley, S. M. 1981. *The new evolutionary timetable*. New York: Basic Books. A short readable review of evolution and origin of species by an advocate of punctuationalism.

Stebbins, G. L. 1982. *Darwin to DNA: Molecules to humanity*. San Francisco: Freeman. A readable introduction to evolution by a major scholar in the field.

White, M. J. D. 1978. *Modes of speciation*. San Francisco: Freeman. A major introduction to the speciation process.

See also Annual Review of Ecology 1984:15 for a series of four major review articles on speciation.

PART III
Conditions for Life

INTRODUCTION

Homeostasis

Most organisms live in a variable environment. An inability to survive and reproduce within a range of environmental conditions means death. They have to maintain a relatively constant internal environment, within the narrow limits required by their cells, organs, and enzyme systems. That calls for some means of regulating the internal environment relative to the external one. Organisms have to regulate their body temperature, acidity, water, and salts in fluids and tissues, to mention a few factors. They have to control how they take in substances for cellular chemical reactions and absorb heat, and how they lose excessive intake and waste products of metabolism. The maintenance of a fairly constant internal environment within tolerance limits is called *homeostasis.*

Homeostasis involves feeding external environmental information into a system, which then responds. An example is temperature regulation in humans. The normal temperature for humans is 37°C (98.6°F); we call such a norm, a *set point.* When the temperature of the environment rises, sensory mechanisms in the skin detect it and send a message to the brain, which acts (involuntarily) on the information and relays the message to the effector mechanisms that increase blood flow to the skin and induce sweating. Water excreted through the skin evaporates, cooling the body. When the environmental temperature falls below a certain point, a similar action in the system takes place, this time reducing blood flow and causing shivering, an involuntary muscular exercise producing more heat. This type of reaction, which halts or reverses movement away from a set point, is called *negative feedback.*

If the environmental temperature becomes extreme, the homeostatic system breaks down. If the environmental temperature becomes too warm, the body is unable to lose heat fast enough to hold the temperature at normal. Body metabolism speeds up, further increasing body temperature, eventually ending in heatstroke or death. If the environmental temperature drops too low, metabolic processes slow down, further decreasing body temperature, eventually resulting in death by freezing. Such a situation, in which feedback reinforces change, driving the system to higher and higher or lower and lower values, is called *positive feedback.*

Organisms have a limited set of tolerances called *homeostatic plateaus,* within which conditions must be maintained. If environmental conditions exceed the operating limits of the system, it goes out of control. Instead of negative feedback governing the system, positive feedback takes over. Movement away from the homeostatic plateau can destroy the system.

Organisms, then, are limited by a number of conditions and often by an interaction among them. One organism may hve a wide range of tolerance for one condition, say salinity, others a narrow range of tolerance. To complicate matters, an organism may have a wide range of tolerance for many conditions and a narrow range for only one. That one narrow range can limit the distribution and fitness of the organism. Further, the maximum and minimum levels of tolerance for a species vary seasonally, geographically, and within the stage of the life cycle. For example, temperatures that are optimal for growth may not be optimal for reproduction. Organisms that exhibit a wide range of tolerance for all environmental influences would be widely distributed, and those that do not would be more restricted.

These relationships are described by two laws. The *law of the minimum* was set forth in 1840 by a German organic chemist, Justus von Liebig, who analyzed the relation between surface soils and plants. He noted that plants require certain kinds and quantities of nutrients (in his experiments supplied by manure) of food materials. If one of these food substances is absent, the plant dies. If it is present in minimal quantities only, the growth of the plant will be minimal.

Subsequent studies over a century disclosed that not only nutrients but also other environmental conditions, such as moisture and temperature, affect the growth of plants. Studies on the physiology and distribution of animals showed that they, too, were limited by food, temperature, water, humidity, and other environmental

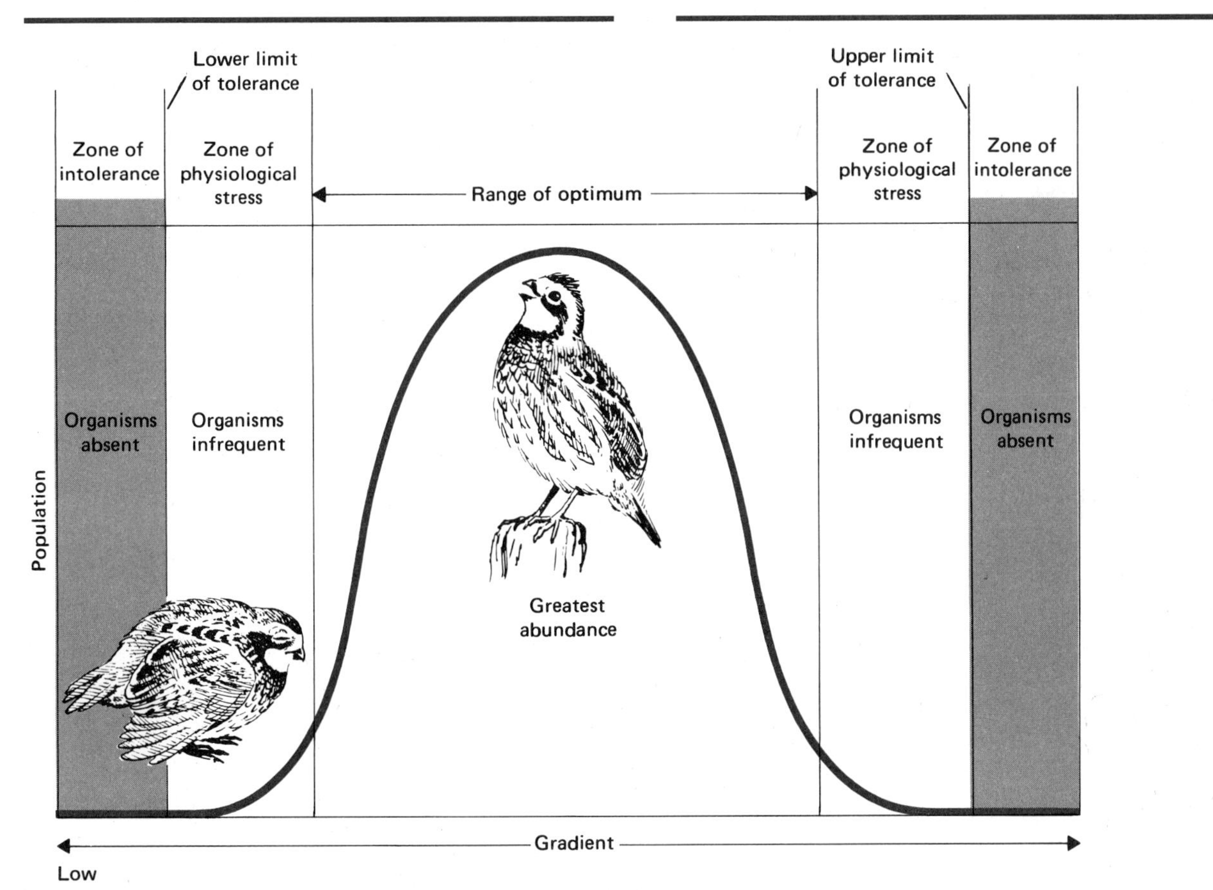

FIGURE III.1 The law of tolerance.

conditions. Not only were plants and animals limited by too little of a substance or condition; too much also limited the presence or fitness of an organism. Organisms, then, live within a range of too much and too little, the limits of tolerance (Figure III.1). This concept of maximum and minimum substances or conditions limiting the presence or success of an organism have been combined into the *law of tolerance.*

CHAPTER 4 Climate

Outline

Objectives

On completion of this chapter, you should be able to:

1. Describe the fate of solar energy reaching Earth.
2. Describe how the atmosphere heats and circulates.
3. Explain adiabatic heating and cooling and how it influences atmospheric movements.
4. Explain the Coriolis effect on atmospheric circulation and ocean currents.
5. Describe stable and unstable air masses and inversions.
6. Explain the origin of local winds and their effects.
7. Describe microclimates and their ecological effects.
8. Describe the differences in the microclimates of north-facing and south-facing slopes and in the microclimates of urban and rural areas.
9. Tell why microclimatic differences occur.

Climate is one of those terms we take for granted without giving thought to their meaning. Often people confuse climate with weather. *Weather* is the temperature, humidity, precipitation, winds, cloudiness, and other atmospheric conditions at a given place and time. *Climate* is the summation of weather conditions over a long period of time, their mean values and variances. Climate can be described locally, regionally, and globally.

EARTH IN THE SUN'S RAYS

Earth is always immersed in sunlight. The planet intercepts the full spectrum of incoming radiation on the outer edge of the atmosphere. The amount of solar radiation striking Earth's atmosphere, measured perpendicular to the sun's rays and at the mean distance of Earth from the sun, is 1.94 cal/cm^2/min. This quantity is known as the *solar constant.*

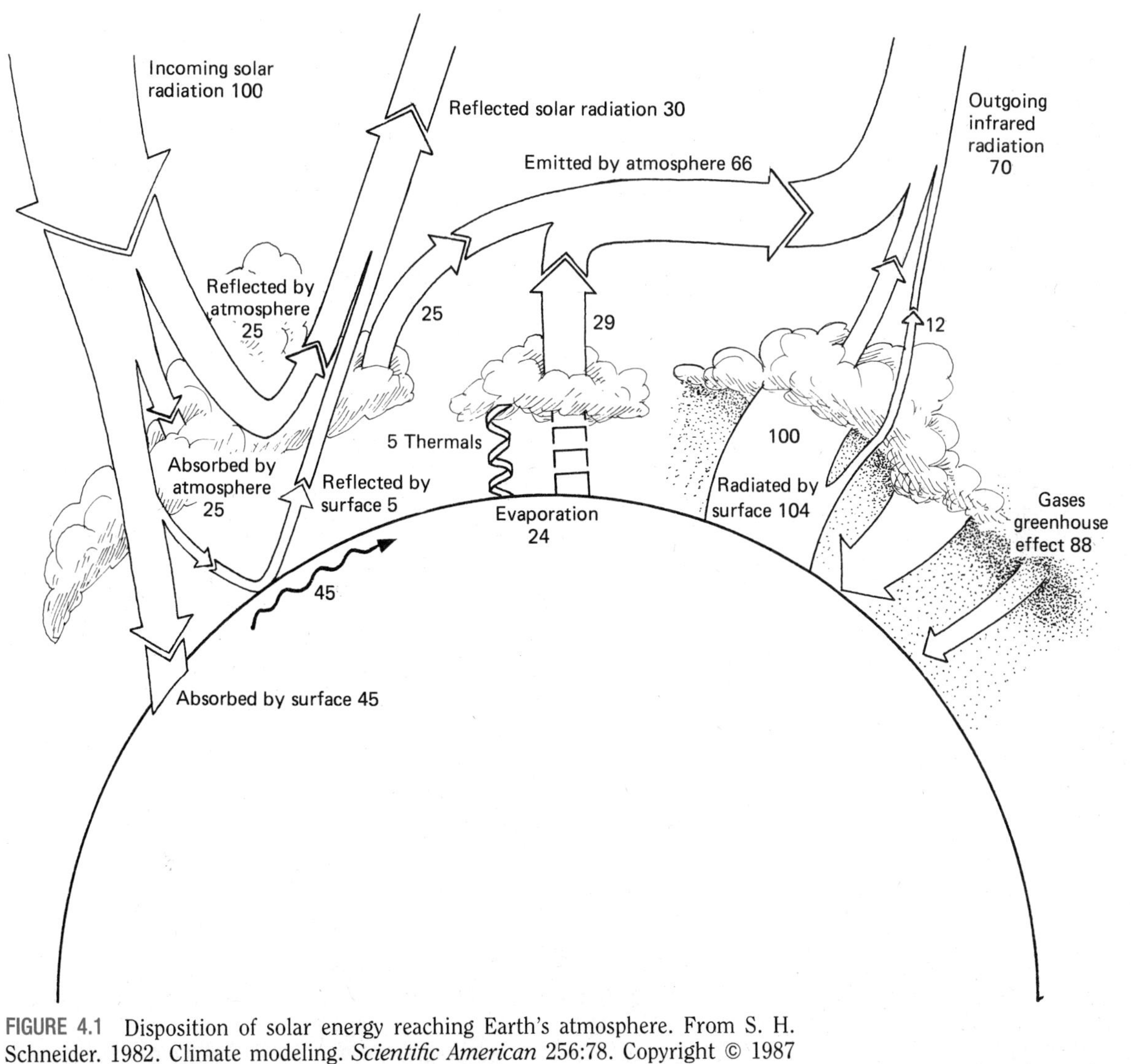

FIGURE 4.1 Disposition of solar energy reaching Earth's atmosphere. From S. H. Schneider. 1982. Climate modeling. *Scientific American* 256:78. Copyright © 1987 by Scientific American, Inc. All rights reserved.

Of the energy intercepted by Earth's atmosphere, considered in terms of 100 units, 30 units are reflected directly back to space, 25 units by the atmosphere, and 5 units by Earth. Twenty-five units are absorbed by the atmosphere, and 45 units are absorbed by Earth's surface (Figure 4.1).

Some of the radiation (29 units) absorbed by Earth is emitted back to the atmosphere as long-wave radiation by thermals and evaporation. The remainder is emitted by the surface as long-wave or infrared radiation. Although short-wave radiation passes through the atmosphere with ease, long-wave radiation cannot readily escape outward. Although the amount of energy flux—rate of flow—from Earth's surface is 104 units, only 4 units pass directly through the atmosphere to outer space. The enveloping blanket of atmospheric gases, especially CO_2, and water vapor trap 100 units. Twelve of these units escape the atmosphere and clouds as outgoing infrared radiation. The atmosphere sends back to Earth 88 units of long-wave radiation. The absorption of outbound long-wave radiation by the atmosphere and its emission back toward Earth is called the *greenhouse effect* (Figure 4.2).

The atmosphere does not exactly act like greenhouse glass. Greenhouse glass permits a relatively normal radiant energy exchange, but it traps a small volume of air beneath the glass, insulating against uneven heat losses. Unlike the atmospheric blanket, the greenhouse does not trap radiant energy. So far, Earth too has maintained a fairly constant temperature, losing to outer space the same amount of energy it receives.

The sun's energy does not reach Earth uniformly. The shape of Earth and its tilt on its axis influence the amount of energy reaching any particular point on Earth. At the equator, the rays of the sun strike Earth vertically. There solar energy is more concentrated and the temperatures are higher. At the poles, rays of the sun intercept Earth's atmosphere at an angle and have to penetrate a deeper blanket of air. More energy is scattered in the atmosphere and reflected back to space. Because energy that reaches the surface is distributed over a wider area, less solar energy per unit of surface is absorbed and temperatures are lower (Figure 4.3).

The elliptic orbit of Earth about the sun and the tilt of Earth's axis to this plane influence the concentration of solar energy and temperatures during the 365-day

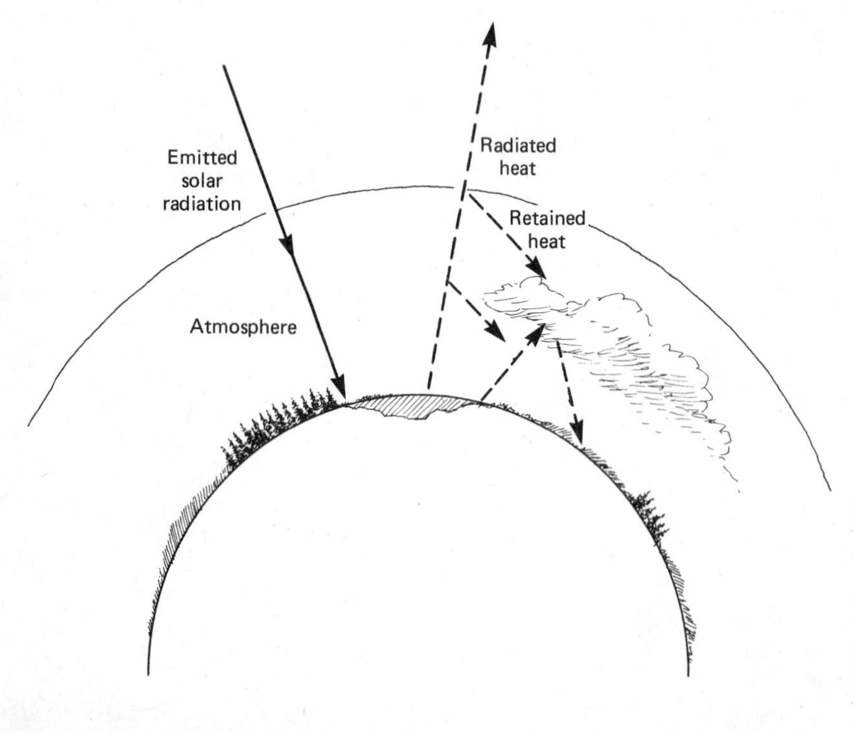

FIGURE 4.2 The greenhouse effect. High-energy, short-wave solar radiation enters Earth's atmosphere, is absorbed by Earth and reradiated back as long-wave radiation. Carbon dioxide and moisture retain much of this heat and reradiate it back to Earth.

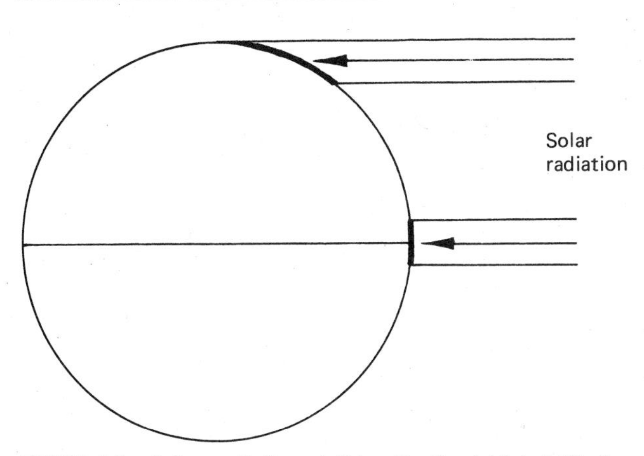

FIGURE 4.3 Solar radiation striking Earth at high latitudes arrives at an oblique angle and spreads over a wider area. Thus it is less intense than energy arriving vertically at the equator.

passage and give us our seasons (Figure 4.4). Earth's surface at all times lies half in the sun's rays and half in shadow, marked by the dividing line, the circle of illumination. The circle of illumination always lies at right angles to the sun. Two times a year, at the time of the vernal and autumnal equinoxes (March 20 or 21 and September 22 or 23), the circle of illumination passes through the poles. At the summer and winter solstices (June 21 or 22 and December 22 or 23), the circle of illumination is tangent to the Arctic and Antarctic circles (Figure 4.5). Thus, at the fall and spring equinoxes, the sun's rays fall directly on the equator, so that at noon the angle of the sun's rays at the equator is 90°. At the North Pole the sun is at the horizon and keeps that position as Earth rotates. At the time of the winter solstice, the angle of the sun's rays at noon is 66.5° at the equator and 90° at the Tropic of Capricorn. The sun remains below the horizon at the North Pole and above the horizon at an angle of 23.5° at the South Pole. At the summer solstice, the position of the sun's rays is 90° at the Tropic of Cancer. At that time of year, the sun remains above the horizon for the entire 24-hour day.

The summer hemisphere receives more solar energy

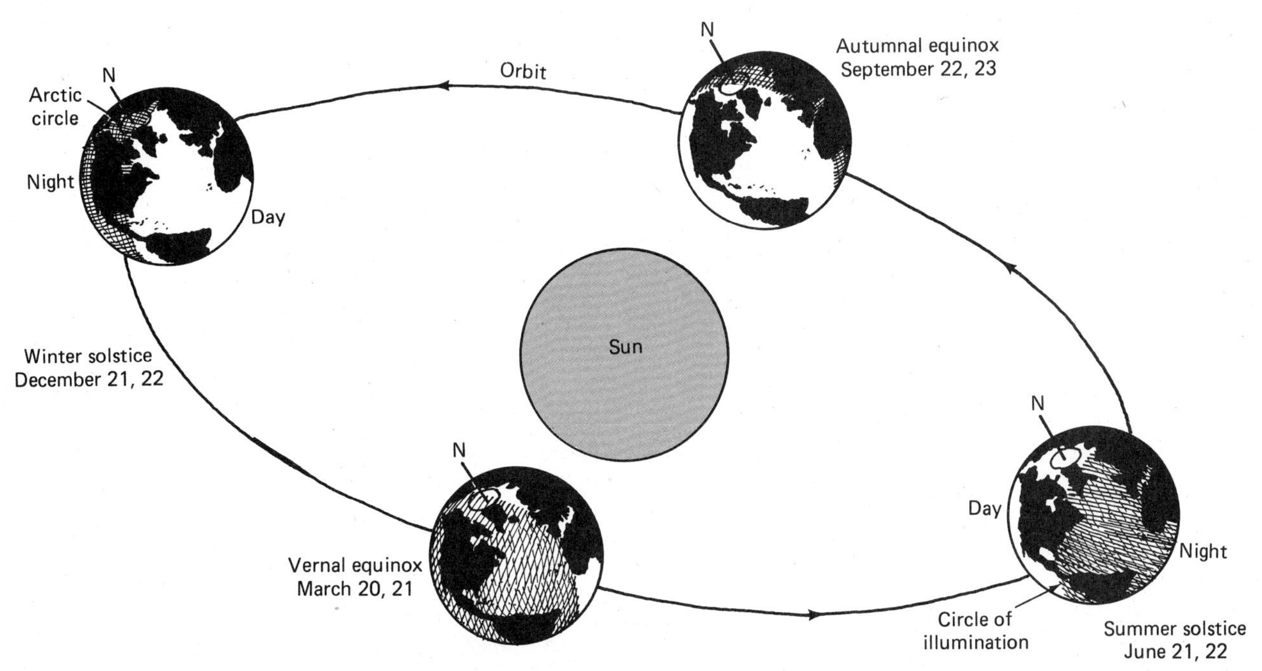

FIGURE 4.4 Because Earth's axis is tilted at about 23° to the plane of its orbit about the sun, the amount of solar radiation reaching any point of Earth's surface varies during the year.

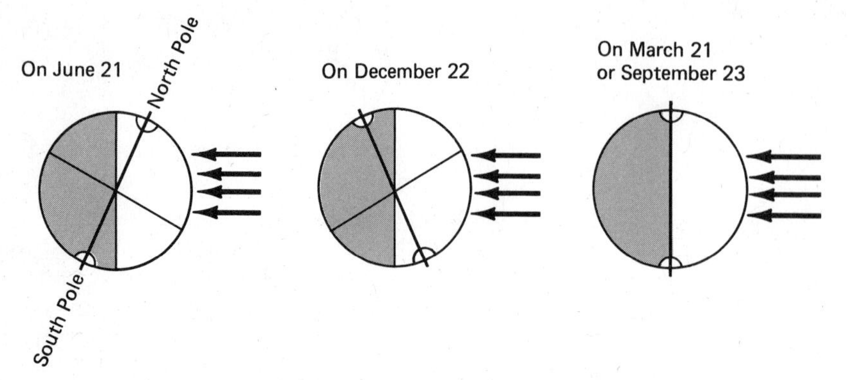

FIGURE 4.5 Angle of the sun at the equinoxes and at the winter solstice.

than the winter hemisphere, but over the year there is more radiation at low latitudes and less at high latitudes. Heat moves from the tropics to high latitudes. Because of this transfer and the steady solar radiation at low latitudes, the average air temperature there remains rather constant. Higher latitudes experience greater variation in temperature because of greater variation in incoming solar radiation.

ATMOSPHERIC MOVEMENT

The atmosphere is constantly in motion. To understand the movement and circulation of air masses, you need to have some appreciation of the behavior of gases. When gases expand, they cool; when they compress, they heat. The change in temperature comes about because of the degree of crowding of the gas molecules. The greater the crowding (compression), the more often molecules collide, raising the temperature. In this process no energy is exchanged between the gas and its environment. Such a process is called *adiabatic.*

When a parcel of warm air rises, it moves into an area of lower atmospheric pressure, expands, and so expends some of its energy. The decrease in the internal energy of the air parcel cools it at a rate of 10°C/1000 m. When it sinks, the air mass warms at the same rate. The rate of decrease in air temperature with height is called the *dry adiabatic lapse rate.* An air mass, however, usually contains some moisture, which condenses as it cools. The heat it releases (latent heat) partially counteracts the cooling of the air. This reduced rate is known as the *moist adiabatic lapse rate,* which averages 6°C/1000 m.

INVERSIONS

If a rising parcel of air cools faster than surrounding air does, it will fall to a level where the surrounding air has the same temperature. When that condition prevails, the atmosphere is said to be *stable* (Figure 4.6). In a layer of air in which the temperature increases with height—that is, the air is warmer above than it is below—atmospheric conditions are very stable, and the condition is called an *inversion.*

If a parcel of rising air cools slower than the surrounding still air, it continues to rise, pulling up air from the ground behind it. That action creates eddies, turbulences, and upward currents, which lead to thunderstorms. Such a condition is termed *unstable.*

Instability of the atmosphere is caused by differential heating of the Earth and its lower atmosphere. By day short-wave solar radiation is absorbed in different amounts over the land, depending upon vegetation, slope, soil, season, and so on. Lower layers of the atmosphere—heated by Earth's surface—rise in small volumes; colder air falls. That turbulence is increased by winds.

After the sun goes down, Earth begins to lose heat. The layer of surface air is cooled, while the air aloft remains near its warmer daytime temperature. In mountainous or hilly country, cold, dense air flows down slopes and gathers in the valleys. The cold air is then trapped beneath a layer of warm air (Figure 4.7).

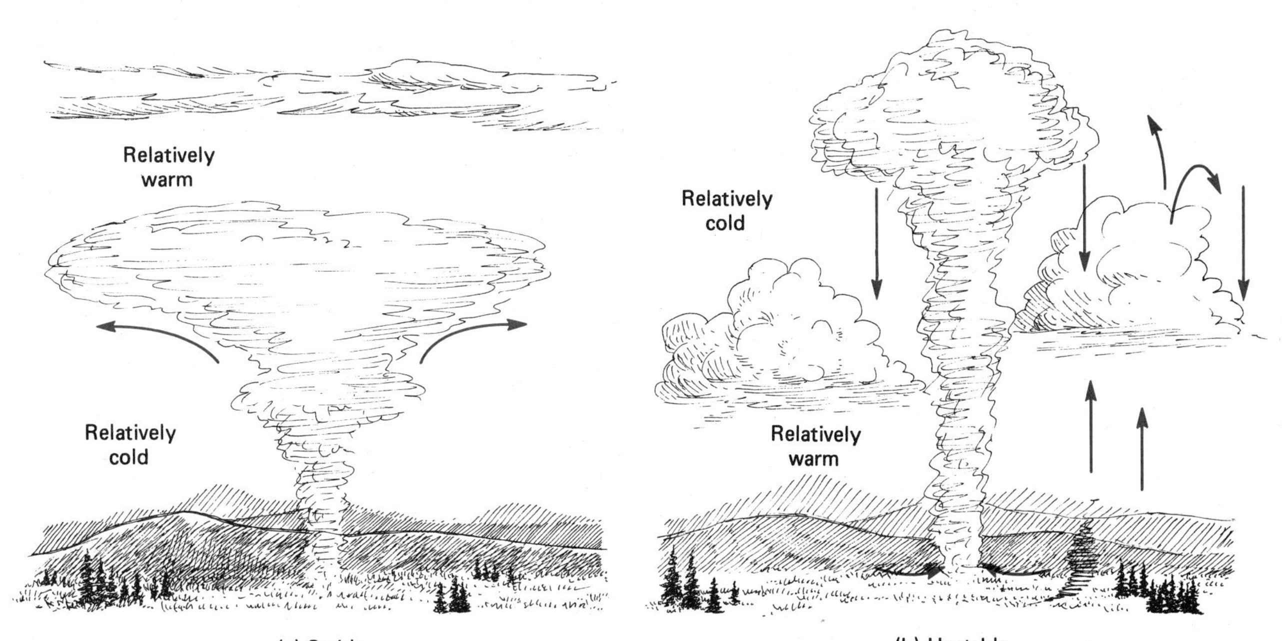

FIGURE 4.6 Stable and unstable air masses as illustrated by rising smoke columns. (a) In a stable air mass heating and mixing are confined to a shallow layer, and the temperature of the air increases rapidly. It cools slower with height than rising air. Vertical motion is damped. In this illustration the smoke column flattens and drifts apart. (b) An unstable air mass cools faster than rising air does. Heating and mixing take place throughout a deep layer. Temperature of the rising air mass is warmer than that of the surrounding air. The air keeps rising until its temperature equals that of the surrounding atmosphere. Such an air mass has upward and downward currents and gusty winds.

Such *radiation inversions* trap impurities. Smoke from industry and other heated pollutants rise until their temperature matches the surrounding air. Then they flatten out and spread horizontally. As pollutants continue to accumulate, they may fill the entire area with smog. Such inversions are most intense if the atmosphere is stable.

Similar but more widespread inversions occur when a high-pressure area stagnates over a region. In a high-pressure area, airflow is clockwise and spreads outward. The air flowing away from the high must be replaced, and the only source of replacement air is from above. Therefore surface high-pressure areas are regions of sinking air movement from aloft, called *subsidence.* When high-level winds slow down, cold air at high levels in the atmosphere tends to sink. The sinking air becomes compressed as it moves downward, and as it warms, it becomes drier. As a result, a layer of warm air develops at a higher level in the atmosphere (Figure 4.8). Rarely reaching the ground, it hangs several hundred to several thousand feet above the ground, forming a *subsidence inversion.* Such inversions tend to prolong the period of stagnation and increase the intensity of pollution. The subsidence inversion that brings our highest concentrations of pollution is often accompanied by lower-level radiation inversions.

Often along the west coast of the United States, and occasionally along the east coast, the warm seasons produce a coastal or *marine inversion.* In this case cool, moist air from the ocean spreads over low land. This layer of cool air, which may vary in depth from a few hundred to several hundred thousand feet, is

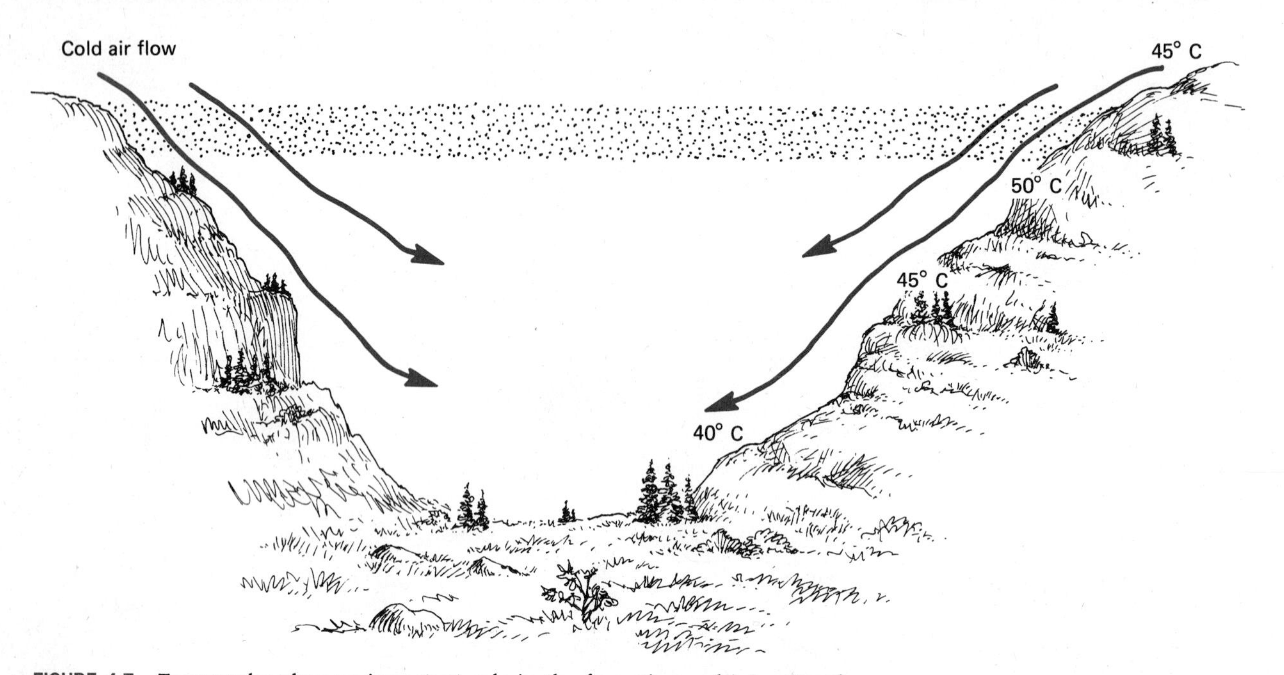

FIGURE 4.7 Topography plays an important role in the formation and intensity of nighttime inversions. At night air cools next to the ground, forming a weak surface inversion in which the temperature increases with height. As cooling continues during the night, the layer of cool air gradually deepens. At the same time, cool air moves downslope. Both cause the inversion to become deeper and stronger. In mountain areas the top of the night inversion is usually below the main ridge. If air is sufficiently cool and moist, fog may form in the valley. Smoke released in such an inversion will rise only until its temperature equals that of the surrounding air. Then the smoke flattens out and spreads horizontally just below the thermal belt.

topped by warmer, drier air, which also traps pollution in the lower layers.

Inversions break up when air close to the ground is heated, causing it to circulate up through the inversion layer, or when a new air mass moves into the area.

GLOBAL CIRCULATION

A similar phenomenon of rising air masses occurs globally on a grander scale. Air heated at equatorial regions rises until it reaches the stratosphere, where temperature no longer decreases with altitude. There the air, whose temperature is the same as or lower than that of the stratosphere, can rise no farther. With more air rising, the air mass is forced to spread north and south toward the poles. As the air masses approach the poles, they cool, become heavier, and sink over the arctic regions. The heavier air then flows toward the equator, replacing the warm air rising over the tropics (Figure 4.9).

If Earth were stationary and without irregular land masses, the atmosphere would circulate just as shown in Figure 4.9. Earth, however, spins on its axis from west to east. The linear velocity of Earth is greatest at the equator—1041 mi/hr or 465 m/sec. Any object on the surface there is moving at the same rate. Because the distance around Earth decreases toward the poles, linear velocity also decreases. At 30° latitude an object on the surface would be moving 403 m/sec; at 60° latitude, 233 m/sec; and at the poles, 0 m/second. How-

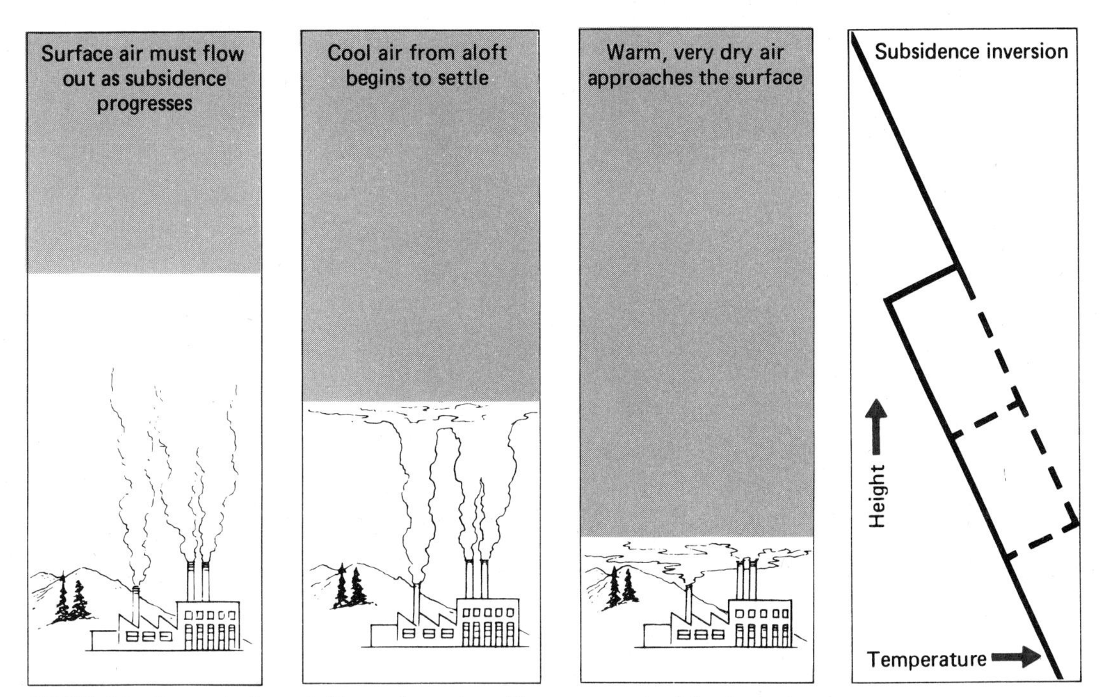

FIGURE 4.8 Descent of a subsidence inversion. The movement of the inversion is traced by successive temperature measurements, shown by a dashed line. The nearly horizontal dashed lines indicate the position of the descending base of the inversion. The solid line indicates temperature of the surrounding air. The temperature lapse rate in the descending layer is nearly dry adiabatic. The bottom surface is marked by a temperature inversion. Two features—temperature inversion and a marked decrease in moisture—identify the base of the descending layer. Below the inversion is an abrupt rise in the moisture content of the air.

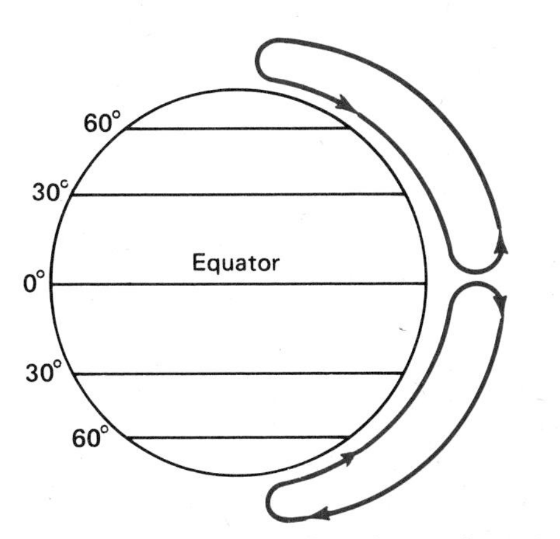

FIGURE 4.9 Circulation of air cells and prevailing winds on an imaginary, nonrotating Earth. Air heated at the equator rises and moves north and south, and upon cooling at the two poles, descends and moves back toward the equator.

ever, objects maintain their momentum unless acted upon by some outside force. If an object moves northward from the equator, it speeds up relative to Earth's rotation and its path appears to deflect to the right or east (Figure 4.10). If the object moves south, it will also be deflected to the right, but the deflection appears to be left or westerly relative to the Northern Hemisphere. That is what happens to parcels of air. The phenomenon is known as the *Coriolis effect,* named after the nineteenth-century French mathematician G. C. Coriolis, who first analyzed it. It is not a simple centripetal force but an apparent effect of a number of forces that act upon objects set in motion on Earth's spherical surface.

The Coriolis effect, then, deflects air movements and prevents a direct, simple flow from the equator to the

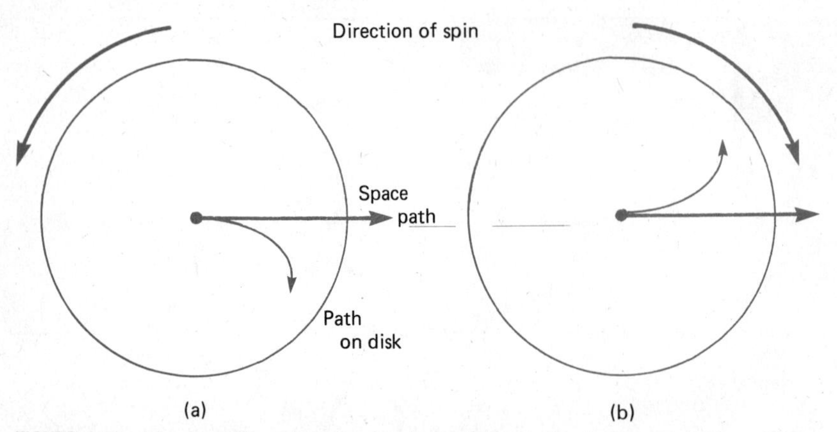

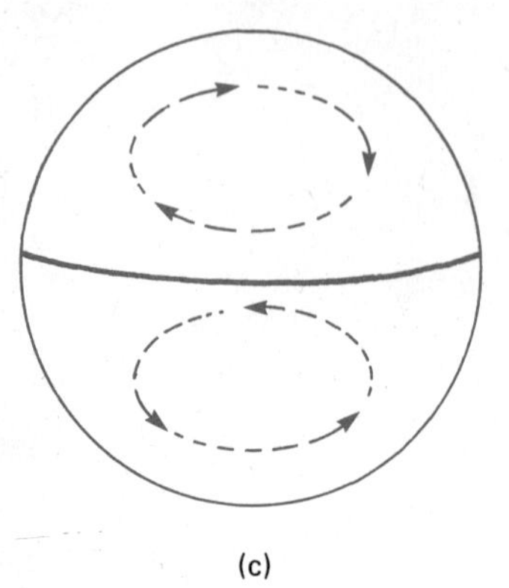

FIGURE 4.10 The Coriolis effect visualized. Earth's rotation is eastward. Viewed from above the North Pole, the sense of Earth's rotation is counterclockwise. Viewed from above the South Pole, the sense of Earth's rotation is clockwise. (a) Consider a disk rotating counterclockwise as representing the Northern Hemisphere. Any object thrown horizontally in a straight path in space will trace on the disk a curved path deflected to the right. (b) Similarly, an object thrown in a straight line on a disk rotating clockwise representing the Southern Hemisphere will be deflected to the left. (c) For this reason, movement of masses of air and water on Earth is deflected to the right in the Northern Hemisphere and to the left in the Southern Hemisphere.

poles. In the Northern Hemisphere airflow is directed to the right and in the Southern Hemisphere to the left. The result is a series of belts of prevailing east winds in the polar regions, the polar easterlies, and near the equator, the easterly trade winds. In the middle latitudes is a region of west wind known as the westerlies. These belts break the simple flow of air toward the equator and the flow aloft to the poles into a series of cells (Figure 4.11).

The flow is divided into three cells in each hemisphere. The air that flows up from the equator forms an equatorial zone of low pressure, a region of calm called the doldrums by sailors. The equatorial air rises, cools, loses its moisture in heavy precipitation, and spreads northward and southward away from the equatorial region. Because the northward flow of air becomes nearly a westerly flow, northward movement is slowed, with air piling up at about 30° north latitude and losing considerable heat by radiation. The air mass sinks, forming a semipermanent high pressure belt encircling Earth, a region of light winds known as the horse latitudes. Air that has descended warms, picks up moisture, and rises again. Some flows northward toward the pole and some southward toward the equator at the surface. The northward-flowing air current turns right to become the prevailing westerlies; the southward-flowing air, also deflected to the right, becomes the northeast trades of the low latitudes. The air aloft gradually moves northward, continues to lose heat, descends at the polar region, gives up additional heat at the surface, and flows southward. The flow of air deflected to the right becomes the polar easterlies. Southward-flowing air meets rising warm air moving toward the poles to produce a semipermanent low pressure area at about 60° north latitude. Similar flows take place in the Southern Hemisphere, but airflow is deflected to the left by the rotating Earth.

Although tropical regions about the equator are always warm, the sun is directly over the equator only during the spring and fall equinoxes. On June 21, summer in the Northern Hemisphere, the sun is directly over the Tropic of Cancer; on December 21, summer in the Southern Hemisphere, it is directly over the Tropic of Capricorn. At these times of the year, these areas on Earth receive the most intense solar radiation, affecting the heating of land, water, and air masses.

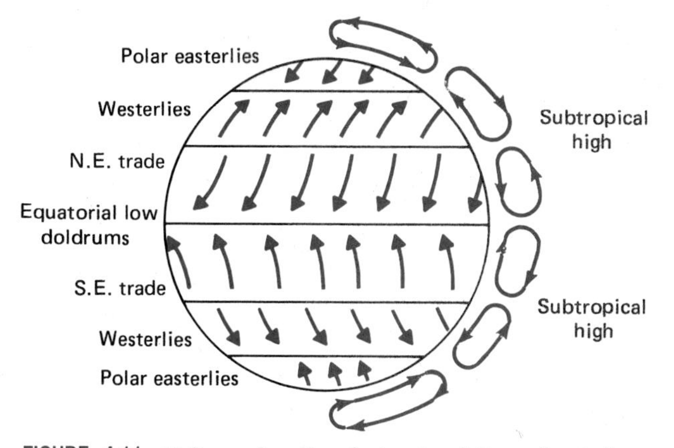

FIGURE 4.11 Belts and cells of air circulation about the rotating Earth.

The junction of rising and falling air masses over the equatorial regions shifts. This movement is known as the *intertropical convergence* (Figure 4.12). The intertropical convergence brings the rainy season to those tropical regions that experience a wet season and a dry season, regions dominated by seasonal tropical forests and savannas.

The interaction of winds and heating produces more or less permanent high-pressure cells known as the subtropical highs in the Atlantic and Pacific oceans; winds and cooling produce low-pressure cells such as the Aleutian and Icelandic lows. The highs are more pronounced during the summer months, the lows during the winter months. Also produced are the monsoon winds, dry winds that blow from continental interiors to the oceans in summer, and winds heavy with moisture that blow from the oceans to the interiors in winter.

Last, there are moving air masses with their cyclonic and anticyclonic frontal systems. These major air circulations are responsible for the changing swirls or cloud patterns seen over Earth from space.

In the Northern Hemisphere, winds in high-pressure cells, called *anticyclones,* flow clockwise and move outward and downward from the center of the system. Because air moves down from high altitudes, highs are characterized by minimum cloudiness and little precipitation. A high-pressure area is surrounded on all sides by low pressure. Low-pressure cells, called *cyclones,* flow counterclockwise. The air moves inward and upward, resulting in cooling, increased relative humidity, and, with adequate moisture, precipitation. In the Southern Hemisphere winds move counterclockwise around a high-pressure system and clockwise around a low-pressure system (Figure 4.13).

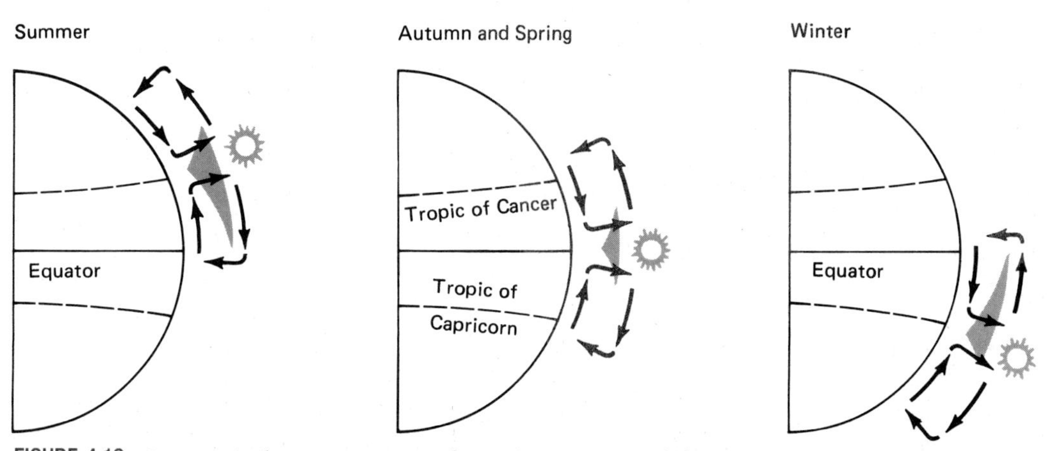

FIGURE 4.12 *Intertropical convergences* produce rainy seasons and dry seasons. As the distance from the equator increases, the longer the dry season and the less the rainfall. These oscillations result from changes in the altitude of the sun between the equinoxes and the summer and winter solstices, diagramed in Figure 4.5.

FIGURE 4.13 High- and low-pressure cells, showing the clockwise and counterclockwise flows of air.

LOCAL WINDS

Winds are a continual but variable influence on the environment. Those near Earth's surface are strongly affected by local landscapes' topography.

Winds that move with the leading edge of the air mass, or frontal system, or that are carried by the general circulation winds aloft are *general winds.* They contrast with *local convective winds* caused by temperature differences within a locality. Among the most familiar are the land and sea breezes along the shores of oceans, large inland lakes, and bays.

In mountainous topography local winds can be exceedingly complex. Differences in the heating of air over mountain slopes and canyon and valley bottoms result in several wind systems. Because of the larger heating surface, air in mountain valleys and canyons becomes warmer during the day. Similarly, the larger cooling area of the valleys causes a reversal of this situation at night. The resulting pressures cause a flow of air upslope by day and downslope by night. Combined valley and upslope winds exit at the ridge tops by day. As the slopes go into the shadows of late afternoon, the slope and valley winds shift direction and become downslope winds.

Local winds affect soil moisture and humidity and have an important bearing on forest fire conditions and local drought situations. The drying action of high, warm winds in winter, when soil moisture is low or unavailable, causes physiological drought. The wind removes humid air about the leaves and increases water loss. Losing more water than they are able to absorb, evergreens, in particular, dry out, and their foliage turns brown.

Plants that normally grow tall become low and spreading when high winds are frequent and regular, a situation characteristic of the timberline. On high, windswept ridges, cushion plants with small, uniform, crowded branches are most common. Because of constant desiccation, cells of plants growing in these places never expand to normal size, and all organs are dwarfed. Terminal branch shoots are killed back by desiccation, by blasting of ice particles, and along the ocean by salt spray. The terminals are replaced by strong laterals, which form a mat close to the ground.

Shallow-rooted trees and trees with brittle woods—such as the willows, cottonwoods, and maples—are thrown or broken by strong winds. Windthrow is most prevalent among trees growing in dense stands that, through logging or natural damage, are suddenly exposed to the full force of the wind.

Hurricanes and other violent windstorms sweeping over forested areas may uproot and break trees across a considerable expanse of land. Hurricanes and strong storms also cause deaths among animals and often carry individuals far from their normal environment and set them down elsewhere.

Winds are important means of dispersal for seeds and small animals such as spiders, mites, and even snails. Wind may also play a secondary role in the distribution of small mammals. The deeper accumulation of litter and snow in areas sheltered from the wind supports more small mammals than do more exposed areas.

RELATIVE HUMIDITY

The moisture content of the air is usually expressed as *relative humidity,* the percentage of moisture in the air relative to the amount of water the air could hold at saturation at the existing temperature. If air is warmed while its moisture content remains constant, the relative humidity drops, because warm air can hold more moisture than cool air. As the relative humidity drops, the *saturation pressure deficit* (the difference between the partial pressure of water at saturation and the prevailing vapor pressure of the air at the same temperature) increases, and increased evaporation takes place.

Relative humidity varies during a 24-hour period. Generally, it is lower by day and higher by night. Relative humidity under a closed forest canopy is higher than on the outside during the day, and it is lower than on the outside during the night.

Over normal surfaces, relative humidity during the day usually increases with height because of the decrease in temperature with height. That contrasts with absolute humidity, which decreases with height. But if a temperature inversion occurs, especially at night, the relative humidity decreases upward to the top of the inversion, then changes little or increases only slightly.

In any one area relative humidity varies widely from one spot to another, depending upon topography. Variations in humidity are most pronounced in mountain country. Low elevations warm up and dry out earlier in the spring than high elevations. Because daytime temperatures decrease with altitude, as does the *dew point* (the temperature at which atmospheric water condenses), relative humidities are greater on the tops of mountains than in the valleys. As nighttime cooling begins, temperature change with altitude is reversed. Cold air rushes downslope and accumulates at the bottom. Through the night, if additional cooling occurs, the air becomes saturated with moisture, and fog or dew forms by morning. Dew is the condensation of moisture on ground surfaces, vegetation, and other objects because the surface is cooler than the air. Fog is a cloud hanging close to the ground. Differences in humidity and in the resulting formation of fog and dew can produce vegetative differences on mountain slopes. They are most pronounced on slopes of mountains along the Pacific coast of North America.

Temperature and wind both exert a considerable influence on evaporation and relative humidity. An increase in air temperature causes convection currents, mixing surface layers with drier air above. Cyclonic disturbances also mix moisture-laden air with drier air above. As a result, the vapor pressure of the air is lowered, and evaporation from the surface increases.

For organisms, seasonal distribution of rainfall is more important than average annual precipitation. A great difference exists between a region receiving 120 cm of rain rather evenly distributed throughout the year and a region in which nearly all of the 120 cm of rain falls within a month or two. In the latter situation, typical of tropical and subtropical climates, organisms must face a long period of drought. An alternation of wet and dry seasons influences the reproduction and activity of organisms as much as light and temperature in the temperate regions.

OCEAN CURRENTS

The movements of Earth and air and solar energy produce ocean currents, the horizontal movements of water around the planet. In the absence of any land masses, oceanic waters could circulate unimpeded around the globe, as the flow of water does around the Antarctic continent. However, land masses divide the ocean into two main bodies, the Atlantic and the Pacific. Both oceans are unbroken from high latitudes, north and south, to the equator; and both are bounded by land masses on either side.

Each ocean is dominated by two great circular water motions, or *gyres,* each centered on a subtropical high pressure area. Within each gyre the current moves clockwise in the Northern Hemisphere and counterclockwise in the Southern Hemisphere (Figure 4.14). The movements of the currents are caused partly by the prevailing winds, the trades or tropical easterlies on the equator side and the prevailing westerlies on the pole side. The two gyres, north and south, are separated in both oceans by an equatorial countercurrent that flows eastward. That current results from the return of the lighter (less dense) surface water piled up on the western side of the ocean basin by the equatorial current.

As the currents flow westward, they become narrower and increase their speed. Deflected by the continental basin, they turn poleward, carrying cold water with them. The two major currents in the Northern Hemisphere are the Gulf Stream in the Atlantic and Kuroshio Current in the Pacific. Their counterparts in the Southern Hemisphere are the Brazil Current in the Atlantic and the Australian Current in the Pacific.

These ocean currents influence the climates of continental land masses, especially in the coastal regions.

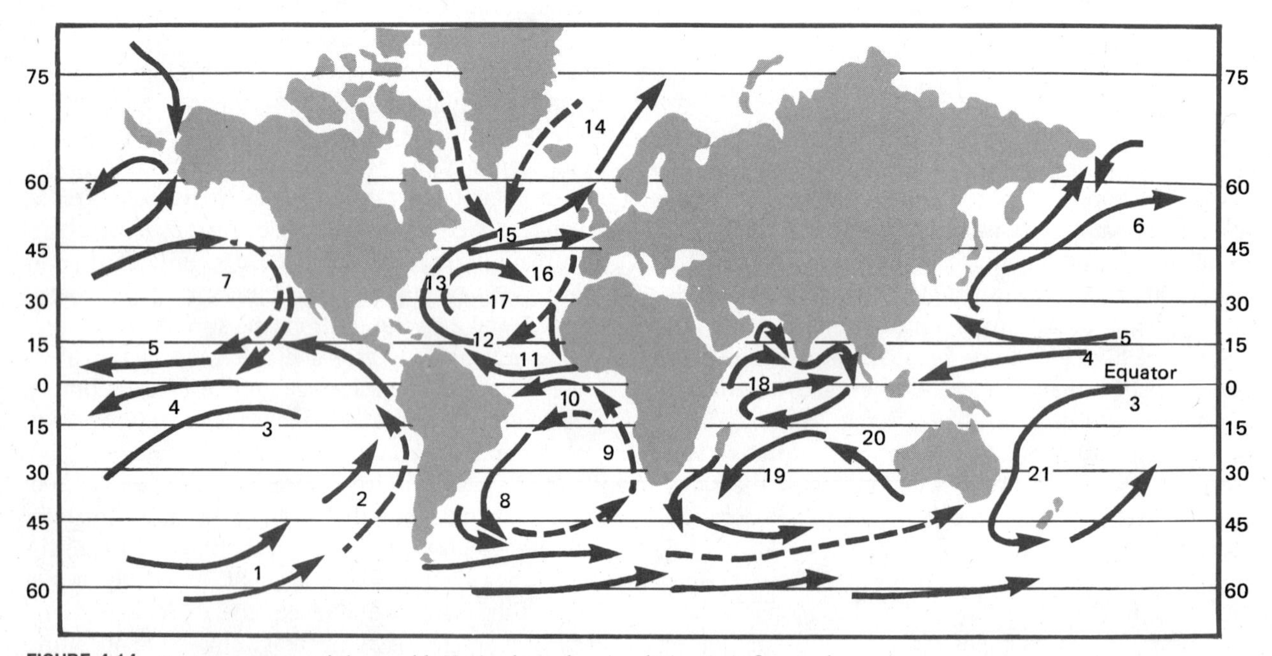

FIGURE 4.14 Ocean currents of the world. Notice how the circulation is influenced by continental land masses and how oceans are interconnected by currents. (1) Antarctic West Wind Drift; (2) Peru Current; (3) South Equatorial Current; (4) Counter Equatorial Current; (5) North Equatorial Current; (6) Kuroshio Current; (7) California Current; (8) Brazil Current; (9) Benguela Current; (10) South Equatorial Current; (11) Guinea Current; (12) North Equatorial Current; (13) Gulf Stream; (14) Norwegian Current; (15) North Atlantic Current; (16) Canaries Drift; (17) Sargasso Sea; (18) Monsoon Drift (summer: east; winter: west); (19) Mozambique Current; (20) West Australian Current; (21) East Australian Current. (Dashed arrows represent cool water.)

The Gulf Stream, for example, carries warm water up along the Atlantic Coast, across to England, then south along the Canary Islands. That warm current moderates the climates of northern North America and England, which otherwise would be much colder. Currents moving toward the equator along the west side of the continents are warmer than adjacent land masses. Air over the water warms, picks up moisture, cools over land, and drops heavy rains, as it does along the coast of British Columbia and the northwestern United States.

MICROCLIMATES

When the weather report states that the temperature is 30° C and the sky is clear, the information may reflect the general weather conditions for the day. However, on the surface of the ground in and beneath the vegetation, on slopes and cliff tops, in crannies and pockets, the climate is quite different. Heat, moisture, air movement, and light all vary radically from one part of the community to another to create a whole range of "little" or "micro" climates. The microclimate, rather than the local climate, defines the conditions under which most organisms live. It is, as David Gates defines it, "the climate in the immediate vicinity of an object or organism."

On a summer afternoon the temperature under calm, clear skies may be 28° C at 2 m (6 ft), the standard height of temperature recording. But on or near the ground—at the 50 mm (2 in.) level—the temperature may be 10° higher; and at sunrise, when the temperature for the 24-hour period is the lowest, the temperature may be 5° lower. Thus, in the middle eastern part of the United States, the temperature near the ground may correspond to the temperature at the 1.83 m level in Florida, 700 miles to the south; and at sunrise the temperature may correspond to the 1.83 m level temperature in southern Canada. Even greater extremes occur above and below the ground surface. In New Jersey, March temperatures about the stolons of clover plants 12 mm (0.5 in.) above the surface of the ground may be 21° C, while below the surface, the temperature about the roots is −1° C. The temperature range within a vertical distance of 89 mm (3.5 in.) is 20° C. Under such climatic extremes, most organisms exist.

The chief reason for the great differences between the ground and the 2 m level is solar radiation. During the day the soil, the active surface, absorbs solar radiation, which comes in short waves as light, and radiates it back as long waves to heat a thin layer of air above (Figure 4.15). Because airflow at ground level is almost nonexistent, the heat radiated from the surface remains close to the ground. Temperatures decrease sharply in the air above this layer and in the soil below. Thus, on a sunny but chilly spring or late winter day, you can walk on muddy ground while the air about you is cold.

Heat absorbed by the ground during the day is reradiated by the ground at night. This heat is partly absorbed by water vapor in the air above. The drier the air, the greater the outgoing heat and the stronger the cooling of the surface of the ground and the vegetation. Eventually, ground and vegetation are cooled to the dew point, and water vapor in the air may condense as dew. After a heavy dew a thin layer of chilled air lies over the surface, the result of rapid absorption of heat in the evaporation of dew.

By altering wind movement, evaporation, moisture, and soil temperatures, vegetation influences or moderates the microclimate of an area, especially near the ground. Temperatures at the ground level under the shade are lower than in places exposed to sun and wind. On fair summer days a dense forest cover can reduce

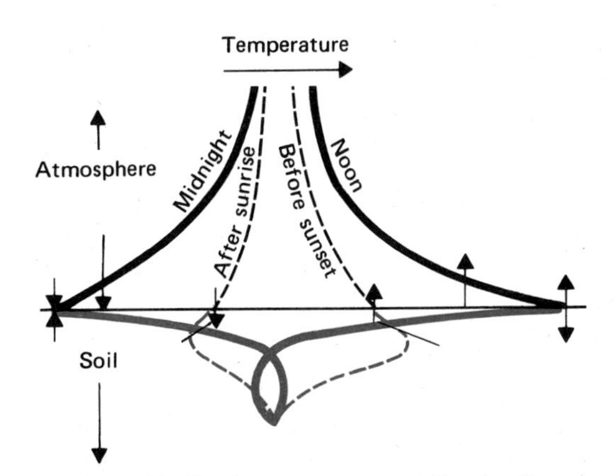

FIGURE 4.15 Idealized temperature profiles in the ground and air for various times of day.

the daily range of temperatures at 25 mm (1 in.) by 7° to 12° C, compared with the temperature in the soils of bare fields.

Vegetation also reduces the steepness of the temperature gradient and influences the height of the active surface, the area that intercepts the maximum quantity of solar insolation. In the absence of vegetation or in the presence of very thin vegetation, temperature increases sharply near the soil; but as the plant cover increases in height and density, the leaves of the plants intercept more solar radiation (Figure 4.16). The plant crowns then become the active surface and raise it above the ground. As a result, daytime temperatures are higher just above the dense crown surface and lowest at the surface of the ground.

Within dense vegetation air movements are reduced to convection and diffusion. In dense grass and low plant cover, complete calm exists at ground level. This calm is an outstanding feature of the microclimate near the ground. It influences both temperature and humidity and creates a favorable environment for insects and other animals.

Humidity differs greatly from the ground up. Because evaporation takes place at the surface of the soil or at the active surface of plant cover, the vapor content (absolute humidity) decreases rapidly from a maximum at the bottom to atmospheric equilibrium above. Relative humidity increases above the surface, because actual vapor content increases only slowly during the day, while the capacity of the heated air over the surface to hold moisture increases rather rapidly.

The greatest microclimatic differences exist between north-facing and south-facing slopes. South-facing slopes receive the most solar energy, which is maximal when the slope grade equals the sun's angle from the zenith point. North-facing slopes receive the least energy, especially when the slope grade equals or exceeds the angle of sun rays.

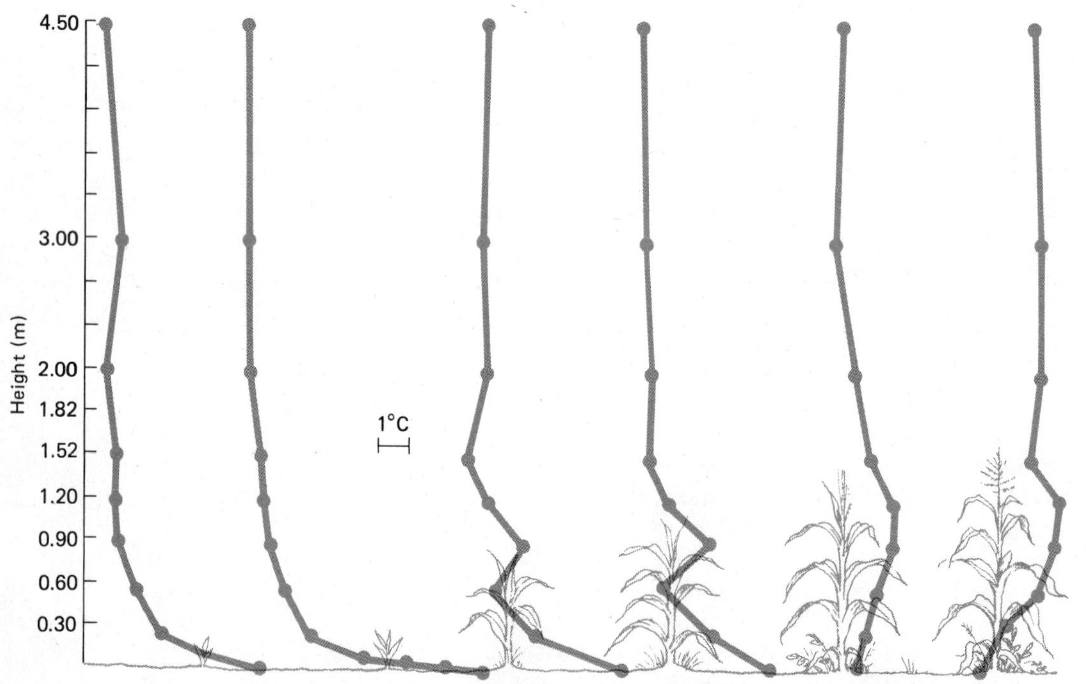

FIGURE 4.16 Vertical temperature gradients at midday in a cornfield from the time of seeding to the time of harvest. Note the increasing height of the active surface.

At latitude north 41° (about central New Jersey and southern Pennsylvania), midday sunlight on a 20° slope is, on the average, 40 percent greater on south-facing slopes than on north-facing slopes during all seasons. This difference has a marked effect on the moisture and heat of the two sites. High temperatures and associated low vapor pressures pull out moisture from the soil and plants. The evaporation rate is often 50 percent higher, the average temperature higher, the soil moisture lower, and the extremes more variable on south-facing slopes. Thus, the microclimate ranges from warm, dry conditions with wide extremes on the south-facing slope to cool, moist, less variable conditions on the north-facing slope. Conditions are driest on the top of south-facing slopes, where air movement is the greatest, and dampest at the bottom of the north-facing slopes (Figure 4.17).

In the central and southern Appalachians, north-facing slopes are steeper and include many minor *microreliefs*—small depressions and benches created largely by the upheaved roots of thrown trees. South-facing slopes are longer and less steep because of long-term downward movements of the soil.

The widest climatic extremes occur in valleys and pockets, areas of convex slopes, and low concave surfaces. These places have much lower temperatures at night, especially in winter; much higher temperatures during the day, especially in summer; and a higher relative humidity. Protected from the circulating influences of the wind, the air becomes stagnant. It is heated by sunlight and cooled by terrestrial radiation, in sharp contrast to the wind-exposed, well-mixed air layers of the upper slopes. In the evening cool air from the uplands flows down the slope into the pockets and valleys to form lakes of cool air. Often when the warm air in the valley comes in contact with the inflowing cold air, the moisture in the warm air condenses as valley fog.

The whole north-south slope climate is the result of a long chain of interactions: solar radiation influences moisture; moisture influences the species of trees and other plants occupying the slopes (Figure 4.18); the species of trees, in turn, influence mineral recycling, which is reflected in the nature and chemistry of the surface soil and the nature of the herbaceous ground cover.

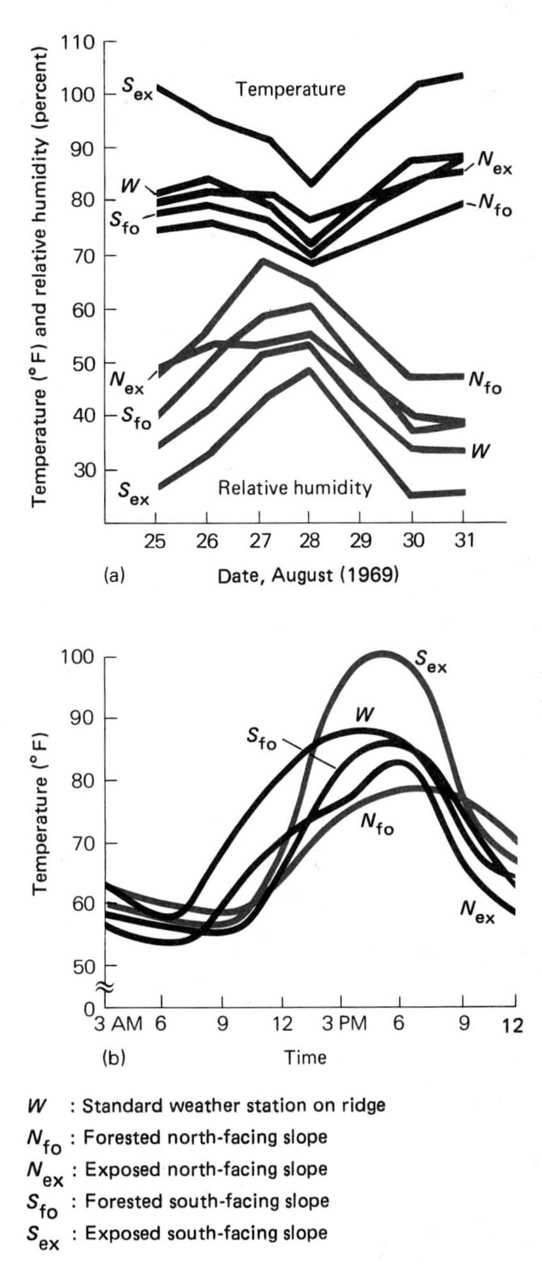

FIGURE 4.17 Temperature and humidity microclimates on a slope in Greer, West Virginia.

CLIMATE MODIFICATION

Ever since humans became an important ecological force they have altered Earth's vegetation and inadver-

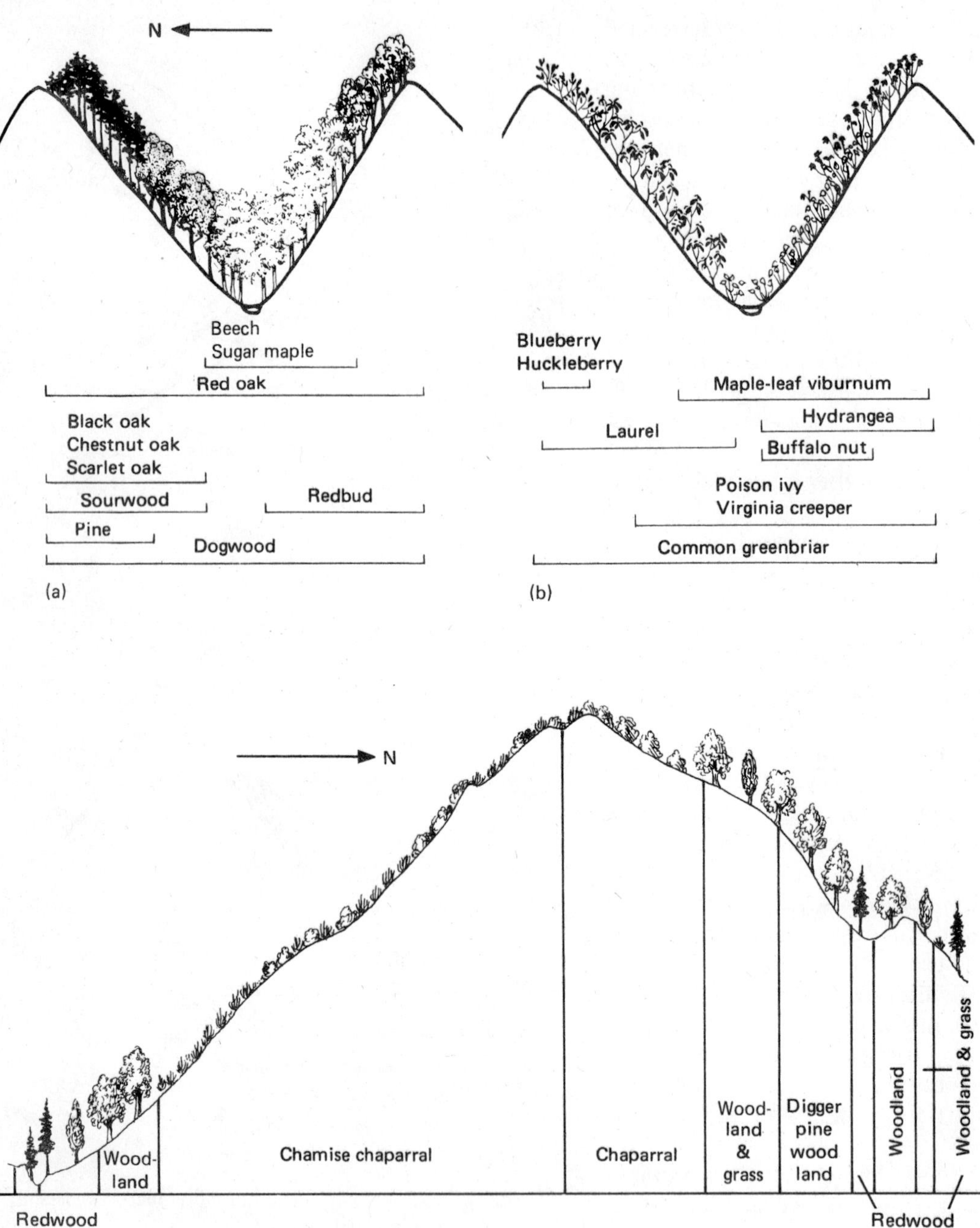

FIGURE 4.18 Influence of microclimate on the distribution of vegetation on north-facing and south-facing slopes. (a, b) Trees and shrubs in the hill country of southwestern West Virginia. (c) The vegetation of the Point Sur area of California. In each case note the similarity of vegetation on the lower parts of the north-facing and south-facing slopes.

tently influenced regional and global climate. Agricultural and grazing practices reduced parts of the world to desert; clearing of dense forest for grazing and crop lands, diversion of rivers, and draining of wetlands have altered heat balance and humidity. Twenty percent of the total area of the continents has been drastically changed, influencing heat and water budgets of Earth. Human impact on climate, however, has been most pronounced since the advent of industrialization. The ever-increasing burning of fossil fuels, massive clearing of forests, especially in tropical regions, and the concentration of populations into large urban complexes continue to inject large amounts of heat, carbon dioxide, and pollutants into the atmosphere (see Chapter 24).

The effect of these human-induced inputs into the atmosphere is to intensify the greenhouse effect, predicted to increase global warming. A major impact of global warming would not necessarily be change in average weather and global temperatures, but the occurrence of extreme events such as summer heat waves, abnormally warm or cold spells in winter, and prolonged droughts. The subtropical monsoonal rain belts would become wetter, and the growing seasons at high latitudes would become longer. Some midlatitudinal areas would experience wetter springtimes, while interior midlatitudinal areas would experience drier midsummer conditions, severely affecting grain production and water supplies. Sea levels could rise as much as 1 m, flooding coastal regions. All these events could influence the distribution and survival of many plants and animals, affect agricultural crop production, and create serious problems for human settlements and economic activity.

On a smaller scale, human impact on climate is exemplified by the pronounced climatic variances in large urban areas. The climate of such areas is a product of urban structure and the density and activity of its occupants. In the urban complex, stone, asphalt, and concrete pavement and buildings, with their high capacity for absorbing and reradiating heat, replace natural vegetation, with its low conductivity. Rainfall on impervious surfaces is drained away as fast as possible, reducing evaporation. Metabolic heat from masses of people and waste heat from buildings, industrial combustion, and vehicles raise the temperature of the surrounding air. Industrial activities, power production, and vehicles pour water vapor, gases, and particulate matter into the atmosphere in great quantities.

The effect of this storage and reradiation of heat is the formation of a *heat dome* about cities, large and small, in which the temperature may be 6° to 8° C higher than in the surrounding countryside. Heat domes are characterized by high temperature gradients. The highest temperatures are associated with areas of highest density and activity, while temperatures decline markedly toward the periphery of the city. Although they are detectable throughout the year, heat domes are most pronounced during the summer and early winter and are more noticeable at night, when heat stored by pavements and buildings is reradiated to the air.

During the summer the buildings and pavement of the inner city absorb and store considerably more heat than the vegetation of the countryside. In cities with narrow streets and tall buildings, the walls radiate heat toward each other instead of toward the sky. At night these structures slowly give off heat stored during the day.

In the winter solar radiation is considerably less because of the low angle of the sun, but heat accumulates from human and animal metabolism, home heating, power generation, industry, and transportation. In fact, the heat contributed by these sources is 2½ times that contributed by solar radiation. This energy reaches and warms the atmosphere directly or indirectly, moderating the winter climate of the city over that of the countryside.

Throughout the year urban areas are blanketed with particulate matter, CO_2, and water vapor. The haze reduces solar radiation reaching the city, which may receive 10 to 20 percent less than the surrounding countryside. At the same time, the blanket of haze absorbs part of the heat radiating upward and reflects it back; part of this heat warms the air, and part warms the ground. The higher the concentration of pollutants, the more intense is the heat island. The particulate matter has other microclimatic effects. Because of the city's low evaporation rate and the lack of vegetation, relative humidity is lower in the city than in surrounding rural areas. However, the particulate matter acts

TABLE 4.1 Climate of the City Compared to the Country

City	Country
Condensation nuclei and particles	10 times more
Gaseous admixtures	5–25 times more
Cloud cover	5–10 percent more
Winter fog	100 percent more
Summer fog	30 percent more
Total precipitation	5–10 percent more
Relative humidity, winter	2 percent less
Relative humidity, summer	8 percent less
Radiation, global	15–20 percent less
Duration of sunshine	5–15 percent less
Annual mean temperature	0.5°–1.0° C more
Annual mean wind speed	20–30 percent less
Calms	5–20 percent more

as condensation nuclei for water vapor in the air, producing fog and haze. Fogs are much more frequent in urban areas than in the country, especially in the winter (Table 4.1).

Summary

Solar radiation—transformed into chemical, thermal, and kinetic energy—is the basis of life on Earth. Variations in heat and Earth's daily rotation produce the prevailing winds, move ocean currents, and influence rainfall patterns.

Daily heating and cooling cause air masses to rise and sink. Under certain conditions the temperature of air masses increases with height rather than decreases. Such an air mass is very stable, creating an inversion, which can trap atmospheric pollutants and hold them close to the ground.

Local winds are produced by temperature changes in air masses along coastal areas and in mountainous regions. Differences in the heating of air over mountain slopes and valley bottoms produce upslope breezes by day and downslope breezes by night. Such local winds can influence microclimatic conditions.

Atmospheric moisture is expressed in terms of relative humidity. Relative humidity is higher at night than during the day and greater at higher elevations, but varies widely with topography, vegetation, wind, and temperature.

The actual climatic conditions under which organisms live vary considerably from one local area to another. These variations, or microclimates, are influenced by topographic differences, height above the ground, vegetative cover, exposure, and other factors. Most pronounced are environmental differences between ground level and upper strata and between cool north-facing and warm south-facing slopes.

Human activities have modified Earth's climate. Of considerable concern is global warming, which could seriously affect natural ecosystems, agriculture and forestry, water supply, and human and social well-being. A strong example of climate modification on a local scale is the microclimate over urban areas. A city is characterized by the presence of a heat dome. Compared to surrounding rural areas, a city has a higher average temperature, particularly at night, more cloudy days, more fog, more precipitation, a lower rate of evaporation, and lower humidity.

Review and Study Questions

1. What is the solar constant? What is the fate of the sun's energy when it reaches Earth's atmosphere?
2. Why does less solar energy reach the poles than the equator?
3. How does Earth's interception of the sun's rays through the year influence the seasons?
4. How does Earth maintain a thermal balance?
5. What is adiabatic heating and cooling? What is the lapse rate?
6. What is the Coriolis effect, and how does it affect atmospheric circulation?
7. What causes ocean currents? What are gyres?
8. When is an air mass stable? Unstable? What is an inversion? A radiational inversion? A subsidence inversion?
9. What are local winds? How are they caused, and what is their effect?
10. What is a microclimate?
11. Explain how temperatures change above and below vegetation during the growing season. How might that influence animal life?
12. Why and how is the microclimate of a north-facing slope different from that of a south-facing slope?

*13. The predicted global warming is a hotly debated

topic internationally, especially because correcting the causes can be economically costly. What will the costs be if we don't take corrective measures? Debate the topic.

Selected References

Gates, D. M. 1962. *Energy exchange in the biosphere.* New York: Harper & Row. A basic advanced reference.

Gates, D. M. 1972. *Man and his environment: Climate.* New York: Harper & Row. An excellent introduction.

Geiger, R. 1965. *Climate near the ground.* Cambridge, MA: Harvard University Press. A classic book on microclimate.

Lee, R. 1978. *Forest microclimatology.* New York: Columbia University Press.

Schneider, S. H. 1989. *Global warming.* San Francisco: Sierra Club Books. A sound discussion of the subject.

Strahler, A. 1971. *The earth sciences.* New York: Harper & Row. Detailed presentation of climate, among other topics.

Study of Man's Impact on Climate (SMIC). 1971. *Inadvertent climate modification.* Cambridge, MA: MIT Press. A good overview.

Woodward, F. I. 1987. *Climate and plant distribution.* Cambridge, England: Cambridge University Press. More technical introduction to climate and its influence on plant distribution.

CHAPTER 5

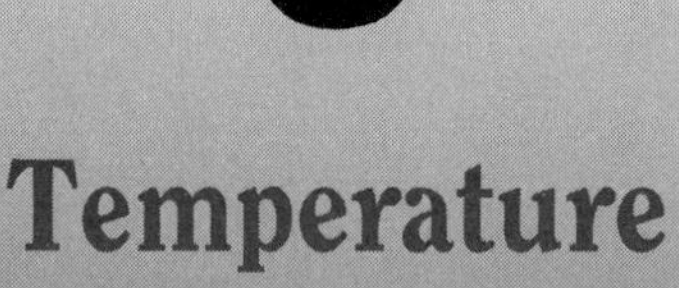

Temperature

Outline

Objectives

On completion of this chapter, you should be able to:

1. Describe the thermal balance of organisms.
2. Distinguish among poikilothermy, homeothermy, and heterothermy.
3. Explain how animal behavior, body structures, and body functions aid in the maintenance of temperature.
4. Explain countercurrent circulation and its adaptive value.
5. Distinguish among hibernation, estivation, torpor, and diapause.
6. Describe how plants meet extremes of heat and cold.
7. Explain the role of temperature in the distribution of plants and animals.

HEAT BALANCE

All organisms live in a thermal environment and exchange energy with it. They absorb solar radiation, which may be direct, diffused from the sky, or reflected from the ground (Figure 4.1). In addition, they absorb infrared thermal radiation from rocks, soil, vegetation, and atmosphere. Organisms produce metabolic heat and lose heat to the environment.

The transfer of heat between organisms and environment occurs in several ways (Figure 5.1). Heat moving by diffusion from a warm solid object to a cool object is *conduction.* How fast heat is moved depends upon the degree of contact between the two objects and their temperature difference. The greater the temperature difference and the thinner the conducting layer, the faster heat moves. *Convection* takes place when air or water moves over an object. Heat transfer by convection is much more rapid than by conduction for a given temperature difference. *Evaporation* depends upon the difference in vapor pressure between the air and the object as well as the resistance of the surface to the loss of moisture. If the humidity of the air is high, little evaporative loss occurs. Another important type of heat transfer is infrared or *thermal radiation.* Radiant energy impinging on an object is absorbed and then converted into heat at the surface of the absorbing object. You experience that type of heat transfer when you stand in front of a fire.

In a general way, when the surrounding temperature is lower than the temperature of the organism, heat is lost to the environment. When the surrounding temperature is higher, heat moves from the environment to the plant or animal. The problem, then, is for the organism to balance heat gains with heat losses.

The heat balance of an organism may be summarized by the following expression:

Heat gain (solar radiation + thermal radiation + food energy storage)
= heat loss (thermal radiation + conduction + convection + evaporation)

THERMAL RELATIONS IN ANIMALS

We divide animals into two broad groups by the way they handle heat (Figure 5.2). Popularly, *poikilotherms* —which means variegated "varied temperatures"—are called "cold-blooded" because they are cool to the touch; actually their body temperatures may get high. *Homeotherms*—which means "same temperature"—are known as "warm-blooded." The body temperature of poikilotherms is more or less in equilibrium with their thermal environment. Their body temperature varies as the environmental temperature varies. Poikilotherms tend to be *ectothermic.* They depend upon the environment as a source of heat to maintain body temperature. Homeotherms regulate their body temperature by physiological means. They tend to be *endothermic,* depending upon internally produced metabolic heat. The two terms, poikilothermy and homeothermy, do not represent two distinct groups of organisms. Rather the two represent extremes in a gradient of physiological responses.

POIKILOTHERMY

Rates of metabolism among poikilotherms are controlled by environmental temperatures. Rising temperatures increase rates of enzymatic activity, which control metabolism and oxidation of carbohydrate reserves. For every 10° C rise in temperature, the rate of metabolism in poikilotherms doubles. Obviously, poikilotherms have an upper limit that they can tolerate. Conversely, when ambient temperatures fall, metabolic activity declines. Most terrestrial poikilotherms are able to maintain a somewhat constant body temperature by behavior. Lizards, for example, may vary their body temperature no more than 4° to 5° C when active, and amphibians 10° C. The range of body temperatures at which poikilotherms carry out their daily activities is called the active temperature range (ACT). Poikilotherms save metabolic energy by adapting their physiological and developmental processes to a limited ACT.

Poikilotherms such as fish and mollusks are completely immersed in water and do not maintain any appreciable difference between their body temperature and the surrounding water. Aquatic poikilotherms are poorly insulated, and water quickly absorbs any metabolic heat. Because seasonal water temperature is relatively stable, fish and aquatic invertebrates maintain a fairly constant seasonal body temperature. They become temperature specialists with a low range of temperature variation.

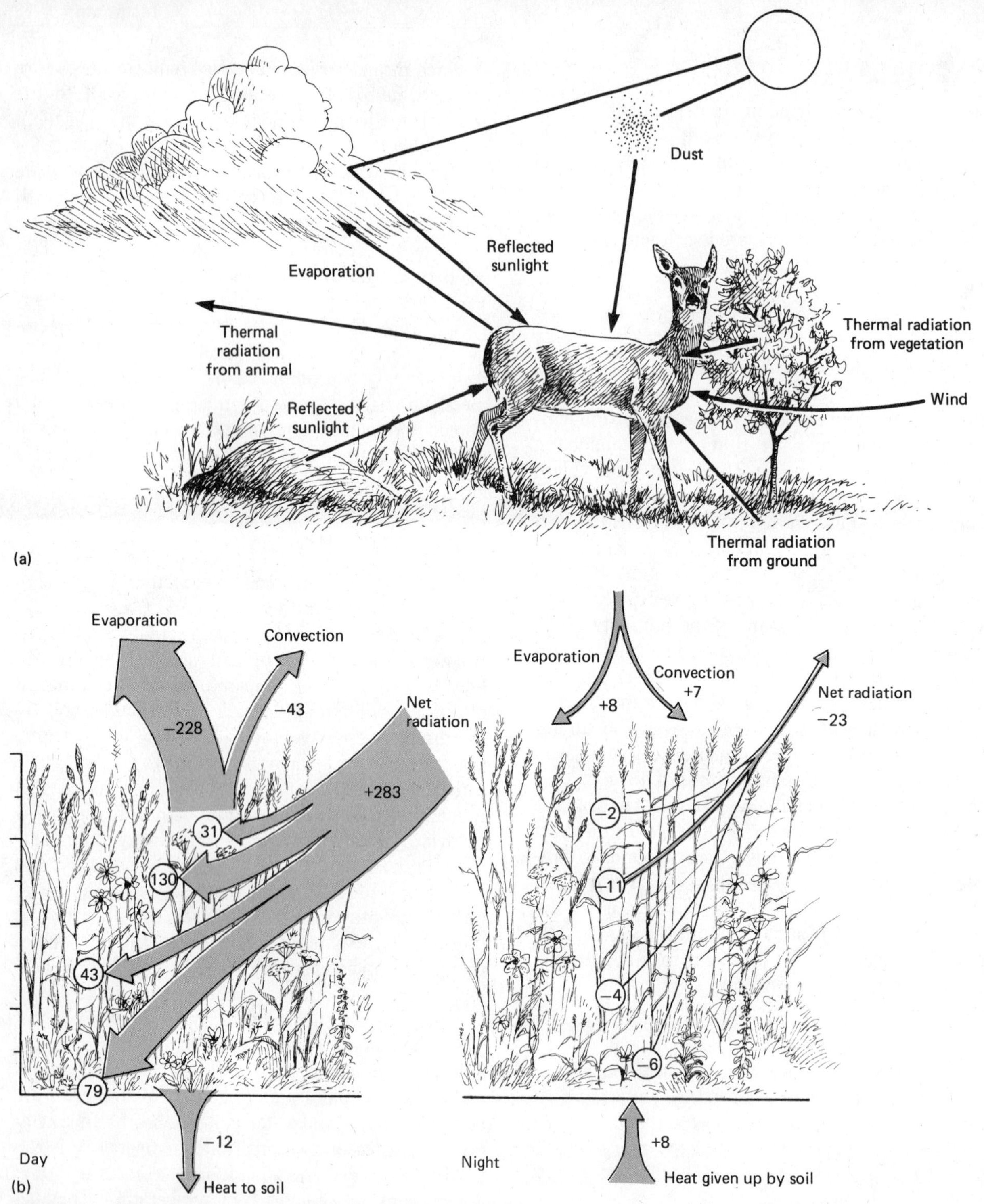

FIGURE 5.1 (a) Exchange of energy between a mammal and its environment. (b) Energy exchange in a meadow on a sunny day (left) and night (right). Net radiation and heat exchange are reversed at night.

(a)

FIGURE 5.2 Representatives of poikilothermic, homeothermic, and heterothermic animals. (a) The poikilothermic long-tailed salamander lives along margins of streams. (b) The homeothermic snowshoe hare inhabits the northern forest regions of North America. (c) The myotis bat undergoes daily torpor and hibernates in winter.

(b)

(c)

Fish and aquatic invertebrates adjust seasonally to changing water temperatures by *acclimatization*: They undergo cellular and enzymatic changes over a period of time. Poikilotherms have an upper and lower limit of tolerance to temperature that varies with species. If they live at the upper end of their tolerable range, poikilotherms will adjust their physiology to that thermal environment, but at the expense of being able to tolerate the lower range. Similarly, during cold temperature periods, the animals adjust to a low temperature that would have been lethal when they were acclimatized to a higher temperature range. Because water temperatures change slowly through the year, aquatic poikilotherms are able to make the adjustment slowly. If they are suddenly subjected to a temperature higher or lower than the one to which they are acclimated, they will experience thermal shock and die.

Poikilotherms have a low metabolic rate and high thermal conductance between body and environment. Most of their energy production—50 to 98 percent—is from anaerobic metabolism, which depletes cellular energy stores and accumulates lactic acid in the tissues. It results in rapid physical exhaustion, often within 3 to 5 minutes.

Nevertheless, poikilothermy has its advantages. Poikilotherms are able to allocate more of their energy intake to biomass rather than to metabolic needs. Because they do not depend upon internally generated body heat, ectotherms can curtail metabolic activity in times of food and water shortage or temperature extremes. Their low energy demands enable poikilotherms to colonize areas of limited food and water. Nor are they restricted to a minimal body size because of metabolic heat loss. That fact permits poikilotherms to exploit resources and habitats unavailable to endotherms.

HOMEOTHERMY

By means of internal heat production, homeotherms maintain a body temperature higher than ambient. A high body temperature enables homeotherms to remain active regardless of environmental temperatures. It is associated with specific enzyme systems designed to operate within a narrow temperature range. A high body temperature may be necessary to maintain the intricate nervous coordination associated with highly developed behavioral patterns.

High body temperatures may have evolved from an inability of large animals to dissipate rapidly the heat produced during periods of high activity. That situation would favor enzyme systems that function at a high temperature, with a set point around 40° C. Otherwise, the animal would have to expend an enormous amount of water for evaporative cooling to ambient temperature. The ability to operate at a high temperature provides homeotherms with greater endurance and the means of remaining active at low environmental temperatures.

Homeotherms have a high metabolic rate and low thermal conductance. Because of efficient cardiovascular and respiratory systems that bring oxygen to the tissues, homeotherms can maintain a high level of aerobic energy production. Thus, they are able to sustain a high level of physical activity without exhaustion. What causes physical exhaustion in homeotherms is the heat generated by activity. Homeotherms are able to function independently of external temperatures and to a point are able to exploit a wider range of thermal environments. They can generate energy rapidly when the situation demands, such as escape from predators and pursuit of prey.

The major disadvantage of homeothermy is the high energetic cost of aerobic metabolism, which leaves a minimum of energy for biomass accumulation. A high resting metabolism is not compatible with a low-energy way of life. As a result, metabolic costs weigh heavily against smaller homeotherms, which places a lower limit on body size.

BODY SIZE

A close relation exists between body size and basal metabolic rate. Basal metabolism is proportional to body mass raised to the ¾ power. As body weight increases, the weight-specific metabolic rate decreases. A doubling of body mass, for example, would increase the basal rate of metabolism by 75 percent. Conversely, as body mass decreases, basal metabolism increases, exponentially with very small body sizes (Figure 5.3). Within any given taxonomic group, small animals have a higher metabolism and require more food per unit of body weight than large ones. Small shrews (*Sorex* spp.), for example, require daily an amount of food (wet weight) equivalent to their own body weight. It is as if a 150-pound human were to need 150 pounds of food daily to stay alive. Thus, small animals are forced to spend most of their time seeking and consuming food.

Part of the reason lies in the ratio of surface area to body mass (Figure 5.4). Heat is lost to the environment in proportion to surface area exposed. All other things being equal, large animals lose less heat to the environment than small ones. To maintain homeostasis, small animals have to burn energy rapidly. In fact, the weight-specific rates of small endotherms rise so rap-

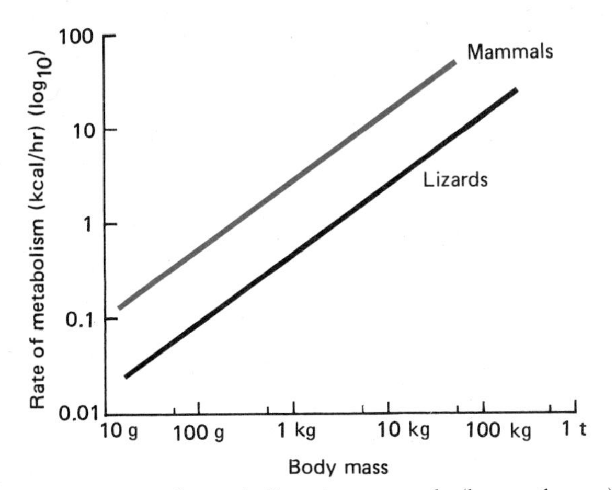

FIGURE 5.3 Basal metabolism in mammals (homeotherms) and standard rate of metabolism (30° C) in lizards (poikilotherms) as a function of body mass. Rate of metabolism is greater in mammals than in lizards of corresponding body size. The slope is close to 0.75 for both homeotherms and poikilotherms. The metabolic rate does not increase in direct proportion to body mass.

idly that below a certain size they could not meet their energy demands. Five grams is about as small as an endotherm can be and still maintain a metabolic heat balance. Few endotherms ever get that small. Because of that problem, most young birds and mammals begin life as ectotherms, depending upon body heat of the parents to maintain their body temperatures. That allows young animals to allocate most of their energy to growth.

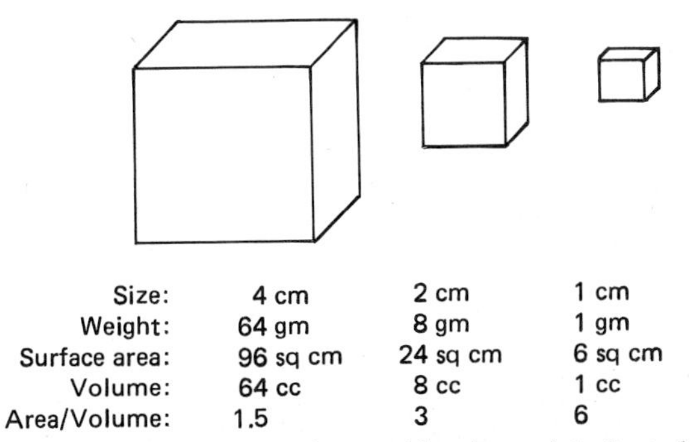

FIGURE 5.4 These three cubes—with edges of 4, 2, and 1 cm—point out the relationship between surface area and volume. A small object has more surface area in proportion to its volume than a large object of similar shape.

Among poikilotherms standard metabolic rates are a function of prevailing environmental temperatures. Because they do not depend on internally generated body heat to maintain tissue temperature, such poikilotherms as reptiles and amphibians have a resting metabolic rate only 10 to 20 percent of that of birds. During periods of inactivity their temperatures drop to ambient, with a reduction in metabolic rate. Ectotherms, then, are able to exploit a world of small body size.

Large poikilotherms, especially large reptiles, retain some of their body heat because of a small body surface to volume ratio. That, possession of scaled skin, and a greater resistance to changes in environmental temperature allow such poikilotherms to acquire a degree of endothermy. From such large ectotherms in the past, endothermism in birds and mammals, both reptilian descendants, probably evolved. It seems likely that mammals evolved from therapsids, partially committed to endothermy because of their large body size and modified skin. Once mammals acquired endothermism, it was retained by smaller types.

ANIMAL HEAT STRATEGIES

To survive and flourish, animals have to maintain a thermal balance with their environment. To accomplish that end both poikilotherms and homeotherms have evolved physical and physiological means to avoid gaining and losing excessive amounts of heat and to reduce the physiological stress of heat and cold.

BEHAVIORAL MECHANISMS

To maintain a tolerable and fairly constant body temperature, terrestrial and amphibious poikilotherms resort to behavioral mechanisms. These animals—including many insects such as butterflies, dragonflies, damselflies, and others—bask in the sun to raise their body temperatures to the level necessary to become highly active (heliothermism). When they become too warm, these animals seek the shade. Amphibians can maintain fairly uniform body temperatures in summer by moving between different thermal environments (Figure 5.5). Reptiles raise or lower their bodies relative to the ground and change body shape to increase or decrease the conduction of heat between themselves and the rocks or soil on which they rest. They may also seek shade or sunlight or burrow into the soil to adjust their temperatures. Thus, the body temperature of poikilotherms does not necessarily follow ambient temperature.

Behavioral means of reducing heat gains and losses are also practiced by homeotherms. In the heat of a summer's day, birds and mammals seek shady places. Desert mammals go underground by day and emerge to become active by night. In winter some mammals, such as rabbits, go underground during periods of inactivity. Larger mammals, such as deer, seek the thermal cover of conifers and rhododendron thickets. Mammals such as flying squirrels and birds such as penguins and bobwhite quail huddle together during periods of cold, reducing individual surface area and conserving body heat.

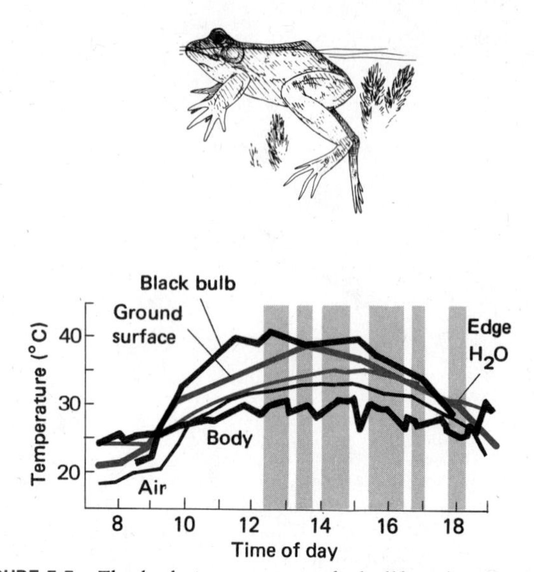

FIGURE 5.5 The body temperature of a bullfrog (*Rana catesbyiana*) measured telemetrically. Dips in the black bulb temperature indicate the effects of cloud cover, convection, or both. Water temperature around the pond's edge varies from one location to another as much as 2° to 3° C. Therefore in shallow water a frog may show a higher body temperature than that recorded for edge water. Note the relative uniformity of temperature the bullfrog maintains by moving in and out of water.

MORPHOLOGICAL MECHANISMS

Insulation

To regulate the exchange of heat between the body and the environment, homeotherms and certain poikilotherms utilize some form of insulation—a covering of fur, feathers, or layers of body fat. Such insulation reduces conductance. The higher the value of the insulation, the lower the conductance. For mammals, fur is a major barrier to heat flow, but its insulation value varies with thickness, which is greater on large mammals than small ones. Small mammals are limited in the amount of fur they can carry, because a thick coat could reduce their ability to move around. Mammals change the thickness of their fur with the season. A heavy coat of winter fur is shed for a much thinner one in summer. Aquatic mammals, especially those of arctic regions, and such arctic and antarctic birds as auklets and penguins have a heavy layer of fat or blubber beneath the skin. Birds reduce conductance by fluffing the feathers and drawing the feet into them, making the body a round feather ball. Some arctic birds, such as ptarmigan, have the tarsi feathered; among most, the tarsi are scaled and act as thermal windows.

Although the major function of insulation is to retain body heat, it may also serve to block absorption of heat by the body. In a hot thermal environment, an animal either has to rid itself of excess body heat (see "Physiological Mechanisms") or prevent heat from being absorbed by the body. One means is to reflect solar radiation by light-colored fur or feathers. Another is to grow a heavy coat of fur to insulate the body against environmental heat, a method employed by large mammals of the desert, notably the camel. Heat is absorbed by outer layers of hair and lost to the environment.

Insulation is also employed by a number of poikilotherms. Moths and bumblebees have a dense, fur-like coat over the thoracic region, which serves to retain the high temperature of flight muscles while the insects are in flight (Figure 5.6). The long, soft hairs of caterpillars, together with changes in body posture, act as selective insulation by reducing convective heat exchange.

PHYSIOLOGICAL MECHANISMS

To maintain a constant body temperature, homeotherms have to employ some physiological means of reducing body heat or increasing it. To become active at relatively low temperatures, some poikilotherms become endothermic, raising body temperature far above ambient.

Continually Active Animals

Metabolism. When ambient temperatures fall to a point beyond which body insulation is no longer effective in reducing heat loss, homeotherms increase basal metabolism. As a last resort the animal may revert to shivering, a form of involuntary muscular activity that increases heat production.

A number of small mammals can increase heat production by nonshivering *thermogenesis*—burning highly vascular brown fatty tissue capable of a high rate of oxygen consumption. Brown fat, found about the head, neck, thorax, and major blood vessels, is prominent in three groups of mammals: cold-acclimated adults, hibernators, and newborn young.

To become active, some poikilotherms become endothermic and raise their body temperatures above ambient by physiological means. Insects such as the sphinx moth and bumblebee raise their body temperatures by increasing their rate of metabolism through rapid contraction of the wing muscles. For example, to fly, the bumblebee has to raise its muscle temperature to between 35° and 40° C.

Evaporative Cooling. Many birds and mammals employ evaporative cooling to reduce the body-heat load. They lose some heat by evaporation of moisture diffused through the skin. Birds and mammals can accelerate evaporative cooling by sweating and panting. Sweating is restricted to certain groups of mammals, particularly horses, pigs, humans, and others that possess sweat glands in the skin. Panting in mammals and gular fluttering in birds increase the movement of air over moist surfaces in the mouth and esophagus.

Hyperthermia. Storing body heat is not exactly a sound option to avoid high thermal environments, because of an animal's limited tolerance for high body temperatures. However, certain mammals—such as the camel, oryx, and some gazelles—do just that. The camel, for example, stores up body heat by day and unloads it by night by conduction and radiation, especially when water is limited. Its temperature can fluctuate from 34° C in the morning to 41° C by late afternoon. By storing body heat these animals reduce the need for evaporative cooling and thus reduce water loss. They decrease thermal gradient between the environment and the body, thereby decreasing the heat flow from the environment. Finally—in the camel, especially—the fur presents a barrier to heat gain.

Cold Tolerance. Many poikilothermic animals of temperate and arctic regions withstand long periods of below-freezing temperatures in winter through supercooling and resistance to freezing.

Supercooling takes place when the body temperature falls below freezing without freezing body fluids. The amount of supercooling that can take place is influenced by the presence of certain solutes in the body. Supercooling is employed by some arctic marine fish, certain insects of temperate and cold climates, and reptiles exposed to occasional cold nights.

Some intertidal invertebrates of high latitudes and certain aquatic insects actually survive the cold by freezing and then thawing out when the temperature moderates. In some, more than 90 percent of the body fluids may become frozen, and the remaining fluids contain highly concentrated solutes. Ice forms outside the shrunken cells, and muscles and organs are distorted. After thawing they quickly resume normal shape.

Other animals, particularly arctic and antarctic fish and many insects, resist freezing because they increase the concentration of solutes in body fluids, particularly glycerol. Glycerol protects against freezing damage, increasing the degree of supercooling. Wood frogs (*Rana sylvatica*), spring peepers (*Hyla crucifer*), and gray tree frogs (*H. versicolor*) can successfully overwinter just beneath the leaf litter because they accumulate glycerol in their body fluids.

Countercurrent Circulation. To conserve heat in a cold environment and to cool vital parts of the body in a hot environment, a number of animals have evolved countercurrent heat exchangers (Figure 5.6a). For ex-

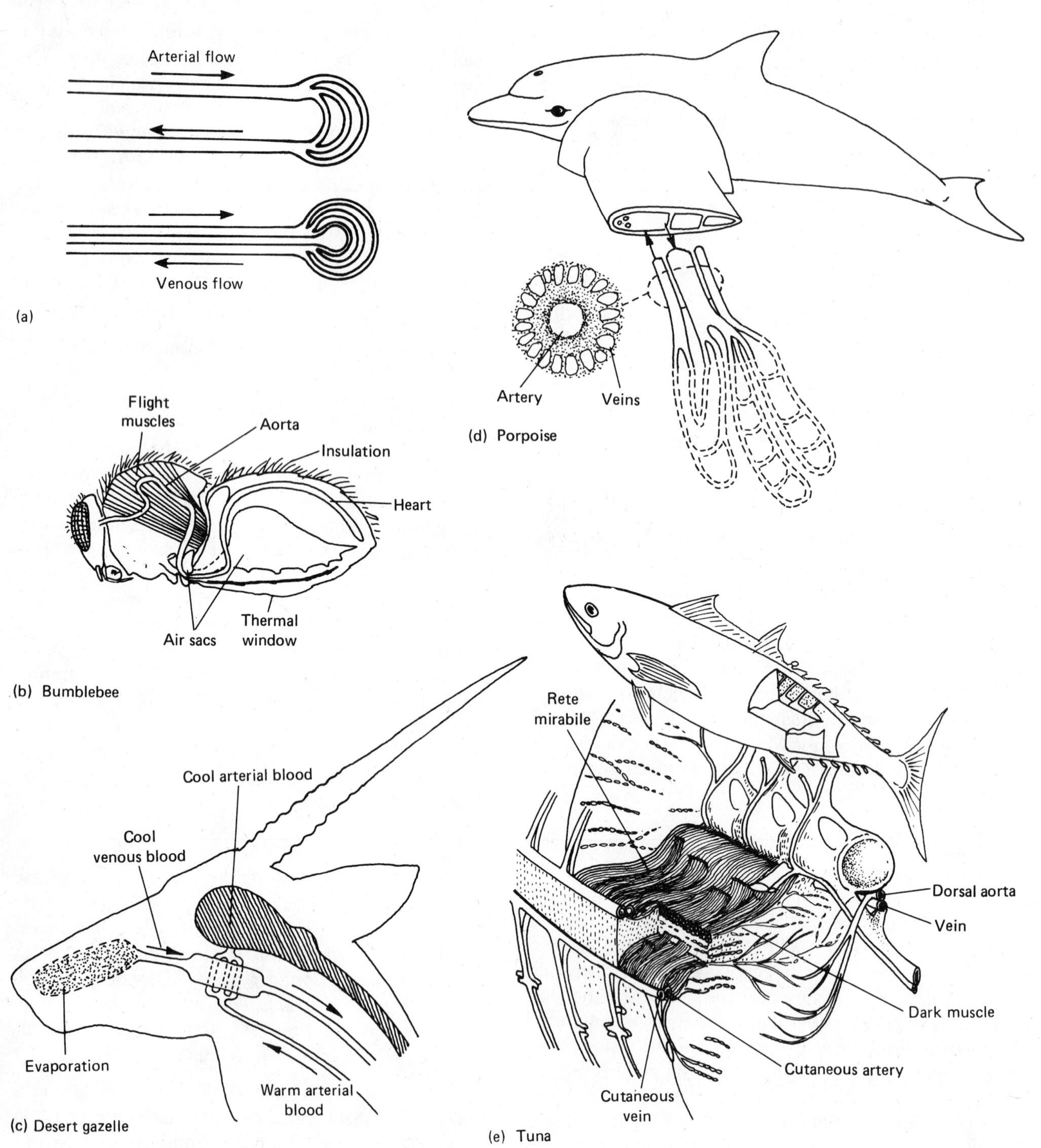
Arterial flow
Venous flow
(a)
Flight muscles
Aorta
Insulation
Heart
Air sacs
Thermal window
(b) Bumblebee
Cool arterial blood
Cool venous blood
Evaporation
Warm arterial blood
(c) Desert gazelle
Artery
Veins
(d) Porpoise
Rete mirabile
Dorsal aorta
Vein
Dark muscle
Cutaneous artery
Cutaneous vein
(e) Tuna

ample, the porpoise, swimming in cold arctic waters, is well insulated with blubber, but it could experience an excessive loss of body heat through its uninsulated flukes and flippers. It maintains its body core temperature by exchanging heat between arterial and venous blood in these structures. Arteries carrying warm blood from the heart to the extremities are completely surrounded by veins (Figure 5.6d). Warm arterial blood loses its heat to the cool venous blood returning to the body core. As a result, blood entering the flippers is cool, so that little body heat is lost to the environment, while blood returning to the deep body is warmed. In warm waters, where the animals need to get rid of excessive body heat, blood bypasses the heat exchanger. Venous blood returns unwarmed through the veins close to the skin's surface to cool the body core. Such vascular arrangements are common in the legs of mammals and birds and the tails of rodents, especially the beaver.

Many animals have arteries and veins divided into a large number of small, parallel, intermingling vessels that form a discrete vascular bundle or net known as a *rete*. In a rete the principle is the same as in the blood vessels of the porpoise's flippers. Blood flows in opposite directions, and a heat exchange takes place.

A countercurrent heat exchange also functions to dissipate heat. The oryx, an African desert antelope exposed to high daytime temperatures, and running African gazelles can experience elevated body temperatures yet keep the highly heat-sensitive brain cool by

FIGURE 5.6 Countercurrent circulation. (a) A generalized model. Arteries close to the veins allow an exchange of heat from the warm arterial blood moving from the body core to returning cool venous blood. (b) Temperature regulation in the bumblebee involves a heavy insulation of pile on the thorax and dorsal side of the abdomen and a lack of pile on the ventral side of the abdomen, which acts as a thermal window. A narrow petiole between the thorax and abdomen and air sacs in the anterior part of the abdomen retard heat flow from the thorax to the abdomen. The heart pumps cool blood forward to the thorax. As blood passes in the aorta through the petiole, it is heated by warm blood flowing back from the thorax. (c) The desert gazelle can keep a cool head in spite of a high body core temperature by means of a rete. Arterial blood passes in small arteries through a pool of venous blood cooled by evaporation as it drains from the nasal region. (d) The porpoise and its relatives, the whales, use flippers and flukes as a temperature-regulating device. Arteries in the appendages are surrounded by several veins. Venous blood returning to the body core is warmed through heat transfer from arterial blood. In turn, arterial blood flowing to the appendages is cooled before reaching the flippers, retaining body heat. (e) Tuna and sharks possess a heat exchanger, or rete, a vascular structure in which roughly equal numbers of very small arteries (about 0.04 mm in diameter) are interspersed with very small veins (about 0.08 mm in diameter). The rete maintains a high temperature in the swimming muscles independent of the temperature of the water in which the fish swims. The swimming muscles of most fish are supplied with blood from a large dorsal aorta that runs along the vertebral column and sends branches out to the periphery. In tuna and sharks, the blood vessels that supply the dark red swimming muscles run along the sides of the fish just beneath the skin. From them arise many parallel fine blood vessels that eventually make up the rete. The cold end of the heat exchange is at the surface of the fish; the warm end is in the swimming muscles. Arterial blood the same temperature as the water flows from the gills in fine arteries into the rete. Arterial blood gains heat from the venous blood coming from the muscles. When venous blood has reached the large veins beneath the skin, it has lost its heat, which has been returned to the muscles by way of the arterial blood. Such a heat exchanger allows tuna and sharks to maintain a muscle temperature as much as 14° C warmer than the water in which the fish swim.

a rete in the head (Figure 5.6c). The external carotid artery passes through a cavernous sinus filled with venous blood cooled by evaporation from the moist mucous membranes of the nasal passages. Arterial blood passing through the cavernous sinus is cooled on the way to the brain, reducing the temperature of the brain to 2° to 3° C lower than the body core. Such a mechanism is also found in dogs and other animals that pant and maintain a high skin temperature. Animals that sweat to regulate temperature have to hold the blood at the temperature required to maintain the brain.

Countercurrent heat exchangers are not restricted to homeotherms. Certain poikilotherms that assume some degree of endothermism employ the same mechanism. The swift, highly predaceous tuna and mackerel sharks possess a rete in the band of dark muscle tissue used for sustained swimming effort (Figure 5.6e). Metabolic heat produced in the muscle warms up the venous blood, which gives up the heat to the adjoining newly oxygenated blood returning from the gills. Such a countercurrent exchange increases the temperature and power of the muscles because warm muscles are able to contract and relax more rapidly. Sharks and tuna maintain relatively constant body temperatures regardless of the water temperature.

The bumblebee, too, uses a countercurrent heat exchanger to maintain the high thoracic temperature necessary for flight and foraging. The heart of the bumblebee is a tube lying close to the dorsal surface of the abdomen. It extends in a ventral loop beneath a large insulating air sac in the anterior abdominal wall (Figure 5.6b). The loop passes through the thin waist and next to the ventral diaphram to the thorax. The loop brings cool blood from the heart into close proximity to warm blood flowing through the ventral diaphram and in spaces surrounding the loop back to the abdomen. Heat from the warm blood flowing back to the abdomen is transferred to cool blood entering the thorax through the loop. Without such heat exchange the bumblebee would lose so much heat to the environment that flight and heat production could no longer continue.

Periodically Inactive Animals

To escape the stress of heat and cold, some animals become dormant or inactive, emerging when environmental conditions are milder. Because the body temperature of inactive poikilotherms tracks that of the environment, these animals have no choice but to become dormant during cold weather.

Diapause. In addition to supercooling, many insects exhibiting frost hardiness enter a resting stage called *diapause,* characterized by a cessation of feeding, growth, mobility, and reproduction. Among many insects diapause is a genetically determined, obligatory resting stage before development can proceed. It is timed mostly through day length (see Chapter 9) and is associated with falling temperatures. Diapause prevents the appearance of a sensitive stage of development at a time when low temperatures would kill individuals. Diapause ends with the lengthening of daylight and the return of warm temperatures.

Hibernation. Diapause is often called hibernation, which it is not. *Hibernation* is the state of winter inactivity in many poikilotherms and small homeotherms. It is a torpor characterized by the cessation of coordinated locomotory movements. Hibernating poikilotherms experience such physiological changes as decreased blood sugar, increased liver glycogen, altered concentration of blood hemoglobin, altered carbon dioxide and oxygen content in the blood, altered muscle tone, and darkened skin. Hibernating homeotherms invoke controlled hypothermia; they relax homeothermic regulation and allow the body temperature to approach ambient temperature. Birds and mammals that sometimes regulate their body temperatures physiologically and sometimes do not are called *heterotherms.*

Entrance into hibernation by homeotherms is a controlled physiological process difficult to explain and difficult to generalize from one species to another. Some hibernators, such as the groundhog (*Marmota monax*), feed heavily in late summer to lay on large fat reserves, from which they will draw energy during hibernation. Others, like the chipmunk, lay up a store of food instead. All hibernators, however, have to acquire a metabolic regulatory mechanism different from that of the active state.

Hibernating homeotherms experience a reduced heart rate, respiration, and total metabolism, and a body temperature below 10° C. Associated with hibernation is a high level of CO_2 in the body and a change in blood acid level. Acidosis affects cellular processes, inhibits breakdown of glucose, lowers the threshold for shivering, and reduces the metabolic rate. Hibernating homeotherms, however, maintain the ability to rewarm spontaneously from the hibernating state using only heat generated from within.

As the animal rouses from hibernation, it hyperventilates without any change in the breathing intervals, CO_2 stores in the body decrease, and blood pH rises. These changes are followed by shivering in the muscles, resulting in high lactic acid production and a rise in metabolism.

Hibernation provides certain selective advantages to small homeotherms. Maintaining a high body temperature during periods of cold is energetically too costly because of a high rate of body heat loss and a scarcity of food. It is easier and far less expensive to allow the body temperature to drop. That eliminates the need to keep warm and to seek scarce food, and it reduces the metabolic costs of maintaining body tissues. Even with periodic arousal, which is metabolically costly, the animal expends less energy coming out of torpor than it would have used had it remained homeothermic and inactive during the same period.

Prolonged Sleep. In contrast, grizzly and black bears do not hibernate but enter a winter sleep. A bear's metabolism during dormancy is 50 to 60 percent of its normal state. Its heartbeat drops from 40 to 10 beats per minute. It metabolizes fat exclusively as a source of energy. Through its winter sleep the bear is capable of coordinated movements if aroused, and the female gives birth to her cubs.

Daily Torpor. *Daily torpor* is the dropping of body temperature to approximately ambient temperature for a part of each day, regardless of season.

Closely related to hibernation, daily torpor is experienced by a number of birds, such as hummingbirds and poorwills, and small mammals, such as bats, pocket mice, kangaroo mice, and even white-footed mice. Such daily torpor is not necessarily associated with scarcity. Rather, it seems to have evolved as a means of reducing energy demands over that part of the day in which the animals are inactive. Nocturnal mammals, such as bats, go into torpor by day, and hummingbirds go into torpor by night. As the animal goes into torpor, its body temperature falls steeply. As the homeothermic response is relaxed, the temperature declines to within a few degrees of ambient. Arousal returns body temperature rapidly to normal as the animal renews its metabolic heat.

Estivation. A dormancy similar to that of hibernation is experienced by some desert mammals, birds, and amphibians during the hottest and driest parts of the year. Such summer dormancy is called *estivation.* It is less easily detected, so less studied, than hibernation.

PLANT ADAPTATIONS TO TEMPERATURE

Plants are fixed in place in their environment. Unlike animals, they cannot move to a more favorable situation. Therefore they experience a range of variable temperatures (Figure 5.7). For example, in early spring in the northeastern United States, temperatures about the stolons of clover plants 3 cm above the ground may be 21° C, while below the surface the temperature of the soil about the roots may be −1° C. Over a distance of only 9 cm, the temperature range is 22° C.

To complicate the picture, an individual plant consists of subpopulations of leaves, buds, and twigs. Some of these plant parts may be fully exposed to the sun, while others are shaded by twigs and leaves above. Because of different exposures to the thermal environment and different tolerances to heat and cold, some leaves, buds, and twigs may succumb to environmental extremes while others and the whole plant still live.

Plant metabolism contributes little to the internal temperature of the plant. Strong sun can increase internal leaf temperatures 10° to 20° C above ambient temperature by day. How much heat a plant gains depends upon the reflectivity of leaves and bark, the orientation of the leaves to incoming radiation, and the size and shape of the leaves. Surfaces perpendicular to the sun's rays absorb more heat than those lying at some other angle. Leaves sharply angled to 70° intercept little midday sun, instead intercepting most of the solar radiation during the morning and evening hours. Because plants lose heat by convection and transpiration, the size and shape of leaves are important. Deeply lobed leaves, like those of some oaks, and small, finely cut leaves, like those of semitropical thorn acacias, lose heat more effectively than broad, unlobed leaves. They expose more surface area per volume of leaf to the air than large leaves. Temperature within an individual leaf may vary because the edges of leaves can cool faster than the central part of the blade. As a result, leaf margins may collect dew or experience frost damage while the midportion of the leaf is unaffected. Stems and twigs may have the same or higher temperature than the surrounding air. Large trunks of trees have a higher temperature than ambient air because of heat storage and low conductance, a feature exploited by cavity-nesting birds and mammals.

METABOLIC ADAPTATIONS

Extremes of both heat and cold kill plants by damaging membranes, inactivating enzymes, and denaturing proteins. Plants have evolved some mechanisms that enable them to deal with both heat and cold. One is the resistance of plant protein to freezing. Plants can tolerate subzero (0° C) temperatures if the temperature decreases slowly, allowing ice to form outside the cell walls. The effect is dehydration, which can be reversed when temperatures rise. If the temperature falls too rapidly, intracellular freezing and frost damage to cell structure and function occur.

Tolerance by plant protoplasm is mostly genetic, varying among species and among geographically separated populations of the same species. In seasonally changing environments, plants develop *frost hardening* through the fall and achieve maximum hardening in winter (Figure 5.8). Plants acquire frost hardness—the turning of sensitive cells into hardy ones—through the formation or addition of protective compounds in the cell. They synthesize and distribute antifreeze substances, such as sugars, amino acids, and nontoxic protective compounds. Once growth starts in the spring this tolerance is quickly lost.

During the growing season some plants avoid frost damage by lowering the freezing point of cell fluids, which is accomplished by increasing certain dissolved substances such as sugars and sugar alcohols. That results in the cell sap's supercooling for short periods of time. *Supercooling* takes place when cell sap is lowered to a temperature somewhat below freezing without its freezing immediately. Further resistance to chilling is obtained by insulation. Some species of arctic and alpine plants and very early blooming species of temperate regions possess hairs that may act as heat traps and prevent cold injury. The interior temperature of cushion-type or ground-hugging leaves may be 20° C higher than the surrounding air.

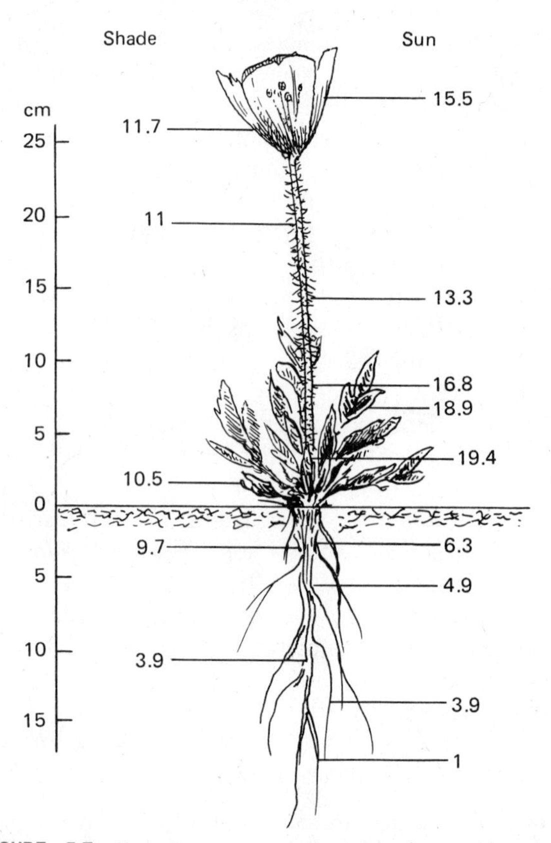

FIGURE 5.7 Temperature variation in the arctic plant *Novosieversia glacialis* on a sunny day with an air temperature of 11.7° C. Note the wide range in temperature experienced by various parts of the plant above and below ground.

Most vascular plants succumb to heat damage at tissue temperatures of 50° to 55° C. Death of a plant by heat results because of a disturbance to nucleic acid and protein metabolism and disruption of membrane function. Paradoxically, heat resistance is highest when resistance to cold is also highest and is lowest during main periods of growth. When exposed to heat, plants can initiate adaptive processes in cells, especially in the midday of summer, when a short-term increase in heat

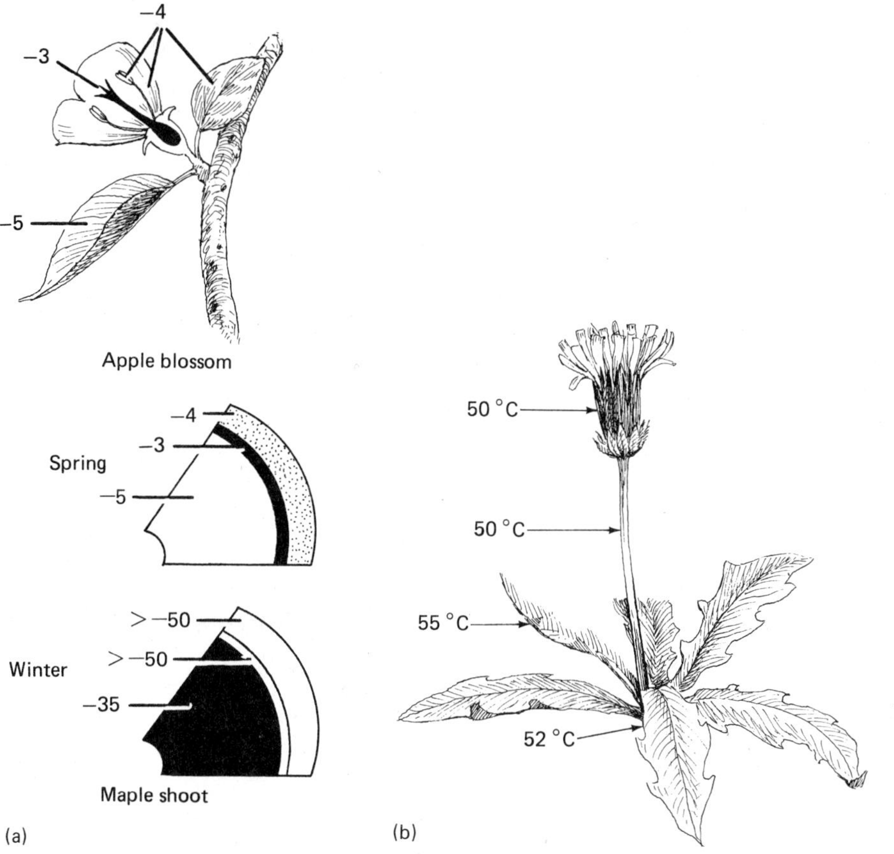

FIGURE 5.8 (a) Frost resistance in an apple blossom in spring and in a maple shoot in spring and winter. The temperature given is for 50 percent damage. The most sensitive tissue is shown in solid black. (b) Average summer heat resistance in an arctic plant, *Sempervivum montanum*. The temperatures in °C are for 50 percent injury. Note that the rosette leaves are more resistant than the shoot and the flower.

resistance takes place. This resistance is lost by evening.

ENDOTHERMISM

A very few plants are endothermic to a degree. Exhibiting some endothermism are certain members of the largely tropical arum lily family, which includes philodendrons as well as the skunk cabbage (*Symplocarpus foetidus*) of eastern North America. The arum lily family is characterized by a fleshy stem of flowers or inflorescence, called a spadix, enclosed by a leafy hood, or spathe.

The most endothermic is the skunk cabbage, which pushes its green and purple spathe and spadix up through the cold mud and snow of early spring. R. M. Knutson measured the temperatures of skunk cabbage spadices over a three-year period. He found that the spadix maintained an internal temperature 15° to 30° C above ambient air temperatures of −15° to +15° C. In doing so, skunk cabbage consumed oxygen at a rate comparable to that of a homeothermic animal the size of a shrew or a hummingbird. Skunk cabbage remained endothermic over the 14-day period of flowering. Throughout that period the temperature of the spadices remained about the same. On colder days and by night, skunk cabbage generated more heat. As long as the temperatures remained above freezing, spadix temperatures remained at a nearly constant 21° to 22° C. Skunk cabbage obtains its energy from the large quantity of respiratory tissue in the spadix as well as a nearly inexhaustible supply of energy stored in the deep, fleshy root. Attracted to the warmth of the spadix and to volatized chemicals that give off the odor of rotting meat are ectothermic insects of early spring, especially flies.

TEMPERATURE AND DISTRIBUTION

Most living organisms have a temperature range outside which they fail to grow or reproduce. Within the favorable range, organisms have an optimum temperature at which they best maintain themselves. The optimum temperature may vary with life stage or activity. The optimum temperature for photosynthesis is lower than that for respiration. The seeds of many plants will not germinate and the eggs and pupae of some insects will not hatch or develop normally until chilled.

Temperatures for some species are so different from those for other species that the animals cannot inhabit the same area. Some organisms, particularly plants growing under suboptimal temperatures, cannot compete with the surrounding growth, a situation that would not exist under optimal conditions.

Generally, the range of a species is limited by the lowest critical temperature in the most vulnerable stage of its life cycle, usually the reproductive stage. A classic example of temperature limitation on animal distribution is found among four species of ranid frogs. The wood frog breeds in late March, when water temperapture is about 10° C. Its eggs can develop at temperatures as low as 2.5° C. Larval stages transform in about 60 days. This frog ranges into Alaska and Labrador, farther north than any other North American amphibian or reptile. The meadow frog breeds in late April, when water temperatures are about 15° C, and the larvae require around 90 days to develop. As a result, the northern limit of its range is southern Canada. The southernmost species of the three, the green frog, does not breed until the water is about 25° C, and the eggs will develop at 33° C, a lethal temperature for the others. Its eggs, however, will not develop at all until the temperature exceeds 11° C. The range of the green frog extends only slightly above the northern boundary of the United States.

Changes in aquatic temperatures brought about by the clearing of forest vegetation along streams and rivers and by the discharge of heated water from power plants can affect the local distribution of species. Many aquatic organisms have a range of temperature optima, one for several different stages of their life cycles. An optimal range for adult survival may not be the optimal range for the development of eggs and larvae. For example, largemouth bass start nest building at 4.4° C and begin to spawn at 8.9° C, although the preferred living temperature is 16° to 17.8° C. Temperature tolerances vary among species. Cold water fish, such as brook trout, and cold water invertebrates, such as stoneflies, are sensitive to warm water. The upper lethal temperature for brook trout acclimatized to warmer water is 15° C, whereas largemouth bass acclimatized to 20° C die at 26.7° C.

Altered temperatures also interfere with the migratory behavior of fish. Anadromous fishes, those that spawn in fresh water and grow to maturity in salt water, are highly sensitive to temperature and have rigid time schedules for upstream and downstream migrations. High temperatures act as a barrier both to spawning by adults and to downstream migration of young. They interfere with the timing of migration or inhibit movements altogether.

Warming water can change the structure and functioning of ecosystems. As temperatures rise, diatoms characteristic of cool water are replaced by green algae, and finally at high temperatures by blue-green algae. Warming temperatures eliminate such heat-sensitive species as trout and stoneflies. If temperatures do not rise too high, these cold water species are replaced by warm-tolerant species such as largemouth bass and carp. Even they will disappear if the water temperature exceeds 29° C.

The broad distribution of vegetation about the world appears to be temperature-controlled by the absolute minimum critical temperature tolerated by various plants for survival. These temperatures range from chilling in the tropics to deep freezing in the boreal region. The minimum temperature tolerated coincides with the temperature at which supercooled water in the plant freezes and causes death. Where minimum temperatures are greater than 15° C, temperature is not limiting, and the region is dominated by broadleaf evergreens if rainfall is adequate. Areas where temperatures range between −1° and 15° C are dominated by broadleaf evergreens resistant to chilling, provided rainfall is adequate. Areas where minimum temperatures range from −15° to 0° C support broadleaf evergreens resistant to freezing through supercooling. Broadleaf deciduous forests dominate where the minimum temperatures range between −40° and 15° C. Regions with minimum temperatures below −40° C are dominated by evergreens and deciduous needleleaf evergreens.

Within this broad classification, individual species exhibit their own distributional limits. For many the northward distribution is related to freezing resistance, controlled by the minimum temperature regularly experienced, especially during the growing season, and the northern limits are controlled by frost. In seasonal climates periods of the lowest temperatures, normally in winter, are not decisive, but cold stress experienced during the growing season is. In contrast boreal species and cold-tolerant species may tolerate warm temperatures, but their southward distribution is limited by their poor competitive ability with less cold-tolerant species they encounter. In arboretums, for example, many boreal trees grow quite well, but they would never survive in the region outside cultivation.

Over the course of Earth's history, shifting climates have influenced the dispersal and distribution of vegetation. During the Pleistocene, the northern hemisphere experienced northward advances and retreats of coniferous and deciduous tree species in response to advances and retreats of the ice sheet. At the height of the Pleistocene glaciation, the global temperature was 5° C colder than at present. As the glacier retreated, spruce moved northward at the rate of about 200 km a century, aided by north-blowing winds and north-flowing rivers. Most species moved northward out of their refugia at the rate of 10 to 40 km per century.

If human-induced global warming occurs as predicted, how will the vegetation respond? If Earth experiences a global warming of 2° C over the next 50 years, northward dispersal rates of plants would have to be on the order of 500 km per century, much too fast for plant species to accomplish. As a result, many plant species would disappear and the nature of ecosystems would change, affecting their associated animal life. The northward warming could result in the disappearance of the arctic tundra, replaced by spruce forest, and the melting of the 2 to 3 m arctic permafrost, bringing about the loss of the wet coastal tundra and its fauna. Shifts in rainfall patterns in the tropics, resulting in little or no rain in the wet season and no rain in the dry season, would impact the flowering and fruiting of tropical plants, directly affecting the animal life dependent upon them and necessary for the pollination and seed dispersal of the plants. The drier tropical and subtropical regions of Africa, Asia, and Australia would experience increased wetness. Central North America would experience much drier conditions than at present, shifting both southwest deserts and grain production northward. Thus, an increase in global temperature would certainly impact regional climates, as well as change the character of natural ecosystems, agriculture, forestry, and water supply.

Summary

All life lives in a thermal environment. Organisms must balance heat inputs and outputs between themselves and their environment. Animal life basically is either poikilothermic or homeothermic. Poikilotherms depend upon the environment as a source of heat and their body temperatures more or less follow ambient temperatures. Homeothermic animals depend upon metabolically produced heat to maintain body temperatures. Thus they maintain a fairly constant body temperature independent of environmental temperatures.

To maintain a thermal balance between the organism and the environment, animals employ a variety of approaches. They may resort to behavioral activities, such as seeking the sun or shade, or they may employ insulating layers of fur, feathers, or fat. Homeotherms and some poikilotherms may increase metabolic rates to increase body heat. They may resort to evaporative cooling or even to hyperthermia to handle excessive heat loads. Some cold-tolerant poikilotherms utilize supercooling, involving the synthesis of glycerol in body fluids to resist freezing in winter. Other animals possess a well-developed countercurrent circulation, which involves exchange of body heat between arterial and venous blood. Such a mechanism allows the animals to retain body heat by reducing heat loss through the extremities or to cool blood flowing to such vital organs as the brain. Some animals enter a state of winter dormancy or hibernation to reduce the high energy costs of staying warm. Hibernation involves a whole rearrangement of metabolic activity to run at a very low level. Heartbeat, breathing, and body temperature are all greatly reduced. A similar slowdown is involved in estivation, or summer dormancy. Torpor involves a much shorter period of metabolic slowdown. Daily torpor does not involve the extensive metabolic changes characteristic of deep hibernation. Often confused with hibernation is diapause, a resting stage in development or growth associated with falling temperatures.

Plant metabolism contributes little to internal plant temperature, but plants have evolved certain mechanisms to resist extremes of heat and cold. They include reflectivity of leaves and bark, leaf size and shape, and orientation of leaves toward the sun. Frost hardening involves the synthesis of protective antifreeze substances in the cells of shoots, buds, roots, and seedlings. Plants are ectothermic, but a few members of the arum family are seasonally endothermic. The warmth of skunk cabbage appears to attract pollinating insects.

Temperature regimes set upper and lower limits within which organisms can grow and reproduce. These temperature ranges influence the distribution of many plants and animals.

Review and Study Questions

1. What is involved in the thermal balance of plants and animals?
2. Distinguish between poikilothermy and homeothermy. What are the selective advantages and disadvantages of each?
3. What are the relationships among body size, metabolic rate, and temperature regulation?
4. What is acclimatization? How does it function among poikilotherms, particularly fish?
5. How might homeothermy have evolved?
6. What are some behavioral mechanisms employed by poikilotherms to maintain a fairly constant body temperature during their season of activity?
7. How can body insulation aid animals to maintain their thermal integrity?
8. How do homeotherms respond when ambient temperatures fall below body temperature? Poikilotherms?
9. How do homeotherms use evaporative cooling?
10. How do some animals use hyperthermia, or buildup of body heat, to reduce the ultimate heat load imposed by the environment?
11. Both plants and animals employ supercooling as a means of tolerating low environmental temperatures. How are they similar?
12. What is countercurrent circulation, and how can it enable some animals to maintain a fairly constant body temperature? Why is it important to some desert mammals and to sharks?
13. Distinguish among hibernation, estivation, torpor, and diapause.
14. What metabolic changes are involved in entrance to hibernation? Why is the black bear not a true hibernator?
15. How do plants build up a tolerance to cold? To heat?
16. Speculate on how endothermy might have evolved in the temperate zone member of the arum family, the skunk cabbage.

*17. Consider a population of fish during winter living in the environs of a power plant discharging heated water in a river. The plant shuts down for three days. What effect would that have on the fish?

*18. Bats hibernate in winter, becoming ectotherms for the season, but female little brown bats (*Myotis lucifugus*) shut off thermoregulation some 18 days before giving birth in late spring and during lactation. What is the advantage of such a physiological change? Consider the bat's summer habitat, energy demands, and daily activity patterns.

Selected References

The following are basic references on plant physiology that deal in part with temperature as well as moisture:

Etherington, J. R. 1982. *Environment and plant ecology,* 2nd ed. New York, Wiley.

Larcher, W. 1980. *Physiological plant ecology,* 2nd ed. New York: Springer-Verlag. Excellent reference on plant ecophysiology.

Long, S. P., and F. I. Woodward, eds. 1988. *Plants and temperature.* Symposia of the Society for Experimental Biology, Number XXXII. Cambridge, England: The Company of Biologists Limited. Advanced and comprehensive.

Strain, B. R., and W. B. Billings, eds. 1975. *Vegetation and the environment.* Vol. 6, *Handbook of vegetation science.* The Hague: W. Junk.

Turner, N. C., and P. J. Kramer, eds. 1980. *Adaptation of plants to water and high temperature stress.* New York: Wiley.

The following are basic references on animal physiology with a strong emphasis on ecophysiology:

Folk, G. E. 1974. *Textbook on environmental physiology,* 2nd ed. Philadelphia: Lea & Febiger.

Heinrich, B. 1979. *Bumblebee economics.* Cambridge, MA: Harvard University Press. Excellent on energetics of bumblebees.

Heinrich, B., ed. 1981. *Insect thermoregulation.* New York: Wiley. Temperature regulation by insects, including heterothermy.

Hill, R. W., and G. A. Wyse. 1989. *Animal physiology.* New York: Harper & Row. A comprehensive text strong on ecophysiology.

Lyman, C. P., A. Malan, J. S. Willis, and L. C. H. Wang. 1982. *Hibernation and torpor in mammals and birds.* New York: Academic Press.

Schmidt-Nielsen, K. 1979. *Animal physiology: Adaptation and environment.* New York: Cambridge University Press. An excellent reference.

CHAPTER 6

Moisture

Outline

Objectives

On completion of this chapter, you should be able to:

1. Explain how the structure of water affects its behavior.
2. Discuss the ecologically important physical properties of water.
3. Explain relative humidity, saturation, pressure deficit, and water potential.
4. Discuss how plants reduce the impact of water stress and respond to excessive moisture.
5. Describe the ways in which animals respond to moisture stress.
6. Describe some means by which plants and animals cope with a saline environment.
7. Discuss the relationship between moisture and plant distribution on a local and regional scale.

Water is essential to all life. Means of obtaining it and conserving it have shaped terrestrial life. Means of living within it have been the overwhelming influence on aquatic life. Because of its enormous importance, water and its properties merit discussion.

THE STRUCTURE OF WATER

Because of the physical arrangement of its hydrogen atoms and its hydrogen bonds, water is a unique substance. A molecule of water consists of one large atom of oxygen and two smaller atoms of hydrogen, bonded together covalently (Figure 6.1a). *Covalence* is the sharing of an electron. Each hydrogen proton shares its single electron with oxygen, leaving unattached two pairs of oxygen electrons (Figure 6.1b). The repulsive force between these two pairs and between them and electrons in the O–H bonds pushes the bonds toward each other. As a result, the water molecule is not symmetrical. Instead of a linear arrangement of the atoms as H–O–H, the arrangement is V-shaped (Figure 6.1a). The two hydrogen atoms are separated by 105°. The shared hydrogen atoms are closer to the oxygen atom than they are to each other. If the molecule were linear, then the two bonds' polarities—the pull they exert—would cancel each other out. Because of the V-shaped arrangement, the side of the molecule on which the H atoms are located is electropositive and the opposite side is electronegative, polarizing the water molecule (Figure 6.1c).

Because of its polarity, each water molecule becomes coupled with its neighboring molecules (Figure 6.1e). The H or positive end of one molecule attracts the negative or oxygen end of the other. The 105° angle of association between the hydrogen atoms encourages an open tetrahedral-like arrangement of the water mole-

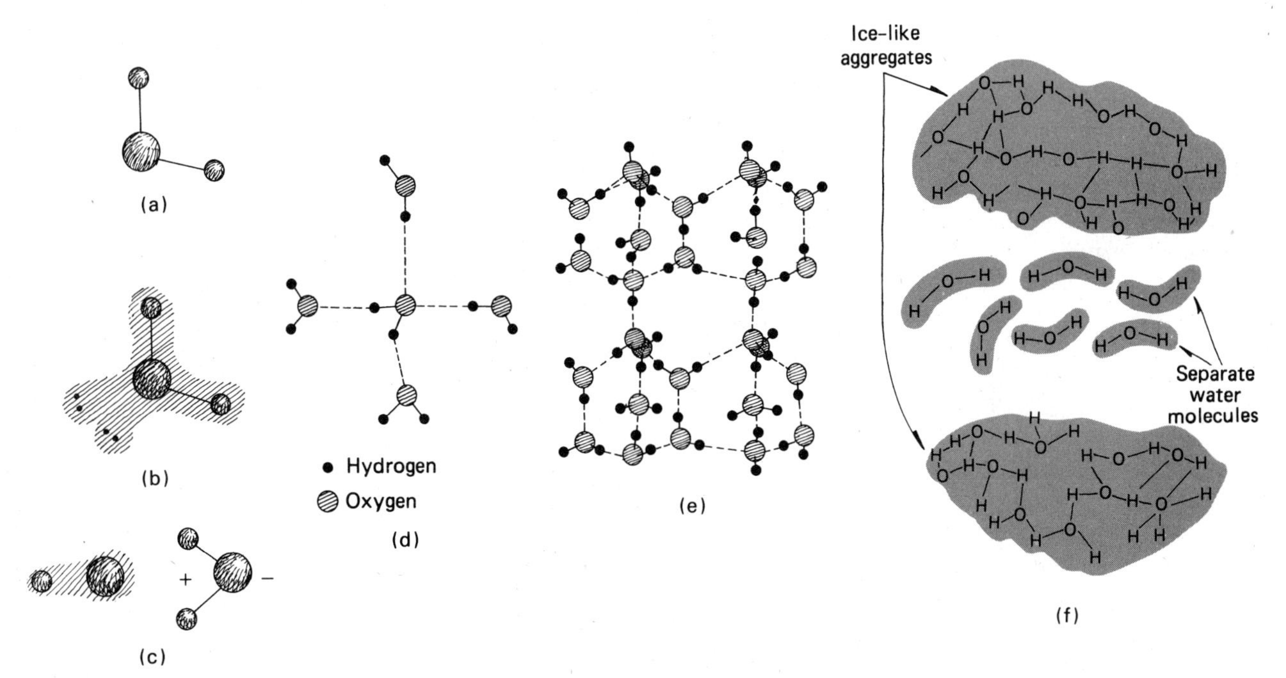

FIGURE 6.1 The structure of water. (a) An isolated water molecule, showing the angular arrangement of the hydrogen atoms; (b) lone pairs of electrons in a water molecule; (c) the polarity of water; (d) hydrogen bonds to one molecule of water in ice; (e) the open lattice structure of ice; (f) the structure of liquid water.

cules. This situation, in which the hydrogen atoms act as connecting links between water molecules, is known as *hydrogen bonding.* The simultaneous bonding of an H atom to two different water molecules accounts for the lattice arrangement of water.

In ice the lattice is complete. Each oxygen atom is hydrogen-bonded to two hydrogen atoms, and each hydrogen atom is covalently bonded to one oxygen atom and is hydrogen-bonded to another oxygen atom (Figure 6.1d). Ultimately each oxygen atom is connected to four other oxygen atoms by hydrogen atoms. One such unit built upon another gives rise to a lattice with large open spaces (Figure 6.1e). Water molecules so structured occupy more space than they would in the liquid form. As a result, water expands upon freezing and ice floats.

As ice melts, the hydrogen bonds break and the lattice work partially collapses. The volume occupied by the water molecules decreases and the density increases, until water achieves its greatest density at 3.98° C. At this temperature, contraction of the molecules brought about by the partial collapse of the lattice structure balances normal thermal expansion of warming molecules. As water is heated, more hydrogen bonds break, converting water to a liquid state, which is a mixture of individual and aggregated molecules (Figure 6.1f). Further heating results in a complete breakdown of the hydrogen bonds, separating aggregates of water molecules into individuals. At this point water enters the vapor state.

Seawater behaves somewhat differently. The density of seawater (salinity of 24.7 parts per thousand, ‰, of dissolved substances)—or rather, its specific gravity relative to that of an equal volume of pure water (whose specific gravity equals 1) at atmospheric pressure—is correlated with salinity. At 0° C the density of seawater with a salinity of 35‰ is 1.028. The lower its temperature, the greater the density of seawater; the higher the temperature, the lower the density. No definite freezing point exists for seawater. Ice crystals begin to form at a point on the temperature scale that varies with salinity. As pure water freezes out, the remaining unfrozen water becomes higher in salinity and lower in its freezing point until finally a solid block of ice crystals and salt is formed.

THE PHYSICAL PROPERTIES OF WATER

SPECIFIC HEAT

Water is capable of storing tremendous quantities of heat energy with a relatively small rise in temperature. It is exceeded in this capacity only by ammonia, liquid hydrogen, and lithium. Thus, water is described as having a high *specific heat,* the number of calories necessary to raise one gram of water one degree centigrade. The specific heat of water is given the value of 1.

Because such great quantities of heat must be absorbed before the temperature of natural waters such as ponds, lakes, and seas rises 1° C, they warm up slowly in the spring and cool off just as slowly in the fall. This fact prevents wide seasonal fluctuations in the temperature of aquatic habitats and moderates the temperatures of local or worldwide environments.

LATENT HEAT

Not only does water have a high specific heat; it also possesses the highest heat of fusion and heat of evaporation, collectively called *latent heat,* of all known substances that are liquid at ordinary temperatures. Large quantities of heat energy must be removed before water can change from a liquid to a solid, and absorbed before ice can be converted to a liquid. It takes approximately 80 calories of heat to convert 1 g of ice at 1° C to a liquid state. The same amount of heat would raise 1 g of water from 0° to 80° C.

Evaporation occurs at the interface between air and water at all ranges of temperature. Here, again, considerable amounts of heat are involved: 536 calories are needed to overcome the attraction between molecules and convert 1 g of water at 100° C into vapor—as much heat as is needed to raise 536 g of water 1° C. When evaporation occurs, the source of thermal energy may be the sun, the water itself, or objects in or around it. Rendered latent at the place of evaporation, the heat involved is returned to actual heat at the point of condensation. Such phenomena play a major role in worldwide weather.

VISCOSITY

The viscosity of water is also high because of the energy in the hydrogen bonds. Imagine liquid flowing through a glass tube. The liquid behaves as if it consisted of a series of parallel concentric layers flowing one over another. The rate of flow is greatest at the center; because of internal friction between layers, the flow decreases toward the sides of the tube. This same phenomenon can be observed along the side of any stream or river with uniform banks. The water along the banks is nearly still, while the current in the center may be swift. This resistance between the layers is called *viscosity.*

This lateral or laminar viscosity is complicated by eddy viscosity, in which water masses pass from one layer to another. Eddies create turbulence both horizontally and vertically. Biologically important, eddy viscosity is many times greater than laminar viscosity.

Viscosity is the source of frictional resistance to objects moving through the water. This resistance is 100 times that of air, so animals must expend considerable muscular energy to move through the water. A mucous coating on fish reduces surface resistance. Streamlining does likewise; in fact, the body forms of some aquatic organisms have evolved under the stresses of viscosity. The faster an aquatic organism moves through the water, the greater the stress placed on the surface and the greater the volume of water that must be displaced in a given time. Replacement of water in the space left behind by the moving animal adds additional drag on the body. An animal streamlined in reverse, with a short, rounded front and a rapidly tapering body, meets the least resistance in the water. The acme of such streamlining is the sperm whale.

SURFACE TENSION

Within all substances particles of the same matter are attracted to one another. Water is no exception. Molecules of water below the surface are surrounded by other molecules. The forces of attraction are the same on all sides. At the water's surface is a different set of conditions. Below is a hemisphere of strongly attractive similar water molecules; above is the much smaller attractive force of the air. Therefore molecules on the surface are drawn downward, and the liquid surface tends to be as small as possible, taut like the rubber of an inflated balloon. This *surface tension* is important in the lives of aquatic organisms.

This skin of water is able to support small objects and animals, such as the water striders and water spiders that run across the pond surface. To other organisms surface tension is a barrier, whether they wish to penetrate the water below or to escape into the air above. For some the surface tension is too great to break; for others it is a trap to avoid while skimming the surface to feed or to lay eggs. If caught in the surface tension, an insect may flounder on the surface. The imagoes of mayflies and caddisflies find surface tension a handicap in their efforts to emerge from the water. Slowed down at the surface, these insects become easy prey for trout.

Surface tension is important in other ways to all life. It is the force that draws liquids through the pores of the soil and the conducting networks of plants. To overcome this force, structural adaptations have evolved that prevent the penetration of water into the tracheal systems of aquatic insects and the stomata and internal air spaces of aquatic plants.

RESPONSE TO MOISTURE

Just as organisms need to maintain a thermal balance with their environment, they need to maintain a water balance. Water taken up by an organism must equal water lost; absorption must equal evaporation, transpiration, and excretion. It is difficult, however, to separate responses of organisms to variations of moisture in the environment from responses to variations in temperature because they are so closely related. Evaporative water loss, for example, is strongly influenced by high temperatures in both plants and animals, especially those that use evaporation as a way of staying cool.

Organisms live in a moisture environment, ranging from environments entirely aquatic to those deficient in moisture, either physically (as in arid regions) or physiologically (as in saline habitats). At one extreme, organisms must conserve water by preventing its loss to the environment. At the other extreme, organisms must get rid of excess water absorbed or prevent an excessive intake of water. Most organisms fall somewhere between these extremes.

PLANT RESPONSE TO MOISTURE

Lack of water is a major selective force in the evolution of plants' ability to cope with moisture stress. A large group—including algae, fungi, lichens, and most mosses—tend to match inner moisture with atmospheric moisture. They possess no protection against water loss. As water becomes less available, the plants dry out. Their cells shrink without disturbing the fine protoplasmic structures within them, and their vital processes gradually become suppressed. When moisture conditions improve, the plants imbibe water, and the cells fill and resume normal functioning. Such plants restrict their growth to moist periods, and their biomass is always small.

Other plants, mostly ferns and seed plants, are somewhat independent of fluctuations of atmospheric moisture. They store water in a vacuole within the cell and have a protective cuticle that slows down evaporation, stomata that regulate transpiration, and an extensive root system to draw water from the soil.

Ferns and seed plants lose water to the atmosphere by an evaporative process. *Evaporation* is the physical process of converting liquid water to vapor. *Transpiration* is the process of water movement through plants to the atmosphere interface. Evaporation at the leaf/atmosphere interface is the force that drives transpiration. The entire process is *evapotranspiration.*

Flooding

At times of heavy, prolonged rainfall and flooding, terrestrial plants have too much water about their roots. Water displaces air in the soil, so the roots lack oxygen.

Some plants of floodplains, swamps, and the wet arctic tundra have evolved structures that enable them to endure flooding. Mangroves have pneumatophores and cypress have root knees that protrude above the water and carry oxygen to the roots. Some grasses and herbaceous plants have hollow tubes leading from the leaves to the roots, through which oxygen diffuses.

Plants intolerant of flooding experience water stress similar to that experienced during drought, because flooding actually inhibits water uptake. Flooded plants show symptoms similar to those of drought—wilting, yellowing and premature coloring of leaves, and leaf fall. These plants also experience metabolic disturbances, including the accumulation of toxic substances and anaerobic metabolites in the roots. Among them is the gas ethylene. Ethylene normally is produced in small amounts and is highly insoluble in water. As ethylene accumulates in the roots, its diffusion from the roots and the diffusion of oxygen into the roots are slowed. In plants somewhat tolerant of flooding, ethylene accumulates to such a high level that it causes the cells of the root cortex to break and separate, forming interconnected vertical chambers that allow some exchange of gases between submerged and better aerated roots. Prolonged flooding for more than half the growing season kills trees.

Trees growing on poorly drained soils may develop shallow, spreading root systems. When their roots contact the water table, they change their direction of growth from downward to horizontal or upward, toward oxygen. Such trees are subject to windthrow and other storm damage and are sensitive to drought and frost. As long as the vapor pressure inside the leaf is greater than the vapor pressure of the atmosphere, such a plant loses water to the environment. It maintains water balance by drawing water from the soil with its roots and conducting it to stems and leaves.

These plants do not give up water easily. They hold it back by a combination of osmotic pressure and turgor pressure of water in the cells pushing outward against the cell walls. The two make up *water potential,* the measure of energy in water. Water tends to move in the direction of low water potential. Thus, water moves from the soil into roots and from leaf to atmosphere. Plants are able to control up to a point the movement of water from leaf to atmosphere. They balance that loss by removing water from the soil as long as it is available. In effect, transpiration dries out the soil.

Drought

The most common form of water stress in plants is water deficiency. It may be a physical drought brought about by the lack of soil water and excessive transpiration, or it may be a physiological drought of winter. In cold and windy weather, plants may be unable to replenish their water supply because of dry or deeply frozen soil. In winter the water ducts of plants are filled with ice. If the sun warms the twigs but the soil is not thawed sufficiently to allow the plants to replace water lost through transpiration, twigs and even the whole plant experience dehydration.

The ability of a plant to withstand water stress depends upon its *drought resistance,* the sum of its drought tolerance and drought avoidance. *Drought tolerance* is the ability of a plant to maintain its physiological activity in spite of a lack of water or to survive drying of its tissues. Few species possess any great degree of drought tolerance during the growing season, mainly because the water potential decreases during drought. Instead, most plants depend upon some type of *drought avoidance,* which involves some interaction between their internal water and that of the environment. Predictably, drought avoidance is most effective among plants of dry habitats and less effective among species of damp sites. Drought avoidance may involve timing or physiological mechanisms.

The first type of avoidance is most common among desert ephemeral plants. Mainly annuals, they germinate, grow, flower, and go to seed only when rains wet the surface soil. Seeds of these plants may lie dormant for years waiting for the right moisture conditions.

The most common approach to drought avoidance among plants is physiological. The usual way is to regulate the stomatal openings in the leaf through which water escapes. A plant responds to drought by partially closing its stomata and opening them for shorter periods of time. In the early period of stress, the plant closes its stomata and reduces transpiration during the hottest part of the day and resumes normal activity in the cooler hours. As the water balance worsens, a plant opens its stomata only in the morning. As drought continues, the plant may keep its stomata closed, and it may fold or roll the leaves to reduce surface area. The plant now transpires through the cuticle, which gives off only a fraction of the water lost through the stomata.

Closing the stomata cuts down on transpiration, but at the same time it reduces the carbon dioxide (CO_2) intake necessary for photosynthesis. As a result, the rate of photosynthesis declines. That, in turn, affects carbon metabolism, with some wide-ranging effects over time. The ability of the leaf to store carbon in a usable form is reduced. The lack of carbon can affect not only the metabolism of the leaf but other organs as well. Carbon reduction is often accompanied by a lowering of the nitrate concentration in the sap, which affects nitrogen metabolism in the plant.

Xeric species (plants adapted to dry conditions) are able to continue photosynthesis at a more negative water potential over a longer period of drought. *Mesic* species (adapted to dampness) close their stomata at a much higher level of leaf water potential. Such action works for a short-term drought, but unless the drought ends quickly, mesic plants may experience metabolic damage.

There are other physiological measures some plants take. One is to reduce transpiration across the cuticle by changing its thickness, its structure, or the chemical composition of the lipids it contains. A plant can increase its water uptake by extending its root system into areas of the soil still untapped. Some plants can increase water storage in cells, or succulence. That mechanism is common among some desert plants. Desert succulents are exemplified by the cacti, which usually lack leaves and carry on photosynthesis in the stem. Succulents obtain their water from the upper layers of soil after it rains and store that water in their cells. They conserve the water by separating the light and CO_2 fixation reactions of photosynthesis (see Chapter 17). That allows the plants to go for a long period without opening the stomata. When succulents do open their stomata, it is at night, when water losses are minimal.

Lacking succulence, some plants of arid and semiarid regions become drought-deciduous, shedding their leaves during the dry season. Some desert shrubs, such as ocotillo (*Fouquieria splendens*), shed their leaves four or five times a year and renew them with each rain. Other plants possess extremely small leaves or lack leaves altogether. Those plants, such as paloverde (*Cercidium floridum*), carry on photosynthesis through green stem tissue rich in chlorophyll. Such plants have the advantage of being able to start photosynthesis immediately when the rains come, without the time lag for the formation of new photosynthetic tissue.

Plants of riverine and streamside habitats and deep depressions in arid and semiarid regions escape drought by tapping a deep, permanent or semi-permanent water supply. Such trees and shrubs, called *phreatophytes* (well plants), may send roots down to a depth of 60 or 80 m to reach water, although 10 to 30 m is more typical. Examples are cottonwood (*Populus fremontii*), willows (*Salix* spp.), salt cedar (*Tamarix* spp.), and mesquite (*Prosopis glandulosa*).

Long periods of dry weather that result in soil drought reduce plant growth and cause the dieback of plants or outright death. Drought-injured plants are vulnerable to outbreaks of insects and are highly susceptible to fire. Drought can also influence the composition of plant communities. For example, during the drought of 1933 to 1939 on the plains, buffalo grass either disappeared entirely from some ranges or was reduced to small, scattered patches. Its more resistant associate, blue grama, was never killed uniformly and persisted. When the drought ended, in 1940 to 1942, buffalo grass responded rapidly to moisture and blue grama declined.

ANIMAL RESPONSE TO MOISTURE

Animals, too, need to maintain a water balance with their environment. Their adaptations to moisture are more complex than those of plants. A more or less universal mechanism, the excretory system, rids the body of excess water or conserves it.

Osmotic pressure pushes water through cell membranes to the side with more salts. Aquatic organisms living in fresh water have more salts in their bodies than the surrounding water. Their problem is to rid themselves of excess water. Protozoans accomplish that by means of contractile vacuoles. Freshwater fish maintain osmotic balance by absorbing and retaining salts in special cells in the body and by producing copious amounts of watery urine. Other animals maintain dilute urine in comparison to body fluids. Aquatic amphibians balance the loss of salts through the skin by absorbing ions directly from the water and by active transport across the skin and gill membranes. Aquatic animals do not drink.

Terrestrial animals have three major means of gaining water and solutes: through drinking, through food, and through production of metabolic water from food. They lose water and solutes through urine, feces, evaporation over the skin, and respiration.

Animals have a variety of strategies for coping with water balance. Amphibians store water from the kidneys in the bladder and, if circumstances demand it, conserve water for metabolism by reabsorbing it through the bladder wall. They can also gain water directly across the skin. Birds and reptiles have a salt gland and a cloaca, a common receptacle for the digestive, urinary, and reproductive tracts. Water in the cloaca is reabsorbed back into the body. Mammals possess kidneys capable of producing urine with high osmotic pressure and inorganic ion concentration. They lose water through the kidneys, respiratory system, and sweat glands.

Like plants, animals of arid environments are faced with a severe problem of water balance. They can solve the problem in one of two ways—either by evading the drought or by avoiding its effects. Animals of semiarid and desert regions may evade drought by leaving the area during the dry season. That is the strategy employed by many of the large African ungulates. Some small animals, such as the spadefoot toad of the southwestern United States, estivate below ground and emerge when the rains return. Some invertebrates, such as the flatworm *Phagocytes vernalis,* which occupies ponds that dry up during the summer, encyst. Other aquatic or semiaquatic animals retreat deep into the soil until they reach groundwater level. Many insects undergo diapause, just as they do when confronted with unfavorable temperatures.

Other animals remain active during the dry season but conserve moisture. One way is to reduce respiratory water loss. Some small desert rodents reduce the temperature of respired air by a countercurrent heat exchange in respiratory passages. Moisture-laden warm air from the lungs passes over the cooled nasal membranes, leaving behind condensed water on the walls of the nasal passages. Dry, warm air inhaled by the rodent is humidified by the evaporation of water on the surface membranes of the nasal passages. An African desert ungulate, the oryx, reduces daytime losses of moisture by becoming hyperthermic (see Chapter 5). A substantial rise in its daytime body temperature reduces evaporative losses. The oryx further reduces water loss by suppressing sweating and by panting only at very high temperatures. Further, the oryx reduces its metabolic rate; by lowering the internal production of calories, the animal reduces the need for evaporative cooling. By night the oryx reduces its nonsweating evaporation across the skin by reducing its metabolic rate below that of the daytime. With a lowered nighttime body temperature, the saturation level for water vapor in exhaled air is lower.

There are other approaches to the problem. Some small desert mammals reduce water loss by remaining in burrows by day and emerging by night. Many desert mammals, from kangaroo rat to camel, produce highly concentrated urine and dry feces and extract water metabolically from the food they eat. In addition, some desert mammals can tolerate a certain degree of dehydration. Desert rabbits may withstand water losses up to 50 percent of their body weight; camels can tolerate a 27 percent loss.

Water balance can have other effects. Moisture influences the speed of development and even the fecundity of some insects. If the air is too dry, the eggs of some locusts and other insects may become quiescent. There is an optimum humidity at which nymphs develop the fastest. Some insects lay more eggs at certain relative humidities than above or below them. Heavy rains and prolonged wet spells cause widespread death among mammals and birds, especially the young, from drowning, exposure, and chilling. Excessive moisture and cloudy weather kill insect nymphs, inhibit insect pollination of plants, and spread parasitic fungi, bacteria, and viruses among both plants and animals.

PROBLEMS OF SALINE ENVIRONMENTS

Saline environments—oceans, salt marshes, estuaries, and alkaline deserts—present special water balance problems. The saltwater environment is a physiological desert, where the concentration of salts outside the bodies of organisms can dehydrate them osmotically. In marine and brackish environments, the problem is to inhibit the loss of water through the body wall by osmosis and prevent an accumulation of salts in the system. Algae and invertebrates get around that problem by possessing body fluids that have the same osmotic pressure as seawater. In a way, that adds to the problem of marine invertebrates because they cannot obtain fresh water through their food. Marine teleost fish absorb water with the salt into the gut. They secrete magnesium and calcium through the kidneys and pass these ions off as a partially crystalline paste. The fish excrete sodium and chlorine by pumping the ions across membranes of special cells in the gills. This pumping process is one type of *active transport*. Salts move against a concentration gradient at the cost of metabolic energy. Sharks and rays retain urea to maintain a slightly higher concentration of salt in the body than in surrounding seawater. Birds of the open sea are able to utilize seawater, for they possess special salt-secreting glands located on the surface of the cranium. Gulls, petrels, and other seabirds excrete from these glands fluids in excess of 5 percent salt. Petrels and tube-nosed swimmers forcibly eject the fluids through the nostrils; other species drip the fluids out of the internal or external nares.

Among marine mammals the kidney is the main route for the elimination of salt. Porpoises have highly developed renal capacities to eliminate salt loads rapidly. In marine mammals the urine has a greater osmotic pressure than blood and seawater (hyperosmotic); the physiology is poorly understood.

Vertebrates in the Arctic and Antarctic have special problems. As seawater freezes, it becomes colder and more salty. The only alternative for most organisms is to increase the solute concentration in the body fluids to lower the body temperature. Some species of fish in the Antarctic possess in their blood a glycoprotein antifreeze substance that enables the animals to exist in temperatures below the freezing point of blood.

Plants inhabiting saline environments are known as *halophytes*. Characteristically, they accumulate high levels of ions within the cells, especially in the leaves. That concentration, which may equal or exceed that of

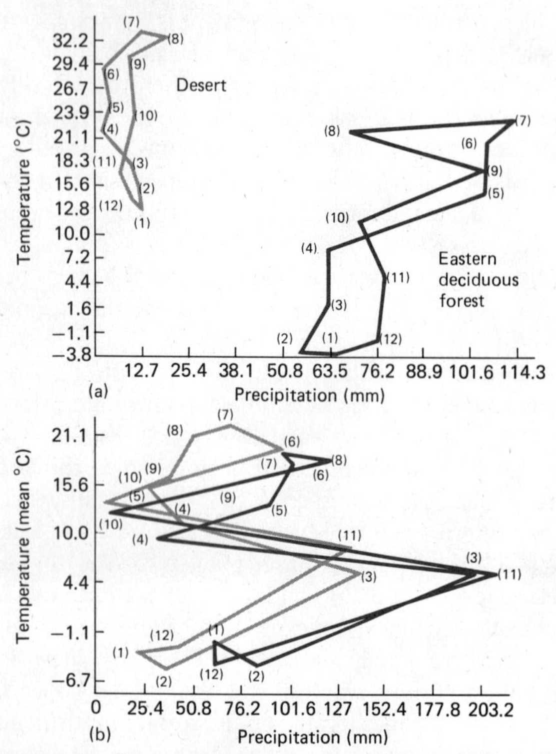

FIGURE 6.2 Temperature-moisture climographs. These 12-sided polygons let us picture and compare moisture conditions. (a) Comparison of the hot, dry desert of western North America and the cool, moist, deciduous forest region of eastern North America. (b) Comparison of conditions on the rainshadow side and the high-rainfall side of the Appalachian Mountains of West Virginia. Parenthetical numbers represent months of the year.

seawater, allows halophytes to maintain a high cell water content in the face of a low external water potential. Some halophytes compensate for an increase in sodium and chlorine uptake with a corresponding dilution of internal solutions with water stored in tissues. In addition, some plants exhibit high internal osmotic pressure, many times that of freshwater and terrestrial plants; possess salt-secreting glands, usually on the leaves, from which the excess salt can be washed away by rainwater; secrete heavy cutin on the leaves; and are succulent. A number of desert plants allow only certain ions to pass across cellular membranes and keep others out. That allows the plants to absorb essential nutrients from the soil and helps maintain internal osmotic pressures higher than those of the surroundings.

Halophytes vary in their degree of tolerance. Some, such as salt marsh hay grass (*Spartina patens*), do best at low salinities. Others, such as salt marsh cord grass (*S. alternifolia*), do best at moderate levels of salinity. A few, such as glassworts (*Salicornia* spp.), are tolerant of high salinities. All halophytes appear to be able to grow in nonsaline environments but are unable to compete with nontolerant species there.

INTERACTION OF TEMPERATURE AND MOISTURE

A close interaction exists between temperature and moisture in terrestrial environments; and the two determine in large measure the climate of a region and the distribution of vegetation. Low-moisture conditions are more extreme when temperatures are high or low. Moisture, in turn, heightens the effects of temperature. Cold is more penetrating and high temperatures are more noticeable when relative humidity is high.

Mean monthly temperatures and relative humidities or precipitation can be plotted on a *climograph,* a composite picture of the climate of an area (Figure 6.2). Twelve monthly dots for the year connected form an irregular polygon, which can be compared with another. In this way climates can be compared much more easily than by tables. Such climographs are useful to contrast one region or one year with another. Often they help determine the suitability of an area for the introduction of exotic animals, particularly game birds.

In North America the distribution of vegetation is influenced more by the pattern of moisture than by temperature. In Europe the pattern of vegetation more nearly follows that of temperature, resulting in broad belts of vegetation running east and west. In North America only in the far north do the vegetation zones (tundra, tiaga, and boreal coniferous forests) stretch in these directions. Below these boreal zones, because available moisture becomes less from east to west, vegetation belts run north and south. Humid regions along the coast support natural forest vegetation. This zone

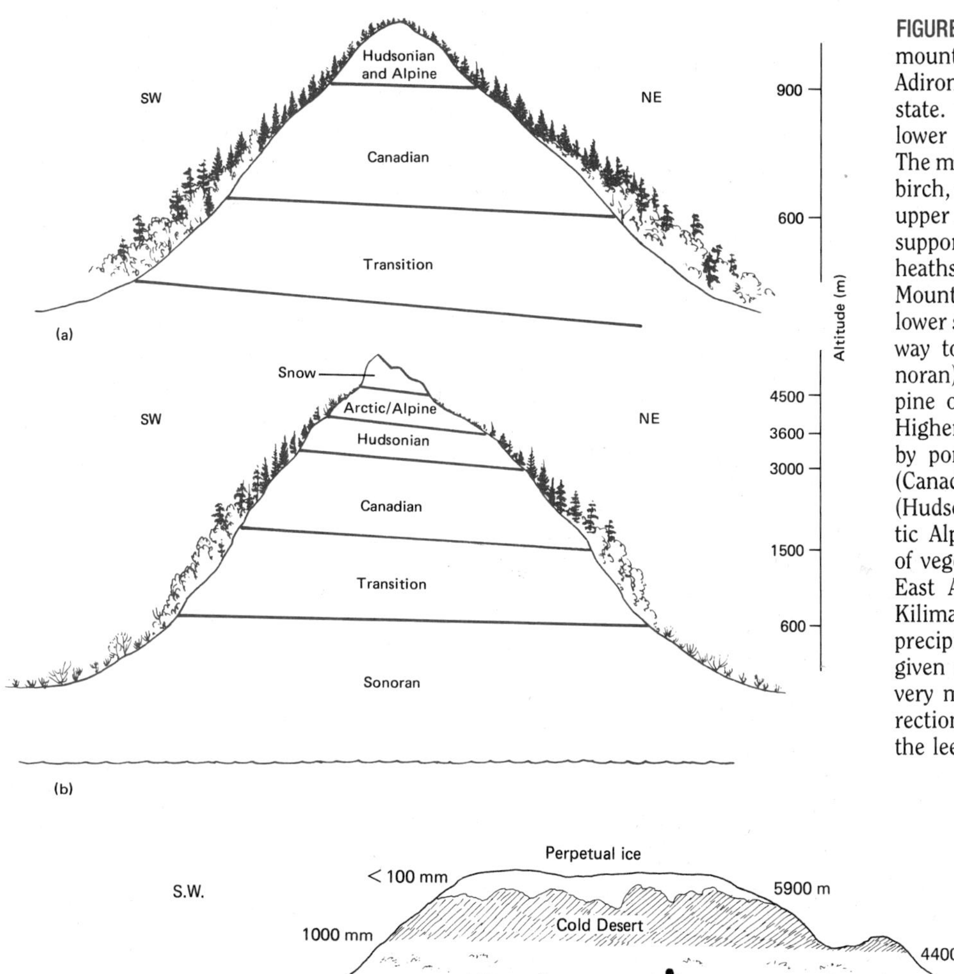

FIGURE 6.3 Altitudinal zonation in mountains. (a) Mount Marcy in the Adirondack mountains of New York state. The Transition forest on the lower slope is northern hardwoods. The mid-slope is Canadian with paper birch, red spruce, and balsam fir. The upper slopes, Hudsonian and Alpine, support dwarf spruce, willow, and heaths. (b) The composite Rocky Mountain situation in which the lower slope supports oak, which gives way to chaparral and junipers (Sonoran), and to oak and lodgepole pine on the mid-slope (Transition). Higher up, these trees are replaced by ponderosa pine and Douglas fir (Canadian), then spruce and fir (Hudsonian), and finally tundra (Arctic Alpine). (c) Altitudinal zonation of vegetation in a tropical region in East Africa, as represented by Mt. Kilimanjaro. Approximate annual precipitation for different altitudes is given (mm). The southwest slope is very moist because it lies in the direction of the southwest monsoon; the leeward slope is much drier.

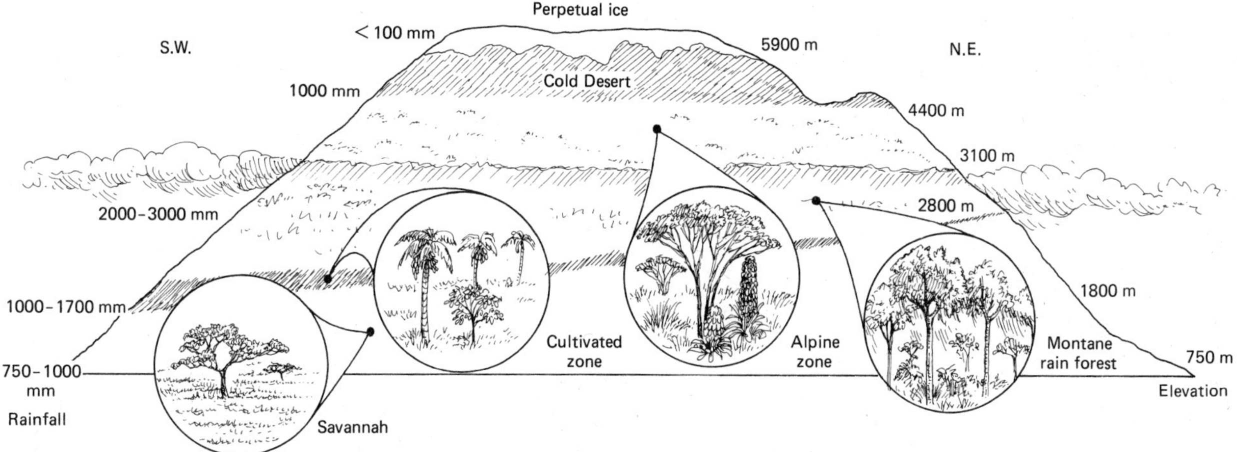

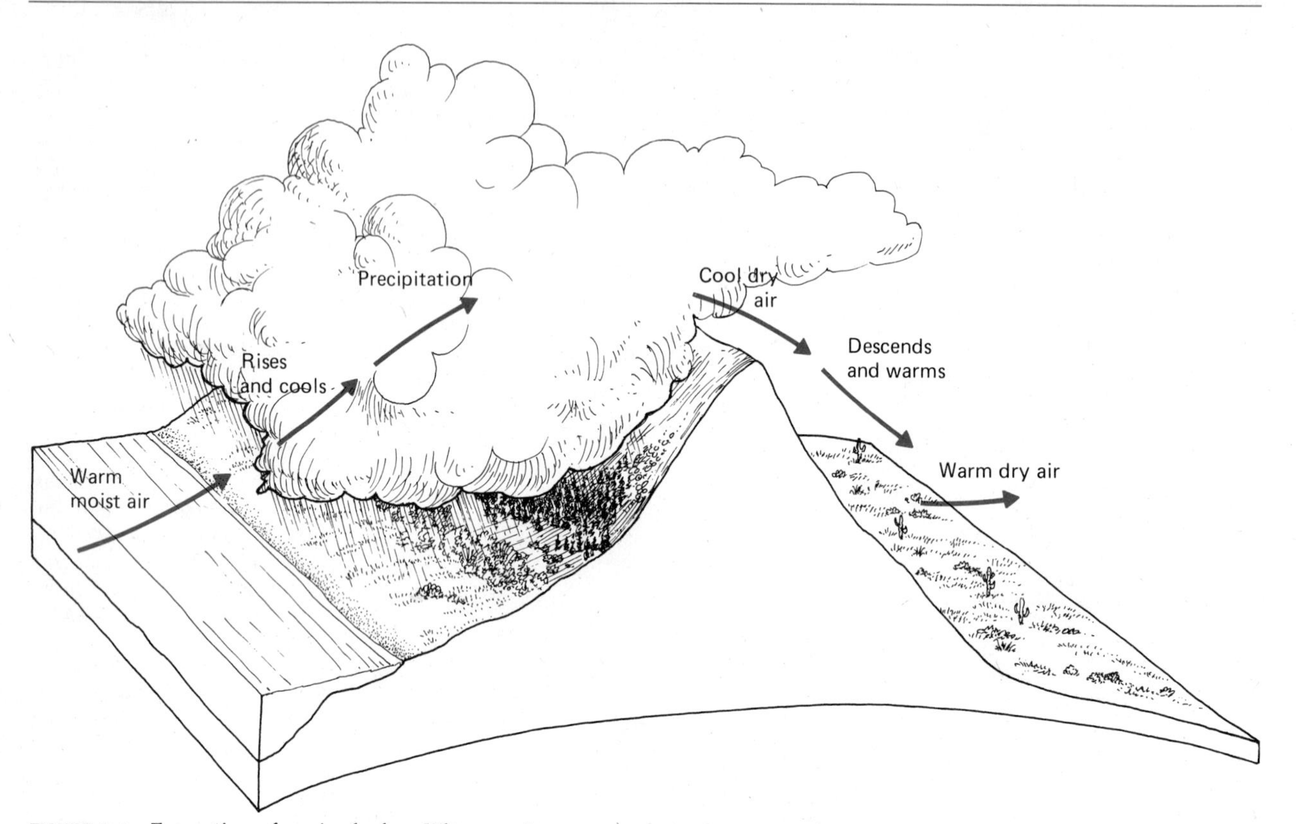

FIGURE 6.4 Formation of a rain shadow. When an air mass encounters a topographical barrier, such as a mountain range, the air mass is forced to go over it. As it rises, the air mass cools and loses its moisture (precipitation) on the windward side. The abundant moisture stimulates heavy vegetational growth. The relatively dry air descending the leeward slope picks up moisture, creating arid to semiarid conditions.

is broadest in the east. West of this region is a subhumid zone where precipitation is less and evaporation is higher. Here the ratio of precipitation to evaporation is about 3:4 and the land supports (or once supported) a tall-grass prairie. Beyond is semiarid country, where the precipitation-evaporation ratio is 1:2. It supports a short-grass prairie. Further west and on the lee of the mountains is desert. On the coastal regions of the Pacific Northwest, influenced by oceanic currents, the climate is mild and wet and supports dense coniferous rain forests.

In mountainous country, vegetation zones reflect temperature and moisture changes on an altitudinal gradient (Figure 6.3). These belts often duplicate the pattern of latitudinal vegetational distribution. In general, belts about the mountain's base reflect the climate of the region. Above is the montane level, which has greater humidity and lower temperature as altitude increases. Here the forest changes from deciduous to coniferous. Further up is a subalpine zone that includes coniferous forest trees adapted to a more rigorous climate than the montane species. Above the subalpine lies the krummholz, a land of stunted trees, followed by the alpine or tundra zone where the climate is cold and grasses, sedges, and small tufted plants replace trees. On the top of the highest mountains is a land of perpetual ice and snow. In the Southern Hemisphere, altitudinal zonation may be more complex. In the Australian Alps, for example, the high latitudes are dominated by a snow gum, a species of eucalyptus that occupies a niche similar to subalpine conifers, before the vegetation gives way to alpine tundra.

Mountains modify regional vegetation by changing the patterns of precipitation. Mountain ranges intercept air flows. As an air mass reaches the mountains, it ascends, cools, becomes saturated (because cold air holds less moisture than warm air), and releases much of its moisture on the windward side. As the cool, dry air descends on the leeward side, it warms and picks up moisture (Figure 6.4). As a result the windward side of a mountain range supports lusher vegetation than the leeward side, where dry, often desertlike conditions exist. Thus in North America the westerly winds that blow over the Sierra Nevada and Rockies drop their moisture on the west-facing slopes that support excellent forest growth, while the eastern sides are desert or semidesert. Similar conditions exist in the Himalayas and the mountain ranges in Europe.

Summary

The physical properties of water make it an ideal medium for life. It is most dense at 4° C, a feature of considerable importance to aquatic ecosystems. The result is a warm upper layer of water in lakes and ponds in summer and floating ice in winter. It has a high specific heat and a high latent heat, which moderate the climate near bodies of water and aid in the heat balance of organisms. Its high viscosity and surface tension are important in the capillary flow of water, especially in the soil and the conducting vessels of plants.

All life responds to moisture. Water stress is a major selective force in the evolution of plants and animals. Some plants have a moisture content matching that of the environment. Others maintain a stable water balance, independent of environmental fluctuations, by a combination of osmotic pressure and turgor pressure, collectively known as water potential, which decreases from the soil to the leaf. Because of water potential, water is able to move from the soil through the plant to the atmosphere.

To maintain water balance during long periods of dry weather, plants depend upon drought resistance—a combination of drought avoidance and drought tolerance. Some plants avoid drought by being ephemeral, such as the desert annuals that germinate and bloom only when rains are sufficient. Other plants reduce transpiration water loss by closing their stomata, increasing leaf thickness, possessing a waxy cuticle, storing water in cells (succulence), or shedding leaves during the dry season. Still other plants extend roots into areas of soil where the water has not been tapped. Some species, called phreatophytes, of riverine habitats, have deep roots that reach a permanent water supply.

Plants intolerant to flooding experience stress similar to that experienced during drought, and many of the symptoms—yellowing of leaves and leaf fall—are the same. Flooded plants also undergo metabolic disturbances and changes in root growth. In extreme cases, flooding can kill trees.

Animals likewise need to maintain a water balance. Usually they do so by some sort of excretory system, ranging from contractile vacuoles in protozoans to kidneys in birds and mammals. Some animals avoid drought by encysting; others go into diapause or estivation. Desert animals reduce water loss by becoming nocturnal, avoiding the heat of day; by producing highly concentrated urine; by using only metabolic water; and by tolerating a certain degree of dehydration.

Organisms inhabiting saline environments live in a physiologically arid region. Their problem is to maintain osmotic balance with the environment by getting rid of excess salt. Plants may accumulate and excrete salt through their leaves or store water in them. Many animals have salt-excreting glands.

Moisture or the lack of it has a major influence on the distribution of plants, both globally and locally. Some are restricted to moist sites, others to dry sites. Temperature and moisture also interact to influence plant distribution and plant communities.

Review and Study Questions

1. How does the physical structure of water influence specific heat, latent heat, viscosity, and surface tension?

2. What is relative humidity? Saturation pressure deficit?
3. Why is relative humidity higher at night than by day? At higher elevation than at lower elevation?
4. Why can you collect moss plants, dry them, put them away in herbarium envelopes, and then some years later moisten the specimens to return them to original condition for study?
5. What is transpiration? Water potential?
6. Distinguish between drought resistance and drought tolerance.
7. In what ways do plants cope with drought and water stress?
8. What are the advantages and disadvantages to a plant of closing its stomata to reduce transpiration?
9. What is succulence? How does it enable a plant to withstand aridity?
10. What are phreatophytes?
11. What problems does flooding create for plants?
12. How do animals of arid regions avoid dehydration?
13. How do plants and animals of a salt marsh face up to the problems of living in a saline environment?
14. In what way does moisture influence the local distribution of plants?
15. How do temperature and moisture interact to influence plant distribution?

*16. Prepare a climograph for your area and different regions in your state. Necessary data are compiled in Federal Weather Bureau monthly summaries available at most libraries.

Selected References

Daubenmire, R. F. 1974. *Plants and environment,* 3rd ed. New York: Wiley.

Feldman, L. J. 1988. The habits of roots. *Bioscience* 38:612–618.

Gilles, R., ed. 1979. *Mechanisms of osmoregulation in animals.* New York: Wiley.

Kozlowski, T. T. 1982. Plant responses to flooding of soil. *Bioscience* 34:162–167.

Kozlowski, T. T. 1984. *Flooding and plant growth.* Orlando, FL: Academic Press.

Lange, O. L., L. Kappen, and E.-D. Schulze, eds. 1976. *Water and plant life: Problems and modern approaches.* Ecological Studies No. 19. New York: Springer-Verlag.

Larcher, W. 1980. *Physiological plant ecology,* 2nd ed. New York: Springer-Verlag.

Lee, R. 1980. *Forest hydrology.* New York: Columbia University Press.

Maloiy, G. M. O., ed. 1979. *Comparative physiology of osmoregulation in animals.* New York: Academic Press.

Schulze, E.-D., R. H. Robichaux, J. Grace, P. W. Rundel, and J. R. Ehleringer. 1987. Plant water balance. *Bioscience* 37:30–37.

Outline

CHAPTER

Light

Objectives

On completion of this chapter, you should be able to:

1. Describe the nature of light as it reaches Earth.
2. Explain the fate of visible light in the plant canopy and in water.
3. Contrast shade-tolerant and shade-intolerant plants.
4. Distinguish among the C_3, C_4, and Crassulacean acid cycles in photosynthesis.
5. Explain the ecological importance of the three pathways.
6. Explain the significance of the leaf area index to photosynthesis.
7. Describe photosynthetic efficiency.

The energy for life comes from visible light. Plants fix some of this light energy through the process of photosynthesis. They store part of this energy as new biomass; the rest they use for metabolic processes.

Light is also an important environmental influence in many ways. The ability or inability of plants to grow in the shade of others affects the nature and structure of plant communities. The structure of plants influences the kind and number of animals in a given area. Light influences the daily and seasonal activities of both animals and plants.

THE NATURE OF LIGHT

Light is solar radiation (see Chapter 4) in the visible range, embracing wavelengths of 0.40 microns to 0.70 microns. This segment of the solar spectrum is known as *photosynthetically active radiation* (PAR).

Light that arrives at Earth's atmosphere is not quite the same light that arrives on Earth's surface (Figure 7.1). The high-level ozone layer absorbs nearly all wavelengths, but especially the violets and blues. Molecules of atmospheric gases scatter the shorter wavelengths, giving a bluish color to the sky and causing Earth to shine in space. Water vapor scatters all wavelengths, so an atmosphere with much water vapor is whitish—thus, the grayish appearance of a cloudy day. Dust scatters long wavelengths to produce reds and yellows in the sky. Because of the scattering of solar radiation by dust and water vapor, part of it reaches Earth as diffuse light from the sky, known as *skylight*.

Light intercepted by Earth is either reflected, absorbed, or transmitted through objects. Of greatest ecological interest is light reaching vegetation. About 6 to 12 percent of photosynthetically active light striking a leaf is reflected. The degree of reflection varies with the nature of the leaf surface. Because green light is most strongly reflected, leaves appear green. Most of the red light is absorbed by chloroplasts in the mesophyll of the leaf and used in photosynthesis. A remaining fraction of light is transmitted through the leaf. How much depends upon its thickness and structure. A leaf may transmit up to 40 percent of the light it receives, but 10 to 20 percent is more usual. Transmitted light is primarily green and far red. Light filtering through a forest is mostly in wavelengths over

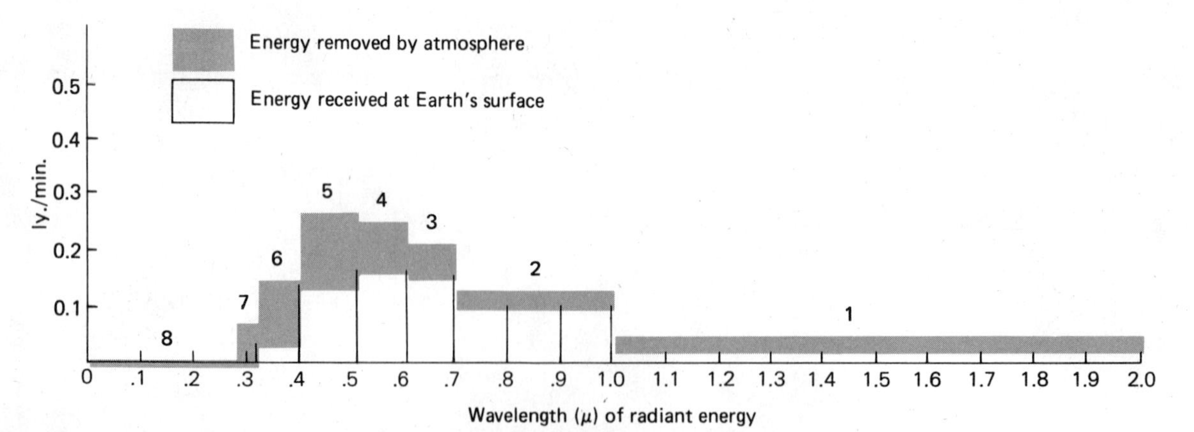

FIGURE 7.1 Energy in the solar spectrum before and after depletion by the atmosphere from a solar altitude of 30°. Figures above the bars indicate (1) near infrared with wavelengths over 1 micron; (2) near infrared, 0.7–1.0 μ; (3–5) visible light; (3) red; (4) green, yellow, and orange; (5) violet and blue; (6–8) ultraviolet. Note the strong reduction in ultraviolet. Nearly all the wavelengths are absorbed by the ozone at high levels. The region of peak energy is shifted toward the red end of the spectrum. Visible light in the blue wavelength is scattered rather than absorbed, producing the blue light of the sky.

0.50 microns. In the depths of a forest, even green light may be extinguished.

LIGHT IN THE PLANT CANOPY

When you walk into a forest in summer, one of the most obvious changes is a decrease in light. If you were able to view the understory of a grassland, you would observe much the same thing. Most of the sunlight that floods open spaces is intercepted by a leafy canopy. The amount of light that penetrates a stand varies with the nature and position of the leaves. Horizontal leaves arranged in layers intercept more light than leaves arranged in an upright position. As the layers of leaves increase, less light reaches the ground. By contrast, leaves arranged at a 45° angle to the perpendicular allow more light to penetrate the canopy.

Only about 1 to 5 percent of the light striking the canopy of a typical temperate hardwood forest reaches the forest floor. In a tropical rain forest, only 0.25 to 2 percent gets through. More light travels through pine stands—about 10 to 15 percent—but densely crowned Norway spruce allow only 2.5 percent of the light in the open to reach the forest floor. Woodlands comprised of trees with relatively open crowns, such as oaks and birches, allow light to filter through. There light dims out gradually, as it does in grasslands. In grasslands most of the light is intercepted by the middle and lower layers (Figure 7.2).

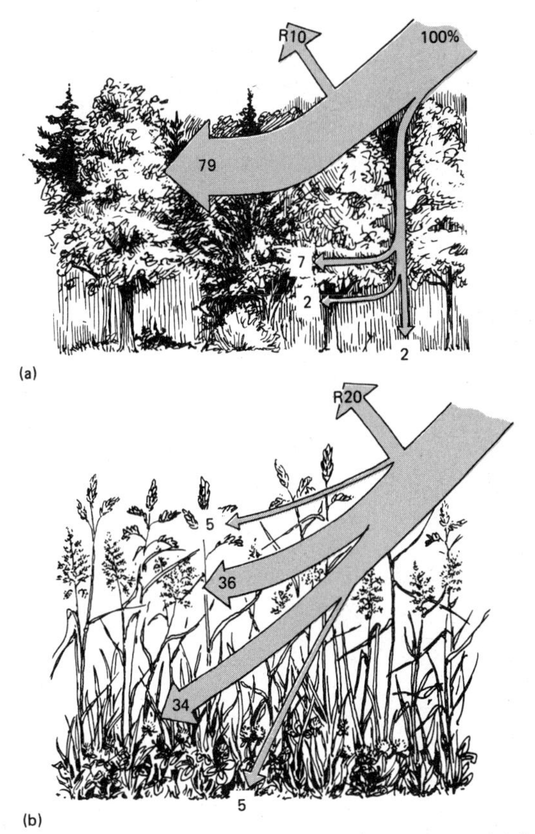

FIGURE 7.2 Thinning of light by the plant canopy. (a) In a boreal mixed forest, 10 percent of the PAR is reflected from the upper crown, and most of the remainder is absorbed within the crown. (b) In a meadow, 20 percent is reflected from the upper surface, while the greatest absorption occurs in the middle and lower regions, where the leaves are most dense. Only 2 to 5 percent of the incident photosynthetically active radiation reaches the ground.

Foliage density, the key to light interception, is expressed as *leaf area index* (LAI). LAI is the percentage of ground area covered by leaves. It is expressed as:

$$\frac{\text{Total leaf area}}{\text{Area of ground}}$$

An LAI of 3, for example, means that a given area of ground is covered by three times that area of leaves arranged in layers.

Some plants have an optimal LAI at which there is minimal shading of one leaf by another. As LAI increases beyond that point, photosynthesis decreases. Other plants do not show this response, primarily because of the angle of the leaf, which influences light interception. Horizontal leaves perpendicular to incoming light intercept the most light, but a number of layers of horizontal leaves reduces the amount of light reaching lower leaves. Leaves growing at an angle require a much higher LAI to intercept the same quantity of light as horizontal leaves. As their LAI increases above a certain value, angled leaves carry on photosynthesis at a faster rate. Such an adaptation in leaf position to obtain maximum photosynthetic rates and still possess a high LAI is characteristic of corn, grasses, beets, turnips, and other row crops.

Individual trees themselves, especially those growing in the open, are affected by the thinning of light through the canopy. Trees with relatively open crowns have leaves distributed within the crown. Light intensity is great enough for interior leaves to survive and carry on photosynthesis. Trees such as cedars and spruce have canopies so dense that no leaves can grow in the deepest part of the crown (Figure 7.3).

Although light in the forest dims down through the canopy, some sunlight does penetrate gaps in the crown and reaches the forest floor at open sky intensity as sunflecks. Ferns and shade-tolerant herbaceous plants of the forest floor intercept these sunflecks. These plants achieve a high rate of photosynthesis in spite of the shade.

SHADE TOLERANCE

Plants are either shade-tolerant or shade-intolerant (shade avoiders). Shade avoiders are plants of open sites receiving full sunlight. They establish themselves rapidly on disturbed sites; can tolerate extreme site conditions; achieve fast growth in the open; and in the case of trees and shrubs, produce seed at an early age and disperse them widely. They carry on photosynthesis more efficiently in full sunlight and have both a high rate of photosynthesis and a high rate of respiration. They rapidly convert photosynthate into growth.

By contrast, shade-tolerant plants are not competitive with shade-intolerant plants in full sun. They have a lower rate of photosynthesis and, perhaps more important, a lower rate of respiration than shade avoiders. For those reasons they grow more slowly in all environments. Many have the ability to carry on photosynthesis at low light intensities as well as the ability to open their stomata in dim light rapidly. That enables the plants to take advantage of increased sunlight from sunflecks over a short period of time. They may live suppressed in a shaded understory for many years, but

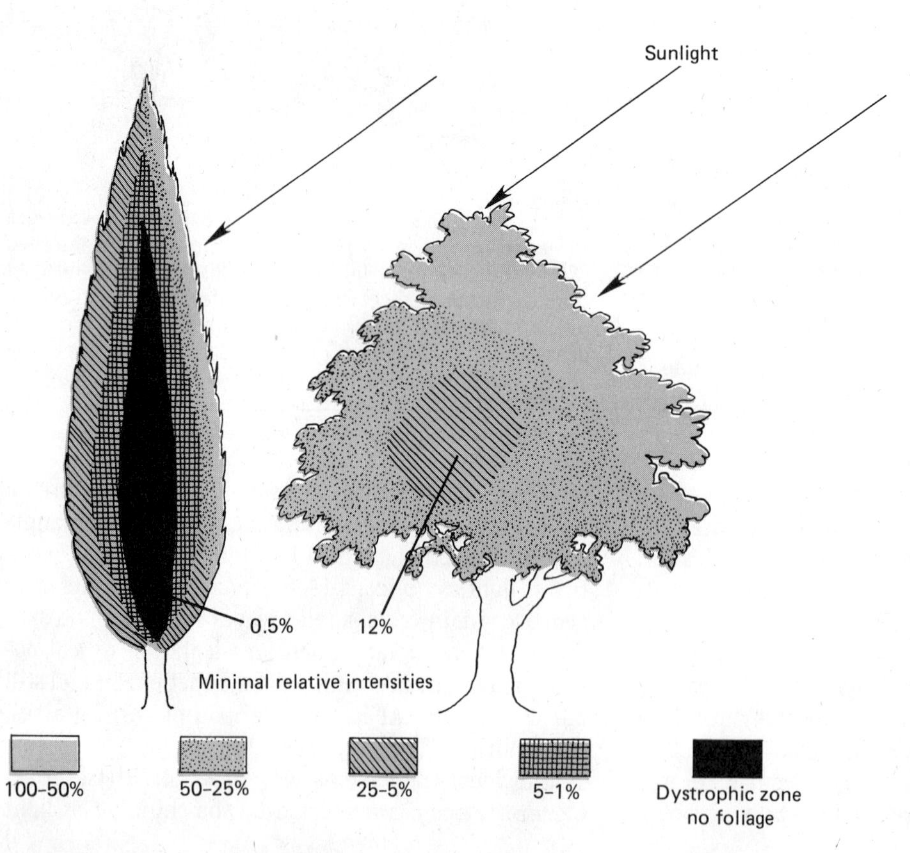

FIGURE 7.3 Reduction of light in a dense crown of cypress (left) and an open crown of olive (right) on a clear summer day. A dense growth of needles in the cypress quickly depletes light to 0.5 percent of illumination in the open. The interior of the tree is too dark to support any foliage. For this reason dense conifers lack interior foliage. The more open crown of the olive, with small reflecting leaves, receives diffuse light strong enough to support leafy branches even within the darkest part of the crown.

when released from overhead shade, they respond immediately with rapid growth.

When forced to grow in the shade, shade avoiders favor stem growth over leaf growth (an effort to reach light), and their leaves are widely spaced, reducing shade on individual leaves. These plants experience a reduction in photosynthesis while still undergoing a high rate of respiration.

Shade tolerance or intolerance, however, may involve more than light. Moisture, too, plays a role. Plants of the understory may experience a shortage of soil moisture resulting from competition with canopy vegetation. Intolerant plants might be able to carry on photosynthesis at a rate higher than required to balance respiration (*light compensation point*), but the rate is not enough to allow roots to expand into deeper soil where moisture is available.

The story of a plant's reaction to shade may be told in its leaves. Plants that show little plasticity in leaf shape in either sun or shade seem to function only under one set of environmental conditions. Other plants are more flexible. They change leaf shape and size in response to light conditions. Sun leaves are deeply lobed and relatively thick, with well-developed support and conduction systems. Shade leaves are either weakly lobed or not lobed, and have a large surface area per unit of weight, a thinner epidermis, fewer stomata, and less support and conduction tissue. In part, their structure reflects less heat, moisture, and physical stress. Suddenly exposed to the sun, shade leaves lose an excessive amount of moisture and intercept too much light, damaging the chloroplasts. The shock kills them. When shaded bark on trunks of trees is exposed to direct sunlight, the light stimulates adventitious buds in the bark to develop into *epicormic branches*.

The amount of light that penetrates a terrestrial stand of vegetation varies with the season (Figure 7.4). In early spring in temperate regions, when leaves are just expanding, 20 to 50 percent of the light may reach the forest floor. Spring flowering plants make use of this flood of light by completing the reproductive phase of their life cycle before the canopy closes. When less than 10 percent of the light reaches the forest floor, flowering is over. In fall, when leaves begin to drop, increased light again reaches the forest floor, and another minor surge of flowering, involving certain goldenrods and asters, takes place. In dense coniferous forests, where light is low throughout the year, only ferns and mosses grow in the dim light.

LIGHT IN WATER

Aquatic environments also experience a reduction in light. As light, particularly sunlight, strikes the surface of water, as much as 10 percent of it at midday is scattered or reflected. In early morning and late afternoon, when the angle of the sun is low, reflection is greatly increased, and little light penetrates. That effect shortens the daylength beneath the surface.

Light that penetrates the water drops off exponentially with increasing depth. Some of it is absorbed and scattered by water molecules as well as by dissolved substances and suspended sediments. In turbid waters or waters supporting a heavy growth of phytoplankton, light may penetrate only a few centimeters below the surface. As water depth increases from 0.1 to 100 m, visible light becomes limited more and more to a narrow band of blue light at wavelengths of about 0.50 microns. That is part of the reason why water of deep, clear lakes looks blue. Eventually, blue light is filtered out, and the remaining green light is poorly absorbed by chlorophyll. In effect—except in the clearest, shallow water—aquatic habitats are shade habitats, and the plants are adapted to grow under reduced light.

PHOTOSYNTHESIS

Photosynthesis is the conversion of CO_2 and H_2O into carbohydrates using the energy of light captured by chlorophyll. It is carried out primarily by chlorophyll-bearing plants, aquatic and terrestrial. Minor contributions are made by photosynthetic bacteria that use hydrogen and various organic compounds as electron donors.

Although photosynthesis is commonly expressed as

$$6CO_2 + 12H_2O + \text{light energy} \rightarrow C_6H_{12}O_6 + O_2$$

it is a complex sequence of reactions (Figure 7.5). Glucose is only an intermediate compound. The final

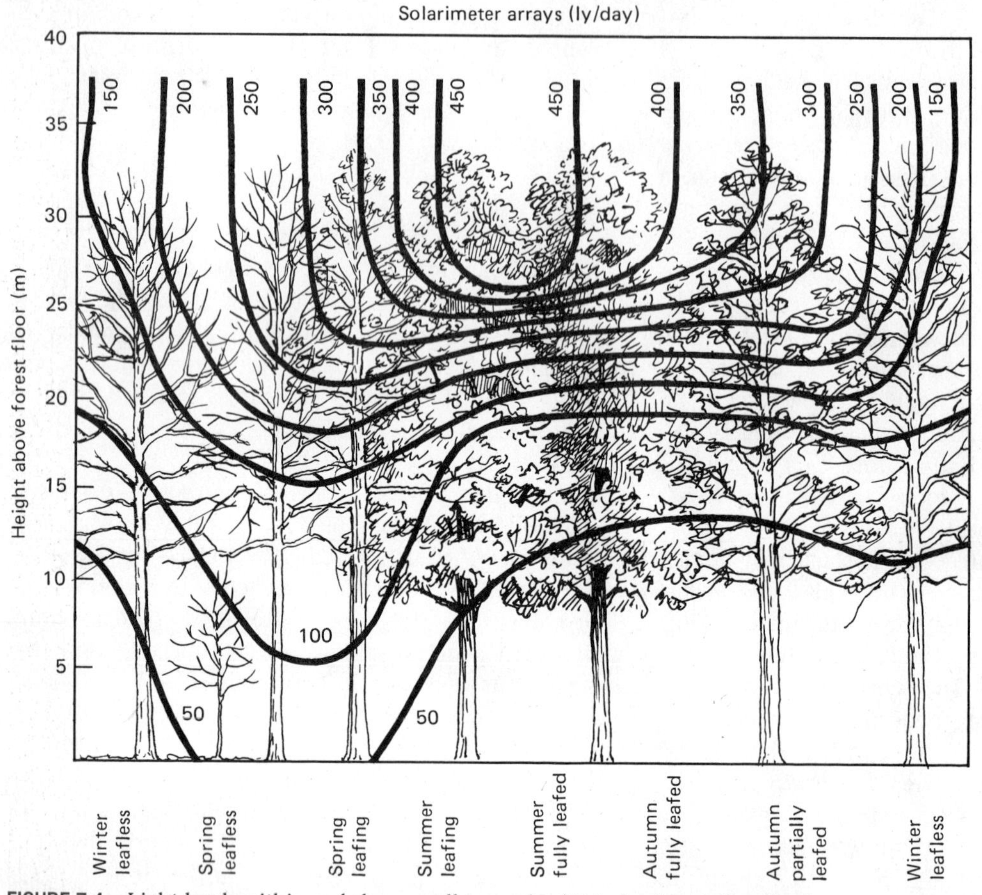

FIGURE 7.4 Light levels within and above a yellow-poplar (*Liriodendron tuliperifera*) over a year. The greatest intensity of solar radiation occurs in midsummer, but the canopy attenuates most of the light, so little more than 10 percent of open light reaches the forest floor. Most illumination reaches the forest floor in spring, when trees are still leafless, the time of spring flowers. The least radiation is received in winter, with lower solar elevations and shorter daylengths. As a result, the amount of solar radiation reaching the forest floor is little more than that of midsummer.

products are more complex sugars, free amino acids, proteins, fats and fatty acids, vitamins, pigments, and coenzymes. All of these substances probably are synthesized in chloroplasts by reactions involving photoelectron transport and photophosphorylation (the addition of one or more phosphate groups to a molecule by light energy). Synthesis of various products takes place in different parts of plants or under different environmental conditions. Mature leaves of some species of plants produce only simple sugars, whereas young shoots and rapidly developing leaves produce fats, proteins, and other products.

This is not an appropriate place to describe in detail the photosynthetic process. You can find complete reviews in various texts on plant physiology and good basic summaries in several outstanding general biology texts. What is of interest here are those features of photosynthesis that relate to ecological functions, es-

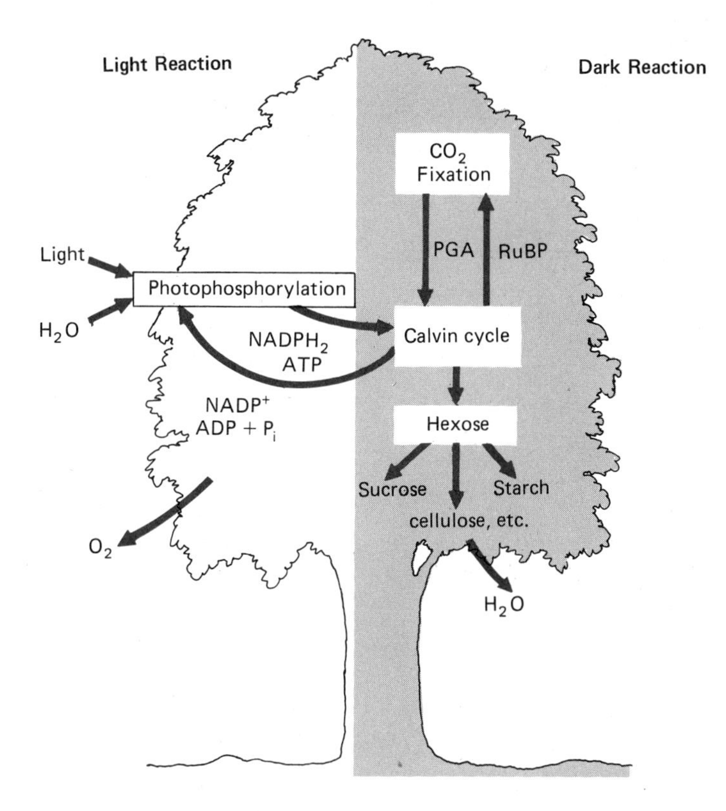

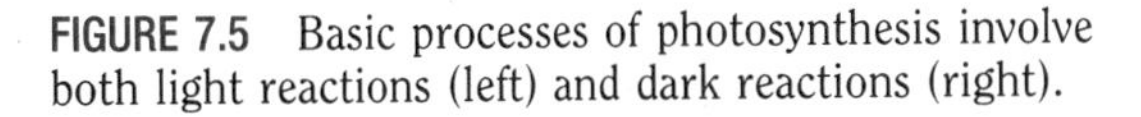
FIGURE 7.5 Basic processes of photosynthesis involve both light reactions (left) and dark reactions (right).

pecially energy transfer and adaptation of plants to various environmental situations.

THE C_3 CYCLE

Photosynthesis consists of two parts: a light reaction and a dark reaction. The light reaction converts light energy into chemical energy and, as the term implies, takes place in light. The dark reaction converts CO_2 into sugars and starches. In spite of its name, the reaction does not require darkness but probably takes place independent of light.

Involved in the light reaction are two pigments or photosystems that absorb light energy (photosystem I, active at 700 nanometers far red wavelength, and photosystem II, active at 680 nanometers); water; and a hydrogen carrier, NADP (nicotinamide adenine dinucleotide phosphate), in the chlorophyll. Ignoring the details of the interactions of the two photosystems and the exchange of energy, what happens is this: The light reaction traps light energy and stores it for a short time in a high-energy phosphate bond in ATP (adenosine triphosphate). Hydrogen, obtained from water, is used to reduce $NADP^+$ to NADPH, and O_2 is given off as a gas:

$$2H_2O + 2NADP^+ \xrightarrow[chlorophyll]{light} 2NADPH + 2H^+ + O_2$$

In the dark reaction, CO_2 enters the process. In the presence of ATP, cells of chloroplasts fix CO_2 into carbohydrates:

$$ATP + NADPH + H^+ + CO_2 \rightarrow$$
$$ADP + P_i^* + NADP^+ + \text{carbohydrate}$$

The reaction, so described, leaves much of the story missing. Involved is a five-carbon compound, ribulose biphosphate, RuBP. It combines with CO_2 to produce a

* P_i is an inorganic phosphate; an abbreviation for the free phosphate group HPO_4^{2-} added directly to ADP.

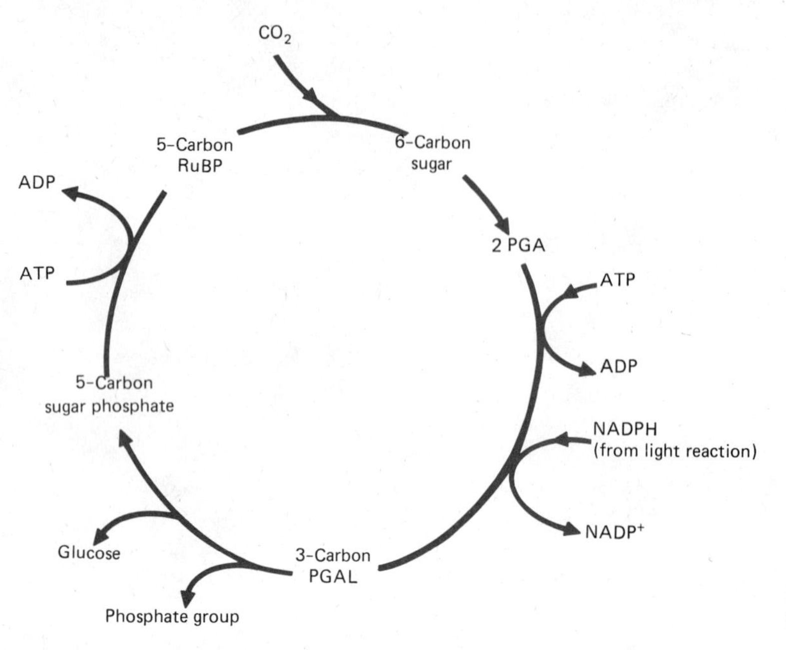

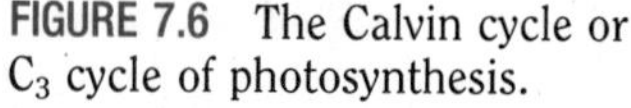

FIGURE 7.6 The Calvin cycle or C_3 cycle of photosynthesis.

six-carbon molecule that soon breaks into two three-carbon molecules of phosphoglyceric acid, PGA (Figure 7.6). Each PGA molecule next receives a second phosphate with a high-energy bond from a molecule of ATP. This new high-energy bond is then broken to supply the energy needed to reduce PGA to PGAL, phosphoglyceraldehyde. This reaction involves an addition of H^+ from NADPH. Some of the three-carbon PGALs are joined to form hexoses, six-carbon sugars. Other PGALs in combination with another ATP are used to reform new five-carbon RuBP, returned to the system to react with CO_2. Thus the light reaction really involves two reactions, a noncyclic one leading to the dark reaction and a cyclic one reforming reactive RuBP. This common form of photosynthetic reaction (often called the Calvin cycle after its discoverer, Melvin Calvin, and his associates) is referred to as the C_3 pathway (Figure 7.7), and the plants that possess it as their sole photosynthetic reaction are called C_3 plants.

RuBP has an affinity not only for CO_2 but for O_2 as well. When RuBP combines with O_2 instead of CO_2, it forms a two-carbon molecule, phosphoglycolate, and one molecule of PGA, instead of two PGAs. The PGA is returned to the cycle, reducing pentose phosphate to form hexose sugars, while phosphoglycolate gives rise to CO_2 and some glucose, but with a considerable expenditure of energy.

This associated pathway, known as *photorespiration,* reduces C_3 efficiency. It ties up RuBP that could be used for fixing CO_2. When O_2 is fixed, energy is lost,

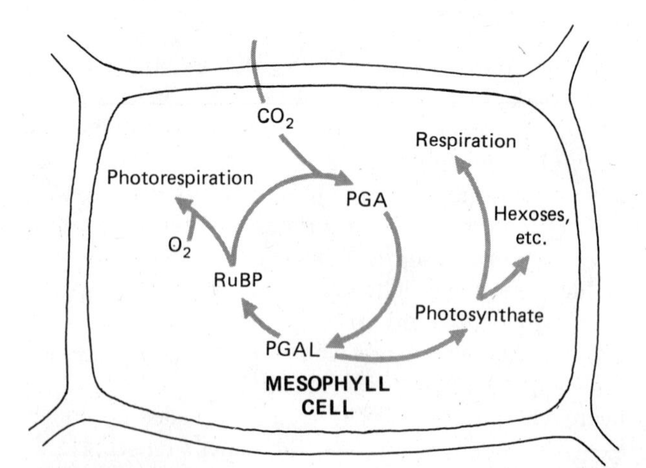

FIGURE 7.7 Basic dark reaction of the C_3 pathway of photosynthesis.

and only half as much PGA is produced. On the average, half the CO_2 fixed in photosynthesis is lost in photorespiration. A rise in temperature increases that loss and decreases the yield of photosynthesis.

Why plants should engage in such an energy-wasting process is not clear. One hypothesis is that photorespiration may be a safety valve for the dissipation of energy in case CO_2 becomes limiting. Under that condition, ATP and $NADP^+$ may become fully charged. To utilize that energy, RuBP fixes oxygen and prevents injury to chlorophyll and a breakdown of the photosynthetic system. When CO_2 is not limiting, the "escape" system still functions to a certain extent.

Photorespiration must not be confused with respiration. Green plants, like animals, undergo typical cellular respiration and generate energy by breaking down carbohydrates, releasing CO_2.

THE C_4 CYCLE

Other plants utilize a different photosynthetic pathway, called C_4. It is facilitated by a plant anatomy different from C_3 plants (Figure 7.8). C_3 plants have vascular bundles surrounded by a layer or sheath of colorless cells, and the cells containing chlorophyll are irregularly distributed throughout the mesophyll of the leaf. Plants exhibiting the C_4 cycle have vascular bundles surrounded by distinctive bundle sheath cells, rich in chlorophyll, mitochondria, and starches. Mesophyll cells are arranged laterally around the bundle sheaths and have few chloroplasts.

C_4 plants first fix CO_2 by acceptor molecules of phosphoenolpyruvate (PEP), a three-carbon compound, to form the four-carbon acids malate and asparate. This reaction takes place in the chlorophyll-containing mesophyll cells. Malic and aspartic acids move to the cells of the bundle sheaths (Figure 7.9), where enzymes break them down to form the three-carbon PGL as in the C_3 cycle. The three-carbon carrier molecules return to the mesophyll cells, where they are converted to PEP to receive more CO_2. Because these plants use a four-carbon molecule in photosynthesis, they are called C_4 plants, and the photosynthetic pathway is the C_4 system.

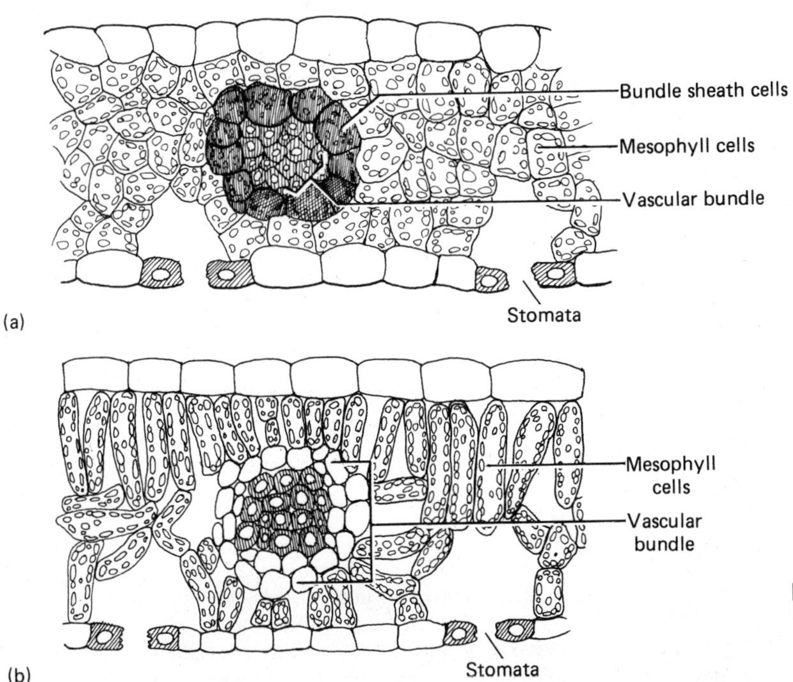

FIGURE 7.8 Comparison of the leaf anatomy of a C_4 plant (a) and a C_3 plant (b). In the C_4 plant, the bundle sheath cells form a ring around the vascular bundle.

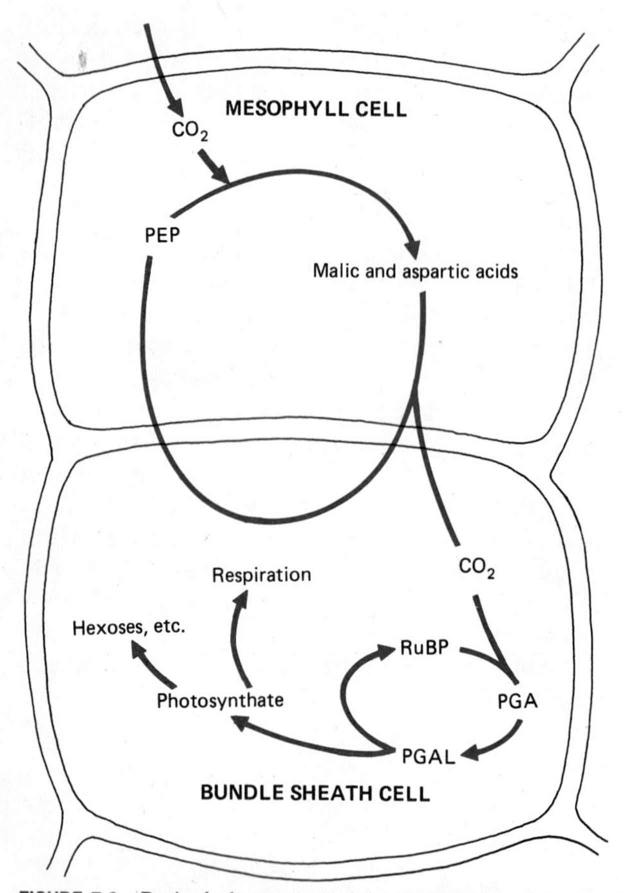

FIGURE 7.9 Basic dark reaction of the C_4 pathway of photosynthesis. Notice the reactions that take place in the mesophyll and bundle sheath cells and the relationship of the Calvin cycle to the C_4 pathway.

The major difference between the C_3 and C_4 pathways is the initial fixation of CO_2. The C_4 pathway involves an extra step, which has certain advantages. By fixing CO_2 in the mesophyll and transporting it to the cells of the bundle sheath, the plants can reduce the concentration of CO_2 in the mesophyll to near zero while maintaining a high concentration of CO_2 in the cells of the bundle sheath. PEP involved in the transport is more reactive with CO_2 than RuBP and thus fixes it more efficiently at low concentrations. As a result, oxygen has no opportunity to inhibit the photosynthetic process. The C_3 cycle, however, requires two additional ATP molecules to fix each molecule of CO_2, but the rate of photosynthesis is twice as fast.

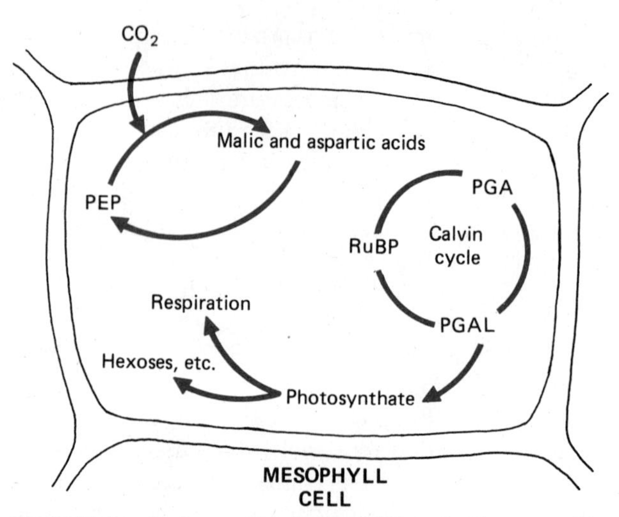

FIGURE 7.10 Basic reaction in the CAM cycle. Compare this cycle with the C_4 cycle (Figure 7.9).

CRASSULACEAN ACID METABOLISM (CAM)

A third type of photosynthetic process is the Crassulacean acid metabolism (CAM) (Figure 7.10). It is common to a small group of desert plants, mostly desert succulents, in the families Cactaceae, Euphorbiaceae, and Crassulaceae. To reduce the loss of water from transpiration, these plants close the stomata by day and

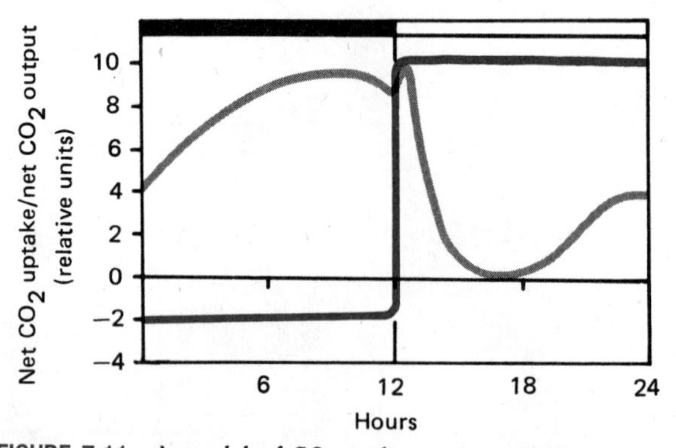

FIGURE 7.11 A model of CO_2 exchange in a CAM plant and a non-CAM plant with 12 hours of daylight and 12 hours of dark. The CAM plant takes in CO_2 during the night, while the non-CAM plant loses it. The non-CAM plant has a high intake of CO_2 during the day. The CAM plant, however, does not experience a negative CO_2 balance.

open them by night to take in CO_2, the reverse of other plants (Figure 7.11). CO_2 is fixed by PEP at night and stored as malate in the vacuoles of cells. By day the stomata close, and malate moves from the vacuole into the cell. CO_2 is removed from malate and, in the presence of light energy, is used to refix CO_2 by the way of the Calvin cycle. CAM is similar to the C_4 pathway except there is no division of the process between different cells.

DIFFERENCES AMONG C_3, C_4, AND CAM PLANTS

Physiologically, C_3 and C_4 plants are different (Table 7.1). C_3 plants have an optimum temperature for photosynthesis of 16° to 25° C (Figure 7.12). The CO_2 compensation point (at which CO_2 uptake is equal to CO_2 output) is in the range of 30 to 70 parts per million CO_2. C_3 plants have a low light-saturation threshold also (Figure 7.13). Photosynthesis is inhibited when light falling on leaves reaches a range of 3000 to 6000 footcandles.

Compared to C_3 plants, C_4 plants have a higher optimum temperature for photosynthesis, 30° to 45° C. They have a lower CO_2 compensation point, 0 to 10 parts per million, and exhibit no photorespiration, thus saving energy for other metabolic processes. They have a high light-saturation threshold; in fact, it is difficult for C_4 plants to reach light saturation, even in full sunlight (10,000 to 12,000 footcandles). The maximum rate of photosynthesis for C_4 plants in full sunlight is 40 to 80 mg CO_2/dm^2 leaf area/hour, whereas the net rate for C_3 plants is 15 to 35 mg CO_2/dm^2 leaf area/hour.

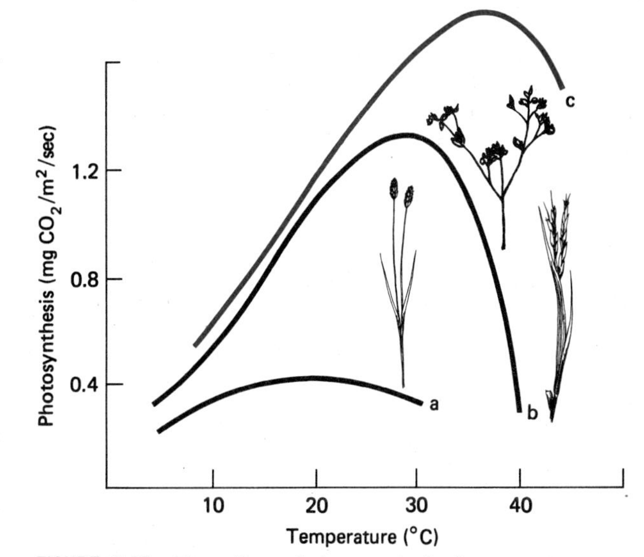

FIGURE 7.12 The effect of changes in leaf temperature on the photosynthetic rates of C_3 and C_4 plants. (a) C_3 plants, represented by the north temperate grass *Sesleria caerulea,* exhibit a decline in the rate of photosynthesis as the temperature of the leaf increases. (b and c) C_4 plants are represented by a north temperate C_4 grass *Spartina anglica* and a C_4 shrub of the North American hot desert, Arizona honeysweet, *Tidestromia oblongifolia.* The C_4 species increase their rate of photosynthesis as the temperature of the leaf increases, up to a certain point.

In spite of the differences, plants cannot be divided arbitrarily into C_3 and C_4 groups. Some are difficult to place on the basis of either CO_2 compensation or pri-

TABLE 7.1 Differences Among C_3, C_4, and CAM Plants

Characteristic	*C_3 Plants*	*C_4 Plants*	*CAM Plants*
Light-saturation point	3000–6000 footcandles	8000–10,000+ footcandles	?
Optimum temperature	16°–25° C	40°–80° C	30°–35° C
CO_2 compensation point	30–70 ppm	0–10 ppm	0–4 ppm
Maximum photosynthetic rate (mg CO_2/dm^2 leaf area/hr)	15–35	30–45	3–13
Maximum growth rate (g/dm²/day)	1	4	0.02
Photorespiration	High	Low	Low
Stomata behavior	Open day, closed night	Open day, closed night	Closed day, open night

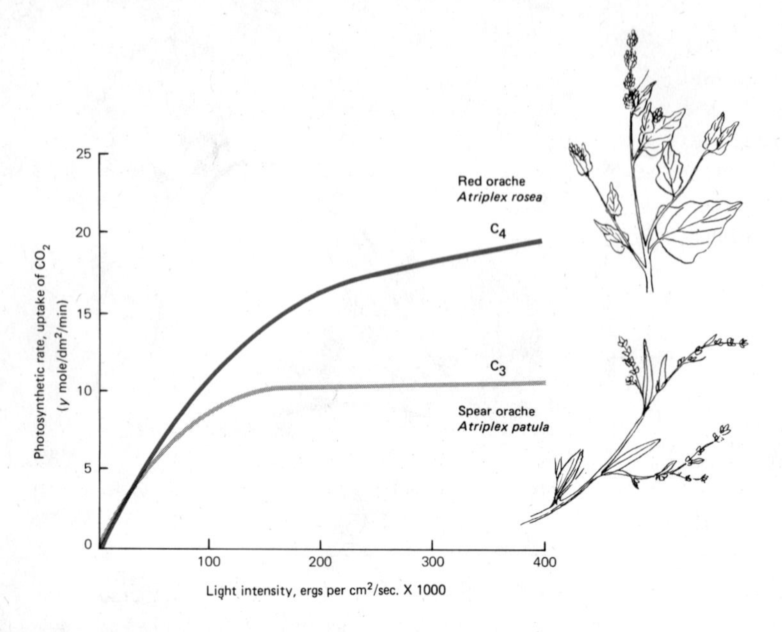

FIGURE 7.13 Effect of changes in light intensity on the photosynthetic rates of C_3 and C_4 plants grown under identical conditions. The C_3 species, spear orache (*Atriplex patula*), exhibits a decline in the rate of photosynthesis as measured by CO_2 uptake as light intensity increases. The C_4 species, red orache (*Atriplex rosea*), shows no such inhibition.

mary products of photosynthesis. Some C_3 species do not become light-saturated at low light levels.

Differences between C_3 and C_4 plants are ecologically important. Because they are more efficient photosynthetically under conditions of high light and temperature, C_4 plants may be more competitive and adjust more easily to environmental stress than C_3 plants. C_4 plants are mostly native to warm and tropical regions, where both light intensity and water are limiting. At low temperatures, the photosynthetic yield of C_3 plants exceeds that of C_4 plants; it matches that of C_4 plants at 25° to 30° C and is decidedly inferior to that of C_4 plants at temperatures over 30° C.

Although C_4 plants are sensitive to low temperatures, no step in the C_4 pathway appears to be unusually sensitive to low temperatures. Features that provide C_4 plants with their advantage at high temperatures (such as stability of proteins and enzymes and transport of photoelectrons) are not necessarily related to the C_4 pathway. The C_4 pathway probably evolved among plants with a tolerance for higher temperatures in various water-limited environments. There selection would favor those plants with an internal mechanism able to concentrate CO_2. Such characteristics, advantageous in warm climates, would be disadvantageous in cool climates. For that reason, the C_4 pathway probably became linked genetically to those characteristics that favor heat tolerance.

C_4 plants, which undoubtedly evolved in tropical climates and then migrated to temperate climates, have the more advanced photosynthetic process. The C_3 pathway is common to all plants, including those with a C_4 pathway. The latter is not found in algae, mosses and lichens, ferns, and more primitive angiosperms. It is known to exist in hundreds of monocotyledonous and dicotyledonous species comprising 100 genera and 10 plant families. Grasses comprise about half of the known C_4 species. In North America most of these species are subtropical in distribution. From Florida to Texas, 65 to 82 percent of the grass species are C_4 plants; in the central plains, 31 to 61 percent; and in the northern part of the continent, 0 to 23 percent. Those with the C_4 pathway are warm-season grasses; the rest are C_3 cool-season grasses, putting on growth and flowering in spring and early summer. Apparently, no C_4 plants grow on the tundra.

CAM plants, also inhabitants of hot climates, not only fix CO_2 from malate, like C_4 plants; they fix CO_2 from the atmosphere as well, depending on environmental conditions. Because CAM plants live in areas where water is deficient, they have to maintain a positive carbon balance, or at least prevent a negative balance, during long dry periods and still maintain a positive water balance. These plants evolved an adaptive variant of the C_4 photosynthetic pathway to achieve the maximum possible carbon gain with a minimum loss of water. They achieve that end at the expense of rapid growth.

EFFICIENCY OF PHOTOSYNTHESIS

Photosynthetic efficiency in converting the energy of the sun to organic matter can be assessed from two viewpoints: (1) the amount of energy required to release a molecule of oxygen and (2) the ratio of calories stored up to light received. In terms of energy input, photosynthesis is a rather inefficient process. To release 1 mole of oxygen and to fix 1 mole of carbon dioxide, a green plant needs an estimated 320 kilocalories of light energy. For each mole of oxygen lost, approximately 120 kilocalories of energy are fixed. That efficiency is approximately 38 percent. Efficiencies calculated for isolated chloroplasts and for some algae amount to 21 to 33 percent.

When calories stored relative to light energy available are considered, efficiency is considerably less. The usable spectrum, 0.4 to 0.7 wavelengths, is only half the total energy incident on vegetation. Highest short-term efficiency measured over a period of weeks of active growth may amount to 12 to 19 percent. When photosynthetic efficiency is computed by year or by growing season, it is still lower.

This low efficiency reflects many limiting factors: nutrients, water, light, temperature, atmospheric gases (especially pollutants), competition with other plants, leaf area, geometry of the vegetative canopy, and the degree to which the plants are preyed upon by herbivores.

Summary

Light, like moisture and temperature, is an important aspect in the lives of plants and animals. As visible light passes through Earth's atmosphere, certain wavelengths, especially violets and blues, are reduced more than others. Light is reduced even further in wavelength and intensity as it is reflected, absorbed, or transmitted by layers of vegetation or by the depth of water. Based on their response to the intensity of light reaching them, we call plants either shade-tolerant or shade-intolerant. Shade-tolerant plants have lower rates of photosynthesis, lower rates of respiration, and physiological characteristics that enable them to grow under conditions of low light. Shade-intolerant plants, the shade avoiders, have higher rates of photosynthesis and respiration, which make them inefficient under low light.

Photosynthesis is the process by which green plants utilize the energy of the sun to convert carbon dioxide and water into carbohydrates. Plants possess one of three different photosynthetic pathways. Most plants use the C_3 pathway, which involves the formation of a three-carbon phosphoglyceric acid used in subsequent reactions. C_4 plants use a four-carbon process in which carbon dioxide reacts to form malic and aspartic acids. The CO_2 first is fixed in these compounds, and then is released to the C_3 cycle.

These two groups of plants, C_3 and C_4, possess structural and physiological differences that are important ecologically. C_4 plants can carry on photosynthesis at higher leaf temperatures and at lower CO_2 concentrations than C_3 plants. C_3 plants are less efficient because of photorespiration, in which plants substitute O_2 for CO_2 in part of the C_3 cycle. C_3 plants are distributed from the arctic to the tropics. C_4 plants are most successful in hot, arid environments. Because of their low CO_2 compensation point, C_4 plants are able to fix CO_2 in spite of the closure of stomata to reduce water loss. Most highly adapted to an arid environment are plants with a third type of photosynthetic pathway, CAM. These plants absorb CO_2 by night, when water loss is reduced, store CO_2 as malic acid in cell vacuoles, and release it during the day to the C_3 cycle.

In spite of the various pathways, the overall efficiency of photosynthesis, measured as a ratio of gross and net photosynthesis to solar radiation, is low. During the growing season, plants achieve an efficiency of roughly 1 to 3 percent or lower if net rather than gross photosynthesis is used in the calculation.

Review and Study Questions

1. What is photosynthetically active radiation?
2. What part of light is reduced as it travels through Earth's atmosphere?
3. What is leaf area index (LAI)? How does it relate to light penetration of the vegetative canopy?
4. Contrast a shade-tolerant plant with a shade avoider.
5. What is the light compensation point?
6. What characterizes the C_3 cycle, the C_4 cycle, and the Crassulacean acid cycle in photosynthesis?
7. What are the C_3, C_4, and CAM pathways of photosynthesis?
8. What is the ecological significance of each?
9. Why is the C_4 pathway considered an evolutionary advance over the C_3 pathway? Under what environmental conditions was it probably selected?
10. How does LAI relate to photosynthetic rates?
11. How efficient is photosynthesis in storing energy?

Selected References

Bainbridge, R., G. C. Evans, and O. Rackham, eds. 1966. *Light as an ecological factor.* Oxford, England: Blackwell Scientific Publications. Dated but still an outstanding reference.

Barberm, J., ed. 1977. *Primary processes of photosynthesis.* Vol. 2. *Topics in photosynthesis.* New York: Elsevier. Primary reactions in the capture of light energy.

Burris, R. H., and C. C. Black, eds. 1976. *CO_2 metabolism and plant productivity.* Baltimore: University Park Press.

Cooper, J. P. 1975. *Photosynthesis and productivity in different environments.* New York: Cambridge University Press.

Govindjee, ed. 1982/1983. *Photosynthesis.* Vols. 1 and 2. Orlando, FL: Academic Press. Comprehensive examination of energy conversion, CO_2 assimilation, and plant productivity.

Grime, J. 1971. *Plant strategies and vegetative processes.* New York: Wiley. Excellent discussion of shade tolerance.

CHAPTER 8

Periodicity

Outline

Objectives

On completion of this chapter, you should be able to:

1. Discuss the role of light in the daily and seasonal cycles of plants and animals.
2. Discuss circadian rhythms and their relation to the biological clock.
3. Discuss how the biological clock functions as a timekeeper.
4. Discuss the relationship between biological clocks and critical daylengths and their role in the annual rhythms in the lives of animals and plants.
5. Describe circadian rhythms, circannual rhythms, tidal rhythms, phase, shift, free-running phenomena, *Zeitgebers,* entrainment, critical daylength, and short-day and long-day organisms.
6. Explain how the biological clock might function in tidal rhythms of marine animals.

One aspect of communities with which everyone is familiar is rhythmicity, the recurrence of daily and seasonal changes. Dawn ends the darkness, and bird song signals its arrival. Butterflies, dragonflies, and bees become conspicuous, hawks seek out prey, and chipmunks and tree squirrels become active. At dusk, light fades and daytime animals retire, the blooms of water lilies and other flowers fold, and animals of the night appear. Foxes, raccoons, flying squirrels, owls, and moths take over niches occupied by others during the day. As the seasons progress, daylength changes, and with it, other conspicuous activities. Spring brings migrant birds and initiates the reproductive cycles of many animals and plants. In fall the trees of temperate regions become dormant, insects and herbaceous plants disappear, summer-resident birds return south, and winter visitors arrive. Underlying these rhythmicities is the daily and annual movement of the Earth relative to the sun.

DAILY PERIODICITY

Because life evolved under the influences of daily and seasonal environmental changes, it is natural that plants and animals would have some rhythm or pattern to their lives that would synchronize them with fluctuations in the environment. For years biologists have been intrigued by the means by which organisms keep their activities in rhythm with the 24-hour day, including such phenomena as the daily pattern of leaf and petal movements in plants, emergence of insects from pupal cases, and sleep and wakefulness of animals (see Figure 8.1). At one time biologists thought that these rhythmicities were entirely exogenous—that is, that the organisms responded only to external stimuli such as light intensity, humidity, temperature, and tides. Laboratory investigations, however, indicate that they are not.

At dusk in the forests of North America, a small squirrel with silky fur and large, black eyes emerges from a tree hole. With a leap the squirrel sails downward in a long, sloping glide, maintaining itself in flight with broad membranes stretched between its outspread legs. Using its tail as rudder and brake, it makes a short, graceful, upward swoop that lands it on the trunk of another tree. It is the flying squirrel, *Glaucomys volans,* perhaps the commonest of all our tree squirrels; but because of its nocturnal habits, this mammal is seldom seen. Unless it is disturbed, the flying squirrel does not come out by day. It emerges into the forest world with the coming of darkness; it returns to its nest with the first light of dawn.

If the flying squirrel is brought indoors and kept under artificial conditions of night and day, the animal will confine its periods of activity to darkness, its periods of inactivity to light. Whether the conditions under which the animal lives are 12 hours of darkness and 12 hours of light or 8 hours of darkness and 16 hours of light, the onset of activity is always shortly after dark. The squirrel's day-to-day activity forms a 24-hour period. This correlation of the onset of activity with the time of sunset suggests that light has a regulatory effect on the activity of the squirrel.

The *photoperiodism* (response to changing light and darkness) exhibited by the squirrel is not quite so simple. There is more to it than the animal's becoming active because darkness has come. If the squirrel is kept in constant darkness, it still maintains a relatively constant rhythm of activity from day to day. However, in the absence of any external time cues, the squirrel's activity rhythm deviates from the 24-hour periodicity exhibited under light-and-dark conditions. The daily cycle under constant darkness varies from 22 hours, 58 minutes to 24 hours, 21 minutes, the average being less than 24 hours (most frequent: 23 hours, 50 minutes and 23 hours, 59 minutes). The length of the period maintained under a given set of conditions is an individual characteristic. Because of the deviation of the average cycle length from 24 hours, each individual squirrel gradually drifts out of phase with the day-night changes of the external world (see Figure 8.2). If the same animals are held under continuous light, a very abnormal condition for a nocturnal animal, the activity cycle is lengthened, probably because the animals, attempting to avoid running in the light, delay the beginning of their activity as much as they can.

CIRCADIAN RHYTHMS

The flying squirrel and many other forms of life studied to date, including humans, all possess a rhythm of activity that under field conditions exhibits a period-

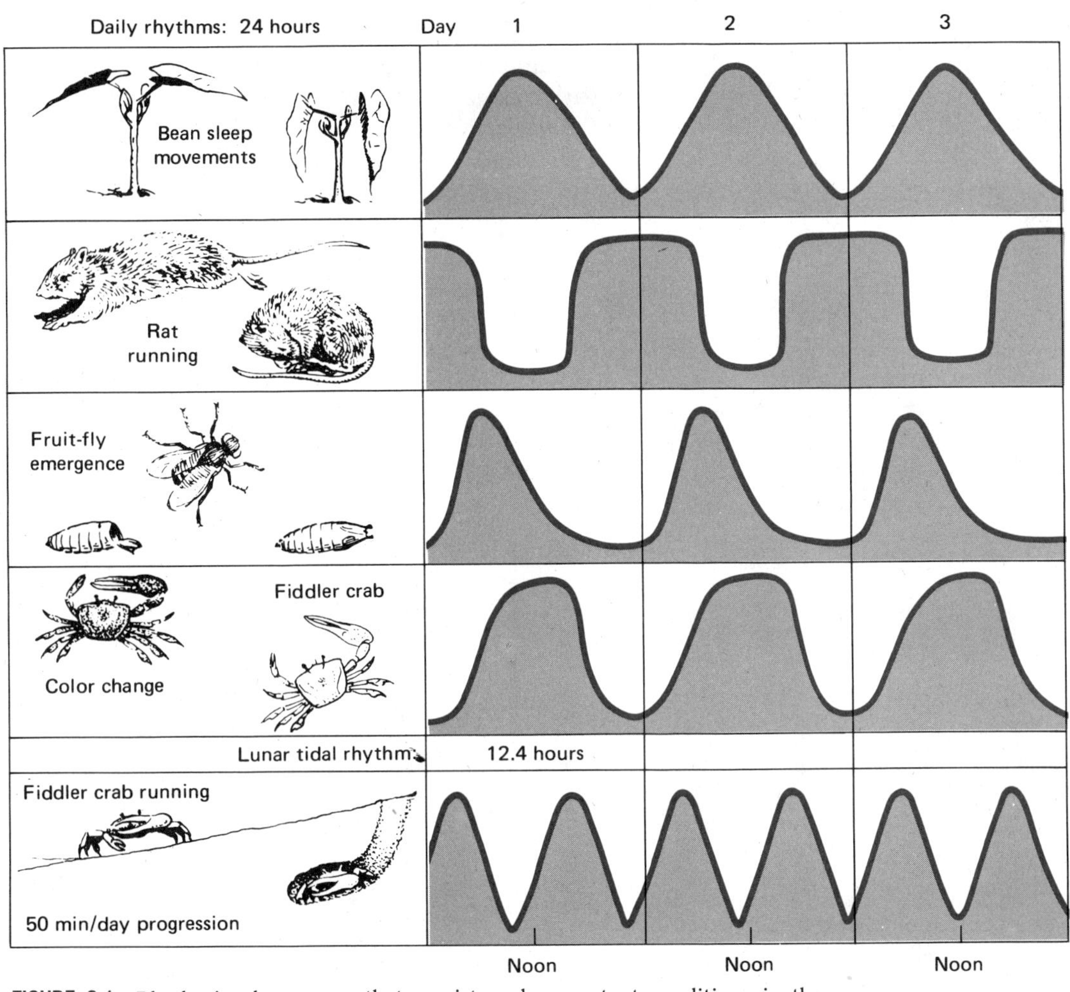

FIGURE 8.1 Rhythmic phenomena that persist under constant conditions in the laboratory. The rhythms relate to external cycles.

icity of 24 hours. Moreover, when these organisms are brought into the laboratory and held under constant conditions of light, darkness, and temperature, away from any external time cues, they still exhibit a rhythm of activity of approximately 24 hours. Because these rhythms approximate but seldom match the periods of the Earth's rotation, they are called *circadian* (from the Latin *circa,* "about," and *dies,* "day"). The period of the circadian rhythm, the number of hours from the beginning of activity on one day to the beginning of activity on the next, is referred to as *free-running.* In other words, it exhibits a self-sustained oscillation under constant conditions.

Thus, many plants and animals are influenced by two periodicities: the internal circadian rhythm of approximately 24 hours and the external environmental rhythm, usually precisely 24 hours. Some environmental "timesetter" must adjust the rhythm to match that of the outside world. The most obvious timekeepers, cues, synchronizers, or *Zeitgebers* are temperature and light. Of the two, light is the master Zeitgeber. It brings the circadian rhythm of many organisms into phase

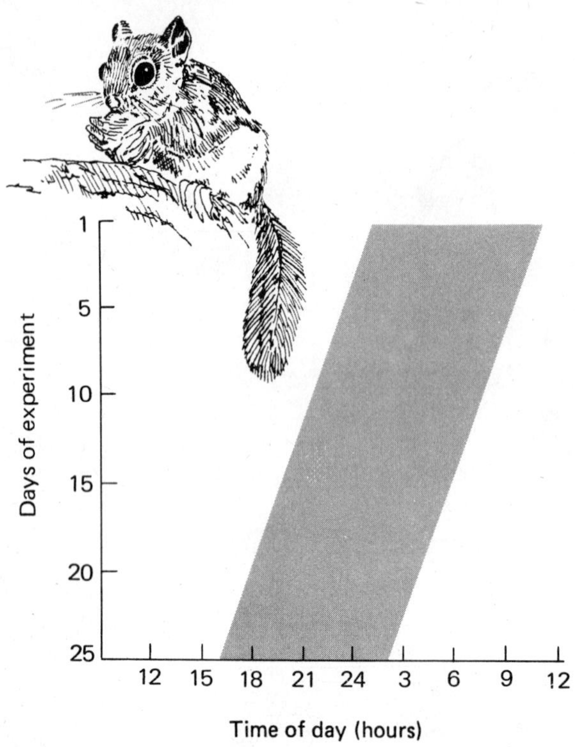

FIGURE 8.2 Drift in phase of activity rhythm of a flying squirrel held in continuous darkness at 20° C for 25 days.

with the 24-hour photoperiod of the external environment. We call this effect *entrainment.*

ENTRAINMENT

The activity rhythm of most organisms shows an entrainment to light-dark cycles. The flying squirrel, both in its natural environment and in artificial day-night schedules, synchronizes its daily cycle of activity to a specific phase of the light-dark cycle. This fact was demonstrated in a series of experiments by P. J. DeCoursey. She held flying squirrels in constant darkness until their circadian rhythms of activity were no longer in phase with the natural environment. Then she subjected the squirrels to a light-dark cycle that was out of phase with their free-running period. If the light period fell in the animals' subjective night, it caused a delay in the subsequent onset of activity. Synchronization took place in a series of stepwise delays until the animals' rhythms were stabilized with the light-dark change (see Figure 8.3). If the light period fell at the subjective dawn or at the end of the dark period (when the animals' activity period was about to end), it caused an advance of activity toward the dusk period. If the light fell in the animals' inactive day phase, it had no effect.

The flying squirrels do not need to be exposed to a whole light-dark cycle to bring about a shift in the phase of the activity rhythm. A single ten-minute light period is sufficient to cause a *phase shift* in locomotory activity, provided that it is given during the squirrel's light-sensitive period.

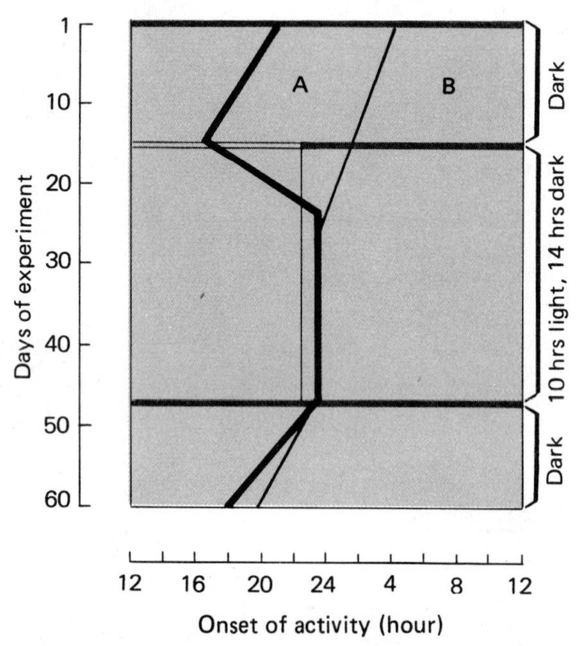

FIGURE 8.3 Synchronization of flying squirrels with a circadian rhythm of less than 24 hours in constant darkness to a cycle of 10 hours light and 14 hours darkness. For squirrel A the rephasing light fell during subjective night; synchronization was accomplished by a stepwise delay. The onset of activity was stabilized shortly after light-dark change. For squirrel B light fell in the subjective day, and the free-running period continued unchanged until the onset drifted up against the "dusk" light change. This obstacle prevented the rhythm from drifting forward by a delaying action of light. When the squirrel was returned to constant darkness, the onset of activity continued the forward drift.

THE BIOLOGICAL CLOCK

That plants and animals have an internal timekeeping mechanism, or a biological clock, has been suspected for a long time. The relative preciseness of the timekeeper has been demonstrated for years. A major question, however, has been the location of the clock in living organisms.

LOCATION

Basically, the biological clock is cellular. In one-celled organisms and plants, the clock seems to be located in individual cells. In multicellular animals the clock is associated with the brain.

Skillful surgical procedures have allowed circadian physiologists to discover the location of the clock in some insects, birds, and mammals. In most insects studied, the clock, including the photoreceptors, is located either in the optic lobes (on the forebrain) or in tissue between the optic lobes and the brain. In the cockroach and cricket, however, the photoreceptor for the entrainment of circadian rhythm of locomotion and stridulation (chirping) is located in the compound eye, but the controlling clock is in the brain. In birds the clock appears to be located in the pineal gland, deep in the lower central part of the brain. In mammals the clock is located in the suprachiasmatic nuclei, a number of specialized cells arranged left and right of the brain just above the crossing of the optic nerves coming from the eyes.

REQUIREMENTS

To function as a timekeeper, the biological clock has to fulfill certain conditions. It has to be an internal mechanism with a natural rhythm of about 24 hours. Its rhythm must have the capability of being changed or reset by signals such as changes in the time of dawn and dusk. The clock has to be able to run continuously in the absence of any environmental timesetter. Also, it has to be able to run the same at all temperatures (that is, it has to be temperature-compensated). Cold temperatures must not slow it down, nor warm temperatures speed it up. If that happened, the block would not keep in phase with environmental time.

MODELS

Several basic models of the biological clock have been proposed. An early one is the oscillating circadian rhythm of sensitivity to light, proposed by E. Bunning in 1960. The cycle or time-measuring process begins with the onset of light or dawn. The first half (12 hours) is light-requiring (*photophilic*) and the second half is dark-requiring (*scotophilic*). Short-day effects are produced when light does not extend into the dark period. Long-day effects are produced when it does (Figure 8.4). A variation is a two-oscillator model in which one

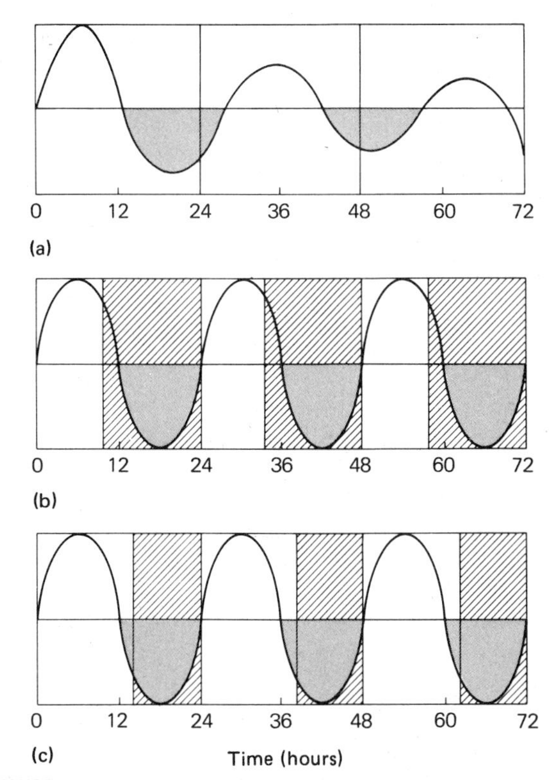

FIGURE 8.4 The Bunning model. Oscillations of the clock cause an alternation of half-cycles with different sensitivities to light. The shaded portion represents the dark-sensitive part of the cycle, the white area the light-sensitive portion. The hatched area represents periods of darkness or night. (a) The free-running clock in continuous light (or continuous dark) tends to drift out of phase with the 24-hour photoperiod. (b) Short-day conditions allow the dark to fall within the white or half-cycle. (c) In the long day, light falls in the dark (shaded) half-cycle.

oscillation is regulated by dawn and the other by dusk. This simple model does not explain all photoperiodic responses, but it is the base upon which other models are built.

Is one master clock driving all other clocks? Is a population of clocks involved? It appears that the biological clock is organized as a hierarchy of oscillators (Figure 8.5). Rhythms influencing various physiological and behavioral phenomena are controlled by "slave" or subservient clocks coupled to a master clock. These clocks or pacemakers may be integrated groups of cells within organs such as the brain, where they have specific timekeeping functions. When the master clock is reset by a light signal, it in turn resets the other clocks. One slave oscillator may take longer to resynchronize than another with a different period. Therefore resetting the master clock can put the subservient clocks out of phase. Such transient disturbances cause the discomforts of jet lag.

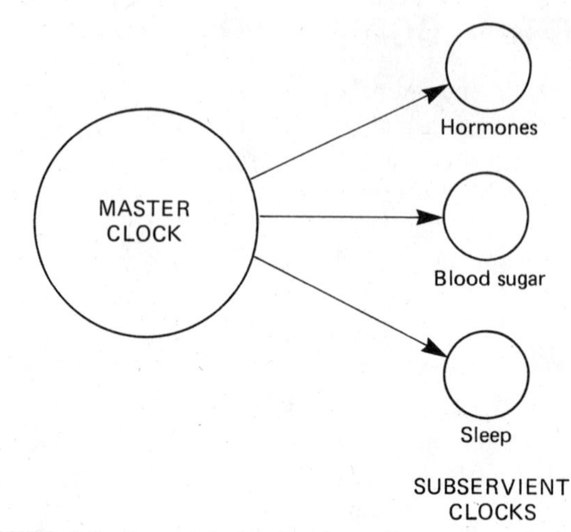

FIGURE 8.5 A model showing how the master clock, entrained to environmental changes in light, in turn resets the subservient clocks that control the rhythms of many physiological activities.

Jet lag is experienced not only by those who fly across a number of time zones. Swing-shift workers for whom night becomes day and day night over a period of weeks suffer from jet lag. Jet lag has a pronounced effect on the well-being and efficiency of those who work night shifts: medical personnel, police, air traffic controllers, pilots, and others. Submarine duty that calls for 6 hours on and 12 hours off causes a phase advance in the biological clock in every cycle; such sailors experience continuous jet lag. Abnormal phase relationships of body temperature and REM sleep are associated with cyclic depression in humans. The biological clock is one feature that reminds us of our close relationship to all other organisms.

FUNCTIONS

The biological clock provides plants and animals with a sense of time, keeping them in phase with their changing daily environment. It also gives some animals, such as the honeybee, a time memory. It enables the insect to return to a known source of nectar and pollen when it is most available. Because its memory involves a circadian rhythm, the bee can remain in the hive for several days during bad weather, and still remember the time of day when a particular group of plants is producing nectar. The biological clock also serves as a time-compensation mechanism. A number of animals orient themselves to the sun and use it like a compass. These animals are able to maintain a sense of direction because their circadian clock compensates for the sun's apparent movement during the day. Such time compensation has been demonstrated in bees, fish, turtles, and birds. Further, the biological clock permits organisms to anticipate and prepare for upcoming daily and seasonal events.

TIDAL RHYTHMS

Along the intertidal marshes the fiddler crabs, *Uca,* so named because of the enormously enlarged claw of the male, which he waves about incessantly, swarm by the hundreds across the exposed mud of salt marshes and mangrove swamps during low tide. As high tide moves in and inundates the marsh, the fiddler crabs retreat to their burrows, where they await the next low tide (Figure 8.6). Other intertidal organisms, from sand beach crustaceans and salt marsh periwinkles to the intertidal fish, the blennies, live under the influence of daily lunar and tidal cycles.

Fiddler crabs brought into the laboratory and held

FIGURE 8.6 A fiddler crab has retreated to its burrow in anticipation of the incoming high tide.

under constant conditions of temperature and light, devoid of tidal cues, exhibit the same tidal rhythmicity in their activity as they would back on the marsh. This tidal rhythm mimics the ebb and flow of tide every 12.4 hours, one-half of the lunar day of 24.8 hours, the interval between successive moonrises (Figure 8.7). Under the same constant conditions, fiddler crabs exhibit a circadian rhythm of color change, dark during the day and light by night.

Is the clock involved unimodal with a 12.4 hour cycle, or bimodal with a 24.8 hour cycle, close to the period of the circadian çlock? Do two clocks operate, a circadian one with a period of 24 hours and a lunar one with a period 24.8 hours? J. D. Palmer and his associates at the University of Massachusetts, based on their experimental evidence, suggest that two independent clocks synchronize tidal activity, one for each of the lunar-daily activity peaks. Further, the same master clock that governs tidal rhythms also drives circadian rhythms. Organisms, even at the cellular level, do not depend on one clock, any more than most of us rely

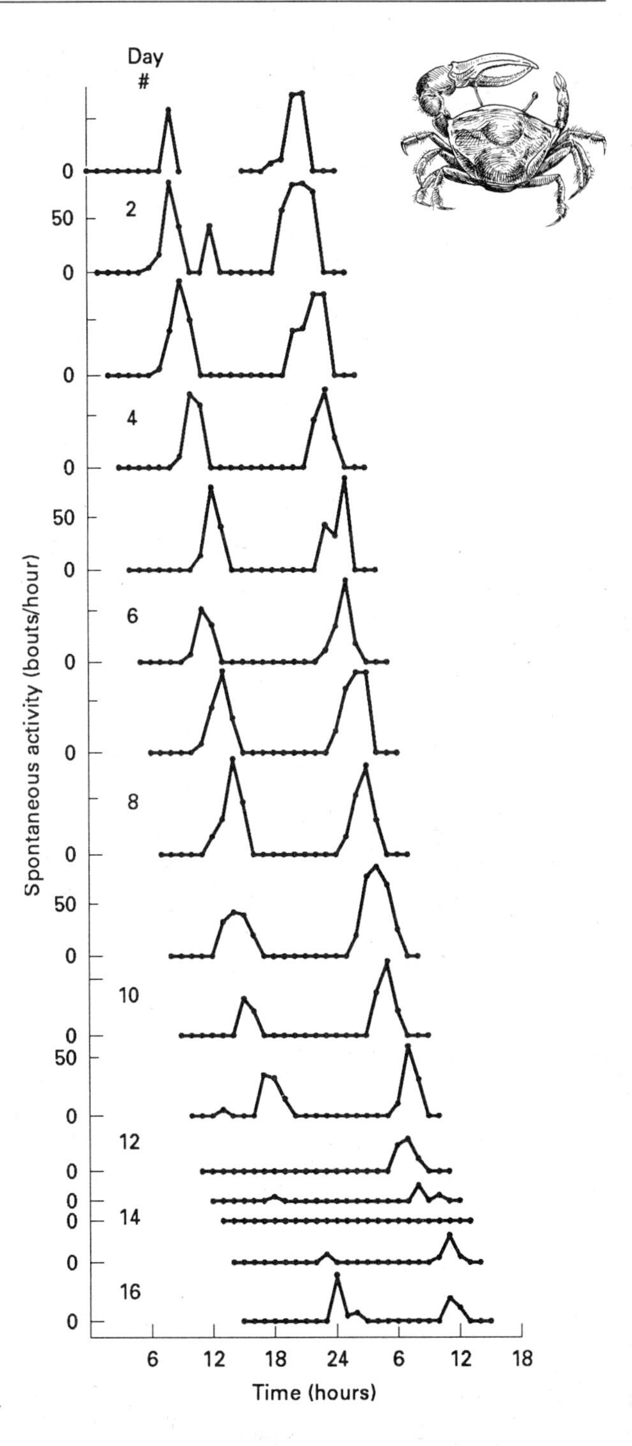

FIGURE 8.7 Tidal rhythm of the first 16 days of a fiddler crab monitored in the laboratory and maintained in a constant light and a constant temperature of 22° C. The fiddler crab alternated bouts of activity and inactivity, as it did when exposed to the tides in its natural habitat. When plotted against the solar day, the peaks move to the right because the lunar day is 51 minutes longer than the solar day.

on one mechanical clock. Organisms have built-in redundancies. This redundancy enables the various clocks to run at different speeds (see Figure 8.5), governing different processes with slightly differing periods.

ANNUAL PERIODICITY

Changes in daily photoperiods over the course of a year allow organisms to time annual events in their life cycles, from flowering and reproduction to food storage, migration, and hibernation. The biological clock picks up signals for seasonal changes in behavior and physiological activity.

CRITICAL DAYLENGTH

The signal is *critical daylength.* When a period of light reaches a certain portion of the 24-hour day, it inhibits or promotes a photoperiodic response. Critical daylength varies among organisms, but it usually falls somewhere between 10 and 14 hours. Through the year plants and animals compare that time span with the actual length of day or night. As soon as the actual daylength or nightlength is greater or lesser than the critical daylength, the organisms respond appropriately. Some organisms are *day neutral*—not affected by daylength but controlled by some other influence, such as rainfall or temperature. Others may be *short-day* or *long-day* organisms (Figure 8.8). Short-day plants, for example, are those whose flowering is stimulated by daylengths shorter than the critical daylength. Long-day plants are those whose flowering is stimulated by daylengths longer than a particular value. The latter usually bloom in late spring and summer.

Because the same period of dark and light occurs two times a year as the days lengthen in spring or grow shorter in fall, the organism could get its signals mixed. For some insects that would be impossible because the sensitive period occurs only once in their developmental phase. In other organisms the situation is different. For them the distinguishing characteristic is the direction from which the critical daylength is approached.

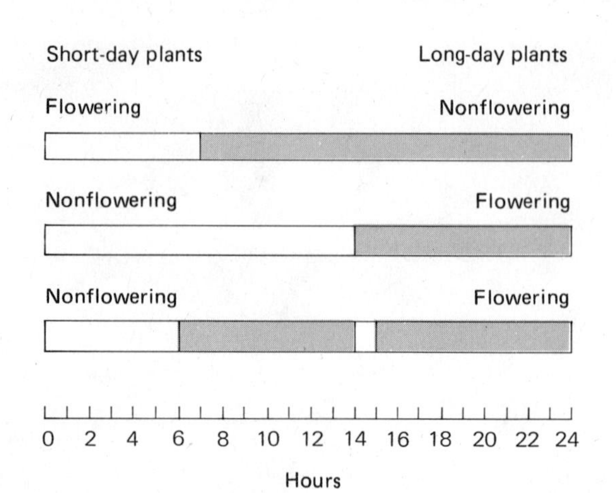

FIGURE 8.8 The time of flowering in long-day and short-day plants is influenced by photoperiod. When plants are held under short-day and long-night conditions, short-day plants are stimulated to flower and long-day plants are inhibited. When daylength is increased, flowering is inhibited in the short-day plants and stimulated in the long-day plants. If the dark period of the short-day and long-day plant is interrupted, each reacts as if it had been exposed to a long day: the long-day plant flowers and the short-day plant does not. In reality, short-day and long-day plants respond not to the length of light but to the length of darkness. The two might more accurately be called long-night and short-night plants.

In one situation the critical daylength is reached as long days move into short and at another time as short days move into long.

DIAPAUSE

In the cotton fields of the southern United States lives the pink cotton bollworm, the larva of a tiny moth. Except for a few hours directly after hatching, the larva spends its life in the flower buds or bolls of cotton. At the fourth larval instar stage, the insect goes into *diapause,* a stage of arrested growth, over winter. The onset of dispause comes in late August; but not until near the autumnal equinox, September 21, when the night becomes equal to or longer than the day, does the number of diapausing larvae sharply increase. In late winter, as the days begin to lengthen, the insect comes out of diapause and continues its growth. The emergence from diapause reaches its maximum right after the spring equinox, when the days are just slightly longer than those that induced diapause.

When the larvae of the pink bollworm were exposed to regimes of light and dark in the laboratory, the insect would go into diapause only when the light phase of the 24-hour day was 13 hours or less. A light period of 13.25 hours prevented them from going into diapause. So precise is the time measurement in the insect that a quarter-hour difference in the light period determines whether or not it goes into diapause. Diapause terminated most rapidly under photoperiods of 14 hours, less rapidly at 16 and 12. Thus, to the pink bollworm, the shortening days of late summer and fall forecast the coming of winter and call for diapause; and the lengthening days of late winter and early spring are the signals for the insect to resume development, pupate, emerge as an adult, and reproduce.

REPRODUCTIVE CYCLES

Experimental work with a number of species of birds has shown that the reproductive cycle is under the control of an outer seasonal rhythm of changing daylengths and an inner physiological response timed by a circadian rhythm. After the breeding season the gonads of birds studied to date regress spontaneously. The *refractory period* begins, a time when light cannot induce gonadal activity. Short days hasten the termination of the refractory period; long days prolong it. After the refractory phase is completed, the *progressive phase* begins in fall and winter. During this period,the birds fatten, they migrate, and their reproductive organs increase in size. This process can be speeded up by exposing the birds to a long-day photoperiod. Completion of the progressive phase brings the birds into the *reproductive* stage. A similar photoperiodic response exists in the cyprinid fish, the minnows.

Seasonal light cycles also influence the breeding cycles of many mammals (see Figure 8.9). For example, the flying squirrel has two peaks of litter production, the first in early spring, usually April in the northeastern United States, and the second in later summer, usually August. To produce litters in April, the flying squirrel must be in breeding condition in January and February.

Investigations of the influence of photoperiod under laboratory conditions on gonadal development in the flying squirrel have shown that descent of the testes into the scrotum (in nonbreeding condition, the testes of the squirrel are held in the body cavity) occurs in January under short-day and long-night conditions. An accelerated increase in daylength hastens the descent. After the squirrel's maximum photoperiod in summer, the testes regress and remain in that condition while photoperiod decreases. If the decrease in photoperiod is unseasonably accelerated, so that critical daylength comes two months early, followed by increased photoperiod, the testes descend two months early. Likewise, the female ovulates when the photoperiod increases from 11 to 15 hours, and ovulation ceases when photoperiod decreases.

OTHER SEASONAL PATTERNS

Numerous seasonal patterns are influenced by photoperiod. One example is the food storage behavior of the flying squirrel. Through a series of laboratory experiments under various controlled light and temperature conditions, I. Muul demonstrated that the food-storing behavior is triggered when the critical daylength reaches about 12 hours in early fall. Such control synchronizes the exploratory and storing behavior of flying squirrels with the ripening of the mast crop—nuts and acorns—and prevents a premature harvest.

Antler growth of deer is also a photoperiodic response. Growth is triggered by the lengthening days of spring, and velvet is shed and antlers harden in the shortening days of fall. As the days grow their shortest in December, deer drop their antlers, their loss brought about by a drop in the male hormone testosterone, the production of which is also photoperiodically controlled. The critical daylength for the stimulation of antler growth is 12.75 hours of daylight.

SEASONAL MORPHS

Photoperiod can also result in seasonal morphs among a number of insects. One example is the Pierid butterfly, the veined, white *Pieris napa,* in western North America. Populations of this butterfly along the Cali-

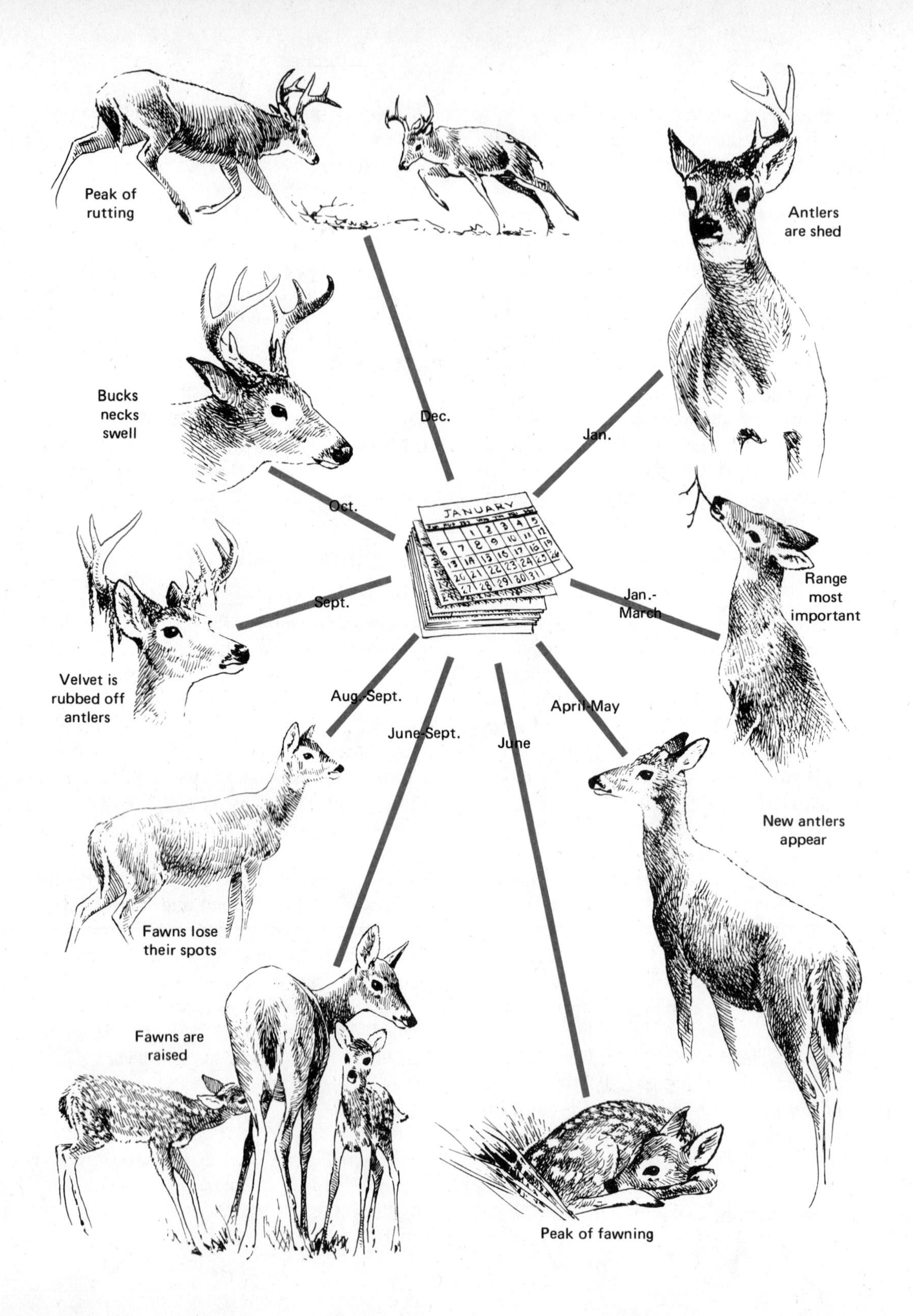

Peak of rutting
Antlers are shed
Bucks necks swell
Dec.
Jan.
Oct.
JANUARY
Range most important
Sept.
Jan.-March
Velvet is rubbed off antlers
Aug.-Sept.
April-May
June-Sept.
June
New antlers appear
Fawns lose their spots
Fawns are raised
Peak of fawning

fornia coast have two generations a year and occur in two phenotypes. One is a dark spring form with heavy black scaling on the veins of the hind wings and the other a lighter summer form almost devoid of black scales. Inland populations have only one generation a year, and all adults are the dark form. Both populations, however, are capable of producing both phenotypes. The dark forms are induced by the short days of autumn and emerge from overwinter diapause pupae. The light summer forms are induced by the long days of summer and emerge from nondiapause pupae. Such photoperiodic responses may be important in the reproductive isolation of closely related populations and the evolution of species.

CIRCANNUAL CLOCKS

All of the activities discussed above exhibit a circannual periodicity, suggesting the existence of a circannual clock. To prove its existence, we need to demonstrate conditions similar to those required of a circadian clock: (1) a free-running rhythm in the absence of environmental cues; (2) a period that approximates but deviates from a 365-day cycle; and (3) temperature independence.

Considerable evidence exists that a circannual clock controlling body weight and hibernation exists in the golden-manteled ground squirrel (*Citellus lateralis*) of western North America. When held for several years under constant temperature regimes and constant artificial days of 12L/12D, the squirrels exhibited circannual rhythms of about a year. Such a rhythm has certain advantages. It enables the animals to prepare well in advance for winter and to rouse in spring without the need of environmental cues to penetrate a deep burrow. It also brings the animals into breeding condition amost immediately after hibernation, so reproduction can take place in the spring.

Most birds show strong circannual rhythms in molt, body weight (fat deposition), testis growth, and nocturnal restlessness. Do we need to invoke a circannual clock in them? The cycles can be explained by seasonal photoperiod response and a circadian clock.

ADAPTIVE VALUE OF PHOTOPERIODICITY

Light and dark may trigger activity rhythms that relate more directly to other aspects of the environment, more significant to the organism than light and dark themselves. The transition from day to night, for example, is accompanied by a rise in humidity and a drop in temperature. Wood lice, centipedes, and millipedes, which lose water rapidly when exposed to dry air, spend the day in a fairly constant environment of darkness and dampness under stones, logs, and leaves. At dusk, when the humidity of the air is more favorable, they emerge. In general, these animals show an increased tendency to escape from light as the length of time they spend in darkness increases. On the other hand, their intensity of response to low humidity decreases with darkness. Thus, they come out at night into places too dry for them during the day; and as light comes, they quickly retreat to their damp hiding places.

Among some animals the biotic rather than the physical aspects of the environment relate to the activity rhythm. Deer undisturbed by humans may be active by day, but when they are hunted and disturbed, they become strongly nocturnal. Predators must relate their feeding activity to the activity rhythms of their prey. Moths and bees must visit flowers when they are open and provide a source of food. The flowers must have a rhythm of opening and closing that coincides with the time when the insects that pollinate them are flying.

The entrainment of the phase of its activity rhythm to a natural light-dark cycle means more to an organism than simply an adjustment to a precise 24-hour period. More important, the entrainment serves to time the activities of plants and animals to a day-night cycle

FIGURE 8.9 Seasonal cycle of the white-tailed deer. The annual cycle is attuned to the decreasing daylength of fall, during which the breeding season begins, and to the lengthening days of spring, when antler growth begins.

in a manner that is adaptive. Over the year photoperiod enables organisms to adjust to seasonal changes in the environment. It keeps the organisms' seasonal activities—flowering, seeding, courtship, mating, reproduction, migration, hibernation—in tune with the rest of the population and the world about them.

Summary

An almost universal feature of life is an internal biological clock, physiological in function. Its basic structure is probably chemical and is involved in the makeup of the cell. It is free-running under constant conditions, with an oscillation or fluctuation that has its own inherent frequency. For most organisms this fluctuation deviates more or less from 24 hours; for that reason it is called a circadian rhythm. Under natural conditions the circadian rhythm is set or entrained to the 24-hour day by external time cues, or Zeitgebers, which synchronize the activity of plants and animals with the environment. Because the most dependable external timesetter is light and dark—day and night—most of the selected species studied so far are entrained to a 24-hour photoperiod. The onset and cessation of activity are usually synchronized with dark and dawn, the response depending upon whether the organisms are diurnal (light-active) or nocturnal (dark-active).

The biological clock synchronizes the activities of plants and animals not only with night and day, but also with the seasons of the year. The possession of a self-sustained rhythm with approximately the same frequency as that of the environment enables organisms to predict such situations as the coming of spring. It brings plants and animals into a reproductive state at a time of year when the probability of survival of offspring is the highest. It synchronizes within a population such activities as mating and migration, dormancy and flowering. Species-specific or population-specific synchronization with the environment is a result of natural selection.

Review and Study Questions

1. What is photoperiodism?
2. What is a circadian rhythm? How does it relate to the 24-hour day? What is meant by the term *free-running*?
3. What is a Zeitgeber? Phase shift? How do the two relate?
4. What is a biological clock? Where is it located?
5. What conditions must a biological clock fulfill to function as a timekeeper?
6. Discuss four functions a biological clock can serve.
7. Why cannot intertidal organisms depend on a solar-day clock alone?
8. What is critical daylength? A long-day organism? A short-day organism?
9. How is diapause influenced by changing daylength?
10. How does photoperiodism influence the seasonal activity of animals?
11. Argue for and against the existence of a circannual clock.
12. Speculate on how photoperiodism could have a role in speciation.
13. What is the adaptive value of photoperiodism?

Selected References

Aschoff, J., S. Daan, and G. A. Gross, eds. 1982. *Vertebrate circadian rhythms*. New York: Springer-Verlag.

Beck, S. D. 1980. *Insect photoperiodism*, 2nd ed. New York: Academic Press. A good review.

Biological Clocks. Special section in *Bioscience*. 1983. *Bioscience* 33:424–457.

Brady, J. 1982. *Biological timekeeping*. New York: Cambridge University Press.

Bunning, E. 1973. *The physiological clock*, 3rd ed. New York: Academic Press. A book by one of the pioneers in the field.

DeCoursey, P. J., ed. 1976. *Biological rhythms in the marine environment*. Columbia: University of South Carolina Press.

Edmunds, L. N. 1988. *Cellular and molecular bases of biological clocks*. New York: Springer-Verlag.

Farner, D. S. 1985. Annual rhythms. *Ann. Rev. Physiol.* 47:65–82.

Gwinner, E. 1986. *Circannual rhythms*. New York: Springer-Verlag.

Johnson, C. H., and J. W. Hastings. 1986. The elusive mechanisms of the circadian clock. *Am. Sci.* 74:29–36.

Moore-Ede, M. C., F. M. Sulzman, and C. A. Fuller. 1982. *The clocks that time us*. Cambridge, MA: Harvard University Press. Circadian rhythms in humans.

Naylor, E. 1985. Tidally rhythmic behavior of marine animals. *Symp. Soc. Exp. Biol.* 39:63–93.

Palmer, J. D. 1976. *An introduction to biological rhythms.* San Diego: Academic Press.
Palmer, J. D. 1990. The rhythmic lives of crabs. *Bioscience* 40:352–358.
Pengelley, E. T., ed. 1974. *Circannual clocks: Annual biological rhythms.* New York: Academic Press.
Saunders, D. S. 1982. *Insect clocks,* 2nd ed. Elmsford, NY: Pergamon Press. An important and interesting reference.
Winfree, A. T. 1986. *The timing of biological clocks.* Scientific American Library 19. New York: W. H. Freeman. A popular but challenging discussion of biological clocks.

CHAPTER

Soil

Outline

Objectives

On completion of this chapter, you should be able to:

1. Provide a general description of soil and its features.
2. Explain how soil develops.
3. Describe the soil profile and the horizons of the major soil orders.
4. Distinguish between mull and mor humus.
5. Contrast calcification, podzolization, laterization, and gleization.
6. Explain the importance of cation exchange in the soil and the role of soil acidity.
7. Define soil texture and describe the sizes and combinations of the various particles that make up soil.
8. Point out the features of the soil as an environment for life.
9. Discuss the selective pressures of soils on plants.
10. Discuss the causes and consequences of soil erosion.

Soil is the foundation upon which all terrestrial life and much aquatic life depend. It is the medium in which plant life is rooted, a reservoir of mineral nutrients needed by plants upon which, in turn, animal life depends. It is the site of decomposition of organic matter and the staging area for the return of minerals to the nutrient cycle. Roots occupy a considerable portion of the soil, to which they tie the vegetation and from which they pump water and minerals in solution needed by plants for photosynthesis and other biogeochemical processes. Vegetation, in turn, influences the development of soil, its chemical and physical properties, and its organic matter content. Thus, soil acts as a pathway between the organic and mineral worlds, a pathway easily obstructed or destroyed by human interference.

A DEFINITION OF SOIL

As familiar as it is, soil is difficult to define. One definition has soil as a natural product formed from weathered rock by the action of climate and living organisms. Another states that soil is a collection of natural bodies of earth that is composed of mineral and organic matter and is capable of supporting plant growth. Such definitions seem inadequate or stilted. Indeed, one eminent soil scientist, a pioneer of modern soil studies, Hans Jenny, will not give an exact definition of soil. In his book *The Soil Resource* (1980: 364), he writes:

> Popularly, soil is the stratum below the vegetation and above hard rock, but questions come quickly to mind. Many soils are bare of plants, temporarily or permanently; or they may be at the bottom of a pond growing cattails. Soil may be shallow or deep, but how deep? Soil may be stony, but surveyors [soil] exclude the larger stones. Most analyses pertain to fine earth only. Some pretend that soil in a flowerpot is not soil, but soil material. It is embarrassing not to be able to agree on what soil is. In this the pedologists are not alone. Biologists cannot agree on a definition of life and philosophers on philosophy.

Of one fact we are sure. Soil is not just an abiotic environment for plants. It is teeming with life—billions of minute animals, bacteria, and fungi. The interaction between the biotic and the abiotic makes the soil a living system.

SOIL DEVELOPMENT

PHYSICAL WEATHERING

Soils begin with the weathering of rocks and their minerals. Exposed to the combined action of water, wind, and temperature, rock surfaces peel and flake away. Water seeps into crevices, freezes, expands, and cracks the rock into smaller pieces. Accompanying this disintegration and continuing long afterward is the decomposition of the minerals themselves. Water and carbon dioxide combine to form carbonic acid, which reacts with calcium and magnesium in the rock to form carbonates. They either accumulate deeper in the rock material or are carrried away, depending on the amount of water passing through. Primary minerals that contain aluminum and silicon, such as feldspar, are converted to secondary minerals, such as clay. As iron is especially reactive with water and oxygen, iron-bearing minerals are prone to rapid decomposition. Iron remains oxidized in the red ferric state or is reduced to the gray ferrous state. Fine particles, especially clays, are shifted or rearranged within the mass by percolating water and on the surface by runoff, wind, or ice.

Eventually, the rock is broken down into loose material; this stuff may remain in place, but more often than not, much of it is lifted, sorted, and carried away. Material transported from one area to another by wind is known as *loess;* that transported by water as *alluvial, lacustrine* (or lake), and *marine* deposits; and that transported by glacial ice as *till.* In a few places soil materials come from accumulated organic matter such as peat. Materials remaining in place are called *residual.*

This mantle of unconsolidated material is called the *regolith.* It may consist of slightly weathered material with fresh primary minerals, or it may be intensely weathered and consist of highly resistant minerals such as quartz. Because of variations in slope, climate, and native vegetation, many different soils can develop in the same regolith. The thickness of the regolith, the kind of rock from which it was formed, and the degree of weathering affect the fertility and water relations of the soil.

BIOLOGICAL WEATHERING

Eventually, plants root on this weathered material. More often than not, intense weathering goes on under some plant cover, particularly in glacial till and water-deposited materials, because they are already favorable sites for some plant growth. Thus, soil development often begins under some influence of plants. They root, draw nutrients from mineral matter, reproduce, and die. Their roots penetrate and further break down the regolith. The plants pump up nutrients from its depths and add them to the surface and so recapture minerals carried deep into the material by weathering processes. Through photosynthesis, plants capture the sun's energy and add a portion of it in the form of organic carbon—approximately 18 billion metric tons, or 1.7×10^{17} kilocalories—to the soil each year. This energy source, the plant debris, enables bacteria, fungi, earthworms, and other soil organisms to colonize the area.

The breakdown of organic debris into humus is accomplished by decomposition and, finally, mineralization. Higher organisms in the soil—millipedes, centipedes, earthworms, mites, springtails, grasshoppers, and others—consume fresh material and leave partially decomposed products in their excreta. This material is further decomposed by microorganisms, the bacteria and fungi, into various compounds of carbohydrates, proteins, lignins, fats, waxes, resins, and ash. These compounds are then broken down into simpler products, such as carbon dioxide, water, minerals, and salts. The latter process is called *mineralization.*

The fraction of organic matter that remains is called *humus.* It is not stable, as it represents a stage in the decomposition of soil organic matter. New humus is being formed as old humus is being destroyed by mineralization. The equilibrium set up between the formation of the new humus and the destruction of the old determines the amount of humus in the soil.

CHEMICAL WEATHERING

The activities of soil organisms, the acids produced by them, and the continual addition of organic matter to mineral matter produce profound changes in the weathered material. Rain falling upon and filtering through the accumulating organic matter picks up acids and minerals in solution, reaches mineral soil, and sets up a chain of complex chemical reactions. They continue in the regolith. Calcium, potassium, sodium, and other mineral elements, soluble salts, and carbonates are carried in solution by percolating water deeper into the soil or are washed away into streams, rivers, and eventually the sea. The greater the rainfall, the more water moves down through the soil and the less moves upward. Thus, high precipitation results in heavy leaching and chemical weathering, particularly in regions of high temperatures. These chemical reactions tend to be localized within the regolith. Organic carbon, for instance, is oxidized near the surface, while free carbonates precipitate deeper in the rock material. Fine particles, especially clays, also move downward.

These localized chemical and physical processes in the parent material result in the development of layers in the soil, called *horizons,* which impart to the soil a distinctive *profile.* Within a horizon a particular property of the soil reaches its maximum intensity, and away from this level it decreases gradually in both directions. Therefore each horizon varies in thickness, color, texture, structure, consistency, porosity, acidity, and composition.

SOIL HORIZONS

In general, soils have four major horizons: an organic, or O, horizon and three mineral horizons—the A, characterized by major organic matter accumulation, by the loss of clay, iron, and aluminum, and by the development of a granular, crumb, or platy structure; the B, characterized by an illuvial concentration of all or any of the silicates, clay, iron, aluminum, and humus, alone or in combination, and by the development of a blocky, prismatic, or columnar structure; and the C, material underlying the two horizons that is either like or unlike the material from which the soil is presumed to have developed. Below all three horizons may lie the R horizon, the consolidated bedrock.

Because the soil profile is essentially a continuum, often there is no clear distinction between one horizon and another. Horizon subdivisions (Figure 9.1) are indicated by arabic numbers—for example, O_1, O_2, A_1, A_2, and so forth; lowercase letters are used to indicate significant qualitative departures from the central concept of each horizon—for example, A_{2g} or B_t.

The O horizon—once designated as L, F, H, or A_0 and A_{00}—is the surface layer, formed or forming above the mineral layer and composed of fresh or partially decomposed organic material, such as that found in temperature forest soils. It is usually absent in cultivated soils. This layer, with the upper part of the A horizon, is the region where life is most abundant. It is subject to the greatest changes in soil temperatures and moisture conditions and contains the most organic carbon. It is the site where most or all decomposition by organisms takes place.

Of all the horizons of the soil, none is more important or ecologically more interesting than the organic horizon. The importance of the organic layer was stressed early in the history of ecology. Darwin, in his famous work "The Formation of Vegetable Matter through the Action of Worms, with Observations on Their Habits" (1881), pointed out the influence of these animals on the soil. At the same time, in 1879 and 1884, the Danish forester P. E. Muller described the existence of two types of humus formation in the temperate forest soil; he called them *mull* and *mor*. He observed differences in their vegetation, soil structure, chemical composition, and fauna. He regarded humus formation as a close interaction among chemical, physical, and biological interactions.

Humus to most of us is the black organic matter we buy in bags at the local garden center to add to flower beds or use as a potting soil. It is a black amorphous substance, 80 to 90 percent humic acid, and it varies in nature depending upon the vegetation from which it derived. Although there are a number of ways to describe humus, the most useful is based on visual features.

Mull humus, neutral to slightly acid, is well mixed with fine mineral matter, which absorbs the humic

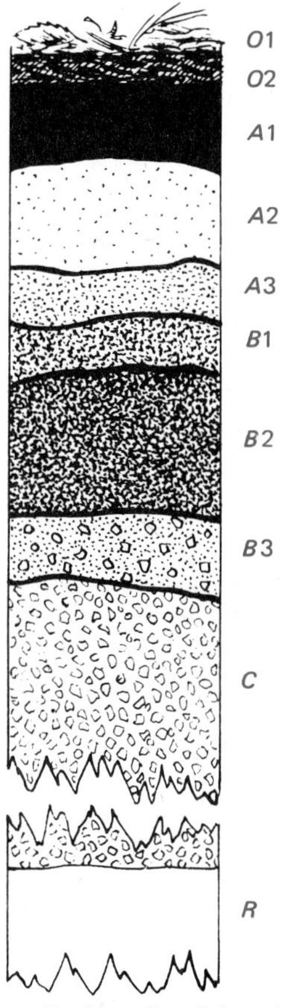

FIGURE 9.1 A generalized profile of the soil. Rarely does any one soil possess all of the horizons shown. O_1—Loose leaves and organic debris. O_2—Organic debris partly decomposed or matted. A_1—A dark-colored horizon with high content of organic matter mixed with mineral matter. The A horizon is the zone of maximum biological activity. A_2—A light-colored horizon of maximum leaching. It is prominent in podzolic soils; faintly developed or absent in chernozemic soils. A_3—Transitional to B, but more like A than B; sometimes absent. B_1—Transitional to B, but more like B than A; sometimes absent. B_2—A deeper-colored horizon of maximum accumulation of clay minerals or of iron and organic matter; maximum development of blocky or prismatic structure or both. B_3—Transitional to C. C—The weathered material, either like or unlike the material from which the soil presumably formed. A gley layer may occur, as well as layers of calcium carbonate, especially in grasslands. R—consolidated bedrock.

substance like a dye. Because of the fragmentation of plant material and the mixing of the material with the mineral soil to which it is inseparably bound, soils with mull humus show no sharp break between the O and the A horizons.

Mor humus has a well-defined, unincorporated, and matted or compacted organic deposit resting on mineral soil, because soil organisms capable of living in its acid condition have little mechanical influence on the soil. The main decomposing agents are acid-producing fungi.

The intermediate or *moder* type of humus is well mixed with mineral matter, but plant residues are transformed into the droppings of small arthropods. These droppings, together with plant fragments and mineral matter, form a loose, netlike structure held together by chains of small droppings that is easily separated. There is a range of intermediate types, the structure of which is influenced by acidity and amount of soil faunal activity.

Finally, there is *skeletal* or raw humus, the one with which you are most familiar, which has not developed into any of the other three types. It looks much like mor.

PROFILE DIFFERENTIATION

The differentiation of the soil profile into horizons and the nature of the soil material—its content and distribution of organic matter, its color, and its chemical and physical characteristics—are influenced over large areas by the combined action of vegetation and its prime determinant, climate (Figure 9.2). The soil beneath native grassland differs from that beneath native forest.

CALCIFICATION

Grassland vegetation developed in the subhumid-to-arid and temperate-to-tropical climates of the world—the plains and prairies of North America, the steppes of Russia, and the veld and savanna of Africa. Dense grass-root systems may extend many feet below the surface. Each year nearly all of the vegetative material above the ground and part of the root system are turned back to the soil as organic residue. Although this material decomposes rapidly the following spring, it is not completely gone before the next cycle of death and decay begins. The humus then becomes mixed with mineral soil by the action of the soil inhabitants, developing a soil high in organic matter. The humus content is greatest at the surface and declines gradually with depth.

Because the amount of rainfall in grassland regions is generally insufficient to remove calcium and magnesium carbonates, they sink only to the average depth reached by the percolating waters. The high calcium content of the surface soil is maintained by grass, which absorbs large quantities from lower horizons and redeposits them on the surface. Likewise, there is little loss of clay from the surface layer. This process of soil development has been called *calcification*.

Soils developed by calcification have a distinct A horizon of great thickness and an indistinct B horizon characterized by an accumulation of calcium carbonate. The A horizon is high in organic matter and nitrogen, even in tropical and subtropical regions.

PODZOLIZATION

Forests are the dominant vegetation in the humid regions. Here the cycle of organic matter accumulation differs from that of the grassland. Only part of the organic matter—leaves, twigs, and some trunks—is turned over annually. Leaves, which are the largest source of organic matter, and the vegetation of the ground layer remain on the surface. Dead roots add little to soil organic matter, because they die over an irregular period and are not concentrated near the surface. Because only the leaves are returned regularly to the soil and much of the mineral matter and energy is tied up in trunks and branches, most of the nutrients turned back to the soil come from annual leaf fall. The amount of nutrient return varies with the species composition of the forest, because trees differ in the nutrient content of their leaves. For example, basswood, quaking aspen, hickories, American elm, and flowering dogwood contain more calcium in their leaves and return more of it to the soil than sugar maple, red

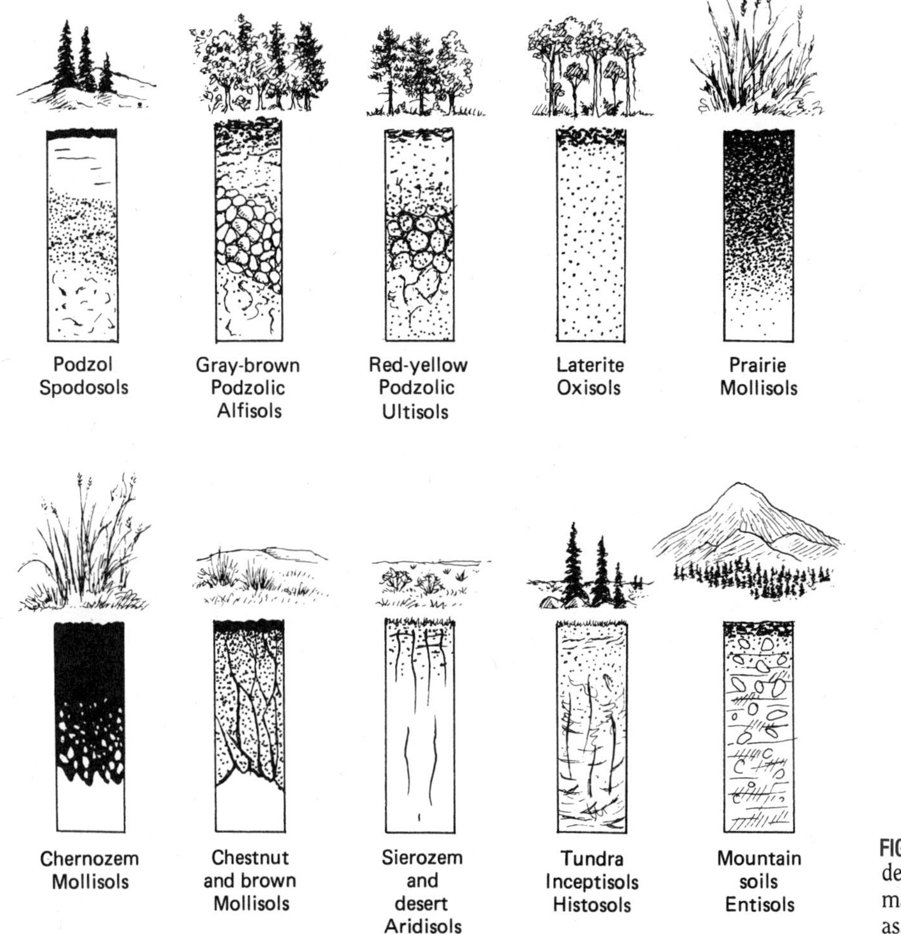

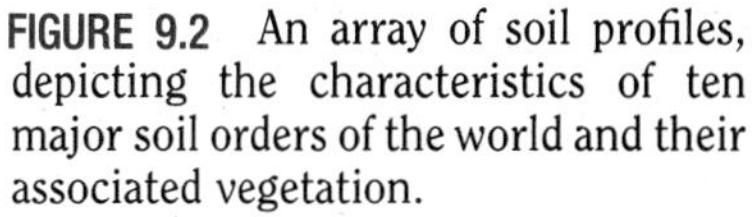

FIGURE 9.2 An array of soil profiles, depicting the characteristics of ten major soil orders of the world and their associated vegetation.

maple, yellow birch, and red oak. The latter return more than beech, red pine, white pine, and hemlock.

Rainfall in forested regions is sufficient to leach away many basic elements, especially calcium, magnesium, potassium, iron, and aluminum. Because trees take up fewer bases (alkaline substances) than grasses, they generally return an insufficient amount to the surface soil to prevent it from becoming acid. The degree of acidity, however, will vary depending on the forest composition and its site. Some forests in the southern Appalachians, particularly those containing yellow poplar and basswood and growing on north and northeast slopes, have a rather high, often neutral, pH in their surface horizons, even though they grow on soils weathered from acid sandstone. Increased acidity may cause the dispersion and downward movement of organic and clay colloids.

A soil developed by this process is called *podzolic* and the process *podzolization,* from the Russian, meaning "ash beneath," referring to the leached horizon of strongly podzolized soils. These soils are characterized by a white A_2 horizon and a brilliant yellow-brown B horizon, the result of accumulations of iron and aluminum compounds and humus. Iron accumulations in some podzolic soils may act as a cement, creating a hardpan layer in the B horizon. This layer, called *ortstein*, impedes the free circulation of air and water.

LATERIZATION

In the humid subtropical and tropical forested regions of the world, where rainfall is heavy and temperatures high, the soil-forming process is much more intense. Because temperatures are uniformly high, the weathering process in these regions is almost entirely chemical, brought about by water and its dissolved substances. The residues from this weathering—bases, silica, alumina, hydrated aluminosilicates, and iron oxides—are freed. Because precipitation usually exceeds evaporation, the water movement is almost continuously downward. With only a small quantity of electrolytes present in the soil water because of continual leaching, silica and aluminum silicates are carried downward, while sesquioxides of aluminum and iron remain behind. The reason is that these sesquioxides are relatively insoluble in pure rainwater, whereas the silicates tend to be precipitated as a gel in solutions containing humic substances and electrolytes. If humic substances are present, they act as protective colloids around iron and aluminum oxides and prevent their precipitation by electrolytes.

The end product of such a process is a soil composed of silicate and hydrous oxides, clays, and residual quartz, deficient in bases, low in plant nutrients, and intensely weathered to great depths. Because of the large amount of iron oxides left, these soils show a variety of reddish colors and generally lack distinct horizons. Below, the profile is unchanged for many feet. The clay has a stable structure, and unless precipitated, the iron is hardened into a cemented laterite. The soil is very pervious to water and is easily penetrated by plant roots. This soil-forming process is termed *laterization* or *latosolization*.

SALINIZATION

Arid and semiarid regions have relatively sparse vegetation. Because of the lack of plant growth, which is limited by low rainfall, there is little organic matter and nitrogen in the soil. Light precipitation results in slightly weathered and slightly leached soils high in plant nutrients. Their horizons are usually faint and thin. In these regions occur areas where soils contain excessive amounts of soluble salts, either from parent material or from the evaporation of water draining in from adjoining land. The infrequent rainwater penetrates the soil, but soon afterward evaporation at the surface draws the salt-laden water upward. The water evaporates, leaving saline and alkaline salts at or near the surface to form a crust, or *caliche*. The soil-forming process producing this crust is called *salinization*.

GLEIZATION

Calcification, podzolization, and laterization are all processes that take place in well-drained soil. Under poorer drainage conditions, a different soil-development process is at work. The slope of the land determines to a considerable extent the amount of rainfall that will enter and pass through the soil, the concentration of erosion materials, the amount of soil moisture, and the height at which the water will stand in the soil. The amount of water that passes through or remains in the soil determines the degree of oxidation and breakdown of soil minerals. In areas where water stays near or at the surface most of the time, iron, because of an inadequate supply of oxygen, is reduced to ferrous compounds, which give a dull gray or bluish color to the horizons.

This process, called *gleization*, may result in compact, structureless horizons. Gley soils are high in organic matter because more is produced than can be broken down by humification, which is greatly reduced because of the absence of soil microorganisms. On gentle to moderate slopes, where drainage conditions are improved, gleization is reduced and occurs deeper in the profile. As a result, the subsoil will show varying degrees of mottling of grays and browns. On hilltops, ridges, and steep slopes where the water table is deep and the soil well drained, the subsoil is reddish to yellowish brown because of the presence of oxidized iron compounds.

SOIL CHARACTERISTICS

CHEMISTRY

Chemical elements in the soil are absorbed in soil particles, dissolved in soil solution, and included in mineral and organic matter. These ions move from soil to plant, from plant to animal, and into the biogeochemical cycle (see Chapter 24). In aquatic systems the ions are dissolved and obey the laws of diffusion and dilute solutions. In soils, ions are limited in their mobility because they are closely held to solid particles of clay and humus.

The key to the availability of nutrients in the soil is the nature of the clay-humus complex. It is made up of platelike particles in the soil called *micelles*. Micelles consist of sheets of tetrahedron and octahedron aluminosilicates (silicate, aluminum, iron combined with oxygen, and hydroxyl ions). The interior of the plates is electrically balanced, but the edges and sides are negatively charged. They attract positive ions, water molecules, and organic substances. The number of negatively charged sites on soil particles that attract positively charged cations is called the *cation exchange capacity* (Figure 9.3). These positively charged ions can be replaced by still other ions in the soil solution.

Cation exchange capacity varies among soils, depending upon the structures of clays and the amount of organic matter or humus. Some clays are made up of octahedron and tetrahedron sheets arranged in large lattices that expand when moist. An extensive surface area, external and internal, allows ion exchanges not only between micelles but within the micelles as well. Other clays have octahedrons and tetrahedrons arranged randomly. They are amorphous clays that do not expand when moist, and cation exchange is of a lesser magnitude. When compared on the basis of weight, soils with a high organic content, and thus organic colloids, have a higher cation exchange capacity, up to four times greater, than inorganic colloids. This difference emphasizes the importance of organic matter to soil fertility.

Negatively charged ions attract cations of Ca^+, Na^+, Mg^+, K^+, and H^+, among others. Some cling to the micelles more strongly than others. H^+ ions are especially tenacious. Hydrogen ions added by rainwater, cationic acids from organic matter, and metabolic acids from roots displace other cations, such as Ca^+. If not taken up by plants, these cations are carried deeper into the soil or are removed altogether through the groundwater and frequently move into aquatic systems. The cation exchange capacity has a pronounced effect on soil fertility and availability of nutrients to plants.

Ions are available to plants only when dissolved in

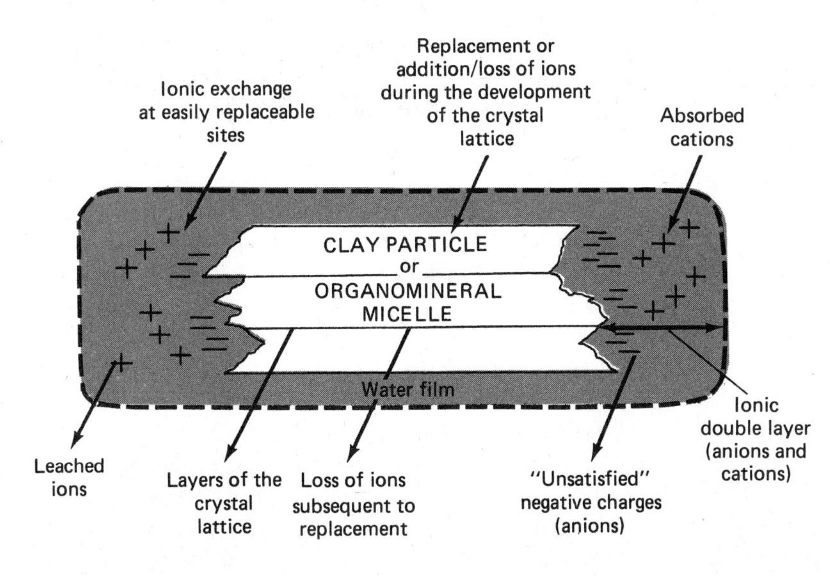

FIGURE 9.3 Cation exchange on a soil particle. Chemical activity is greatest at the edges. Weakly adsorbed cations may be exchanged with incoming ions and lost through leaching. More strongly adsorbed cations remain firmly attached to negative charges at active surfaces. In humid regions the most abundant exchangeable minerals are, in order, H and Ca, Mg, K, and Na; in arid regions they are, in order, Ca and Mg, Na and K, and H.

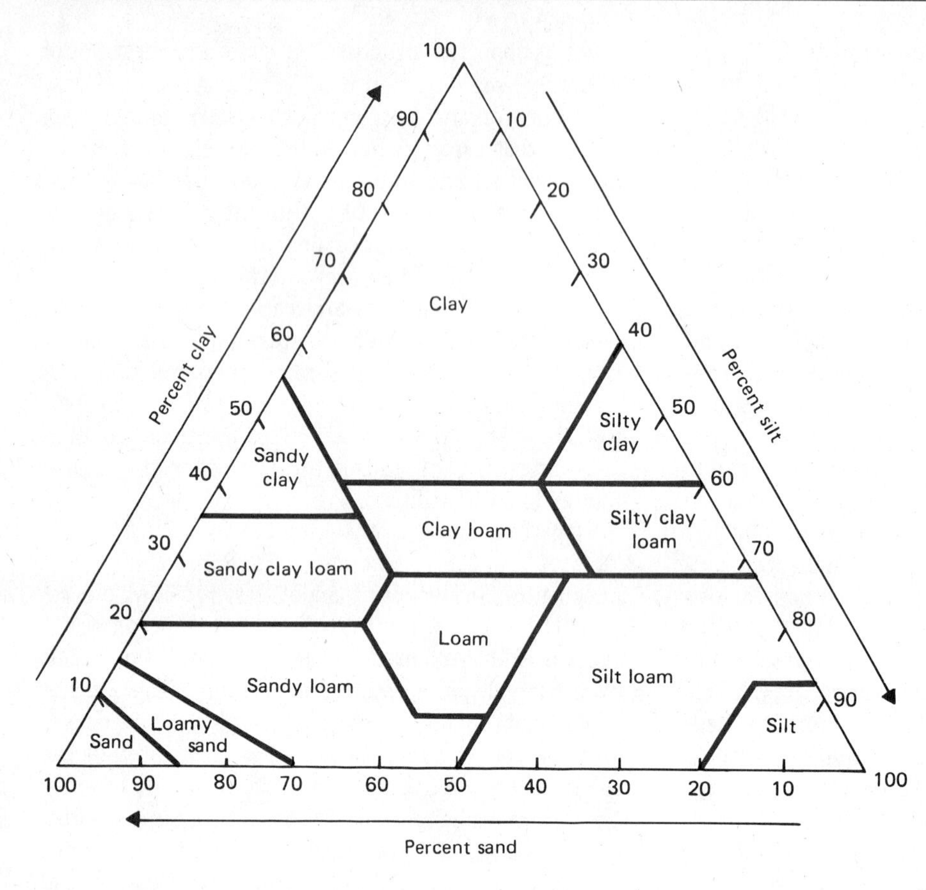

FIGURE 9.4 The percentages of clay (below 0.002 mm), silt (0.002 to 0.05 mm), and sand (0.05 to 2.0 mm) in the basic soil textural classes.

soil solution. Ions in soil solution maintain an equilibrium with ions adsorbed on micelles, which, in turn, maintain an equilibrium with absorbed ions in the micelles. As plants remove ions from the soil solution in the vicinity of the roots, other ions diffuse to the region. That, in turn, enhances the release of ions from the micelles.

Acidity is one of the most familiar of all chemical conditions in the soil. Typically, soils range between a very acid pH of 3 and a pH of 8, strongly alkaline. Soils just over a pH of 7 (neutral) are considered basic and those of 6.6 or less acid.

Soil acidity has a pronounced effect on nutrient availability. As soil acidity increases, the proportion of exchangeable Al^+ increases, and Ca^+, K^+, Na^+, and other nutrients decrease. Such changes bring about not only nutrient deprivation but also aluminum toxicity. Harmful effects of low pH in both soil and aquatic environments are due not so much to the acid as to the toxic Al^+ and Fe^+ ions released in soil and water.

TEXTURE AND STRUCTURE

Differences among soils and among horizons within a soil are primarily reflected in texture, arrangement, and color. The *texture* of a soil is determined by the proportion of different-size soil particles (Figure 9.4).

Texture is partly inherited from parent material and partly a result of the soil-forming process. Particles are classified on the basis of size into gravel, sand, silt, and clay. Gravel consists of particles larger than 2.0 mm. Sand ranges from 0.05 to 2.0 mm, is easily seen, and feels gritty. Silt consists of particles from 0.002 to 0.05 mm in diameter, which can scarcely be seen by the naked eye and feel and look like flour. Clay particles, too small to be seen under an ordinary microscope, are colloidal in nature. Clay controls the most important properties of soils, including plasticity and exchange of ions between soil particles and soil solution. Most soils are a mixture of these various particles.

Soil particles are held together in clusters or shapes of various sizes, called *aggregates* or *peds*. The arrangement of these aggregates is called *soil structure*. There are many types of soil structure. Soil aggregates may be classified as granular, crumblike, platelike, blocky, subangular, prismatic, or columnar (Figure 9.5). Structureless soil can be either single-grained or massive. Soil aggregates tend to become larger with increasing depth. Structure is influenced by texture, plants growing on the soil, other soil organisms, and the soil's chemical status.

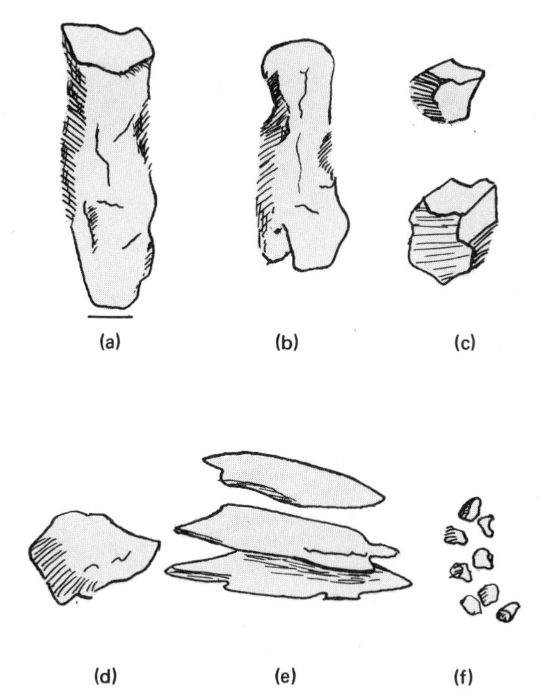

FIGURE 9.5 Some types of soil structure. (a) Prismatic, (b) columnar, (c) angular blocky, (d) subangular blocky, (e) platelike, (f) granular.

MOISTURE

Water that enters the soil will percolate or seep into its pore system. After a heavy rainfall most of the pore space will be filled with water; if there is an excess to drain away gravitationally, the soil is *saturated* (Figure 9.6). If all the pores are filled with water, but no water drains away because of internal capillary forces, the soil is at *field capacity*. Field capacity represents the maximum amount of water a soil can hold at one-third atmosphere after gravitational water has drained away. It varies considerably with the texture of the soil. Highly porous sandy soils have low field capacity, whereas fine-textured clay and humic soils have a high field capacity. The portion of water held between soil particles by capillary forces is *capillary water*.

As soil loses capillary water through evaporation and withdrawal by plants, it reaches a level where no further water is available to plants. At this level, called the *wilting point*, water is tightly held against colloidal surfaces, even though up to 25 percent of the pore space may contain water. As the soil dries even further, the only moisture remaining is *hygroscopic water*, that portion which adheres to the soil particles as a thin film and is unavailable to plants.

A close relationship exists between the amount of water in the soil and the amount of air in the soil. Highly saturated soils are deficient in oxygen, which creates anaerobic conditions (see Chapter 6). Soils with maximum aeration would cause wilting in plants.

MAJOR ORDERS OF SOILS

The combination of climate, vegetation, soil material, and time produces unique soils, the smallest repetitive unit of which is called a *pedon*. Over the years, attempts have been made to classify soils, from the most local types to great soil groups and orders. Although a detailed classification of soils is extremely important to professional soil scientists and ecologists, to understand vegetational distribution and to determine suitability of soils for agriculture, urban and suburban development, road construction, landfills, the susceptibility of soils to erosion, and more, a broad classification of higher types or orders has the greatest general interest. Eleven main orders are described in Table 9.1;

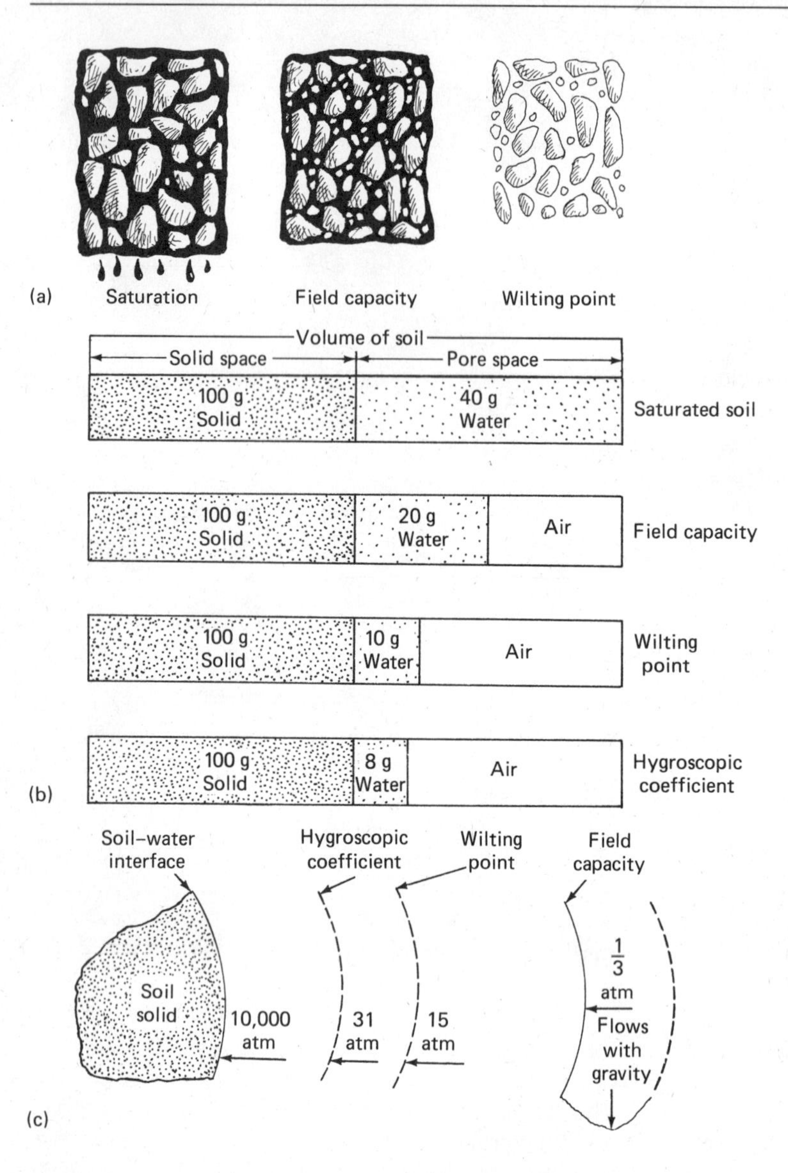

FIGURE 9.6 The proportions of solid matter, water, and air at different moisture levels. (a) Soils at saturation, field capacity, and wilting point. (b) Bar graphs illustrating the proportions of solid, water, and air at the three levels, plus the hygroscopic coefficient. (c) The relationship between moisture levels and bars of atmospheric pressure.

ten associated great soil groups are illustrated in Figure 9.2. These orders correspond roughly with broad climatic zones of Earth (Figure 9.7).

SOIL AS AN ENVIRONMENT

The soil is a radically different environment for life than the one above the surface; yet the essential requirements do not differ. Like animals that live outside the soil, soil fauna require living space, oxygen, food, and water.

The soil in general possesses several outstanding characteristics as a medium for life. It is relatively stable, both chemically and structurally. Climate is far more constant in the soil than above it. The atmosphere remains saturated, or nearly so, until soil moisture drops below a critical point. The soil affords a refuge from high and low extremes in temperature, wind, evaporation, light, and dryness, permitting soil

TABLE 9.1 The Eleven Major Soil Orders

Order	*Derivation and Meaning*	*Description*	*Approximate Equivalents*
Entisol	Coined from *recent*	Dominance of mineral soil materials; absence of distinct horizons; found on floodplains	Alluvial soils, azonal soils, regosol, lithosol
Vertisol	Latin *verto*, "inverted"	Dark clay soils that exhibit wide, deep cracks when dry	Grumusols
Inceptisol	Latin, *inceptum*, "beginning"	Texture finer than loamy sand; little translocation of clay; often shallow; moderate development of horizons	Brown forest soil, sol brun acide, acide, humic gley, weak podzols
Aridisol	Latin, *aridus*, "arid"	Dry for extended periods; low in humus, high in base content; may have carbonate, gypsum, and clay horizons	Sierozems, red desert soils, solonchak
Mollisol	Latin, *mollis*, "soft"	Surface horizons dark brown to black with soft consistency; rich in bases; soils of semihumid regions	Chestnut, chernozem, prairie; some brown and brown forest and associated humic gleys
Spodosol	Greek, *spodos*, "ashy"	Light gray, whitish A_2 horizon on top of a black and reddish B horizon high in extractable iron and aluminum	Podzol, brown podzolic soils
Alfisol	Coined from *Al* and *Fe*	Shallow penetration of humus, translocation of clay; well-developed horizons	Gray-brown podzolic, gray wooded soils, noncalcic brown soils, some planisols
Ultisol	Latin, *ultimus*, "last"	Intensely leached; strong clay translocation, low base content; humid, warm climate	Red-yellow podzolic, red-brown laterite, some latisols
Oxisol	French, *oxidé*, "oxidized"	Highly weathered soils; red, yellow, or gray; rich in kaolinite, iron oxides, and often humus; in tropics and subtropics	Laterites, latosols
Histosol	Greek, *histos*, "organic"	High content of organic matter	Bog soils, muck
Andisol	Japanese, *ando, an* "black," *do* "soil"	Developed from volcanic ejecta; not highly weathered; upper layers dark colored; low bulk density	None, a new Order

fauna to make relatively easy adjustments to unfavorable conditions.

On the other hand, soil has low penetrability. Movement is greatly hampered. Except to such channeling species as earthworms, soil pore space is important, for it determines the size of the living space, the humidity, and the available air.

The variability of these conditions creates a diversity of habitats, reflected in the diversity of species found in the soil (Figure 9.8). The number of different species found in the soil is enormous, representing practically every invertebrate phylum. There are 250 species of Protozoa alone in English soils. The number of species of soil animals exclusive of Protozoa found in a variety of habitats in Germany varies from 68 to 203. In the soil of a beech wood in Austria live at least 110 species of beetles, 229 species of mites, and 46 species of snails and slugs. E. C. Williams counted 294 species of soil animals, exclusive of Protozoa, in the Panama rain forest.

Only a part of the soil litter is available to most soil animals as living space. Spaces between the surface litter, cavities walled off by soil aggregates, pore spaces between individual soil particles, and root channels and fissures are all potential habitats. Most of the soil fauna are limited to pores and cavities larger than themselves. Large species of mites inhabit loose soils with a crumb structure, in contrast to smaller forms inhabiting compact soils. Larger soil species are confined to upper layers, where the soil interstices are largest.

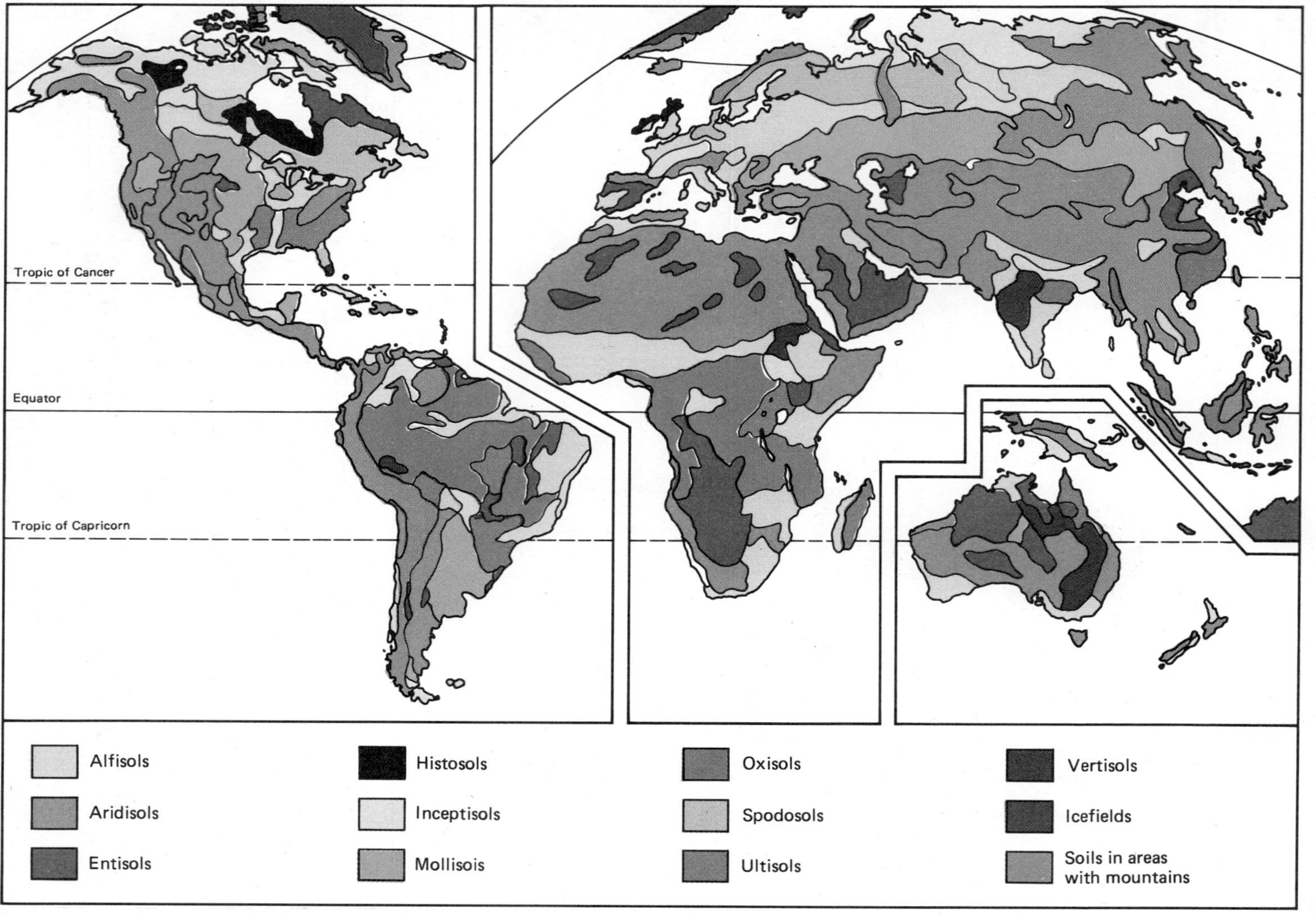

FIGURE 9.7 The world distribution of major soil orders.

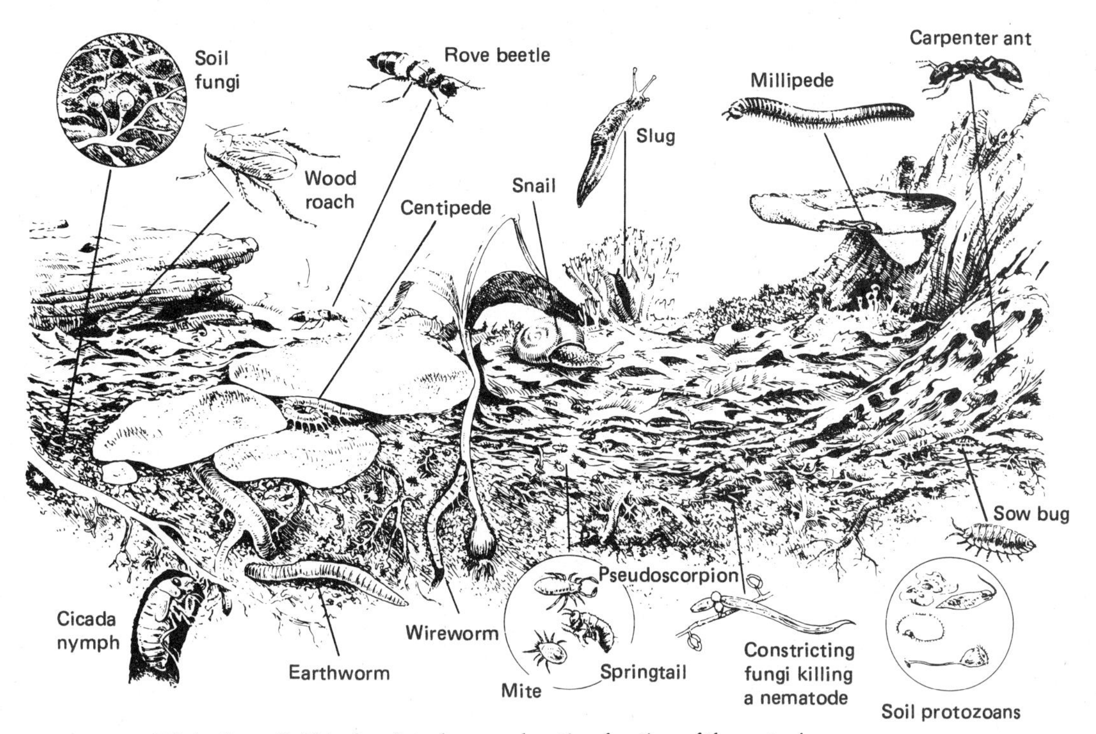

FIGURE 9.8 Life in the soil. This drawing shows only a tiny fraction of the organisms that inhabit soil and litter. Note the fruiting bodies of fungi, which furnish food for soil-associated organisms.

Water in the spaces is essential, because the majority of soil fauna is active only in water. Soil water is usually present as a thin film coating the surface of soil particles. This film contains, among other things, bacteria, unicellular algae, protozoa, rotifers, and nematodes. Most of them are restricted in their movements by the thickness and shape of the water film in which they live. Nematodes are less restricted, for they can distort the water film by means of muscular movements and thus bridge the intervening air spaces. If the water film dries up, these species encyst or enter a dormant state. Millipedes and centipedes, on the other hand, are highly susceptible to desiccation and avoid it by burrowing deeper into the soil.

Excess water and lack of aeration are detrimental to many soil animals. Excessive moisture, which typically occurs after heavy rains, is often disastrous to soil inhabitants. Air spaces become flooded with deoxygenated water, producing a zone of oxygen shortage for soil inhabitants. If earthworms cannot evade this zone by digging deeper, they are forced to the surface, where they die from ultraviolet radiation. The snowflea (a collembola) comes to the surface in the spring to avoid excess soil water from melting snow. Many small species and immature stages of larger species of centipedes and millipedes may be completely immobilized by a film of water and unable to overcome the surface tension imprisoning them. Adults of many species of these organisms possess a waterproof cuticle that enables them to survive some temporary flooding.

VEGETATION AND SOILS

A close relationship exists between vegetation and soils. The vegetation both reflects the soil and influences its development. If you can identify certain plants, you often can tell soils at a glance.

Two contrasting vegetational communities exist on high-lime (or limestone) and low-lime (or acid) soils. In general, limestone soils support plants that have a high uptake of calcium, need an adequate supply of phosphorus and magnesium, and are intolerant of aluminum ions, which inhibit the uptake of phosphorus. Plants of acid soils cannot tolerate a high level of calcium, which inhibits their uptake of potassium and iron. The vegetation of acid soils includes such plants as heaths, laurels (*Kalmia*), blueberries (*Vaccinium*), oaks (*Quercus*), and chestnuts (*Castanea*). Limestone soils in eastern North America support such plants as hickory, red cedar (*Juniperus virginiana*), maples, walking ferns (*Camptosorus*), spleenwort (*Asplenium*), and blazing star (*Liatris*). Limestone soils, especially where outcrops are common and the soils shallow, are xeric sites, because crevices and channels in the rocks carry the water far below the surface.

A strong division between the plant communities of limestone and dolomitic soils and those of sandstone or noncarbonate soils exists in the semiarid, subalpine White Mountains of eastern California. This discontinuity is most pronounced along the boundaries of the two geological contacts. The noncarbonate soils are dominated by sagebrush (*Artemisia tridentata*) and silvery lupine (*Lupinus argenteus*). The carbonate soils are dominated by bristlecone pine (*Pinus aristata*), limber pine (*Pinus flexilis*), phlox (*Phlox covillei*), and locoweed (*Astragalus kentrophyta*).

In the Pacific Northwest and northern California are large areas of serpentine soils derived from green-colored rocks such as igneous peridotite and metamorphic serpentine rocks. Such soils are low in calcium and high in nickel, magnesium, and chromium, in that order. The effects of these elements—modified by topography, elevation, and water availability—are conspicuous where plants on serpentine soils grow next to plants on nonserpentine soils. Serpentine areas are characterized by stunted vegetation and an unusual flora, which includes endemics—plants that can grow nowhere else. A typical serpentine community is Jeffrey pine (*Pinus jeffreyi*)—grass woodland, which contrasts strongly with adjacent stands of Douglas-fir. At high elevations the serpentine community is dominated by grasses and at lower elevations, where the moisture conditions are better, by a greater diversity of species. Other areas of serpentine soils occur in Quebec, where tundra contrasts with taiga on nonserpentine soils; in northern California, where chaparral or serpentine contrasts with oak woodland; in Appalachian regions; and in Cuba, where savanna and shrub contrast with tropical forests.

SOIL EROSION

W. C. Lowdermilk, assistant chief of the U.S. Soil Conservation Service in the 1930s, wrote: "In a larger sense a nation writes its record on the land . . . —a record that is easy to read by those who understand the simple language of the land." Past civilizations and nations have left a sorry record. The impoverished lands today—the Near East, Egypt, Syria, Lebanon, North Africa, the Mediterranean—all once supported flourishing agriculture, extensive forests, and great empires. Because of overgrazing, deforestation, and poor land use, these ancient empires based on agriculture are gone, the remains buried beneath soil eroded from the hills.

In spite of past experience, soil erosion still is the world's major environmental problem. Worldwide loss of topsoil amounts to about 23 billion tons a year, and it is increasing as human population growth forces agriculture to move into marginal, easily erodable land. Each year the Soviet Union abandons an estimated one-half billion hectares of cropland because of erosion; dust storms in North Africa carry up to 100 million tons of soil annually out to the Atlantic, speeding up desertification of that part of Africa; and the Ganges in India and the Yellow River in China carry billions of tons of soil to the ocean, as does the Mississippi.

Soil erosion was severe in the United States during the early 1930s, when plowing of the Great Plains and drought resulted in the Dust Bowl and unvaried cropping of cotton and tobacco gullied most of the South. Alarmed, President Franklin D. Roosevelt and Congress established the Soil Conservation Service, which took massive action. Solutions involved planting windbreaks in the Midwest, grassland restoration, fallowing, crop rotation instead of monoculture, terracing, contour farming, and gully control. Such approaches were working until after World War II, when family farms declined and corporate interests with large land holdings took over agricultural production. Corporate farmers tore out windbreaks, terraces, fencerows, and other soil-holding features that interfered with the use of large-scale machinery and abandoned crop rotation for monocultural production of corn, soybeans, and cotton. Again erosion grew severe as soil was treated as an economic input, not as a renewable resource.

Soil protected by a vegetative cover maintains its integrity. Vegetation breaks the force of the wind and disperses raindrops, breaking their force. Rain trickles slowly through the litter, infiltrating the soil. If rainfall exceeds the soil's capacity to absorb it, the excess runs across the surface, but its movement will be slowed.

Stripped of its protective vegetation and litter by plowing, logging, grazing, road building, and urban construction, soil is removed by wind and water. As the upper layers of humus-charged, granular, highly absorptive soil are removed, humus-deficient, less stable, less absorptive, and highly erodable layers beneath are exposed. If the subsoil is clay, it absorbs water so slowly that heavy rains produce an abrasive and rapid runoff.

Rain falling on bare ground hammers the soil's surface, breaking down soil aggregates and bringing fine particles to the surface to create a seal. Unable to infiltrate the soil, the water moves across the surface as runoff, carrying soil with it. The least conspicuous type of water erosion is *sheet erosion*, a more or less even and cumulative removal of soil over a field. Sheet erosion can be aggravated by soil compaction, caused by heavy agricultural equipment traveling over fields and overuse of recreational areas and campsites by outdoor recreationists. When runoff tends to concentrate in streamlets, its cutting force is increased; rills and gullies carry runoff and soil away even more rapidly. *Gully erosion* often begins with furrows running up and down hill, in wheel ruts formed by all-terrain and off-road vehicles and construction equipment, on recreational hiking trails, livestock trails, and logging skid trails.

Bare soil, finely divided, loose and dry, as it often is after tillage, is ripe for wind erosion. The forward velocity of the wind well above the soil's surface is much higher than near the surface, where it approaches zero. Just above the surface, wind movement meets surface irregularities, producing eddies with an upward velocity two to three times the forward velocity near the surface. The eddies dislodge the most erodable grains of soil occupying the most exposed positions on the surface and move them along a short distance on the surface. Suddenly the grains shoot upward, to a height influenced by the size and density of the particles and the nature of the surface. When the particles strike the ground again, they rebound and dislodge still other particles, forcing them to jump into the air. The impact of the particles on the surface initiates still another movement, downhill *surface creep*. Meanwhile fine particles of dust, bounced by larger particles, rise high enough to be picked up by the wind and carried on as dust clouds. Often dust particles are lifted high into the atmosphere and carried for hundreds of miles.

Erosion by wind and water carries away organic layers, exposes the subsoil, depletes nutrients, deposits soil elsewhere, increases runoff, and causes land ruin and abandonment. Land abandoned because of soil mismanagement is usually so degraded that natural vege-

TABLE 9.2 Effects of Erosion on Soil Structure, Nutrient Availability (Phosphorus), and Productivity of Maimi, a Fine Loamy Soil in Indiana

Erosion Phase	*Clay (%)*	*Organic Matter (%)*	*Available Water (%)*	*Phosphorus (kg/ha)*	*Corn (kg/ha)*	*Soybeans (kg/ha)*
Slight	15.4	1.89	16.10	106.4	9227	3026
Moderate	18.1	1.64	11.47	96.5	8788	2825
Severe	22.1	1.51	4.76	76.4	7658	2690

FIGURE 9.9 Soil eroded from upstream watersheds is filling in the upper end of this reservoir in California.

tation has difficulty colonizing it. Erosion becomes progressively worse unless extreme measures are taken to restore vegetation.

Effects of erosion are felt both on site and off. Erosion on agricultural and forest lands reduces soil organic matter and increases clay content (Table 9.2). It reduces water-holding capacity of the soil, intensifying drought conditions in dry weather and accelerating erosion in wet weather. Erosion degrades soil structure and reduces plant nutrients and plant rooting depth, depressing crop yield. Studies have shown that the loss of one inch (2.5 cm) of topsoil reduces corn and wheat yields by 6 percent. Costly, energy-demanding chemical fertilizers mask the real effects of soil erosion on the inherent fertility of the soil. The economic cost of soil erosion amounts to close to a billion dollars annually, including cost of erosion control and loss in crop production.

The cost of off-site effects may be more than twice that of on-site effects. Soil eroded by wind and water has to go somewhere. Soil erosion carries sediment into rivers, clogging navigation and requiring dredging; it fills in reservoirs and hydroelectric dams, shortening their lives (Figure 9.9); it contributes to water pollution and reduction in water quality. Wind erosion contributes significantly to particulate air pollution, illness, and damage to machinery.

All forms of soil erosion destroy the integrity of ecosystems and ecological cycles. They raise the cost of food, fostering hunger and famine. For humans the ecological consequences could well be social disorder and degradation of life. There is a saying, "Poor soils make poor people." It could not be truer.

Summary

Soils are the base for terrestrial ecosystems. Soil is the site of the decomposition of organic matter and of the return of mineral elements to the nutrient cycle. It is the home of animal life, the anchoring medium for plants, and their source of water and nutrients. Soil begins with the weathering of rocks and minerals, which involves the leaching out and carrying away of mineral matter. Its development is guided by slope, climate, original material, and native vegetation. Plants rooted in the weathering material further break down the substratum, pump up nutrients from its depths, and add all-important organic material. This material, through decomposition and mineralization, is converted into humus, an unstable product that is continuously formed and destroyed by mineralization.

As a result of the weathering process, accumulation and breakdown of organic matter, and the leaching of mineral matter, four horizons or layers form in the soil: the O, or organic, layer; the A horizon, characterized by accumulation of organic matter and loss of clay and mineral matter; the B horizon, in which mineral matter accumulates; and C, the underlying material. These horizons divide into subhorizons.

Of all the horizons none is more important than the humus layer, which plays a dominant role in the life and distribution of plants and animals, in the maintenance of soil fertility, and in much of the soil-forming process. Humus is usually grouped into three types: mor, characteristic of acid habitats, whose chief decomposing agents are fungi; mull, characteristic of deciduous and mixed woodlands, whose chief decomposing agents are bacteria; and intermediate types, which are highly modified by the action of soil animals.

Profile development is influenced over large areas by vegetation and climate. In grassland regions the chief soil-forming process is calcification, in which calcium accumulates at the average depth reached by percolating water. In forest regions podzolization—the leaching of calcium, magnesium, iron, and aluminum from the upper horizons and the retention of silica—takes place. In tropical regions laterization, in which silica is leached and iron and aluminum oxides are retained in the upper horizons, is the major soil-forming process. In semiarid regions the soil-forming process involves salinization. The soils contain an excessive amount of soluble salts that are drawn to the surface by the evaporation of water from infrequent rains. This process results in a crust of alkali salts. Gleization takes place in poorly drained soils. Organic matter decomposes slowly, and iron is reduced to the ferrous state.

Soils and horizons differ in texture, color, structure, moisture, and chemistry. Each combination of climate, vegetation, soil material, slope, and time results in a unique soil, of which the smallest repetitive unit is the pedon. Soils can be classified into eleven major orders, based on major soil-forming processes.

Worldwide, soils continue to be depleted by erosion, in which topsoil loss exceeds formation. Deforestation, poor agricultural practices, urbanization, road building, and other disturbances expose the soil to the erosive forces of water and wind. Soil erosion destroys natural ecosystems and farming ecosystems and fills rivers, streams, lakes, reservoirs, and navigation ways with silt. Wind erosion carries soil far from its source in clouds of dust and increases particulate air pollution. Soil erosion in all its forms impoverishes regions and nations, reduces food production, and causes extensive economic losses.

Review and Study Questions

1. Why is it difficult to define soil?
2. What role does physical and chemical weathering play in soil development?
3. What is the role of plant and animal life?
4. What is mineralization? Humus? Do the two relate?
5. Distinguish between mull and mor humus. What is their ecological significance?
6. Compare calcification, podzolization, and laterization.
7. What is gleization? Under what conditions does it take place?
8. What is a micelle? How do micelles relate to cation exchange capacity?
9. What is cation exchange capacity? How does it affect nutrient uptake? Soil acidification?
10. What is soil texture? How does it relate to soil particle size?
11. How do soil particles relate to soil structure?
12. What characterizes the soil as an environment for life?
13. In what way is soil, especially its acidity and alkalinity, a selective force in the evolution of plants?

*14. From your local or state soil conservation office obtain a soil survey for your area. Note the pattern of soil distribution. What are the local soil series? Relate local soils to agricultural development, urban development, soil erosion problems, and forest distribution.

*15. What is causing erosion in your local area? Notice the degree of siltation in streams, rivers, dams, and lakes after heavy rains. What efforts are being made to reduce erosion? Your district and state Soil Conservation Service offices can help.

*16. Look up the various techniques of reducing soil erosion. Which are best, and why?

Selected References

Brady, N. C. 1990. *The nature and properties of soils*, 10th ed. New York: Macmillan. A basic text.

Brown, L. R. 1984. Conserving soil. Pages 53–73 in L. Strake, ed., *State of the world 1984*. New York: W. W. Norton.

Brown, L. R., and E. C. Wolf. 1984. *Soil erosion: Quiet crisis in the world economy*. Washington, D.C.: Worldwatch Institute.

Eyre, S. R. 1968. *Vegetation and soils: A world view*. Chicago: Aldine. Survey of world soils.

Farb, P. 1959. *The living earth*. New York: Harper & Row. An excellent popular discussion of soil.

Furley, P. A., and W. N. Newey. 1983. *Geography of the biosphere*. London: Butterworths. See Chapters 4 and 9 on soils.

Jenny, H. 1980. *The soil resource*. New York: Springer-Verlag. A definitive reference source.

Lutz, H. J., and R. F. Chandler. 1946. *Forest soils*. New York: Wiley. Dated in places, but still an excellent reference.

Sears, P. B. 1947. *Deserts on the march*. Norman: University of Oklahoma Press. A classic, written after the Dust Bowl. It still holds true.

Worster, D. 1979. *The dust bowl*. New York: Oxford University Press. An in-depth look at the Dust Bowl, past and present.

PART IV
Population Ecology

INTRODUCTION

Looking at Populations

How do we, as individuals, perceive the world around us? Most see a friend, a tree in the forest, a flower in the field, a squirrel in the park, or a bird nesting in the backyard as an individual. Rarely do we see each as part of a larger unit, a group of similar, interacting individuals of the same kind living in the same place at the same time. Such a group makes up a *population*.

A population has a life of its own. It has special ways to allot resources such as food, space, and mates and to regulate growth. Animals may cooperate in hunting, nest building, or defense. Plants and animals interact as they mate and rear offspring. What happens to an individual in a population affects the lives of the others.

Populations have unique features. They have an age structure, density, and distribution in time and space. They exhibit a birthrate, a mortality rate, and a growth rate. They respond in their own ways to competition within and without, to predation, and to other pressures.

Populations affect one another—limiting, helping one another, and even evolving together. The relations of one population with another influence the structure and function of whole ecosystems.

Population ecology developed around animals. Only recently have ecologists begun to look at plant populations. Although traditional approaches used in the study of animal populations can be applied to them, plant populations do not lend themselves to the same demographic analyses (nor do some of the invertebrates, such as corals and hydroids, that grow in colonies). The reasons lie in the nature of plants themselves.

Consider a tree. The tree appears as one unit; but if you look up into it, you will realize that the tree is a collection of subpopulations: buds, twigs, shoots, and leaves above ground and roots and their extensions below ground. These subunits make up *metapopulations* that have their own demography: birthrates, death rates, and growth rates. The birth and death of these individual subunits determine not only the rate of growth but also the form the plant will take.

A tree or shrub grown from a seed is an individual. That individual, with its own genetic characteristics, is termed the *genet*. It reproduces sexually, producing a new generation from seeds. If it is cut, new buds may develop along the root collar and send up sprouts or coppice growth. Although the original stem is dead, the genet lives on through new vegetative growth. Trees such as black locust and aspen and shrubs such as sumac grow underground, adding root extensions that send up new shoots or suckers. New metapopulations, known as *ramets*, develop in an area surrounding the original parent genet. These clones may cover a considerable area and look like individuals of different ages. Because they are reiterations of the parent stem, the genet may be very old although its trees are young. As a result, it is hard to tell the life spans of plants, especially trees, shrubs, and some perennial herbs. Such characteristics of plants complicate studies of their demography.

CHAPTER 10

Density, Distribution, and Age

Outline

Objectives

On completion of this chapter, you should be able to:

1. Define population density, crude density, and ecological density.
2. Describe the types of population distribution.
3. Discuss problems and methods of determining the density of a population.
4. Describe the age structure of a population and stable age distribution.
5. Discuss the significance of different types of age pyramids.
6. Discuss ways in which the ages of plants and animals can be determined.
7. Contrast age structure of plants with that of animals.

Three attributes of a population most obvious to a casual observer are density, distribution, and age structure. You are aware of the change in density and distribution of human populations from the concentrated masses of the city to the sparse populations of the countryside. You notice the years when gypsy moths seem to be stripping trees everywhere. You observe young rabbits, speckle-breasted young robins, deer fawns, and the full array of age classes in human populations, which emphasize age structure.

DENSITY

When you get caught in traffic going home, make your way through crowded streets, or enjoy the freedom of lightly traveled roads, you are aware of population density. You respond to that density in some way. You may seek out the crowds, or you may escape to the peaceful countryside or the quiet of your house or apartment. Then again, you may become frustrated at the delays high densities impose on you.

Individuals in natural populations are also affected by density in some way. Trees in crowded stands may grow more slowly, and some may succumb to a lack of water, nutrients, and light, unequally shared. Scarce food may be denied to smaller or less aggressive individuals in a mammalian population; and access to nest sites may be denied to some birds because not enough sites exist to meet the demand. Having too few individuals in a population may reduce their chance of finding a mate or inhibit the performance of behavioral activities essential to the welfare of the population. Density of a population affects the spread of diseases and parasites and influences the risks of individuals succumbing to predation. In affecting the welfare of individuals in a population, density in part controls its birthrates, mortality rates, and growth.

However important it may be, density is an elusive characteristic, difficult to define and hard to determine. Density can be characterized as the number of individuals per unit of space—as so many per square mile, per hectare, or per square meter. That measure is *crude density*. But individuals in a population do not occupy all the space within a unit, because not all of it is suitable habitat. A biologist might estimate the number of deer living in a square-mile area. The deer, however, might avoid half the area because of human habitation, land use, and lack of cover and food. Goldenrods inhabiting old fields grow in scattered groups or clumps because of soil conditions and competition from other old field plants.

No matter how uniform a habitat may appear, it is usually patchy, because of microdifferences in light, moisture, temperature, or exposure, to mention a few physical conditions, or because of the lack of sites available for colonization. Each organism occupies only those areas that can adequately meet its basic requirements.

To account for those conditions or effects, the density of organisms should refer to the amount of area available as living space. That measure would be *ecological density*. Ecological densities are rarely estimated because what portions of the habitat represent living space are difficult to determine. In a Wisconsin study, the densities of bobwhite quail were expressed as the number of birds per mile of hedgerow rather than as birds per acre.

DISTRIBUTION

How organisms are distributed over space has an important bearing on density. Individuals of a population may be distributed randomly, uniformly, or in clumps (Figure 10.1). Individuals of a population are distributed randomly if the position of each is independent of the other. Forest trees of the canopy layer are frequently spaced at random. So are some invertebrates of the forest floor, particularly spiders, and the clam *Mulinia lateralis* of the intertidal mud flats of the northeastern coast of North America.

By contrast, individuals distributed uniformly are more or less evenly spaced. In the animal world uniform distribution usually results from some form of intraspecific competition, such as territoriality. Uniform distribution happens among plants when severe competition exists for crown or root space, as among some forest trees, or for moisture, as among desert plants.

The most common type is clumped (sometimes called contagious) distribution, which consists of scat-

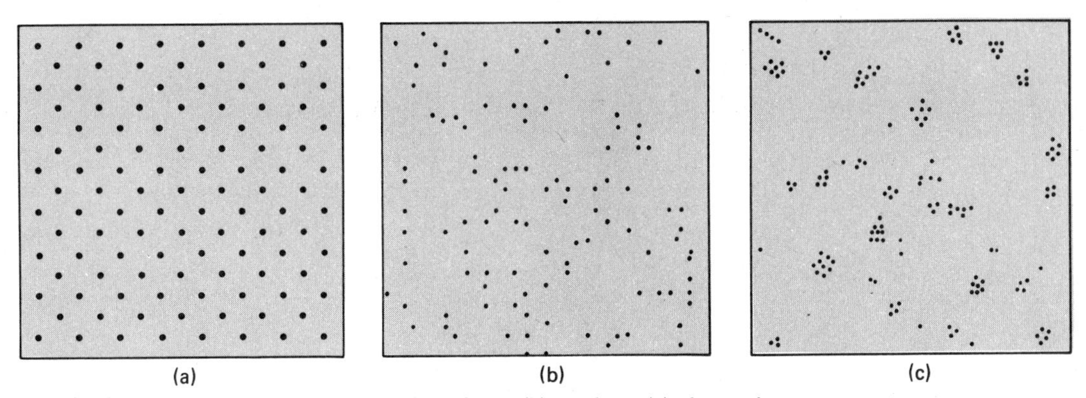

FIGURE 10.1 Patterns of distribution: (a) uniform, (b) random, (c) clumped.

tered groups. Clumping is the result of responses to habitat differences, daily and seasonal weather changes, reproductive patterns, and social behavior. The distribution of human populations is clumped or aggregated because of social behavior, economics, and geography. There are various degrees and types of clumping. Groups may be randomly or nonrandomly distributed over an area. Aggregations may range from small ones to a single centralized group. If environmental conditions encourage it, populations may be concentrated in long bands or strips along some feature of the landscape such as a river, leaving the rest of the area unoccupied.

Populations are not uniformly distributed over a region. Just as the local distribution of a population may be defined by the spacing of individuals, the distribution of species is defined on a larger scale by the spacing of populations. These populations may be concentrated in clusters within a given region. Regional distributions of populations make up the *range* of a species (Figure 10.2). The boundaries of a species' range are not fixed, but fluctuating. Habitat changes, competition, predation, climatic changes—all can influence the extent of a species' range. It may expand one year, contract another. The many variables that influence the distribution of a species over its range and the regional adaptations of a species are the subject matter of *biogeography.*

Organisms in populations are not only dispersed in space; they are also distributed in time. Temporal distribution may be circadian (related to daily changes of daylight and dark), lunar, tidal, and seasonal (see Chapter 8). Distribution may also involve longer periods, including annual fluctuations, successional changes, and evolutionary changes.

DETERMINING DENSITY AND DISTRIBUTION

Density and dispersion distribution of individuals in a population are almost inseparable. With uneven dispersion, a change of boundaries can change the level of density (Figure 10.3).

Fortunately, ecologists rarely need to know the exact abundance or density of a population. (*Abundance* is the number of individuals in a given area, in contrast to *density,* which is the number expressed per unit area.) Both can be determined only by a direct count of all individuals in a population. Except for unusual habitat situations, such as antelope living on an open plain or waterfowl concentrated in a marsh, direct counts are either extremely difficult or impossible to obtain. Even if possible, direct counts may be too time-consuming or too expensive.

For these reasons ecologists use other means of obtaining information on population density. One is sampling, in which the area of study is divided into subunits in which animals or plants are counted in a prescribed manner. From the sample data is determined the mean density of the unit sampled. The mean is then multiplied by the total number of sampled and

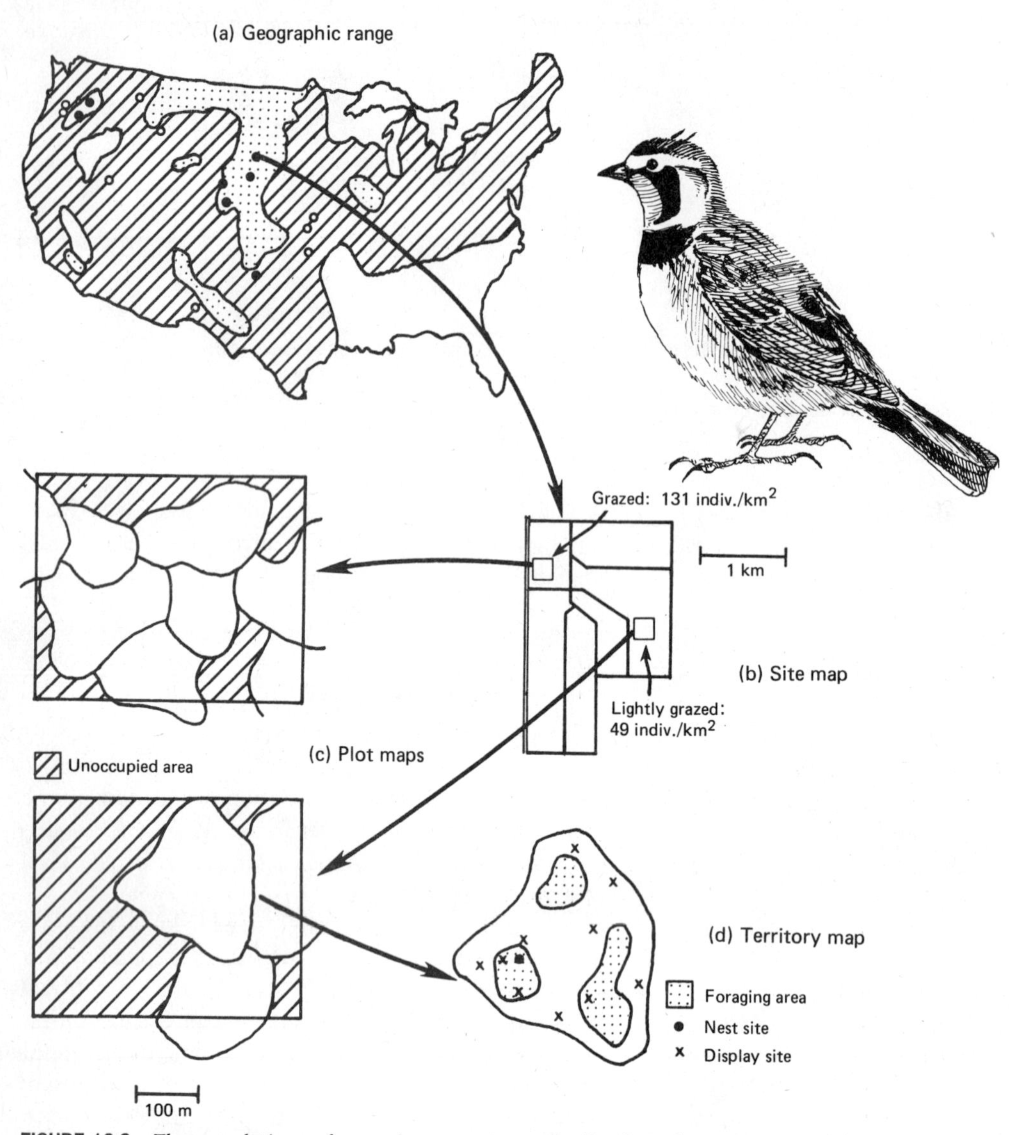

FIGURE 10.2 The populations of a species are not equally distributed over its range. (a) The range map of the horned lark (*Eremophila alpestris*). Dotted areas show highest density. (b) Within a region distribution is influenced by the availability of habitats. (c) Within a given habitat the bird's distribution is influenced by territorial behavior. (d) Within each territory certain parts may be used for different purposes, such as singing and nesting.

unsampled areas to arrive at the estimated population. Confidence limits can be calculated on the estimate. Sampling is used most widely in the study of populations of plants and sessile (attached) animals because no movement in and out of the area is involved. Sampling is also important in the study of enclosed fish populations, as in lakes and ponds.

For most ecological work with animals, indices of

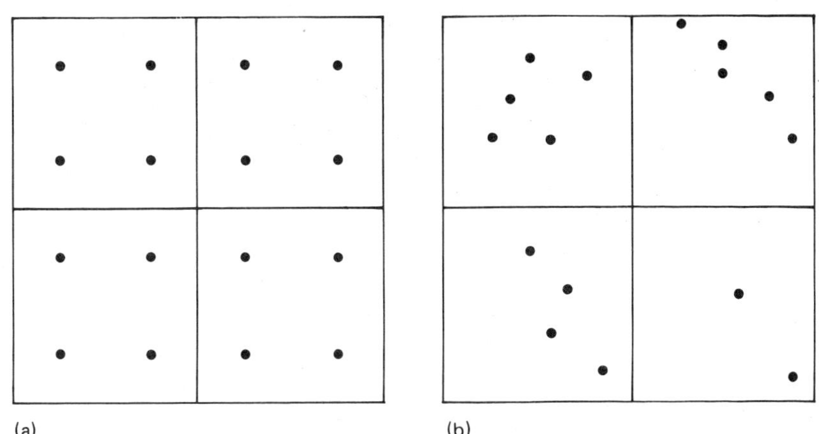

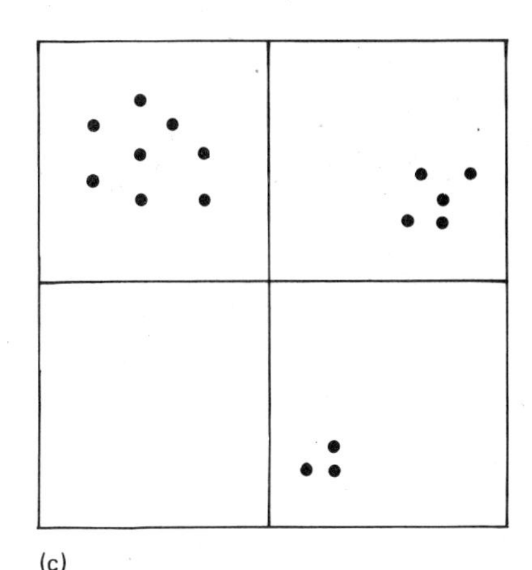

FIGURE 10.3 Crude density for each of the subpopulations within the three large squares is 16. If each area is divided into four sampling units, density estimates would be quite different, depending upon which sample was selected. In the uniform population (a) any one sampling unit would give a correct estimate of the population: 4 × 4 (to explode the population to the full area) = 16. For the random population (b) the estimates would be 20, 20, 8, and 16. In the clumped population (c) the estimates would be 32, 20, 12, and 0. Thus any estimates of crude density should indicate how individuals of a population are distributed.

relative abundance may be sufficient (Figure 10.4). You might count the number of mourning doves seen along a prescribed length of road traveled year after year, or you might note the number of drumming ruffed grouse heard along a stretch of trail, or you might record the number of tracks crossing dusty roads. The results can be converted to numbers of individuals per kilometer or per hour. Such counts cannot stand alone; but if you have a series of such index figures collected from the same area over a period of years, you can follow trends in density or abundance. Or you can obtain counts from different areas during the same year to compare numbers from one habitat to another. Most population data on birds and mammals are based on indices of relative abundance rather than on direct counts.

Getting some idea of how individuals of a population are dispersed also involves sampling the population. One method uses sample plots and compares the data to some mathematical distribution such as a Poisson, variance-to-mean ratio, or chi-square goodness-of-fit. Other methods involve measuring distances between individuals or distances to individuals from random points located in the habitat and treating the data to appropriate analysis.

AGE STRUCTURE

Unless a population consists of seasonal breeders with nonoverlapping generations, such as annual plants and animals, it will be characterized by a certain age structure that influences considerably the birthrates and death rates of a population.

Populations can be divided into three ecological periods: prereproductive, reproductive, and postreproductive. In plants the prereproductive period is usually termed the juvenile period. The relative length of each period depends largely on the life history of the organism. Among annual species the length of the prereproductive period has little significant effect on the potential rate of population growth. In longer-lived plants and animals, the length of the prereproductive period has a pronounced effect on the population's rate of growth. Organisms with a short prereproductive period often increase rapidly with a short span between

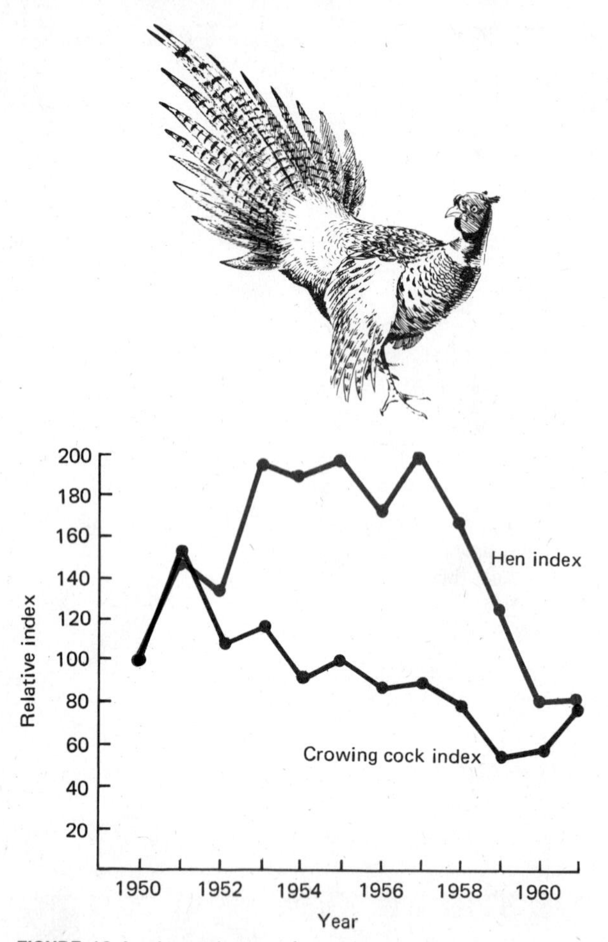

FIGURE 10.4 An estimate of population abundance. Density trends in a ring-necked pheasant (*Phasianus colchicus*) population in Wisconsin were followed by counting spring crowing over a prescribed route. The hen index is obtained by multiplying the average number of calls per stop in the crowing cock counts by the number of hens observed per cock during the preceding winter.

generations. Organisms with a long prereproductive period generally increase more slowly and have a long span of time between generations.

Theoretically, all continuously breeding populations tend toward a *stable age distribution;* that is, the distribution of individuals in the various age classes reaches and remains at a fixed distribution when the population is increasing at a constant rate and when the age-specific birthrates and age-specific mortality rates remain constant. When the population reaches a constant size in which deaths equal births, then we call the fixed proportion of individuals in its age classes a *stationary age distribution.* Few natural populations ever reach stable age distributions, because annual variations in environmental conditions, emigration, and immigration influence mortality and natality. Changes in age structure in a population of elk are illustrated in Figure 10.5. This changing age structure reflects sporadic and occasionally heavy human predation, winter mortality, and predation on calves which, in turn, affected fecundity.

Any influence that causes age ratios to shift because of changes in age-specific death rates affects the population birthrate. In populations in which life expectancy for the oldest age classes is reduced, a higher portion falls into the reproductive class, automatically increasing the birthrate. Conversely, if life is extended, a greater portion of the population falls into the postreproductive class, thus reducing the birthrate. Rapidly growing populations are usually characterized by declining death rates, especially in the very young age classes, which inflates the younger age groups. Declining or stabilized populations are characterized by lower birthrates with fewer young to rise into the reproductive age classes and by a larger portion in the older age classes.

ANIMAL POPULATIONS

The age structure of a population represents the ratio of the various age classes in a population to each other at a given time. To determine age structure we need some means of obtaining the ages of members of a population. For humans this task is not a problem, but it is with wild populations. Age data for wild animals can be obtained in a number of ways, the method varying with the species. The most accurate, but most difficult, method is to mark young individuals in a population and to follow their survival through time. Such a method, however, requires both a large number of marked individuals and, of course, time. For this reason other, less accurate methods are employed. These methods include an examination of a representative sample of carcasses for the wear and replacement of teeth in deer and other ungulates; the growth rings

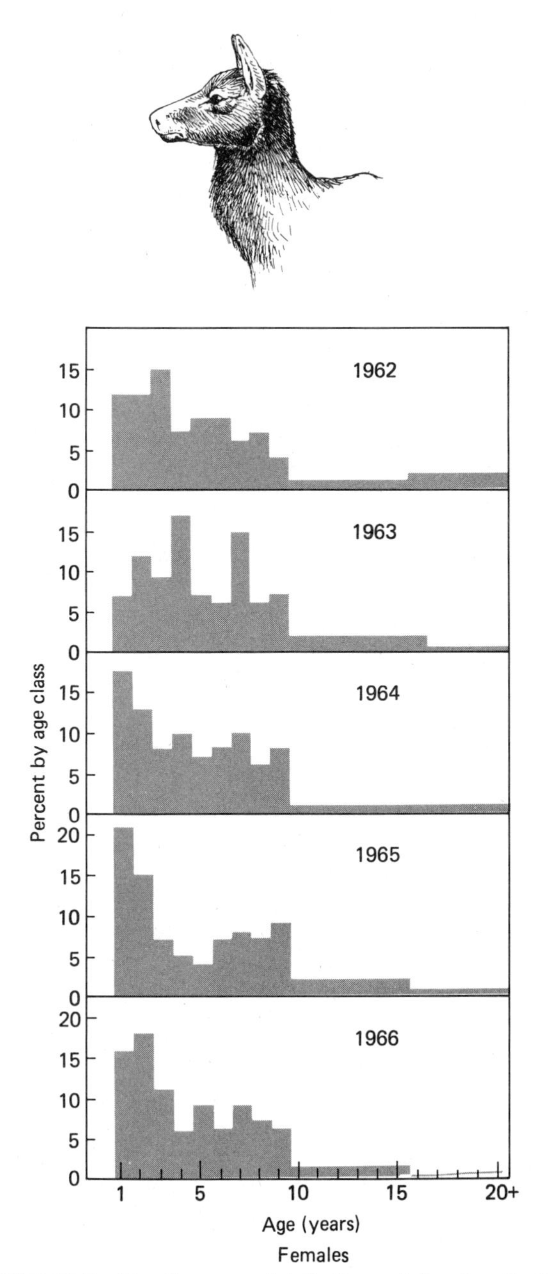

FIGURE 10.5 Standing age distribution of females in the northern Yellowstone elk (*Cervus elaphus*) from 1962 to 1966. Note how the lower age classes move into and influence the older age classes. These age pyramids suggest an unstable population, but statistically there is little significant difference among 1964, 1965, and 1966.

in the cementum of teeth of carnivores and ungulates; the annual growth rings in the horns of mountain sheep; and, in rabbits, the weight of eye lens, which increases with age. Among birds, aging methods involve observations of plumage changes and wear, which can only separate juveniles from subadults (in some species) and adults.

Age structure is visualized best by means of age pyramids (Figure 10.6) that compare the percentage of one age group in a population to other age groups. As the population changes with time, the number of individuals and, thus, the ratio in each age class change. Growing populations are characterized by a large number of young, which expands the base of the age pyramid. This large class of young eventually moves up into the reproductive age classes. If the young are as prolific as the parents, the young age class will expand further. Declining populations are characterized by fewer individuals being added to the population and a higher proportion of the individuals in the population moving into the older age classes. With fewer young, fewer individuals will enter the reproductive age classes, further depressing the population. In this way the age structure changes as a population increases, decreases, or remains the same.

Although such generalizations are applicable to human populations because of their long life span and low juvenile mortality, age structure is much less predictive of population growth in wild animal populations. Population growth for wild animals also depends on the close relationship between animals and their resources, both habitat and food, and the competitive relationships among individuals for those resources. For example, a normally increasing population of a wild species should have an increasing number of young; a decreasing population, a decreasing number of young. But within this framework there can be a number of variations. A population may be decreasing, yet show an increasing percentage of young. Another population might remain static with an increasing percentage of young. Both situations result when too many adults are hunted, leaving few to carry on reproduction. Only by correlating such information as density, mortality, and reproduction with changing age distribution can age structure provide some insight into the dynamics of wild populations.

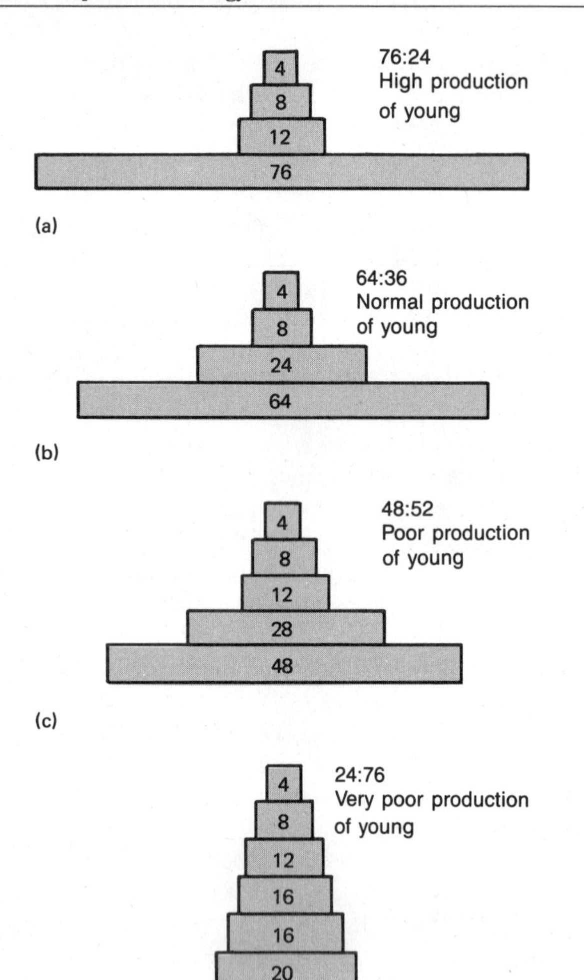

FIGURE 10.6 Theoretical age pyramids, especially applicable to mammals and birds. (a) Growing populations are characterized by a large number of young. The age pyramid of a growing population has a broad base and a narrow, pinched top, indicating that increasing numbers of young may enter the reproductive age and increase all age classes through time. (b) A growing population normally held just below carrying capacity (such as rabbits) should have a ratio of about two-thirds young age class to one-third adults. (c) If a population is declining, then fewer individuals are being added to the reproductive and other age classes in the population and a higher proportion may be in the older age classes. (d) If the population is neither growing nor declining, then the number of individuals in each age class tends to remain the same (stationary age distribution). The pyramid of such a population is a narrow one with a narrow base of young.

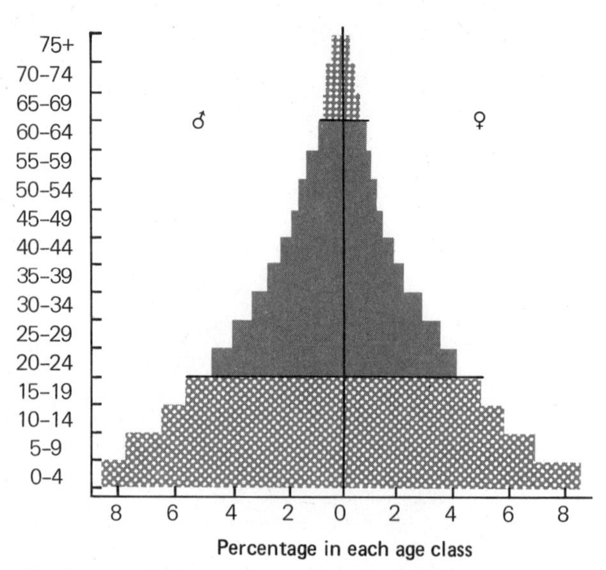

FIGURE 10.7 The age pyramid for India in 1970 shows a rapidly expanding population. Note the broad base of young that will enter the reproductive age classes.

Because age structure of human populations more accurately reflects population dynamics than those of wild animals, age structure is best illustrated by age pyramids of human populations. The age pyramid for India in Figure 10.7 has a broad base, reflecting inflated young age groups, and a narrow top, reflecting relatively few individuals in old age classes. Such a pyramid indicates a youthful population in which increasing numbers will enter the reproductive period. Figure 10.8, depicting an age pyramid for the United States, illustrates a population with a large reproductive age group, but declining young. This pyramid reflects a decreased birthrate brought about by birth control and predicts a smaller reproductive age class later on. The rate of population growth in such a population is slow. The pyramid for Sweden in Figure 10.9 is narrow. The ratio between one age class and another is about the same. It suggests that the population is neither growing nor declining. It has reached zero population growth.

Although age structure may not be a reliable predictor of future population growth, it does provide insight into current population problems. For example, the increase in individuals in the older age classes in the United States portends social and economic changes and problems as the population ages. Coun-

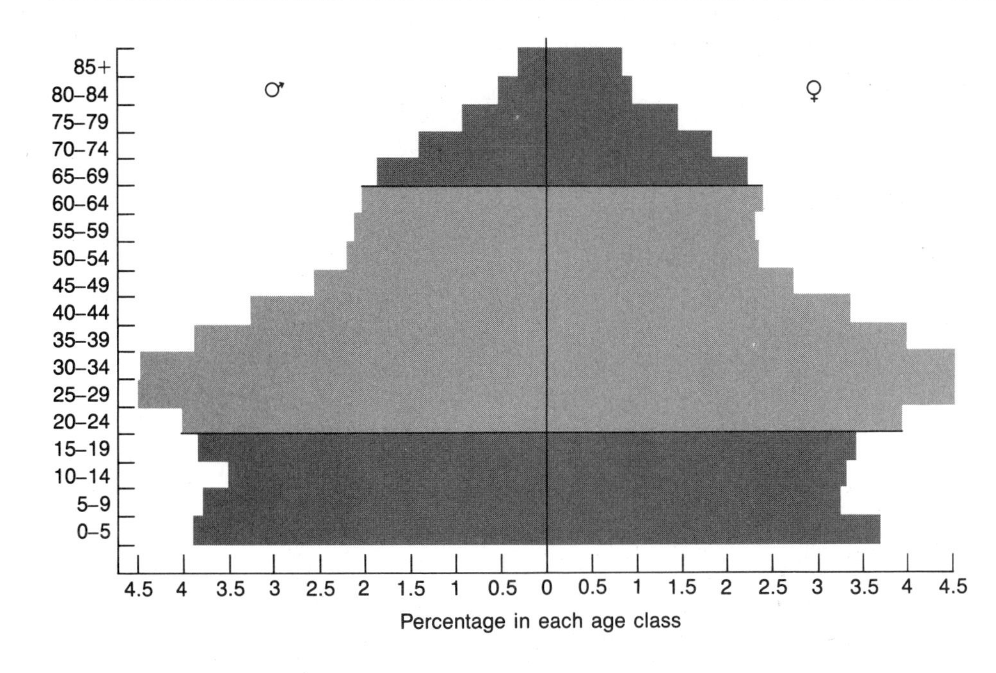

FIGURE 10.8 The age pyramid for the United States in 1988 exhibits a constrictive shape. The youngest age classes are no longer numerically the largest. This type of pyramid reflects declining fertility, which, if it persists, will in time further distort the age structure.

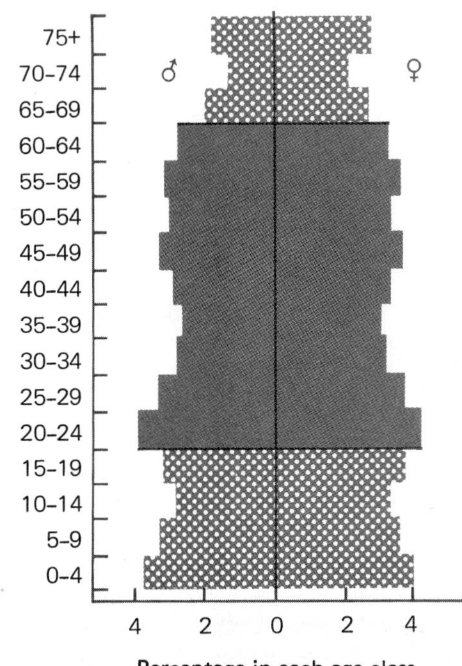

FIGURE 10.9 The age pyramid for Sweden in 1970 is characteristic of a population that is approaching zero growth.

tries faced with famine, political turmoil, and aggressive behavior typically have a high proportion of their population in the younger age classes. Analysis of age structure provides insights into developing population policies and decisions that have important ecological as well as social and political consequences.

PLANT POPULATIONS

Studies of the age structure of plant populations are few (Figure 10.10). The reasons are in part because demographic techniques have been late in coming to the study of plant populations and in part because of the difficulty of determining age when plants reproduce both sexually and asexually.

A modification of age structure has been used by foresters for years as one guide to timber management. They employed size (dbh, diameter at breast height) as an indicator of age on the logical assumption that diameter increases with age. The greater the diameter, the older the tree. Such assumptions, foresters discovered, were valid for dominant canopy trees; but with their growth suppressed by lack of light, moisture, or

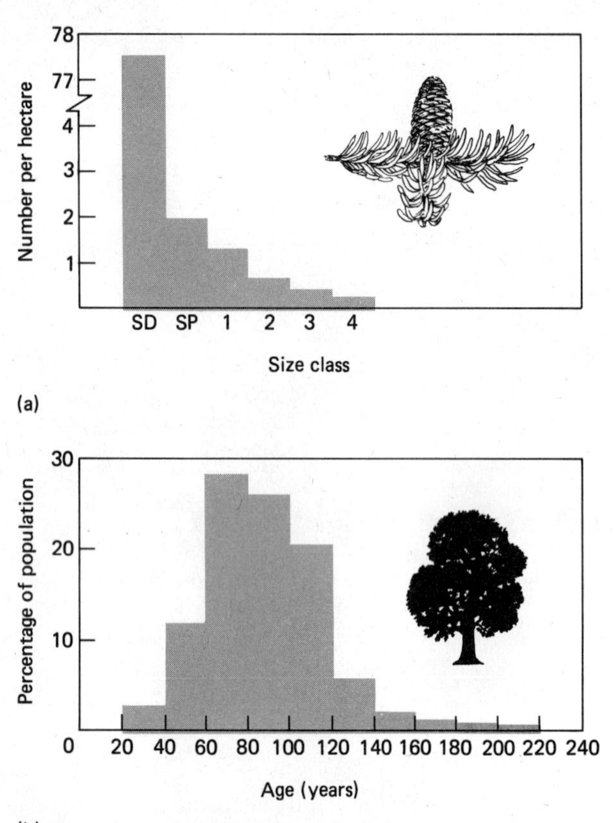

FIGURE 10.10 Age pyramids for two plant populations. (a) Balsam fir (*Abies balsamea*) stand on Lake Superior in Ontario. Size classes are: seedlings (SD), 1 to 3 cm diameter at breast height (dbh); saplings (SP) greater than 3 cm and less than or equal to 9 cm; and trees broken down in 7.62 cm diameter classes. This stand is young, with most of the population in younger age classes. The fewest trees are 30 cm (12 inches) dbh. (b) An oak (*Quercus*) forest in Sussex, England, is dominated by trees in the 60 to 100 year age class. There has been no recruitment for 20 years.

nutrients, understory trees and even so-called seedlings and saplings added little to their diameter size. Although their smaller diameter suggested youth, the small trees often were the same age as their large dominant associates.

Trees can be aged approximately by counting annual growth rings. Attempts to age nonwoody plants, especially those with corms that seem to exhibit rings of annual growth, have not been successful. The most accurate method of determining the age structure of relatively short-lived herbaceous plants is to mark individual seedlings and follow them through their lifetimes.

Because the growth and survival of juvenile trees and seedling reproduction are often inhibited by the competitive effects of the dominant overstory trees, the distribution of age classes in many forests is nonstable (Figure 10.10a). One or two age classes dominate the site until they die or are removed, allowing new young age classes to develop. There is a disproportionately higher number of old individuals, and when they succumb, there is a high local recruitment of young trees that in turn will dominate the site for years (Figure 10.10b).

Summary

A population is a group of individuals of the same kind living in the same place at the same time. The size of a population in relation to the area it occupies is its density. Populations and individuals are distributed in some kind of pattern over the landscape. Some are uniformly distributed, some are randomly distributed, but most are clumped in aggregations.

Individuals making up the population may be divided into three ecological periods: prereproductive, reproductive, and postreproductive. These periods and the age groups comprising them may be represented visually as age pyramids.

The distribution of individuals within each age group influences considerably the birthrate, mortality rate, and population growth. A large number of young about to enter the reproductive age suggests a potentially increasing population, whereas a high proportion of individuals in the postreproductive age classes suggests a zero or declining population growth. Populations tend toward a stable age distribution, in which the proportion of individuals in each age class remains the same, as long as growth continues at a constant rate. When deaths equal births and the proportion of individuals in the population remains constant, the population has arrived at a stationary age distribution.

Review and Study Questions

1. Distinguish between ecological density and crude density.
2. What are the three ways in which populations may be dispersed across the landscape?
3. What are the three age periods into which individuals in a population may be divided? Discuss their significance to a population.
4. What is meant by stable age distribution?
5. Contrast age structure in animal and plant populations. Explain the difference.

*6. Age pyramids, which represent the past of a population, often are used to predict its future. Think of some drawbacks to using present age structure to forecast future population developments. What can age pyramids tell us?

Selected References

Blower, J. G., L. M. Cook, and J. A. Bishop. 1981. *Estimating the size of animal populations.* London: George Allen & Unwin.

Brower, J. E., and J. H. Zar. 1982. *Field and laboratory methods for general ecology,* 2nd ed. Dubuque, IA: William C. Brown.

Caughley, G. 1977. *Analysis of vertebrate populations.* New York: Wiley.

Cox, G. W. 1985. *A laboratory manual of general ecology.* Dubuque, IA: William C. Brown.

Krebs, C. J. 1989. *Ecological methods.* New York: Harper & Row.

Mueller-Dombois, D., and H. Ellenberg. 1974. *Aims and methods of vegetation ecology.* New York: Wiley.

Nielsen, L. A., and D. L. Johnson, eds. 1983. *Fishery techniques.* Bethesda, MD: American Fisheries Society.

Schemnitz, S. D., ed. 1980. *Wildlife management techniques manual.* Washington, D.C.: The Wildlife Society.

Smith, R. L. 1990. *Student resource manual for ecology and field biology,* 4th ed. New York: Harper & Row.

Southwood, T. R. E. 1978. *Ecological methods.* London: Chapman and Hall.

CHAPTER 11

Mortality, Natality, and Survivorship

Outline

Objectives

On completion of this chapter, you should be able to:

1. Express mortality as the probability of dying or of surviving.
2. Understand the construction of a life table and the derivation and meaning of life expectancy.
3. Distinguish among several kinds of life tables.
4. Plot mortality and survivorship curves.
5. Distinguish among different types of survivorship curves.
6. Explain how net reproductive rates are determined.

Demography—the study of populations—mostly involves statistical review of the arrivals and departures of individual members. Some arrive by immigration or depart by emigration, but most comings and goings involve births (natality) and deaths (mortality). The difference between the two processes determines the growth or decline of a population.

MORTALITY

Mortality, which begins even in the uterus and the egg, is usually expressed as the probability of dying, or the mortality rate. It is obtained by dividing the number of individuals that died during a given interval of time by the number alive at the beginning of the period: $q = d_t/N_t$. The complement of the probability of dying is the probability of survival, the number of survivors at the end of a given time interval divided by the number alive at the beginning of the interval. Because the number of survivors is more important to a population than the number dying, mortality is better expressed either as the probability of surviving or as life expectancy, the average number of years to be lived in the future by members of the population.

To obtain a clear and systematic picture of mortality and survival, we can construct a *life table.* The life table, first developed by students of human populations and widely used by life insurance companies, is simply an account book of deaths. It consists of a series of columns headed by standard notations, each of which describes mortality relations within a population when age is considered. It begins with a group or cohort of a certain size, usually 1000 at birth or hatching (see Tables 11.1 through 11.4). The cohort of 1000 is obtained by converting data collected in the field to the equivalent numbers had the starting density of the cohort been 1000.

The columns include x, the units of age; l_x, the number of organisms in a cohort that survive to age x, $x + 1$, and so on; and d_x, the number or fraction of a cohort that dies during the age interval x, $x + 1$. The column d_x can be summed to give the number dying over a particular period of time. If l_x and d_x are converted into proportions—that is, if the number of organisms that died during the interval x, $x + 1$ is divided by the number of organisms alive at the beginning of age x, the result is q, the age-specific mortality rate. The q_x column, however, cannot be summed to give the overall mortality rate at any specific age.

Two other columns are L_x, the average years lived by all individuals in each age category, and T_x, the number of time units left for all individuals to live from age x onward. These two columns are used to calculate e_x, the life expectancy at the end of each interval. The values for L_x are obtained by summing the number alive at the age intervals x and $x + 1$ and dividing the sum by 2. T_x is calculated by summing all the values of L_x from the bottom of the table to the top. Life expectancy, e_x, is obtained by dividing T_x for a particular age class x by the l_x value for that age class.

VERTEBRATE LIFE TABLES

At one time data for life tables could be obtained only for laboratory animals and humans. As census methods and age determination techniques became more refined, sufficient data for at least an approximate life table could be acquired for some other species. It is difficult to obtain information on mortality and survival in wild animals. Mortality can be estimated by determining the ages at death of a large number of animals born at the same time. This procedure could involve the marking or banding of a considerable number of animals. Such a method provides information for the d_x column. We also can record the ages at death of animals marked at birth but not necessarily born during the same season or year. Data from several years and several cohorts are pooled to provide the information for the d_x column. Another approach is to determine the age of death of a representative sample of carcasses of the species concerned. Such information also goes into the d_x column. Recording the ages of death of a sample of a population wiped out by some catastrophe could provide data for an l_x series. Life tables derived from the aging of animals taken during a hunting season provide information for the l_x column because the sample came from a living population, but the data are biased in favor of older age classes, especially if the data are collected between breeding seasons.

TABLE 11.1 Dynamic Life Table for Red Deer Hinds on the Isle of Rhum, 1957

x *Age* *(Years)*	l_x *Survivors at Beginning of Age Class* x	d_x *Deaths*	*1000* q_x *Mortality Rate/1000*	e_x *Further Expectation of Life (Years)*
1	1000	0	0	4.35
2	1000	61	61.0	3.35
3	939	185	197.0	2.53
4	754	249	330.2	2.03
5	505	200	396.0	1.79
6	305	119	390.1	1.63
7	186	54	290.3	1.35
8	132	107	810.5	0.70
9	25	25	1000.0	0.50

There are two kinds of life tables. One is a cohort or *dynamic life table,* recording the fate of a group of animals all born at the same time. An example is Table 11.1. It is based on the number of red deer born in 1957, obtained from census data (the young are distinguishable from adults), and from samples of deer that died from 1957 to 1966.

The other type is the *time-specific life table,* in which the mortality of each age class in a given population is recorded over a year. It is constructed from a sample of animals of each class taken in proportion to their numbers in a population. It involves the assumption that the birth and death rates are constant and that the population is stationary. An example of a time-specific life table is Table 11.2, which is constructed from data on the age distribution of a population sampled in one year, 1957. The l_x schedule shows survival from year 1 rather than from birth.

TABLE 11.2 Time-Specific Life Table for Red Deer Hinds on the Isle of Rhum

x	l_x	d_x	*1000* q_x	L_x	T_x	e_x
1	1000	137	137.0	931.5	5188.0	5.19
2	863	85	97.3	820.5	4256.5	4.94
3	778	84	107.8	736.0	3436.0	4.42
4	694	84	120.8	652.0	2700.0	3.89
5	610	84	137.4	568.0	2048.0	3.36
6	526	84	159.3	484.0	1480.0	2.82
7	442	85	189.5	399.5	996.0	2.26
8	357	176	501.6	269.0	596.5	1.67
9	181	122	672.7	120.0	327.5	1.82
10	59	8	141.2	55.0	207.5	3.54
11	51	9	164.6	46.5	152.5	3.00
12	42	8	197.5	38.0	106.0	2.55
13	34	9	246.8	29.5	68.0	2.03
14	25	8	328.8	21.0	38.5	1.56
15	17	8	492.4	13.0	17.5	1.06
16	9	9	1000.0	4.5	4.5	0.50

INSECT LIFE TABLES

The life tables described so far are typical of long-lived animals in which generations overlap and in which different ages are alive at the same time. However, a tremendous number of animals are annual species. They have one breeding season, and generations do not overlap, so all individuals belong to the same age class. The l_x values can be obtained by observing a natural population over its annual season and estimating the size of the population at each time of observation. For many insects the l_x value can be obtained by estimating the size of the surviving population from egg to adult. If records are kept of weather, abundance of predators and parasites, and the occurrence of disease, deaths from various causes can also be estimated.

Table 11.3 represents the fate of a cohort from a single egg mass. The age interval, or x column, indi-

TABLE 11.3 Life Table of a Sparse Gypsy Moth Population in Southeastern New York

x	l_x	d_{xf}	d_x	*100* q_x
Eggs	450	Parasites	67.5	15
		Other	67.5	15
		Total	135.0	30
Instars I-III	315	Dispersion, etc.	157.5	50
Instars IV-VI	157.5	Parasites	7.9	5
		Disease	7.9	5
		Other	118.1	75
		Total	133.9	85
Pre-pupae	23.6	Desiccation, etc.	0.7	3
Pupae	22.9	Vertebrate predators	4.6	20
		Other	2.3	10
		Total	6.9	30
Adults	16.0		5.6	35
Generation	—	—	439.6	97.69

cates life history stages, which are of unequal duration. The l_x column indicates the number of survivors at each stage. The d_x column gives a breakdown of deaths by causes in each stage; d_{xf} is the cause. In this particular population dispersion and predation account for most of the losses. Note that no life expectancy is calculated because there is none. All the adult population will die in late summer.

PLANT LIFE TABLES

Mortality in plants is beginning to receive considerable conceptual treatment; but mortality and survivorship in plants are not easily condensed in life tables. To begin with, age is difficult to determine. Mortality of individuals usually stimulates growth of the survivors, increasing their biomass and the size of the *metapopulation*—buds, leaves, and stems. Seedlings make up a large numerical proportion of individuals but an extremely small proportion of the biomass and the metapopulation. In a plant population, size rather than age may be more important. Further, it is difficult to separate and even identify individuals. Are the plants really *genets* (genetic individuals) or *ramets* (asexual clones)? The "parent" plant may die, yet live on in sprouts and root suckers. The plant demographer has to deal with mortality (and natality) on two levels, the genet and the metapopulation.

The life table approach in the study of plant demography is most useful in studying (1) seedling mortality and survival, (2) population dynamics of perennial plants whose individuals can be marked as seedlings, and (3) life cycles of annual plants. An excellent example of the third type is Table 11.4. The time of seed formation is considered the initial point in the life cycle. The l_x column indicates the number of plants alive at the beginning of each stage and the d_x column the number dying. The L_x column gives the mean number of plants alive during the life cycle and the T_x column the total number of plants remaining to the members of the population at the beginning of each life cycle stage. Life expectancy of these annual plants dropped rapidly in the seed stages and returned to a high level after seedling establishment. Although individuals that became established had a good chance of surviving, the high early mortality resulted in a low mean life expectancy.

Another approach to the life table of plants is the yield table developed by foresters (Table 11.5). Like the life table for vertebrates, the yield table considers age classes and number of trees in each class, with additional columns giving diameter, basal area, and vol-

TABLE 11.4 Life Table for a Natural Population of *Sedum smallii*

x	D_x	A_x	A'_x	l_x	d_x	$1000q_x$	L_x	T_x	e_x
Seed produced	4	0–4	−100	1000	160	160	920	4436	4.4
Available	1	4–5	− 10	840	630	750	525	756	0.9
Germinated	1	5–6	+ 13	210	177	843	122	230	1.1
Established	2	6–8	+ 35	33	9	273	28	109	3.3
Rosettes	2	8–10	+ 81	24	10	417	19	52	2.2
Mature plants	2	10–12	+126	14	14	1000	7	14	1.0

ume. Yield tables chart the mortality of trees by the reduced number of individuals in each age class, but as the numbers decline, basal area and biomass increase. Mortality does not necessarily reflect a declining population but rather a maturing one. Like life tables, yield tables are not constant for a species; they are constructed for different site classes based on different environmental conditions.

MORTALITY AND SURVIVORSHIP CURVES

From the life table two kinds of curves can be plotted: mortality curves, based on the q_x column, and survivorship curves, based on the l_x column.

TABLE 11.5 Yield Tables for Douglas-Fir on Fully Stocked Hectare, Total Stand

	Site Index 200		
Age (Years)	*Trees Per (Ha)*	*Av. dbh (cm)*	*Basal Area (m^2)*
20	1427	14	23
30	875	23	35
40	600	30	46
50	440	38	53
60	345	46	58
70	282	53	62
80	242	59	66
90	210	65	69
100	187	70	72
110	172	75	75
120	157	80	77
130	147	83	79
140	137	87	81
150	127	91	83
160	120	94	85

MORTALITY CURVES

Mortality curves (Figure 11.1) plot mortality rates directly in terms of $1000q_x$ against age. They consist of two parts: (1) the juvenile phase, in which the rate of mortality is high; and (2) the postjuvenile phase, in which the rate decreases as age increases until mortality reaches some low point, after which it increases again. For mammals a roughly J-shaped curve results. For plants the mortality curve may assume a number of patterns, depending upon the type of plant, annual or perennial, and the method used to plot the data.

SURVIVORSHIP CURVES

Survivorship curves may be plotted in a number of ways. The usual method is to plot the logarithmic number of survivors, the l_x column, against time, with the time interval on the horizontal coordinate and survivorship on the vertical coordinate. Another method is to plot survivorship against time intervals scaled as percentage deviations from mean length of life. That method allows the direct comparison of survivorship curves of organisms that have very different life spans.

The validity of survivorship curves depends upon the validity of the life table. Life tables, and so survivorship curves, are not typical of some standard population; instead, they depict the nature of a population at particular places and times under different environmental conditions. For this reason survivorship curves are useful for comparing the population of one time, area, or sex with the population of another.

Survivorship curves fall into three general types (Figure 11.2). If mortality rates are extremely high in early life—as in oysters, fish, many invertebrates, and some plants—the curve is concave, called Type III. If

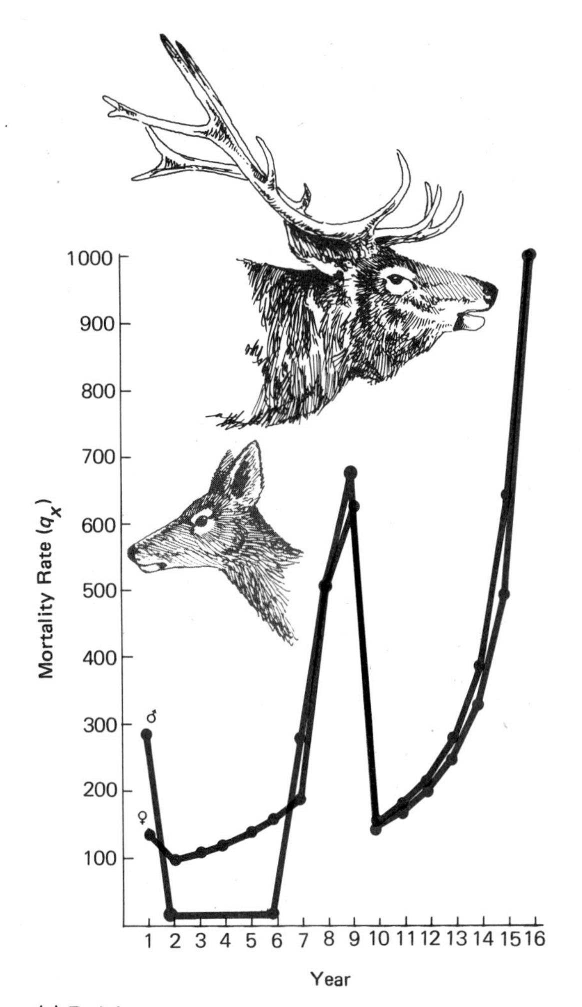

(a) Red deer

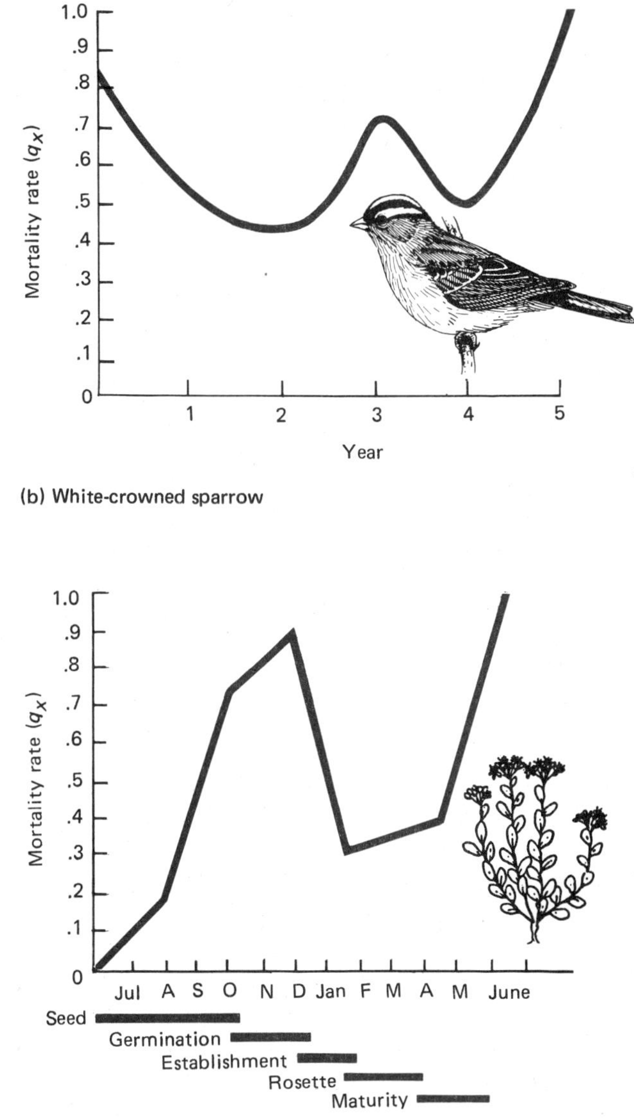

(b) White-crowned sparrow

(c) Sedum

FIGURE 11.1 Examples of mortality curves. All are roughly J-shaped. The break in the curves results from a sharp rise in mortality for the red deer (a) and white-crowned sparrow (b). Sedum (c) has two J-shaped curves, one for seeds through germination and another for the established plants through youth (the rosette phase) and maturity.

mortality rates are constant at all ages, the survivorship curve will be linear, or Type II. Such a curve is characteristic of the adult stages of birds, rodents, and reptiles as well as many perennial plants. When individuals tend to live out their physiological life spans and when there is a high degree of survival throughout the species life span followed by heavy mortality at the end, the curve is strongly convex, or Type I. Such a curve is typical of humans and other mammals and some plants.

The generalized survivorship curves are models to which survivorship of a species can be compared (Fig-

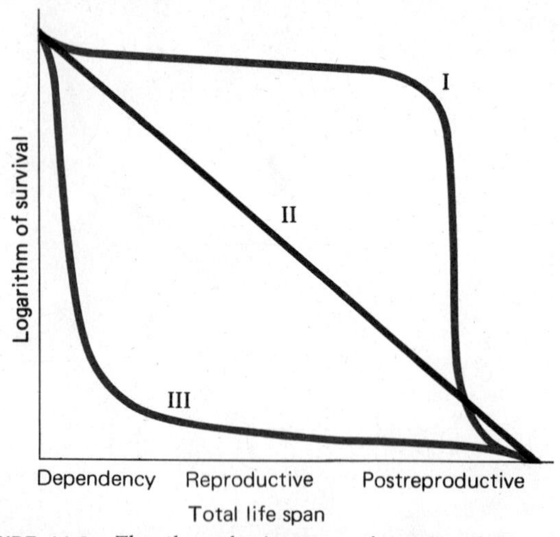

FIGURE 11.2 The three basic types of survivorship curves. Type I, curve for organisms living out the physiological life span of the species; Type II, curve for organisms in which the rate of change in mortality is fairly constant at all age levels, a more or less uniform percentage decrease in the number that survive; Type III, curve for organisms with a high mortality early in life.

ures 11.3 and 11.4). Most survivorship curves are intermediate between models.

Mortality in subunits or metapopulations of plants—the buds, leaves, shoots, twigs, and roots—may also be analyzed by the use of survivorship curves (Figure 11.5). Knowledge of leaf mortality is important in estimating plant production (see Chapter 22). Failure to allow for the "birth" of new leaves and the "death" or turnover of mature leaves can make estimates too low.

NATALITY

The greatest influence on population increase is usually natality, the production of new individuals. Natality may be expressed as either a crude birthrate or a specific birthrate. If the number of births in a given period, such as a year, is divided by the estimated population at the midpoint of the period and the result multiplied by 1000, the figure represents the *crude birthrate*, expressed as births per 1000 population per unit of time. A more precise way of expressing the birthrate is the number of births per female of age x per unit of time, because reproductive success varies with age. If females of reproductive age are divided arbitrarily into age classes and the number of births for each age class is tabulated, we get an *age-specific schedule of births*. Because population increases are a function of the female, the age-specific birth schedule can be modified further by determining only the mean number of females born in each female age group. This information is known as the *gross reproductive rate*. It contrasts with the *net reproductive rate, R*, the number of females left during a lifetime by a newborn female or the mean number of females born in each female age group. Because it is calculated by multiplying the gross reproductive rate, m_x, by the survival of each class, l_x, it includes adjustments for mortality in females in each age group.

How net reproductive rates are determined is demonstrated in the fecundity table for white-crowned sparrows (Table 11.6). The fecundity table uses the survivorship column, l_x, from the life table and an m_x column, the mean number of females born to females in each age group. To calculate the net reproductive rate, the l_x values are converted to a proportionality, in this case by dividing each age value by 1000 females of the birth year. Age 0 (x column) females lay no eggs; therefore, their m_x value is 0. The m_x value for a female aged 1 year is 3.142, and the m_x values increase with age. To adjust for mortality, the m_x values are multiplied by the corresponding l_x, or survivorship, values. The resulting value, $l_x m_x$, gives the mean number of females born in each age group adjusted for survivorship.

Thus, for age class 1 the m_x value is 3.142, but when adjusted for survival, the value drops to 0.525, and for age 5 the m_x value of 4.0 drops to 0.024, reflecting poor survival of the adult females. If you multiply the proportion of females living from age x to age t, l_t/l_x, by m_x, you obtain the **reproductive value** of each age group, v_x. Note that in spite of their lower m_x value, females age 1 have the highest reproductive value. When the adjusted m_x values, $l_x m_x$, are summed over all ages at which reproduction occurs, the sum represents the number of females that will be left during a lifetime by a newborn female, or R_0. If the R_0 value is 1, the female has replaced herself. If the value is less

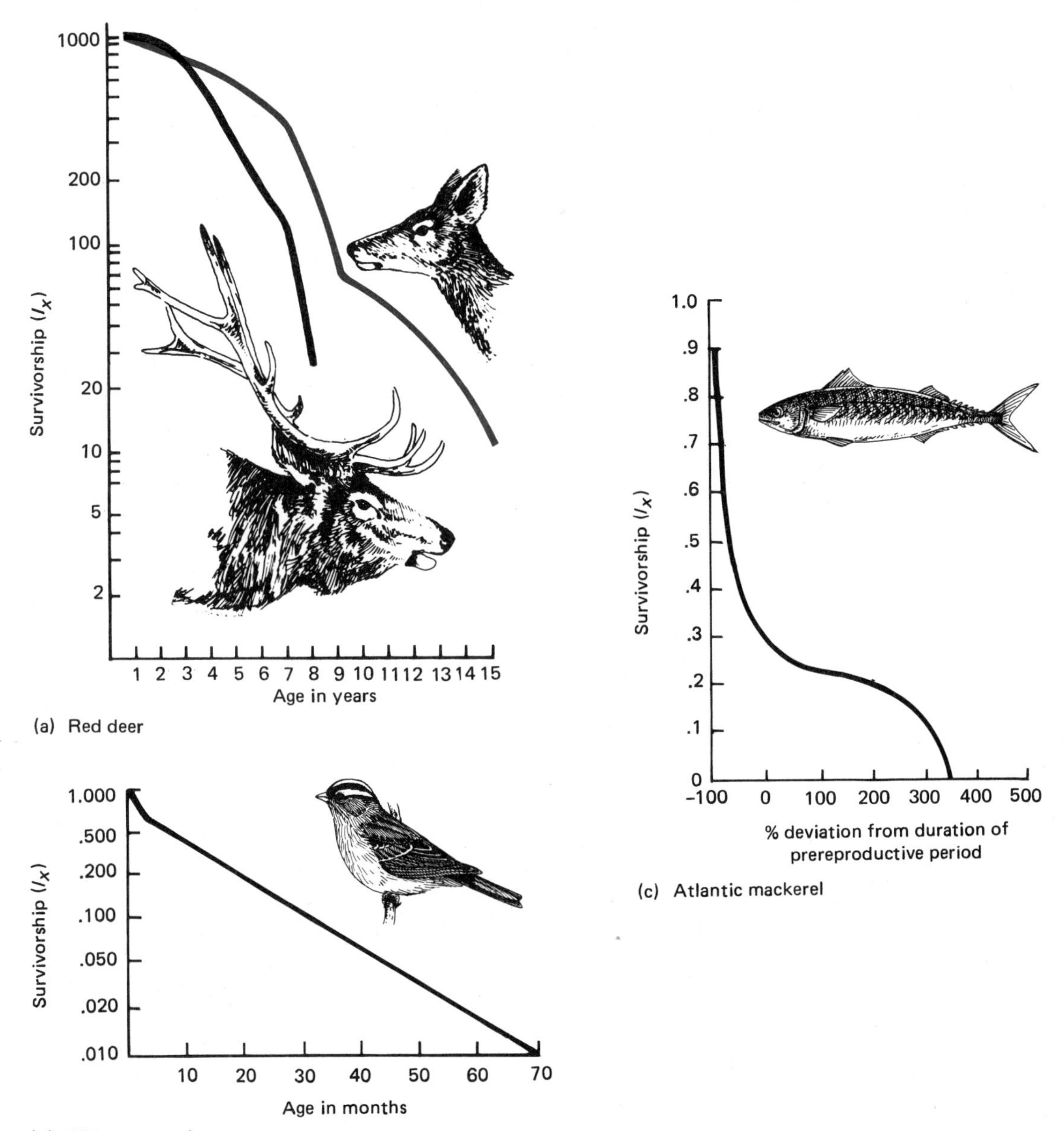

FIGURE 11.3 Examples of survivorship curves for animals. (a) The red deer has a Type I form, tending to approach physiological longevity. Note the greater survivorship of the females. (b) The white-crowned sparrow possesses a Type II curve, seemingly typical for birds. (c) The Atlantic mackerel exhibits a Type III curve.

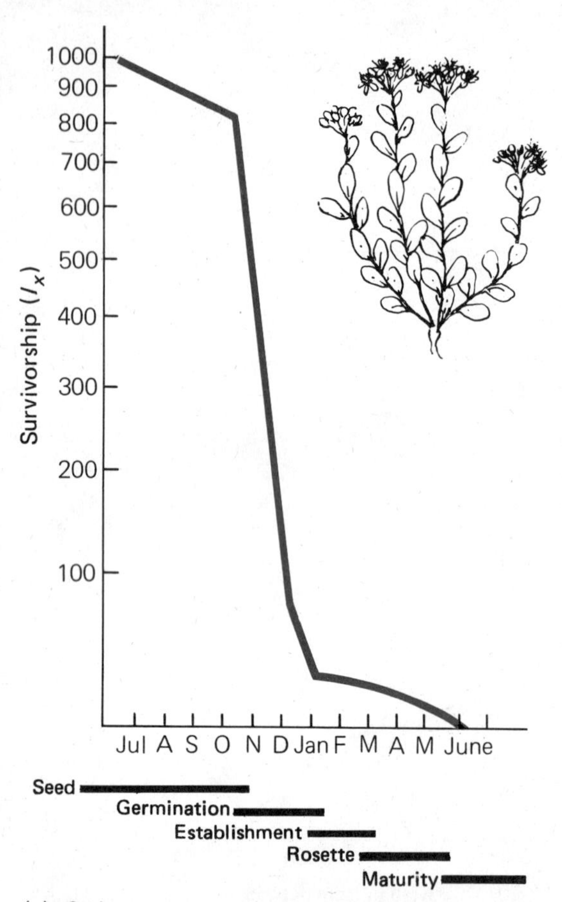

(a) *Sedum*

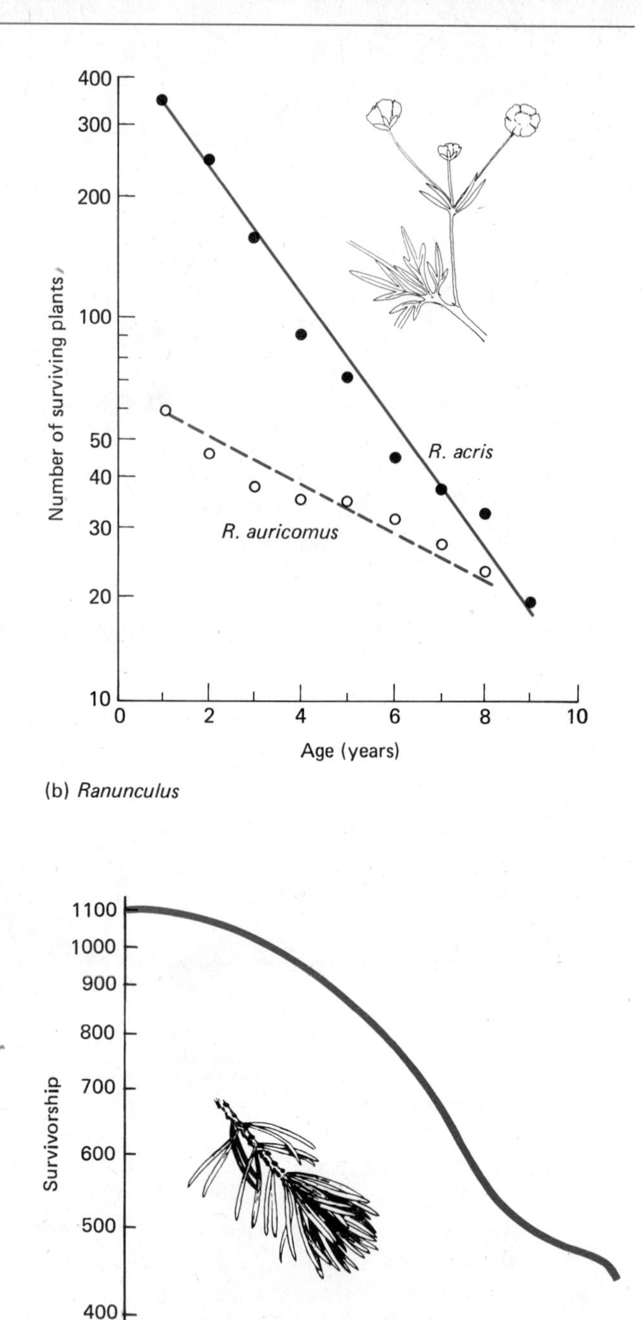

(c) *Pinus banksiana*

FIGURE 11.4 Survivorship curves for some plants. (a) *Sedum smallii* suggests elements of a Type I curve; (b) two species of buttercup (*Ranunculus*) exhibit a Type II curve; and (c) jack pine (*Pinus banksiana*) has a Type I curve.

than 1, the female is not replacing herself. If the value is much over 1, she is leaving additional offspring behind.

Another fecundity table is Table 11.7. Reproduction in the example of a red deer begins with the 3-year-olds and increases until age 7, when a decline in mean number of births begins. For a red deer population with the survivorship indicated, the net reproductive rate is 1.316. Fecundity in the white-crowned sparrows (Table 11.6) increases with age; the net reproductive rate is 1.042.

Natality in plants, like mortality, is perplexing be-

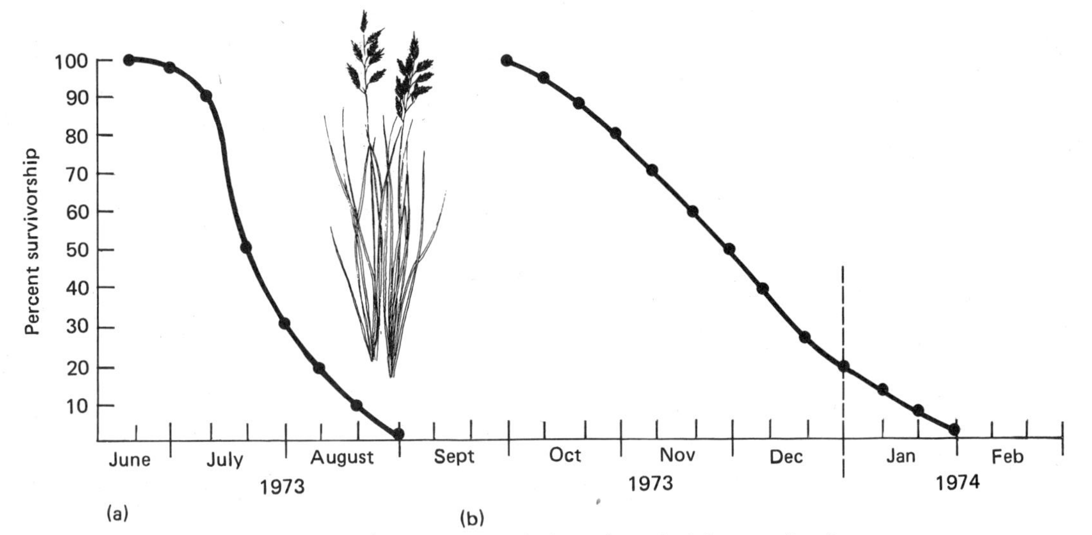

FIGURE 11.5 Survivorship curves for a metapopulation of marked leaves of red fescue (*Festuca rubra*) at two different times of year, (a) summer and (b) late summer extending into winter.

cause plants reproduce both sexually and asexually. If you consider only the genet, or genetic individual, then natality is restricted to sexual reproduction. Involved are two separate populations, seeds and seedlings, and two separate processes, the production of seeds and the germination of seeds. Except for annuals and biennials, which have one reproductive effort resulting in the death of the parent plant, seed production by individual plants is hard to estimate. Woody plants and other perennials, even within a population, vary in longevity and in seed production over a period of years as well as in the ability of the seeds to germinate.

The formal equivalent of birth in plants is germination. Before "birth," seeds usually undergo a varying period of dormancy, often necessary before they can

TABLE 11.6 Fecundity Table for Female White-Crowned Sparrows

x	l_x	m_x	l_xm_x	xl_xm_x*	v_x
0	1.000	0	0	0	1.042
1	0.167	3.142	0.525	0.525	6.234
2	.083	3.333	.277	0.554	6.221
3	.048	3.556	.171	0.513	4.994
4	.012	3.750	.045	0.180	5.750
5	.016	4.000	.024	0.120	4.000
		$R_0 =$	1.042		
		Total weighted age		1.892	

* Data to be used in Chapter 12.

TABLE 11.7 Fecundity Table for Red Deer

x	l_x	m_x	l_xm_x	xl_xm_x
1	1.000	0	0	0
2	.863	0	0	0
3	.778	0.311	0.242	0.726
4	.694	.278	.193	.772
5	.610	.308	.134	.667
6	.526	.400	.210	1.26
7	.442	.476	.210	1.47
8	.357	.358	.128	1.024
9	.181	.447	.081	0.729
10	.059	.289	.017	.170
11	.051	.283	.014	.154
12	.042	.285	.012	.144
13	.034	.283	.010	.130
14	.025	.282	.007	.098
15	.017	.285	.005	.075
16	.009	.284	.003	.048
		$R_0 =$	1.316	
		Total weighted age		7.467

sprout. The seeds of some plants remain dormant for years, buried in soil or mud as a seed bank, until they are exposed to conditions conducive to germination. Once the seed has germinated, the seedling is subject to mortality. Thus, the plant population at all times consists of two parts, one growing and producing seeds and the other stored as seeds in a dormant state.

Among many plants, though, genetic individuals make up only a fraction of the plant population. Also involved are ecologic individuals, the ramets, and populations of buds, leaves, shoots, and sprouts, some connected to the "parent" plant and others living independently. Do asexually reproduced plants represent natality? Genetically they are part of the original individuals, even though they eventually may produce their own seeds. Can you count them as new individuals? Also, does the production of new growth by a plant—the development of buds, the growth of new shoots, the unfolding of new leaves—represent natality at the level of the metapopulation, just as the death of buds and leaves represents mortality? Both have a pronounced influence on the plant itself. Such questions are yet to be resolved.

Summary

Population size is influenced by the number of individuals added to the group by births and immigration and by the number leaving by death and emigration. The difference between the two determines the growth and decline of populations.

Mortality, concentrated in the young and the old, is the greatest reducer of populations. Mortality and its complement, survivorship, are best analyzed by means of a life table, an age-specific summary of mortality operating on a population.

From the life table we can derive both mortality curves and survivorship curves. They are useful for comparing demographic trends within a population and among populations living under different environmental conditions and for comparing survivorship among various species. In general, mortality curves assume a J shape. Survivorship curves fall into one of three major types: Type I, in which survival of young is low; Type II, in which mortality, and thus survivorship, is constant through all ages; and Type III, in which individuals tend to live out their physiological life spans. Survivorship curves follow similar patterns in both plants and animals.

Survivorship and mortality in plants are complicated by the survivorship and mortality of parts of plants. The metapopulation of leaves, twigs, and clones also exhibits its own demographic characteristics.

Birth has the greatest influence on population increase. Like deaths, births are age-specific. Certain age classes contribute more to the population than others. Reproductive values of various age groups as related to their survival can be determined from fecundity tables derived from the life table.

Like mortality, natality in plants is complicated by vegetative reproduction and sexual reproduction through seed production and germination. In general, the latter are considered true natality because they involve the production of genets, or genetic individuals. Again, clones, leaves, buds, and twigs exhibit their own "birth" rates.

Review and Study Questions

1. What are mortality, natality, survivorship, and fecundity? How are they interrelated?
*2. What life history characteristics influence birthrates? (See R. L. Smith, *Ecology and field biology*, 4th ed.)
3. What is a life table? What information is needed for its construction?
4. What are the advantages and weaknesses of a life table in a study of population dynamics?
5. Distinguish between a mortality curve and a survivorship curve. From which columns of the life table are they derived?
6. Why are birth and natality more difficult to study in plants than in animals?

Selected References

Caughley, G. 1977. *Analysis of vertebrate populations*. New York: Wiley.

Harper, J. L. 1977. *The population biology of plants.* London: Academic Press. The major reference on plant population biology.

Krebs, C. 1985. *Ecology: The experimental analysis of distribution and abundance.* New York: Harper & Row. See Chapter 11.

Smith, R. L. 1990. *Ecology and field biology,* 4th ed. New York: Harper & Row. See Chapter 14.

Also see Suggested References at the end of Chapter 10.

CHAPTER 12

Population Growth

Outline

Objectives

On completion of this chapter, you should be able to:

1. Distinguish between exponential and logistic growth.
2. Understand the relationship between net reproductive rate, finite rate of increase, and annual rate of increase.
3. Explain what is meant by carrying capacity and its relationship to the logistic growth curve.
4. Discuss cycles and irregular fluctuations in populations.
5. Discuss reasons why populations become extinct.

Mortality and natality are the two major forces influencing population growth. If births exceed deaths, the population increases. If births equal deaths, the population remains the same. If deaths exceed births, the population is headed for extinction. Two additional influences on population growth are immigration, an influx of new individuals into a population, and emigration, the dispersal of individuals from a population.

GROWTH POTENTIAL

The rate at which populations change can be estimated from R_0, the net reproductive rate, determined from the fecundity table (Chapter 11). R_0 is the expected number of female offspring an average newborn female will produce during her lifetime, assuming discrete generations.

Many populations, including invertebrates, plants, and all vertebrates, have overlapping generations. The parental generation continues to contribute to population growth, although at a reduced rate, while their offspring are reproducing. In this case we have to convert generation time T to mean cohort generation time T_c by adding the product of age (x) and the number of expected offspring produced per age ($l_x m_x$) to give a total weighted age, as we did in Tables 11.6 and 11.7. This total weighted age divided by R_0, the net reproductive rate, gives us the *mean cohort generation time,* T_c. For the red deer the mean generation time is 5.67 years, and for the white-crowned sparrow it is 1.8 years. The red deer produce 1.32 offspring in an average of 5.67 years and the white-crowned sparrows produce 1.04 offspring in an average of 1.8 years.

Because it is much more useful to compute population growth by year than by generation, we convert R_0 to an *annual finite rate of increase,* designated by the Greek letter lambda (λ):

$$\lambda = R_0^{1/T_c}$$

For the red deer:

$$\lambda = 1.32^{1/5.67}$$
$$\lambda = 1.32^{0.176} = 1.05$$

For the white-crowned sparrow the annual rate of increase is 1.02 (Table 12.1). Notice that the annual rate of increase is lower than the generation rate of increase.

TABLE 12.1 Growth Rate Parameters

Parameter	*Red Deer*	*White-crowned Sparrow*
T_c	5.67	1.8
R_0	1.316	1.042
$R = \lambda$	1.05	1.023
r	0.048	0.022

When $R = \lambda = 1$, females are replacing themselves, and the population remains the same. When λ is greater than 1, the population increases. How fast it increases is revealed by how much λ exceeds 1. When λ is less than 1, the population is declining. The populations of red deer and white-crowned sparrows, as depicted by the life tables, are growing slowly.

EXPONENTIAL GROWTH

The general equation for population growth is $N_t = N_0\lambda$, in which N_t is the population size at some given time in the future, N_0 is the initial population, and λ is the annual rate of increase. To determine the population growth of the red deer, as an example, you might start with an initial population of $N_0 = 100$ and $\lambda = 1.05$.

The equation can be stated as $N_{t+1} = N_t\lambda$, or more concisely, $N_t = N_0\lambda^t$, in which λ is raised to the power of the appropriate time interval. For example, $N_2 = N_0\lambda^2$. Some growth rates are given in Table 12.2.

If you lack sufficient life table data to construct a fecundity table, you can estimate annual rate of increase, λ, from the ratio of numbers at successive time intervals, provided you can obtain sufficient census data:

$$\lambda = \frac{N_{t+1}}{N_t}$$

For example, a mule deer herd in Colorado over three years had annual populations of 10,449, 10,702, and 11,153. Lambda in this case would be $N_2/N_1 = 10{,}702/10{,}449 = 1.02$; and $N_3/N_2 = 11{,}153/10{,}702 = 1.04$.

The equation $N_t = N_0\lambda^t$ describes a population that grows exponentially, like compound interest. Such

TABLE 12.2 Exponential Growth of Hypothetical Populations ($N_t = N_0R_0^t$; $N = 100$)

Year	$\lambda = 1.51$	$\lambda = 1.32$	$\lambda = 1.04$	$\lambda = 0.887$
0	100	100	100	100
1	151	132	104	89
2	228	174	108	79
3	344	230	112	70
4	519	304	117	62
5	785	400	122	55
6	1185	529	126	48
7	1789	689	132	43
8	2702	922	136	38
9	4081	1216	142	34
10	6162	1606	148	30

growth can occur when λ is greater than 1, the environment remains constant, and there are extra resources. The examples of exponential growth in Table 12.2 are plotted in Figure 12.1. Note how the shapes of the curves vary with the value of λ. The closer λ comes to unity, the slower the growth. The population with value of λ = 1.04 is barely replacing itself, and the population with the value below 1 is declining rather than increasing exponentially.

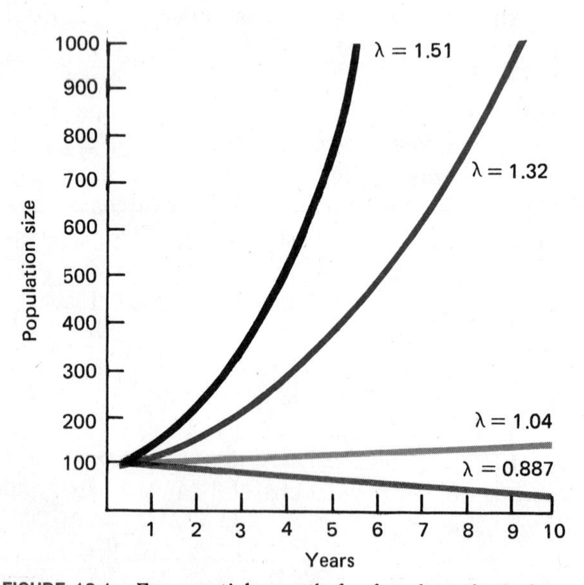

FIGURE 12.1 Exponential growth for four hypothetical populations with different values of λ.

These growth curves suggest several features of population growth. It is influenced by heredity and by life history features, such as age at the beginning of reproduction, the number of young produced, survival of young, and length of the reproductive period.

A population may increase at an exponential rate until it overshoots the ability of the environment to support it. Then the population declines sharply from starvation, disease, or emigration. From a low point the population may recover to undergo another phase of exponential growth, it may decline to extinction, or it may recover and fluctuate about some level far below the high level once reached.

The J-shaped or exponential curve is characteristic of some invertebrate and vertebrate populations introduced into a new or unfilled environment. An example of an exponential growth curve is the rise and decline of a human population in southwestern West Virginia (Figure 12.2). Growth between 1830 and 1950 was exponential. Between 1950 and 1960 the population declined dramatically; and except for a rise between 1970 and 1980, the decline continued to 1990. The decline produced a curve typical of a population that exceeds the carrying capacity of the environment. Growth stops abruptly and declines sharply in the face of environmental deterioration. As coal resources decline and mined land is left unfit for other uses, there is a wave of emigration.

Globally, human populations are experiencing exponential growth. Some Western nations have zero growth or modest population decline; but most Third

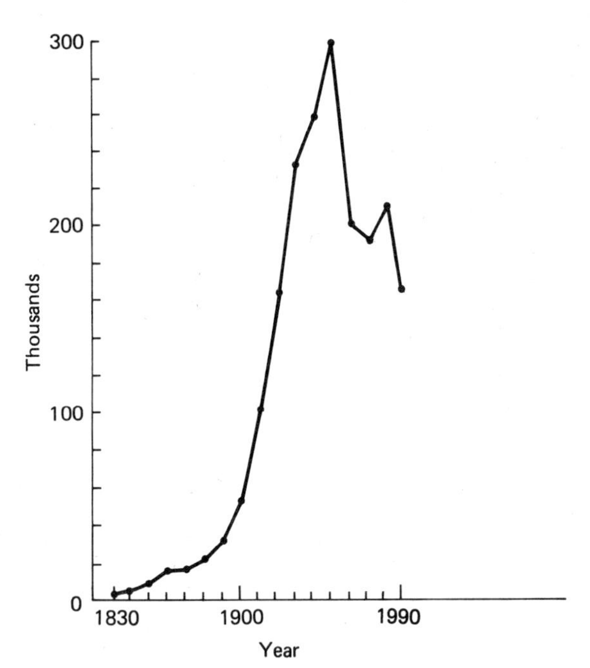

FIGURE 12.2 Population growth curve for five southwestern West Virginia counties, whose economy depends upon coal. Note the exponential-type growth, spurred by immigration, followed by a sharp population decline.

World countries in Africa, Asia, and Latin America are experiencing growth rates of 1.8 to 2.9 percent. Already the environment is unable to support such huge populations. Ultimately in one manner or another—birth control, war, famine, disease, or mass migration—they will decline sharply, because no environment can support sustained exponential growth.

RATE OF INCREASE

The finite rate of increase, lambda, can be expressed as the *rate of increase, r,* which is more commonly used in population studies. The rate of increase is obtained by taking the natural log of λ. Thus for the red deer $r = \log_n 1.05 = 0.048$; and for the white-crowned sparrow $r = 0.022$. Lambda, then, is often expressed as e^r where e is the base of natural logarithm, 2.71828. For the red deer $e^r = 2.71828^{0.048} = 1.049$ or 1.05.

The determination of r by this method results in an approximation of the actual rate of increase only. To determine the value of r_m, the intrinsic rate of increase, requires a more complex mathematical calculation.

The rate of increase depends upon the exponential rate at which a population grows if it has a stable age distribution appropriate to current life table parameters of age, survival, and fecundity. It also depends upon mean fecundity and mean survival at each age in the population. Because age structure is seldom stable and fecundity and survival vary over time, r, like λ, reflects the past and not the present.

Nevertheless, the use of r has advantages. When population growth is measured as r, it has the same value as an equivalent rate of population decrease. Consider the declining population with $\lambda = 0.887$. If the population were increasing at the same rate, then the value of λ would be 1.127. It is hard to see that $\lambda = 0.887$ is the inverse of $\lambda = 1.127$. It is easy to see the connection between $r = +0.120$ and $r = -0.120$. Thus, r allows a direct comparison of rates and converts easily from one form to another. In addition, r allows the computation of doubling time.

Doubling time is the time required for a population to double its size from a given base. If $N_t/N_0 = 2$, then $e^{rt} = 2$, $rt = \log_n 2 = 0.6931$. Therefore doubling time is $0.6931/r$. The doubling time for the red deer population is 0.6931/0.049 or 14 years. For the white-crowned sparrow it is 31 years. The same method is used to deterine the doubling time of human populations.

With some of the mystery removed from r, we will use the term from this point on. To present exponential growth in terms of r, you would use the equation $N_t = N_0 e^{rt}$. Thus for the red deer at time $t = 4$, assuming an initial population at N_0 as 100:

$$N_4 = 100\,(2.71828^{0.048 \times 4})$$
$$= 100\,(2.71828^{0.192}) = 121$$

LOGISTIC GROWTH

For populations in the real world, the environment is not constant and resources are not unlimited. As the density of a population increases, competition among its members for available resources also increases. With fewer resources to share and with an unequal distribution of those resources available, mortality increases, fecundity decreases, or both occur. As a result, population growth declines with increasing density, eventually reaching a level at which population growth ceases. That level is called *carrying capacity,* or K. Theoretically, at K the population is in equilibrium with its resources or environment. In other words, population growth is density-dependent, in contrast to exponential growth, which is independent of population density.

Inhibitions on the growth of a population by competition among its members for available resources can be described mathematically by taking the exponential equation $N_t = N_0e^{rt}$ and adding to it a variable to account for the effects of density. The braking effect of population growth can be described by:

$$\frac{K - N}{K}$$

In this formula as N approaches K, the value of the variable decreases toward 0.

Then:

$$\frac{dN}{dt} = rN\left(\frac{K - N}{K}\right) = rN\left(1 - \frac{N}{K}\right)$$

in which dN/dt represents the instantaneous rate of change in population density N, K is carrying capacity and $(K - N)/K$ is the unutilized opportunity for population growth. As the population grows, this unutilized opportunity declines. This equation describes a logistic, sigmoidal, or S-shaped growth curve (Figure 12.3).

As an example of the logistic growth rate consider a hypothetical population with a starting size of 100 and a rate of increase r of 0.412 ($\lambda = 1.51$). It will increase like this:

Year	Size
0	100
1	134
2	173
3	214
4	253
5	289
7	342
9	372
12	391
16	398
20	399.6
25	399.9
36	400

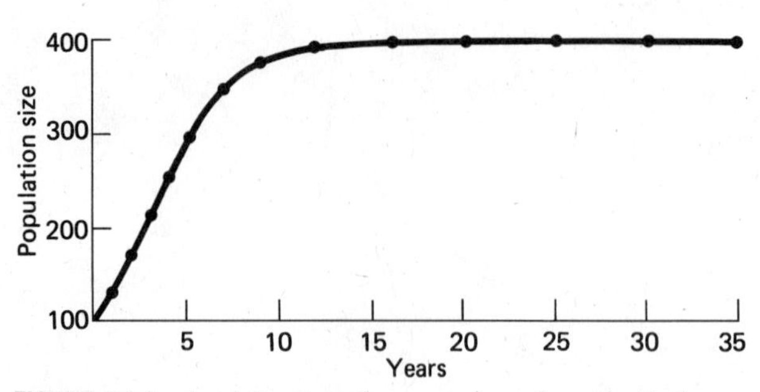

FIGURE 12.3 Logistic growth curve for a hypothetical population.

Note how the rate of increase is slow at first, accelerates, and then slows. The point in the logistic growth curve where growth is maximal is $K/2$, known as the *inflection point*. From this point on, population growth slows. As population density approaches carrying capacity, N approaches K and the rate declines.

The logistic growth curve is theoretical, a mathematical model of how populations might grow under favorable conditions. Although natural populations might appear to grow logistically, they rarely do.

For example, consider the growth of a white population of European descent in Monroe County, West Virginia, which has always had a stable economic base of agriculture and small industry. It was settled by Europeans in the early 1700s and was well established in 1800, when the first U.S. Census was taken. The population grew most rapidly between 1800 and 1850, so those years provide the data to estimate r. The population reached 13,200 in 1900 and has fluctuated about that number since that year. The rate of increase r was calculated as 0.074 and K was set at 13,200. Although the actual growth curve of the population mimics the logistic curve, the logistic curve rose much more steeply than the actual growth curve and predicted that the population would reach K around 1870, 30 years before it actually did (Figure 12.4).

The reasons for nonconformity are obvious. The age structure was not stable, birthrates and death rates varied from census period to census period; immigration and emigration were common to the population. The most surprising feature of the population is its relative stability after reaching K.

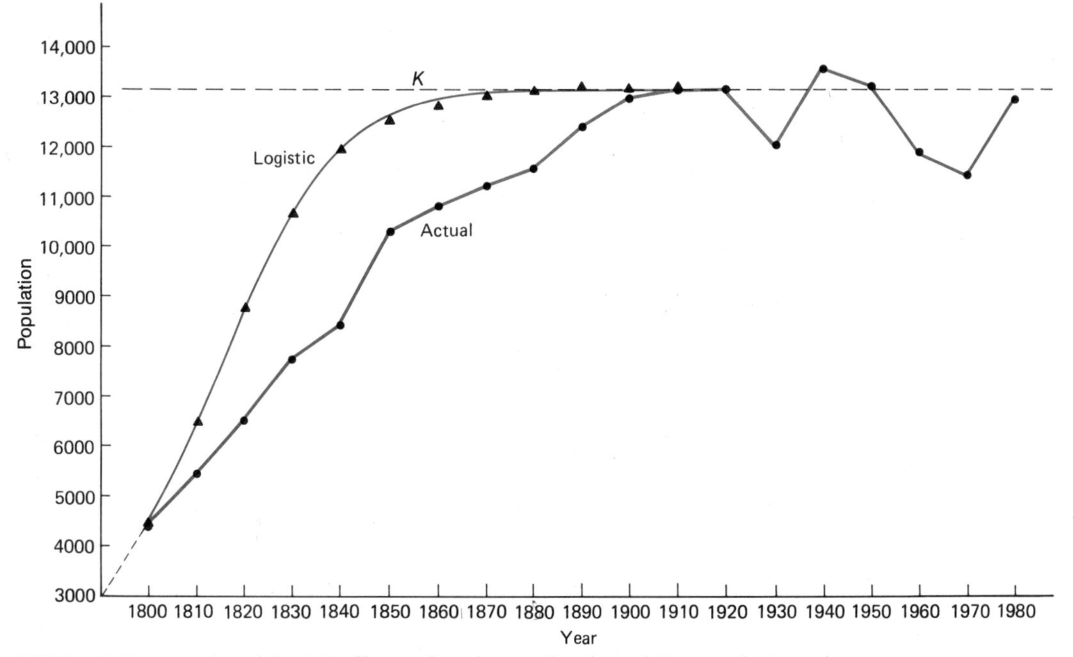

FIGURE 12.4 Actual and logistically predicted growth of a white population of European descent in Monroe County, West Virginia.

FLUCTUATIONS

The logistic equation suggests that populations function as systems regulated by feedback. Growth results from positive feedback (illustrated by the exponential curve) and slows with the negative feedback of competition and dwindling resources. As the population approaches the upper limit of the environment, it theoretically responds instantaneously, as density-dependent reactions set in. Rarely does such feedback work as smoothly in practice as the equation suggests. Often adjustments lag, and available resources may be sufficient to allow the population to overshoot equilibrium. Unable to sustain itself on the available resources, the population then drops, but not before it has altered resource availability to future generations. The density of the previous generation and the recovery of resources, especially the food supply, build a time lag into population recovery.

Time lags result in fluctuations in the population. The population may fluctuate widely without any reference to equilibrium size. Such populations may be influenced by some powerful outside force such as weather or by some chaotic changes inherent in the population. A population may fluctuate about the equilibrium level, K (derived as the mean size of the population over time), rising and falling between upper and lower limits. Some populations oscillate between high and low points in a manner more regular than we would expect to occur by chance. These fluctuations are called cycles.

The two most common oscillation intervals in animal populations are 9 to 10 years, typified by the lynx and snowshoe hare, and 3 to 4 years, typified by lemmings (Figure 12.5). These cyclic fluctuations are confined largely to simpler ecosystems, such as the northern coniferous forest and tundra. Cycles in the snowshoe hare, in part, involve an interaction between the hare and its overwinter food supply, mostly small aspen twigs. Overutilization of the food supply by a growing population of hares reduces the ability of the plants to recover from excessive pruning. Ultimately,

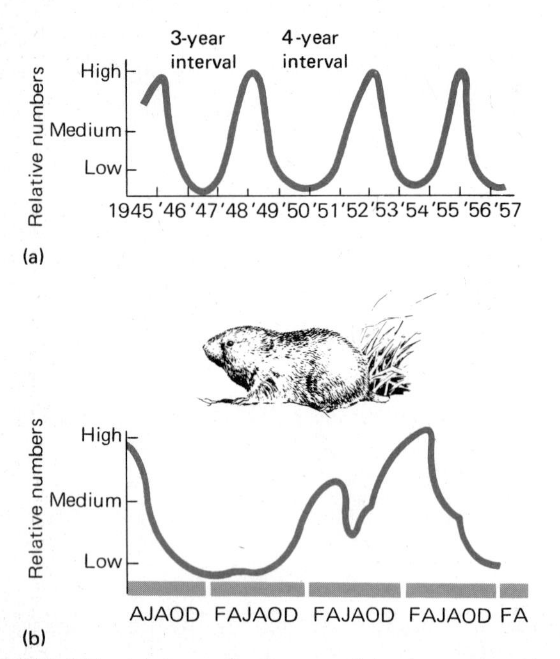

FIGURE 12.5 (a) Generalized curve of the four-year cycle of the brown lemming (*Lemmus sibiricus*) near Barrow, Alaska. (b) Generalized curve of a single four-year oscillation, showing subordinate fluctuations. The letters stand for months April, July, August, October, December, and February.

the decreased plant growth triggers an overwinter food shortage, resulting in a heavy mortality of hares. While the hare population is low, the vegetation recovers, stimulating a resurgence of the hare population and initiating another cycle.

EXTINCTION

As the growth curves suggest, population growth is maximum when there are neither too many nor too few individuals. This *Allee effect* is shown in Figure 12.6. When the population falls below or exceeds these points, the rate of increase declines. Increasing sparseness is associated with a reduction in the rate of increase. The population may become so low that r becomes negative and the population declines toward extinction.

The heath hen is a classic example. Formerly abundant in New England, this eastern form of prairie chicken was driven eastward by excessive hunting to

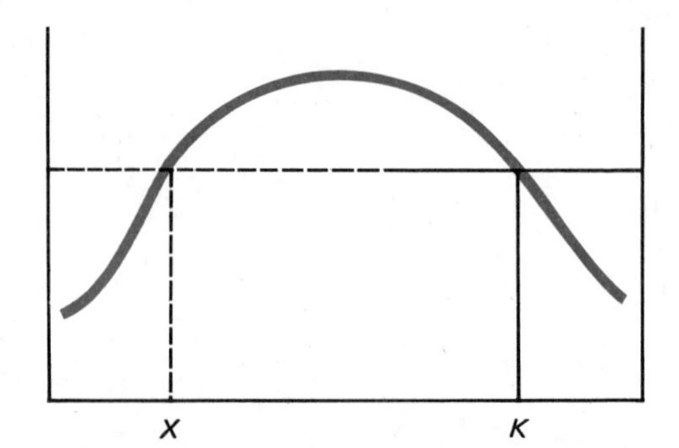

FIGURE 12.6 The Allee effect. Population growth is highest when populations are at moderate densities. Populations decline when their size exceeds some upper limit, K, or falls below some lower limit, X. In both situations r is negative.

Martha's Vineyard, off the Massachusetts coast, and to the pine barrens of New Jersey. By 1880 it was restricted to Martha's Vineyard. Two hundred birds made up the total population in 1890. Conservation measures increased the population to 2000 by 1917; but that winter a fire, gales, cold weather, and excessive predation by goshawks reduced the population to 50. The number of birds rose slightly by 1920, then declined to extinction in 1925. The last bird died in 1932.

There are several causes for the decline of sparse populations. When just a few animals are present, the females of reproductive age may have a small chance of meeting a male in the same reproductive condition. Many females remain unfertilized, reducing average fecundity. A small population faces the prospect of an increased death rate. The fewer the animals, the greater may be the individual's chances of succumbing to predation. Also, small populations may not be large enough to stimulate social behavior necessary for successful reproductive activity.

Extinction is a natural process, albeit a selective one. Species differ in their probability of extinction, a probability dependent in part on the characteristics of the species rather than wholly on random factors. Some of the qualities of a species that favor a high rate of extinction are large body size, small or restricted geographic range, habitat or food specialization, lack of genetic plasticity, and loss of alternate prey species among carnivores.

Extinctions are not spread evenly over Earth's history. Most extinctions have clustered in geologically brief periods of time (less than several tens of millions of years). One occurred in the late Permian, 225 million years ago, when 90 percent of the shallow-water marine invertebrates disappeared. Another occurred in the Cretaceous, 65 to 125 million years ago, when the dinosaurs vanished. That extinction was perhaps brought about by some extraterrestrial influence such as asteroids striking Earth, interrupting oceanic circulations, influencing climatic conditions, and accompanying turbulent volcanic and mountain-building activity.

One of the great extinctions of mammalian life took place during the Pleistocene, when such species as the woolly mammoth, giant deer, mastodon, and giant sloth vanished from Earth. Some students of the Quaternary believe that climatological changes brought about by the advance and retreat of the ice sheet caused the extinctions. Others argue that the disappearance of certain large mammals was caused by overkill by Pleistocene hunters, especially in North America, as human populations swept through North and South America between 11,550 and 10,000 years ago. Perhaps the large grazing herbivores could not withstand the combined predatory pressure of the humans and the large carnivores.

We usually think of extinction as taking place simultaneously over the full range of a species. Actually, it begins with isolated local extinctions caused when conditions deteriorate. Eventually, one local extinction added to another sums to total extinction.

The most important cause of extinction today is habitat alteration, which is a local phenomenon. Cutting and clearing away of forest, drainage and filling of wetlands, conversion of grasslands to croplands, construction of highways and industrial complexes, and urbanization and suburbanization greatly reduce available habitat for many species. When a habitat is destroyed, its unique plant life is eliminated, and the animals must either adapt to changed conditions or seek a new place to settle.

Because of the rapidity of habitat destruction, no evolutionary time exists for a species to adapt to changed conditions. Forced to leave, the dispossessed usually find the remaining habitats filled and face competition from others of their own kind or from different species. Restricted to marginal habitats, the animals may persist for a while as nonreproducing members of a population or succumb to predation or starvation. As the habitat becomes more and more fragmented, the animals are broken down into small, isolated, "island" populations, out of contact with other populations of their species. As a result, genetic variations in the isolated populations are reduced, making members of those populations less adaptable to environmental changes (see Chapter 2). The maintenance of local populations often depends heavily on the immigration of new individuals. As distance between local populations, or "islands," increases and as the size of the local populations declines, immigration becomes impossible. As the local population falls below some minimum level, it may become extinct simply through random fluctuations (see Chapter 2). At the current rate of habitat change by humans, you will see mass extinctions in your lifetime.

Summary

Population growth depends on differences between additions through births and immigration and removals through death and emigration. When additions exceed removals, the population increases. The difference between the two (when measured as an instantaneous rate) is r, the rate of increase. It is derived from the net reproductive rate, R_0, determined from a fecundity table or from the change in population size from one period to another. Both take into account age and survivorship.

In an unlimited environment, populations expand geometrically or exponentially, described by a J-shaped curve. Such growth may occur when a population is introduced into an unfilled habitat.

Because resources are limited, geometric growth cannot be sustained indefinitely. Population growth eventually slows and arrives at some point of equilibrium with the environment, the carrying capacity (termed K).

However, natural populations rarely achieve a stable level, but fluctuate about some mean. Some fluctuations have peaks and lows that occur more regularly than we would expect by chance. The two most common intervals are 3 to 4 years, as in lemmings, and 9 to 10 years, as in snowshoe hares.

When removals from a population exceed additions,

populations may decline to extinction. Extinction is a natural process taking place over long periods of time. Old species disappear and new species arise. When populations are very small, chance events alone can lead to extinction. At the present time, extinctions have accelerated at an alarming rate as human populations have expanded, destroying habitats and contaminating the environment.

Review and Study Questions

1. Contrast exponential and logistic population growth.
2. Contrast R, net reproductive rate, with λ, the finite rate of increase, and with r, rate of increase. How does R relate to r?
3. How might you determine the carrying capacity of a given habitat if the population tended to fluctuate rather than arrive and remain at some upper level as the logistic equation assumes? Do fluctuations necessarily relate to changes in carrying capacity?

*4. Discuss extinction as a process. Relate that process to some of our current endangered species, such as the California condor, whooping crane, spotted owl, and desert tortoise. For information consult the annual *Audubon Wildlife Report* (Academic Press, Orlando, FL).

5. How does habitat fragmentation, destruction, and deterioration relate to viable population size and extinction? Relate Chapter 2 to this chapter.

*6. Apply the concepts of this chapter to the world problem of human population growth. Review the population growth rates of some less developed countries. For information consult the *Population Bulletin* published by the Population Reference Bureau.

Selected References

Begon, M., and M. Mortimer. 1986. *Population ecology: A unified study of plants and animals.* Sunderland, MA: Sinauer Associates.

Clark, W. C., and R. E. Munn. 1986. *Sustainable development of the biosphere*. Cambridge, England: Cambridge University Press. Good source on population growth and the environment.

Fitzgerald, S. 1989. *International wildlife trade: Whose business is it?* Washington, DC: World Wildlife Fund. A documentary on why species go extinct.

Frankel, O. H., and M. E. Soulé. 1981. *Conservation and evolution*. Cambridge, England: Cambridge University Press.

Kaufman, L., and K. Mallory, eds. 1986. *The last extinction*. Cambridge, MA: MIT Press. Excellent overview.

Shaw, J. 1985. *Introduction to wildlife management*. New York: McGraw-Hill. Discusses several concepts of carrying capacity.

Terbrough, J. 1988. *Where have all the birds gone?* Princeton, NJ: Princeton University Press. Effects of habitat fragmentation on neotropical birds.

Wilson, E. O., and W. H. Bossert. 1971. *A primer of population biology*. Sunderland, MA: Sinauer Associates.

See also references at the ends of Chapters 2, 10, and 11.

CHAPTER 13

Population Regulation

Outline

Objectives

On completion of this chapter, you should be able to:

1. Define competition and distinguish between scramble and contest competition.
2. Discuss the effects of intraspecific competition on growth and reproduction, on density and biomass in plants, and on density-dependent mortality and natality.
3. Discuss the significance of dispersal.
4. Describe social hierarchy and explain how it may function in population regulation.
5. Discuss territoriality and its possible role in population regulation.

No population continues to grow indefinitely. Even those exhibiting exponential growth ultimately confront the limits of the environment. Most populations, however, do not behave in an exponential fashion. As the density of a population increases, interactions among members of the population and the availability of resources result in increased mortality, reduced natality, or both. If the population drops below the density the environment is able to support, mortality decreases, natality increases, and the population grows (Figure 13.1).

Implicit in the concept of population regulation is density dependence. Density-dependent effects influence a population in proportion to its size. At some low density there is no interaction. Above that point the larger the population becomes, the greater is the proportion of individuals affected. Density-dependent mechanisms act largely through environmental shortages and competition for resources. If the effects of a particular influence do not change proportionately with population density, or if the proportion of individuals affected is the same at any density, then the influence is density-independent.

INTRASPECIFIC COMPETITION

Population regulation, in part, involves competition among individuals of the same species for environmental resources. Competition results only when a needed resource is in short supply relative to the number seeking it. As long as resources are abundant enough to allow each individual a sufficient amount for survival and reproduction, no competition exists. When resources are insufficient to satisfy adequately the needs of all individuals, the means by which they are allocated has a marked influence on the welfare of the population.

TYPES OF COMPETITION

When resources are limited, a population may exhibit one of two responses. All the individuals can share the resources equally with none of them obtaining enough for growth or reproduction, provided the population density remains high or competition remains intense. Such competition is called *scramble.* Or some individuals can claim enough resources for maintenance and reproduction while denying others shares of the resources. That type of competition is called *contest.* Generally, under the stress of limited resources, a species will exhibit only one type of competition. Some are scramble species and others are contest species. Still others vary according to the stage in the life cycle. For example, the larval stages of some insects experience scramble competition and the adult stages face contest competition.

The outcome of scramble and contest competition differs. Scramble competition can produce chaotic oscillations in the population over time, and it limits the average density of the population below that which the resources could support if an adequate amount of resources were supplied only to part of the population. For that reason, scramble competition can waste resources relative to population growth. In contest competition the deleterious effects of limited resources are confined to a fraction of the population, the unsuccessful individuals. That eliminates or greatly reduces the wastage of resources, permits the maintenance of a relatively high population density, and maintains some numerical constancy.

Both scramble and contest types of competition have a threshold level below which no competition takes place and all individuals survive or maintain fitness. Above that threshold, scramble competition results in little production of offspring at best and could reduce the population to zero. Once a population experiencing contest-type competition passes that threshold, a fraction of the individuals will get all the resources they need to survive and produce offspring. The remaining individuals will get less than they need and produce no offspring or die. Thus, in scramble competition most individuals get less than they need, whereas in contest-type competition only a portion of the population experiences that fate.

CONSEQUENCES

Because the intensity of intraspecific competition is density-dependent or density-proportional, it rises slowly. It involves no sudden thresholds; it increases gradually, at first affecting just the quality of life, later affecting individual survival and reproduction.

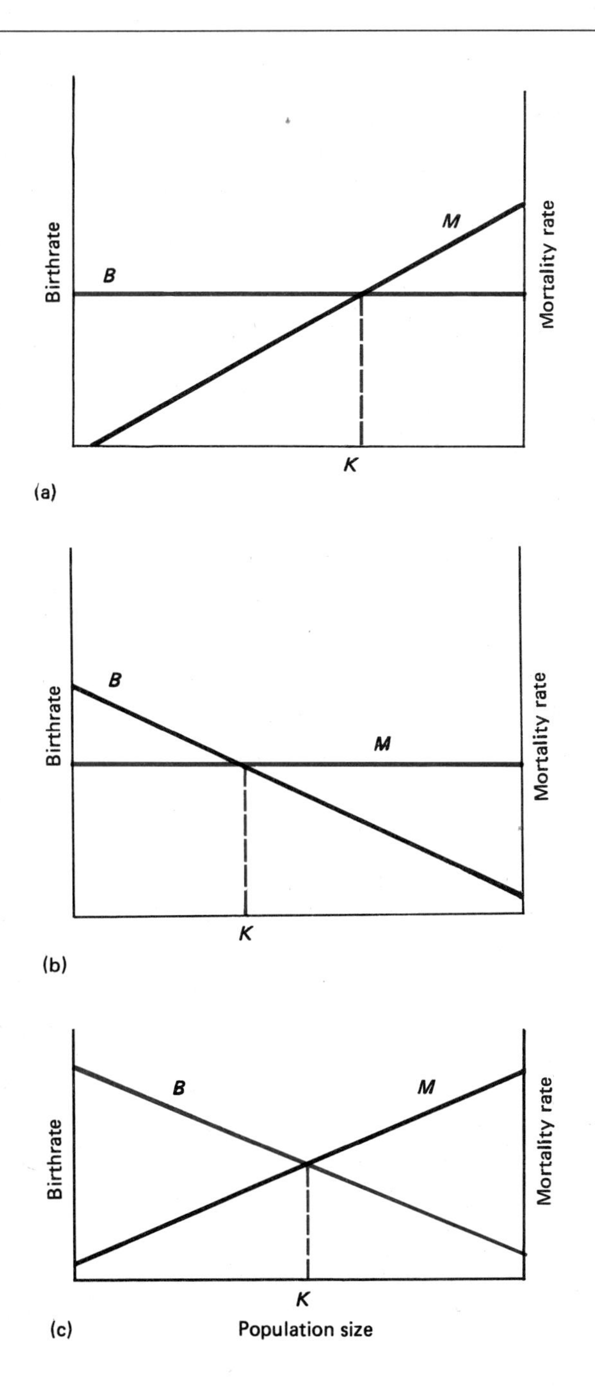

Retarded Growth and Delayed Reproduction

As population density increases toward a point at which resources are insufficient, individuals in scramble competition reduce their intake of food. That slows the rate of growth and inhibits reproduction. Examples of this inverse relationship between denstiy and rate of body growth may be found among populations of poikilothermic vertebrates. Frog larvae reared experimentally at high densities failed to grow normally (Figure 13.2a). They experienced slower growth, required a longer time to reach the size at which transformation from the tadpole stage takes place, and had a lower probability of completing metamorphosis. Those that did reach threshold size were smaller than those living in less dense populations. Fish living in overstocked ponds exhibit a similar response to density (Figure 13.2b). Bluegills (*Lepomis macrochirus*), for example, normally grow to the size of a dessert plate; but in overstocked and underharvested farm ponds, they rarely grow beyond the size of a half dollar and never reproduce.

Like fish and tadpoles, plants, too, may respond to density in a scramble fashion through reduction in individual growth. Botanists call it *phenotypic plasticity*. Individuals adjust their growth form, size, shape, number of leaves, flowers, and production of seeds in a scramble fashion to the limited resources available. In some plants, reproduction by seed at high densities is almost nonexistent; it is solely by vegetative means.

FIGURE 13.1 Population regulation, if it is to function, requires density-dependent birthrates (*B*) and mortality rates (*M*). In (a) the birthrate is independent of population density, as indicated by the straight line. It remains unchanged as the population increases. But in (a) the death rate increases as population increases. As long as the *B* exceeds the *M*, the population increases toward carrying capacity (*K*). At *K* the population reaches equilibrium, maintained by increasing mortality. In (b) the situation is reversed. *M* is independent of population density, but *B* declines as density increases until the population reaches *K*. At that point equilibrium is maintained by a decreased birthrate. In (c) both *B* and *M* are density-dependent, and the population reaches equilibrium when the birthrate equals the death rate. Fluctuations in each will tend to hold the population at or near the equilibrium point and influence population density. If the birthrate increases, then the mortality rate increases.

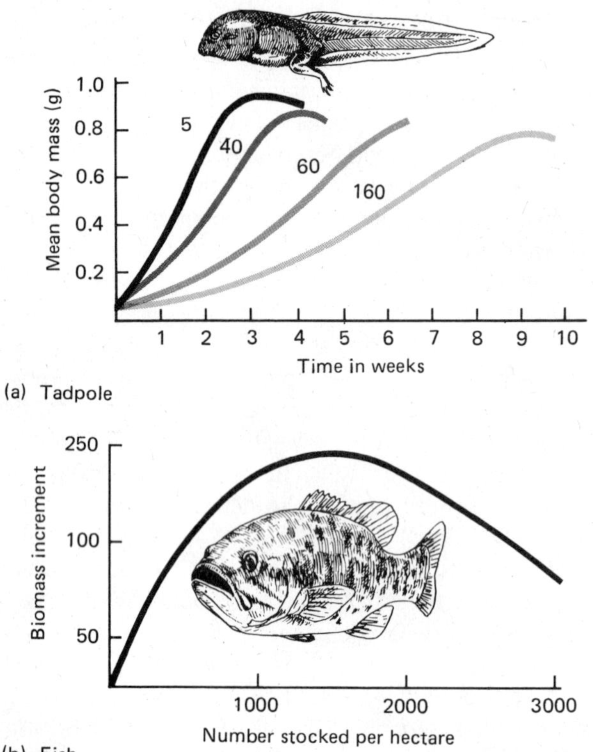

FIGURE 13.2 Two examples of the effect of population density on the growth of individuals in a population. (a) Influence of density on the growth rate of the tadpole *Rana tigrina.* Note how rapidly individual growth rates decline as density increases from 5 to 160 individuals confined in the same space. (b) Growth of fish (bass) as influenced by population density. Growth declines rapidly with density above 1500/ha. Both graphs reflect the scramble type of competition.

As the density of plants increases, the number of vegetative offspring also declines. For example, at high densities the genets of perennial ryegrass (*Lolium perenne*) produce few tillers, and their weight is low. At low densities the genets produce many tillers, and their weight is high.

Decrease in Density and Increase in Biomass

A close relationship exists between plant density and growth of individual plants as measured by biomass accumulation. Up to a certain point, all plant seedlings exhibit an increase in biomass. Then as their size increases, plants interfere with each other, competing for the same resources—light, moisture, nutrients, space. Plants initially respond to competition and increasing density in a plastic manner, modifying their form and size. Up to a point, plants are successful. Beyond that point the population begins to experience density-dependent mortality. Response takes place the earliest in dense seedling populations, but in time it also occurs in less dense populations as plant size increases. As biomass of individual plants increases, fewer plants per unit area are needed for competition to develop. As a result, plants starting at different densities converge toward some common value, with decreases through time.

As mortality thins the population, more biomass accumulates in the remaining individuals. Thus an old growth forest has a high biomass but fewer individuals than a younger stand (see Table 11.5).

Density-Dependent Mortality—The African Buffalo

Among some animals, large mammals in particular, the effects of density become prominent as the population approaches carrying capacity. To demonstrate density dependence in such a population you need to know something about the level of the population. Is it still expanding? Is it at some equilibrium level? One way to determine that is to reduce a population and allow it to expand again.

That is what happened to the African buffalo (*Syncerus caffer*), studied in detail on the Serengeti by A. R. E. Sinclair. The African buffalo (not to be confused with a bison) is a large bovine ungulate of the African savanna. Its food is grass, preferably the protein-rich leaves. In 1895, rinderpest, a measlelike disease of cattle, spread from domestic cattle to the African buffalo and wildebeest (*Connochaetes taurinus*). Herds were decimated in the early 1900s. Veterinarians eventually eliminated the disease in cattle, which reduced its incidence in wild bovines. The disease continued to cause a high mortality among juveniles as late as 1964. Released from heavy juvenile mortality, the African buffalo expanded dramatically, reaching an apparent equilibrium density in the 1970s.

The critical time of year for the buffalo population is the dry season, when food may become scarce. Rainfall determines the productivity of grass. The greater the rainfall, the more vigorously the grass grows, increasing the amount of forage available in the dry season. Equilibrium density varies with mean annual rainfall; the greater the rainfall, the greater the density of buffalo (Figure 13.3).

During the wet season on the African savanna, food is abundant, but during the dry season, the quality of food declines as the grasses dry. The buffalo become more selective, seeking green leaves, moving to the moist riverine habitat, breaking up into smaller units, and utilizing different areas. As the dry season progresses, buffalo become less selective, consuming dry leaves and stems that they would otherwise have rejected. As food quantity and quality decline, competition in a scramble fashion becomes keener. Individual buffalo feeding in any one area reduce the food available to neighboring animals. The more buffalo present, the less food there is available for each individual. Eventually, the quality of food as measured by its available protein drops below maintenance level, and the animals use up their fat reserves. Undernourished and lacking the protein intake necessary to maintain immunity to disease and parasites they normally harbor, adults become ill. Older animals are the most vulnerable. The number dying depends upon the rapidity with which the adults use up their energy reserves before the coming of the rainy season and new growth. If the next season sees more rainfall and that rainfall extends sporadically into the dry season, the mortality of adults the following year is reduced.

Thus, mortality of adults, as influenced by rainfall and poor resources, determines equilibrium density and regulates the population (Figure 13.4). By contrast, juvenile mortality appears to be density-independent. It results from one or several randomly fluctuating environmental variables and causes fluctuations in the population. Density-dependent adult mortality compensates for the disturbances and dampens fluctuations.

The examples suggest that populations do respond to increasing numbers in a density-dependent fashion through intraspecific competition. The timing of the response depends upon the nature of the population. Among larger mammals, such as the American bison, with a long life span and low reproduction, regulating mechanisms do not function until the population nears carrying capacity (Figure 13.5). Among organisms characterized by high reproduction and short life spans, response to density may occur much earlier.

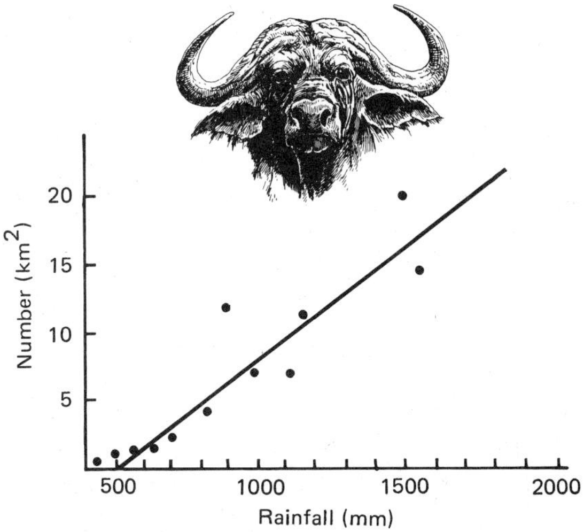

Number (km^2)
20
15
10
5
500
1000
1500
2000
Rainfall (mm)

FIGURE 13.3 The relationship between rainfall and African buffalo abundance.

MECHANISMS OF REGULATION

By what means does intraspecific competition regulate populations? Stress, dispersal, and behavior all may play a part.

STRESS

As a population reaches a high density, individual living space becomes restricted. Often aggressive contacts among individuals increase. One hypothesis of population regulation is that increased crowding and social contact result in stress. Such stress triggers hyperactivation of the system that controls the endocrine glands. Profound hormonal changes result in a

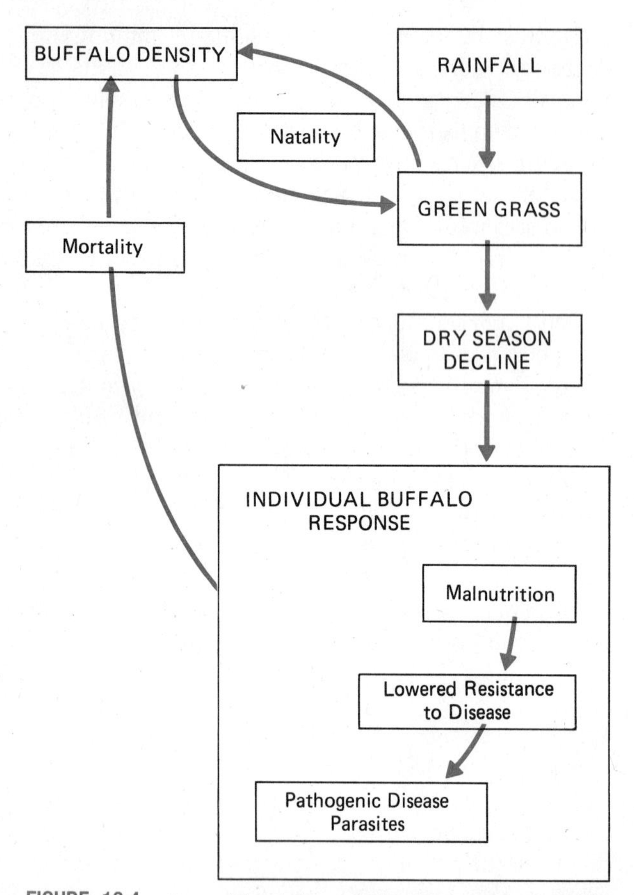

FIGURE 13.4 A model of the relationship among rainfall, forage production, population density, mortality, and natality in the African buffalo. The dry season causes a decline in quantity and quality of forage, resulting in malnutrition and lowered resistance to disease. Increased mortality of adults results in population decline.

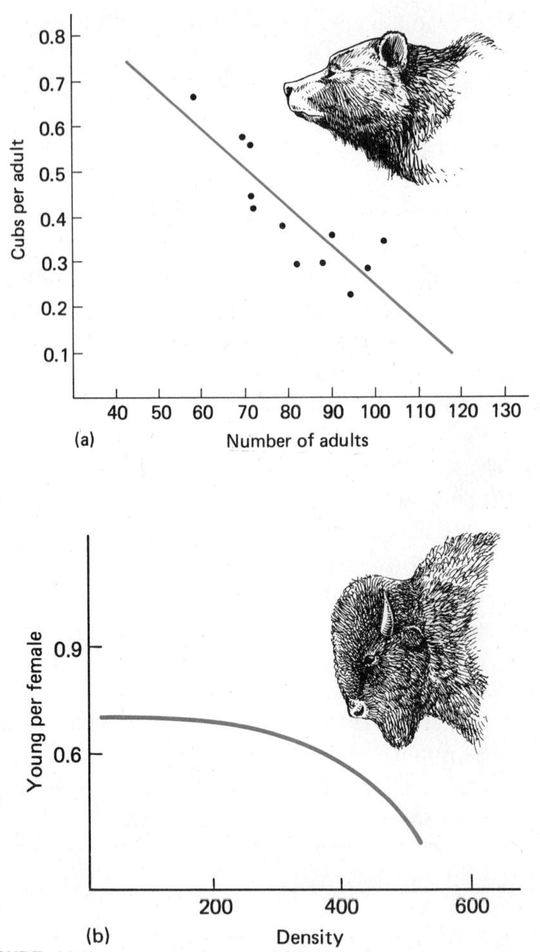

FIGURE 13.5 An example of linear and nonlinear density-dependent change in a large mammal population. (a) A linear density-dependent decrease in the number of young produced per adult female grizzly bear as the population increases. (b) The birthrate of the American bison, expressed as young per adult female, begins to decline at a certain density and declines more sharply as the density increases.

suppression of growth, a curtailment of reproductive functions, and delayed sexual activity among individuals in the population. These hormonal changes may also suppress the immune system and break down white blood cells, increasing vulnerability to disease. Social stress among pregnant females may increase intrauterine mortality and cause inadequate lactation, stunting nurslings. Thus, stress results in decreased births and increased infant mortality.

Such population-regulating effects have been confirmed in confined laboratory populations of several species of mice and to a lesser degree in enclosed wild populations of woodchucks (*Marmota monax*) and Old World rabbits (*Oryctolagus cuniculus*). K. Myers and his associates experimented with rabbits at several densities in different living spaces within confined areas of their natural habitat. Those living in the smallest space, in spite of a decline in numbers, suffered the most debilitating effects. Sexual and aggressive behavior increased, reproduction declined, fat about the kidneys decreased, and kidneys showed inflammation and pitting on the surface. Weights of liver and spleen decreased, and adrenal size increased.

Young rabbits born to stressed mothers were stunted in all body proportions and in organs. As adults they exhibited such behavioral and physiological aberrations as a high rate of aggressive and sexual activity and large adrenal glands relative to body weight. In contrast, rabbits from low- to medium-density populations showed excellent health and survival.

Pheromones, or chemical releasers, present in the urine of adult rodents can encourage or inhibit reproduction. One study involved wild female house mice (*Mus musculus*) living in high-density and low-density populations confined to grassy areas within a highway cloverleaf. Urine from females of a high-density population was absorbed onto filter paper. The paper was placed with juvenile wild female mice held individually in laboratory cages. Urine from females in low-density populations was placed with other juvenile test females. Juvenile females exposed to urine from high-density populations experienced delayed puberty, whereas females exposed to urine from low-density populations did not. The results suggest that pheromones present in the urine of adult females in high-density populations may delay puberty and help to slow further population growth. Juvenile female house mice exposed to urine of dominant adult males accelerate the onset of puberty.

The response of plants to stress is influenced by their adaptiveness. For example, some plants are adapted to shade. Under the light-intercepting canopy of other plants or similar low-light conditions, shade-adapted plants will respond by growing slowly, conserving energy by reducing the rate of photosynthesis, and forgoing the production of flowers and seed. In effect, they wait for a time when sufficient light will stimulate rapid growth. Plants adapted to other forms of stress follow a similar pattern. Plants adapted to open light respond to low light intensities by growing rapidly in height (which under some conditions will get them up into the light). However, they develop thin cell walls, which reduce the supporting ability of the stems, resulting in weak, spindly individuals, and they are highly vulnerable to fungal infections, which can be fatal.

Under conditions of low nutrients, low moisture, or other environmental stress, some groups of plants will use up their nutrient reserves and fail to produce seeds. That limits their ability to respond to improving environmental conditions. In contrast, ruderal plants, those adapted to persistent and severe disturbance, respond to stress by producing seeds at the expense of vegetative development. Individual plants are small and poorly developed; yet the number of seeds relative to individual plant biomass is high. Seeds of such plants can survive buried in the soil for long periods of time. They are able to germinate rapidly when a disturbance exposes the seeds to light and fluctuating daily temperatures. Examples of such plants are annual weeds.

Although such responses to stress do influence individual fitness among plants and the maintenance and expansion of populations, there is little evidence that stress in plants acts in any sort of regulatory manner.

DISPERSAL

Instead of coping with stress (to state the situation somewhat anthropomorphically), some animals run away from a bad situation. They seek vacant habitats. Although dispersal is most apparent when population density if high, it is a relatively constant phenomenon. Some individuals leave the parent population whether it is crowded or not. However, there is no hard and fast rule about who disperses.

When a lack of resources forces out individuals, usually subadults driven out by adult aggression, the odds are that they will perish, although a few may arrive at some suitable area and settle down. Such dispersal results from overpopulation; it has little influence on population regulation.

More important to population regulation is dispersal that takes place when the population density is low or increasing, but well before the population reaches a density at which food and cover are overexploited. Individuals who participate are not a random selection of the population but ones that are in good condition, belong to any sex or age group, have a good chance of survival, and show a high probability of settling in a new area. Some evidence exists that such individuals are genetically predisposed to disperse.

Such individuals can maximize their fitness only if they leave their birthplace. When intraspecific competition at the home place is intense, dispersers can locate in habitats where resources are more accessible, breeding sites are more available, and competition is less. Further, the individual reduces the risk of inbreeding (see Chapter 2). At the same time dispersers also incur the risks. They are living in unfamiliar terrain; their hybrid young, produced by outcrossing with wholly unrelated individuals, may not be well adapted to the environment.

Dispersers, according to recent studies, travel no further than necessary. How far they go depends on the density of the surrounding populations, and the availability of suitable habitat. Generally, individuals travel in a straight line from their birthplace or make a number of exploratory forays before leaving, and then occupy the first uncontested site they locate. How well they fare in their new location depends upon the quality of habitat. Those that settle into optimum habitat experience high survival and reproduction, whereas frustrated dispersers experience low survival.

Does dispersal actually serve to regulate a population? Although dispersal is positively correlated with population density, no relationship exists between the proportion of the population leaving and population increase or decrease. Dispersal may not function as a regulatory mechanism, but it contributes strongly to population expansion, aids in the persistence of local populations, and functions as a source of natural selection by sorting out phenotypes and genotypes.

Successful dispersal of plants may depend upon the frequency of dispersal. The more often the plant has a heavy seed crop, the more often it will have new propagules available for colonization, even though seed losses will be enormous. Regardless of the means of dispersal, most of the seeds are dropped near the parent plant. Because they furnish a concentrated source of food, seeds near the parent plant usually are subject to more intense predation by seed-eating animals than are seeds scattered some distance away. Thus, the probability of seed survival increases with the distance from the parent plant.

BEHAVIOR

Intraspecific competition expresses itself in social behavior as the degree of tolerance between individuals of the same species. Social behavior appears to be a mechanism that limits the number of animals that can live in a particular habitat, have access to the food supply, and engage in reproductive activities.

To prove that social behavior limits populations in a density-dependent fashion, we have to show that (1) a substantial portion of the population consists of surplus animals that do not breed because they either die or attempt to breed and fail; (2) such individuals are prevented from breeding by dominant individuals; (3) nonbreeding individuals are capable of breeding if dominant individuals are removed; and (4) breeding animals are not completely utilizing food and space.

Social Hierarchy

Many species of animals live in groups with some kind of social organization based on intraspecific aggressiveness and intolerance and on the dominance of one individual over another. Two opposing forces are at work. One is mutual attraction; the other is a negative reaction against crowding, the need for personal space.

Each individual occupies a position in the group based on dominance and submissiveness. In its simplest form the group includes an alpha individual dominant over all others, a beta individual dominant over all but the alpha, and finally an omega, an individual subordinate to all others. Individuals settle social rank by fighting, bluffing, and threatening at initial encounters between pairs of individuals or at a series of such encounters. Once social rank is established, it is maintained by habitual subordination of those in lower positions, reinforced by threats and occasional punishment meted out by those of higher rank. Such organization results in social harmony, stabilizing and formulating intraspecific competitive relationships and resolving disputes with a minimum of fighting and wasted energy.

Social dominance may play a role in population regulation if it affects reproduction and survival in a density-dependent manner. An example is the wolf. Wolves live in small groups of 6 to 12 or more individuals, called packs. The pack is an extended kin group consisting of a mated pair, one or more juveniles from the previous year who do not become sexually mature until the second year, and several related, nonbreeding adults.

The pack has two social hierarchies, one headed by an alpha female and the other headed by an alpha male, the leader of the pack to whom all other members defer. Below the alpha male is the beta male, closely related, often a full brother, who has to defend his position against pressure from males below (Figure 13.6).

Mating within the pack is rigidly controlled. The alpha male (occasionally the beta male) mates with the alpha female. She prevents lower-ranking females from mating with the alpha and other males, while the alpha male inhibits mating attempts by other males. Thus, each pack has one reproducing pair and one litter of pups each year. They are reared cooperatively by all members of the pack.

The level of the wolf population in a region is governed by the size of the packs, which hold exclusive areas (see "Territoriality"). Regulation of the size of the pack is achieved by events within the pack that influence the amount of food available to each wolf. The food supply itself does not affect births and deaths, but the social structure that leads to an unequal distribution of food does. The reproducing pair, the alpha female and the alpha male, has priority for food; they, in effect, are independent of the food supply. The subdominant animals, male and female, with little reproductive potential, are affected most seriously. At high densities the alpha female will expel other adult females from the pack. Other individuals may leave voluntarily. Unless these animals have the opportunity to settle successfully in new territory and form a new pack, they fail to survive (see "Dispersal").

The social pack, then, becomes important in population regulation. As the number of wolves increases, the size of the pack increases. Individuals are expelled or leave, and the birthrate declines because most sexually mature females do not reproduce. Overall, the percentage of reproducing females declines. When the population of wolves is low, sexually mature females and males leave the pack, settle in unoccupied habitat, and establish their own packs (see "Dispersal") with one reproducing female. As a result, nearly every sexually mature female reproduces, and the wolf population increases. However, at very low densities, females may have difficulty locating males to establish a pack and so fail to reproduce or even survive.

Territoriality

A more complex form of grouping is *territoriality*, a situation in which an individual animal defends an exclusive area that is not shared with rivals (Figure 13.7). Defense involves definitive behavioral patterns: song and call, intimidation displays such as spreading of wings and tail in birds and baring of fangs in mammals, attack and chase, and scent markings that evoke escape and avoidance in rivals. As a result, territorial individuals tend to occur in more or less regular patterns of distribution.

Territoriality and social hierarchy represent degrees of manifestation of the same basic patterns of dominance. It is difficult to draw a sharp line between the two. A gradient in behavior exists from no social organization at one end to group territoriality at the other. Under one set of environmental conditions—a certain amount of space, population density, and season of the year—individuals of a species may be highly territorial, whereas under another set of conditions, territoriality might disappear into social hierarchy.

For example, the dragonfly *Leucorrhini arubicunda*, like other dragonflies, maintains a territory along the edge of a pond. When the population density is low, individuals are spaced 3 to 7 meters apart and aggressive interaction is frequent. As density increases, the level of territorial defense and attachment to the site decreases. Thus territoriality can be considered a spatial organization of dominance hierarchies in which the individual holds the highest rank in the hierarchy in its own territory or center of activity.

Why should a cardinal or a wolf pack defend a territory? The proximate reasons vary. For some it is the acquisition and protection of a needed resource such as food, nesting site, a mating area, or a mate. The ultimate reason is an increased probability of survival and improved reproductive success—in short, increased fitness. By defending a territory, the individual

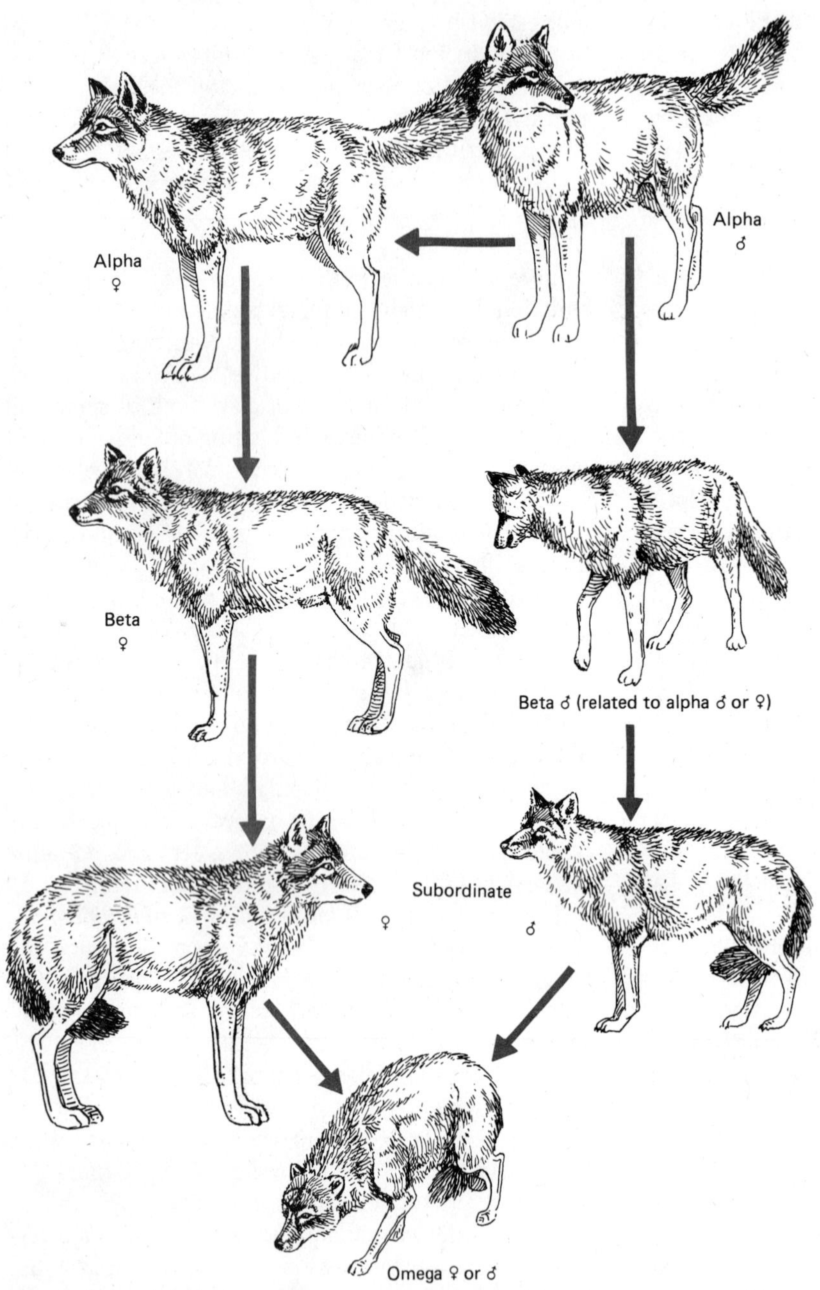

FIGURE 13.6 Social hierarchy in a small wolf pack. Note the two separate hierarchies, one male and one female, with one individual occupying the omega position. The alpha male is the dominant individual in the pack.

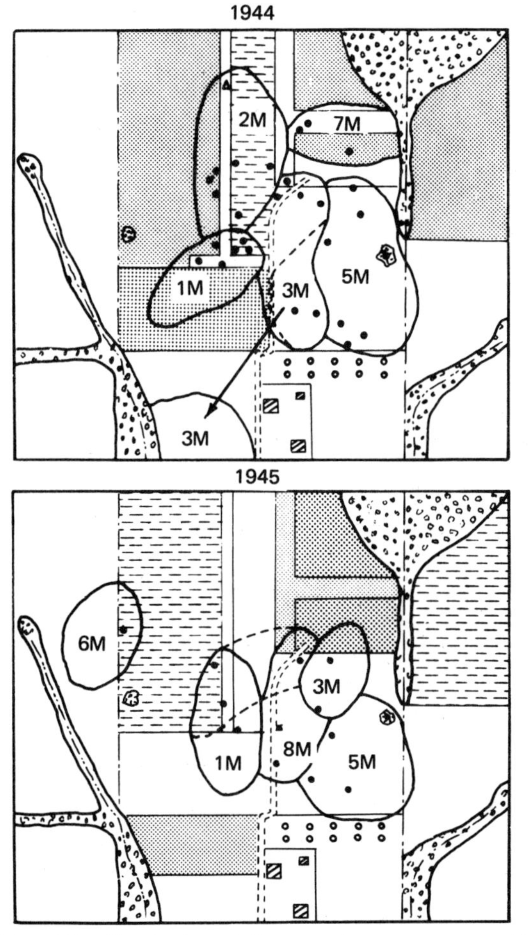

FIGURE 13.7 Territories of the grasshopper sparrow (*Ammodramus savannarum*), determined by observations of banded birds. Dots indicate song perches. Note how they are distributed near the territorial boundaries. Shaded areas represent crop fields. Dashed lines indicate territory shifts prior to second nesting. Note the return of the same males to nearly the same territorial area the second year. Such behavior is termed *philopatry*.

forces others into suboptimal habitat, reducing their fitness and increasing the proportion of its own offspring in the population.

Defending a territory is costly in energy and time and can interfere with courtship, mating, feeding, and rearing young. Situations exist in which a territory is economically defendable and in which it is not. It is worth defending a predictable resource not highly dispersed. If resources are unpredictable and patchily distributed, it may be advantageous not to defend.

Associated with ownership of territory is size. As the size of a territory increases, the cost of territorial defense increases. In general, territory size tends to be no larger than required, varying from year to year (Figure 13.7) and from locality to locality, depending upon resources and number of animals seeking space. Brewer's sparrow, a sagebrush bird of the Great Basin in western United States (*Spizella breweri*), achieves a maximum territory size at moderate population densities. It does not expand beyond that at lower densities, despite available habitat, nor does it seem to have a lower size limit at very high densities. The bird's associate, the sage sparrow (*Amphispiza belli*), expands its territory rapidly at low densities with no upper limit, but apparently has a minimal size at high densities, when the habitat seems to be saturated.

For some animals, birds in particular, it is not the size of the territory that counts, but its quality. Some males, perhaps the most aggressive, are the most successful in claiming the best territories, usually measured by some features of vegetation that make them superior nesting sites. Less successful males occupy suboptimal territories, and some males secure no territories at all. The most successful males are assured of a mate, whereas a male bird settled on a poor territory may be unable to attract one.

The number of territorial owners a habitat can support is determined by the total area available divided by the minimum size of the territory. When the available area is filled, excess animals are ejected or denied access. These individuals make up a floating reserve. Such a floating reserve of potential breeding adults has been described for a number of species, including the red grouse (*Lagopus lagopus*) in Scotland, Australian magpie (*Gymnorhina tibicen*), Cassin's auklet (*Ptychoramphus aleuticus*), and the white-crowned sparrow (*Zonotrichia leucophrys*) of California. Studies of a banded white-crowned sparrow population indicated a nonbreeding surplus of potential breeding individuals. In fact, 24 percent of the territorial holders entered the population two to five years after banding, and 25 percent of the nestlings that acquired territories did so two to five years after their hatching. Territory holders that disappeared during the breeding season were replaced almost immediately.

Few data exist on the social organization of the floaters. Floaters may form flocks with a dominance hierarchy on areas not occupied by territory holders, as do red grouse and Australian magpies. They may live singly off the territories, as white-crowned sparrows do; or accepted as nonrivals, they may spend much time on the breeding territories of others, as do the rufous-collared sparrows (*Zonotrichia capensis*) of Costa Rica.

Territoriality can function as a mechanism of population regulation. If all pairs that settle on an area get a territory, then territoriality results only in the spacing out of a population. No regulation of population size results. By contrast, if territorial size has a lower limit, then the number of pairs that can settle on an area is limited, and individuals that fail have to leave. In such a situation territoriality might regulate the population, but only if an excess of nonterritorial males and females of reproductive age live in the area, as is the case for the red grouse and white-crowned and rufous-collared sparrows. Then reproduction is limited by territoriality, and density-dependent population regulation results. Surplus individuals are limited or excluded from breeding, and the population will not increase beyond some upper limit set by the number of territories available.

Home Range

Territoriality refers specifically to an area defended during some portion of the year. Differing from territory is *home range*, land over which an animal normally roams during the year or life. It is often but not necessarily associated with aggressive behavior. Among some species territory and home range are the same. Among others only a part or core area may be defended, or the area may not be defended at all. The two sexes

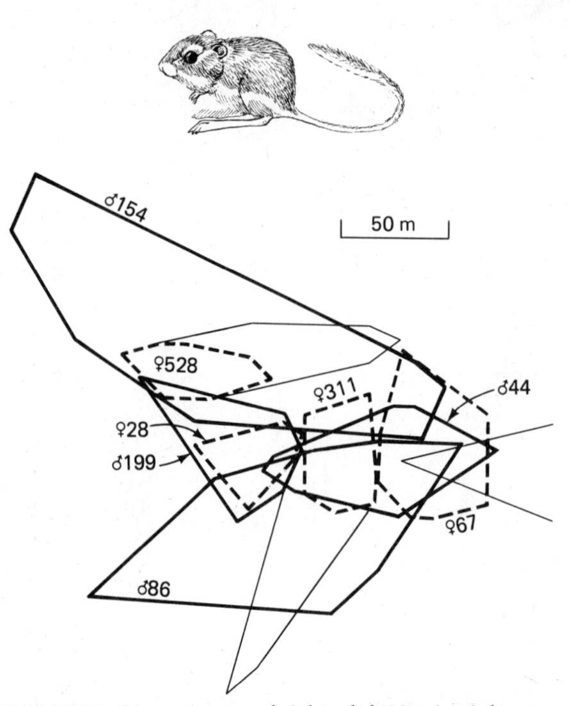

FIGURE 13.8 Home ranges of eight adult Merriam's kangaroo rats (*Dipodomys merriami*) in Arizona. Males are shown by solid lines, females by dashed lines. Dotted lines show excursions outside of the usual home ranges. Note the overlap of all home ranges, the greater home range size of the males, and the inclusion of female home ranges within the home ranges of males.

may have the same or different home ranges (Figure 13.8), which may overlap. The home range of a male may embrace those of several females.

The home range is highly variable, even within a species and among individuals. Seldom is a home area rigid in its use, its size, or its establishment. The home range may be compact and continuous, or it may be broken into discontinuous parts reached by trails. Irregularities in spatial and temporal distribution of food produce corresponding irregularities in home range and frequency of visitation.

Overall size of the home range is influenced by food resources, mode of food gathering, body size, and metabolic needs. In general, carnivorous animals require a larger home range than herbivorous and omnivorous animals of the same size. Males and adults have larger

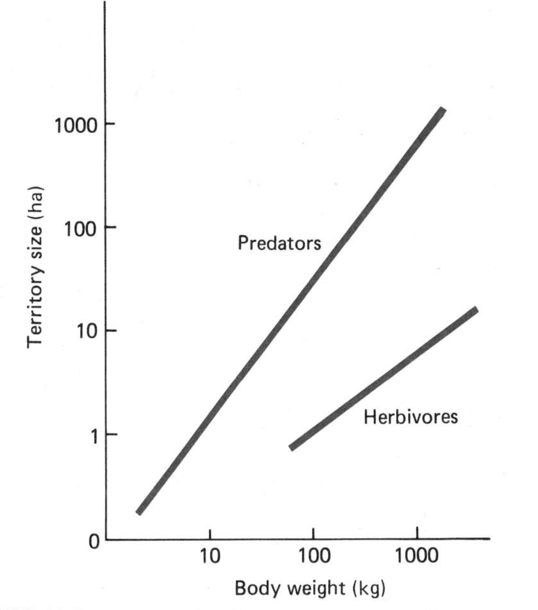

FIGURE 13.9 Relationship between the size of home range and the body weight of North American mammals.

TABLE 13.1 Sizes of Home Ranges and Territories

Species	*Home Range*	*Diet*
White-footed mouse	0.32–4.26 ha	O
Kangaroo rat	6.67 ha m 1.66 ha f	O
Snowshoe hare	5.8 ha	H
White-tailed deer	142 ha m 45 ha f	H
Skunk	17–38 ha	O
Red fox	259–777 La	O
Fisher	30.9 km^2 m 16.3 km^2 m	C
Timber wolf (pack)	1404 km^2	C
Bobcat	95.7 km^2 m 31.2 km^2 f	C
Mountain lion	39–78 km^2 m 13–65 km^2 f	C
Grizzly bear	147 km^2 f + yrlings	O
Ovenbird	0.2–0.6 ha	
Song sparrow	0.2–0.6 ha	
Meadowlark	1.2–6.0 ha	
Black-capped chickadee	3.4–6.8 ha	
Downy woodpecker	4 ha	
Pileated woodpecker	121 ha	
Red-tailed hawk	1.3–7.8 km^2	
Golden eagle	49–154 ha	

m = male; f = female; H = herbivore; O = omnivore; C = carnivore.

home ranges than females and subadults. Weight alone is sufficient to account for differences within a species without invoking any competitive interactions. The home range of herbivores and omnivores increases at a nearly constant rate as body weight increases; among carnivores, home range increases as body weight increases (Figure 13.9). The more concentrated the food supply, the smaller is the home range. Home ranges of some mammals and the center of activity within them appear to be in flux, determined by such variables as the location of food supply and reproductive condition of neighboring males and females.

The degree of aggressive behavior among individuals may limit the size of the home range. Dominant individuals may hold the largest home range, with some overlap, while subdominant individuals occupy home areas within those of dominant individuals. The dominant animal is able to range more freely over a larger area and has access to more resources than subdominant individuals. Usually a male can control highly desirable locations, especially in relation to food and females, and force subdominant animals to occupy the area that is left. Restriction of activities to a home area also confers security. Animals become familiar with the location of food, shelter, and escape cover.

The size of home ranges and of territories is of more than academic interest. It is an indicator of how much habitat is necessary to support viable populations of various species of animals. In general, animals with small home ranges or territories can be contained in a smaller landscape than large mammals and predatory birds, the populations of which may require thousands of square miles (Table 13.1). Management of such expansive habitats is a major ecological and economic problem.

SPACE CAPTURE BY PLANTS

Plants obviously are not territorial in the same sense that animals are territorial; but plants can capture and hold on to space for a period of time. This phenomenon in the plant world can be considered analogous to territoriality in animals, especially if you accept the alternative definition of territoriality—individual organisms spaced out more than we would expect from a random occupancy of suitable habitat.

Plants from dandelions to trees do capture a certain amount of space and defend it over time from intraspecific and interspecific competitors. They exclude other individuals from the same and smaller size classes. When a dandelion plant spreads its rosette of leaves on the ground, it effectively eliminates all other plants from the area covered by its leaves. The faster-growing trees in a forest achieve a height dominance over others, allowing them to expand their canopies over the tops of others. Plants with expansive crowns or rosettes of leaves intercept light, eliminating competition and restricting occupancy of the ground below to shade-tolerant species. Because the root systems of trees more or less mirror in the ground the expanse of the canopy, dominant trees are also in a superior competitive position for nutrients and moisture. Because of their longevity, some plants, especially trees, occupy space over a long period of time, preventing invasion by other individuals of the same or other species. Plants successful in capturing space increase their fitness at the expense of others. The survival rate of a few adults is high; others are eliminated.

Plants may also capture space by the release of organic toxins that reduce competition for light, nutrients, and space. A variety of phenolic compounds released by roots and by leaves due to litterfall and rainwater throughfall accumulate in the soil and inhibit the germination of seeds and the growth of other plants, herbaceous and woody (see "Allelopathy," Chapter 15).

Summary

Populations do not increase indefinitely. As resources become less available to an increasing number of individuals, birthrates decrease, mortality increases, and population growth slows. If population declines, mortality decreases, births increase, and population growth speeds up. Between positive and negative feedback, the population arrives at some form of regulation.

Regulation involves intraspecific competition. Competition occurs when resources are in short supply relative to the number seeking them. Competition can take two forms: scramble and contest. In scramble competition, resources are shared equally by all individuals in the population, none of which receives a sufficient amount for growth and reproduction. In contest competition, sufficient resources for growth and reproduction are claimed by dominant individuals, who deny them to others. The latter produce no offspring or perish.

Intraspecific competition can result in decreased or retarded growth of individuals, decreased density and increased biomass in individual plants, delayed reproduction, and density-dependent mortality in animals brought about by malnutrition, decline in immunity to diseases, and parasites.

Responses to intraspecific competition may take a number of forms. One is increased stress brought about by crowding, fighting, lack of food or space, or attempts to acquire mates. Stress may result in delayed reproduction, abnormal behavior, increased adrenal activity, reduced growth and production of flowers and seeds, and reduced ability to resist disease and parasitic infections.

Increased stress, however, might lead to dispersal. Dispersal is a constant phenomenon in populations at presaturation levels. It apparently is not a random selection of individuals; rather, the individuals seem to be genetically programmed to disperse. Many end up in submarginal habitats, and some successfully occupy new or unfilled habitat. At saturation levels dispersal is a response to overcrowding, and the dispersers are surplus to the population. There is no strong indication that dispersal results in population regulation, but dispersal is important in population expansion.

Intraspecific competition may be expressed as social behavior, a social organization based on the dominance of one individual over another. Dominant individuals secure most of the resources, and in times of resource shortages, the effects are borne by subdominant individuals.

One form of social hierarchy is territoriality, in which a defended or exclusive area is maintained by an individual or a group. Animals defend territories by songs, calls, displays, chemical scents, and fighting. Because of the costs incurred in territorial defense, owners can afford the benefits derived only if their increased fitness exceeds costs. Territoriality is a form of contest competition in which a certain portion of the population is excluded from reproduction. These nonreproducing individuals act as a floating reserve of potential breeders, available to replace losses of territorial holders. In such a manner, territoriality can act as a population regulating mechanism.

Review and Study Questions

1. What is competition? Distinguish between scramble and contest competition and discuss the importance of each in population regulation.
2. Explain how density-dependent effects might influence population growth and size.
3. How might stress influence population growth in mammals? In plants? How do plants respond to stress?
4. How is dominance expressed in social hierarchy? In what way can it function as a population-regulating mechanism?
5. What is territoriality, and how does it relate to social hierarchy?

*6. How can territorial behavior increase the fitness of the male holder of the territory? The female? Speculate on the idea that the female may choose the territory and take the male that goes with it rather than choose the male. For a start, consult "The evolutionary effects of mate selection" by W. A. Searcy, 1982, *Annual Review of Ecology and Systematics*, 13:57–85.

7. How can territoriality regulate population? What is the role of the floater?
8. Comment on the remark, "If this wood is cut, the animals will just move elsewhere?" What is the probable fate of the displaced individuals, and why?

*9. Look up the home ranges of such large mammals as the elephant, rhino, grizzly bear, and others and then discuss the problem of maintaining viable populations on such limited areas as national parks.

10. What are some problems associated with the release of animals into unfilled habitat? Partially filled habitat? This question is central to the reestablishment of new populations or the introduction of new males into a small population to increase genetic diversity (see Chapter 2).

Selected References

Chepko-Sade, B. D., and Z. T. Halpin. 1987. *Mammalian dispersal patterns.* Chicago: University of Chicago Press.

Gaines, M. S., and L. R. McClenaghan, Jr. 1980. Dispersal in small mammals. *Annual Review of Ecology and Systematics* 11:163–169.

Greenwood, P. J., and P. H. Harvey. 1982. The natal and breeding dispersal in birds. *Annual Review of Ecology and Systematics,* 13:1–21.

Grime, J. P. 1979. *Plant strategies and vegetative processes.* New York: Wiley. Role of stress in plant populations.

Harper, J. L. 1977. *Population biology of plants.* New York: Academic Press.

Houston, D. B. 1982. *The northern Yellowstone elk: Ecology and management.* New York: Macmillan. Excellent population study. The role of density-dependent mortality is discussed.

Krebs, J., and N. B. Davies, eds. 1984. *Behavioural ecology: An evolutionary approach,* 2nd ed. Oxford, England: Blackwell Scientific Publications. Chapters on aspects of territoriality and related topics.

Mech, L. D. 1970. *The wolf: The ecology and behavior of an endangered species.* Garden City, NY: Doubleday.

Myers, K., C. S. Hale, R. Mykytowycz, and R. L. Hughes. 1971. The effects of varying density and space on sociality and health in animals. Pp. 148–187 in A. H. Esser, ed., *Behavior and environment: The use of space by animals and men.* New York: Plenum.

Nice, M. M. 1937. *Studies in the life history of the song sparrow I.* Transactions of the Linnaean Society of New York. IV. Reprint edition. New York: Dover Publications. A classic study in territoriality.

Sinclair, A. R. E. 1977. *The African buffalo: A study of resource limitations of populations.* Chicago: University of Chicago Press.

Smith, S. M. 1978. The "underworld" in a territorial adaptive strategy for floaters. *American Naturalist* 112:570–582. Excellent study on behavior of floaters.

Tamarin, R. H., ed. 1978. *Population regulation.* Benchmark Papers. Stroudsburg, PA: Dowden, Hutchinson, and Ross. Collection of important papers on the topic.

Zimen, E. 1981. *The wolf: A species in danger.* New York: Delacourt Press. Excellent behavioral study of the European wolf.

CHAPTER 14

Life-History Patterns

Outline

Objectives

On completion of this chapter, you should be able to:

1. Discuss the importance of energy allocation to reproduction.
2. Compare semelparity with iteroparity.
3. Discuss the relationship between the number of young and parental investment in them, and the ecological significance of the relationship.
4. Show the relationship among size, age, and fecundity and discuss possible reasons.
5. Compare *r*-selection with *K*-selection and discuss the values and shortcomings of the concepts.
6. Discuss the various modes of sexual reproduction, including sexual selection and mating systems.
7. Define a mating system and distinguish among monogamy, polygamy, polygyny, and polyandry.

Ecology in recent years has acquired certain anthropomorphic terms, such as *strategy* (life-history strategy, reproductive strategy, optimal foraging strategy, mating strategy) and *tactics* (reproductive tactics, foraging tactics). The use of these terms in ecology is unfortunate in some ways, because they imply conscious, detailed planning on the part of living organisms toward a desired end or goal—that oak trees and warblers consciously set out a course in their life cycles. Nothing could be further from the real situation.

The term *life-history strategy* in ecology means the selective processes involved in achieving fitness by living organisms. Such processes involve, among other things, fecundity and survivorship; physiological adaptations; modes of reproduction; age at reproduction; number of eggs, young, or seeds produced; parental care; means of avoiding environmental extremes; size; and time to maturity. Success among individuals and species is measured only in terms of successful offspring. How this success is achieved becomes the organism's life-history strategy or pattern. This pattern is the outcome of evolution and represents the best that natural selection could provide given the gene pool available.

An ideal situation for any organism is to reproduce shortly after birth, produce a large number of highly adapted offspring, live for a long period of time, reproduce frequently, lavish parental care on the offspring, and devote the maximum amount of energy to reproduction. Realistically, any organism has to compromise considerably with the ideal to achieve optimal fitness. The various means by which that goal is achieved are the subject of this chapter.

REPRODUCTIVE EFFORT

ALLOCATION OF ENERGY

Energy captured by an organism (see Chapter 22), like income used to run a household or business, must be allocated to certain essential uses. Some must go to growth, to maintenance, to the acquisition of food, to defending territory and escaping predators; and some has to go to reproduction. To achieve optimal fitness, an organism has to budget its energy carefully. How it budgets that energy and time in reproduction makes up the organism's *reproductive effort*.

The more energy an organism spends on reproduction, the less it has to budget for growth and maintenance. As a result, the individual may grow more slowly to the next age or fail to survive. For example, reproducing females of the terrestrial isopod *Armadillidium vulgare* have a lower rate of growth than nonreproducing females. Nonreproducing females devote as much energy to growth as reproducing females devote to both growth and reproduction.

How much energy organisms invest in reproduction varies. Energy costs involve not only the weight of the progeny but also the amount of energy expended in rearing the young. Herbaceous perennials may invest between 15 and 25 percent of annual net production in new plants, including vegetative propagation. Wild annuals (single-stage reproducers) expend 15 to 30 percent; most grain crops, 25 to 36 percent; and corn and barley, 35 to 40 percent. The lizard *Lacerta vivipara* invests 7 to 9 percent of its annual energy assimilation in reproduction. The female Allegheny Mountain salamander, *Desmognathus ochrophaeus,* spends 48 percent of its annual energy flow on reproduction, including energy stored in eggs and energy costs of brooding.

TIMING OF REPRODUCTION

If an organism is to contribute its maximum to future generations, it has to balance the profits of immediate reproductive investment against the costs of future prospects, including fecundity and its own survival, a trade-off between present progeny and future offspring.

One strategy is to invest all energy into growth, development, and storage and then expend the energy in one massive reproductive effort and die. Such a reproductive strategy is employed by most insects and other invertebrates, some species of salmon, annual plant, and some bamboos. Certain bamboos delay flowering for 100 to 120 years, produce one massive crop of seeds, and die. These organisms sacrifice future prospects by expending all their energy in one suicidal act of reproduction. That mode of reproduction is *semelparity.*

Another strategy is to produce fewer offspring in a series of separate reproductive events, after which the organism survives to reproduce again. That strategy is called *iteroparity*. For an iteroparous organism the problem becomes one of timing reproduction—early in life or at a later age. Early reproduction reduces survivorship and the potential for later reproduction. Later reproduction increases growth of organisms and improves survivorship but reduces fecundity. Energy expended in repeated reproduction weighs against future prospects of fecundity and survival. Mammals, for example, experience increased mortality during and after each reproduction (Figure 14.1).

Most organisms make a trade-off between present progeny and future offspring and survival. The nature of these trade-offs varies from species to species and within species. If an individual grows rapidly while investing no energy in reproduction and is short-lived as an adult, its reproductive strategy will probably be a single reproductive effort. Any delays in the production of offspring reduce the probability that the individual will survive to reproduce. If an organism is longer-lived and its chances for survival as an adult are high, it can reproduce more than once throughout a lifetime. Under these conditions the strategy is usually delayed reproduction. However, some semelparous organisms exhibit delayed reproduction. Mayflies may spend several years as larvae before emerging to the surface of the water for an adult life of several days devoted to reproduction. Certain Pacific salmon do not reproduce until they are 3 to 4 years old. Periodical cicadas spend 13 to 17 years below ground developing before they emerge as adults to stage an outstanding exhibition of a single-term reproductive effort.

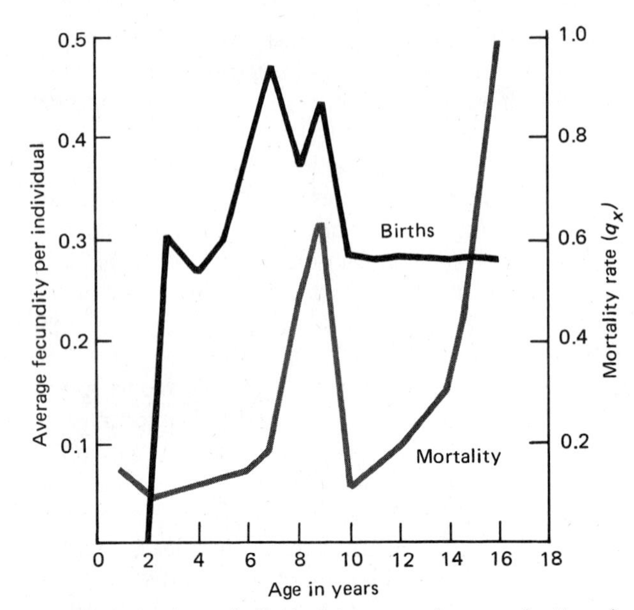

FIGURE 14.1 As an individual increases its reproductive effort, it is likely to reduce its survivorship. In red deer, for example, an increased rate of birth results in an increased rate of mortality. Energy expended in repeated reproduction works against the potential for survival.

PARENTAL INVESTMENT AND NUMBER OF YOUNG

The number of offspring produced in each reproductive effort may range from many small ones to a single large individual. The number produced relates to the amount of parental investment per individual offspring. If a parent produces a large number of young, it can afford only minimal investment per individual. In such cases, animals provide no parental care, and plants store a minimal amount of energy in seeds. Such organisms usually inhabit disturbed sites or unpredictable environments, or places such as the open ocean, where opportunities for parental care are minimal or difficult at best. By dividing energy allocated for reproduction among as many young as possible, parents increase the chances that some young will successfully settle somewhere. By such a strategy parents increase their own fitness but decrease the fitness of the young.

Parents that produce few young are able to expend more energy on each individual. The amount of energy will vary, of course, depending upon the number of young produced, their size, and their stage of maturity when born. Some organisms expend less energy during incubation or gestation and invest more energy after birth of the young. The young are born or hatched in a helpless condition and require considerable parental care. Such animals are called *altricial*. Other animals have longer incubation periods or longer gestation periods, so the young are born in an advanced stage of development and are able to forage for themselves shortly after birth. Such young are called *precocial*.

Examples are gallinaceous birds and ungulate mammals. Plants may produce relatively few large seeds with large amounts of stored energy for the germinating seedling. Examples are acorns, walnuts, hickory nuts, and coconuts. In such cases, parents decrease their own fitness but increase the fitness of their young.

The number and size of young depend upon the selective pressures under which the organism evolved or to which it must adjust. Among plants, short-lived annuals have numerous small seeds. High production of numerous offspring ensures that some will survive and germinate the next growing season. Optimum seed size and number for perennial plants relate to dispersal ability, colonizing ability, and the need to escape predation. Plants that colonize disturbed or unpredictable environments may have small, windblown seeds that can be carried great distances. Because a minimal amount of energy is invested in each seed, these plants can afford heavy losses of seeds to ensure that some successfully survive and reproduce. Plants subject to heavy predation may have small seeds that provide less attractive food packages. Again, an abundance of seeds is an insurance that some will escape predation (see Chapter 16).

Plants associated with more stable environments may produce fewer and larger seeds, with a large store of energy that the seedling can use to get established. Such plants may invest a considerable amount of energy in the production of toxins or heavy seed coats to reduce predation. Some plants may adjust their production of seeds to meet current environmental conditions. Such adjustments may involve a genetic polymorphism. For example, the common weed purslane speedwell, *Veronica peregrina,* a plant of moist soils, growing in an environment where moisture conditions are optimal produces few but heavy seeds. The same species growing where moisture is limited and competition from grass is keen grows taller and produces lighter seeds that are better able to disperse. In other cases the adjustment may involve a plastic response, as with the annual *Polygonum cascadense.* Plants growing on harsh, open environments allocate proportionately more of their energy to reproduction than those growing in more moderate habitats.

Animals exhibit similar adjustments. "Annual" species, such as insects that overwinter in the egg stage, produce enormous numbers of small eggs. Birds such as quail and mammals such as rabbits and mice that are subject to heavy predation produce large numbers of young.

One way by which animals can apportion energy to reproduction is through the adjustment of clutch size and litter size. David Lack proposed that clutch size in birds evolved to equal the average largest number of young the parents can feed. Thus, clutch size is an adaptation to food supply. Temperate species, he argued, have larger clutches because increasing daylength provides the parents a longer time to forage for food to support large broods. In the tropics, where daylength does not change, foraging time is limited.

Martin Cody modified Lack's ideas by proposing that clutch size results from different allocations of energy to egg production, avoidance of predators, and competitive ability. In temperate regions periodic local climatic catastrophes can hold a population below carrying capacity. Natural selection then favors a high rate of increase and thus larger clutches on the average than in tropical regions. The predictability of the tropical climate makes the maintenance of carrying capacity more important than increased production of young. More energy is expended in avoiding predation and in meeting competition.

These theories are supported by field examples. Birds in temperate regions do have larger clutch sizes than those in the tropics (Figure 14.2). Mammals at higher latitudes have larger litters than those at lower latitudes. Lizards living at lower latitudes have smaller clutches and higher reproductive success than those living at higher latitudes. Even among insects, tropical species produce fewer eggs and fewer clutches than their temperate-region counterparts.

Within a given region, production of young may reflect abundance of food, and the optimal clutch size may not be the average for the region. The optimal size may be more flexible, a response to available resources. T. M. Spight and J. Emlen provided two marine snails, *Thais lamellosa* and *T. emarginata,* with an increased food supply. Adult *T. lamellosa* increased their average egg clutch sizes from 930 to 1428, and *T. emarginata* spawned more times during the year. Likewise, within local populations, individual pairs of European magpies (*Pica pica*) have variable clutch sizes

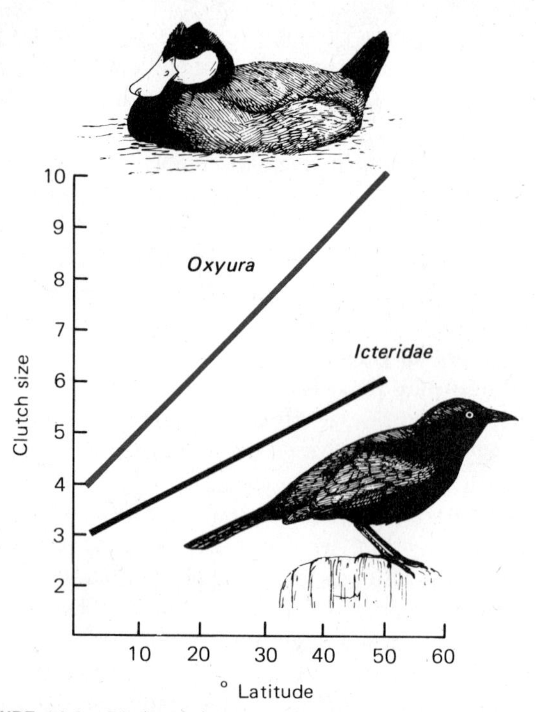

FIGURE 14.2 Birds of the same family tend to have smaller clutches at low latitudes than at high latitudes. Clutch size of the family Icteridae (blackbirds) and for the genus *Oxyura* (ruddy ducks) of the family Anatidae illustrates this trend.

during the same year. Large clutches and high adult survival are associated with high territorial quality. Average clutch size, optimal for the birds, goes with average territories. Birds able to acquire the highest-quality territories have the largest clutches.

AGE, SIZE, AND FECUNDITY

A relationship exists between fecundity and size and age. Among plants and poikilothermic animals, in particular, fecundity increases with size and age. Among homoiotherms the relationship is not so clear-cut. Age at maturity and seed production vary among trees. Some, like Virginia pine (*Pinus virginiana*), will start to produce seed at 5 to 8 years when growing in open stands. In dense stands Virginia pine may delay seed production for 50 years. Quaking aspen *(Populus tremuloides)* produces seeds at the age of 20 years and white oak not until the age of 50 years. Acorn production varies with size. Trees 40 cm in diameter at breast height produce around 700 acorns; those 60 to 66 cm produce 2000 or more acorns.

Perennial herbaceous plants and even annuals exhibit size and fecundity relationships. Among the *Plantago* species, for example, annuals have a higher reproductive effort and seed output per unit of leaf area than perennial species; but perennials with a larger average leaf area produce a greater weight of seeds per plant. Perennials delay flowering until they have attained a sufficiently large leaf area. Flowering plants are larger than nonflowering ones. Annuals show no such relationship between leaf area and reproductive effort, which seems to be independent of plant size once reproduction starts; but size differences among annuals do result in differences in seed production. Small plants produce few seeds, even though the plants themselves may be contributing the same proportionate share to reproductive efforts. Annual jewelweeds (*Impatiens* spp.) growing in optimal environments produce both outcrossed and self-fertilized flowers, the latter produced late in the season (see "Mating Strategies"). Jewelweeds growing in stressed environments produce only a few self-fertilized flowers and seed early in the season.

Similar patterns exist among poikilothermic animals. Fecundity in fish increases with size, which, in turn, increases with age. Because early fecundity reduces both growth and later reproductive success, there is a selective advantage to delaying sexual maturation until more body growth is achieved. Gizzard shad (*Dorosoma cepedianum*) reproducing at 2 years of age produce 59,000 eggs. Those delaying reproduction until the third year produce 379,000 eggs. Egg production then declines to about 215,000 eggs in later years. Among these fish only about 15 percent spawn at 2 years of age; 80 percent breed at 3 years.

Delayed maturation is characteristic of fish and other poikilotherms. Among 104 species of freshwater fish in the United States, only 15 percent mature the first year, and 21 percent the second year. Sixty-four percent of all species mature at 3 years of age or older. By contrast, 55 percent of 171 species of placental mammals reach reproductive maturity at 1 year of age, 20 percent the second year. Only 25 percent mature at the age of 3 or older.

An increase in fecundity is associated with a decrease in body mass and an increase in the rate of metabolism. Such relationships exist over a wide range of unicellular and multicellular organisms, including poikilotherms and homoiotherms. A close correlation exists among fecundity, metabolism, and food among mammals. Mammals with a high rate of metabolism have higher fecundity than mammals with a low rate of metabolism. Because basal metabolism in mammals increases with a decrease in size, small mammals have a higher metabolic rate than large ones.

Metabolic activity apparently is also related to quantity and quality of food in space and time. Mammals that feed on tree leaves, dried fruits and seeds, and insects have lower rates of metabolism than grazing ungulates, most rabbits and hares, microtine rodents, and carnivores. A seed-eating rodent with the same body size as a grazing rodent would have a lower rate of metabolism and lower fecundity. Unlike some poikilotherms, mammals cannot divert energy from maintenance to reproduction. They all have certain basic metabolic costs needed to maintain homeostasis, which cannot be changed. Energy above that is available for reproduction. All mammals, then, probably have as high a rate of increase as their metabolisms will support.

r-SELECTION AND *K*-SELECTION

The review of reproductive strategies suggests two points: (1) Species living in harsh or unpredictable environments in which mortality is largely independent of population density allocate more energy to reproduction and less to growth, maintenance, and adjustments to the environment. (2) Species living in a stable or more predictable environment in which mortality results from density-related factors and competition is keen allocate more energy to nonreproductive activities. The former are known as ***r-strategists,*** because environmental conditions keep growth of such populations on the rising part of the logistic curve. In such situations, genotypes with a high *r* are favored. The latter are called ***K-strategists*** because they are able to maintain their densest populations at equilibrium (asymptote) or carrying capacity.

According to this concept, *K*-strategists and *r*-strategists are under different selection pressures. *r*-selection favors those genotypes that confer the highest possible intrinsic rate of increase; rapid development; early maturity; small body size; short life span; early, even single reproductive effort; a large number of offspring; and minimal parental care. As a result, *r*-strategists typically are opportunists; they have the ability to colonize temporary or disturbed habitats where competition is minimal. They have means of wide dispersal and respond quickly to disturbance to the population. *K*-selection favors genotypes that confer the ability to cope with physical and biotic pressures, to tolerate relatively high population densities; they have delayed reproduction, slower development, larger body size, relatively long life span, repeated reproduction, and few offspring. Animals care for young; plants produce seed with stored food. *K*-strategists are resource-limited, and the population remains at or near carrying capacity. These qualities, along with the lack of a means of wide dispersal, make *K*-strategists poor colonists.

The original intent of the theory of *r*-selection and *K*-selection was to serve as a model of density-dependent natural selection. Within that context it is a useful concept, if you recognize its limitations. Frequently the concept is used to distinguish equilibrium (*K*) from nonequilibrium (*r*) populations without considering age structure. Traits favored in populations with low adult mortality include longer life span, slower development, delayed reproduction, and high investment in young, all *K*-traits. In an increasing, nonequilibrium population, *r*-traits, including faster development, earlier reproduction, and perhaps less investment in young, may predominate. At one stage in its life history, a population may exhibit *r*-traits and at another time *K*-traits.

Another problem is that density dependence is the heart of the concept. *K*-species or equilibrium species are assumed to be at a density-dependent carrying capacity. However, fluctuating environments can influence *K* by changing the resource base, by increasing or decreasing the patchiness of the environment, and by influencing density-independent mortality, which in turn can select for some particular life-history trait more related to *r* than to *K*.

r-selection and *K*-selection can be considered as poles of a continuum from *r* to *K*. Such a view tempts a classification of species as either *r*-selected, *K*-selected, or somewhere in between, but we can hardly compare elephants with mice. The concept of *r*-selection and *K*-selection is most useful when applied to individuals within a population or to populations within a species. Under certain conditions individuals or populations will exhibit *r*-selected traits or *K*-selected traits. Meadow mice living in situations where dispersal can take place, reducing the population below carrying capacity, easily exhibit characteristics of *r*-selection, whereas those living under conditions in which there is no place to disperse assume *K*-selected characteristics.

Because of the difficulty of forcing plants into *r* and *K* categories, J. Grime has proposed the idea of ruderal, or *R*-strategists; competitive, or *C*-strategists; and stress-tolerant, or *S*-strategists. *R*-strategists reproduce early in life, possess high fecundity, experience lethal reproduction (semelparity), occupy uncertain or disturbed habitats, and have well-dispersed seeds. *C*-strategists and *S*-strategists occupy more stable environments and are relatively long-lived, often drastically reducing the opportunity for seedling establishment, resulting in high juvenile survival. Beyond these characteristics, the two have evolved quite different life-history strategies. *C*-strategists reproduce early and repeatedly, utilizing an annual expenditure of energy stored prior to seed production. *S*-strategists have delayed maturity, intermittent reproductive activity, and long-term energy storage.

MATING STRATEGIES

Sexual reproduction is common to most multicellular organisms. Even those that rely primarily on asexual or vegetative reproduction revert, even if infrequently, to sexual reproduction. Sexual reproduction allows the gene pool to become mixed, increasing the genetic variability necessary to meet changing selective pressures and to prevent an accumulation of harmful mutations. However, for the individual, sexual reproduction is expensive. Each individual can contribute only one-half of its genes to the next generation. The success of that contribution depends upon a member of the opposite sex. For that reason each individual must acquire the best possible mate.

TYPES OF SEXUAL REPRODUCTION

Sexual reproduction can take a variety of forms. The most familiar is separate male and female individuals. Plants with that characteristic are called *dioecious,* and examples are holly trees and stinging nettle *(Urtica)*. Some organisms may possess both male and female sex organs. They may be *monoecious* or *hermaphroditic.* There is a difference. Hermaphrodites have both male and female organs in the same individual. Among plants that means flowers contain both stamens and ovules. Among animals individuals possess both testes and ovaries, a condition common to a number of invertebrates such as earthworms. Monoecy is restricted to plants, in which the individual possesses separate male and female flowers, such as birch *(Betula)* and aspen *(Populus)* trees. Some hermaphrodites are simultaneous, others sequential. The latter type are one sex when young and develop into the opposite sex when mature.

Such sex reversal seems to be stimulated by a social change involving sex ratio in the population. Sex reversal among several species of marine fish can be initiated by the removal of one or more individuals of the other sex. Among some coral fish, removal of females from a social group stimulates an equal number of males to change sex and become females. Among other species, removal of males stimulates a one-to-one replacement of males by sex-reversing females. Some mollusks and bivalves undergo sex change, but the change is exclusively from male to female. Among these organisms, male size has little effect on breeding success, but the most fecund females are large. So in the presence of many males, a small individual could improve his fitness by switching to and growing rapidly as a female.

In general, a sex change is favorable if differences in reproductive success result from differences in size in both sexes and if the individual can increase his reproductive value if he switches sex. Under conditions of intense male competition, in which only the largest males mate, it may pay for the small male to become a small female, because all females breed, even though female fecundity increases with size. A male can breed only when large and a successful competitor. When male competition is less intense, it may pay for the male to start life as a male because all males breed, and then switch to the female sex when he is larger and older. Such sex changes, in some species at least, involve hormonal factors associated with the central nervous system or the gonads.

Plants also exhibit gender change. One such plant is jack-in-the-pulpit, *Arisaema triphyllum*, a clonal woodland herb whose genet is a perennial corm and whose ramet is a single annual shoot. Jack-in-the pulpit produces staminate (male) flowers one year, an asexual vegetative shoot the next, and a carpellate (female) or monoecious shoot the next. Over its life span a jack-in-the pulpit may produce both genders as well as a nonsexual vegetative shoot, but in no particular sequence. Usually, an asexual stage follows a gender change. Gender changes in some plants are normally stimulated by environmental changes in moisture and light. Gender change in jack-in-the-pulpit appears to be triggered by an excessive drain on its photosynthate by female flowers. If the plant is to survive, one carpellate flowering could not follow another. To avoid death, the plant reduces its reproductive effort the next year by changing its gender or becoming vegetative.

Mating in hermaphroditic and monoecious species can involve either outcrossing, in which genetic variability is assured, or self-fertilization, which involves no genetic change. Apparently, most hermaphroditic animals are not self-fertilized. Earthworms mate with other earthworms, for example. Some animal hermaphrodites are completely self-sterile. A few, like certain land snails, are self-fertilized. Many hermaphroditic plants are self-compatible but have evolved different means to prevent self-fertilization. Anthers and pistils may mature at different times, or the pistil may extend well above the stamens. Other plants have evolved more effective means. One is a genetic mechanism that prevents the growth of a pollen tube down the style of the same individual. Other hermaphroditic species may be divided into two or three morphologically different types. Pollination can take place only between members of different types.

Although many hermaphroditic organisms possess mechanisms to reduce self-fertilization, the capability of self-fertilization in hermaphrodites does carry certain advantages, especially among plants. A single self-fertilized individual is able to colonize a new habitat and then reproduce itself, establishing a new population. Other hermaphrodites produce self-fertilized flowers under stressed conditions, ensuring a new generation. Jewelweed, for example, under normal and optimal environmental conditions produces cross-pollinated flowers. Under adverse environmental conditions or after the cross-fertilized flowers have set seed, jewelweed (as well as violets and other species) produces tiny, self-fertilized flowers that never open. They have vestigial petals, no nectar, and few pollen grains and remain in a green, budlike stage. Thus, if outcrossing fails or never develops, the plants have ensured a next generation by self-fertilization.

SEXUAL SELECTION

Possessing an optimal reproductive strategy is of little consequence unless the organism has some means not only of bringing the sexes together but also of enabling individuals to choose a partner with the greatest possible fitness. That choice involves some form of *sexual selection*. Sexual selection consists of members of one sex competing among themselves to mate with members of the other sex (intrasex competition) and members of the other sex (usually female) showing preference for those that win. Supposedly, males (and in some situations females) that win the intrasex competition are the fittest, and those who choose them ensure their own fitness.

How does a female select a mate with the greatest fitness? That is still a largely unanswered question. Actually, the female has little to go on. She might select a winner from among males that bested others in combat—as in bighorn sheep, elk, and seals—or ritualized display. She might base her choice on intensity of courtship display. Whatever the situation, the selection process comes down to salesmanship on the part of the male and sales resistance on the part of the female. The female may attempt to elicit as much courtship behavior as possible from a potential mate. Females, for example, may force males to display in groups, as in prairie chickens and in some frogs and toads. They usually choose the larger and probably older males. In those species in which males monopolize a group of females, such as elk and seals, the females accept the dominant male; but even in that situation the females have some choice. Protestations by female elephant seals over the attention of a dominant male may attract other large males nearby who may attempt to dislodge the male from the group. Such behavior ensures that the females will mate only with the highest-ranking male.

Often females choose males who offer the highest-quality territories. In that situation the question is whether the female selects the male and accepts the territory that goes with him or whether she selects the territory and accepts the male that goes with it. By choosing a male with a high-quality territory, the female can best assure her own fitness.

Both male and female seek to assure their own maximum fitness; but what increases male fitness is not necessarily what improves female fitness. Sperm are cheap. With little investment involved, males should mate with as many females as possible to achieve maximum fitness. Because females invest considerably more in reproduction, it is to their advantage to be more selective in choosing a mate. In general, they should refuse to mate with a promiscuous male or one with several mates. Instead, they should favor one-to-one relationships in which the male helps to care for the young. For males such a relationship carries certain risks. The male must guard the female from other males to ensure that the offspring he helps to raise are really his own. Females are fairly sure that the offspring they raise are their own.

T. H. Clutton-Brock and his associates have spent many years studying the ecology and behavior of red deer on the Isle of Rhum off the coast of Scotland. They were able to follow the reproductive success of a number of stags and hinds. For the stags, differences in lifetime reproductive success resulted from differences in harem size, time the stags were able to hold a harem, quality of the rutting area, and longevity. The first two factors, related to body size and condition, and thus to fighting ability, were most important. Over 50 percent of the stags over four years of age failed to breed, and of the successful stags, only 5 percent sired more than four calves a year. Reproductive activity of the males rarely exceeded four years, the length of time a stag could hold a harem.

Although we might expect that successful polygamous males would exhibit greater lifetime reproductive success than the hinds, that was not the case. Lifetime reproductive success of hinds was influenced mostly by the quality of the home range and the size of the matrilineal group, with reproductive success greatest in the smaller groups. Success among hinds was much less variable than among stags. This lesser variability related in part to the longer reproductive period of hinds and the consistent breeding success of certain hinds. By contrast, reproductive success in males had pronounced age variations. Such differences between the two sexes led Clutton-Brock to suggest that in polygamous mammals we systematically overestimate the strength of sexual selection.

Sexual selection is considered mostly an attribute of animals. Although Darwin recognized such selection in plants, we have difficulty imagining how plants would engage in sexual selection. Mostly, sexual selection in plants is associated with the evolution of dioecy from largely hermaphroditic plants. It may involve largely intrasexual competition among hermaphroditic flowers for the dispersal of pollen. Selective advantage would go to those with the largest flowers that provided the most pollen and nectar. The more energy these flowers put into pollen, the less energy is available for seed production by the female component. If some hermaphrodites carry a mutant gene for partial female sterility, all their energy could be directed toward pollen production. Such flowers would have a selective advantage over normal hermaphrodites. At the same time, male-sterile flowers would not waste energy on pollen production. They would have a selective advantage over normal and female-sterile hermaphrodites. They could channel most of their resources to seed production, not to pollen. Ultimately, that type of sexual selection could lead to the evolution of dioecy.

Fitness of flowering plants depends upon male pollen reaching the female stigma. Males are under selective pressures of delivering pollen, reducing wastage of pollen, and thus reducing allocation of resources to pollen production. Females receive an array of pollen from different males. Genetic quality of pollen may influence the growth of pollen tubes to the ovule. Pollen carried from different plants by insects may result in multiple paternity. In the wild radish, for example, pollen from a number of male plants may fertilize different ovules in one female plant. That situation increases genetic diversity in the female's offspring as well as size of the seed head. Thus fertilization by pollen from a number of plants increases the fitness of that female over others receiving all their pollen from one male.

Sexual selection in both plants and animals involves major selective processes acting on both sexes during the whole reproductive process. The role of intrasexual competition among male animals, however, is more easily observed and better understood than female sexual selection. We still do not know what criteria females use to make the important choice of the fittest male. We still have difficulty separating intramale competition for mates from female selection of males.

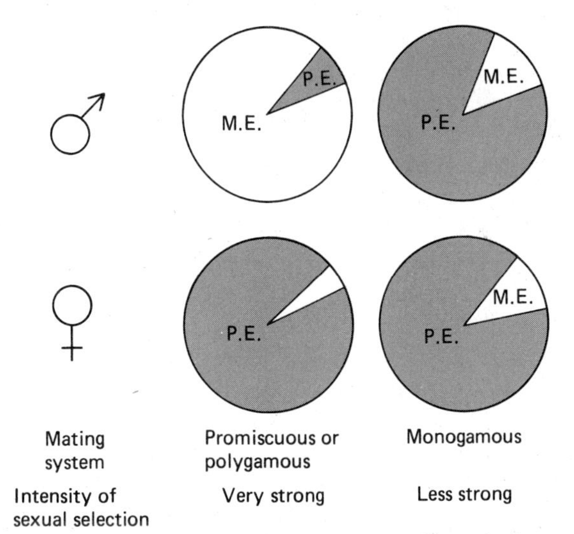

FIGURE 14.3 Reproductive effort, represented by a circle, can be partitioned into parental effort (shaded area) and mating effort (open area). In general, males put more effort into mating than females, but in monogamous species parental efforts are nearly equal.

MATING SYSTEMS

The behavioral mechanisms and the social organization involved in obtaining a mate are called a *mating system*. A mating system includes such aspects as the number of mates acquired, the manner in which they are acquired, the nature of the pair bond, and the pattern of parental care provided by each sex (Figure 14.3).

The structure of mating systems runs from monogamy through many variations of polygamy. *Monogamy* is the formation of a pair bond between one male and one female, and both parents typically care for the young. Monogamy is prevalent among insectivorous birds, carnivorous birds and mammals, certain herbivorous mammals such as beaver and muskrats, gibbons, and humans. *Polygamy* is the acquisition by an individual of two or more mates, none of which is mated to other individuals. A pair bond exists between the individual and each mate. There are two kinds of polygamy. One is *polygyny*, in which an individual male gains control of or access to two or more females. In this mating system, males put most effort into mating and females into parental care. The other is the less common *polyandry*, in which a female gains control of or access to two or more males. In this system, found in some shore birds, males invest most effort in parental care and females in mating and laying eggs. Mating systems in which one male has several short pair bonds with different females in sequence, as in grouse and white-tailed deer, are called *polybrachygyny*. Finally, there is *promiscuity*, in which males and females copulate with one or many of the opposite sex but form no pair bonds. True promiscuity is rare or nonexistent in the natural world. The various mating systems have been classified by Emlen and Oring according to the means by which individuals gain access to the limited sex, male or female.

Summary

Life-history strategy includes all the selective processes used by an organism to achieve optimal fitness. These processes include, among others, fecundity, survival, mode of reproduction, age at reproduction, parental care, number of young produced, and time to maturity—all related to the amount of energy allocated to reproduction.

To achieve optimal fitness, an organism has to balance immediate reproductive efforts against the costs of future prospects. At one extreme is semelparity, in which organisms expend all their energy in one lethal reproductive effort. At the other end is iteroparity, in which organisms produce fewer young at one time but have repeated reproduction through a lifetime.

The number of young produced relates to the amount of parental investment. Organisms that produce a large number of offspring have a minimal investment in each offspring. They can afford to send a large number into the world with a chance that a few will survive. By so doing, they increase parental fitness but decrease the fitness of the young. Organisms that produce few young invest considerably more in each individual, providing them with a greater ability to survive. Such organisms increase the fitness of the young at the expense of the fitness of the parents.

A large number of young is characteristic of annual plants, short-lived mammals, insects, and semelparous species. Few young are characteristic of long-lived species. Iteroparous species may adjust the number of young in response to environmental conditions and the availability of resources. In general, clutch and litter sizes increase from the tropics toward the poles.

A direct relationship between size and fecundity exists among plants and poikilotherms. The larger the size, the more young are produced. Mammals exhibit a relationship between fecundity and metabolism. Mammals with a high rate of metabolism have high fecundity. Because basal metabolism decreases with size, small mammals have both high metabolism and a high fecundity.

A review of reproductive strategies suggests two types of selection: r-selection and K-selection. r-selection favors genotypes with a high rate of increase, rapid development, early maturity, small body size, short life span, large numbers of offspring, minimal parental investment in individual young, and minimal parental care. K-selection favors genotypes that tolerate high population densities, delayed reproduction, slow development, large body size, long life, fewer offspring, large parental investment in individual young, parental care in animals, and large seeds with stored food in plants.

Several mating strategies are associated with sexual reproduction. Sexual reproduction commonly involves individual males and individual females. Plants with separate males and females are called dioecious. Organisms possessing both male and female sex organs may be monoecious or hermaphroditic. Hermaphrodites have both male and female organs in the same individual or flower. Monoecy, restricted to plants, involves separate male and female flowers on the same individual.

An important component of mating strategy is sexual selection. In general, males compete among themselves for the opportunity to mate with the other sex, and females show a strong preference for the winner. Supposedly, males that win intrasexual competition are the most fit. By choosing among those males, females ensure their own fitness. There is some evidence for sexual selection among plants. It appears to be associated with the evolution of dioecy from largely male-dominated hermaphroditic flowers for the dispersal of pollen.

Review and Study Questions

1. What is meant by energy allocation? How does it affect reproduction and growth?
2. Compare semelparity with iteroparity. What environmental conditions favor semelparity?
3. Discuss the range of iteroparity from a few young at a late age to many small litters at an early age. What are the advantages and disadvantages of reproduction at an early age?
4. Contrast parental investment in many young and few young. Under what conditions should each of the investments be made?
5. Compare altricial and precocial young. How do parental investments in each differ?
6. How might plants and animals apportion energy to reproduction according to resource availability?
7. What is the relationship between size and fecundity in plants and poikilothermic animals? Does the size relationship hold among homeotherms, especially mammals? Explain.
8. What is the relationship between fecundity and metabolism?
9. What is the difference between r-selected and K-selected organisms?

10. What are some weaknesses of the above concept? In what situations is it useful?
11. What is dioecy? What is the difference between a monoecious and a hermophroditic organism?
12. What are some selective advantages of hermaphroditism?
13. What is sexual selection? In what ways does selection differ between males and females?
14. How might sexual selection evolve in plants? For more information see Willson 1983, Willson and Burley, 1983.
15. Define a mating system and distinguish among monogamy, polygamy, polygyny, and polyandry.

Selected References

Bajema, C. J., ed. 1984. *Evolution by sexual selection theory.* Benchmark Papers in Systemic and Evolutionary Biology. New York: Scientific and Academic Editions.

Boyce, M. S. 1984. Restitution of *r*- and *K*-selection as a mode of density-dependent natural selection. *Annual Review of Ecology and Systematics,* 15:427–447.

Clutton-Brock, T. H., F. E. Guinness, and S. D. Albion. 1982. *Red deer: Behavior and ecology of two sexes.* Chicago: University of Chicago Press. An outstanding study of reproductive behavior and success based on a long-term study of known individuals.

Cody, M. L. 1966. A general theory of clutch size. *Evolution,* 20:174–184.

Ellstrand, N. C. 1984. Multiple paternity within the fruits of wild radish *Raphanus sativus. American Naturalist,* 123:819–828.

Emlen, S. T., and L. W. Oring. 1977. Ecology, sexual selection, and the evolution of mating systems. *Science,* 197: 215–223.

Freeman, C. L., K. T. Harper, and E. L. Charnov. 1980. Sex change in plants: Old and new observations and new hypotheses. *Oecologica,* 47:222–232.

Grime, J. P. 1979. *Plant strategies and vegetative processes.* New York: Wiley.

Gubernich, D. J., and P. H. Klopher, eds. 1981. *Parental care in mammals.* New York: Plenum.

Krebs, J. R., and N. B. Davies. 1981. *An introduction to behavioral ecology.* Oxford: Blackwell Scientific Publications. Excellent discussions of topics covered in this chapter.

Krebs, J. R., and N. B. Davies, eds. 1984. *Behavioral ecology: An evolutionary approach,* 2nd ed. Oxford: Blackwell Scientific Publications.

Lack, D. 1954. *The natural regulation of animal numbers.* Oxford: Oxford University Press (Clarendon Press).

McGregor, P. K., J. R. Krebs, and C. M. Perrins. 1981. Song repertoires and lifetime reproductive success in the great tit (*Parus major*). *American Naturalist,* 118:49–59. A study of reproductive success of male and female birds.

Primack, R. B. 1979. Reproductive effort in annual and perennial species of *Plantago* (Plantaginaceae). *American Naturalist,* 114:51–62.

Shapiro, D. Y. 1980. Serial sex changes after simultaneous removal of males from social groups of coral reef fish. *Science,* 209:1136–1137.

Spight, T. M., and J. Emlen. 1976. Clutch size of two marine snails with a changing food supply. *Ecology,* 57:1162–1176.

Thornhill, R., and J. Alcock. 1983. *The evolution of insect mating systems.* Cambridge, MA: Harvard University Press.

Warner, R. R. 1988. Sex change and size-advantage model. *Trends in Ecology and Evolution,* 3:133–136.

Wasser, S. K., ed. 1983. *Social behavior of female vertebrates.* New York: Academic Press.

Willson, M. F. 1983. *Plant reproductive ecology.* New York: Wiley.

Willson, M. F., and N. Burley. 1983. *Mate choice in plants: Tactics, mechanisms, and consequences.* Princeton, NJ: Princeton University Press.

Wilson, E. O. 1980. *Sociobiology: The abridged edition.* Cambridge, MA: Harvard University Press.

Outline

CHAPTER 15 Interspecific Competition

Objectives

On completion of this chapter, you should be able to:

1. Describe the various types of relationships between species.
2. Contrast interference and exploitation competition.
3. Describe four theoretical outcomes of interspecific competition.
4. Explain how potentially competing species may coexist.
5. Explain resource partitioning.
6. Define the niche and how it changes.

Although the most intense relationships exist between individuals of the same species, they do not live apart from individuals of other species. Living in close association, different species experience some sort of relationship. They may compete with one another for some shared resource, such as food, light, space, or moisture. One may depend upon the other as a source of food. Others may in some way mutually aid each other. Such interrelations can be beneficial or detrimental to either or both species involved, or they may have no effect at all.

RELATIONS BETWEEN SPECIES

If we designate a positive effect of one species on another as a +, a detrimental effect as a −, and no effect as a 0, we can express nine different ways in which a population may interact (Table 15.1). When neither of the two populations affects the other, the relationship is (00), or neutral. If the two populations mutually benefit each other, the interaction is (+ +), or positive. If the relationship is mutually detrimental, it is negative (− −). The remaining interactions are unequal.

When one species maintains or provides a condition necessary for the welfare of another but does not affect its own well-being by doing so, the interaction (0+), (0+) or (+0), is called *commensalism.* An example is an epiphytic plant growing on the trunk or limb of a tree. The tree provides support only, and the epiphyte gets its nutrients through its aerial roots. The interaction (0−) or (−0), in which one species reduces or adversely affects the population of another but the affected species has no influence, positive or negative, on the first, is called *amensalism.* Amensalism is difficult to demonstrate in the laboratory, and its role in nature is questioned. It may be considered a form of competition.

TABLE 15.1 Interspecific Relationships

	+	0	−
+	+ +	+ 0	+ −
0	0 +	0 0	0 −
−	− +	− 0	− −

+ +	Both benefit from the interrelationship: *mutualism.*
+ 0 0 +	Fitness of one increases; the other remains neutral: *commensalism.*
− + + −	Fitness of one is reduced; fitness of the other is increased: *predation* and *parasitism.*
0 − − 0	Fitness of one declines; fitness of the other is unchanged: *amensalism.*
− −	Fitness of both declines: *competition.*
0 0	No influence on each other: *neutralism.*

Ecologically more important is that relationship which benefits both populations (+ +), called *mutualism.* If the relationship is not essential for the survival of either population, it is *nonobligatory mutualism.* If the relationship is essential for survival of both populations, it is *obligatory mutualism.*

Mutualism and commensalism sometimes are collectively termed *symbiosis,* a word of Greek origin, meaning simply "living together." The term was introduced by the German botanist A. de Bary in 1897 to refer to both beneficial and adverse relationships between species, including parasites and their hosts. However, today we often equate the word with mutualism. Strong suggestions have been made to use the term in its original historical meaning to avoid confusion. Part of the argument for such broad usage is the lack of clear-cut distinctions along the continuum from parasitism to mutualism. In fact, the fungi-algae associations in lichens, once the standard example of mutualism, turn out to be controlled parasitism, in which the fungus is actually an obligatory parasite of the alga.

Other relationships of great importance to population ecologists are *competition* (− −), which is detrimental to populations of both species—each inhibits the other—and predation and parasitism, (− +) or (+ −), in which the population of one species benefits at the expense of another. *Predation* involves the killing and consumption of prey. *Parasitism* involves one organism's feeding on another, where the prey or host is rarely killed outright. The two, parasite and host, live together for some time. The host survives, although its fitness is reduced. If it does succumb, it is because in a weakened condition it is unable to resist other infections. It is not to a parasite's selective advantage to kill its host, for by doing so, it eliminates its own habitat and energy source.

A special type of predation and parasitism combined is *parasitoidism.* Parasitoids include certain wasps (Hymenoptera) and flies (Diptera) that lay their eggs in or on a single host individual, like a parasite. When the egg hatches, the larva develops on or inside the host, consuming it as the larva grows. By the time the larva arrives at the pupal stage, the host has been nearly consumed and killed. Thus the parasitoid acts as a predator.

INTERSPECIFIC COMPETITION

The relationship in which the populations of both associated species are affected adversely (− −) is *interspecific competition.* Interspecific competition, like intraspecific competition, involves the seeking of a resource in short supply, but two or more species are involved. Both types of competition may take place simultaneously. Gray squirrels, for example, may be competing among themselves for acorns during a poor crop year. At the same time, white-footed mice, white-tailed deer, wild turkeys, and blue jays may be vying for the same crop. Because of competition, individuals within a species may be forced to broaden the base of their foraging efforts, and populations of various species may be forced to turn away from acorns to foods less in demand. Thus intraspecific competition selects for a broadening of the resource base, or generalization, whereas interspecific competition favors a reduction of the resource base and specialization—points to be explored later.

Like intraspecific competition, interspecific competition comes in two forms, interference and exploitation. *Interference competition,* which equates with contest competition, involves direct or aggressive interactions between competitors. *Exploitative competition,* which is similar to scramble competition, depletes the resource to a level where it is of little value to either population.

The concept of interspecific competition is one of the cornerstones of evolutionary ecology. Competition, the struggle to survive, the survival of the fittest, was one of the foundations on which Darwin based his idea of natural selection. Because it is advantageous for a species to avoid it, competition has been regarded as the major force behind species divergence and specialization. Although the concept had a strong influence on ecological thinking, it is one of the least known and most controversial areas of ecology.

OUTCOMES

In the early part of the twentieth century, two mathematicians, Alfred Lotka and Vittora Volterra, independently arrived at mathematical expressions to describe the relationship between two species utilizing the same resource. The equations are modifications of the logistic growth equation (Chapter 12), one for each species. To these pairs of logistic equations Lotka and Volterra added competition coefficients, constants that describe the degree to which species 1 inhibits the growth of species 2 and species 2 inhibits the growth of species 1. These equations consider both intra- and interspecific competition:

$$\frac{dN_1}{dt} = r_1N_1\left(\frac{K_1 - N_1 - \alpha_{12}N_2}{K_1}\right)$$

$$\frac{dN_2}{dt} = r_2N_2\left(\frac{K_2 - N_2 - \alpha_{21}N_1}{K_2}\right)$$

where K_1 and K_2 are the carrying capacities for species 1 and species 2, respectively; α_{12} is a constant representing the inhibitory effect of species 2 on species 1; and α_{21} is the inhibitory effect of species 1 on species 2.

Initially, each species grows at its own intrinsic rate of increase, and each experiences intraspecific competition. In the presence of interspecific competition, the outcome is influenced by the relative values of K_1, K_2, α_{12}, and α_{21} (Figure 15.1). In the absence of either species, no interspecific competition exists, and population growth is regulated by intraspecific competition. In the presence of competing species, species 1 decreases the carrying capacity, K, for species 2 at a certain rate, and species 2 decreases the carrying capacity for species 1 at a certain rate, because each has to share limited resources with the other.

Depending upon the combination of values for the Ks and the alphas, the Lotka-Volterra equations predict four different potential outcomes. In two situations one species wins out over the other. In one case species 1 inhibits further increase in species 2 while it can still increase itself, and species 2 dies out. In the other case species 2 inhibits further increase in species 1 while continuing to increase itself, and species 1 eventually disappears. In the third situation each species when abundant inhibits the growth of other species more than it inhibits its own growth. The outcome depends upon which species initially is the most abundant. The two species may coexist for a while in an unstable equilibrium. In real-life situations the outcome may depend upon which of the two species has the competitive advantage in the face of environmental change over time. In the fourth situation the two species achieve a stable equilibrium and coexist. In that situation neither population reaches its own carrying capacity, and individuals of each species inhibit their own population growth more than they inhibit the population growth of the other species.

EVIDENCE FROM THE LABORATORY

The theoretical Lotka-Volterra equations stimulated studies of competition in the laboratory, where under controlled conditions, the outcome is more easily determined. One of the first to study competition experimentally was the Russian biologist G. F. Gause. When he reared two species of protozoan, *Paramecium aurelia* and *P. caudatum,* together in tubes containing fixed amounts of bacterial food, *P. caudatum* died out. Its competitor had a higher rate of increase and tolerated higher population densities. In another experiment, Gause reared the loser, *P. caudatum,* with another species, *P. bursaria.* These two coexisted, because *P. caudatum* fed on bacteria suspended in solution, whereas *P. bursaria* confined its feeding to bacteria on the bottom of the tubes.

Thomas Park at the University of Chicago in the 1940s and 1950s conducted a number of classic competition experiments with laboratory populations of flour beetles. He found that the outcome of competition between two species, *Tribolium castaneum* and *T. confusum,* depended upon temperature, humidity, and fluctuations in the total number of eggs, larvae, pupae, and adults. Often the final outcome of competition was not determined for generations.

In the 1980s David Tilman of the University of Michigan and his associates grew laboratory populations of two species of diatoms, *Asterionella formosa* and *Synedra ulna,* both of which require silica for the formation of cell walls. They monitored not only population growth and decline but also the consumption and level of silica. When grown alone in a liquid medium to which the resource (silica) was continually added, both species kept silica at a low level. When grown together, *S. ulna* took silica to a level below which *A. formosa* could not survive and reproduce (Figure 15.2). In this experiment, *Synedra* competitively excluded *Asterionella* by reducing resource availability, driving it to extinction.

In these laboratory experiments the patterns were similar. The species grew exponentially at low densities. As their densities increased, population growth declined, and each began to influence the other. Under these conditions one species had a greater fitness than the other. A point was reached when one species declined toward extinction, while the population of the other species continued to grow.

THE COMPETITIVE EXCLUSION PRINCIPLE

In three of the four situations predicted by the Lotka-Volterra equations, one species wins and the other loses, eventually driven to extinction, provided the interaction proceeds to saturation level, or *K*. Evidence provided by laboratory experiments tends to support the mathematical models. In competitive situations if one species produces enough individuals to prevent the population increase of another, it can reduce that population to extinction or exclude it from the area.

These observations led to the concept called Gause's principle, although the idea was far from original with him. More recently called the *competitive exclusion principle,* it states that complete competitors cannot coexist. Basically, if two noninterbreeding populations possess exactly the same ecological requirements, and if they live in exactly the same place, and if population A increases the least bit faster than population B, then A will eventually occupy the area completely and B will become extinct.

The competitive exclusion principle assumes that competitors remain genetically unchanged, that immigrants from areas with different environmental conditions do not move into the population of the losing species, and that environmental conditions remain

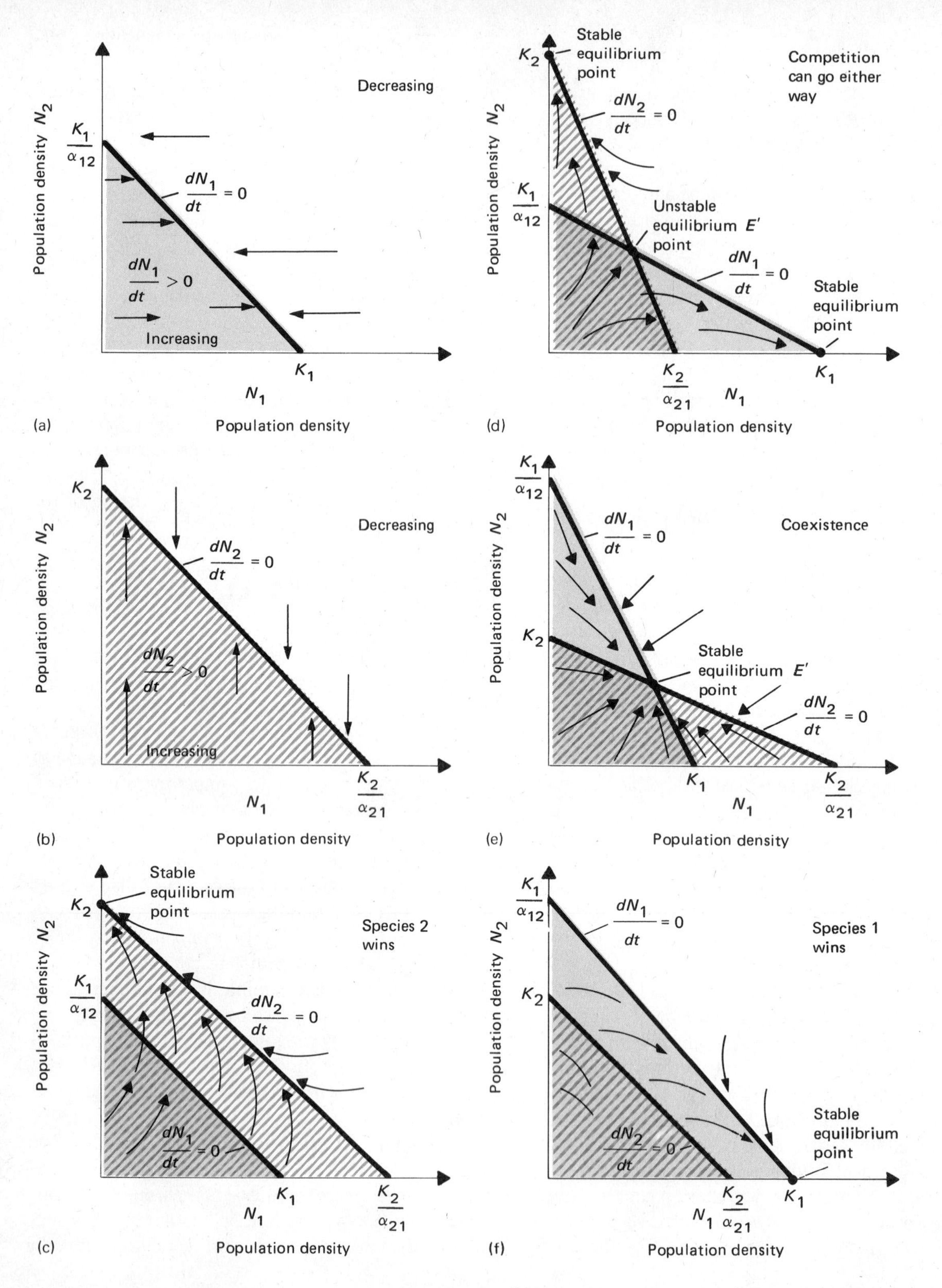

(a)
Decreasing
$\frac{K_1}{\alpha_{12}}$
$\frac{dN_1}{dt} = 0$
$\frac{dN_1}{dt} > 0$
Increasing
K_1
N_1
Population density
Population density N_2
(b)
K_2
Decreasing
$\frac{dN_2}{dt} = 0$
$\frac{dN_2}{dt} > 0$
Increasing
$\frac{K_2}{\alpha_{21}}$
N_1
Population density
Population density N_2
(c)
Stable equilibrium point
K_2
Species 2 wins
$\frac{K_1}{\alpha_{12}}$
$\frac{dN_2}{dt} = 0$
$\frac{dN_1}{dt} = 0$
K_1
$\frac{K_2}{\alpha_{21}}$
N_1
Population density
Population density N_2
(d)
Stable equilibrium point
K_2
Competition can go either way
$\frac{dN_2}{dt} = 0$
$\frac{K_1}{\alpha_{12}}$
Unstable equilibrium E' point
$\frac{dN_1}{dt} = 0$
Stable equilibrium point
$\frac{K_2}{\alpha_{21}}$
K_1
N_1
Population density
Population density N_2
(e)
$\frac{K_1}{\alpha_{12}}$
$\frac{dN_1}{dt} = 0$
Coexistence
K_2
Stable equilibrium E' point
$\frac{dN_2}{dt} = 0$
K_1
$\frac{K_2}{\alpha_{21}}$
N_1
Population density
Population density N_2
(f)
$\frac{K_1}{\alpha_{12}}$
$\frac{dN_1}{dt} = 0$
Species 1 wins
K_2
$\frac{dN_2}{dt} = 0$
Stable equilibrium point
$\frac{K_2}{\alpha_{21}}$
K_1
N_1
Population density
Population density N_2

constant. Such conditions rarely exist. The idea of competitive exclusion, however, has stimulated a more critical look at competitive relationships in natural situations, including studies to determine what ecological conditions are necessary for coexistence among species.

EVIDENCE FROM THE FIELD

Demonstrating interspecific competition in laboratory "bottles" is one thing; demonstrating competition under natural conditions is another. In the field, we have no control over the environment; we have difficulty knowing whether populations are at or below carrying capacity; we lack full knowledge of the life-history requirements or the subtle differences between species. Further, the competitive outcome among associated species may have been settled through evolutionary time. What we observe in the field may be the outcome of competition in the past rather than current competition.

Competitive Exclusion

Few clear-cut examples of competitive exclusion have been experimentally demonstrated. These experiments involve interactions between native species. The best examples are nonexperimental ones involving the impact of introduced species on native fauna. The aggressive European starling and the English house sparrow compete strongly with the cavity-nesting bluebird and flicker for nesting sites and drive the native species away. (A sharp reduction in the populations of the two invaders has considerably lessened the competitive interactions.) An introduced, highly competitive fish, the carp (*Cyprinus carpo*), can occupy virtually every type of freshwater habitat and easily outcompete other fishes, such as bass and trout, for both space and food. Uprooting and consuming aquatic plants, they eliminate spawning cover for fish such as pike and perch, cause nesting fish to desert their nests, and muddy the water so fishes such as sunfish cannot see to forage. Competitive exclusion, then, invokes more than competition for a limited resource.

An example of exclusion brought about by physiological tolerance, aggressive behavior, and restriction to habitats in which one organism has the competitive advantage involves small mammals. On the eastern slope of the Sierra Nevadas live four species of chipmunks: the alpine chipmunk (*Eutamias alpinus*), the

FIGURE 15.1 Models of competition between two species. In (a) and (b) populations of species 1 and 2 in absence of competition will increase in size and come to equilibrium at some point along the diagonal line or isocline. The line represents all equilibrium conditions ($dN/dt = 0$) at which the population just maintains itself ($r = 0$). In the shaded areas below the line r is positive and the population increases (as indicated by the arrows). Above the isocline r is negative and the population decreases (as indicated by the arrows). In (a) the intercepts of the N_1 isocline are K_1 and K_1/α and the intercepts of the N_2 isocline are K_2 and K_2/α_{21}. In (c) species 1 and 2 are competitive. Because the isocline, the zero growth curve of species 2, falls outside the isocline of species 1, species 2 wins, leading to the extinction of species 1, and the stable equilibrium point is at K_2 ($N_2 = K_2, N = 0$). In (f) the situation is reversed and species 1 wins, leading to the exclusion of species 2. Equilibrium is at K_2 ($N_1 = K_1, N_2, = 0$). In (d) and (e) the isoclines cross. Each species, depending upon the circumstances, is able to inhibit the growth of the other. In (e) neither species can exclude the other. Each by intraspecific competition inhibits the growth of its own population more than it inhibits the growth of the other population. There is one equilibrium point where the lines cross and both species coexist at densities below their respective carrying capacities. In (d) each species inhibits the growth of the other more than it inhibits the growth of its own population. Three possible equilibrium points exist. Two are stable: $N_1 = K_1, N_2 = 0; N_2 = K_2, N_1 = 0$. At these two points one species excludes the other. The other equilibrium point where the two lines intersect is unstable. Which species wins often depends upon the initial proportion of the two species or changing environmental conditions.

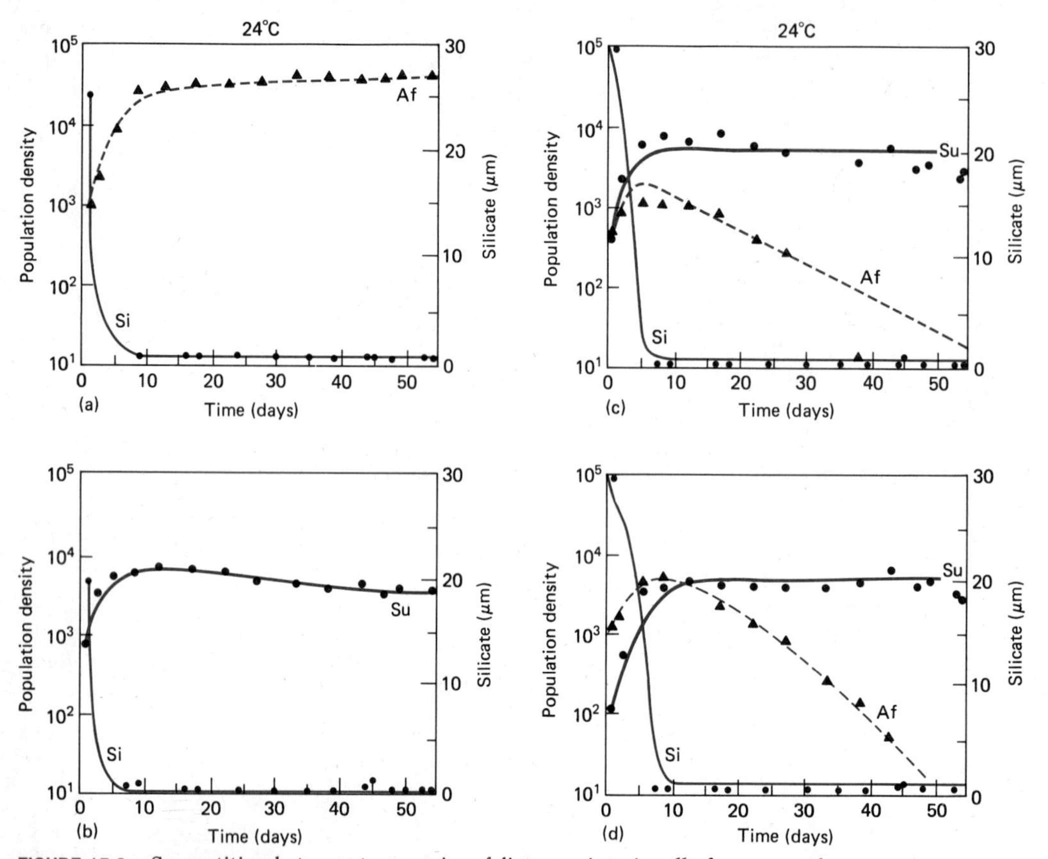

FIGURE 15.2 Competition between two species of diatoms, *Asterionella formosa* and *Synedra ulna,* for silica. (a) and (b) When grown alone in a culture flask, both species reach a stable population that keeps silica at a constant low level. *Synedra* draws silica lower. (c) and (d) When the two are grown together, *Synedra* drives *Asterionella* to extinction. *Synedra* reduces the silica level to a point below which the other species cannot exist.

lodgepole pine chipmunk (*E. speciosus*), the yellow-pine chipmunk (*E. amoenus*), and the least chipmunk (*E. minimus*), at least three of which have strongly overlapping food requirements.

Each species occupies a different altitudinal zone (Figure 15.3). The line of contact is determined partly by interspecific aggression. The upper range of the least chipmunk is determined by aggressive exclusion from the dominant yellow-pine chipmunk. Although the least chipmunk is capable of occupying a full range of habitats from sagebrush desert to alpine fell fields, it is restricted in the Sierras to sagebrush habitat. It is physiologically more capable of handling heat stress than the others, enabling it to inhabit extremely hot, dry sagebrush. If the yellow-pine chipmunk is removed from its habitat, the least chipmunk moves into the vacated open pine woods; but if the least chipmunk is removed from the sagebrush, the yellow-pine chipmunk does not invade the habitat. The aggressive behavior of the lodgepole pine chipmunk in turn determines the upper limit of the yellow-pine chipmunk. The lodgepole pine chipmunk is restricted to shaded forest habitat because it is the most vulnerable to heat stress. Most aggressive of the three, the lodgepole pine

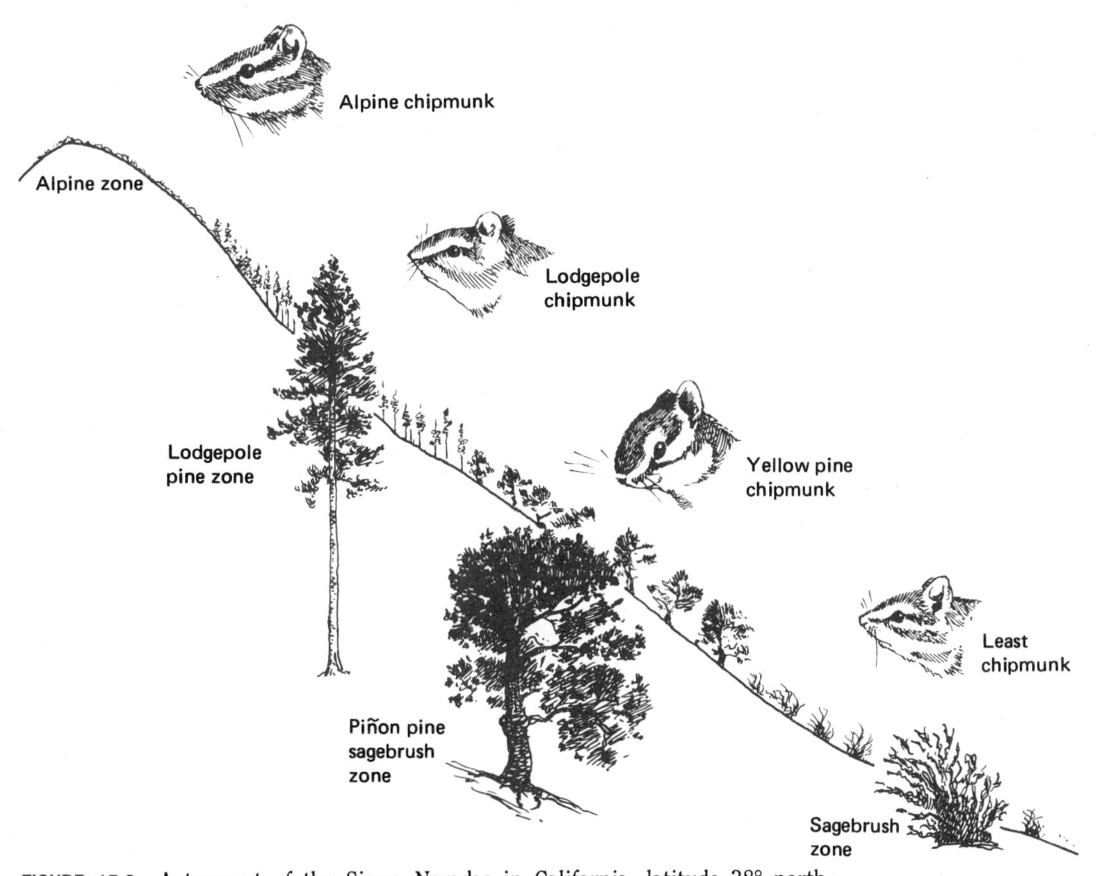

FIGURE 15.3 A transect of the Sierra Nevadas in California, latitude 38° north, showing vegetational zonation and altitudinal range of four species of chipmunks (*Eutamias*) that inhabit the east slope.

chipmunk may limit the downslope range of the alpine chipmunk. Thus, the range of one species, the least chipmunk (two if you include the alpine chipmunk), is determined through aggressive exclusion by another species and its ability to survive and reproduce on habitat hostile to the other species. The other two species are restricted to certain habitats by their own physiology and by aggressive exclusion.

Under competitive situations in the past, aggressive behavior was probably selected for in the lodgepole pine and yellow-pine chipmunks because of a seasonally limited, often patchy food supply that can be cached for winter use and economically defended. Aggressiveness was probably not selected for in the least chipmunk because such activity would not be metabolically feasible in the hot sagebrush desert outside the physiological range of the other species. Instead, selection was for tolerance of heat stress.

The competitive outcome among some herbivores may be influenced by plant defenses (see Chapter 13). Lambs-quarter (*Chenopodium*) is attacked by two phloem-feeding aphids that never actually come in contact with each other, although they share the same resource. One species, *Hayhurstia atriplicis,* feeds above ground, where it forms leaf galls on the plant. The other species, *Pemphigus betae,* feeds underground on the roots. *Pemphigus* has no apparent effect on its host. *Hayhurstia,* however, can reduce plant biomass of the host by an average of 54 percent and seed set by an average of 60 percent. Some lambs-quarters are resistant to leaf galling by *Hayhurstia;* others are not. In their experimental studies N. A. Moran and T. G. Whigham found that on susceptible plants, *Hayhurstia* reduced the number of *Pemphigus* by an average of 91 percent, and often eliminated them entirely. Thus on susceptible plants, *Pemphigus* experiences competitive exclusion (Figure 15.4). On plants resistant to galling, colonies of *Hayhurstia* were smaller and did not affect *Pemphigus* infesting the roots of the same plant. Although *Hayhurstia* had a negative effect on *Pemphigus,* the latter had no measurable impact on *Hayhurstia,* which suggests amensalism.

In this example (and others), competitive exclusion does not fit the theoretical framework described by the Lotka-Volterra equations. The competing species do not occupy the same habitat and increase simultaneously in it, so that exclusion results from superior fitness of one over the other, as demonstrated in laboratory populations. Instead, exclusion is determined quickly by physiological tolerances and by aggressive behavior.

Unstable Equilibrium

An unstable equilibrium usually results in one species winning over another. Under natural conditions the species that wins more often than not is determined by the local environment.

Unstable equilibrium can be observed more easily among plants than among animals. From the time seedlings germinate and develop, demand for growing space, light, moisture, and nutrients increases. Plants that most effectively utilize resources in short supply have the best chance for survival. Species of plants whose roots are in the same horizon have to compete for limited moisture and nutrients. In western North America the shallow-rooted annual cheat grass *Bromus tectorum* grows early in spring and often reduces moisture to the point where slow-growing annuals, perennials, and even shrubs are unable to withstand the competition. In drier regions plants that develop roots rapidly after germination have the competitive advantage. Weaker plants are overtopped and eventually crowded out by more vigorous and aggressive species.

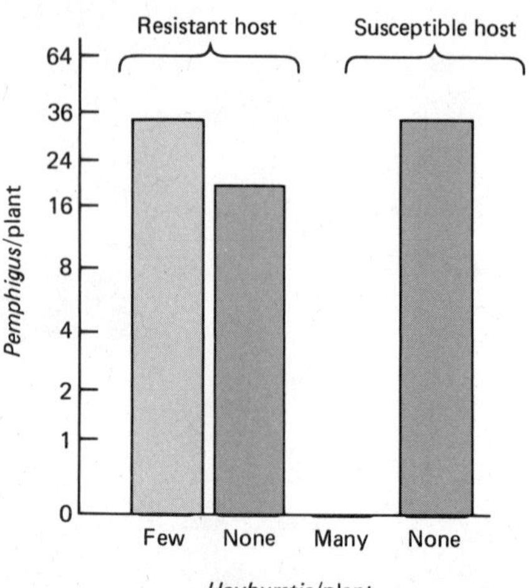

FIGURE 15.4 Exclusion of *Pemphigus betae* by *Hayhurstia atriplicis* in a screen cage experiment. The same competitive exclusion carries through to natural stands.

Under optimum conditions some plants will dominate and suppress their associates. During moist years the prairie grass little bluestem dominates the grassland, but during dry years it almost disappears. Then two associates, side-oats grama and blue grama, grasses of the same genus associated with each other, become dominant.

Coexistence

Coexistence results when competing species share resources although such competition reduces the fitness of both. Among animals, especially birds, coexistence may involve interspecific territoriality, in which the area is divided among the competing species, each behaving toward the other as if both or all were the same species.

Coexistence can exist among different taxonomic groups. An example is the competitive relationship between rodents and ants in the desert scrub of Arizona. Both groups feed on seeds of desert plants; both overlap considerably in the size of seeds they consume. Production of those seeds is strongly influenced by desert rainfall. A strong correlation between the number of species of rodents and ants and mean annual rainfall suggests that ants and rodents may compete for the limited amount of seeds produced.

J. Brown and D. W. Davidson undertook field experiments to determine to what extent rodents and ants competed for a limited seed supply. They established six experimental plots. From two of the plots they removed rodents by trapping and excluded immigrants by fencing. From another two plots they eliminated the ants by repeated applications of an insecticide. The remaining two plots, from which both ants and rodents were removed, were the controls. They discovered that ants and rodents did indeed compete. Removal of rodents resulted in an increase of ants, who consumed as many seeds alone as rodents and ants combined did previously. Removal of ants had a similar effect on rodents. Where both ants and rodents were removed, the amount of seeds increased. Thus, both ants and rodents depressed each other's population and perhaps the fitness of the seed plants as well.

Allelopathy

A form of competition among plants only is *allelopathy,* the production and release of chemical substances by one species that inhibit the growth of other species. These substances may range from acids and bases to relatively simple organic compounds that reduce competition for nutrients, light, and space. Produced in profusion in natural communities as secondary substances, most compounds remain innocuous, but a few may influence community structure. For example, broomsedge (*Andropogon virginicus*) produces chemicals that inhibit the invasion of old fields by shrubs and thus maintains its dominance. The allelopathic effects of goldenrods (*Solidago*), asters (*Aster*), and certain grasses prevent tree regeneration in glades.

J. L. Harper, the English plant ecologist, suggests that toxic interactions in higher plants may not be as common as supposed for two reasons. First, higher plants rapidly evolve tolerances to such environmental toxins as zinc, nickel, and copper and have developed tolerances to herbicides. Second, complex organic molecules are broken down by the action of soil microbes, and plant toxins probably experience the same fate.

RESOURCE PARTITIONING

Observations of a number of species sharing the same habitat suggest that they coexist by partitioning available resources. Animals may utilize different kinds and sizes of food, feed at different times, or forage in different areas. Plants may occupy a different position on a soil moisture gradient, require different proportions of nutrients, or have different tolerances for light and shade. Each species exploits a portion of the resources unavailable to others. Such partitioning is regarded as an outcome of interspecific competition.

The Theory

Consider species A, which, in the absence of any competitor, utilizes a range of different-sized food items. Picture that utilization as a bell-shaped curve on a graph, with food as the ordinate and fitness as the abscissa. Most individuals feed about the optimum. Individuals at either tail feed on larger or smaller food items. As population size increases, the range of food taken may increase as intraspecific competition forces some individuals to seek food at the two extremes. Such intraspecific competition fosters increasing genetic variability in the population.

Now allow a second species, B, to enter the area. When its resource use curve is superimposed on the curve of species A, B shows considerable overlap. Selective pressure from interspecific competition forces both species A and species B to narrow their range of resource use. Natural selection will favor those individuals living in areas of minimal or no overlap. Ultimately, the two species will narrow their ranges of resource use. They will diverge, moving to the left and the right on the graph. Direct interspecific competition will be reduced, and the two species will coexist. Thus, whereas intraspecific competition favors expansion of the resource base, interspecific competition narrows it. The populations have to arrive at some balance between the two.

Enter the third species, C, which utilizes a food resource in the area of overlap between A and B (Figure 15.5). Because it can make optimal use of resources in that part of the gradient, it will force A and B to restrict further the range of resources they use. The result is a partitioning of resources along the resource gradient. Such a group of functionally similar species whose members interact strongly with one another but weakly with the remainder of the community is called a *guild.*

Field Examples

Any number of intensive field studies turn up examples of presumed resource partitioning. One example involves three species of annuals growing together in a

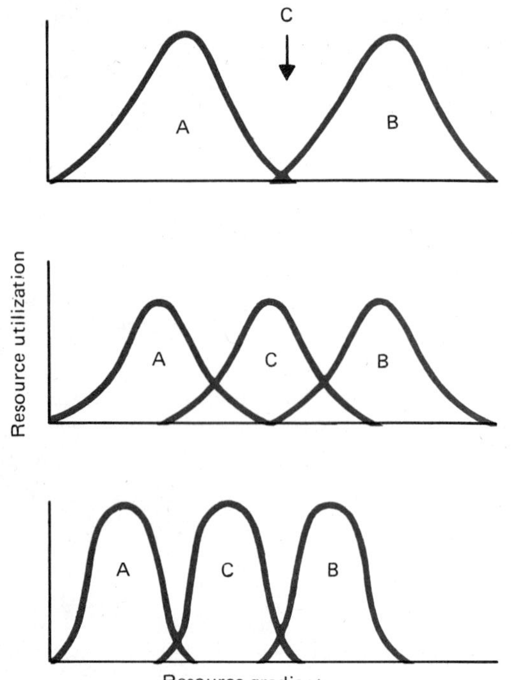

FIGURE 15.5 Theoretical resource gradient utilized by three competing species, A, B, and C. A and B share the resource gradient with minimal overlap. A third species, C, whose optimal resource utilization lies between A and B, comes in. In response to selection pressures, A and B narrow their range of resource utilization to the optimum, and C utilizes that portion of the resource used at less than an optimal level by A and B.

field on prairie soil abandoned one year after plowing. Each plant exploits a different part of the soil resource (Figure 15.6). Bristly foxtail (*Setaria faberi*) has a fibrous, shallow root system that exploits a variable supply of moisture. It possesses the ability to recover rapidly from water stress, to take up water rapidly after a rain, and to carry on a high rate of photosynthesis even when partially wilted. Indian mallow (*Abutilon theophrasti*) has a sparsely branched taproot extending to intermediate depths where moisture is adequate during the early part of the growing season but is less available later on. That plant is able to carry on photosynthesis at a low water potential. The third species, smartweed (*Polygonum pensylvanicum*), possesses a taproot that is moderately branched in the upper soil layer and develops mostly below the rooting zone of other species, where it has a continuous supply of moisture.

Another example, in which circumstantial evidence strongly supports resource partitioning, involves guilds of seed-eating rodents on three major deserts of western North America (Great Basin, Mojave, and Sonoran). The guild in each of these deserts includes pocket mice and kangaroo rats of different body sizes (Figure 15.7). Species of similar body size occur together less frequently than would be expected on the basis of chance, and rarely do so locally. Differences in body size allow each to utilize different seed sizes and to forage in different microhabitats. Behavioral differences, too, are involved. The quadrapedal pocket mouse *Perognathus penicillatus* feeds in dense vegetation and under shrub canopy, where it is protected from predators. The much swifter bipedal kangaroo rat *Dipodomys merriami* feeds on larger seeds in more open areas.

Such patterns of resource use are examples of interspecific competition in action. By dividing the resources in some manner, each species avoids direct competition with the other(s). Such competition may be taking place currently, or it may have occurred in the past. It is regarded as instrumental in directing the evolutionary divergence of species.

THE NICHE

Almost inseparable from the concept of interspecific competition is the concept of the niche. *Niche* is one of those terms in ecology that defies a rigorous definition. Ecologists seem to know what it means, but attempts to put that meaning down on paper satisfactorily are elusive.

The word *niche* in everyday terms means a recess in a wall where you place something, usually an ornamental object; or it is a place or position in life suitable for a person. In ecology it means an organism's place and function in the environment. Or does it?

One of the first to propose the idea of the niche in ecology was Joseph Grinnell, a California ornithologist. He suggested that the niche be regarded as a subdivision of an environment occupied by a species. Essentially, Grinnell was describing the habitat of a species. Charles Elton, in his classic book *Animal Ecology* (1927), considered the niche as the fundamental role of the organism in the community—what it does, its relationship to its food and its enemies. Basically, this

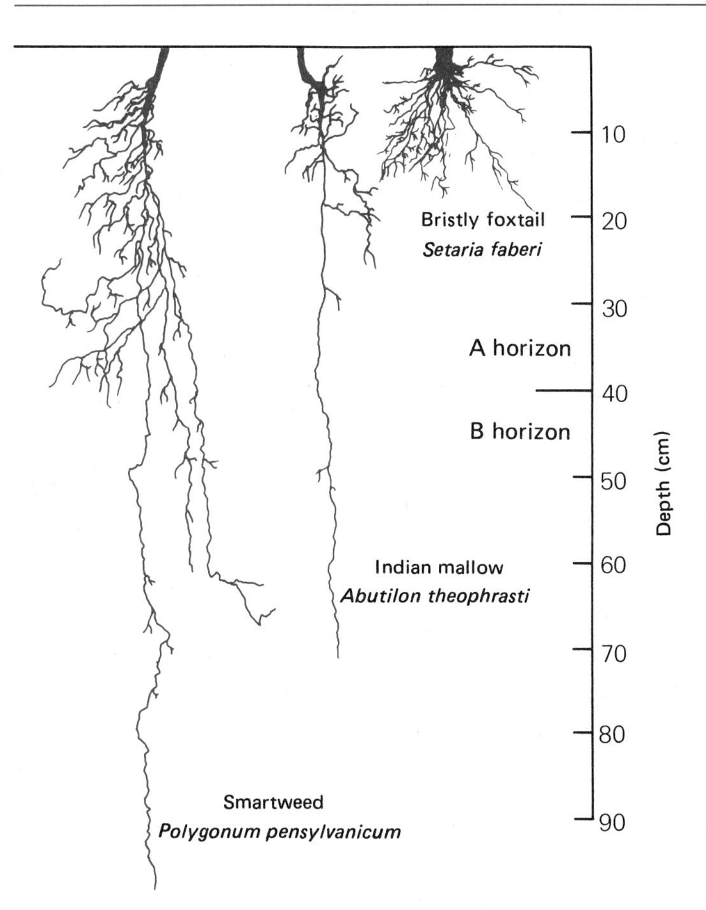

FIGURE 15.6 Partitioning of the prairie soil resource at different levels by three species of annuals in a field one year after disturbance.

concept stresses the occupational status of the species. Other definitions are variations on the same theme. One considers the habitat as the animal's address and the niche as its occupation. Another considers the niche only as a functional position, whereas the habitat and the niche combined comprise the *ecotope.* Yet another definition regards the niche as embracing all the ways in which a given individual, population, or species conforms to its environment.

The definition that most closely links the niche to competition is the one proposed by G. E. Hutchinson. It is based on the competitive exclusion principle. According to Hutchinson, an organism's environment consists of many physical and biological variables, each of which can be considered as a point in a multidimensional space. Hutchinson called that space the *hypervolume.*

We can visualize the multidimensional niche to a certain extent by creating a three-dimensional one. Consider three niche-related variables for a hypothetical organism: food size, foraging height, and humidity (Figure 15.8). Suppose the animal can consume only a certain range of food size. Food size, then, is one dimension of its niche. Add foraging height, the area to which it is limited for seeking food. If we graph that on a second axis and enclose the space, we have a rectangle, representing a two-dimensional niche. Suppose, too, that the animal can survive and reproduce only within a certain range of humidity. Humidity can be plotted on a third axis. By enclosing that space, we come up with a volume, a three-dimensional niche. Of course, more than three variables, biotic and abiotic, influence a species' or an individual's fitness. A number of these niche dimensions, n—difficult to visualize and impossible to graph—make up the n-dimensional hypervolume that would be a species' niche. An individual or a species free from the interference of another could occupy the full hypervolume or range of variables to

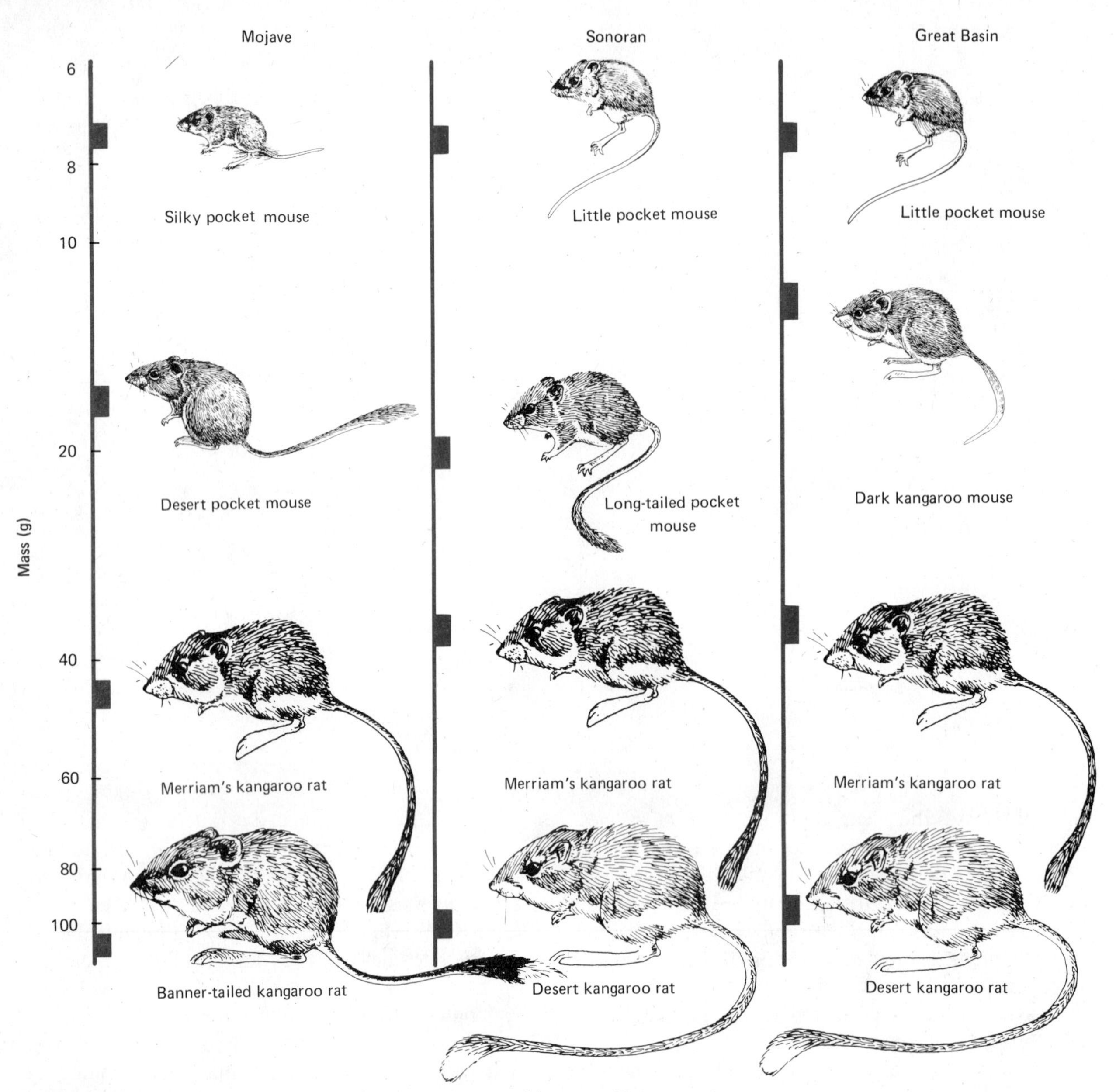

FIGURE 15.7 Resource partitioning, perhaps an outcome of interspecific competition, is illustrated by guilds of rodents in the major desert regions of North America: the Mojave, the Sonoran, and the Great Basin. Size distribution of the four species is similar, even though the identity of the species differs among the deserts. Sizes are plotted on a logarithmic scale so that equal spacing indicates equal ratios. Body mass in each community ranges from 7 to 100 g, and the minimum body size ratio between species exceeds 1.75.

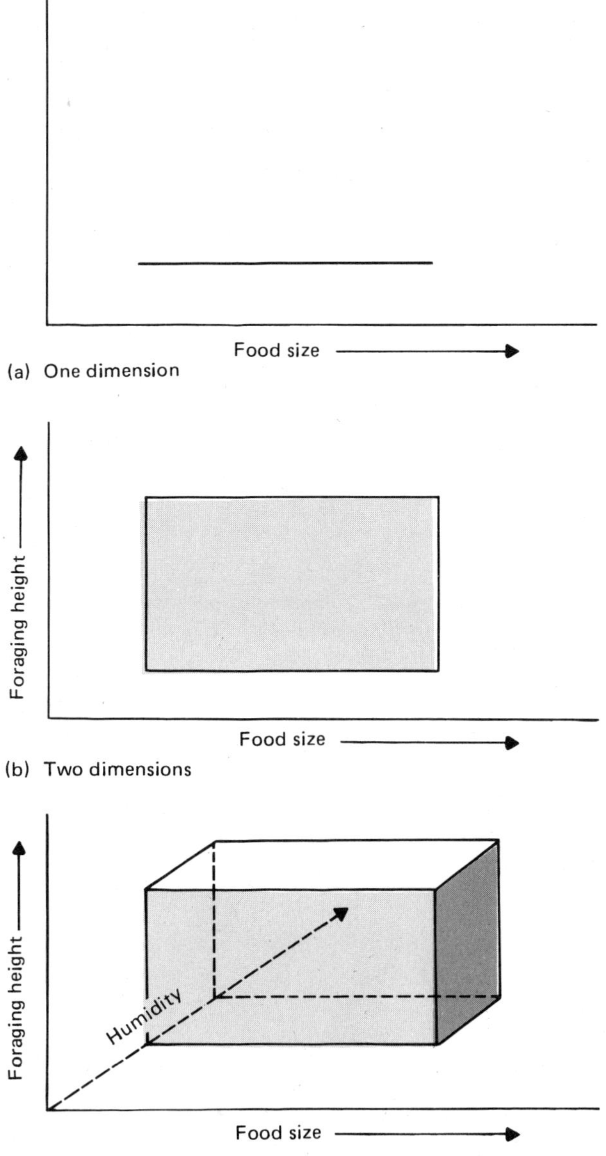

FIGURE 15.8 An illustration of niche dimension. Assume three elements comprising the hypothetical organism's niche: food size, foraging height, and humidity. Graph (a) represents a one-dimensional niche involving food size. In graph (b) a second dimension has been added, foraging height. By enclosing that space, we obtain a two-dimensional niche. Now suppose the organism can survive and reproduce only within a certain range of humidity, graphed as a third axis. Enclosing all those points, we arrive at (c) a three-dimensional niche space, or volume, for the organism.

which it is adapted. That is the idealized *fundamental niche* of a species.

The fundamental niche assumes the absence of competitors, but this case is rare. Competitive relationships force the species to constrict a portion of the fundamental niche it could potentially occupy. In those parts its fitness might be reduced to zero. The conditions under which an organism actually exists in any given situation are its *realized niche* (Figure 15.9). Like the fundamental niche, the realized niche is an abstraction. In their studies, ecologists usually confine themselves to one or two niche dimensions, such as a feeding niche, a space niche, or a tolerance niche.

P. D. Putwain and J. L. Harper studied the population dynamics of two species of dock, *Rumex acetosa* and *R. acetosella,* each growing in hill grasslands in North Wales. *R. acetosa* grew in a grassland community dominated by velvet grass (*Holcus lanatus*) and red and sheep fescues (*Festuca rubra* and *F. ovina*). *R. acetosella* grew in a community dominated by sheep fescue and bedstraw (*Galium saxatile*). To determine interference and niches of the two docks, Putwain and Harper treated the flora with specific herbicides to remove selectively in different plots (1) grasses and (2) nongrasses, except *Rumex* species. All species except *R. acetosella* spread rapidly after the grasses were removed. *R. acetosella* increased only after both grasses and nongrasses were removed. The niches of the two

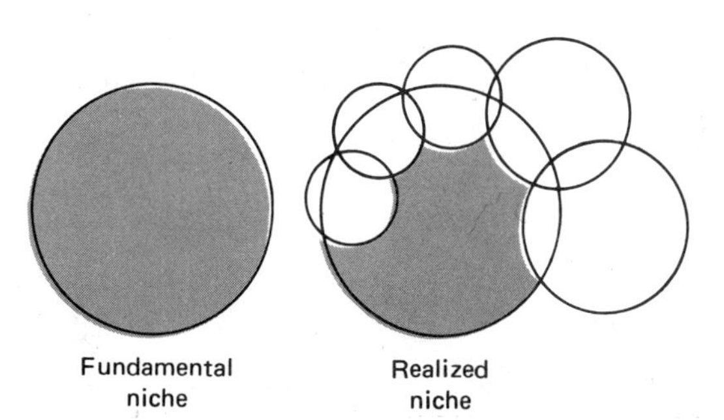

FIGURE 15.9 Fundamental and realized niches. The fundamental niche of a species represents the full range of environmental conditions, biological and physical, under which the species can successfully exist. However, under pressure of superior competitors, the species may be completely displaced from parts of its fundamental niche and forced to retreat to that part to which it is most highly adapted, its realized niche.

plants are diagrammed and explained in Figure 15.10. The presumed fundamental niche of *R. acetosella* (R) overlaps the fundamental niches of both grasses and nongrasses. Only when these competitors are eliminated does *R. acetosella* realize its fundamental niche. *R. acetosa,* however, overlaps only with the grasses, and only their removal is necessary to permit that dock to expand.

Note that the niches of seedlings differ from those of the mature plant. The fundamental and realized niches of an organism can change with its growth and development. Insects with a complex life cycle may occupy one niche as a larva and an entirely different niche as an adult. Other organisms change food and cover requirements as they grow larger.

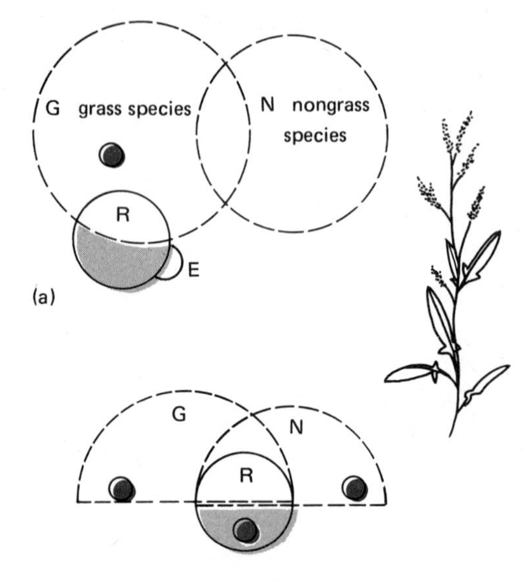

FIGURE 15.10 Niches of competing species of dock in mixed grassland swards: (a) *Rumex acetosa* and (b) *R. acetosella.* The fundamental niches of grass species (G) and nongrass species (N) overlap. The fundamental niche of *Rumex* species (R) is shown as a continuous line, and the realized niche is shaded. E is that part of the fundamental niche of *R. acetosa* that is expressed in the presence of nongrass species only and does not overlap the fundamental niches of G and N. The fundamental niches of seedlings—the small, dark-colored circles—are contained within those of mature plants.

NICHE OVERLAP

Niche overlap occurs when two or more organisms use a portion of the same resource, such as food, simultaneously. The amount of niche overlap (Figure 15.11) is assumed to be proportional to the degree of competition for that resource. With little or no competition, niches may be adjacent with no overlap, or they may be disjunct. At the other extreme, according to competition theory, under intense competition, the fundamental niche of one species may be completely

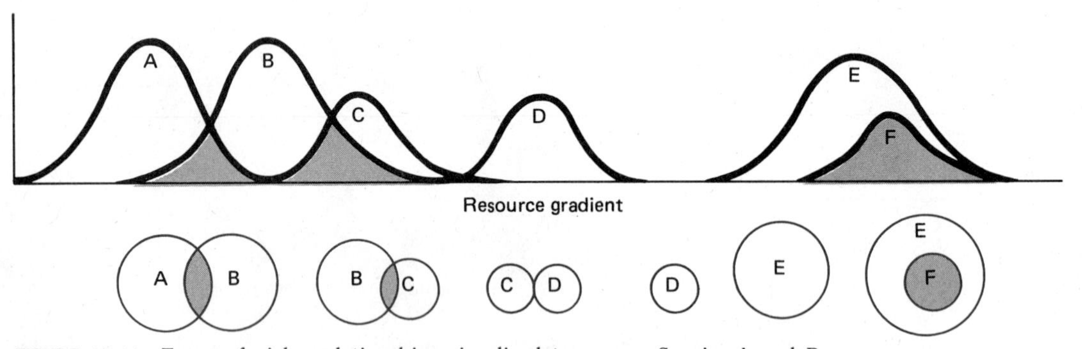

FIGURE 15.11 Types of niche relationships visualized two ways. Species A and B have overlapping niches of equal breadth but are competitive at opposite ends of the resource gradient. B and C have overlapping niches of unequal breadth. Species C shares a greater proportion of its niche with B than B does with C. (In this example, however, B shares its niche also with A at the other end). C and D occupy adjacent niches with little possibility of competition. D and E occupy disjunct niches, and no competition exists. Species F has a niche contained within the niche of E. If F is superior to E competitively, it persists, and E shares that part of its niche with F. (Compare this with the seedlings in Figure 15.10.)

within or correspond exactly to another, as in the case of the *Rumex* seedling. In such instances there can be two outcomes. If the niche of species 1 contains the niche of species 2 and species 1 is competitively superior, species 2 will be eliminated entirely. If species 2 is competitively superior, it will eliminate species 1 from that part of the niche space that species 2 occupies. The two species then coexist within the same fundamental niche.

When fundamental niches overlap, some niche space is shared and some is exclusive, enabling the two species to coexist. However, niche overlap does not mean high competitive interaction. In fact, the reverse may be true. Competition involves a resource in short supply. Extensive niche overlap may indicate that little competition exists and resources are abundant.

For simplicity, niche overlap is usually considered as one- or two-dimensional. In reality, of course, a niche involves the utilization of many types of resources: food, a place to feed, cover, space, and so on. Rarely do two or more species possess exactly the same niche involving all requirements. Species may show overlap on one gradient but not on another. The total competitive interactions may be less than the competition or niche overlap suggested on one gradient alone (Figure 15.12).

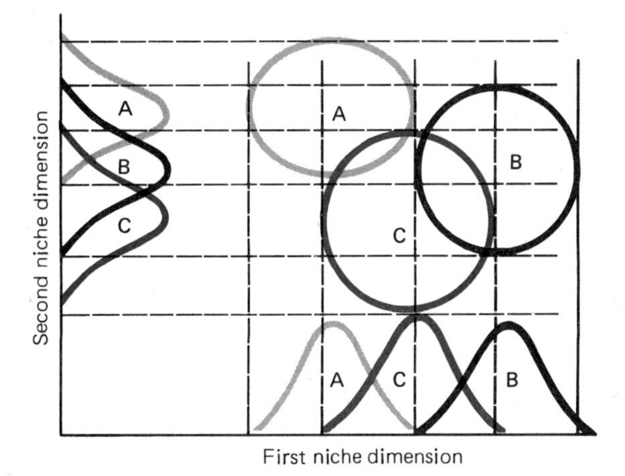

FIGURE 15.12 Niche relationship based on two gradients. When several niche dimensions are considered, niche overlap may be reduced considerably. On resource gradient 1, A and B exhibit no overlap, and on resource gradient 2, they overlap equally and in an opposite way. When both niches are considered (circles), A and B do not overlap. C on resource gradient 2 overlaps equally with B and very little with A. On resource gradient 1, C overlaps with both A and B. When both gradients are considered, C overlaps mostly with B and little with A.

Differences in niches are not solely a characteristic of various species. Division of feeding space, food size, and other niche components may exist between sexes of the same species. The male red-eyed vireo (*Vireo olivaceus*), for example, obtains its insect food in the upper canopy, the female in the lower canopy and near the ground. The feeding areas overlap only about 35 percent (Figure 15.13). Likewise, a pronounced difference in bill size allows the male Arizona woodpecker (*Dendrocopus arizonae*) to forage on the trunk and the female to hunt on the branches.

NICHE WIDTH

When we plot the range of resources—for example, food size utilized by an animal or soil moisture conditions occupied by a plant—the length of the axis intercepted by the curve represents niche width. Theoretically, *niche width* (also called niche breadth and niche size) is the extent of the hypervolume occupied by the realized niche. A more practical definition is the sum total of the different resources exploited by an organism. Measurement of a niche usually involves the measure of some ecological variable such as food size or habitat space (Figure 15.14).

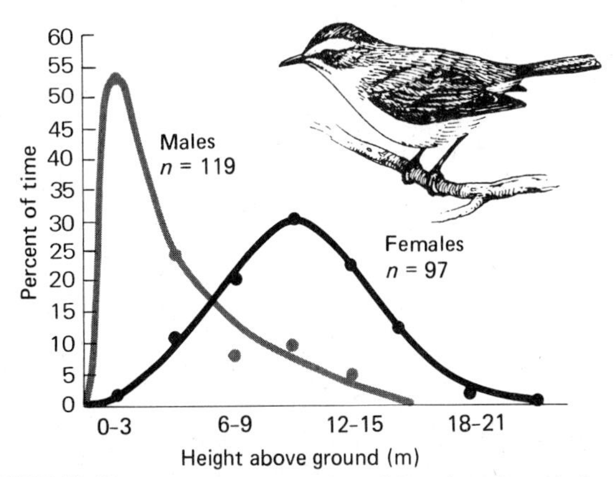

FIGURE 15.13 Separation of male and female red-eyed vireos (*Vireo olivaceus*) by height of foraging. Mean height for males is 11 m; mean height for females is 4 m.

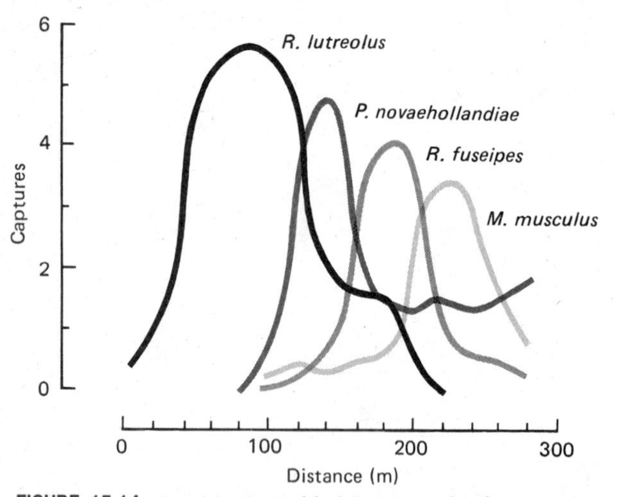

FIGURE 15.14 Partitioning of habitat space by four species of murid rodents in Australia. The numbers of individuals captured are plotted against distance—a transect across a study area with increasing elevation. The transect followed topography and thus a moisture gradient. Both, in turn, affected structure and composition of vegetation, creating habitat differences. Note that in spite of niche overlap, the abundance of each species is highest at some point on the gradient where the populations of the others are lower.

Niche widths are usually described as narrow or broad. The wider the niche, the more generalized the species is considered to be. The narrower the niche, the more specialized is the species. Generalist species have broad niches and sacrifice efficiency in the use of a narrow range of resources for the ability to use a wide range of resources. As competitors they are superior to specialists if resources are somewhat undependable. Specialists equipped to exploit a specific set of resources occupy narrow niches. As competitors they are superior to generalists if resources are dependable and renewable. A dependable resource supply is closely partitioned among specialists with low interspecific overlap. If resource availability is variable, generalist species are subject to invasion and close packing with other species during periods of resource abundance.

NICHE COMPRESSION AND COMPETITIVE RELEASE

If a community consisting of a number of species with broad niches is invaded by a number of competitors, intense competition along points on the resource gradient may force the original occupants to restrict their use of space, range of foods, or other resource-oriented activities. This change is *niche compression.*

Conversely, if interspecific competition is reduced, a species may expand its realized niche by utilizing space or a resource previously unavailable to it. Niche expansion in response to reduced interspecific competition is *competitive release.* Competitive release may occur when a species invades an island that is free of potential competitors, moves into habitats it never occupied on the mainland, and increases its abundance. Such expansion may also follow when a competing species is removed from a community, allowing remaining species to move into microhabitats they previously could not occupy.

NICHE SHIFT

Associated with compression and release is another response, *niche shift.* Niche shift is the adoption of new behavior by two or more competing populations to reduce interspecific competition. The shift may be a short-term ecological response or a long-term evolutionary response involving basic behavioral or morphological traits.

E. E. Werner and J. A. Hall demonstrated niche shift in three competing species of sunfish (Centrarchidae): the bluegill (*Lepomis maerochirus*), the pumpkinseed (*L. gibbosus*) and the green sunfish (*L. cynellus*). When the three species were stocked in separate experimental ponds, the food and habitat preferences were similar, and their average growth rate increased. The three preferred the emergent and submerged vegetation about the edges of the ponds and the large food they found there. When the three species were stocked together in equal densities, they occupied the vegetated zone about the edges. As food resources declined, the bluegill and pumpkinseed left, leaving behind the green sunfish, a more efficient forager, in the vegetation. Bluegill and pumpkinseeds turned their foraging efforts to bottom invertebrates, mainly Chironominae. The pumpkinseed—with its short, widely spaced gill rakers that do not become fouled when the fish sorts through the bottom sediments—was better able to exploit that food supply. The bluegill—with long, fine gill rakers that retain small prey—then shifted to foraging in the open-water column.

The fish also exhibited shifts in habitat utilization. When bluegill and green sunfish were confined together in equal densities in a pond, the green sunfish eventually caused the bluegill to shift to open water; but in the absence of green sunfish, bluegills invaded the dense vegetation. Bluegill and pumpkinseed have a dietary overlap of 50 to 55 percent. When the two are stocked together in a pond, the bluegill again is ultimately forced to the open-water habitat. In the end, however, the bluegill may be the superior competitor. Young sunfish of all three species feed on zooplankton in open water. This places the young of green sunfish and pumpkinseeds in direct competition for small-sized food with both adult and young bluegill. A more efficient open-water forager, the bluegill probably affects the recruitment of the other two species. Possibly its generalized diet and its ability to shift its habitat as the situation requires account for the bluegill's being the most common sunfish in ponds.

Summary

Relations among species may be positive (+), or beneficial; negative (−), or detrimental; or neutral (0). There are nine possible interactions: (00), neutral (of little ecological interest); (+ +), in which both populations benefit each other (mutualism); (− −), in which both populations are affected adversely (competition); (0+) or (+0), in which one population benefits and the other is unaffected (commensalism); (0−) or (−0), in which one population is harmed and the other is unaffected (amensalism); and (+ −) or (− +), in which one population benefits and the other is harmed (predation, parasitism).

Interspecific competition—the seeking of a resource in short supply by individuals of two or more species, reducing the fitness of both—may be one of two kinds: interference and exploitation. Exploitative competition depletes resources to a level of little value to either population. Interference involves aggressive interactions. A particular form of interference competition is allelopathy, the secretion of chemical substances that inhibit the growth of other organisms.

As described by the Lotka-Volterra equations, four outcomes of interspecific competition are possible. Species 1 may win over species 2; species 2 may win over species 1. Both of these outcomes represent competitive exclusion. A third possibility is unstable equilibrium, in which the potential winner is the one most abundant at the outset. A final possible outcome is stable equilibrium, in which the two species coexist, but at lower population levels than if each existed in the absence of the other.

The competitive exclusion principle—two species with exactly the same ecological requirements cannot coexist—has conceptual difficulties. It has, however, stimulated critical examinations of other competitive relationships, especially how species coexist and how resources are partitioned. Groups of functionally similar species that share a spectrum of resources and interact strongly with each other are termed *guilds*.

Closely associated with the concept of interspecific competition is the concept of the niche. Basically, a niche is the functional role of an organism in the community. That role might be constrained by interspecific competition.

In the absence of competition, an organism occupies its fundamental niche. In the presence of interspecific competition, the fundamental niche is reduced to a realized niche, the conditions under which an organism actually exists. When two different organisms use a portion of the same resource, such as food, their niches are said to overlap. Overlap may or may not indicate competitive interaction.

The range of resources used by an organism suggests its niche width. Species with broad niches are considered generalist species, whereas those with narrow niches are considered specialists. Niche compression results when competition forces an organism to restrict its type of food or constrict its habitat. In the absence of competition, the organism may expand its niche and experience competitive release. Organisms may also undergo niche shift by changing their behavior to reduce interspecfic competition.

Review and Study Questions

1. Describe possible relationships among species, using +, increase in fitness; −, decrease in fitness; and 0, no effect.
2. Define commensalism, amensalism, predation, parasitism, and parasitoidism.

3. Distinguish symbiosis from mutualism, obligatory and nonobligatory.
4. Define interspecific competition and distinguish between interference and exploitation.
5. Based on the Lotka-Volterra competition equations, describe four outcomes of competition.
6. What is the competitive exclusion principle? What is the problem with the concept?
7. What conditions often determine the outcome of unstable equilibrium under natural situations?
8. What permits coexistence among competing species?
9. What is allelopathy? How might it be considered a form of competition?
10. What is resource partitioning? What are some weaknesses of the concept?
11. Define niche.
12. Distinguish between a fundamental niche and a realized niche.
13. What is niche overlap? Niche width? How is niche width usually measured?
14. What is meant by niche compression, competitive release, and niche shift?

Selected References

American Naturalist. 1983. Interspecific competition papers, Volume 122, No. 5 (November). A group of controversial articles on interspecific competition.

Brown, J. C., O. J. Reichman, and D. W. Davidson. 1979. Granivory in desert ecosystems. *Annual Review of Ecology and Systematics* 10:210–227.

Diamond, J., and T. Case, eds. 1985. *Community ecology.* New York: Harper & Row. Advanced; many papers on competition.

Hutchinson, G. E. 1978. *An introduction to population ecology.* New Haven, CT: Yale University Press. Builds a strong case for interspecific competition.

May, R. M., ed. 1981. *Theoretical ecology: Principles and applications,* 2nd ed. Sunderland, MA: Sinauer Associates. Advanced and informative.

Pianka, E. R. 1988. *Evolutionary ecology,* 4th ed. New York: Harper & Row. Good discussion on competition.

Pielou, E. C. 1974. *Population and community ecology.* New York: Gordon & Breach.

Rice, E. L. 1984. *Allelopathy,* 2nd ed. Orlando, FL: Academic Press.

Schoener, T. W. 1982. The controversy over interspecific competition. *American Scientist* 70:586–595.

Strong, D. R., Jr., D. Simberloff, L. G. Abele, and A. B. Thistle, eds. 1983. *Ecological communities: Conceptual evidence and the issues.* Princeton, NJ: Princeton University Press.

Werner, E. E., and J. D. Hall. 1976. Niche shifts in sunfishes: Experimental evidence and significance. *Science* 191:404–406.

Werner, E. E., and J. D. Hall. 1979. Foraging efficiency and habitat switching in competing sunfish. *Ecology* 60:256–264.

Wiens, J. A. 1977. On competition and variable environments. *American Scientist* 65:590–597.

CHAPTER 16

Predation

Outline

Objectives

On completion of this chapter, you should be able to:

1. Define predation and distinguish its forms.
2. Sketch the Lotka-Volterra model of predation and its weaknesses.
3. Explain functional response and its three types, numerical response, and total response.
4. Define search image and switching and their importance to predation.
5. Explain the concept of optimal foraging.
6. Describe plant-herbivore and herbivore-carnivore systems.

The poet who described nature as "red in tooth and claw" was thinking of predation. Most of us associate predation with a hawk taking a mouse, a wolf killing a deer. That is a narrow view. A fly laying its eggs on a caterpillar to develop there at the expense of the victim is exhibiting a form of predation called *parasitoidism.* The parasitoid attacks the host (the prey) indirectly by laying its eggs on the host's body. When the eggs hatch, the larvae feed on the host, slowly killing it. True *parasitism* involves a predator living on or within the host. The parasite is physiologically dependent upon the host and seldom kills the host outright (which, in effect, would destroy both its habitat and its food). A deer feeding on woody shrubs and grass and a mouse eating a seed are practicing forms of predation called *herbivory.* Seed consumption is a form of outright predation. Grazing on plants without killing them is a form of parasitism. A special form of predation is *cannibalism,* in which the predator and the prey are the same species. Thus, predation in its broadest sense can be defined as one living organism feeding on another.

Predation in its actions and reactions is more than just a transfer of energy. It represents a direct and often complex interaction of two or more species, of the eaters and the eaten. The numbers of some predators may depend upon the abundance of their prey, and the population of the prey may be controlled by its predators. Each can influence the fitness of the other and favor new adaptations.

THEORY OF PREDATION

In the 1920s Lotka and Volterra extended their models of population growth and competition to predation. Independently they proposed mathematical statements to express the relationship between predator and prey populations. They provided one equation for the prey population and another for the predator population.

The growth equation for the prey population involving the maximum rate of increase per individual and the removal of the prey from the population by the predator is:

$$\frac{dN_1}{dt} = r_1N_1 - PN_1N_2$$

where N_1 = density of the prey population
r_1 = intrinsic rate of increase in the absence of predation
N_2 = density of the predatory population
P = coefficient of predation
N_1N_2 = probability of an encounter between predator and prey

The growth equation for the predator is influenced by the density of the prey population:

$$\frac{dN_2}{dt} = P_2N_1N_2 - d_2N_2$$

where P_2 = coefficient expressing effectiveness of the predator
d_2 = density-independent mortality rate of the predator

The underlying assumptions of the Lotka-Volterra equations are that the prey population grows exponentially and that reproduction in the predator population is a function of the number of prey consumed.

These equations show that as a single predator population increases, the single prey population decreases to a point at which the trend is reversed. The prey increases, followed by an increase in the predator population. The two populations rise and fall, oscillating (Figure 16.1).

About a decade later A. H. Nicholson, an ecologist, and W. Bailey, an engineer and mathematician, developed a mathematical model for a host-parasitoid relationship. That model predicted increasingly violent oscillations in single-predator–single-prey populations living together in a limited area with all the external conditions constant.

Both models state that as the predator population increases, it will consume a progressively larger number of prey until the prey population begins to decline. In turn, the abundance of prey influences the reproductive rate of the predator population. In time, the number of predators overshoots the availability of prey, and the predator population declines to a point where the reproduction of prey more than balances its losses through predation. The prey then increases, followed by an increase in the population of predators. The cycle may continue indefinitely. The prey is never quite de-

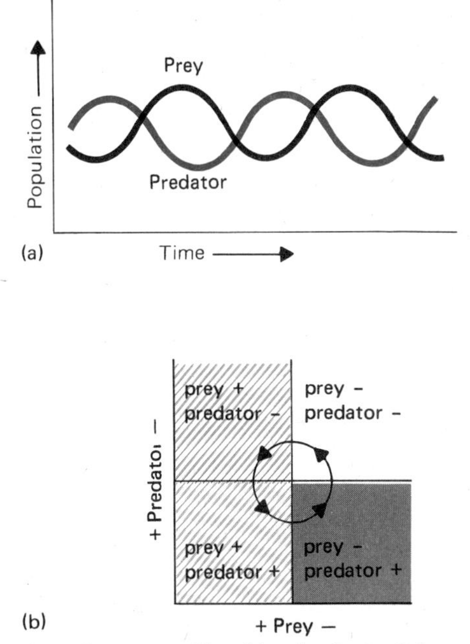

FIGURE 16.1 Pattern predicted by the Lotka-Volterra model of predator-prey interactions. (a) The abundance of each population is plotted as a function of time. An increase of prey is followed by an increase of predators. (b) The zero growth curves of predator and prey are straight and intersect at right angles. A minus sign indicates population decline and a plus sign population increase. Predators increase up to the horizontal line. Prey increase up to the vertical line. The Lotka-Volterra model assumes that reproduction of predators is a function of the number of prey consumed.

stroyed by the predator; the predator never completely dies out.

Both equations are simplistic. They overemphasize the influence of predators on prey populations. They ignore genetic changes, stress, emigration, aggression, the availability of cover, the difficulty of locating prey as it becomes scarcer, and other parameters. The continuing appeal of the Lotka-Volterra equations to population ecologists lies in the straightforward mathematical descriptions.

EXPERIMENTS

The biologist G. F. Gause attempted to demonstrate the Lotka-Volterra predation model experimentally, just as he did the competition models. He reared together under constant environmental conditions a prey population, *Paramecium caudatum,* whose populations grow exponentially, and a predator, the ciliate *Didinium nasutum.* Because *D. nasutum*'s production rate is independent of food intake, these experiments did not properly test the Lotka-Volterra model of predation. However, Gause did demonstrate that in a laboratory habitat the predator *Didinium* exterminated its prey, regardless of the density of the two populations. After the prey was destroyed, the predators died of starvation. Only by introducing prey periodically to the medium was Gause able to prevent it from dying out. In that manner, he was able to maintain the two populations together and produce regular fluctuations in both. The predator-prey relations were overexploitation and annihilation unless immigration took place from other prey populations.

In another experiment Gause introduced sediment in the floor of the tube habitat. There the prey could escape from the predator. When predators eliminated the prey from the clear medium, they died from lack of food. The *Paramecium* that took refuge in the sediment continued to multiply and eventually took over the medium.

Over 20 years later, in a different type of experiment, C. Huffaker attempted to learn whether an adequately large and complex laboratory environment could be established in which a predator-prey system would not be self-exterminating. Involved were the six-spotted mite *Eotetranychus sexmaculatus* and a predatory mite, *Typhlodromus occidentalis.* Whole oranges, placed on a tray among a number of rubber balls the same size, provided food and cover for the spotted mite. Such an arrangement permitted the experimenter to control both the total food resource available and the pattern of dispersion by covering the oranges with paper and sealing wax to whatever degree desired and by changing the general distribution of oranges among the rubber balls. Huffaker could manipulate conditions to simulate a simple environment in which the food of the herbivore was concentrated or a complex environment in which the food was widely dispersed, partially blocked by barriers, and in which refuges were lacking. In both situations the two species at first found plenty of food available for population growth. Density of predators increased as the prey population increased. In the environment where the food was concentrated and dispersion of the prey population was minimal, predators readily found prey, quickly responded to changes in prey density, and were able to destroy the prey rapidly. In fact, the system was self-annihilative. In the environment where the primary food supply and prey were dispersed, predator and prey went through two oscillations before the predators died out. The prey recovered slowly.

Several important conclusions resulted from the study. First, predators cannot survive where their prey population is low. Second, a self-sustaining predator-prey population cannot be maintained without immigration of prey. Third, the complexity of prey dispersal and predator searching relationships, combined with a period of time for the prey population to recover from the effects of predation and to repopulate the areas, has more influence on the period of oscillation than the intensity of predation.

FUNCTIONAL RESPONSE

Models of interactions between predator and prey suggest two distinct responses of the predator to changes in prey density. One is for the individual predator to eat more prey as the prey population increases or take them sooner. The other is for predators to become more numerous through increased reproduction or immigration. The former is called a *functional response* and the latter a *numerical response.*

The idea of a functional response was introduced by the English entomologist M. E. Solomon in 1949 and was explored in detail by C. S. Holling a decade later. The basis of a functional response is that a predator (or parasite) will take or affect more prey as the density of prey increases. Holling classified functional responses into three types (Figure 16.2).

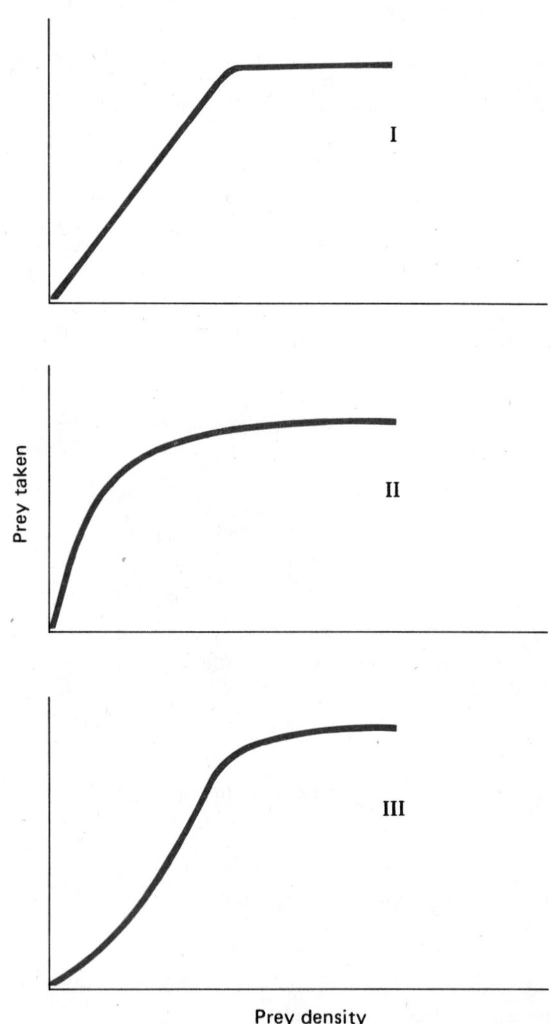

FIGURE 16.2 Three types of functional response curves. In Type I, the number of prey taken per predator increases linearly to a maximum as prey density increases. In Type II, the number of prey taken rises at a decreasing rate to a maximum value. In Type III, the number of prey taken is low at first, then increases in a sigmoid fashion approaching an upper limit.

In a Type I response the number of prey affected per predator increases in a linear fashion to a maximum as prey density increases. Predators of any given abundance take a fixed number of prey during the time they are in contact, usually enough to satiate themselves. Trout feeding on an evening hatch of mayflies is an example. Type I produces density-independent mortality up to satiation.

In a Type II response the number of prey affected rises at a decreasing rate toward a maximum value. A dominant component of a Type II response is handling time for the predator. Handling time includes time spent pursuing the prey, subduing, eating, and digesting it. A tiger or boa constrictor, for example, takes a long handling time. Because handling time limits the amount of prey that the predator can process per unit of time, the number of prey taken per unit of time slows to a plateau while the number of prey is still increasing. For that reason a Type II functional response rarely acts as a stabilizing force on a prey population, unless the prey occurs in patches.

The plateau in Type II may be influenced by an aggregative response of predators to patches of high density. Patches of profitable foraging may attract other predators to the area (Figure 16.3). When one or two members of a predator species discover and begin to feed on the prey item, other members of the population observe and join them. As the number of predators increases on the patch, they interfere with each other. That increases each predator's search time, reduces the efficiency of predation, and encourages predators to leave the crowded area.

Type III functional response is more complex than Type II. In Type III the number of prey taken is low at first, then increases in a sigmoid fashion, approaching a plateau at which the attack rate remains constant. Type III functional response can potentially stabilize a prey population, because the attack rate varies with prey density. At low densities the attack rate is negligible, but as prey density increases (as indicated by the upward sweep of the curve), predatory pressure increases in a density-dependent fashion.

Type II responses involve varying densities of a single prey species. Type III responses invariably involve two or more prey species. Predators take most or all of the individuals that are in excess of a certain minimum number, determined, perhaps, by the availability of prey cover and the prey's social behavior. The population level at which the predator no longer finds it profitable to hunt the prey species has been called the *threshold of security* by P. Errington (Figure 16.4).

Type III responses have been called compensatory because as prey numbers increase above the threshold, surplus animals become vulnerable to predation through intraspecific competition. Below the threshold of security, the prey species compensates for its losses through increased reproduction and greater survival of young. Below the threshold of security, functional response of the predator is low; above the threshold, functional response is marked. An example of that type of predation is detailed by Errington in his notable long-term study of the muskrat. Adult muskrat (*Ondatra zibethicus*) well established on breeding territories and occupying an optimal habitat were largely free from predation by mink (*Mustela vison*) and had the greatest fitness. Animals excluded from the territory or forced to occupy submarginal habitats were highly vul-

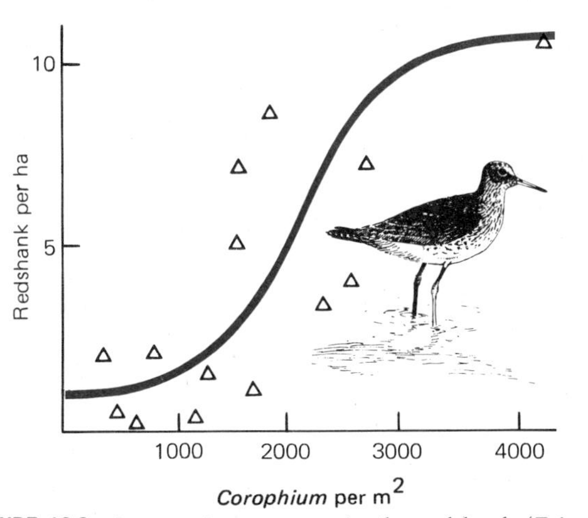

FIGURE 16.3 Aggregative response in the redshank (*Tringa totanus*). The curve plots the density of the predator (the redshank) in relation to the average density of arthropod prey (*Corophium*).

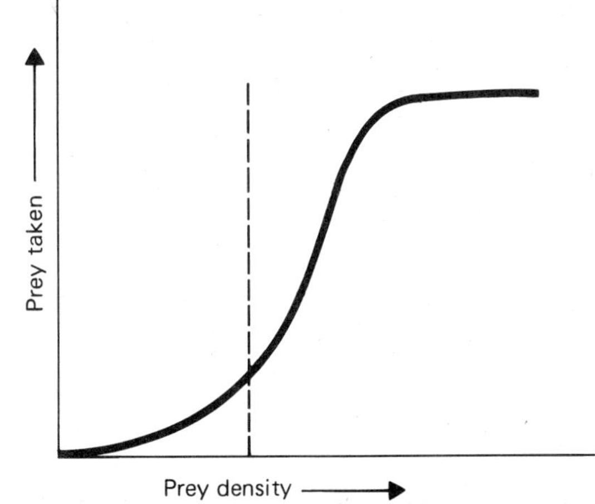

FIGURE 16.4 Compensatory predation as illustrated by a functional response curve. There is no response to the left of the vertical line, which represents the threshold of security for the prey.

nerable to predation or, if they were not, usually failed to reproduce. In either case, these animals could be considered surplus. Their demise did not affect the overall fitness of the prey population.

Search Image

Another reason for the sigmoidal shape of the Type III functional response curve may be the *search image.* According to the search image hypothesis, when a new prey species appears in an area, its risk of becoming prey is low. The predator has not yet acquired a search image—a way to recognize it. Once the predator has secured a palatable item of prey, it is easier to locate others of the same kind. The more adept and successful the predator becomes at securing a particular prey item, the longer and more intensely it concentrates on that item. In time the number of the prey species becomes so reduced or its population so dispersed that encounters between it and the predator lessen. The search image for that species begins to wane, and the predator reacts to another species.

Switching

The Type III functional response curve is characteristic of situations embracing more than one prey species. It assumes a predator capable of choice and an alternate prey. Although a predator may have a strong preference for a certain prey, it can turn to an alternate, more abundant prey species that provides more profitable hunting. If rodents, for example, are more abundant than rabbits and quail, foxes and hawks will concentrate on rodents. That idea was advanced in 1933 by Aldo Leopold in his book *Game Management,* in which he describes alternate prey species as buffer species because they stand between the predator on the one hand and game species on the other. If the population of the buffer prey is low, the predators will turn to game species; foxes and hawks will concentrate on rabbits and quail.

Turning to alternate, more abundant prey has been termed *switching.* In switching, the predator concentrates a disproportionate amount of feeding on the more abundant species and pays little attention to the rarer species. As the relative abundance of the two prey species changes, the predator turns its attention to the less common prey (Figure 16.5).

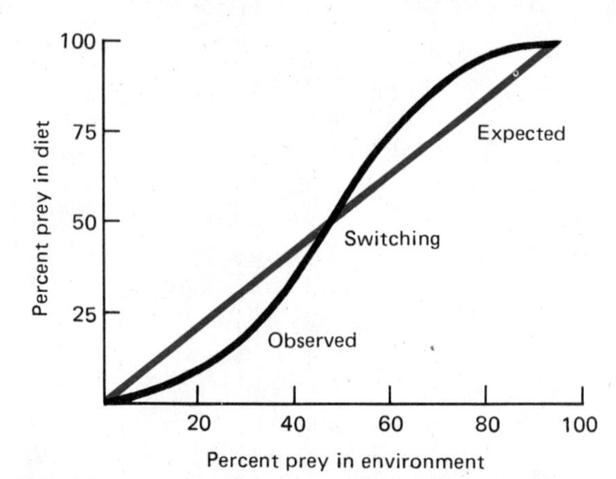

FIGURE 16.5 A model of switching. The color line represents a constant preference with no switch. The black line represents the prey actually taken. Up to a point the percent of prey taken is less than the percent in the environment. Switching occurs at the point where the lines cross.

At what point in prey abundance a predator switches depends considerably on the threshold of security for the prey species involved. The threshold of security may be much lower for a highly palatable prey. A predator may hunt longer and harder for a palatable species before it turns to a more abundant, less palatable alternate prey. Conversely, the predator may turn from the less palatable species at a much higher level of abundance than it would from a more palatable species. For the palatable prey that situation is maladaptive. It is at a selective disadvantage when associated with the less palatable prey. Such selective pressure favors temporal or spatial isolation of the more palatable from the less palatable species, or it favors mimicry.

In spite of switching, some predators deliberately seek out certain prey, no matter how scarce. Predators in a California grassland exhibited a distinct preference for meadow voles over harvest mice and other rodents even though alternate prey was more abundant. In fact, the abundance of alternate prey apparently enables carnivores to maintain a sufficiently high population and to obtain sufficient energy to permit continued predatory pressure on the preferred species. Among herbivores, deer exhibit a pronounced preference for certain species of browse plants. Meadow mice often concentrate on seeds of preferred grasses, even though the seeds of other species are more abundant.

NUMERICAL RESPONSE

As the density of prey increases, the number of predators may also increase. Numerical response takes three basic forms (Figure 16.6): (1) direct response, in which the number of predators in a given area increases as the prey density increases; (2) no response, in which the predator population remains proportionately the same; and (3) inverse response. An example of the last is the response of parasites to an increasing density of budworms in Canada. Parasites increased sharply at first, then declined rapidly as the pupal density of the budworm increased.

Most numerical responses involve an increase by reproductive effort. Because reproduction does require a certain minimal time, a lag exists between an increase of a prey population and a numerical response by a predator population. For example, the population of great horned owl (*Bubo bubo*) in a 160 square kilometer area in Alberta, Canada, increased over a three-year period, 1966 to 1969, from 10 birds to 18 as the population of its prey, the snowshoe hare, increased sevenfold. The portion of owls nesting increased from 20 to 100 percent as the biomass of snowshoe hare in the owls' diets increased from 23 to 50 percent. In that situation, nutrition, controlled by the amount of prey available, influenced fitness. If food is limited because of low prey density, fitness is necessarily low, and positive numerical response is low. With increasing prey density, fitness increases, and the numerical response is proportionately higher (Figure 16.7).

Numerical response may involve both an aggregative response and an increase in fitness. An example can be found among the "fugitive" warblers of northern forests, especially the Tennessee (*Vermivora peregrina*), Cape May (*Dendroica tigrina*), and bay-breasted (*D. castanea*) warblers, whose abundance is dictated by an outbreak of spruce budworm. During such periods, populations of the bay-breasted warblers have increased from 10 to 120 pairs per 40 hectares, and Cape May and bay-breasted warblers have larger clutches than associated warbler species. In fact, Cape May and possibly bay-breasted warblers apparently depend upon occasional outbreaks of spruce budworm for their contin-

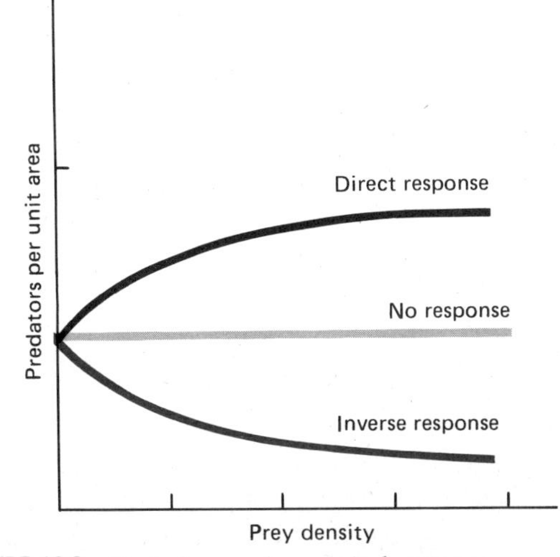

FIGURE 16.6 Basic forms of numerical response.

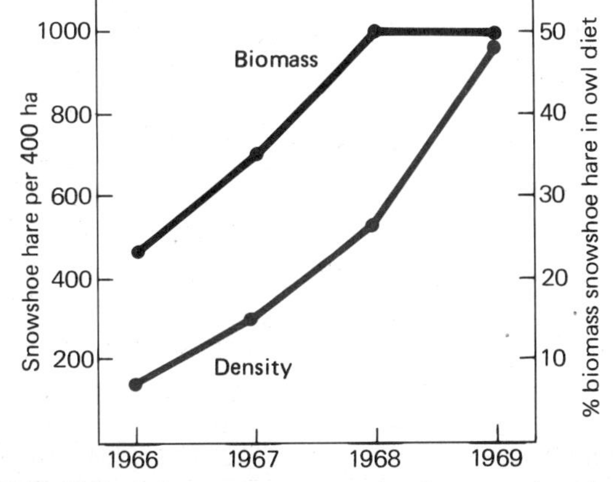

FIGURE 16.7 Numerical response in the snowshoe hare–great horned owl system.

ued existence. At those times, these two species are able to increase more rapidly than other warblers because of extra-large clutches. Between outbreaks, they are reduced in numbers and even become extinct locally.

TOTAL RESPONSE

Functional and numerical responses may be combined to give a total response, and predation may be plotted as a percentage (Figure 16.8). If that is done, predation falls into two types: (1) the percentage of predation declines continuously as prey density increases, and (2) the percentage of predation rises initially and then declines. The latter results in a dome-shaped curve produced by the sigmoid (Type III) functional response curve to prey density and by direct numerical response.

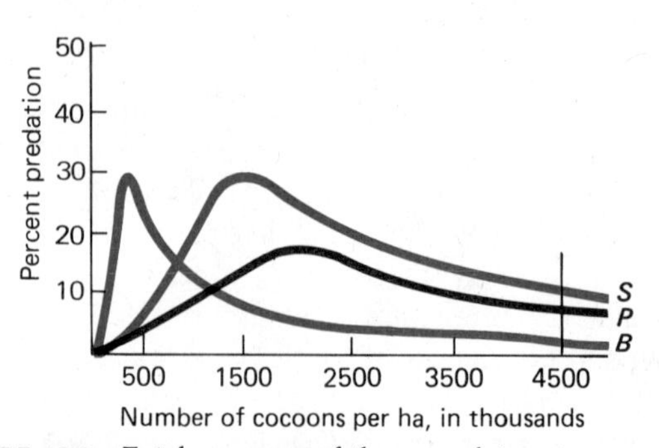

FIGURE 16.8 Total response of three predators to prey density expressed as a percentage of predation to prey density. Total response includes both functional response and numerical response. S = *Sorex* shrew; B = *Blarina* shrew; P = *Peromyscus* mouse.

CANNIBALISM

A special form of predation is cannibalism, euphemistically called intraspecific predation. Cannibalism is killing and eating an individual of the same species. Cannibalism is more widespread and important in the animal kingdom than many ecologists admit, probably because of the moral significance attached to it. In reality, cannibalism is common to a wide range of animals, both aquatic and terrestrial, from protozoans and rotifers through centipedes, mites, and insects to frogs, toads, fish, birds, and mammals, including humans. Interestingly, about 50 percent of terrestrial cannibals, mostly insects, are normally herbivorous species, the ones most apt to encounter a shortage of protein. In freshwater habitats the bulk of cannibalistic species are predaceous, as they all are in marine ecosystems.

Cannibalism has been associated with stressed populations, particularly those facing starvation. Although some animals do not become cannibalistic until other food runs out, others do so when alternative foods decline and individuals are malnourished. Other conditions that may promote cannibalism are (1) crowded conditions or dense populations, even when food is adequate; (2) stress, especially when induced by low social rank; and (3) the presence of vulnerable individuals—such as nestlings, eggs, or runty individuals—even though food resources are adequate.

Whatever the cause, the intensity of cannibalism is influenced by local conditions and by the nature of local populations. In general, cannibalism fluctuates greatly over both long and short periods of time. Among predaceous fish such as walleyes (*Stizostedion vitreum*) and insects such as freshwater back swimmers (*Notonecta hoffmanni*), cannibalism is most prevalent in summer, which coincides with a sudden decrease in normal prey and a reduction of spatial refuges for the young. Those receiving the brunt of cannibalism are the small and young, preyed upon by older and larger individuals.

Even at low rates, cannibalism can produce demographic effects. Three percent cannibalism in the diet could account for 88 percent of mortality among the young of walleyes, 23 to 46 percent among eggs and chicks of herring gulls, 8 percent of young Belding's ground squirrels, and 25 percent of lion cubs. If a large proportion of either a population or a vulnerable age class is eaten, it can result in violent fluctuations in recruitment.

Cannibalism can rapidly decrease the number of intraspecific competitors as food becomes scarce. It decreases as resources become more available to survivors and as vulnerable individuals become scarcer or harder to find. By reducing intraspecific competition at times of resource shortages, cannibalism may actually reduce the probability of local extinction of a population, improve conditions for survivors, and enhance their growth, fecundity, and fitness. Conversely, cannibalism can be at a selective disadvantage if survivors become too aggressive and destroy their own progeny and genotype completely, or reduce their genotype faster than genotypes of conspecific competitors.

OPTIMAL FORAGING

Have you ever watched a robin hopping across a lawn? It flies in, lands, hops along for a short distance as if sizing up the situation, and stops. Then it moves on deliberately in a series of irregular paths across the grass. The bird pauses every few feet, stares ahead as if in deep concentration or cocks its head toward the ground. It either moves on or crouches low as if to brace itself. It pecks quickly in the ground and pulls out an earthworm. On occasion the earthworm pulls back, and in the tug-of-war the worm wins. The robin does not push the action. It lets go and hops to another spot to repeat the food-seeking activity.

The behavior of the robin can be divided into four parts. The robin had to decide where to hunt for food. Once on the area, it had to search for palatable items of food. Having located some potential food, the robin had to decide whether to pursue it or not. If it began the pursuit by pecking in the ground, then the robin had to attempt a capture, in which it might or might not be successful. By capturing the earthworm, the robin earned some units of energy. The robin needed to gain more energy than it expended in its round of foraging activity.

The problem facing the robin and all other animals is securing sufficient energy to maintain itself, feed its young, and lay up fat for migration or winter, yet not spend too much time doing so. Time, in the parlance of the robin, is energy. To spend too much time in securing energy without a sufficient return for the effort results in a type of personal bankruptcy—a loss of fitness. If the robin fails to find suitable and sufficient food in the area of lawn it is searching, it leaves to search elsewhere. If successful, the bird probably will return until the spot is no longer an economical place to feed.

The means animals employ to secure food are termed their *foraging strategy.* Of particular theoretical interest to ecologists is the optimal foraging strategy, against which actual foraging strategy can be measured. An optimal foraging strategy is one that provides the maximum net rate of energy gain, endowing the animal with the greatest fitness. A foraging strategy involves two separate but related components. One is optimal diet; the other is optimal foraging efficiency. Ecologists have come up with certain rules (hypotheses) for each.

OPTIMAL DIET

When the robin searches the lawn for food, it should, by optimal foraging theory, select only those items that provide the greatest energy return for the energy expended. That places an upper and lower limit on the size of food items accepted as well as the degree of palatability. If a food item is too large, it requires too much time to handle; if too small, it does not deliver enough energy to cover the costs of capture. Our robin should reject or ignore less valuable items such as small beetles and caterpillars and give preference to smaller and medium-sized earthworms. These more valuable food items would be classified as preferred food, ones taken out of proportion to their availability relative to all other food items. Our robin, according to optimality theory, should take the most valuable food items first. When those items have been depleted, the robin should turn its attention to the next most valuable food items. Eventually, the animal should expand its diet to include the poorer foods when the discovery of high-value foods falls below a certain rate.

Although such a theory makes practical sense, it is difficult to test in the field. Not only would we have to know exactly what items the animals were consuming; we would also have to know the relative availability of the food items in the habitat. Some studies have been done under controlled conditions. E. E. Werner and J. D. Hall presented a group of ten bluegill sunfish (*Lepomis macrochirus*) with three sizes of *Daphnia* in a large aquarium. After a period of foraging, the fish were killed and their stomachs examined to determine the number and size of *Daphnia* taken. When the density of the prey presented was low, the fish consumed

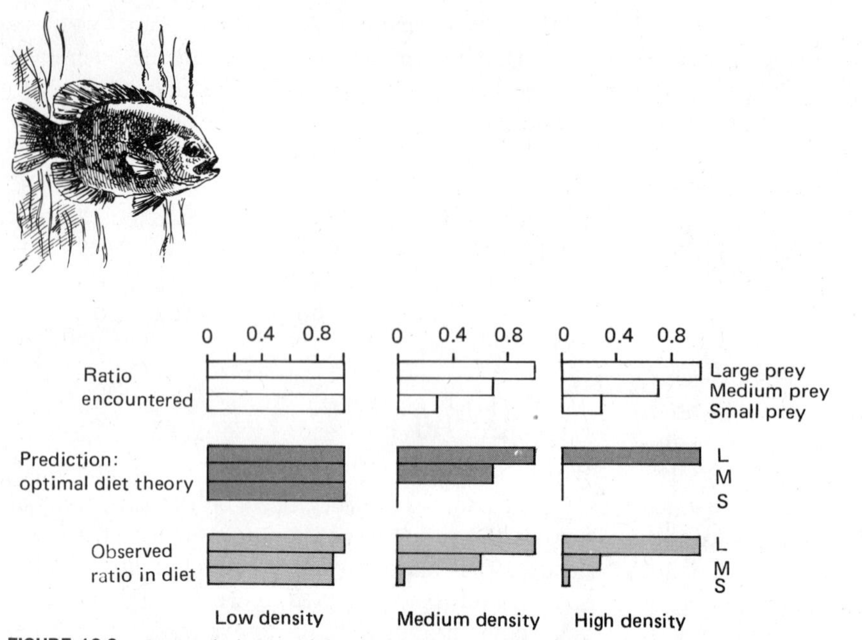

FIGURE 16.9 Optimal choice of diet in bluegill sunfish preying on different sizes of *Daphnia*. The histograms show for each size class at three different densities the ratio of encounter between predator and prey, the prediction of optimal ratios in the diet, and the observed ratios in the diet. Note the preference of the bluegill for large prey.

the three sizes according to the frequency encountered. They took the first prey encountered (Figure 16.9). When the prey population was dense, the fish were highly selective and took the largest prey items. When presented with an intermediate number of prey, the fish consumed the two largest size classes. The results of these feeding trials support optimal foraging theory.

Field studies provide some insight into how animals forage under natural conditions. N. Davies studied the feeding behavior of the pied wagtail (*Motacilla alba*) and the yellow wagtail (*M. flava*) in a pasture field near Oxford, England. The birds fed on various dung flies and beetles attracted to cattle droppings. Prey of several sizes were available: large, medium, and small flies and small beetles. The wagtails showed a decided preference for medium-sized prey (Figure 16.10). The size of prey selected corresponded to the optimum-size prey the birds could handle profitably (Figure 16.11). The birds ignored small sizes. Easy-to-handle small prey did not return sufficient energy, and large sizes required too much time and effort.

OPTIMAL FORAGING EFFICIENCY

The robin lives in a heterogeneous, or patchy, environment. It is made up of clumps of trees, shrubby thickets or plantings, and open lawns. The robin could forage in all these patches of vegetation, utilizing all food items it came across, but it does not. Rather, the bird concentrates its foraging activities in patches of open lawn. Moreover, the robin may restrict its searching to certain parts of the lawns where the environment is more favorable for earthworms, until the earthworms are depleted or have moved deeper in the soil in response to a drying surface. Then the robin is forced to turn its attention to other patches and other foods, such as ground beetles, flies, sowbugs, snails, and millipedes.

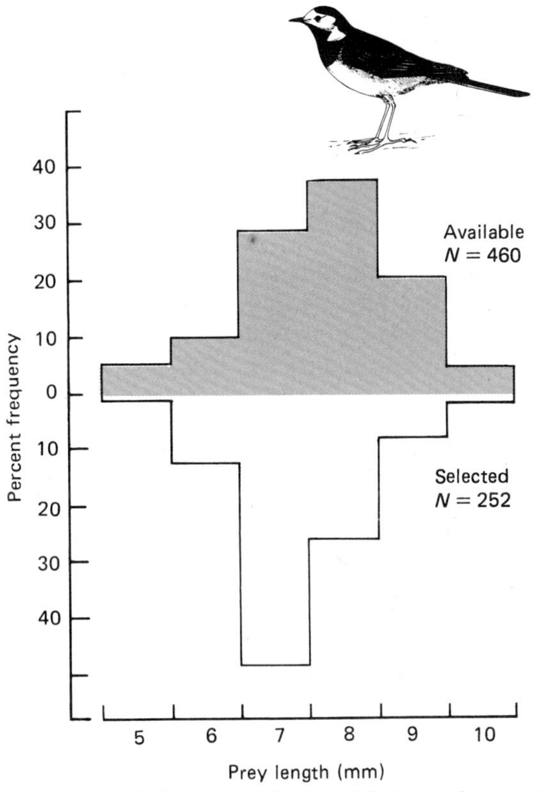

FIGURE 16.10 Pied wagtails show a definite preference for medium-sized prey, which are taken in disproportionate amounts compared to sizes of prey available in the environment.

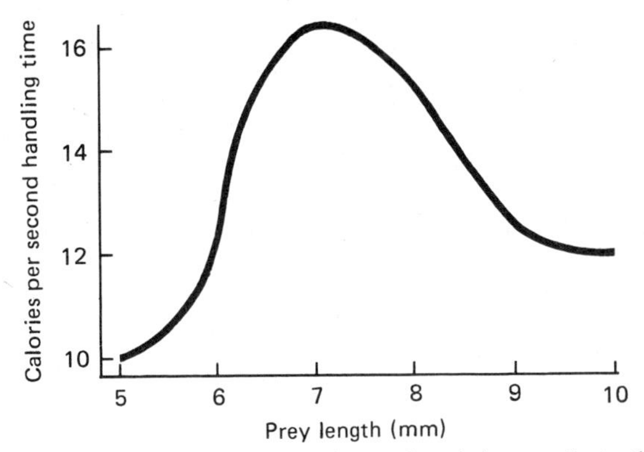

FIGURE 16.11 The prey size chosen by pied wagtails is the optimal size for providing maximum energy per handling time. Small sizes provide too few calories. Large sizes require too much handling time.

The robin has been following the rules of optimal foraging: (1) concentrate foraging activity in the most productive patches; (2) stay with those patches until their profitability falls to a level equal to the average for the foraging area as a whole (Figure 16.12); (3) leave the patch once it has been reduced to the level of average productivity; (4) ignore patches of low productivity.

How well animals go by these rules has also been the object of experimentation, both in the laboratory and in the field. S. F. Hubbard and R. M. Cook studied the foraging behavior of a parasitoid, the ichneumon wasp *Memeritis canescens,* in a laboratory arena containing patches of the host, larvae of *Ephestia cantella.* Hosts in densities of 64, 32, 16, 8, and 4 were placed in petri dishes filled to the brim with plaster of paris, the hosts' substrate. Space between the larvae was filled with wheat bran. One result of the experiment showed that all patches were depleted of larvae to a common level of host abundance. The richest patches suffered the greatest depletion. As exploitation continued, the amount of time spent by the wasp in patches of highest density declined, and the proportion of time spent in the next richest patches increased.

R. Zack and J. B. Falls studied the foraging strategy of the ovenbird (*Seiurus aurocapillus*), a ground-nesting warbler, under seminatural conditions. They exposed captive ovenbirds held individually in natural outdoor pens in typical woodland habitat to a food supply of mealworms presented at four constant locations but interchanged prey densities. The birds concentrated their efforts in areas of high profitability and took a higher percentage of prey available in the high density sites. Although the ovenbirds did not always visit every patch, they always visited the densest ones. The birds quit searching for food after they encountered one or more profitable patches and revisited patches of high prey density. The ovenbirds learned rapidly to find patches of prey; they chose feeding sites nonrandomly, and avoided areas of no food and patches visited previously.

These and other observational and experimental studies support the hypothesis of optimal foraging up

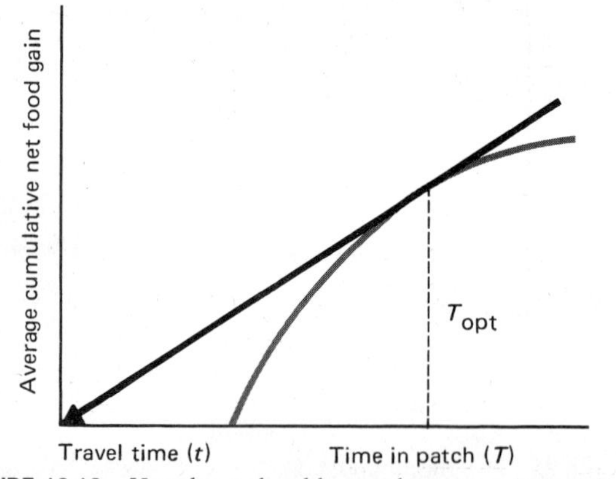

FIGURE 16.12 How long should a predator remain in a habitat patch seeking food? This graph provides a theoretical answer. Time spent in travel and time spent in a habitat patch are plotted against net gain in food, which declines as available food is depleted. The curve represents the cumulative amount of food harvested relative to time in the patch. The straight line represents average food intake per unit of time for the habitat as a whole. Where the line touches the curve, the predator has reached average cumulative net food gain for the habitat as a whole. Beyond this point net food gain declines below average. It is no longer profitable to remain. Thus, the point represents the optimal time (T_{opt}) for the predator to seek a more profitable food patch.

to a point. Where the theory breaks down is in the expectation that animals will select patches in order of their profits and take only optimal food items and ignore the rest. Such choices may be characteristic of animals foraging in a stable laboratory environment. It is not necessarily the way animals behave in the wild.

Being opportunists, animals will take some less than optimal food items (by ecologists' standards) upon discovery, and they may quit foraging in a patch before food items are reduced to some minimal level. Nor will they pass up profitable patches because they do not meet some theoretical expectation. Animals quickly learn where food is and where food is not, and they do not waste time on a patch after it is depleted. Foragers, however, will stay with a patch as long as the rate of replenishment exceeds the rate of depletion. Some animals are highly restricted in their choices. Sedentary animals such as corals, barnacles, and blackfly larvae, filter feeders all, have to take what food flows past them. Others have severely restricted foraging patterns that limit their choices of food.

In spite of its weaknesses, the concept of optimal foraging is valuable. It provides a way to compare how animals do forage, select, and harvest their food.

PREDATOR-PREY SYSTEMS

The usual approach in ecology is to consider predation on plants and predation on animals as two separate entities. They are not separate. Predator-prey interactions at one level influence interactions at other levels. We cannot really understand a herbivore-carnivore system without understanding plants and their herbivores; nor can we understand plant-herbivore relations without understanding predator-herbivore relationships. All three are related, but we are just beginning to understand these interrelations.

PLANT-HERBIVORE SYSTEMS

A deer pulling tender young twigs from a sapling, horses clipping grass in a pasture, caterpillars voraciously chewing tree leaves, bark beetles feeding on cambium, chipmunks eating acorns—all are instances of predation on plants. Evidence of such herbivory occurs to some extent on the leaves and fruit of almost any plant you examine.

If you measure the amount of biomass actually consumed by herbivores, it may be small, perhaps 6 to 10 percent, except in years of major insect outbreaks or in the presence of an overabundance of deer. Consumption, though, is not a good measure of the importance of herbivory. Grazing on plants has a subtle impact on the fitness of both plants and herbivores.

Effects on Plant Fitness

Removal of plant tissue—leaf, bark, stems, roots, sap—affects a plant's fitness and its ability to survive, even though it may not be killed outright. Loss of foliage and subsequent loss of roots decrease plant biomass, reduce the vigor of the plant, place it at a competitive disadvantage with respect to surrounding vegetation, and lower its reproductive effort and fitness. That is especially true in the juvenile stage, when the plant is most vulnerable and least competitive.

Although a plant may be able to compensate for the loss of leaves by increasing photosynthetic assimilation in the remaining leaves, it may be adversely affected by the loss of nutrients, depending upon the age of the tissues removed. Young leaves are dependent structures, importers and consumers of nutrients drawn from reserves in roots and other plant tissues. As the leaf matures, it becomes a net exporter of nutrients, reaching its peak before senescence sets in. Grazing herbivores such as sawfly and gypsy moth larvae, deer, and rabbits concentrate on more palatable, more nutritious leaves. They tend to reject older leaves because they are less palatable, are high in lignin, and often contain secondary compounds (tannin, for example). If grazers concentrate on young leaves, they remove considerable quantities of nutrients.

Plants respond to defoliation with a flush of new growth that drains nutrients from reserves that otherwise would have gone into growth and reproduction. Defoliation also draws on the plants' chemical defenses, a costly response. Often the withdrawal of nutrients and phenols from roots exposes the roots to attack by root fungi while the plant marshals its defenses in the canopy. If defoliation of trees is complete, as often happens during an outbreak of gypsy moths (*Porthetria dispar*) or fall cankerworms (*Alsophila pometaria*), replacement growth differs from the primary canopy removed. The leaves are smaller, and the total canopy area may be reduced by as much as 30 to 60 percent. Defoliation in a subsequent year may cause an even further reduction in leaf size and number. Some trees may end up with only 29 to 40 percent of the original leaf area to produce food in a shortened growing season (Figure 16.13).

FIGURE 16.13 Outbreaks of the defoliating gypsy moth exert a severe stress on trees, especially its primary food, oaks. Note that an unpalatable tree, the black cherry, escaped defoliation.

Severe defoliation and subsequent regrowth alter the tree physiologically. Growth regulators controlling bud dormancy are changed when leaves are removed. The plant uses up reserve food to maintain living tissues until new leaves are formed. Buds for the next year's growth are late in forming. The refoliated tree is out of phase with the season, and the drain on nutrient reserves adversely affects the tree over winter because twigs and tissues are immature at the onset of cold weather. Such weakened trees are more vulnerable to insect and disease attack the next year. Also, because nutrient reserves are used for regrowth and maintenance, no resources are available for reproduction. Defoliation of coniferous species results in death.

Some plant predators, such as aphids, do not consume tissue directly but tap plant juices instead, especially on new growth and young leaves. Sap suckers can decrease growth rates and biomass of woody plants by 25 percent.

Damage to the cambium and the growing tip (apical meristem) may be more important in some plants. Deer, rabbits, mice, and bark-burrowing insects feed on those parts, often killing the plant or changing its growth form.

Moderate grazing, even in a forest canopy, can have a stimulating effect, increasing biomass production, but at some cost to vigor and at the expense of nutrients stored in roots. The degree of stimulation depends upon the nature of the plant, nutrient supply, and moisture. In general, the biomass of grass is increased by grazing up to a point. Then biomass production declines. Adverse effects are greatest when new growth is developing. Defoliation then, as in deciduous trees, results in a loss of biomass, decreased growth, and delayed maturity. Grasses, however, are well adapted to grazing, and most benefit from it. Because the meristem is placed close to the ground, the older rather than the more expensive young tissue is consumed first. Grazing stimulates production by removing older tissue functioning at a lower rate of photosynthesis, reducing the rate of leaf aging, thus prolonging active photosynthetic production and increasing the light intensity on underlying young leaves, among other benefits. Some grasses can maintain their fitness only under the pressures of grazing, even though defoliation reduces sexual reproduction. Exclusion studies show that in the absence of grazing, some dominant grasses disappear.

Plant Responses

Herbivory is a two-way street. Plants, although seemingly passive in the process, have a pronounced effect on the fitness of herbivores. For herbivores it is not the quantity of food that is critical—usually enough biomass is available—but the quality. Because of the complex digestive process needed to break down plant cellulose and convert plant tissue into animal flesh, high-quality forage rich in nitrogen is necessary. Without that herbivores can starve to death on a full stomach. Low-quality foods are tough, woody, fibrous, and indigestible. High-quality foods are young, soft, and green or are storage organs such as roots, tubers, and seeds. Most food is low-quality, and herbivores are forced to live on such resources experience high mortality or reproductive failure. Added to the problem of quality is the chore of overcoming the various defenses of plants that make the food unavailable, hardly digestible, unpalatable, or even toxic.

Secondary chemical products are a first line of defense. Plants accumulate an array of toxic proteins—lectins and protease inhibitors, alkaloids, cyanogenic glycosides, cyanolipids, digestive-inhibiting polyphenols, terpenes, and tannins. These secondary products, poisonous to the plant itself, may be stored in vacuoles within cells and released only when the cells are broken, or they may be stored and secreted by epidermal glands to function as a contact poison or volatile inhibitor.

Production and storage of such metabolites are expensive to a plant and require a trade-off between defense and reproductive effort. Their evolution reflects the relationship between plant availability as prey or host and the presence of herbivores. The mode of defense varies with the nature of the plants, which can be divided into two broad groups.

One group consists of large, conspicuous, usually long-lived, woody plants available to herbivores. They possess the most expensive type of defense—quantitative. Such a defense, not easily mobilized, is the most effective against herbivore specialists. Involved are mostly tannins and resins concentrated near the surface tissues in leaves, in bark, and in seeds. They form indigestible complexes with leaf proteins, reduce the rate of assimilation of dietary nitrogen, reduce the ability of microorganisms to break down leaf proteins in herbivore digestive systems, and lower palatability. The problem with such a defense is the slow response. After defoliation by gypsy moths, oaks a year later increase the phenolic and tannin content of leaves and increase their toughness.

The other group contains short-lived plants, mostly annuals and perennials. These plants employ a qualitative defense, involving plant secondary substances such as cyanogenic compounds and alkaloids that interfere with metabolism. These substances can be synthesized quickly at little cost, are effective at low concentrations, and are readily transported to the site of the attack. They can be shuttled about the plant from growing tips to leaves to stems, roots, and seeds. They can be transferred from seed to seedling. These substances protect mostly against generalist herbivores.

A few herbivores have developed ways of breaching these chemical defenses. Some insects can absorb or metabolically detoxify these chemical substances, and even store the plant poisons and use them in their own defense, as the larvae of monarch butterflies do, or in the production of pheromones. A few beetles and certain caterpillars cut circular trenches in leaves before feeding, stopping the flow of chemical defenses to the leaf area on which they will feed; others cut veins on milkweeds and other latex-producing plants, blocking the flow of latex to the intended feeding site.

Plants may also employ the least costly form of defense, structures that make attack by herbivores difficult, if not impossible. Tough leaves, hairy leaves, thorns, and spines evolved early in the history of the plant, when it was subject to even greater predatory pressure.

HERBIVORE-CARNIVORE SYSTEMS

Herbivory supports carnivory. Unlike herbivores, carnivores are not faced with a lack of quality in their food. The quality is there—all highly proteinaceous and easily digestible—but quantity is frequently lacking. That fact dictates a somewhat different relationship between eater and eaten. Numbers of prey become important. Fitness of the predator depends upon its ability to capture prey; and fitness of the prey depends upon its ability both to elude predators and to overcome plant defenses. That combination puts a squeeze on herbivores.

Some stability exists beween predator and prey. Rarely do predators exterminate their prey. On occasion the prey population may outstrip the ability of the predator to contain it. At times predators can reduce or depress prey populations. Over time, however, predator and prey populations exhibit stability in their fluctuations. With prey species that possess a population regulation mechanism such as territoriality, predators take mostly the surplus population. If prey species are not self-limiting, as appears to be the case with ungulates, and predator populations are, as is the case with the wolf, then the predator may tend to control the populations of the prey. In other cases, the prey and predator populations may be so interrelated that their numbers undergo well-defined cyclic changes over time, as exemplified by the snowshoe hare and the lynx (discussed on page 250).

The relationship between predator and prey is influenced considerably by prey defenses and the ability of predators to overcome them. The predator does not succeed in every encounter, as models of predation assume. Prey employ different means of defense. Chemical defense is widespread among many groups of animals. Some species of fish release alarm pheromones, inducing fright reactions in members of the same and related species. Arthropods, amphibians, and snakes employ odorous secretions to repel predators. Many arthropods possess toxic secondary substances, which they have taken from plants and stored in their own bodies. Other arthropods and venomous snakes, frogs, and toads synthesize their own poisons.

Prey have evolved a number of other defensive mechanisms. Cryptic coloration includes coloration, patterns, shapes, postures, movements, and other behaviors that tend to hide the prey species from predators. Flashing coloration, such as extremely visible color patches, may distract and disorient predators. Warning coloration and mimicry signal that animals possess pronounced toxicity and other chemical defenses. Yellow and black coloration of many bees and wasps, for example, serves notice of danger to would-be predators. All their predators, however, must have some experience with the prey before they learn to associate the coloration with unpalatability or pain.

Some animals associated with inedible species evolve a similar mimetic or false warning coloration (Figure 16.14). The mimic, a palatable species, resembles the inedible model. Among North American butterflies, for example, the palatable viceroy butterfly mimics the monarch, most of which are distasteful to birds. (Most of the viceroy's relatives are blue-black in color.) Once the predator has learned to avoid the model, it avoids the mimic also. The greater the proportion of mimics to the model, the longer the learning time of the predator, because it will take many mimics before encountering the model. A less common type of mimicry involves both unpalatable models and unpalatable mimics. The pooling of numbers between the model and the mimic reduces the losses of each, because the predators learn quickly without having to handle both species.

Other animals employ armor as a defense. Clams, armadillos, turtles, and numerous beetles all withdraw into their armor coat or shells when danger approaches. Another type of defense is exhibited by porcupines and echidnas, whose quills (modified hairs) discourage predators.

Some animal defenses are behavioral. One is alarm

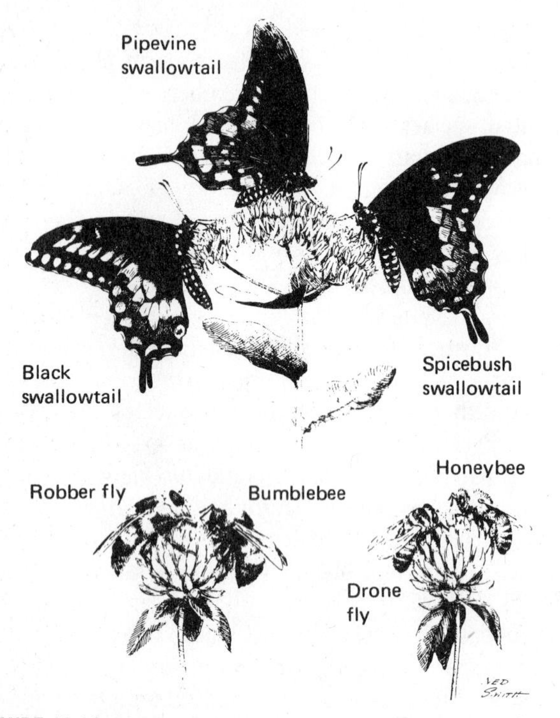

FIGURE 16.14 Mimicry in insects. One model, the distasteful pipevine swallowtail, has as its mimics the black swallowtail and the spicebush swallowtail. The black female tiger swallowtail is a third mimic. All these butterflies are found in the same habitat. The robber fly illustrates aggressive mimicry. It is a mimic of the bumblebee, on which it preys. The drone fly is a mimic of the honeybee.

calls when a predator is sighted. Because high-pitched alarm calls are not species-specific, they are recognized by a wide range of animals close by. Alarm cells often bring in numbers of potential prey that mob the predator. Mobbing may involve harassment at a safe distance or direct attack. Distraction displays, most common among birds, direct the attention of the predator away from the prey.

For some prey, living in groups is the simplest form of defense. The more the prey are congregated, the less likely is a predator to attack, and the less is the chance that any one individual will be taken. By maintaining a tight, cohesive group, prey make it difficult for any predator to obtain a victim.

A more subtle form of defense is timing of reproduction so that most of the offspring are produced in a short period of time. Then prey is so abundant that the predator becomes satiated, allowing a percentage of the reproduction to escape. Such is the strategy employed by ungulates like the caribou.

As prey have evolved ways of avoiding predators, predators necessarily evolved better ways of hunting and capturing prey. Predators have three general methods of hunting: ambush, stalking, and pursuit. Ambush hunting involves lying in wait for prey to come along. This method is typical among some frogs, alligators, lizards, and certain insects. Ambush hunting has a low frequency of success, but it requires minimal energy. Stalking, typical of herons and some cats, is a deliberate form of hunting with a quick attack. The predator's search time may be great, but pursuit time is minimal. Pursuit hunting, typical of many hawks, lions, and wolves, involves minimal search time, because the predator usually knows the location of prey, but pursuit time is generally great. Searchers spend more time and energy to encounter prey. Pursuers, theoretically, spend more to capture and handle prey.

Predators, like their prey, may utilize cryptic coloration to blend into the background or break up their outlines. Predators can deceive by resembling the prey. Robber flies (*Mallophora bomboides*) mimic their prey, bees. The female of certain species of fireflies imitates the mating flashes of other species, attracting males of those species, which she promptly kills and eats. Predators may also employ chemical poisons, as do shrews and rattlesnakes, and group attack on large prey, as do lions and wolves.

THREE-WAY INTERACTIONS

Relationships among vegetation, herbivores, and predators can be manipulated easily enough on paper. How do they work out in the field under natural conditions? An example is the three-level interaction of vegetation, snowshoe hare (*Lepus americanus*), and lynx (*Felix lynx*) (Figure 16.15).

The snowshoe hare inhabits the northern forest, where in winter it feeds on the buds of conifers and twigs of aspen, alder, and willow, termed browse. Browse consists largely of stems less than 3 mm in diameter, young growth with a concentration of nutrients. The hare-vegetation interaction becomes critical when essential browse falls below that needed to

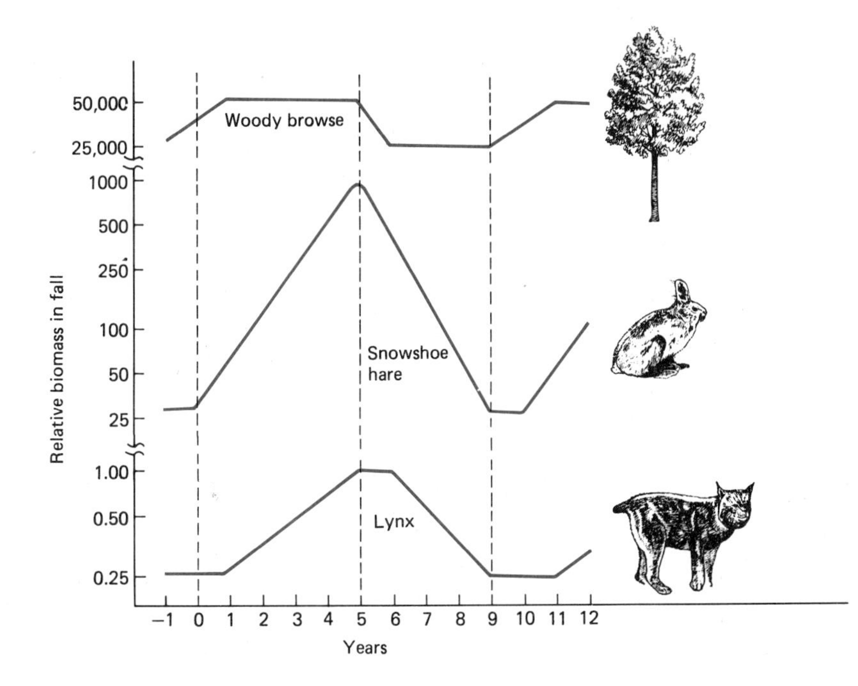

FIGURE 16.15 A graphic model of the vegetation-herbivore-predator cycle involving woody vegetation, snowshoe hare, and lynx. Note the time lag between the cycle of vegetation recovery, the growth and decline of the snowshoe hare population, and the rise and fall of the predator population.

support the population overwinter, approximately 300 g per individual per day of stems 3 to 4 mm in diameter. Excessive browsing and girdling during population increase reduce subsequent increases of woody growth, bringing about a food shortage, which causes a high winter mortality of juvenile hares and lowered reproduction the following summer.

The decline in hares is also related to the chemical defenses of the alders, birches, and some willows, which strongly influence the selection of winter forage among many subarctic browsing vertebrates. Hares avoid these more nutritious plants, especially juvenile growth and buds, because they contain more resin and phenolic glucosides than mature twigs. For example a toxic phenol (pinsylvin methyl ether) in the catkins, buds, and young twigs of green alder (*Alnus crispus*) exceeds three times the concentration in mature twigs and greatly reduces palatability.

The decline in hares elevates the predator-hare ratio, intensifying predation on the hare that extends the period of decline and drives the population to a level below which the habitat could support. The low hare population allows the vegetation to recover and to accumulate biomass. It also causes a sharp decline in predator populations (mostly lynx), who undergo early loss of young. With a decline of predators and a growing abundance of winter foods, the hare population begins to rise sharply, starting another cycle.

Summary

Predation is the consumption of one living organism by another, a relationship in which one organism benefits at the other's expense. However, there is a close interaction between predator and prey. Each can influence the fitness of the other. In its broadest sense predation includes herbivory, parasitism, and cannibalism.

The interaction between predator and prey involves a functional response and a numerical response. A functional response is one in which the predator takes more prey as the density of its prey increases. A numerical response is one in which the number of predators increases as the prey increases.

There are three types of functional responses: Type I, in which the number of prey affected increases in a linear fashion to a maximum as the prey increases; Type II, in which the number of prey affected increases at a decreasing rate toward a maximum value; and Type

III, in which the number of prey taken increases in a sigmoidal fashion as the density of prey increases.

Type III response may involve both a search image and switching. Predators develop a search image for a particular prey as that item becomes more abundant. Switching takes place when a predator turns to an alternate, more abundant prey. Of the three types of functional response, only Type III is important as a population-regulation mechanism.

A particular form of predation is cannibalism, or intraspecific predation. It often but not necessarily occurs when a population is under stress. Cannibalism is particularly common among insects and fish, in which predatory adults feed on grazing young. Cannibalism can become an important mechanism of population control.

Central to the study of predation is the concept of optimal foraging, a strategy that obtains for the predator a maximum rate of net energy gain. There is a break-even point above which foraging is profitable and below which it is not. An optimal diet includes the most efficient size of prey for both handling and net energy return. Optimal foraging efficiency involves the concentration of activity in the most profitable patches of prey and the abandonment of those patches when they are reduced to the average profitability of the area as a whole.

Predatory relationships involve two related systems: plant-herbivore and herbivore-carnivore. Herbivory affects plant fitness by reducing the amount of photosynthate and the ability to produce more. Plants respond by trying to replace losses. Failure to do so means a loss of fitness or even death. At the same time plants affect herbivore fitness by denying them palatable or digestible food. Herbivore-carnivore relations introduce mutual fluctuations.

In response to the selective pressures of predation, prey species, both plant and animal, have evolved measures of defense. Plants' chemical defense involves secondary metabolic products such as alkaloids, phenolics, and cyanogenic glycosides. Chemical defense is most successful against generalist herbivores. Certain specialists breach the chemical defense and detoxify the secretions or sequester the toxins in their own tissues as a defense against predators. Chemical defense in animals usually involves a distasteful or toxic secretion that repels, warns, or inhibits a would-be attacker. Other defensive measures in animals include armor, cryptic and warning coloration, mimicry, and behavior such as distraction displays and alarm calls. Predators have evolved strategies of their own.

Review and Study Questions

1. Define parasitism, parasite, herbivory, and cannibalism.
2. What are the basic features and assumptions of the Lotka-Volterra equations for predation?
3. What is a functional response to predation? A numerical response?
4. Distinguish among Type I, Type II, and Type III functional responses. Why are Type I and Type II not important in population regulation?
5. What is meant by search image? Switching? How do they relate to Type III responses?
6. Speculate about why cannibalism is widespread among animals. What triggers cannibalism, and what are its demographic effects?
7. What is optimal foraging strategy? An optimal diet?
8. How is foraging influenced by a patchy environment?
9. How do plants defend themselves against herbivores? Include two approaches to chemical defense.
10. What effect do herbivores have on plant fitness?
11. How is predator satiation a prey defense? Discuss such a defense from an individual prey's point of view.
12. Discuss the role of alarm calls, mobbing, and distraction displays in animal defense. From an individual point of view, are they altruistic? (Refer to Chapter 2.)
13. Contrast ambush with pursuit and stalking. Evaluate costs and benefits. Under what conditions might each have evolved?
14. Relate the methods of hunting to such defensive and offensive measures as cryptic coloration.

Selected References

Curio, E. 1976. *The ethology of predation.* New York: Springer-Verlag. An interesting treatment of the behavioral side of predation.

Errington, P. L. 1946. Predation and vertebrate populations. *Quarterly Review of Biology* 22:144–177, 221–245. A classic study of predation.

Errington, P. L. 1963. *Muskrat populations.* Ames: Iowa State University Press. Study of a prey species and its predators.

Fox, L. R. 1975. Cannibalism in natural populations. *Annual Review of Ecology and Systematics* 6:87–106. A review of the topic.

Holling, C. S. 1959. The components of predation as revealed by a study of small mammal predation of the European pine sawfly. *Canadian Entomologist* 91:293–320. Classic study on functional response.

Holling, C. S. 1966. The functional response of invertebrate predators to prey density. *Memoirs Entomological Society of Canada* 48:1–86. Further study of functional response.

Hubbard, S. F., and R. M. Cook. 1978. Optimal foraging by parasitoid wasps. *Journal of Animal Ecology* 47:593–604.

Huffaker, C. B. 1958. Experimental studies on predation. Dispersion factors and predator-prey oscillations. *Hilgardia* 27:343–383. Laboratory study of predation.

Krebs, J. R., and N. B. Davies, eds. 1984. *Behavioural ecology: An evolutionary approach,* 2nd ed. Oxford, England: Blackwell Scientific Publications. A theoretical approach.

Kruuk, H. 1972. *The spotted hyena.* Chicago: University of Chicago Press. An excellent study of a predator and its prey.

Polis, G. 1981. The evolution and dynamics of intraspecific predation. *Annual Review of Ecology and Systematics* 12:225–251. Further study of cannibalism.

Rosenthal, G. A., and D. H. Janzen, eds. 1979. *Herbivores: Their interactions with secondary plant metabolites.* New York: Academic Press. Chemical defense by plants reviewed.

Stephens, D. W., and J. W. Krebs. 1987. *Foraging theory.* Princeton, NJ: University of Princeton Press. Current state of foraging theory.

Taylor, R. J. 1984. *Predation.* New York: Chapman and Hall. An overview of predator-prey dynamics.

Zack, R., and J. B. Falls. 1976. Ovenbird (Aves: Parulidae) hunting behavior in a patchy environment: An experimental study. *Canadian Journal of Zoology* 54:1863–1879.

CHAPTER 17

Parasitism and Mutualism

Outline

Objectives

On completion of this chapter, you should be able to:

1. Define parasitism and the types of parasitism.
2. Discuss the modes of transmission and the effects of parasitism on hosts.
3. Discuss social parasitism.
4. Explain mutualism and types of mutualism.
5. Discuss the role of mutualism in ecosystems.
6. Relate coevolution to parasite-host systems, seed dispersal, plant pollination, and obligate mutualisms.

In Chapter 16 we suggested a loose evolutionary relationship between predator and prey. This connection is closer between parasite and host and between mutualists. A relationship in which two interacting populations appear to place reciprocal selection pressures on each other is called *coevolution*. Any evolutionary change in one member changes the selective forces acting on the other. They play a game of adaptation and counteradaptation.

Not all species in coevolutionary relationships grew up together over time. Most coevolutionary responses appear to be general. A trait evolves in one or more species in response to a trait or a suite of traits in several other species. For example, many plants have adapted to pollination by traveling insects. Such generalized relationships, in which genetic changes are not closely coupled, are known as *diffuse coevolution*. Paired interactions between specific parasites and their hosts or between certain mutualists are another story.

PARASITISM AND DISEASE

Although parasites and pathogens are extremely important in interspecific relationships, ecologists have not yet given them enough attention. The effects are dramatic when they are introduced into host populations that have not evolved defenses against them. Then diseases sweep through and decimate the population. Examples are the outbreaks of rinderpest in African ungulates, myxomatosis in European rabbits, and blight in the American chestnut. In most natural populations, though, living in association with normal parasites and diseases, the effects of parasitism are elusive. Bodies of victims are quickly processed by detrivores and scavengers, eliminating traces of disease.

TYPES OF PARASITES

Parasitism is a condition in which two organisms live together but one derives its nourishment at the expense of another. Parasites, strictly speaking, draw nourishment from the tissues of their larger hosts, a case of the weak attacking the strong. Typically, parasites do not kill their hosts as predators do, although the host may die from secondary infection or suffer stunted growth, emaciation, or sterility.

Parasites include viruses, many bacteria, fungi, and an array of invertebrate taxonomic groups including the arthropods. A heavy load of parasites is considered an infection, and the outcome of an infection is a disease. A *disease* is any state in the condition of a plant or animal that deviates from normal. Not all parasites are agents of disease. Many are mutualistic and essential to the well-being of the host. Parasitic bacteria in the rumen of ungulates and the gut of termites are essential for digestion.

Parasites exhibit a tremendous diversity in adaptations to exploit their hosts. Parasites may be plants or animals, and they may parasitize plants or animals or both. They may live on the outside of the host (ectoparasites) or within its body (endoparasites). Some are full-time parasites, others only part time. Part-time parasites may be parasitic as adults and free-living as larvae or the reverse. Parasites have developed numerous ways to gain entrance to their hosts, even to the point of using several hosts as dispersal agents. They have evolved various means and degrees of mobility, ranging from free-swimming ciliated forms to ones totally dependent upon other organisms for transport. They have developed diverse ways of securing themselves to the host to maintain position and means of counteracting the biochemical hazards of living inside a host. Parasites may be restricted to one species or genus of host or a few. A number of parasites of birds, especially certain tapeworms, live only in one particular order. Some parasites may live their entire life cycles on one host, while others require more than one host.

Ecologically, parasites are distinguished by size, as microparasites or macroparasites. *Microparasites* include the viruses, bacteria, and protozoans. They are characterized by small size and a short generation time. They develop and multiply rapidly within the host and tend to induce immunity to reinfection in hosts that survive initial infections. The duration of the infection is short relative to the expected life span of the host. Transmission from host to host is direct, although some other species may serve as a carrier or *vector*.

Macroparasites are relatively large. They include flatworms, acanthocephalans, roundworms, flukes, lice, fleas, ticks, mites, fungi, rusts, smuts, dodders, broomrape, and mistletoe. Macroparasites have a comparatively long generation time, and direct multiplication in the host is rare. The immune response macroparasites stimulate is of short duration and depends upon the number of parasites in the host. Macroparasites are persistent with continual reinfection. They may spread by direct transmission from host to host or by indirect transmission, involving intermediate hosts and carriers.

HOSTS AS HABITATS

Hosts are homes for parasites, and they have exploited every conceivable habitat on and within them. Parasites live on the skin within the protective cover of feathers and hair. Some burrow beneath the skin. They live in the bloodstream, in the heart, brain, digestive tract, liver, spleen, mucosal lining of the stomach, spinal cord and brain, in nasal tracts and lungs, in the gonads and in the bladder, in pancreas, eyes, gills of fish, and muscle tissue, to mention some sites among many. Parasites of insects live on the legs, on upper and lower body surfaces, and even the mouthparts.

Parasites within the host colonize different sites in organ systems. Coccidian protozoans of the genus *Elmeria* occupy different regions: one in the duodenum, another in the lower duodenum and upper small intestine, a third in the lower small intestine, and a fourth in the caecum and rectum. There are similar divisions in the sharing of a particular organ system by closely related parasites in many animals.

Plant parasites, too, divide up the habitat. Some live on the roots and stems; others penetrate the roots and bark to live in the woody tissue beneath. Some live at the root collar where the plant emerges from the soil. Others live within the leaves, on young leaves, on mature leaves, or on flowers, pollen, or fruits.

A major problem for parasites, especially parasites of animals, is gaining access to and escaping from the host. Parasites of the alimentary tract enter the host orally and escape through the rectum, a path used by other parasites as well. Parasites of the lungs enter orally or penetrate the skin and travel to the lungs by way of the pulmonary system. They escape mainly by being coughed up and swallowed into the alimentary tract. Liver parasites exploiting one of the richest habitats in the animal body, arrive there by way of the circulatory system, bile duct, and hepatic portal systems and escape through the same routes. Parasites of the urogenital system enter orally, travel through the gut to the site of infection, exit by the urinary system. Blood parasites enter and escape through the skin, but always with the aid of some obliging vector such as mosquitos and ticks. Parasites that end up in the muscle tissues, where they usually exist in capsules, reach a blind end. For them the only way out is for their host to be killed and consumed by a predator.

MODES OF TRANSMISSION

To survive and multiply, parasites have to escape from one host and locate another, something they cannot do at will. Parasites can escape only during an infective stage, when they must make contact with the host. The infective stage is essential.

All parasites reach a stage in their life cycles within the host when they can develop no further. The *definitive host* is the one in which the parasite becomes an adult and reaches maturity. All others are *intermediate hosts,* which harbor some developmental phase. Parasites may require one, two, or even three intermediate hosts. Each infective stage can develop only when it is independent of the definitive host, and it can continue its development only if it can find another intermediate host or its definitive host. For this reason, many parasites employ animals as intermediate hosts to aid in the location of a definitive host or adapt to the host's habits. Thus the population dynamics of a parasite population is closely tied to the population dynamics of the host.

Direct Transmission

Direct transmission is the transfer of a parasite from one host to another by direct contact or through a carrier, with no intermediate stages in a secondary host. Typically, microparasites are transmitted directly.

For example, a bacterial disease of increasing importance to humans in recent years is Lyme disease, which involves wild animal vectors. Named Lyme disease because the first noted occurrence was at Lyme, Connecticut, the disease involves a bacterial spirochete,

Barretica bungdorferi, which lives in the bloodstream of vertebrates from birds and mice to deer and humans. The spirochete depends upon the deer tick (*Ixodes damnini* in eastern North America and *I. pacificus* in western North America) for transmission from one host to another. The tick is small; the larva is the size of a grain of pepper, the nymph the size of a poppy seed, and the adult the size of a sesame seed. White-footed mice, chipmunks, white-tailed deer, and ground-dwelling birds are reservoirs for the spirochete. The bacterium depends upon the tick as its way of gaining new hosts. Tick larvae hatch in spring from eggs laid by the adult the previous fall and seek out a small host on which to feed. If that host happens to be infected with bacteria, the larva picks it up and transfers it to the next host. The following spring the tick larva molts to become a nymph that seeks its blood meal from mice or chipmunks and carries the infection with it. Having fed, the nymph molts to become an adult. The male and female move to white-tailed deer, feed, mate, and renew the two-year life cycle. Some of the adult ticks attach themselves to humans hiking or working in the woods and fields. If they carry the spirochete, the ticks transmit it to humans. Unless treated, the infection, which may produce such visible symptoms as a bull's-eye rash, causes flu-like respiratory congestion, pain in the joints and muscles, and permanent damage to brain, central nervous system, and heart.

Microparasites that infect plants also are transmitted directly. In the presence of suitable hosts, spores resting in the soil germinate and penetrate the roots. Wind-carried spores come to rest on and infect leaves of the plant. The devastating Dutch elm disease is carried from tree to tree by spore-carrying elm beetles.

Many important macroparasites of animals and plants move from infected to uninfected host by direct contact. Parasitic nematodes (*Ascaris*) live in the digestive tracts of mammals. Female roundworms lay thousands of eggs in the gut of the host, which are expelled with the feces. If they are swallowed by the host of the correct species, the eggs hatch in the intestines of the host, bore their way into the blood vessels, and come to rest in the lungs. From here they ascend to the mouth, usually by causing the host to cough, and are swallowed again to reach the stomach, where they mature and enter the intestines.

The most important external debilitating parasites of birds and mammals are spread by direct contact. They include lice, mites that cause mange, ticks, fleas, and botfly larvae. Many of these parasites lay their eggs directly on the host, but fleas lay their eggs and their larvae hatch in the nests and bedding of the host (even in shag rugs), and eventually leap onto nearby hosts.

Some fungal parasites of plants spread through root grafts. For example, an important fungal infection of white pine (*Pinus strobus*), *Fomes annosus,* spreads rapidly through pure stands of the tree when roots of one tree become grafted onto the roots of a neighbor.

A number of flowering plant macroparasites spread by direct transmission, too. One group is the *holoparasites,* plants that lack chlorophyll and draw their water, nutrients, and carbon from the roots of host plants. Notable among them are members of the broomrape family, Orobanchaceae. Two are the familiar squawroot (*Conopholis americana*), which parasitizes the roots of oaks (Figure 17.1), and beechdrops (*Epifagus virginiana*), which parasitizes mostly the roots of beech trees.

Another group is the *hemiparasites.* They are photosynthetic, but they draw water and nutrients from their host plant. The most familiar hemiparasites are mistletoes, whose sticky seeds attached to limbs send out roots that embrace the limb and enter the sapwood. Mistletoe can reduce growth of its host.

Indirect Transmission

Many parasites, both plant and animal, utilize indirect transmission, spending different stages of the life cycle with different hosts. The brainworm (*Parelaphostrongylus tenuis*) (in spite of its name the parasite is a lungworm) of the white-tailed deer has as its intermediate host during its larval stage a snail or slug that lives in the grass (Figure 17.2). The deer picks up the infected snail while grazing. In the deer's stomach the larvae leave the snail, puncture the deer's stomach wall, enter the abdominal membranes, and by way of the spinal cord reach spaces surrounding the brain. Here the worms mate and produce eggs. Eggs and larvae pass through the bloodstream to the lungs, where the larvae break into the air sacs and are coughed up, swallowed, and passed out through the feces. The larvae are ingested by the snail, where they continue to de-

FIGURE 17.1 A holoparasite common to the deciduous forest of eastern North America is squawroot, a member of the broomrape family, Orobanchaceae. Squawroot is parasitic on the roots of oaks.

velop to the infective stage. There is little evidence that brainworms affect the health of white-tailed deer. Some deer may suffer lung damage, especially if they are infected with other species of lungworms. However, many native and exotic ungulates lacking a natural resistance to brainworms are adversely affected.

Indirect transmission among plant macroparasites is uncommon except among rusts. The wind carries infective stages from primary and intermediate hosts. Examples are white-pine blister rust and wheat rust, both of which have intermediate stages on shrubs of the genus *Ribes*.

Transmission from host to host is the key to parasitic existence. It can take place only with the dispersal of an infective stage independent of the definitive host. Parasites requiring more than one host reach only a certain stage in their life cycles in each, and they can complete the cycle only if they can infect another host. The brainworm of deer, for example, has to locate a snail as an intermediate host to continue its development to an infective stage that can be transmitted back to the deer. For this reason many animal parasites exploit the feeding habits of the definitive host and adapt to the habits of the intermediate hosts. The brainworm of the deer exploits the snail's habit of crawling up grass stems, where it risks being eaten along with the grass by a grazing deer. Unless the snail is swallowed by the deer, the parasite will perish.

Transmission of parasites is further complicated by the patchy or clumped distribution of hosts. Few members of the host population harbor major parasite loads, acting as reservoirs of infection. If uninfected hosts are widely scattered or intermixed among populations of other species, the probability of the parasite or its carrier coming into contact with susceptible individuals is low. In other words, transmission is highly dependent on both the density of hosts and the distance between parasite and potential hosts. Transmission of parasites is most successful when the population of potential hosts is dense, particularly if the parasites depend upon direct contacts among hosts. The rapid spread of viral or bacterial disease in dense animal

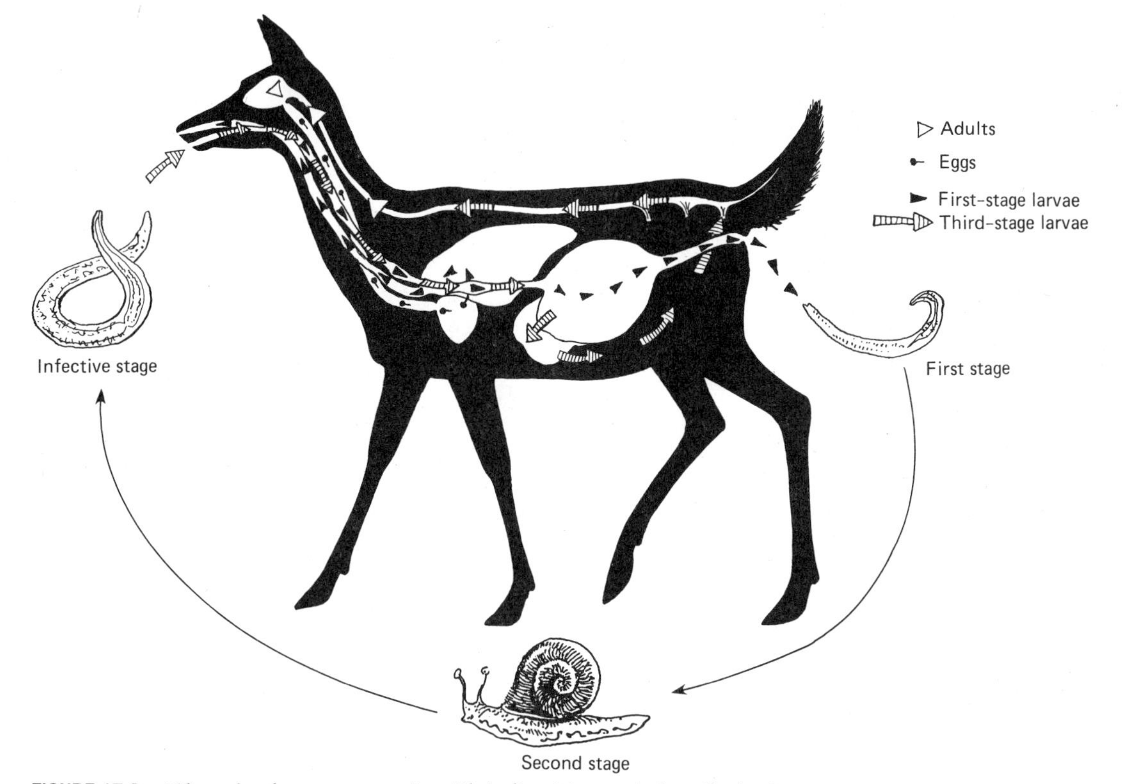

FIGURE 17.2 Life cycle of a macroparasite with indirect transmission, the brainworm (actually a lungworm) *Parelaphostrongylus tenuis* in white-tailed deer (but transmissible to moose and elk). Adult worms in the definitive host, the white-tailed deer, release eggs into the venous circulation. The eggs reach the lungs as emboli (undissolved material). In the lungs the eggs hatch and develop into first-stage larvae. The larvae move into air spaces in the lungs and pass up the bronchial tubes. Subsequently, the larvae are swallowed and pass in the feces. Outside, the first-stage larvae invade the foot of terrestrial snails that move across the feces. The larvae grow in the snail, molt twice, and give rise to the third and infective stage. Deer acquire the third stage by ingesting the infected snails on vegetation. In the alimentary canal the larvae leave the tissues of the snail and penetrate the gastrointestinal wall, cross the peritoneal cavity, and follow the lumbar and other nerves to the vertebral canal, a journey of about ten days. The migrating third-stage larvae enter the dorsal horns of the gray matter of the brain, where they develop for about a month. Forty days after infection, the subadult worms enter the spinal subdural space. Here they mature and migrate to the cranium, where a new cycle begins.

populations is called an *epizootic* (an *epidemic* in humans).

REGULATION BY PARASITES

The impact of parasites on host populations depends in part on the mode of transmission and the density and dispersion of the host population. Microparasites, dependent for the most part on direct transmission, require a high host density to persist. For them ideal hosts live in groups or herds. To persist, these parasites need a long-lived infective stage that does not ensure long-term immunity in the host population. Immunity reduces parasite populations, if it does not eliminate them. An example of a parasite in wild populations that does not confer long-term immunity is rabies; one that does confer immunity to animals that survive the disease is distemper.

Indirect transmission, typical of macroparasites, is more complex. To persist, it requires a highly effective transmission stage, which often involves a close association with food webs. Parasites with indirect transmission exist at low population levels; but because of efficient transmission, they do well and persist for a long time in low population densities of hosts. The longevity of each parasitic stage varies in different hosts. Longevity is high in the definitive host and much less in the intermediate host.

To prove that a parasite regulates a host population, we must show that it increases the host's death rate or decreases its reproductive capability. Such effects are most evident when parasites are introduced into a population with no evolved defenses. In such cases the disease may be density-independent and can reduce populations, exterminate them locally, or restrict their distribution. For example, the fungus *Endothia parasitica* spread rapidly through the American chestnut (*Castanea dentata*). Once a major commercial timber species prominent in the eastern deciduous forest, it has all but vanished.

Such extreme cases are hardly regulatory. What about directly transmitted endemic diseases maintained in the population by a small reservoir of infected carrier individuals? Outbreaks of these diseases appear to occur when the density of the host population is high, and they tend to sharply reduce host populations. Examples are distemper in raccoons and rabies in foxes, both significant in controlling their host populations.

The distribution of macroparasites, especially those with indirect transmission, is highly clumped or overdispersed (Figure 17.3). Some individuals in the host population carry a higher burden of parasites than others. These individuals are the ones that are most likely to succumb to parasite-induced mortality, suffer reduced reproductive rates, or both (Figure 17.4). Such parasitically induced deaths often are caused not directly by parasites but indirectly by secondary infections. Herds of bighorn sheep (*Ovis canadensis*) in western North America may be infected with up to seven different species of lungworms (Nematoda). Highest infections occur in the spring when the lambs are born. Heavy lungworm infestations in the young

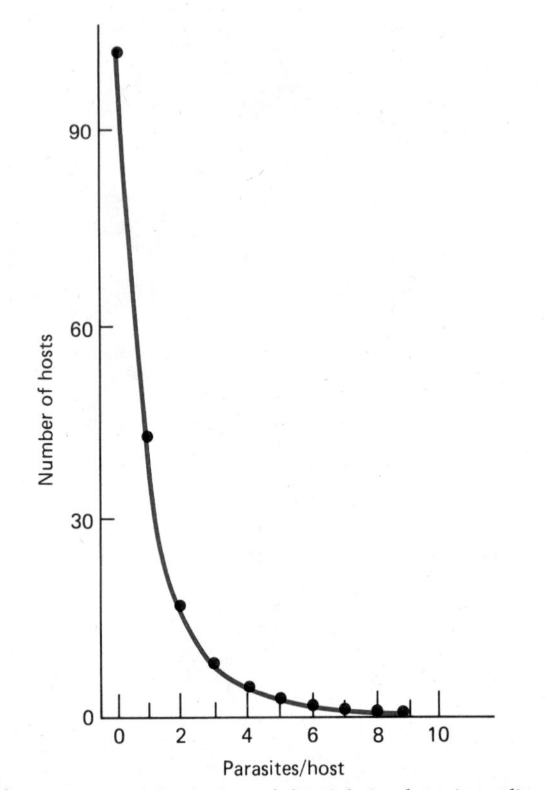

FIGURE 17.3 Overdispersion of the tick *Ixodes trianguliceps* (Birula) on a population of the European field mouse *Apodemus sylvaticus.* Most of the individuals in the host population carry no ticks. A few individuals carry most of the parasite load.

encourage a secondary infection, pneumonia, which kills the lambs. Such infections tend to stabilize or sharply reduce mountain sheep populations by reducing reproductive success. Wildlife biologists increase lamb survival by treating pregnant ewes with an antihelminthic drug disguised in fermenting apple mash.

Host populations may show the combined effects of infestations by different species of ectoparasites and endoparasites, a case of diffuse parasitism. The eastern cottontail rabbit (*Sylvilagus floridanus*) has been declining over parts of its range, in spite of no apparent changes in its habitat. Field evidence seems to point to the role of multiple parasites in holding a local population of rabbits at a level well below the carrying capacity of the habitat.

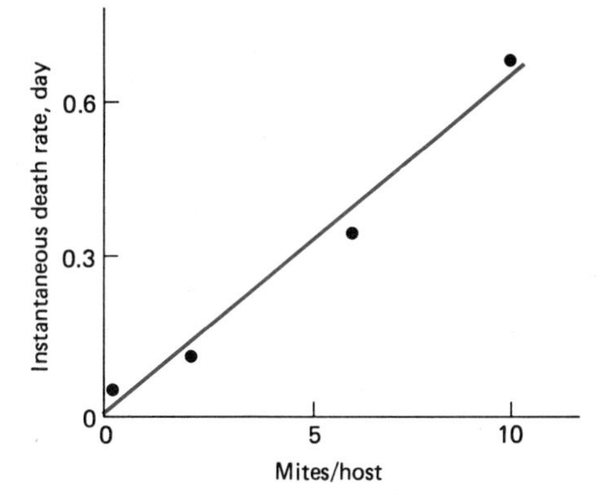

FIGURE 17.4 Effects of parasite density on a host on the rate of host mortality. This example shows the influence of parasitism by the mite *Hydrophantes tenuabilis* on the water measurer *Hydrometra myrae,* a common hemipteran on ponds.

EVOLUTIONARY RESPONSES

The relationship of parasites and their hosts involves an evolutionary response on both sides. As the host builds up defenses against parasitic infections, the parasite evolves milder strains. The parasite gains no advantage if it kills its host. A dead host means dead parasites. A high virulence usually results in a high mortality of hosts. Low virulence ensures a long duration of infection. As an outcome both host and parasite evolve a mutual tolerance.

An example of coevolutionary adaptation and counteradaptation is the interaction of the European rabbit and the viral infection myxomatosis. To control the introduced rabbit, the Australian government introduced the rabbit's viral parasite into the population. The first epizootic of myxomatosis was fatal to 97 to 99 percent of the rabbits. The second resulted in a mortality of 85 to 95 percent; the third, 40 to 60 percent. The effect on the rabbit population was less severe with each succeeding epizootic, suggesting that the two populations were adjusting to each other.

In this adjustment, attenuated genetic strains of virus, intermediate in virulence, tended to replace highly virulent strains. Too high a virulence killed off the host; too low a virulence allowed the rabbits to recover before the virus could be transmitted to another host. Also, a passive immunity to myxomatosis was conferred upon the young born to immune does. Finally, a genetic strain arose in the rabbit population with intrinsic resistance to the disease.

The transmission of myxomatosis depends upon *Aedes* and *Anopheles* mosquitoes, which feed only on living animals. Rabbits infected with the more virulent strain lived for a shorter period than those infected with a less virulent strain. Because the latter live for a longer period, the mosquitoes have access to that virus for a longer time. That fact gives the less virulent strain a competitive advantage over the more virulent.

In those regions where the less virulent strains have the competitive advantage, the rabbits are more abundant because fewer die. That means more total virus is present in those regions than in comparable areas where the more virulent strains exist. Thus, the virus with the greatest rate of increase and density within the rabbit population is not the one with the selective advantage. Instead, the virus whose demands are balanced against supply has the greatest survival value.

SOCIAL PARASITISM

Another form of parasitic relationship is social, in which one organism is dependent on the social structure of another. Social parasitism may be temporary or permanent, facultative (can either lead or not lead a parasitic life), or obligatory (necessary for species survival), within a species or between species. Four forms of social parasitism can be defined.

The first is a temporary facultative parasitism within a species. That type is well developed among ants and wasps. For example, a newly mated queen of the wasp genus *Polistes* or *Vespa* will attack an established colony of her own species and displace the resident egg-carrying queen. Intraspecific nest parasitism is another example. Parasitic females lay eggs in nests of host females of the same species. This practice is well developed among ants and wasps and among certain groups of birds, especially waterfowl (Anatidae). Among these waterfowl are black-bellied tree ducks *(Dendrocygna autumnalis),* goldeneyes *(Bucephala clangula)*, and wood ducks *(Aix sponsa)*, hole-nesting species. The host female responds to parasitism by reducing the size of her own clutch by the number added. The earlier the parasitic female lays her eggs in her host's nest, the greater will be her share of the clutch. Such brood parasitism may have evolved among those waterfowl because suitable nest sites are scarce, nests are easy to locate, and ducks do not defend their nests. Just as important, female waterfowl have a strong affinity for their places of birth. That fact makes it likely that the host and the parasite are related genetically.

A second type of social parasitism is temporary facultative parasitism between species. An example can be found within the formicine ant genus *Lasius*. A newly mated queen of the species *L. reginae* will enter the nest of a host species, *L. alienus,* and kill its queen. The *alienus* workers will care for the *reginae* queen and her brood. In time the *alienus* workers, deprived of their own queen and thus replacements, die out, and the colony then consists of *reginae* workers.

A third type of social parasitism is temporary obligatory parasitism between species. Although common in ants, the most outstanding examples are obligatory egg or brood parasitism in birds. Brood parasitism has been carried to the ultimate by the cowbirds and Old World cuckoos, both of which have lost the art of nest building, incubation, and caring for the young. They pass off these duties to the host species by laying eggs in their nests. The brown-headed cowbird of North America removes one egg from the nest of the intended victim, usually the day she is to lay, and the next day lays one of her own as a replacement. Some host birds counter by ejecting the egg from the nest. Others hatch the egg and rear the young cowbird, usually to the detriment of their own offspring. The host's young may be pushed from the nest or die from lack of food because of the more aggressive nature and larger size of the young cowbird.

A fourth type of social parasitism is permanent obligatory parasitism between species. The parasitic form spends its entire life cycle in the nest of the host. That type of social parasitism is common among ants and wasps. In most cases the species are workerless, and queens have lost the ability to build nests and care for the young. The queen gains entrance to the nest of the host and either dominates the host queen or kills her outright and takes over the colony.

MUTUALISM

Mutualism is a positive, reciprocal relationship at the individual or population level between two different species. Out of this relationship, most obvious at the individual level, both species enhance their survival, growth, or fitness. Evidence suggests that mutualism is more a reciprocal exploitation than a cooperative effort.

OBLIGATORY SYMBIOTIC MUTUALISM

Mutualism may be symbiotic or nonsymbiotic, facultative or obligatory. *Symbiotic* refers to two organisms living together in close association. Parasites and their hosts are a form of symbiosis. In symbiotic mutualism individuals interact physically and their relationship is obligatory. At least one member of the pair cannot live without the other.

Some forms of the relationship are so permanent and obligatory that the distinction between the two interacting populations becomes blurred. A good example is mycorrhizae, a mutualistic relation of plant roots with fungi. So important is this mutualism to the growth of forest trees and the functioning of forest ecosystems that it is the object of expanding research activity. Common to many trees of temperate and tropical forests, one form is *endomycorrhizae.* The roots are infected by mycelia—a mass of fine fungal filaments—from the soil. They penetrate the cells of the host and form a finely bunched network called an arbuscle. The mycelia act as extended roots for the plant, drawing in phosphorus at distances beyond those reached by the roots and root hairs, but do not change the shape or structure of the root. Another form, *ectomycorrhizae,* produces shortened and thickened roots that suggest coral (Figure 17.5). The threads of fungus work between the root cells. Outside the root they develop into a network that functions as extended root hairs.

Mycorrhizae, especially important in nutrient-poor soils, aid in the decomposition of litter and translocation of nutrients, especially nitrogen and phosphorus, from the soil into root tissue. Mycorrhizae increase the ability of roots to absorb nutrients, provide selective ion accumulation and absorption, produce growth regulators, mobilize nutrients in infertile soil, and make available certain nutrients bound up in silicate minerals. In addition, mycorrhizae reduce susceptibility of their hosts to invasion of pathogens by utilizing root carbohydrates and other chemicals attractive to pathogens. They provide a physical barrier to pathogens and stimulate the roots to elaborate chemical inhibitory substances. Without ectomycorrhizae, certain groups of trees, especially conifers, oaks, and birches, could not become established, survive, and grow. In fact many mycorrhizae are specific to certain trees, such as pines and Douglas-fir. In return, the roots of the host provide support and a constant supply of nutrients in a nutrient-poor environment.

The association between the two can be tenuous. Any alteration in the availability of light or nutrients for the host creates a deficiency of carbohydrates and thiamine for the fungi. Interruption of photosynthesis stops fruiting by mycorrhizae.

Mycorrhizal interactions extend beyond the fungi and roots. Although some mycorrhizae have aboveground fruiting bodies—the familiar mushrooms whose spores are released to the air—others have belowground fruiting bodies, known as truffles. These mycorrhizae depend on small mammals, especially voles, to disperse their spores. Small mammals are able to detect underground truffles by smell. They dig up the truffles, eat them, and defecate, spreading the mycorrhizal spores across the forest floor. Mycorrhizal spores cannot germinate until they come in contact with tree roots. Thus, a three-way obligatory relationship exists (Figure 17.5). The tree depends upon mycorrhizae for nutrient uptake from the soil. The mycorrhizae depend upon tree roots for an energy source and upon small mammals for dispersal of spores. Small mammals obtain a significant portion of their food from truffles.

OBLIGATORY NONSYMBIOTIC MUTUALISM

The relationship between the mycorrhizal spores and small mammals is nonsymbiotic. The mutualists live apart yet are dependent upon one another, a common relationship. The approximately 900 species of tropical figs (*Ficus*) have a complex obligatory relationship with pollinating agaonid fig wasps. The wasps lay their eggs in the developing seeds, upon which the larvae feed. Figs experience 44 to 77 percent seed mortality, a high cost for pollination.

Other such obligatory relationships involve shelter, protection against predators, and reproduction. Some of the most interesting exist between ants and plants. Attine ants carry a slow-growing fungus that cannot survive without them. A species of Central American ants lives in the swollen thorns of acacia (*Acacia* spp.), from which they derive shelter and a balanced and almost complete diet for all stages of development. In turn the ants protect the plants from herbivores. At the least disturbance the ants swarm out of their shelters, emitting repulsive odors and attacking the intruder until it is driven away. Neither the ants nor the acacias can survive in the absence of the other.

FACULTATIVE MUTUALISM

Most mutualisms are nonobligatory and facultative, at least on one side. Such mutualisms are diffuse, involving interactions among guilds of species. Such mutualisms promote seed dispersal and pollination; the benefits are spread over many plants, pollinators, and seed dispersers.

Seed Dispersal

Plants with seeds too heavy to be dispersed by wind depend upon animals such as jays, squirrels, and ants to carry the seeds some distance from the parent plant and deposit them in sites favorable for seedling establishment. Some seed-dispersing animals upon which the plant depends may be seed predators, consuming the seeds for their own nutrition. Plants depending on such animals must produce a tremendous number of seeds over their reproductive lifetimes and sacrifice most of them to ensure that a few will survive, come to rest on a suitable site, and germinate.

In the deserts of southwestern United States, in the sclerophyllous shrublands of Australia, and in the deciduous forests of eastern North America a number of herbaceous plants, including many violets (*Viola* spp.),

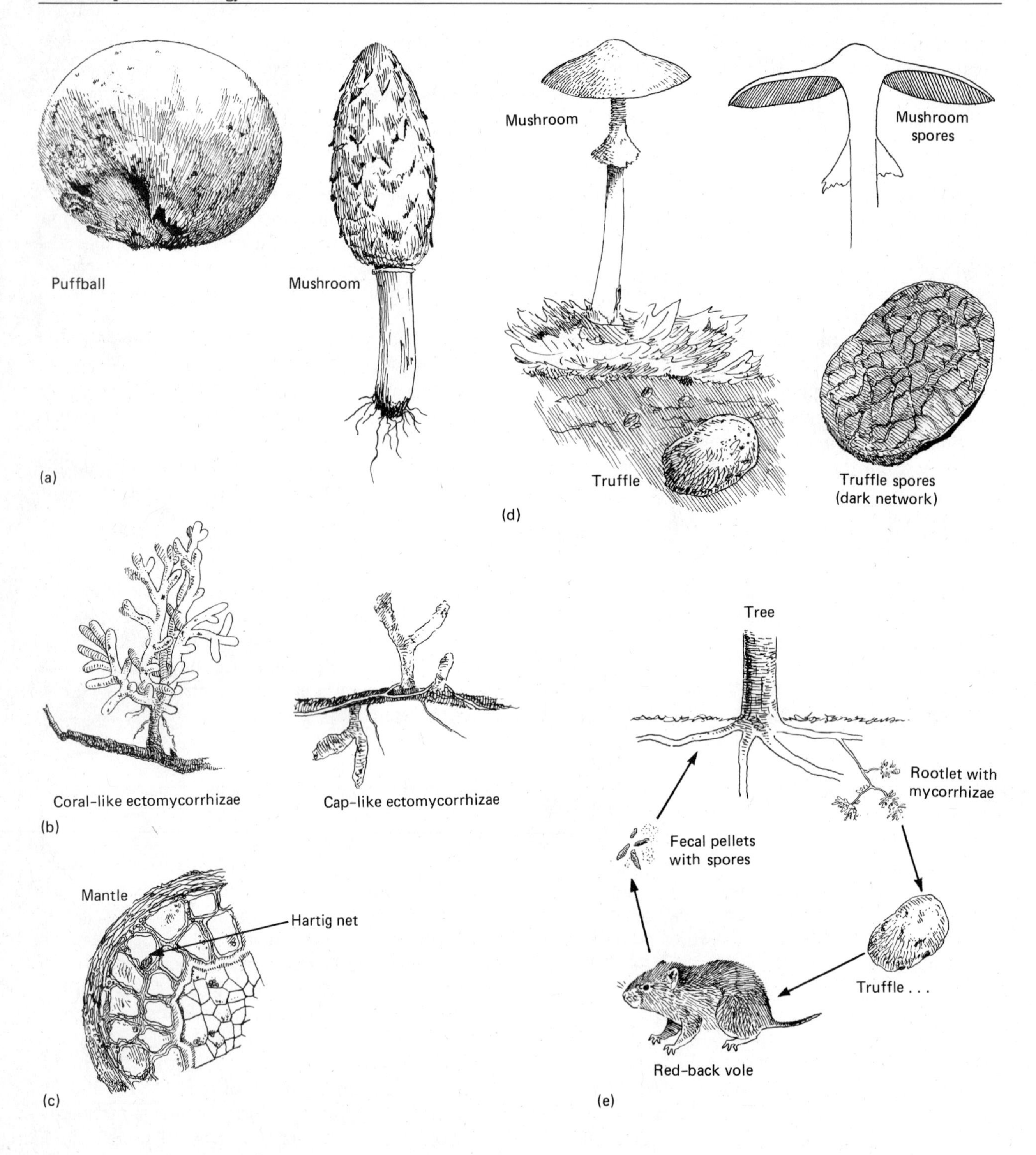
Puffball
Mushroom
(a)
Mushroom
Mushroom spores
Truffle
Truffle spores (dark network)
(d)
Coral-like ectomycorrhizae
Cap-like ectomycorrhizae
(b)
Mantle
Hartig net
(c)
Tree
Rootlet with mycorrhizae
Fecal pellets with spores
Truffle . . .
Red-back vole
(e)

depend upon ants to disperse their seeds. Such plants, called *myrmecochores,* have an ant-attracting food body on the seed coat called an elaiosome. Appearing as shiny tissue on the seed coat, the elaiosome contains lipids and sterols essential to certain physiological functions in insects. The ants carry seeds to their nests, where they sever the elaiosome and eat it or feed it to their larvae. The ants discard the intact seed within abandoned galleries of the nest. Ant nests, richer in nitrogen and phosphorous than surrounding soil, provide a fine substrate for seedlings. Further, by removing seeds far from the parent plant, the ants significantly reduce rodent predation on them.

Plants have an alternative approach for seed dispersal: to enclose the seed in a nutritious fruit attractive to fruit-eating animals, the *frugivores.* Frugivores are not seed predators. They consume only the endocarp surrounding the seed and, with some exceptions, do not impair the vitality of the seed. Most frugivores do not depend exclusively on fruits, which are deficient in certain nutrients, such as protein, and are only seasonally available.

To use frugivorous animals as agents of dispersal, plants must attract them at the right time. Plants discourage the consumption of unripe fruit by cryptic coloration, such as green fruit among green leaves, and by unpalatable texture, repellent substances, and hard outer coats. When seeds mature, plants attract fruit-eating animals by presenting attractive odors, softening the texture, improving succulence, increasing sugar and oils, and "flagging" their fruits with colors—red, black, blue, yellow, white—to catch frugivores' attention.

Most plants have fruits that can be exploited by a large number of dispersal agents. Such plants opt for quantity dispersal, the scattering of a large number of seeds with the chance that a diversity of consumers will drop some seeds in a favorable site.

Such a strategy is typical of, but not exclusive to, plants of the temperate regions, where fruit-eating birds and mammals rarely specialize in one kind of fruit and do not depend exclusively on fruit for sustenance. The fruits are usually succulent and rich in sugars and organic acids and contain small seeds that pass through the digestive tracts unharmed. To accomplish this passage such plants have evolved seeds with hard coats resistant to digestive enzymes. Seeds of some plants may not germinate unless they have been conditioned or scarified by passage through the digestive tract. Large numbers of small seeds may be so dispersed, but few are deposited on suitable sites. The length of time such seeds take within the digestive tracts of some small birds is no more than 30 minutes, so the distance dispersed depends on how far and where the birds go after eating. Such dispersal is a lottery in the truest sense (Figure 17.6).

FIGURE 17.5 Mutualism between tree roots and fungi. (a) Mushrooms and puffballs are fruiting bodies of fungi. The vegetative parts are the white, threadlike hyphae weaving through the dark organic litter. Follow the hyphae carefully, and you may find them threading among the rootlets of trees. (b) Some rootlets support caplike, nodular, or corallike growths, brown, black, white, red, or yellow. They are known as mycorrhizae or fungus roots. Shown are ectomycorrhizae, so named because most of their growth is outside the root. (c) Ectomycorrhizae form a mantle of fungi about the tips of the rootlets. Hyphae invade the tissues of rootlets and then grow between the cells. This growth is known as the Hartig net. (d) There are two general types of mycorrhizae: those whose fruiting bodies are among the familiar mushrooms and puffballs and those whose fruiting bodies, called truffles, are below ground. (Not all fungi that produce mushrooms and puffballs are mycorrhizal.) Ectomycorrhizal fungi spread by spores produced by fruiting bodies. Aboveground mushrooms discharge their spores to the air. Belowground truffles depend upon mammals to disperse them. (e) Mammals digest the fleshy part of the truffle, but the spores become concentrated in fecal pellets. Thus mammals deposit spores in nonforested areas. The lack of ectomycorrhizae is one reason why forest trees often fail to invade open areas.

FIGURE 17.6 The fruits of the serviceberry, *Amelanchier canadensis,* which range in color from green to red to black depending upon the degree of ripening. This midsummer fruit is a choice food dispersed by many birds and mammals, including the black bear.

In temperate regions fruits ripen in early and late summer, when the young are no longer dependent on highly proteinaceous food and both adults and young can turn their attention to a growing abundance of fruits. Such timing of ripening makes fruits available in early summer for resident animals, especially birds, and in late summer and fall for migrant birds.

In tropical forests 50 to 75 percent of the tree species produce fleshy fruits whose seeds are dispersed by animals. Rarely are these frugivores obligates of the fruits on which they feed. Exceptions include the oilbirds and a large number of tropical fruit-eating bats. The others consume many different fruits. Dispersers of the seeds of one plant are also dispersers for others, for several reasons. Fruits vary widely in their nutritional value; by eating a variety of them, frugivores tend to balance their diets. Also, plants have few means to restrict consumption of their fruits to specific frugivores. However, because fruits are of various sizes, shapes, colors, aromas, nutrient contents, and palatability, some are consumed chiefly by mammals and others by birds.

Pollination

The "goal" of seed dispersal by plants is to get the seeds away from the parent plant to some site favorable for seedling establishment. The goal in pollination is much more specific and direct. The plant must transfer its pollen from the anthers of one plant to the stigma of a conspecific.

Some plants simply disperse their pollen to the wind. This method works well and costs little when the plants grow in large homogeneous stands, as grasses and pine trees do. Wind dispersal is unreliable when conspecifics are scattered individually or in patches across a field or forest. These plants depend upon animals for pollen transfer, mostly insects, with some assistance from nectivorous birds and bats.

Like frugivores, nectivorous animals visit the plants to exploit a source of food, not to pollinate. With a few exceptions, the nectivores are generalists; like the frugivores they find little advantage in specializing, except as temporary facultatives. Further, because of the short flowering of each species, often shorter than the availability of fruits, nectivores depend on a progression of flowering plants through the season. Nectivores cannot afford to commit themselves to one flower, but they do concentrate on one species while its flowers are available.

Plants are the ones that have to specialize, enticing certain animals by color and fragrance, dusting them with pollen, and then rewarding them with a rich source of food: sugar-rich nectar, protein-rich pollen, and fat-rich oils. Providing such rewards is expensive for plants. Nectar and oils are of no value to the plant except as attractants for potential pollinators. They represent energy that the plant otherwise might expend in growth.

Many species of plants, such as blackberries, elderberries, cherries, and goldenrods, are generalists themselves. They flower profusely and provide a glut of nectar that attracts a diversity of pollen-carrying insects, from bees and flies to beetles. Other plants are more selective, screening their visitors to ensure some efficiency in pollen transfer. These plants have long corrollas, allowing access only to insects and hummingbirds with long tongues and bills and keeping out small insects that eat nectar but do not outcross the plants. Some, such as the closed gentian, have petals that only large bees can pry open. Orchids, whose individuals are scattered widely through their habitats, have evolved a variety of precise mechanisms in pollen transfer and reception so that pollen is not lost when the insect visits flowers of other species.

In addition to nectar, some plants provide oil as a floral reward. Families including Iridaceae, Orchidaceae, Scrophulariaceae, Concurbitaceae, Solanaceae, and Primulaceae, mostly in neotropical savannas and forest, have specialized oil-secreting organs, called *elaiophores*. Epithelial elaiophores consist of small areas of secretory epidermal cells beneath a protective cuticle on the petals, in which secreted lipids accumulate. Trichome elaiophores are made up of hundreds to thousands of glands that secrete lipids in a thin film of oil exposed to the air, although in some plants the lipids may be protected within deep floral spurs. The oil flowers are visited by highly specialized bees in four families that use the energy-rich floral oils in place of or along with pollen as provisions for developing larvae. These bees possess modified structures designed for mopping up, storing, and transporting oil to the nest.

DEFENSIVE MUTUALISM

A major problem for many livestock producers is the toxic effects on cattle of certain grasses, particularly perennial ryegrass and tall fescue. These grasses are infected by certain fungal endophytes that live inside plant tissues. The fungi (Clavicipitaceae, Ascomycetes) produce physiologically active alkaloids in the tissue of the host grasses. The alkaloids, which impart a bitter taste to the grass, are toxic to grazing mammals, particularly domestic animals, and a number of insect herbivores. In mammals the alkaloids constrict small blood vessels to the brain, causing convulsions, tremors, stupor, gangrene of the extremities, and death. At the same time these fungi seem to stimulate plant growth and seed production. This symbiotic relationship suggests a defensive mutualism. The fungi defend the host plant against grazing.

There are costs to the plant. The fungal infection causes sterility in the host plant by inhibiting flowering or aborting seeds. Some plants have a few counteradaptations that restore fertility; but in most plants the loss of sexual reproduction is balanced by the greater vegetative growth of infected plants and enhanced growth in the absence of herbivores.

POPULATION EFFECTS OF MUTUALISM

Mutualism is easily appreciated at the individual level. We grasp the interaction between an ectomycorrhizal fungus and an oak; we count the acorns dispersed by squirrels and jays, and measure the cost of dispersal in terms of seeds consumed. Mutualism improves the fitness of the fungus, the oak, and the seed predator. What are the consequences at the level of the population?

Defining the population consequences is considerably more difficult for mutualism than for predation and parasitism. The relationship is more difficult to model. Mutualism exists at the population level only if the growth rate of species A increases with the increasing density of species B, and vice versa.

The question is most relevant to nonsymbiotic mutualists. For obligatory symbiotic mutualists the relationship is straightforward: Remove species A and the population of species B no longer exists. If ectomycorrhizal spores fail to infect the rootlets of young pine, they will not develop; and if the young pine invading a nutrient-poor old field fails to acquire its mycorrhizal symbiont, it will not grow well. (Foresters and nursery workers must inoculate nursery-grown pine seedlings with the appropriate mycorrhizae.)

For nonsymbionts, obligatory or facultative, the effect on populations may be limited by the extent to which one species benefits another and by the proportion of each other's life-history cycle in which the mutualistic relationship is involved. An example of the population consequences of mutualism is provided by a demographic analysis of an ant-seed mutualism by F. M. Hanzawa, A. J. Beattie, and D. C. Culver. It involves a guild of ants and golden corydalis (*Corydalis aurea*), an annual or biennial widely distributed in open or disturbed sites in northeastern and western continental United States, Canada, and Alaska. The three compared the survivorship of both seeds and plants, and fecundity, reproduction, and growth rates of two seed cohorts of the plant, one relocated to an ant nest by undisturbed ant foragers and a control cohort of equal numbers hand-planted near each nest. The ant-handled cohort had significantly higher survivorship than the control. The ant-handled cohort produced 90 percent more seeds than the control cohort, its net reproductive rate R_0 was 8.0 and that of the control 4.2. The finite rate of increase of the ant-handled cohort was 2.83 per year, compared to 2.05 per year for the control. The ant-handled cohort experienced greater reproductive success not because of any great difference in fecundity but because of a significantly higher survival to reproductive age. The ant nests were superior microsites.

This study makes the point that close mutualistic relationships affect populations in complex ways essential to the integrity of ecosystems; yet most are still unknown and unappreciated. This fact serves as a warning for us to go slow in determining what organisms we will tolerate and what organisms we will not. We do not know what we are getting rid of. Foresters consider the southern fox squirrel a seed-eating pest in loblolly pine stands in the southern United States. Only recently, though, studies have pointed out that the fox squirrel feeds heavily on the truffles of the mycorrhizal fungus *Elaphomyces granulatus,* associated with longleaf pine, and serves as its dispersal agent. Thus the squirrel, far from being a pest of longleaf pine, is essential for the pine's well-being. Most people consider ants and other insects as pests and if given the choice would eradicate them, not appreciating their importance in the natural world.

EVOLUTION OF MUTUALISM

Symbiotic mutualism might have evolved from predator-prey, parasite-host, or commensal relationships. Initially, one member of the relationship increases the stability of the resource level for the second. In time, energy benefits accrue to the second member, and perhaps its activities begin to improve the fitness of the first. For example, a host tolerant of a parasitic infection may begin to exploit the relationship. In time, the two exploit each other, each for its own benefit, as in the mycorrhizae-plant mutualisms. Selection then favors mutual interaction to the point that the two become totally dependent on each other. At the extreme the two function as one individual, as algae and fungi do in lichens.

Among nonsymbiotic mutualists, the relationship may have begun with exploitation. In plant-pollinator relationships, birds and insects came to plants to feed on pollen and nectar. As a result of this exploitation, these animals carried pollen to other similar plants. Such plants experienced improved fitness and ultimately began to exploit the visitors as a means of dispersing pollen. Selection then favored the development of mechanisms to maintain the relationship, such as sugar-rich nectar to keep the pollinators coming.

Summary

Many species have an evolutionary effect on each other. In response to selective pressures of predation, prey evolve means of defense and predators improve their hunting efficiency. A still closer evolutionary relationship exists between parasites and their hosts and between mutualists.

Parasitism is a situation in which two organisms live together, but one derives its nourishment at the expense of the other. A parasitic infection can result in disease, a condition that deviates from normal well-being. Parasites may be divided into microparasites and macroparasites. Microparasites include the viruses, bacteria, and protozoa. They are small in size, have a short generation time, multiply rapidly in the host, tend to produce immunity, and spread by direct transmission. They are usually associated with dense populations of the host. Macroparasites, relatively large in size, include parasitic worms, lice, ticks, fleas, rusts, smuts, fungi, and other forms. They have a long generation time, rarely multiply directly in the host, are persistent with continual reinfection, and spread by both direct and indirect transmission.

Hosts are the habitat of parasites. The problem of parasites is to enter and escape the host. The life cycles of parasites revolve about these two problems. The adult stages live in the definitive host, from which they escape by direct contact with other hosts or by means of organisms that transmit them. These carriers become intermediate hosts of some developmental or infective stage of the parasite. Transmission from definitive to intermediate to definitive host is considered indirect. Indirect transmission often involves the food chain.

Transmission of parasites, direct or indirect, is complicated by patchy or clumped distribution of the host. As a result parasites become overdispersed. A great load of parasites is carried by a few individuals, while most remain free of infection. A heavy parasitic load can increase mortality and decrease fecundity of the host population. Under certain conditions parasitism regulates population. Because the death of a host does not benefit a parasite, which depends on its host for both food and shelter, natural selection favors less virulent forms that can live in the host without killing it. Hosts and parasites develop a mutual tolerance with a low-grade widespread infection.

Another type of parasitism is social parasitism. It may be temporary or permanent, facultative or obligatory. A common example is brood parasitism among some species of birds, ants, and wasps.

Mutualism is a positive reciprocal relationship between two species. It may have evolved from predator-prey, host-parasite, or commensal relationships. Mutualism may be symbiotic (living together) or nonsymbiotic and obligatory or nonobligatory. Obligatory symbiotic mutualists are physically dependent on each other, one usually living within the tissues of the other. Obligatory nonsymbiotic mutualists depend upon each other, but they live apart. Nonobligatory facultative mutualism includes interaction between plants and guilds of species involved in seed dispersal and pollination. In exchange for dispersal of pollen and seeds, plants reward animals with food—fruit, nectar, and oil. To reduce wastage of pollen some plants possess morphological structures that permit only certain animals to reach the nectar.

The natural world involves innumerable mutualistic relations, few of which we know or understand. The disruption of mutualistic relations, whether obligatory or facultative, can have adverse effects on populations or whole ecosystems.

Review and Study Questions

1. Define parasitism. Why is it more reasonable to classify parasites by size than by taxonomic group?
2. Distinguish between definitive and intermediate hosts. Look up examples beyond those mentioned in the text.
3. Examine the transmission of some common parasitic diseases in humans, such as malaria, Rocky Mountain spotted fever, giardiasis, and trichinosis. How might they be controlled?
4. How might a patchy or clumped distribution of hosts affect the spread of parasites?
5. What is social parasitism? Why might nest parasitism have evolved?
6. What is mutualism? Look at some examples critically. Are they mutualistic? Are they cases of reciprocal exploitation?
7. Distinguish among symbiosis, obligatory symbiotic mutualism, and obligatory nonsymbiotic mutualism.
8. Is fruit predation a chancy way of distributing seeds? Why?

*9. Explore examples of how parasitic infections have altered populations, ecosystems, even human history. You might consider the chestnut blight's effect on North American deciduous forests; the night mosquito, bird pox, and the fate of the Hawaiian honeycreepers; rats, fleas, and plague in Europe; or rinderpest in African wild and domestic ungulates, among many others.

Selected References

Bacon, P. J. 1985. *Population dynamics of rabies in wildlife.* New York: Academic Press.

Boucher, D. H., ed. 1985. *The biology of mutualism.* Oxford, England: Oxford University Press. An excellent reference.

Boucher, D. H., S. James, and H. D. Keller. 1982. The ecology of mutualism. *Annual Review of Ecology and Systematics* 13:315–347.

Burdon, J. J. 1987. *Diseases and plant population biology.* Cambridge, England: Cambridge University Press. Effect of diseases on individual plants and populations.

Edwards, M. A., and U. McDonald. 1982. *Animal diseases in relation to animal conservation.* New York: Academic Press.

Effects of disease on wild animal populations and human interactions.

Futuyma, D. J., and M. Slatkin, eds. 1983. *Coevolution.* Sunderland, MA.: Sinauer. Papers on processes and consequences of coevolution.

Hanzawa, F. M., A. J. Beattie, and D. C. Culver. 1988. Directed dispersal: Demographic analysis of an ant-seed mutualism. *American Naturalist* 131:1–13.

Howe, H. F., and L. C. Westley. 1988. *Ecological relationships of plants and animals.* New York: Oxford University Press. An outstanding presentation of herbivory, pollination, and other interrelationships.

May, R. M. 1983. Parasitic infections as regulators of animal populations. *American Scientist* 71:36–45.

Real, L., ed. 1983. *Pollination biology.* Orlando, FL: Academic Press. Recent research in pollination biology.

Wilson, E. O. 1975. *Sociobiology: The new synthesis.* Cambridge, MA: Harvard University Press. See Chapter 17, "Social Symbiosis," for a detailed discussion of social parasitism.

CHAPTER 18

Human Control of Natural Populations

Outline

Objectives

On completion of this chapter you should be able to:

1. Explain the nature of pests and weeds.
2. Discuss the various methods of and ecological aspects of pest and weed control.
3. Describe integrated pest management.
4. Discuss the concept and application of sustained yield in management of exploited natural populations.
5. Explain the efforts needed to restore endangered species.

Humans have always had a close relationship, positive and negative, with plants and animals. From the first we depended on them for food, clothing, shelter, and tools. In turn we were prey to large carnivores and hosts for an array of parasites. Our early technological advances enabled a much greater exploitation of animals, driving some of them, such as the Pleistocene mammals, to extinction. As our culture advanced, we domesticated certain plants and animals, allowing us to form larger, more interdependent social units. Now some animal species became competitors of or threats to domestic plants and animals. Some species of plants invaded the new habitats provided by agricultural fields and competed with crop plants.

Growing human populations and cultural changes demanded greater exploitation of resources. Forests were leveled to provide building materials for cities, ships, and armies. Large mammals killed for food decreased in widening distances about centers of population. Other animals, unable to exist in altered or diminished habitats, disappeared.

Out of this dependency–enmity relationship grew a cultural relationship with plants and animals. Mammals such as the cave bear, the wolf, and the cat and plants such as the oak and lily became religious symbols or objects of worship. Even amid high technology this symbolic relationship persists in our use of animals as national and state symbols and as names and mascots of athletic teams. Plants and animals also have helped us develop drama, dance, art, and medicine.

In spite of this complex relationship, only recently has a segment of human society realized that natural populations with which we share Earth are declining rapidly from overexploitation, decimated habitats, and poisoned environments. For economic, aesthetic, and moral reasons, and for the sake of long-term human survival, these people have begun managing and conserving wild populations. They take three approaches: reducing populations of destructive organisms, maintaining the populations of exploited plants and animals, and increasing the populations of others in danger of extinction.

REDUCING POPULATIONS

Over most of our existence, we have had to contend with unwanted plants and animals that interfere with our well-being. What constitutes a pest or weed depends on your viewpoint and values.

PESTS AND WEEDS

Pests are animals that humans consider undesirable, a classification that varies with time, place, circumstance, and individual attitudes. Some animals are obvious pests. Certain mice, rats, cockroaches, fleas, mites, lice, and mosquitoes have long been camp followers of humans, adapting quickly to food and shelter offered by human habitation and cultural practices. With the development of agriculture, large predators that threatened domestic animals, herbivorous mammals, large and small, that invaded fields and gardens, and grain-feeding birds all became pests.

Plants also became competitors with humans. Wild plants in fields and gardens compete with crops for light, space, and nutrients, reducing yields and becoming, in human terms, weeds. A *weed* is a plant growing anywhere not desired, a plant out of place. Violets growing in a thicket are wildflowers; violets growing in a lawn may be weeds (depending on the lawn owner's value judgment). Electric companies regard woody plants and trees growing in power line rights-of-way as weeds; to a naturalist, the same growth is wildlife habitat. Some foresters consider any trees that have no commercial value as weed trees; to a naturalist those trees may be beautiful or important for wildlife.

We can look upon pests as occupying a continuum from r to K. Our most familiar and obnoxious pests, such as houseflies, cockroaches, fleas, mites, and scale insects, possess certain r characteristics (see Chapter 14). They have a high rate of increase, small size, and a high rate of dispersal; they seek new and open habitats, adapt well to conditions provided by humans, and spread rapidly in homogeneous habitats where both shelter and food are abundant.

Common weeds are much the same. The most persistent and abundant weeds are easily dispersed, colonize highly disturbed sites, persist a long time in the soil as seeds, respond quickly to disturbed sites, and are tough and resilient. Good examples are ragweeds and dandelion.

Other animals and plants become pests or weeds under different sets of conditions. They have low reproductive rates, occupy more specialized habitats, and need more specialized resources. Often they become pests or weeds because humans move into their habitats. The tsetse fly of Africa, a *K* species, became a pest only when humans moved their cattle into its natural range. Elephants and deer became pests when humans usurped their habitat for agriculture and settlements. Certain trees become "weeds" when humans change the composition of forests and fields.

Most pests and weeds, however, fall somewhere along a continuum between these two. They show a mix of *r* and *K* characteristics. Normally they are regulated by natural predators and parasites. Native mice and rats fall into this category, as well as certain insect pests, such as the spruce budworm and the codling moth, whose larvae feed on apples.

MEANS OF CONTROL

Since the advent of agriculture, humans have been concerned with pest control. After centuries, it is evident that we cannot eradicate pests. The best we can do is control their numbers. Control should reduce pests to a level below which they will not cause economic injury (Figure 18.1), a point at which costs of control are less than or at best equal to the net increase in value derived from such control. Even so, many people initiate control measures when no strong economic values are involved, such as spraying insecticides and herbicides on lawns. Control measures vary with the life history and effect of the pest or weed.

Chemical Control

One means of control is chemical. The ancient Sumerians used sulfur to combat crop pests, and the Chinese as long ago as 3000 B.C. used substances derived from plants as insecticides. By the early 1800s such chemicals as Paris green, Bordeaux mixture, and arsenic were commonly used to combat insect and fungal pests. The major chemical weapons, however, appeared after World War II with the development of organic insecticides (containing carbon). The original impetus for their development was combatting insect vectors of human disease, especially in tropical areas. The initial success of these insecticides encouraged their rapid use in agriculture, for which the chemical industry provided an arsenal of over 500,000 pest products.

These organic chemicals, with varying degrees of toxicity and persistence, are either synthetic (of human manufacture) or botanic (derived from plants). Major groups of synthetic pesticides include the chlorinated hydrocarbons, organophosphates, and carbamates. All are broad spectrum insecticides and in one manner or another disrupt the nervous system. Chlorinated hydrocarbons are fat soluble (see Chapter 25) and environmentally persistent, accumulate in food chains, and are toxic to a wide range of organisms, invertebrate and vertebrate. They include such pesticides as DDT, eldrin, lindane, chlordane, and mirex.

Organophosphates, relatives to nerve gas, are much more toxic to birds and mammals, including humans, and are water soluble, leachable into groundwater. Because they are less persistent, organophosphates are applied more frequently. Some are systemic, taken up by plants and translocated to leaves and stems where they are toxic to leaf-eating and sap-sucking insects. Organophosphates include malathion and parathion.

The carbamates, which include sevin, are derived from carbamate acids. They are more narrow in the spectrum of insects they kill, have low persistence, and are highly toxic to vertebrates.

Another synthetic, diflubenezuron, interferes with the formation of exoskeletons in molting insect larvae. Relatively nonpersistent, it is used widely in gypsy moth control. The pesticide is relatively nonselective, affecting all lepidopteran caterpillars developing at the time of spraying. If spray drift reaches streams, the insecticide kills stream invertebrates. Its effects on soil invertebrates are not well understood.

Prominent among the botanicals are the pyrethrins, toxicants extracted from the flower heads of chrysanthemum plants. They act much like DDT, affecting the nervous system, but they are among the safest and least persistent pesticides. They are commonly used in household sprays. Another important botanical is nicotine, derived from tobacco, and rotenone, obtained from the roots of certain tropical legumes. Rotenone is highly toxic to fish and certain insects and slightly toxic to mammals. It is widely used to reclaim lakes for game fish.

Organic pesticides used to control weeds are called herbicides. They fall into three classes based on their effects on plants. Contact herbicides, such as atrazine, kill foliage by interfering with photosynthesis. Systemic herbicides, such as 2,4-D and 2,4,5-T, are absorbed by

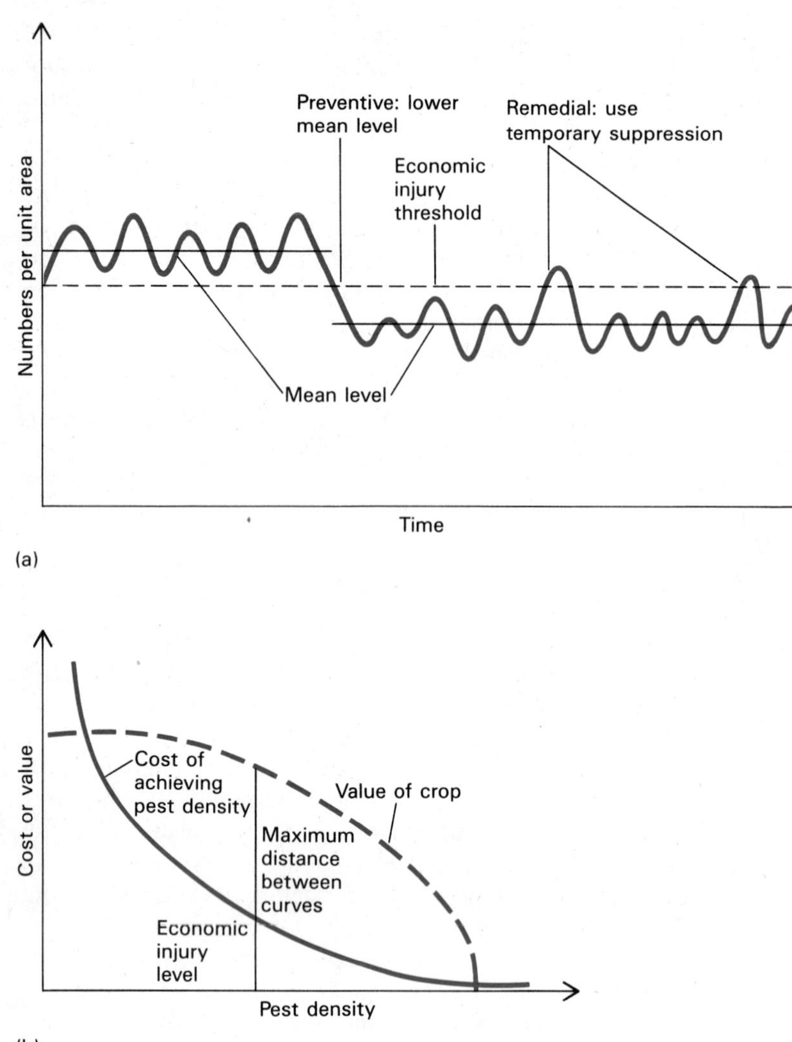

FIGURE 18.1 (a) Economic injury threshold suggests when pest control measures should be taken. When the mean level of a pest population exceeds the threshold, strong preventive measures should be taken to reduce the population below the economic injury level. Pest populations maintained at a mean below the threshold will need only temporary suppression measures to prevent an outbreak. (b) The economic injury level is that point at which the value of the crop exceeds the cost of controlling pest density. Attempting to hold pest density at too low a level exceeds the value of the crop.

plants and overstimulate growth hormones. The plants grow faster than they can obtain necessary nutrients and die. Soil sterilants, such as Dowpow, kill soil microorganisms necessary for certain plants to grow. Although designed to kill plants, many herbicides are extremely toxic to humans, especially 2,4-D and 2,4,5-T, the two components of the notorious Agent Orange. These herbicides contain dioxins that have been linked with birth defects and cancers, including leukemia.

Ease of application, effectiveness in small dosages, low cost, and toxicity gave these insecticides the appearance of a panacea to pest problems. They became, as the entomologist Paul Debach described them, ecological narcotics. However, their toxicity and nonselectivity ended in numerous ecological disasters. Instead of solving or controlling the pest problem, the pesticides compounded the problem by killing natural predators as well as the pests. With their natural predators eliminated and an abundant food supply available, the surviving pests resurged to even greater infestations

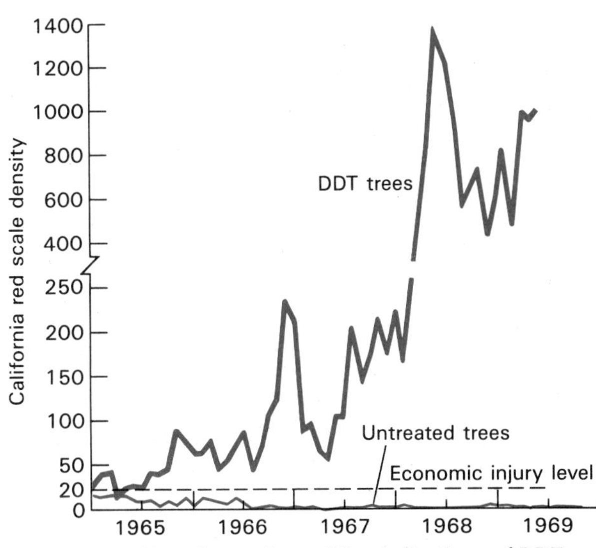

FIGURE 18.2 Experimental monthly applications of DDT on scale-infested lemon trees reduced not only the pest but also its natural enemies. Loss of natural enemies allowed the scale to resurge dramatically, whereas natural predators and parasites kept the scale in check on the unsprayed trees.

(Figure 18.2). Loss of natural predators did more. It released other insect pests that heretofore had been held in check by predation. Now their populations increased dramatically, adding new pests to the situation.

There are numerous worldwide examples. One involves the boll weevil in the cotton fields of the Rio Grande Valley in Texas. For 15 years the insect was held in check with chlorinated hydrocarbons. Then in the late 1950s, the weevil achieved resistance to the insecticide. When cotton growers switched to organophosphate insecticides, two other pests emerged, the bollworm and tobacco budworm. By 1962 they, too, had become resistant to organophosphates, so the growers switched over to carbamates, mostly methyl parathion. By 1968 this insecticide, too, had lost its effectiveness. The original pest, the boll weevil, faded as bollworm and especially the tobacco budworm became the major pests. Today the tobacco budworm is resistant to all registered pesticides and cotton growing has collapsed in northeastern Mexico and southern Texas.

This example points out one of the major failings of chemical pesticides. Through the process of natural selection (see Chapter 2), insect pests evolve a resistance to the pesticide. As one pesticide replaces another, the pests acquire a resistance to them all. By 1988 over 1600 insect pests worldwide had developed resistance to pesticides. Some species, notably houseflies, certain mosquitoes, Colorado potato beetles, cotton bollworms, cattle ticks, and spider mites, have overcome the toxic effects of every pesticide to which they have been exposed. Nearly 50 species of weeds are herbicide resistant. Insect pests need only about five years to evolve pesticide resistance; their predators do so much more slowly.

Biological Control

Farmers observed long ago that natural enemies of pests act as controls. As early as 300 A.D. the Chinese were introducing predaceous ants into their citrus orchards to control leaf-eating caterpillars and boring beetles. In the late Renaissance, naturalists made important observations on insect parasitism. In the 1800s entomologists made important advances in understanding the role of biological control of both insect pests and weeds.

Insect pests have their own array of enemies in their natural habitats. When an animal or a plant is introduced, intentionally or unintentionally, into a new habitat outside its natural range to which it can adapt, it leaves its enemies behind. Freed from predation and finding an abundance of resources, the species quickly becomes a pest or a weed. This fact has led to the search for suitable natural enemies to introduce into populations of pests, in some instances with considerable success.

First entomologists need to discover the major predators of the insect involved. Next they need to collect a great number of them, along with a sufficient number of prey to keep the predators alive until the shipment arrives at the laboratory. In the laboratory the entomologists must rear large numbers of the predators to release in infested areas. Following release, they must observe and evaluate the predators in their new habitat. Some may not adapt to the new environment; they may fail to reproduce; and they may not be as effective as they were back in their native habitat. All in all, finding a biological control agent for pests takes trial and error.

A classic success story is the case of the cottony-cushion scale *Icerya purchasi* in California. The U.S. Department of Agriculture sent an entomologist, Albert Koebele, to Australia, the scale's native habitat. Among the few scales in the orange groves of the Murray River Valley he found a ladybug, *Vedalia cardinalis,* feeding on large female *Icerya.* Koebele brought *Vedalia* beetles to California in 1887. Their explosive multiplication allowed the release of thousands of beetles in California orange groves. By the end of 1889, the scale was no longer a problem. The vedalia beetle controlled the scale until orchardists introduced DDT into the groves. DDT killed the beetle and the scale rebounded. Subsequently spraying has been eliminated and the beetle has been reintroduced.

Biological control has also been used on weeds. One outstanding example is the reduction of prickly pear (*Opuntia* spp.) in Australia by the introduced moth *Cactoblastis cactorum,* a native of Argentina. Another is the control of Klamath weed or Saint-John's-wort (*Hypericum perforatum*). A native of Eurasia and northern Africa, this noxious weed was introduced in 1900 into California along the Klamath River. An aggressive colonizer of overgrazed pastures and toxic to cattle and sheep, it soon occupied over 800,000 hectares in the northwestern United States. Among the 600 species of insects that feed on the plant in its native habitat was the leaf-eating beetle *Chrysolina quadrigema.* Introduced in 1945, this beetle has reduced Klamath weed to a rare roadside weed in California.

Not all predators are suitable agents of control. An insect predator has to be easy to culture commercially. Its life cycle must synchronize with that of the pest. It has to be ecologically compatible with the system into which it is introduced; in other words, the predator itself should not become a pest. Most important, it has to have outstanding ability to search out prey and to disperse widely.

Genetic Resistance

Numerous wild plants and animals have evolved their own defenses against natural enemies (see Chapter 16). These defenses include allelopathic effects against plant competitors and toxins in plant and animal tissue that inhibit or suppress predation. One approach to pest control, then, is to breed for genetic resistance. This tool has been used successfully in crop plants such as corn, wheat, and rice. The process involves the crossing of cultivars with wild relatives to capture the resistant gene into the gene pool of the cultivated plant. Such an approach requires considerable effort in locating, recognizing, and studying wild relatives. As more and more of our natural communities are being destroyed—most crop plants are tropical in origin—we are losing the genetic storehouse on which we could draw.

The latest technique to increase plant resistance to weeds and insect pests is genetic engineering, particularly the use of recombinant DNA material to introduce the gene with the desired trait into the plant. This technique involves the transfer of a single gene that confers resistance to viruses or to herbicides, or enables gene coding for endotoxins, poisons that inhibit the feeding activity of insect enemies.

Genetic engineering for pest control has its dangers. Plants with engineered traits could evolve new genotypes with somewhat different life histories or physiological traits. Engineered crop plants might transfer genes over long distances by hybridizing with related plants. Many crops, such as celery, asparagus, and carrot, have weedy relatives with high reproductive output and efficient seed dispersal. If these weedy relatives acquired the engineered gene, then the gene could spread through the range of the plants. This transfer would create weeds against which current herbicides would be ineffective, and insect-resistant plants could speed the evolution of even more resistant insect pests.

Another genetic approach to insect suppression is sterile male release. Sterile males reared in great numbers in the laboratory are introduced in sufficiently large numbers to ensure they will be involved in a high proportion of the matings in the field. If the number of sterile males is kept high as the population of the pest declines, the proportion of sterile to fertile matings increases. If the population of the pest is initially high, insecticides may be used to reduce the population before sterile males are released. Such a method works only if certain conditions are met: The pest population must be fairly isolated and not subject to immigration of wild males or emigration of sterile males; genetically different subpopulations must not be involved; and no genetic changes must occur in the reared population.

An example is the screwworm control program. The screwworm is the larva of the blowfly *Cochliomyia hominivorax* that lays its eggs in open wounds of warm-blooded animals. The larvae enter the wound and feed on the flesh of the animal. The screwworm became a major livestock pest in the southeastern United States and Texas and an eradication effort began in the 1950s involving the release of factory-reared sterile males. The program was highly successful until 1972, when a major outbreak occurred. The cause was a genetic difference between the sterile males released and the wild type. In 1977 the defective factory strain was replaced and control was regained.

Mechanical Methods

Another approach to pest control is mechanical. For centuries farmers have used fencing and other barriers not only to keep livestock in but to deter predators and to prevent herbivores such as deer and rabbits from feeding on crops. Sticky traps wrapped around trees prevent caterpillars from climbing up to the crown. Traps baited with pheromones and sticky paper capture males of specific insect pests such as gypsy moths. Light traps, although effective, are indiscriminate, killing beneficial insects along with the pest. Various methods of trapping are used to catch mammalian and avian pests. Cultivation, hoeing, and hand weeding are typical methods of eliminating competitive weeds from fields and gardens.

Cultural Methods

Homogeneous habitats provide the opportunity for large outbreaks of pests. Large fields and extensive stands of forest trees of the same and closely related species, such as southern pines and balsam fir, provide huge areas of abundant food and cover. Spruce budworm and southern pine beetle, for example, have swept across hundreds of thousands of acres of forests. One deterrent for rapid spread of pests is the creation of patchy environments in which homogeneous stands are broken up by other types of vegetation, such as interplanted row crops, interspersion of hardwoods with conifers, and hedgerows. These patches not only scatter the food supply, checking the spread of the pest and breaking its population into smaller units more vulnerable to predation, but also harbor enemies of the pests.

Integrated Pest Management

Chemical, biological, and other methods cannot stand alone. Entomologists now have developed a holistic approach to pest control called *integrated pest management.* Integrated pest management considers the biological, ecological, economic, social, and even aesthetic aspects of pest control and mixes techniques. The objective of IPM is to meet the pest not at the point of a major outbreak but at the time when the size of the population is easiest to control. Managers rely first on natural mortality caused by weather and natural enemies with as little disruption of the natural system as possible, so long as numbers hold below the economic injury level (Figure 18.1).

Successful integrated pest management requires the knowledge of the population ecology of each pest, its associated species, and the host species. It involves considerable field work. Monitoring the pest species and its natural enemies by such techniques as counting eggs and trapping adults, scientists determine the necessity, timing, and intensity of control measures. These measures may involve minimal chemical spraying at appropriate times or cultural techniques such as thinning and salvage operations in an affected forest. IPM makes minimal use of chemicals to reduce the development of genetic resistance to pesticides. The control methods must be adjusted from one location to another.

The intensity of control or no control is based on the degree of damage sustainable, the costs of control, and benefits to be derived. Managers make a series of decisions (Figure 18.3): number and activity of the pest at designed spots; potential effects of pest control on the resources; cost effectiveness; the degree and acceptability of environmental and social consequences; the best preventive or suppressive tactics; and success after treatment. Integrated pest management has been employed in the control of spruce budworms in Canada and the southern pine beetle in the United States. It is used successfully in apple orchards of the United States and Canada, in cotton fields of Texas, and in alfalfa-growing regions in the United States.

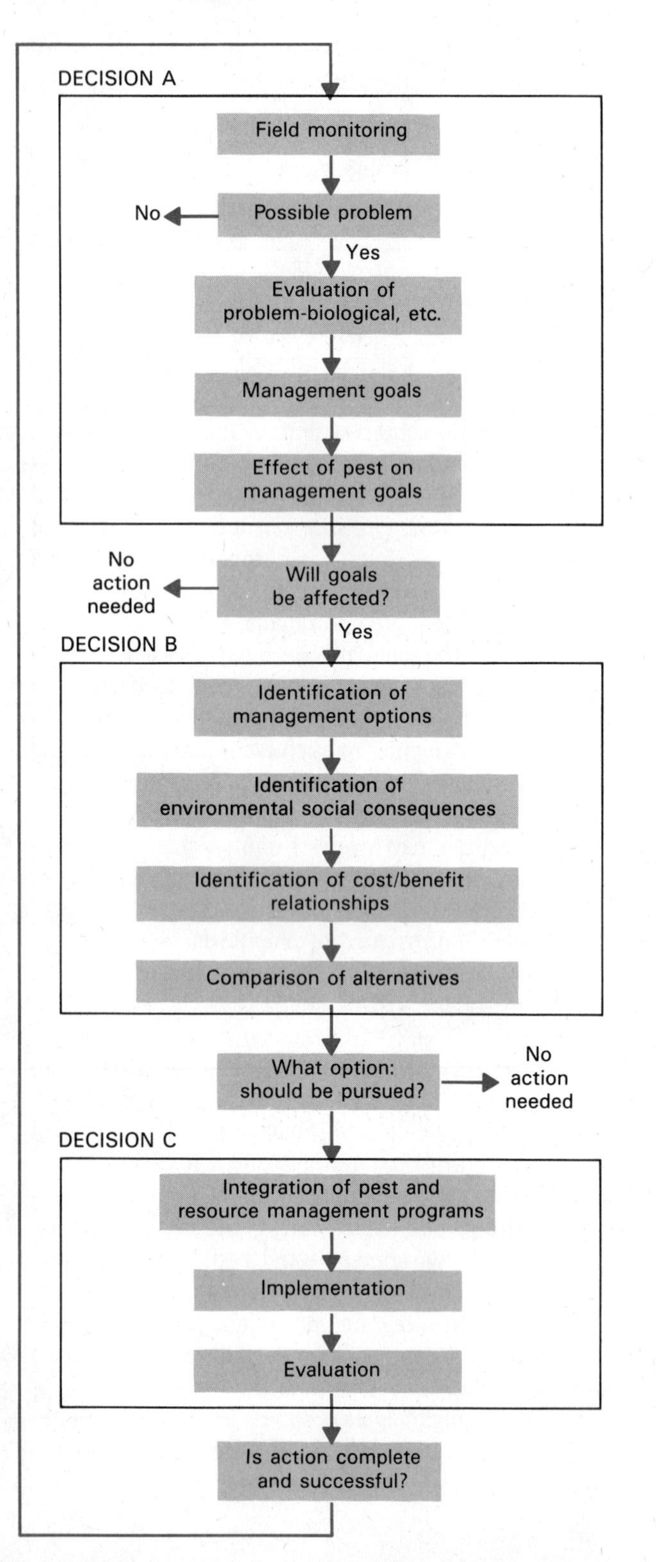

MAINTAINING EXPLOITED POPULATIONS

For thousands of years humans have been exploiting plant and animal populations with little regard for the effects of our actions. Populations of many of these organisms have been able to absorb the effects of exploitation; others have been exterminated or driven to the brink of extinction.

Not until the late 1800s did we make any effort to manage exploited populations to ensure their continuance. A major stimulus was the wide fluctuation in the catch of fishes in the North Sea, which caused parallel fluctuations in commercial income. Debates raged over human impact on the fisheries. Some argued that removal of fish had no effect on reproduction; others that it did. Not until a Danish fishery biologist, C. D. J. Petersen, developed the tagging and mark-recapture technique of estimating population size were biologists able to make some assessment of fish stocks. This technique was augmented by methods for aging fish and by egg surveys. All these studies suggested that overfishing indeed was the culprit.

The real answer surfaced after World War I. During the war, fishing in the North Sea had stopped. After the war, fisherman experienced sizable increases in their catches. Fishery biologists suggested that after fishing removed accumulated stock, population size and catches would stabilize. The size of the population would be determined by the amount and size of fishes caught. In 1931 E. S. Russell of Great Britain presented this model of fishery yield:

$S_2 = S_1 + (A + G) - (M + C)$ where
S_2 = preharvest population, time 2
S_1 = post-harvest population, time 1
A = weight of fish large enough to be caught
G = increase in growth of these fish plus other fish already in the harvested population
M = mortality and decrease in weight
C = fish caught

If $(A + G)$ were kept equal to $(M + C)$, then the population would remain stable.

FIGURE 18.3 A decision-making model of integrated pest management. Note the inputs of biological, social, and environmental information.

The Norwegian fishery biologist Johannes Hjort added a new twist. He stated that the largest sustained catch could be obtained at the precise point where overfishing began. Fishing should cease at that point. Although this theory described scientific fishing, it was not rational fishing, argued the industry. "Rational" fishing was based on profit—an argument still with us today.

SUSTAINED YIELD

Harvesting at a level that will ensure a similar yield over and over without forcing the population into decline is called *sustained yield.* The crop removed per unit time is equal to production per unit time.

In a stable environment largely undisturbed by humans, species populations tend to be dominated by large old individuals. When humans exploit such a population, those individuals are first to go. To compensate, the population exhibits an increased growth rate, reduced age at sexual maturity, increased reproductive effort, and reduced mortality of small members of the population. As the harvest of the species declines, the exploiters are forced to change their techniques and increase the intensity of harvesting. If the stock is overexploited, the age classes become too young to carry on reproduction and the population collapses. The species' niche is taken over by unexploited, highly competitive, and closely sympatric or introduced species. The objective of sustained yield is to avoid such a collapse.

The rate of exploitation and sustained yield of a population are clearly dependent on the rate of increase, r. Sustained yield does not imply holding a population at ecological carrying capacity, K, for at that level the rate of increase is zero. A population stable in the absence of harvest can be managed under sustained yield only by manipulating the population to increase r. One way is to increase the carrying capacity by increasing available resources, fecundity, and survival. The usual way is to lower the density by removing a certain number and then stabilizing the population. Within limits the lower the density of the population is below the carrying capacity, the higher is the rate of increase. The rate of harvest should equal the rate of increase to hold the reproductive population stable at some desired lower density.

Sustained yield is not a particular value for a given population. There may be a number of sustained yield values corresponding to different population levels and different management techniques. The level of sustained yield at which the population declines if exceeded is known as *maximum sustained yield* (MSY) (Figure 18.4). Maximum sustained yield implies the removal of all production over and above that needed to replace the amount harvested. Harvest or removal takes the population down to a level at which the amount removed can be replaced by the remaining stock by the next harvest period. In practice the maximum sustained yield is quite variable. If fluctuations in the environment and the breeding stock are not taken into account, the margin between a level that stabilizes the population and one that drives it to extinction is dangerously thin.

An alternative to MSY is *optimal sustained yield* (OSY). It is more complex because it takes into account both biological and sociological aspects. Much more conservative than MSY, it has built-in safety factors, such as reducing yield below MSY and not insisting on any particular proportion of the population to harvest. Optimal sustained yield, however, can be subject to political and social pressure for increased cropping.

The higher the rate of increase of a population, the higher will be the rate of harvest that produces the maximum amount of harvestable biomass. Species characterized by scramble competition (r-strategists) lose much of their production to a high density-independent mortality. In a population influenced by density-dependent variables, such as climate or temperature, the management objective is to reduce wastage by taking all individuals that otherwise would be lost to natural mortality.

Such a population, however, is difficult to manage because the stock can be severely depleted unless there is repeated reproduction. An example is the Pacific sardine, a species in which there is little relationship between size of breeding stock and number of progeny. Exploitation of the Pacific sardine population in the 1940s and 1950s shifted the age structure of the population to younger age classes. Prior to exploitation, 77 percent of the reproduction was distributed among

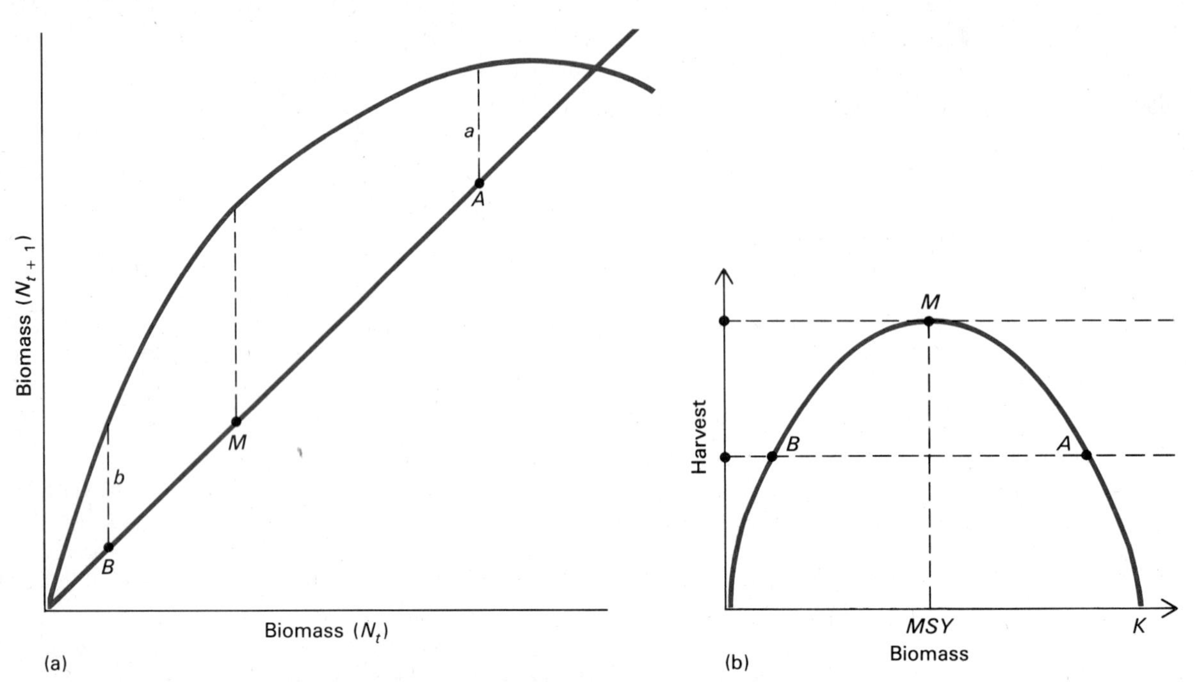

FIGURE 18.4 (a) A sustained yield model for an idealized population of K-strategists harvested under three regimes. The 45° line represents the replacement level of the stock. Where the curve and the diagonal intercept is the point at which reproduction balances losses. In the first regime the population is harvested down from a steady state to a size $N_t = A$. The dashed line a represents the number that must be harvested or replaced each year to hold the population size stable at $N_t = A$. In the second regime the population is reduced to $N_t = B$ and a sustained yield represented by line b could be harvested each year to hold the population stable at $N_t = B$. The yield in this case is a large proportion of a small population. Maximum sustained yield is obtained at the population size where the diagonal line and the curve have maximum separation, $N_t = M$. (b) A parabolic recruitment curve illustrates the concept in a different fashion. Multiple sustained yield is approximately ½K, represented at M on the curve. Points A and B are two different equilibrium levels. At A the stable equilibrium is at a relatively high density, much above MSY. B is an unstable equilibrium point much below MSY. Harvesting to the left of MSY can drive the population to extinction.

the first five-year age classes. In the exploited population 77 percent was associated with the first two years of life. The population approached that of a single-stage reproduction subject to pronounced oscillations (Figure 18.5). Two consecutive years of environmentally induced reproductive failures resulted in a collapse of the population from which the species never recovered.

In populations characterized by density-dependent regulation (K-strategists), the maximum rate of harvest depends on age structure, frequency of harvest, numbers left behind after harvest, and fluctuations both in fecundity and in the environment. It also depends on the density of the population to be harvested and the rate of harvest needed to stabilize the density at that level.

Several rules are applicable to exploitation. To obtain a croppable surplus the population first must be reduced below a steady density. For each density to

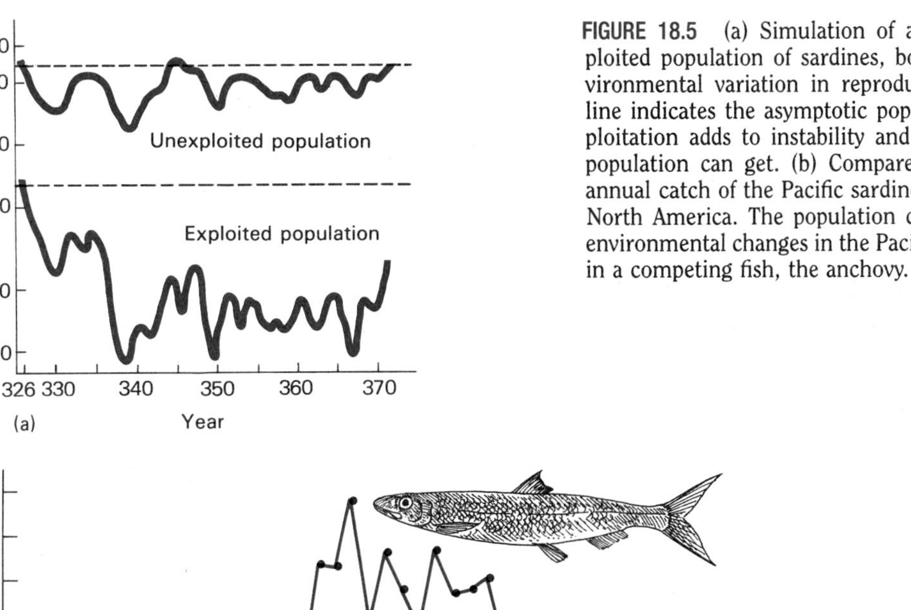

FIGURE 18.5 (a) Simulation of an exploited and an unexploited population of sardines, both subject to random environmental variation in reproductive success. The dotted line indicates the asymptotic population size. Note how exploitation adds to instability and how dangerously low the population can get. (b) Compare this simulation with the annual catch of the Pacific sardine along the Pacific coast of North America. The population collapsed from overfishing, environmental changes in the Pacific Ocean, and the increase in a competing fish, the anchovy.

which the population is reduced, there is an appropriate sustained yield. There are two levels of density from which a given sustained yield can be obtained, but maximum sustained yield can be harvested at one density only (Figure 18.4). If a constant or fixed number is removed from the population each year, the population will decline to and stabilize at the upper population density for which that number is sustained yield. If that number exceeds the maximum sustained yield, the population declines to extinction. If a constant percentage of the standing crop of each year is removed, the population will decline and stabilize at a level at equilibrium with the rate of harvesting. This level may be above or below that generating maximum sustained yield.

Overexploited populations exhibit symptoms of impending disaster. Exploiters experience decreased catch per unit effort as well as the decreasing catch of related species. There is a decreasing proportion of females pregnant, due both to sparse populations and to a high proportion of nonreproducing young. The species fails to increase its numbers rapidly after harvest. A change in productivity relative to age and age-specific survival shows that the ability of the population to replace harvested individuals has been impaired. An outstanding example is the blue whale (Figure 18.6). After 1860 the blue whale became the most commercially important species of whale. Catches in the Antarctic peaked in 1931 at 30,000 animals and declined to under 2,000 in 1963. For the last 40 years of their exploitation the average age of the blue whale caught in the Antarctic was 6 years, mostly immature females or females carrying their first calf. The species is near extinction.

Sustained yield is applied also to animals hunted for

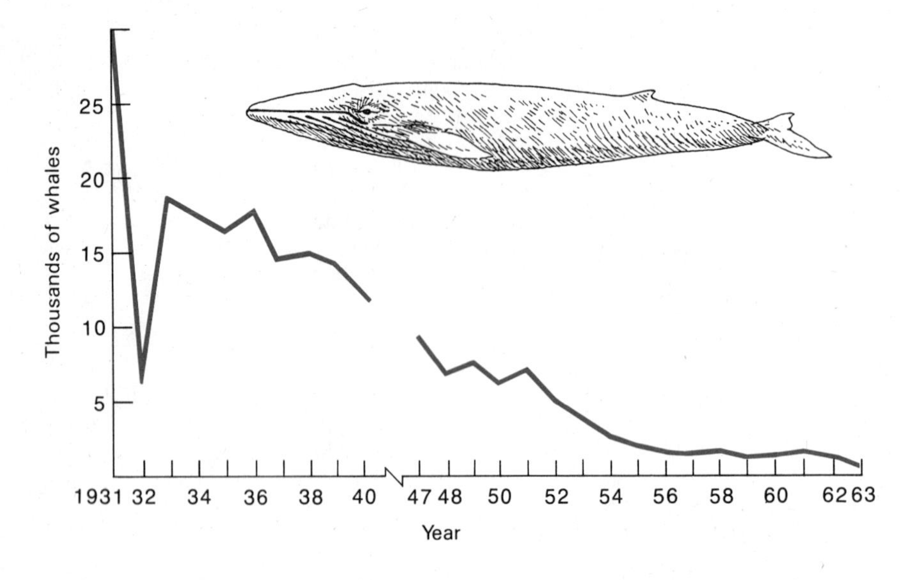

FIGURE 18.6 Annual catch of blue whales in the Antarctic, seasons 1930–1931 through 1962–1963 (war years omitted). Note the rapid decline in the population.

sport. Most game animals are characterized by contest competition and are largely, but not always, density regulated. Wildlife managers assume that regulated hunting take replaces natural mortality. If the surplus were not harvested, the animals would succumb to disease, exposure, predators, and the like. Thus for each individual removed by hunting, mortality is reduced by one. Supposedly population stability and sustained yield can be maintained if the population is harvested at the rate represented by the percentage of young of the year, or in some cases the percentage of year-old animals in the population.

This assumption is weak. If hunting mortality and natural mortality replace each other and if the rate of harvest applied to the population equals the rate of mortality in the absence of hunting, then the only cause of death in the post-young population must be hunting. Similarly, if the nonhunted population is below carrying capacity, no adult should die until K is reached. Obviously other mortality does take place among hunted populations (deer killed on highways, for example). Part of the hunting mortality adds to rather than replaces nonhunting mortality.

PROBLEMS WITH SUSTAINED YIELD

Although sustained yield management, particularly maximum sustained yield, looks good on paper, it depends too heavily on a simplistic equation. Maximum sustained yield views management as a numbers game. The number of biomass removed theoretically is replaced by an equal number of new recruits to the harvestable component. No proof of this theory exists. The usual approach to MSY fails to consider adequately: size and age classes, differential rates of growth among them, sex ratio, survival, reproduction, and environmental uncertainties, all data difficult to obtain. To add to the problem, the resource may be common property. One nation cannot control the policy of another in open water. To attempt MSY without full information is to balance on the edge of catastrophe (see Figure 18.4).

Several approaches to management of exploitable populations are in current use. One is the *fixed quota,* in which a certain percentage based on MSY estimates is removed each harvest period. Harvesting is supposed to match recruitment. Often used in fisheries, such an approach is risky because a fixed quota can easily drive a population to commercial if not actual extinction. Overharvest combined with environmental changes have been responsible for the demise of fisheries such as the Pacific sardine, Peruvian anchovy, and Atlantic halibut.

A second approach is *harvest effort,* often used in establishing seasons for sport hunting and fishing. The number of animals killed is manipulated by controlling

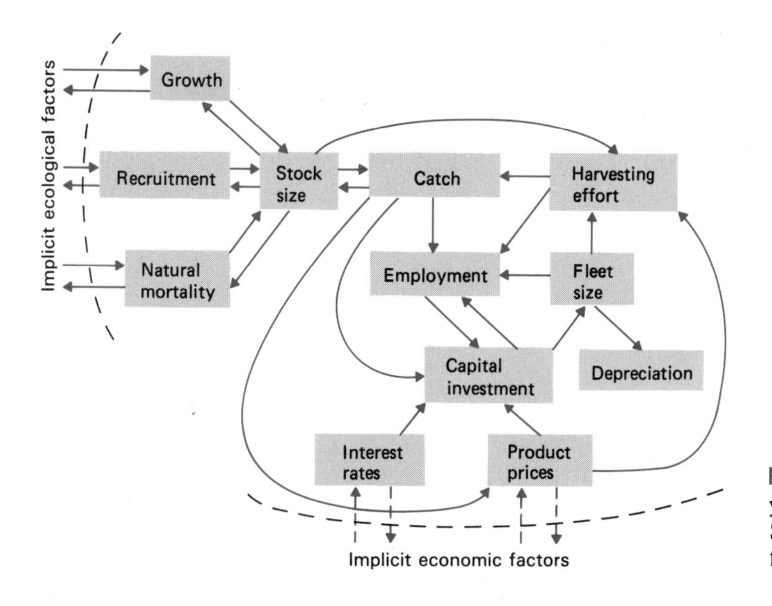

FIGURE 18.7 The relationship between sustained yield models and economic models of harvesting. Sustained yield models ignore powerful economic factors.

hunting effort: the number of hunters in the field, the number of days of hunting (season length), and the bag limit. This approach has been more successful than the fixed quota.

A third approach is the *dynamic pool model.* It assumes that natural mortality is concentrated in early life. It regulates type of equipment (such as size-selective gill-nets that sort out age classes), efficiency of equipment, and season of exploitation. Few dynamic pool models have been developed, let alone put into practice. A general weakness of the model is the inability to estimate natural mortality accurately.

All three models have a major flaw. They fail to incorporate the most important component of population exploitation, economics (Figure 18.7). Once exploitation of a natural resource becomes a commercial enterprise, the pressure is on to increase that exploitation to maintain the economic infrastructure built upon it. Attempts to reduce the rate of exploitation meet strong opposition. It is argued that reduction will result in unemployment and industrial bankruptcy—that in fact, the harvest effort should increase. This, of course, is a short-sighted argument, because in the long run the resource will be depleted and economic collapse will occur. That fact is written across the United States in abandoned fish-processing plants, rusting fishing fleets, and deserted logging towns. With conservative exploitation on lower economic and biological scales, the resource could still be exploited.

The history of the Great Lakes fishery, as exemplified in Lake Erie, illustrates the point well. Before the War of 1812, the lake, whose shores were lightly settled, held an abundance of whitefish, lake trout, blue pike, sauger, and herring. After the war, settlements increased rapidly and so did the exploitation of the fish. A subsistence fishery grew into a thriving industry by 1820. For the next 70 years rapid improvement in transportation, fishing boats, gear, and techniques and an expanding market increased the average rate of the catch 20 percent a year. By 1890 this rapid growth in catch decreased as stocks became depleted. However, increased intensity of fishing, further improvement of equipment, and heavy capital input maintained the size of the catch until the late 1950s (Figure 18.8).

The first fish to go was the lake sturgeon, netted and burned on the shore as a destroyer of fish nets. After the near destruction of the lake sturgeon, fishermen turned to lake trout and whitefish. Lack of regulation, incentives to increase production during World War I, and the introduction of the highly efficient gill net that was extremely wasteful in its catch increased fishing intensity. In 1950 the development of the nylon gill net, which could be left in the water a much longer period of time, resulted in the highest intensity fishing

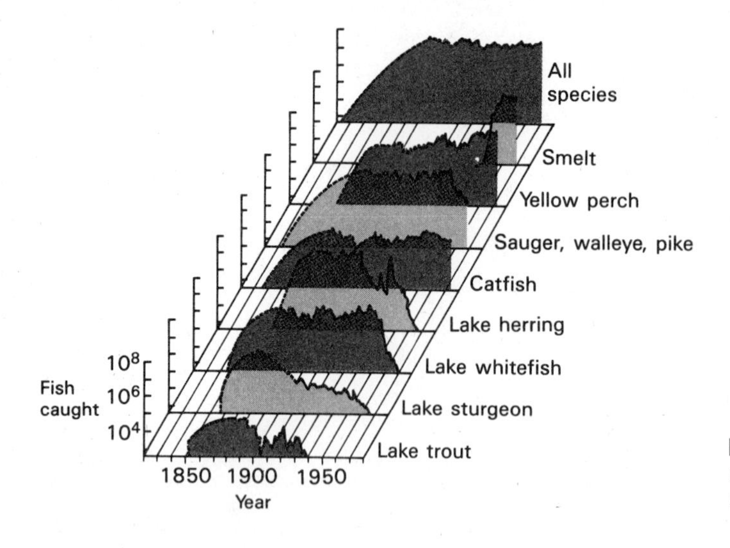

FIGURE 18.8 Annual catches of selected species, and all species combined, by the Lake Erie commercial fishery since 1820. The vertical scale is logarithmic.

ever in Lake Erie for walleye, blue pike, and yellow perch. By 1960 the stocks of walleye and blue pike had been commercially depleted.

Adding to the stress and depletion of stocks was pollution in the lake. Industrial, agricultural, and urban wastes, toxic pollutants and biocides, and runoff from shore developments eliminated or greatly reduced phytoplankton that supported fish life, and stimulated the growth of blue-green algae. Fallout of dead organic matter increased the oxygen demand in the bottom sediments by decomposer organisms. Oxygen-demanding benthic life, such as certain mayflies, gave way to species with low oxygen tolerance, such as bloodworms and chironomids.

An invasion of the rainbow smelt brought about a collapse of Lake Erie fisheries. Young smelt feed on plankton and crustaceans, the old smelt on small fish. In turn the smelt are prey for lake trout, blue pike, walleye, and sauger. With the decimation of those predatory species, smelt increased rapidly and became the predators of young remnant stocks of sauger, blue pike, walleye, and lake herring. Had the invasion of rainbow smelt been halted and had fishing been regulated, the preferred commercial species would still be in harvestable numbers.

The history of the Great Lakes fishery is a microcosm of the history of marine fisheries. Certain stocks of fish, such as Pacific sardine, Atlantic halibut, and Peruvian anchovy, have been exploited to commercial extinction, causing both ecological and economic damage. The sad plight of the whales is another example of decreased catches followed by increased hunting intensity, made possible by greater capital input and technological advances such as factory ships and fleets of hunters. In spite of warnings of overharvesting, the marketplace and short-term profits dictated the take.

Although some international regulations, including a quota system for commercial species of fish, are in effect today, compliance is strictly voluntary. The system does not work. The problem is that traditional fisheries management considers stocks of individual species as single biological units rather than components of a larger ecological system. Each stock is managed to bring in a maximum economic return, overlooking the need to leave behind a certain portion to continue its ecological role of predator or prey. This attitude encourages a tremendous discard problem, euphemistically called "bypass."

Bypass includes all species of fish and other marine life pulled up in the nets that have no economic value to the fisherman concerned. This bypass can include commercially valuable species. Not just nontarget fish are discarded. Taken in the nets in United States waters alone each year are nearly 1 million seabirds, more than 6,000 sea lions and seals, over 10,000 small whales, and thousands of endangered sea turtles. Overall the world's marine fisheries annually catch and throw overboard 12 to 20 billion tons of sea life, about 10 to 20 percent of their total catch of fish and shellfish. Few of the discards survive. Shrimp fisheries remove billions of tons of live fish—including at least 115 species—along with turtles. More than 21 pounds of fish are landed for every pound of shrimp. Up to 70 percent of discards from shrimp fishing are commercially recognized groundfish species, such as croaker, red snapper, and grouper.

The ecological effects of such discard can be enormous. Because much of the bypass consists of juvenile or undersized fish of commercial species, the practice can seriously affect the future of those fisheries. Because the biology and ecology of many species of fish are poorly understood, their removal or reduction can interfere with predator–prey interactions and the dynamics of interspecific competition, modifying ecosystem structure and function. Tons of dead carcasses fall to the bottom to enrich the benthic detrital-based food chains, enhancing nutrient release and energy turnover. Such enrichment could change benthic community structure and function.

Economics also distorts the management of some game species. In too many instances biologists have emphasized the increase of recreational opportunities for hunters rather than the welfare of the species involved. This is evident in the failure to reduce waterfowl bag limits when the population drops, in late-season hunting of grouse well after breeding males have established their spring territories, and in the reluctance to restrict moose hunting in Alaska. Hunting mortality is additive to predatory losses. Rather than restricting human exploitation of moose populations, we kill wolves to reduce natural predatory loss.

Reluctance to reduce seasons or to tighten bag limits relates in part to conservation funding. Most wildlife programs directly or indirectly depend on hunting license revenue. Reduction in such revenue lowers income to wildlife agencies. This loss has a two-edged effect: It may reduce pressure on some hunted species, but it also reduces money for wildlife habitat restoration and acquisition (which benefits all species, including endangered ones) and other programs essential to wildlife welfare. Conservation should be funded by other means.

Management of exploitable populations is too much crisis management. Rarely are steps taken to rescue a species until it is endangered. Then it is protected and expensive recovery programs are initiated. The simpler solution is to manage the population judiciously in the first place.

SUSTAINED YIELD IN FORESTRY

Forests have not fared much better than wildlife. They were destroyed rapidly over the centuries to clear land for agriculture and to supply building materials and firewood. Growing shortages of wood stimulated some form of forest management in Great Britain back in 1600s. Scientific forest management evolved in Europe in France, Germany, and Switzerland in the 1800s. The practices of forest management were introduced into North America by Gifford Pinchot in 1892 on the Biltmore Estate in North Carolina. The seemingly endless forests, however, did not encourage forest management until exploitative logging, land clearing, and fires had decimated forests east of the Mississippi. The same fate was happening to western forests until much of the acreage was set aside by the federal government as national forests and parks.

The goal of sustained yield in forestry is to achieve a balance between net growth and harvest. To accomplish this the mature or unexploited forest must be cut to stimulate regeneration, much in the same manner as an unexploited fish or wildlife population must be reduced to increase r. Management proceeds in a different time frame, because regenerating a forest requires decades. The time frame for the next harvest depends on the desired wood product. Pulp wood and posts and poles require only a short-term rotation of 30 to 40 years; sawlogs require 65 to 100 years. Sustained-yield forestry works best on large acreages where blocks of timber of different age classes can be maintained.

Sustained yield in forestry is much more sophisticated than in wildlife and fisheries for the simple reason that it is easier to inventory, measure growth and potential yield of biomass, and manipulate populations or stands of trees. To achieve this end, foresters have an array of silvicultural and harvesting techniques, from clearcutting to selection cutting, with many variations in between. In recent years foresters and forest ecologists have been giving considerable attention to nutrient cycling and nutrient conservation in managed forests.

The concept of sustained yield is weli ingrained in forestry and is practiced to some degree by large commercial timber companies and federal and state forestry agencies. But timber-cutting practices in some national forests, particularly in the Pacific Northwest and on the Tongras National Forest in Alaska, hardly qualify as sustained yield management when below-cost timber sales are mandated by the government to meet politically determined harvest quotas. Such cuttings are forest mining and not harvesting. Sustained yield management has hardly filtered down to smaller parcels of private forest land, in which poor cutting practices with little concern for regeneration and species composition has left behind forests of poor quality and understocked stands.

The problem of sustained-yield forestry, like that of sustained-yield fisheries, is its strong economic focus on the resource with little concern for the overall ecosystem. Once an old-growth forest is cut, that particular ecosystem will not return. A carefully managed stand of trees, often reduced to one or several species, is not a forest in an ecological sense. Rarely will regenerated forest ever reach an old-growth stage again, because by the time the trees reach economic or financial maturity—based on the type of rotation—they are harvested again. Economic maturity is not the same as ecological maturity.

RESTORING POPULATIONS

When species arrive at the brink of extinction, we undertake valiant efforts to bring them back, when we should have never allowed them to arrive at that condition in the first place. Since North America (and Africa and Australia as well) was first settled by Europeans, its wildlife has been decimated by market hunting, wanton killing, and habitat destruction. Species such as the heath hen, passenger pigeon, and Carolina parakeet became extinct. In the early 1900s, a number of individuals made strenuous efforts to halt the destruction of wildlife. One effort was the passage of the Lacey Act of 1900, which prohibited interstate transportation of wildlife, dead or alive. Another was the Migratory Bird Act of 1913 that gave international protection to migratory birds and ended market hunting for waterfowl. By the 1930s, when the United States was suffering from drought and severe soil erosion, resulting in the Dust Bowl, wildlife was still in serious trouble. Then wildlife restoration efforts began in earnest with the passage of the Pittman-Robertson or Federal Aid in Wildlife Restoration Act, financed by an excise tax on shooting equipment, which supported wildlife research and habitat restoration.

EARLY RESTORATIONS

Comeback of such species as the white-tailed deer, pronghorn antelope, and wild turkey was achieved by a series of actions that allowed low populations to expand. One was strict protection from hunting followed by highly regulated hunting seasons once the populations were well on the way to recovery. States and the federal government set aside refuges and reserves to protect both animals and habitat. They reintroduced wild individuals taken from pockets of abundance to areas of scarcity and empty habitats. What made early restoration efforts most successful, however, was the availability of large areas of empty habitat. During the Depression of the 1930s farmland and rangeland were abandoned and allowed to revert to natural vegetation. Devastated forest lands were growing back, providing outstanding food and cover for white-tailed deer and other species. So successful were restoration efforts that some of the species are approaching a pest status. White-tailed deer are probably more abundant now than at the time of settlement. They are so numerous that they are killed on highways, invade suburban gardens, and destroy agricultural crops.

The wild turkey is a good example of restoration of a species from the brink of extinction (Figure 18.9). Originally the turkey's range included all or parts of the 39 states and extended into Ontario, Mexico, and Central America. By the mid-1800s the species had been eliminated from the northeastern United States and by 1900 from the midwestern states. In 1949 only small populations of eastern wild turkey survived on

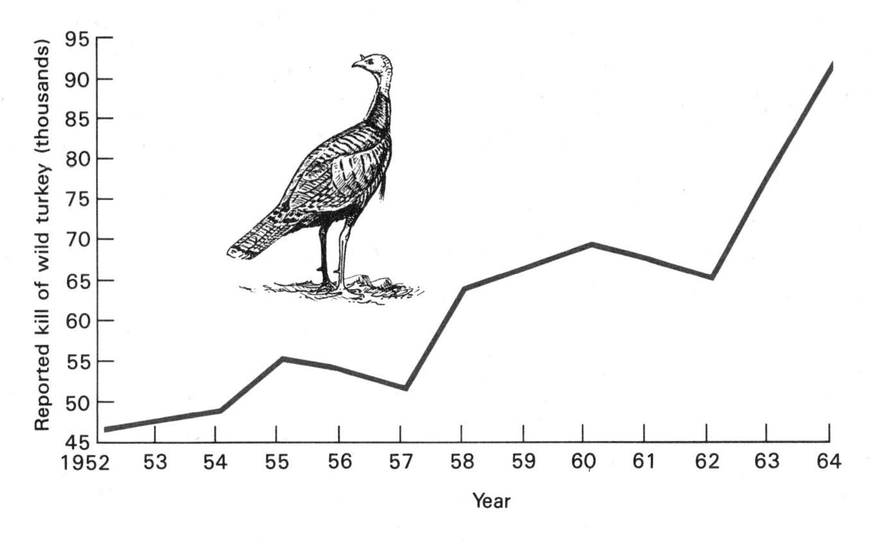

FIGURE 18.9 Growth of the wild turkey population in the United States following restoration through 1964, as derived from hunter harvest data. Note the sharp increase in growth after 1962.

about 12 percent of their ancestral range, most in remote areas of the Appalachian Mountains. Intensive studies of the biology and ecology of the species, financed by P-R money, provided the informational base for restoration efforts. In the western United States restoration efforts involved transplanting small groups of live-trapped wild turkeys from areas of abundance, mostly New Mexico, to suitable habitat in other states, with outstanding results. In South Dakota, for example, the original 29 birds reintroduced in 1948–1951 had increased to 7000 birds by 1960.

Capturing wild turkeys in the West was easily accomplished by baited walk-in traps, but they did not work well in the eastern forest. Some states, notably Missouri and Pennsylvania, resorted to game farm turkeys produced by mating domestic turkey hens with wild gobblers. Such crosses were so maladapted genetically—nesting too early and lacking wildness—that the releases were failures and, worse, hybridized with wild stock and introduced the fatal blackhead disease. At last the cannon-projected net trap (Figure 18.10), intended primarily for capturing waterfowl, proved highly successful in capturing social groups of wild turkeys. Large-scale restoration efforts were aided by changes in forest management that improved turkey habitat, greater public support, continued intensive studies of the bird, and the wild turkey's unforeseen ability to adapt to habitats previously thought unsuitable. Today the wild turkey population is about three million birds, and like the white-tailed deer, flocks have invaded and settled in wooded suburban areas. Successful restoration, then, involved protection including strict law enforcement, public concern and cooperation, adequate financing, scientific studies of the bird's biology, ecology, and behavior, protected reserves in which the population could recover, large areas of suitable habitat, and the adaptability of the species.

Today innumerable species worldwide are facing extinction, and the problem of their restoration is much different and more difficult. Today species are declining from numerous reasons not faced earlier: toxic chemicals in the environment, increasing human populations demanding more of Earth's resources and space, illegal traffic in wildlife, and rapid loss and fragmentation of habitat (Chapter 19) by land development. These fragments either hold small isolated populations of the species or are too small for occupancy. Small isolated populations are subject to many problems and dangers. Concentrated in small pockets, many species are more easily reduced by poaching or predation. Isolated from others of the same species, the populations may be too small for successful mating and reproduction and become vulnerable to inbreeding and loss of heterozygosity (Chapter 14). If the population does expand, the surplus has no available habitat to colonize, a situation the opposite of early restoration efforts.

FIGURE 18.10 A cannon net in action capturing a group of wild turkeys at a bait site. This group was used as transplant stock for an empty habitat.

Elephants restricted to parks and reserves in Asia and Africa and the bison of Yellowstone National Park overpopulate the available habitat and move outside park boundaries, where they come into conflict with human interests.

CONSERVING THREATENED SPECIES TODAY

It is no longer possible for most species to exist as they did in presettlement days or even in the recent past. Their survival depends on management at varying levels of intensity to maintain their populations and protect their habitats. We have a wide range of management options.

Habitat Restoration

Loss of habitat is a major cause of a species' decline. As habitat shrinks or is broken up, the population is compressed into smaller and smaller units.

Habitat restoration involves planting, protection, and maintenance. Eroded lands can be restored to grassland, cutover forest land replanted to trees, and drained wetlands refilled and replanted with aquatic vegetation. Some species require early ephemeral stages of succession (see Chapter 20), which must be maintained by such techniques as cutting and burning. The endangered Kirtland warbler (*Dendroica kirtlandii*), for example, requires early-stage stands of jack pine. Such stands can be maintained only by periodic burning of blocks of overaged pine to stimulate regeneration of young stands. In such cases the management objective is to keep blocks of vegetation in various stages of succession to ensure continuance of acceptable habitat.

Habitat Protection

Small populations cannot increase unless available habitat is protected as reserves against development, agricultural land clearing, and logging. Some of our most endangered species occupy natural reserves, many of which are too small to support a viable population.

One approach is to increase the size of the reserve to provide habitat for an expanding population or to embrace the entire ecosystem. Most reserves do not. For example, the Yellowstone Park is only part of the

Yellowstone ecosystem that includes the surrounding national forests subject to logging and grazing. The bison herd in the park has expanded beyond the capacity of the park itself to support it, especially in winter. Part of the bison problem could be solved if nearby national forest lands open to grazing by private domestic livestock were allotted to buffalo instead.

The success of many reserves is threatened by human activities just outside park boundaries: logging, grazing, agriculture, housing, and recreational development. To protect the reserves, concentric buffer areas of land use should encircle the core area. Land use should be minimal about the boundary and increase with distance.

Many otherwise excellent reserves and parks are compromised because of excessive human activity within park boundaries. Intensive recreational development, trails, and human incursions into the back country that bring people in direct contact with large species of wildlife decrease the value of the reserve and further endanger the species. The grizzly bear in our national parks is a good example. A certain segment of society would prefer to see the grizzly bear exterminated to make back country travel safe for humans.

Large reserves provide the greatest protection and support the largest population, although even the largest reserves may be too small for certain large carnivores and herbivores. Associated with large reserves should be smaller reserves. Although isolated and surrounded by other types of land use, small reserves do hold populations of species that represent different samples of the gene pool. Subject to somewhat different selection pressures and to some degree of genetic drift, these smaller populations help maintain genetic diversity in the species population.

Population integrity and stability can be enhanced among fragmented populations by corridors connecting one habitat patch with another. On a small scale, hedgerows between woodlands act as corridors. On a much larger scale, greenbelts in cities and suburbs, plantings along major highways, and vegetation along rivers and streams provide not only travel corridors for wild animals but also dispersal routes for woody plants. Of particular value are corridors of woody vegetation along rivers and streams, because they provide water, food, and cover for a great diversity of species.

Population Control

If restoration is successful, the species will reach a point where its population exceeds the ability of the habitat to support it. Overpopulation results in deterioration of habitat and a scarcity of food, impairing the ability of habitat to support the population and destroying or damaging habitat for associated species. As the habitat is damaged beyond the ability to support a viable population, the species declines.

Prevention of such damage may require cropping, the trapping and removal of the excess individuals. These surplus individuals provide transplant stock for depleted habitats. For example, the highly productive white rhino (*Diceros simus*) population in the Umfolozi Game Reserve in South Africa has supplied a number of animals for reintroduction into other reserves.

Cropping requires some knowledge of the interrelations and function of the species in the system. Are apparent overpopulation problems simply a part of a natural plant–herbivore cycle, or does the situation require human intervention? For example, elephants, which feed on woody browse, are compressed into parks and reserves where they are converting tree savannas to grassland that cannot provide sufficient food for them. The outcome is starvation of elephants. When the elephant population declines, the vegetation slowly returns. The question is whether to allow such a cycle to proceed or to step in, reduce the elephant herd, and try to achieve some stability between elephants and trees. It is difficult to answer because of the longevity of elephants and the time required for trees to regenerate. The decision affects not only the elephants but the entire ecosystem.

Reintroductions

If the habitat within the natural range of the species is available but the species itself is locally extinct, then the habitat can be recolonized by introducing a core population. Animals or plants for reintroduction may be trapped from wild populations, as was the wild turkey, or captive-bred. Critical to introduction is the size of the initial or founding population (see Chapter 14), which must be large enough to be a cohesive social or reproductive unit.

Introduction of individuals from captive-bred populations to the wild requires special attention and effort. Animals need prerelease and postrelease conditioning in the acquisition of food, finding of shelter, interaction with conspecifics, and fear and avoidance of humans. Numerous attempts have met with failure. Some successful reintroductions include the whooping crane (*Grus americana*), masked bobwhite (*Colinus virginiana*), and Hawaiian goose or nene (*Branta sandvicensis*) among the birds and the American bison and its relative the European wisent (*Bison bonasus*) among the mammals. Recent efforts involve the peregrine falcon (*Falco peregrinus*), the golden lion tamarin (*Leontopithecus rosalia*), and the Arabian oryx (*Oryx leucoryx*). The outcomes of these reintroductions are inconclusive. Problems with reintroductions of captive-bred individuals include high costs, logistical difficulties, and shortage of habitat, and the questionable ability of the individuals to adapt to the wild.

Both wild and captive-bred individuals may be translocated to build up the numbers of individuals already present and to introduce new genetic material into populations. Such translocations must be done carefully. Not only is there danger of introducing disease, but the new individuals must be integrated into the social and breeding structure of the native population to achieve desired success.

Just as important is checking the genetic backgrounds of the transplanted animals to ensure they are adapted to their new environment. If not, these individuals can weaken the resident stock. Attempts to increase depleted populations of the northern bobwhite quail in Pennsylvania and New England with birds of southern origin resulted in the death of both the introduced birds and their "hybrid" offspring during cold winters to which they were not adapted. The northern populations went with them.

Captive Breeding

It is an unfortunate commentary on human treatment of Earth that the only hope for many species of wild things is safety in captivity. Only a small number of a few endangered species can be reared in captivity, but for some species it is the only way they can increase initially (Figure 18.11). Captive propagation, however, has its problems, including small population size, potential inbreeding, incompatibility of captive individuals for mating, and lack of social interaction. Captive propagation programs cannot be carried out indefinitely. After a number of generations, depending on population size, the captive stock will begin to show signs of inbreeding depression and domestication, so it is important to consider opportunities of reintroduction as soon as feasible.

FIGURE 18.11 A captive breeding program is the only means by which some highly endangered species, such as the Sumatran rhino, can be saved. This mother and young, the first to be born in captivity, are part of a captive breeding program in Malaysia.

Protection and Education

Once any species is compressed into an island of habitat, it becomes more vulnerable to poaching and predation. Waterfowl forced to nest in small wetlands surrounded by agricultural land are easy marks for such nest predators as skunks, raccoons, and opossums, all highly adaptable to an agricultural landscape. Under such conditions protection from predators may be necessary. Larger mammals such as bear, bighorn sheep, rhino, and elephant concentrated in a limited area become easy marks for poachers and unscrupulous trophy hunters. Such activity may support the local economy.

Reintroductions need the long-term protection of governments, and especially of nearby communities. In most places it is necessary to demonstrate tangible economic benefits to be derived locally from parks and reserves and protected wildlife, so local people collaborate with the program. Protection must be accompanied by education about the species and its habitat, from international to local levels. Education is absolutely necessary to gain support at the local level. Education at the national and international levels should be aimed at alerting and informing consumers not to buy endangered and threatened species as pets, or ornamental plants and products made from endangered and threatened species. Without such support reintroduction and conservation programs could fail.

Ultimately, the fate of wild creatures and the ecosystems of which they are a part rests with controlling the explosive growth of human populations. As the number of dominant mammal on Earth increases, diversity of life decreases. Eventually we will be left with only those plants and animals that are tolerant of or share the human-dominated landscape. The outstanding ecosystems—the tropical forests, old-growth forest ecosystems, savannas and grasslands, even marine environments and their unique forms of plants and animals—will exist only as memories preserved in books and films.

Summary

Humans interact in three ways with other natural populations: reducing pest populations, maintaining stability and productivity of exploited populations, and increasing or at least maintaining populations of threatened and endangered species. The aim of reducing pests, defined as undesirable animals, and weeds, defined as plants growing anywhere not desired, is to hold the populations below a level at which they cause economic injury. There are a number of control methods. One is the use of organic chemical pesticides and herbicides. Their use involves serious risks of evolution of resistant strains of pests and weeds and of toxic environmental pollution. Biological controls based on natural predators and parasites are effective in keeping some pests below the economic injury level. Loss of natural enemies through the use of chemical pesticides allows a pest population to increase dramatically. Breeding for genetic resistance to pests and weeds is another important and effective tool. The latest is the use of transplanted genes to produce plants that are resistant to viral diseases and insect attack. Such genetic engineering, however, involves a risk that the gene may be transferred to wild, weedy relatives that could create highly resistant strains of weeds and pests. In some situations mechanical controls such as trapping pests and cultivating the ground to kill weeds are effective. A cultural approach creates habitat diversity or a patchy environment, which reduces availability of food and makes dispersal of pests more difficult. The latest approach is integrated pest management. More a philosophy than a specific strategy, it combines in appropriate ways chemical, biological, genetic, mechanical, and cultural control measures. Based on careful monitoring, it weighs environmental and social benefits and costs.

If plant and animal populations exploited for human use are to be continuously productive, they must be managed for sustained yield: The yield per unit time should equal production per unit time. Such an approach to management necessarily differs between K-selected and r-selected species. Based on the logistic equation, sustained yield models fail to take into consideration all aspects of population dynamics, including natural mortality and environmental uncertainty. Sustained yield management considers the resource as a single biological unit, which is managed for maximum economic return, rather than as a component of a larger ecological system. Economic considerations too often outweigh biological considerations, resulting in the eventual commercial and even biological extinction of species. Economic interference in sustained yield also encourages a wastage of marine life, from juvenile fish to dolphins caught in large nets along with commercial species. Euphemistically called bypass, this enormous tonnage of wasted marine life results in ecological disturbance to marine ecosystems and to the overall fishery resource. This emphasis on single species management extends to wildlife management when one species such as moose is favored at the expense of another, often a predatory species.

Species experiencing declining populations require strong efforts at restoration if they are not to become locally or globally extinct. Because the loss of habitat is the major cause of the decline of threatened and endangered species, a major restoration effort involves the protection, restoration, and management of habitats. Habitats of wild creatures are being fragmented and isolated, resulting in small isolated populations. These populations may be too small to remain viable. If they do reproduce successfully, their numbers may exceed the ability of the habitat fragment to support them. In such situations we can use the surplus to recolonize empty habitats. The future of threatened populations depends on the establishment of protected parks and reserves. Surrounded by human land developments, small reserves should be connected by corridors that allow movement and dispersal of individuals among habitat patches. Otherwise we may have to transfer individuals among reserves to reduce inbreeding and genetic drift. In extreme cases of highly endangered species, individuals may have to be captured from the wild and held in captive breeding programs to save the species from extinction. All attempts at restoration and protection of threatened species must be accompanied by protection from poaching and by an educational program to gain support.

Review and Study Questions

1. What three major interactions do we have with natural populations?
2. Define a pest and a weed. Why are the two terms value-laden?
3. Under what conditions do plants and animals become weeds and pests?
4. What are the major control measures used on pests and weeds? What are the ecological advantages and disadvantages of each?
5. What are the major synthetic organic pesticides and what are their environmental impacts?
6. What are the characteristics of an effective biological control agent?
7. Why are most mechanical approaches to pest control ineffective?
8. What are the ecological dangers of plants genetically engineered to resist pests?
9. How can a patchy environment reduce pest outbreaks?
10. What is integrated pest management? How can it channel public input into a local or regional pest control program?
11. Explain sustained yield. What is the difference between maximum sustained yield and optimal sustained yield?

*12. Report on the effects of modern commercial ocean fishing, especially with huge drift nets, on the fishery resource and on other ocean life. Can the ocean withstand such intensive exploitation?

*13. What indirect effect do you have on ocean life when you enjoy an "all you can eat" shrimp dinner?

*14. Why is there obvious unconcern about the destructive harvesting of the ocean when we show strong concern about destruction of tropical forests?

15. How do economic interests block sustained yield practices? Apply your answer to commercial fishing and cutting of old-growth forests in the Pacific northwest.

*16. Comment on the statement that a stand of trees is not necessarily a forest.

17. What approaches must be taken to conserve and restore threatened species? Why is habitat a key element?

*18. The management and recovery of endangered species are often fraught with controversy. Report on the capture of all remaining California condors to undertake a captive breeding program, or on approaches to restoring the black-footed ferret and the golden lion tamarin.

Selected References

Baden, J. A., and D. Leal. 1990. *The Yellowstone primer.* San Francisco: Pacific Institute for Public Policy. An excellent review of the problems of Yellowstone Park, a microcosm of problems facing all parks and reserves.

Bricklemyer, E. C., Jr., S. Ludicello, and H. J. Hartmann. 1989. Discarded catch in U.S. commercial marine fisheries. In *Audubon Wildlife Report 1989/1990.* San Diego, CA: Academic Press, pp. 259–295. Excellent review of current fishery regulations and the discard problem.

Caughley, G. 1976. Wildlife management and the dynamics of ungulate populations. *Applied biology* 1:183–246. Good introduction to sustained yield in wildlife management.

Debach, P. 1974. *Biological control by natural enemies.* London: Cambridge University Press. Somewhat dated but an excellent introduction.

DiSilvestro, R. L. 1989. *The endangered kingdom: The struggle to save America's wildlife.* New York: Wiley. Case histories of restoration and endangerment of selected wildlife species, as well as an assessment of wildlife management.

Dover, M. J., and B. A. Croft. 1986. Pesticide resistance and public policy. *Bioscience* 36:78–85. Discusses pesticide resistance and effective pest control.

Ellstrand, N. C., and C. A. Hoffman. Hybridization as an avenue of escape for engineered genes. *Bioscience* 40:438–442.

Fitzgerald, S. 1989. *International wildlife trade: Whose business is it?* Washington, DC: World Wildlife Fund. Why endangered species need protection; a review of illegal trade in wildlife species.

Gambell, R. 1976. Population biology and the management of whales. *Applied biology* 1:237–343. A short history of whaling, population dynamics of species, and failure of attempts to apply sustained yield management.

Hall, C. A. S., C. J. Cleveland, and R. Kaufman. 1986. *Energy and resources quality: The ecology of the economic process.* Economics of exploitation.

Harris, L. D. 1984. *The fragmented forest.* Chicago: University of Chicago Press. Integrates island biogeography and forest management in the preservation of biotic diversity.

Hoffman, C. A. 1990. Ecological risks of genetic engineering of crop plants. *Bioscience* 40:434–436.

Huffaker, C. B., and R. L. Rabb, eds. 1984. *Ecological entomology.* New York: Wiley. Excellent chapters on evolutionary process in insects, natural control of insects, and application of ecology to insect population management.

Kleiman, D. G. 1989. Reintroduction of captive mammals for conservation. *Bioscience* 39:152–161. Authoritative discussion of guidelines for introducing endangered species into the wild.

Regier, H. A., and G. L. Baskerville. 1986. Sustainable development of regional ecosystems degraded by exploitive development. In W. E. Clark and R. E. Munn (eds.), *Sustainable development of the bisophere.* Cambridge, England: Cambridge University Press, pp. 74–103. Focuses on redevelopment of degraded forestry and fishery resource systems.

Regier, H. A., and W. L. Hartman. 1973. Lake Erie's fish community: 150 years of cultural stress. *Science* 180:1248–1255. History of Lake Erie's fisheries.

Small, G. 1976. *The blue whale.* New York: Columbia University Press. The authoritative, depressing history of the blue whale.

Soulé, M. E. (ed.) 1986. *Conservation biology: The science of scarcity and diversity.* Considers various aspects of conserving endangered populations.

Soulé, M. E., and B. A. Wilcox, eds. 1980. *Conservation biology: An evolutionary-ecological perspective.* Sunderland, MA: Sinauer Associates. Excellent overview of the problems of conserving small populations, including captive propagation.

Thatcher, R. C., J. L. Searcy, J. E. Coster, and G. D. Hertel, eds. 1986. *The southern pine beetle.* U.S.D.A. Forest Service Science and Education Tech. Bull. 1631. Washington, DC: U.S. Department of Agriculture. A colorful, accessible explanation of all aspects of IPM for the southern pine beetle.

Trefethen, J. B. 1975. *An American crusade for wildlife.* New York: Winchester Press. An excellent history of restoration of wildlife and the politics involved.

Walters, C. 1986. *Adaptive management of renewable resources.* New York: Macmillan. An advanced text, but presents challenging approaches to managing renewable resources.

PART V
The Community

INTRODUCTION

The Community Defined

Plants and animals do not live alone as separate entities. They share the same environments and habitats, and interact with one another in various ways. This collection of plant and animal populations interacting directly or indirectly is a *community.* Such a definition embraces the idea of the community in its broadest sense.

Other definitions of a community, more restrictive in their meaning, are commonly used. Zoologists may apply the term community to a restricted assemblage of species, such as a bird community or a mammal community of a particular forest or grassland. Botanists use the term *association* for a plant community possessing a definitive floristic composition. Ecologists may recognize and contrast communities as *heterotrophic* and *autotrophic.* Forests and grasslands are examples of autotrophic communities, which require only the energy of the sun. Assemblages of organisms that inhabit such microhabitats as a fallen log, a tiny pool of water in a tree hollow, or a cave are examples of heterotrophic communities, which are dependent on the autotrophic community for their energy source.

A community necessarily implies some form of organization. How communities are organized is a subject of considerable research; we may never learn exactly how. The intensity and degree of interaction among the species involve herbivory, predation, interspecific competition, parasitism, and mutualism. These interactions can determine the presence and abundance of species. One way of looking at the community is to group interacting species according to functional relationships; feeding relations as reflected in patterns of eating and being eaten; the food web; and guilds, groups of species that share a common resource. Another way of characterizing a community is to examine the way these interrelationships are reflected in species distribution and species diversity.

The community is also characterized by attributes relating to patterns in space and time. All communities exhibit some form of spatial arrangement both vertically and horizontally, and are influenced by the growth form and structure of vegetation, from trees to mosses. Plant structure influences the spatial distribution of associated animal life. Spatial distribution of species is influenced by variations in environmental conditions such as soil, moisture, light, and nutrients and by relationships among the species. For this reason species composition or membership varies greatly among communities. Variations in community composition over time involve diurnal and seasonal changes and, over a much longer period of time, successional changes.

Ecologists find it difficult to draw a sharp line between population ecology and community ecology, because so many aspects of the community relate directly to population interactions. They also find it difficult to draw a line between the community and the ecosystem, because the community is the biotic compartment of the ecosystem. These observations emphasize the interrelatedness of all parts of the biosphere.

CHAPTER 19

Community Organization and Structure: Spatial Patterns

Outline

Objectives

On completion of this chapter, you should be able to:

1. Explain how species dominance influences community structure.
2. Discuss the concept of species diversity.
3. Explain how vertical structure and horizontal patterns of vegetation influence the community.
4. Distinguish between edge and ecotone and discuss the edge effect.
5. Explain the theory of island biogeography.
6. Relate island biogeography theory to habitat fragmentation, species equilibrium, and species management.

TWO VIEWS OF COMMUNITY

The nature of the community has been the object of study and dispute for years. Is a community such as an oak-hickory forest a real entity that is definable, describable, and constant from one stand of oak-hickory to another? Or is it an abstraction, different populations that we group together because they have similar environmental requirements? Such questions are still being debated.

The composition of any one community is determined in part by the species that happen to be distributed on the area and can survive its environmental conditions. Seeds of many plants may be carried in by wind and animals, but only those that are adapted to grow in the habitat where they are deposited and are capable of overcoming competition of species already present will take root and thrive. One adapted species may colonize an area and prevent others equally adapted from entering. Wind direction and velocity, size of the seed crop, disease, and insect and rodent damage all influence the establishment of vegetation. The exact species that settle an area and the number of individual species that succeed seldom if ever are repeated in any two places or times. The element of chance is heavily involved. Nevertheless, there is a certain pattern, with more or less similar groups recurring from place to place. Only a small group of species is potentially dominant because a limited number are well adapted to the general climate and soils of the region they occupy.

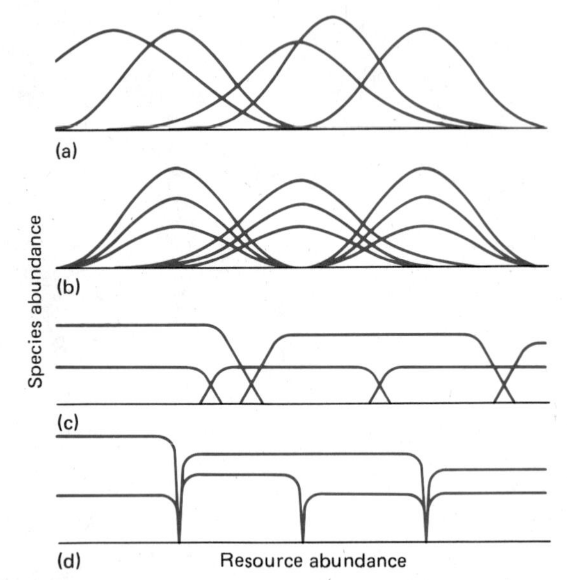

FIGURE 19.1 Four models of species distribution along environmental gradients. (a) The abundance of one species on an environmental gradient is independent of the others. Thus, the association of several species along the gradient changes with the response of the individual species to that gradient. (b) The abundance of one species is associated with that of another. The two or more species are always found in association with each other. (c) The distribution of one species is independent of another on an environmental gradient, but the abundance and distribution of each species are sharply restricted at some point on the gradient by interspecific competition. (d) The distribution of species is sharply restricted by a change in some environmental variable.

Two opposite views of natural communities exist. One regards communities as distinct natural units or *associations*. The distribution and abundance of a species in a community are determined by its interaction with other species in the same community. Species making up the community typically fall into discrete groups. Groups of stands similar to one another form associations. Stands of one association are clearly distinct from stands of other associations.

For practical reasons of study and description, the idea of distinct, definable communities has advantages. However, general observations confirm that species comprising a community do not associate exclusively with one another. Rather, each species appears to be distributed in its own way, according to its own response to varying environmental conditions. Some organisms will succeed only in certain environmental situations and tend to be confined to certain habitats. They have a restricted distribution over an environmental gradient (Figure 19.1). Others are more tolerant and occupy a wider distribution on an environmental gradient such as moisture, temperature, soil, slope position, and the like. This sequence of communities showing a gradual change in composition is called a *continuum*. A community on such a gradient can be described as a discrete area in the continuum. Each community is somewhat different from its neigh-

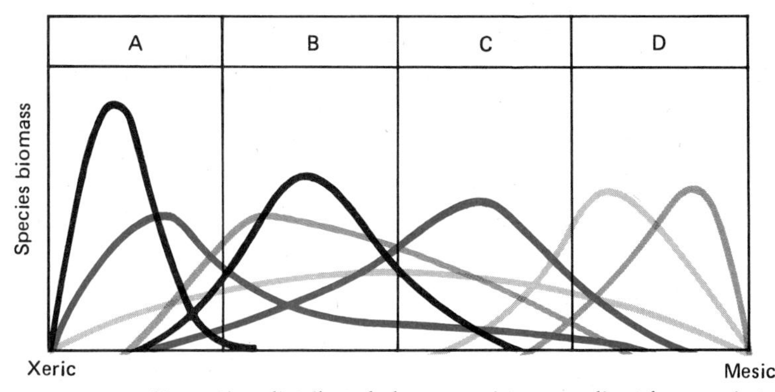

FIGURE 19.2 Vegetation distributed along a moisture gradient from xeric to mesic. Each species responds in its own individual way to moisture; yet overlap in response allows a number of species to associate with each other. The nature of the community, its dominants and associated species, depends on the point at which community boundaries are placed. Each demarked community here is characterized by its own dominants, although some species are shared with other communities. The communities at either end of the gradient are distinct, although they share one ubiquitous species. Shifting community boundaries would result in changes in community composition and dominants.

bor, the difference increasing roughly as the distance between them increases.

The gradient approach emphasizes the species rather than the community itself as the essential unit in the analysis of interrelationships and distribution. Species respond independently to the biotic environment according to their own genetic characteristics. They are not bound into groups of associates that must appear together. Instead, when species populations are plotted along an environmental gradient, bell-shaped curves overlap in a heterogeneous fashion (Figure 19.2). In this view the community is regarded as a collection of populations of species existing under similar environmental conditions.

GROWTH FORMS AND LIFE FORMS

The form and structure of terrestrial communities can be characterized by the nature of the vegetation. The

TABLE 19.1 Raunkiaer's Life Forms

Phanerophytes (Gr. *phaneros,* "visible"). Perennial buds carried well up in the air and exposed to varying climatic conditions. Trees and shrubs over 25 cm; typical of moist, warm environments.

Chamaephytes (Gr. *chamai,* "on the ground"). Perennial shoots or buds on the surface of the ground to about 25 cm above the surface. Buds receive protection from fallen leaves and snow cover. Plants typical of cool, dry climates.

Hemicryptophytes (Gr. *krypos,* "hidden"). Perennial buds at the surface of the ground, where they are protected by soil and leaves. Many plants are characterized by rosette leaves and are characteristic of cold, moist climates.

Cryptophytes. Perennial buds buried in the ground on a bulb or rhizome, where they are protected from freezing and drying. Plants are typical of cold, moist climates.

Therophytes (Gr. *theros,* "summer"). Annuals, with complete life cycle from seed to seed in one season. Plants survive unfavorable periods as seeds and are typical of deserts and grasslands.

Epiphytes. Plants growing on other plants; roots up in the air.

plants may be tall or short, evergreen or deciduous, herbaceous or woody. Such characteristics can be used to describe growth forms. Thus, one might speak of shrubs, trees, and herbs and further subdivide the categories into needle-leaf evergreens, broadleaf evergreens, evergreen sclerophylls (small, tough, evergreen leaves, as in chamise), broadleaf deciduous, thorn trees and shrubs, dwarf shrubs, ferns, grasses, forbs, and lichens.

Perhaps a more useful system is the one designed in 1903 by the Danish botanist Christen Raunkiaer. Instead of considering the plants' growth form, he classified plant life by the relation of the embryonic or meristemic tissues that remain inactive over winter or a dry period (perennating tissue) to their height aboveground. Such perennating tissue includes buds, bulbs, tubers, roots, and seeds. Raunkiaer recognized five principal life forms, which are summarized in Table 19.1 and Figure 19.3. All the species in a region or community can be grouped into these five classes and the ratio among them expressed as a percentage, providing a life form spectrum of the area that reflects the plants' adaptations to the environment, particularly climate (see Table 19.2 and Figure 19.4). A community with a high percentage of perennating tissue well aboveground (phanerophytes) would be characteristic of warm climates. A community with most of its plants chamaephytes and hemicryptophytes would be characteristic of cold climates, and a community dominated by therophytes would be characteristic of deserts.

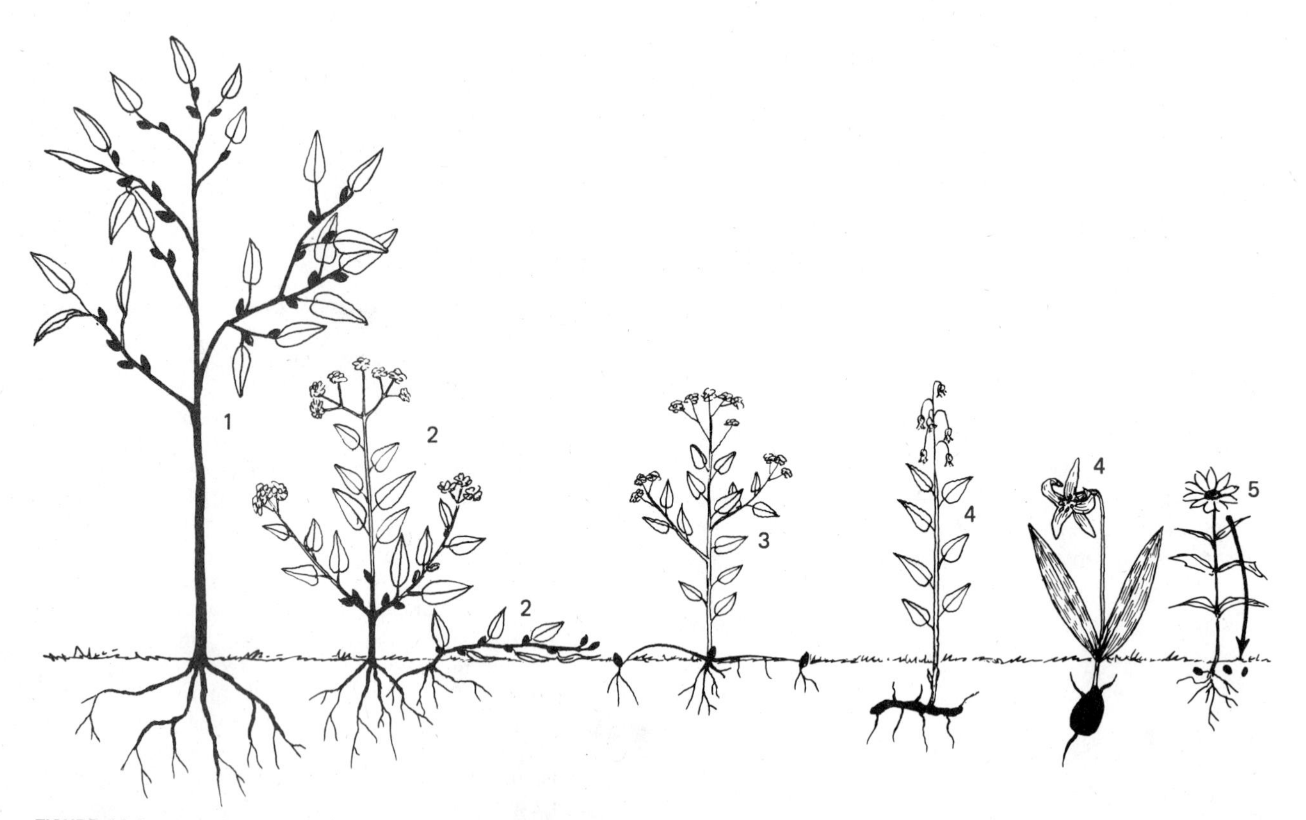

FIGURE 19.3 Raunkiaer's life form. (1) phanerophytes; (2) chamaephytes; (3) hemicryptophytes; (4) geophytes (cryptophytes); (5) therophytes. The parts of the plant that die back are unshaded; the persistent parts with buds (or seeds in the case of therophytes) are dark.

TABLE 19.2 Life Form Spectra of Major Ecosystems

Community	*Ph*	*Ch*	*He*	*Cr*	*Th*
Arctic tundra	1	23	61	13	2
Temperate deciduous forest	15	2	49	13	12
Subtropical forest	34	23	10	5	15
Rain forest	54	6	12	3	16
Desert	26	7	18	7	42

SPECIES DOMINANCE

In a general way, the nature of communities is controlled either by physical or abiotic conditions such as substrate, the lack of moisture, and wave action or by some biological mechanism. Biologically controlled communities are often influenced by a single species or by a group of species that modify the environment. These organisms are called *dominants.*

It is not easy to describe a dominant or to determine the dominant species (see Box 19.1). The dominants in a community may be the most numerous, possess the highest biomass, preempt the most space, make the largest contribution to energy flow or mineral cycling, or by some other means control or influence the rest of the community.

Some ecologists have given the dominant role to those organisms that are numerically superior, but abundance alone is not sufficient. A species of plant, for example, can be widely distributed over the area and yet exert little influence on the community as a whole. In a forest the small or understory trees can be numerically superior; yet the community is controlled

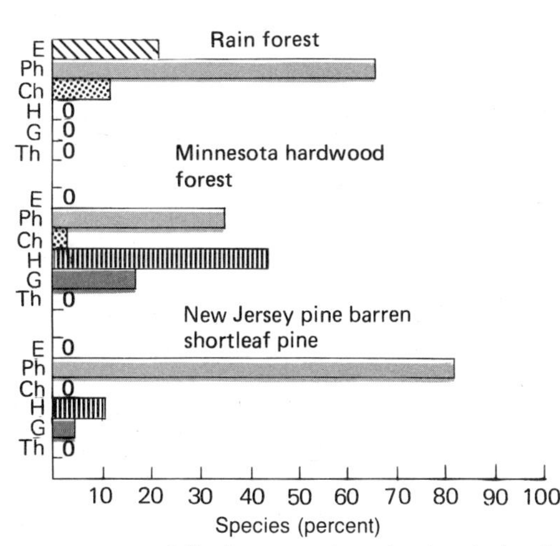

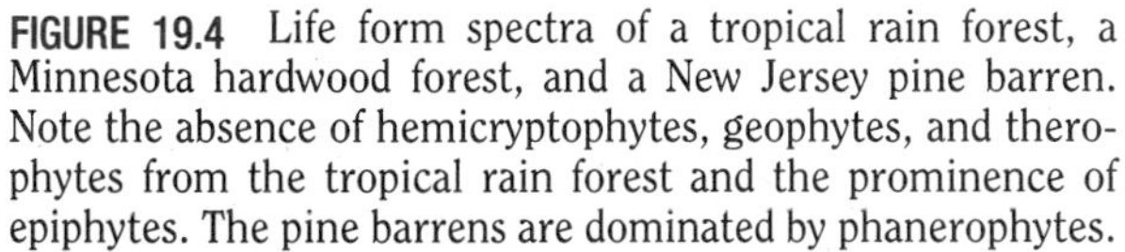

FIGURE 19.4 Life form spectra of a tropical rain forest, a Minnesota hardwood forest, and a New Jersey pine barren. Note the absence of hemicryptophytes, geophytes, and therophytes from the tropical rain forest and the prominence of epiphytes. The pine barrens are dominated by phanerophytes.

BOX 19.1

Some Measures of Dominance

1. Dominance = $\dfrac{\text{basal area or aerial coverage, species A}}{\text{area sampled}}$

2. Relative dominance = $\dfrac{\text{basal area or coverage, species A}}{\text{total basal area or coverage, all species}}$

3. Relative density = $\dfrac{\text{total individuals, species A}}{\text{total individuals, all species}}$

4. Frequency = $\dfrac{\text{intervals or points where species A occurs}}{\text{total number of sample plots or points}}$

5. Relative frequency = $\dfrac{\text{frequency value, species A}}{\text{total frequency values, all species}}$

6. Importance value = relative frequency + relative dominance + relative density

All the above results may be multiplied by 100.

Simpson's index of dominance:

$$\text{dominance} = \frac{\Sigma n_i(n_i - 1)}{N(N-1)}$$

where N = total number of individuals of all species

n_i = total number of individuals of species A

by a few large trees that overshadow the smaller ones. In such a situation the dominant organisms are not those with the greatest numbers but those that have the greatest biomass or that preempt most of the canopy space and thus control the distribution of light. Ecologists measure such dominants by biomass or basal area. In other cases the dominant organism may be scarce yet by its activity control the nature of the community. The predatory starfish *Piaster*, for example, preys on several associated species and reduces competitive interaction among them, so they are able to coexist. If a predator is removed, a number of prey species disappears and one becomes dominant. In effect, the predator controls the nature of the community and must be regarded as the dominant.

The concept of dominance has certain implications. In the first place, the dominant species may not be the most essential species in the community from the standpoint of energy flow and nutrient cycling, although this is often the case. Dominant species achieve their status by occupying niche space that might potentially be occupied by other species in the community. For example, when the American chestnut was eliminated by blight from oak-chestnut forests, the chestnut's position was taken over by other oaks and hickories.

Although dominants frequently shape populations of other levels, dominance necessarily relates to species occupying the same level. If a species or small group of species is to achieve dominance, it must relate to a total population of species, all of which have similar ecological requirements. One or several become dominants because they are able to exploit the range of environmental requirements more efficiently than other species in the same level. The subordinate species exist because they are able to occupy a niche or por-

TABLE 19.3 Structure of One Mature Deciduous Forest in West Virginia

Species	*Number*	*Percentage of Stand*
Yellow-poplar (*Liriodendron tulipifera*)	76	29.7
White oak (*Quercus alba*)	36	14.1
Black oak (*Quercus velutina*)	17	6.6
Sugar maple (*Acer saccharum*)	14	5.4
Red maple (*Acer rubrum*)	14	5.4
American beech (*Fagus grandiflora*)	13	5.1
Sassafras (*Sassafras albidum*)	12	4.7
Red oak (*Quercus rubra*)	12	4.7
Mockernut hickory (*Carya tomentosa*)	11	4.3
Black cherry (*Prunus serotina*)	11	4.3
Slippery elm (*Ulmus rubra*)	10	3.9
Shagbark hickory (*Carya ovata*)	7	2.7
Bitternut hickory (*Carya cordiformis*)	5	2.0
Pignut hickory (*Carya glabra*)	3	1.2
Flowering dogwood (*Cornus florida*)	3	1.2
White ash (*Fraxinus americana*)	2	.8
Hornbeam (*Carpinus caroliniana*)	2	.8
Cucumber magnolia (*Magnolia grandiflora*)	2	.8
American elm (*Ulmus americana*)	1	.39
Black walnut (*Juglans nigra*)	1	.39
Black maple (*Acer nigra*)	1	.39
Black locust (*Robinia pseudoacacia*)	1	.39
Sourwood (*Oxydendrum arboreum*)	1	.39
Tree of heaven (*Ailanthus altissima*)	1	.39
	256	100.00

tions of it that the dominants cannot effectively occupy. Dominant organisms, then, are generalists with a wide range of physiological tolerances. The subdominants tend to be more specialized in their environmental requirements and more limited in their physiological tolerances.

SPECIES DIVERSITY

Species diversity, simply defined, is the number of species in a community. The greater the number, the greater the diversity of species. The concept of species diversity has been expanded of late into biodiversity, a term that is finding its way into popular literature. *Biodiversity* refers to the variety of life forms, of the ecological functions they perform, and of the genetic variations they contain. Biodiversity can be best understood and appreciated if we view it from the community level within the context of species diversity.

MEASURING DIVERSITY

Among the array of species that make up the community, few are abundant, and most are rare. You can discover this characteristic for yourself by counting all the individuals of different species of plants or other organisms in a number of sample plots and determining what percentage of each makes up the whole, or relative abundance.

As an example, Table 19.3 on the facing page presents the structure of a mature woodland consisting of 24 species of trees over 4 inches dbh. Two trees, yellow-poplar and white oak, made up nearly 44 percent of the stand. The four next abundant trees—black oak, sugar maple, red maple, and American beech—each made up a little over 5 percent of the stand. Nine species ranged from 1.2 to 4.7 percent of the stand, while the nine remaining species as a group represented about 5 percent of the stand. Another woodland sample presents a somewhat different picture (Table 19.4). This community consists of ten species, of which two, yellow-poplar and sassafras, make up almost 84 percent of the stand. Both forest stands illustrate the pattern of a few common species associated with many rare ones.

These two tables illustrate two other characteristics of a community—species richness or abundance and the evenness of distribution of individuals among the species. The stand described in Table 19.3 is richer in species than the stand in Table 19.4, and the evenness with which the individuals are distributed among the species is greater in the first stand than in the second.

These two parameters, species richness and species evenness, are useful in measuring species diversity. A community that contains a few individuals of many species will have a higher diversity than will a community containing the same number of individuals but with most of them confined to a few species. For example, a community with ten species of ten individuals each has a higher diversity than a community also with

TABLE 19.4 Structure of Second Deciduous Forest in West Virginia

Species	*Number*	*Percentage of Stand*
Yellow-poplar (*Liriodendron tulipifera*)	122	44.5
Sassafras (*Sassafras albidum*)	107	39.0
Black cherry (*Prunus serotina*)	12	4.4
Cucumber magnolia (*Magnolia grandiflora*)	11	4.0
Red maple (*Acer rubrum*)	10	3.6
Red oak (*Quercus rubra*)	8	2.9
Butternut (*Juglans cinerea*)	1	.4
Shagbark hickory (*Carya ovata*)	1	.4
American beech (*Fagus grandiflora*)	1	.4
Sugar maple (*Acer saccharum*)	1	.4
	174	100.0

ten species but with the 100 individuals apportioned 90, 1, 1, 1, 1, 1, 1, 1, 1, 1, 1.

To quantify species diversity, several indexes have been proposed (Box 19.2). The most widely used is the Shannon index, which has been adapted from communication or information theory:

$$H = -\sum_{i=1}^{s} (p_i)(\log_2 p_i)$$

where H = diversity of species
s = number of species
p_i = proportion of individuals of the total sample belonging to the ith species

The index takes into consideration the number as well as the relative abundance of species. When the diversity index, A, is calculated for the two woodlands, the first, as described in Table 19.3, has a diversity index of 3.59; the second, described in Table 19.4, has a diversity index of 1.87. This index is obtained using $\log_2$, usually employed in the Shannon formula. Diversity calculated using $\log_n$ is 2.49 and 1.30, respectively.

The two components, species richness and evenness, can be separated. The simplest determination of species richness is to count the number of species. In the first woodland that is 24 and in the other, 10. To determine evenness, you first have to calculate H_{max}, what H would be if all species in the community had an equal number of individuals. This number can be calculated by:

$$H_{max} = l_n S$$

where l_n = natural log and S = number of species.

For the first woodland H_{max} is 3.18 and for the second, 2.30.

Evenness (J) is determined by:

$$J = H/H_{max}$$

Evenness of the first woodland is 0.78 and of the second, 0.57. The first woodland has a more even distribution of species than the second.

Up to this point we have considered measures of diversity within a community, or *alpha diversity.* Diversity between communities is *beta diversity.* It can be calculated by such techniques as coefficients of community, percent similarity, distance measures, and others. Two examples are given in Box 19.3. A third type is *gamma diversity,* which describes diversity on a regional basis, including species replacement over large geographical regions.

BOX 19.2

Indexes of Diversity

The Shannon index of diversity is only one of a number of diversity indexes. Based on information theory, it measures the degree of uncertainty. If diversity is low, then the certainty of picking a particular species at random is high. If diversity is high, then it is difficult to predict the identity of a randomly picked individual. High diversity means high uncertainty.

Another common index is Simpson's. It takes a different approach—the number of times we would have to take pairs of individuals at random to find a pair of the same species. This index of diversity is the inverse of Simpson's dominance index (see Box 19.1):

$$\text{diversity} = \frac{N(N-1)}{\Sigma n_i(n_i - 1)}$$

or

$$1 - \frac{\Sigma n_i(n_i - 1)}{N(N-1)}$$

Thus, in a collection of species high dominance means low diversity.

The Shannon and Simpson indexes take into consideration both the richness and evenness of species. A much simpler index of diversity that does not take evenness into account is Margalef's:

$$\text{diversity} = (s - 1)/\log N$$

where s is the number of species and N is the number of individuals. Such an index does not express the differences among communities having the same s and N, so it is much less useful.

BOX 19.3

Community Similarity

A number of methods are available for measuring similarity of communities. The one most often recommended is Morisita's index, based on Simpson's index of dominance. However, here are two simpler approaches: Sorensen's *coefficient of community* and *percent similarity.*

To find the coefficient of community, apply the equation

$$CC = \frac{2c}{s_1 + s_2}$$

where c = number of species common to both communities and s_1 and s_2 = number of species in communities 1 and 2.

For the woodland examples:

$$s_1 = 24 \text{ species}$$
$$s_2 = 10 \text{ species}$$
$$c = 9 \text{ species}$$
$$CC = \frac{2(9)}{24 + 10} = \frac{18}{34} = 52.9$$

Coefficient of community does not consider the relative abundance of species. It is most useful when the major interest is the presence or absence of species.

To calculate percent similarity (PS), first tabulate species abundance in each community as a percentage. Then add the lowest percentage for each species that the communities have in common. For the two woodlands, 15 species are exclusive to one community or the other. The lowest percentage for those 15 species is 0, and they need not be added in.

$$PS = 29.7 + 0.4 + 3.6 + 0.4 + 4.7 + 2.9 + 4.4 + 0.4 + 0.39 = 46.89$$

Percent similarity does consider relative abundance of various species in each community.

THEORIES OF DIVERSITY

Species diversity within and among communities involves three components: space, time, and feeding. All in some manner relate to niche differentiation.

Diversity increases as vertical and horizontal stratification patterns increase, providing more microhabitats to exploit and niches to fill. Animals are able to partition habitats among them; the finer the partitioning, the more kinds of organisms can coexist. Plant diversity is influenced by changes in soil type, drainage, nutrient status, elevation, and the like.

Changing temporal use of habitats also increases diversity. Seasonal changes are most pronounced. Spring flowers give way to summer- and then fall-blooming species. Migrant summer nesting birds are replaced by winter migrants. Species active by day are replaced by nocturnal animals as darkness comes. Such temporal changes and replacements greatly increase the total species diversity in a community. They are often overlooked in determining species diversity.

Subtle differences in feeding habits also encourage diversity. Among groups of ungulates, for example, some are browsers, feeding on woody vegetation, while others feed on herbaceous plants. Among those some feed on young growth; others on more mature growth. Some feed on grasses; others consume herbaceous plants. The degree of specialization is reflected in diversity.

Although temporal, spatial, and trophic components account for differences in diversity, certain questions go unanswered. Why, for example, is species diversity higher in tropical regions than in temperate and arctic regions? That question in particular has intrigued ecologists for years. They have come up with several theories or hypotheses, mostly difficult, if not impossible, to test.

Perhaps the oldest is the *evolutionary time* theory, which dates back to Alfred Wallace (1878). According to this theory, the tropical regions, in the words of Wallace, are "a more ancient world than that represented by the temperate zone," and compared to northern latitudes, relatively undisturbed by glaciation. For that reason, tropical regions have had more time for the evolution of plants and animals. Temperate regions have not experienced sufficient time for species to diverge, adapt to, or occupy completely the changed environment.

Related to the time theory is the *climatic stability* theory. In an unstable climate, species would develop tolerances sufficiently broad to allow them to adjust to and survive in a wide variation in the physical environment, food supply, and the like. Such species, in effect, would occupy broad niches. On the other hand, a stable climatic environment to which they need not constantly respond allows species to adapt to a variety of microclimatic habitats and to specialize in their feeding habits. In effect, they occupy smaller niches.

A climate may be variable over time, but predictable. According to the *climatic predictability* theory, species have evolved ways to take advantage of seasonally predictable variations in climate and depend on those variations in their life cycles. Plants, for example, put on seasonal growth, flower, and fruit during favorable periods of the year and go dormant during the winter or dry season. Desert annuals germinate and bloom only during periods of adequate rainfall. Animals migrate to a more favorable climate when conditions become severe or food and water become scarce, or they enter hibernation or estivation and return or become active when environmental conditions become favorable again. This ability to specialize on predictable environmental conditions and temporal changes in food and other resources results in increased species diversity.

The *heterogeneous environment* hypothesis holds that the more complex the structure of the community, the more potential niches it possesses. That allows a greater opportunity for speciation among organisms to exploit those niches. Thus, a tropical rain forest, with its complex vertical structure, provides many more niches and is able to support many more species than a temperate forest, grassland, or arctic tundra. Variations in altitude and topography, locally or regionally, provide additional habitats, adding further to diversity.

The *productivity* hypothesis states that the more resources available in the form of nutrients, plants, or prey species, the more species are able to specialize. The tropical rain forests, with a long growing season and a large variety of plant species, have a high primary production. For that reason they are able to support many more animal species than temperature or arctic regions, with their much lower productivity. The more energy available in a usable form for organisms, the more species the ecosystem can support. Such an argument, however, does not quite hold for plants growing on nutrient-rich sites. Increased nutrient availability results in a reduction in plant species diversity and the dominance of a few species, even though primary production is high. This is true both in grasslands and Costa Rican forests, ranging from dry forests to rain forests. However, a high primary production that is both stable and predictable allows the coexistence of more species than would be possible under a less predictable set of conditions. More availability of energy allows a greater specialization across a gradient of resources.

The production hypothesis relates to the *competition* hypothesis. In a more variable environment, the major selection forces come from the physical environment. In a more stable environment such as the tropics, selection forces are largely biotic, especially intraspecific and interspecific competition. Competition favors specialization, resulting in smaller niches.

Another hypothesis involving interpopulation relationships is the *predation* theory. It holds that a random or selective removal of prey species by a predator reduces the level of competition among them. That allows more species to coexist locally than would do so in the absence of predation, because populations of competitors are held low enough to prevent any one from becoming dominant.

DIVERSITY GRADIENTS

Species diversity can be used not only to compare communities or habitats within a given region but also to examine global ecosystems. For these reasons species diversity falls into global gradients. Traveling north from the tropics to the Arctic, we find the numbers of species of plants and animals decreasing on a latitudinal gradient. Species of nesting birds are much more numerous in Central America than they are in Newfoundland. The same pattern exists among mammals, fish, lizards, and trees.

Diversity is not restricted to a latitudinal gradient. In oceans, species diversity increases from the conti-

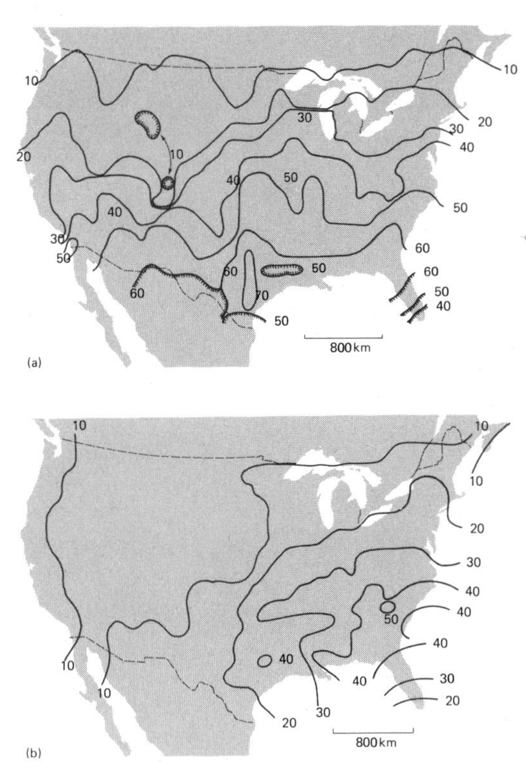

FIGURE 19.5 Pronounced latitudinal variations occur among reptiles and amphibians. (a) Being poikilothermic and endothermic, reptiles have their greatest density in hot desert regions and lower latitudes of North America. (b) Being not only poikilothermic but also highly sensitive to moisture conditions, amphibians reach their greatest diversity in the central Appalachians, then decrease northward and southward as well as westward. Species numbers are lowest in the dry and cold regions of the continent.

nental shelf, where food is abundant but the environment is changeable, to the deep water, where food is less abundant but the environment is more stable. Mountain areas generally support more species than flat lands, peninsulas have fewer species than adjoining continental areas, and islands, small or remote, have fewer species than large islands or those near continental land masses. From east to west in North America, the number of species of land birds and mammals increases. This east-west gradient relates to an increased diversity of the environment both horizontally and altitudinally. Eastern North America has more uniform topography and climate, so it holds fewer species than western North America. However, because of more favorable moisture conditions, amphibians are more abundant and diverse in eastern North America than in the western part of the continent, whereas reptiles are more diverse in the hot, arid regions of the west (Figure 19.5).

VERTICAL STRUCTURE

A distinctive feature of a community is vertical structure (see Figure 19.6), physical and biological. It is determined largely by the life form of the plants—their size, branching, and leaves—which, in turn, influences and is influenced by the vertical gradient of light. The vertical structure of the plant community provides the physical structure in which many forms of animal life are adapted to live. A well-developed forest ecosystem, for example, has several layers of vegetation. From top to bottom, they are the *canopy,* the *understory,* the *shrub layer,* the *herb* or *ground layer,* and the *forest floor.* We could even continue down into the root layer and soil strata.

The canopy, which is the primary site of energy fixation, has a major influence on the rest of the forest. If it is fairly open, considerable sunlight will reach the lower layers and the shrub and the understory tree strata will be well developed. If the canopy is closed, the shrub and the understory trees and even the herbaceous layers will be poorly developed.

The understory consists of tall shrubs such as witch hobble, understory trees such as dogwood and hornbeam, and younger trees; some are the same as those in the crown, while others are of different species. Species that are unable to tolerate shade and competition will die; others will eventually reach the canopy after some of the older trees die or are harvested.

The shrub layer differs with the type of forest. In oak forests on south-facing slopes, blueberries are most characteristic; in the moist cove forests grow buffalo nut, hydrangea, and rhododendron. In the northern hardwood forests, witch hobble, maple-leaf viburnum, and striped maple are common.

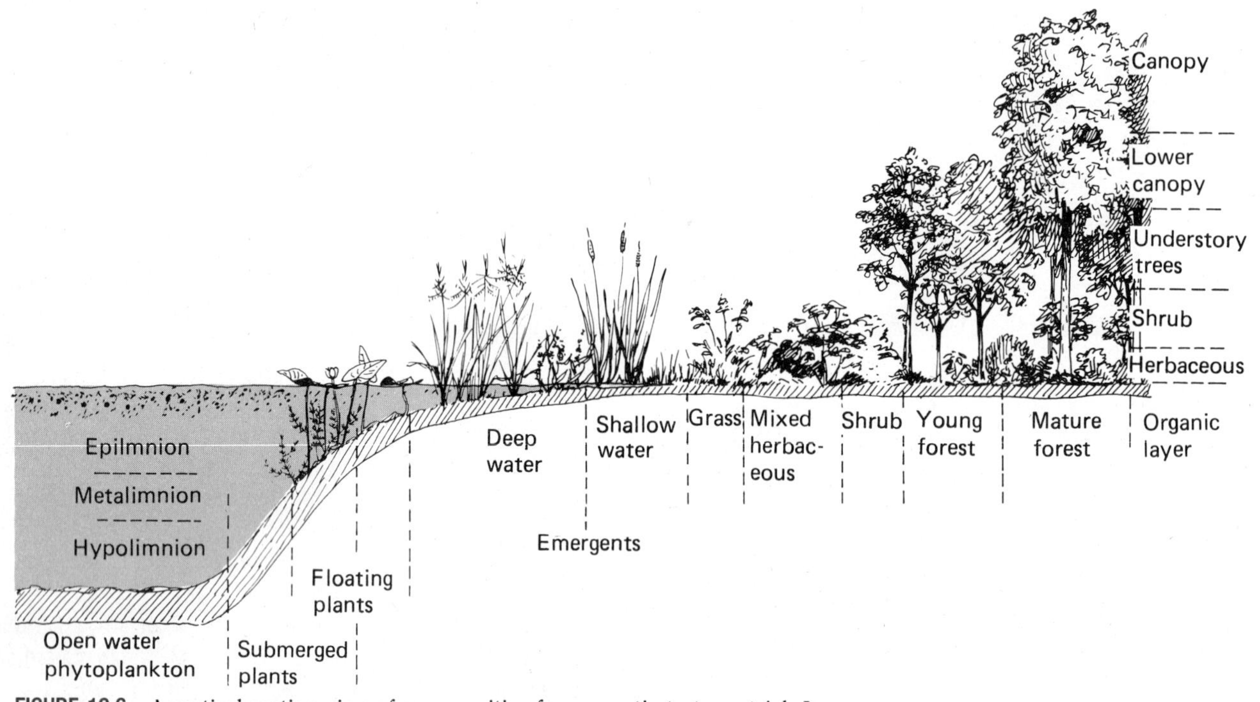

FIGURE 19.6 A vertical section view of communities from aquatic to terrestrial. In both the zone of decomposition and regeneration is the bottom stratum and the zone of energy fixation is the upper stratum. From aquatic to terrestrial types, stratification and the complexity of the community become greater. Stratification in aquatic communities is largely physical, influenced by gradients of oxygen, temperature, and light. Stratification in terrestrial communities is largely biological. Dominant vegetation affects the physical structure of the community and the microclimatic conditions of temperature, moisture, and light. Because the forest has four to five strata, it can support a greater diversity of life than a grassland with two strata. Floating and emergent aquatic plant communities can support a greater diversity of life than open water.

The nature of the herb layer depends on the soil moisture conditions, the slope position, the density of the overstory, and the aspect of the slope, all of which vary from place to place through the forest.

The final layer, the forest floor, is the site where the important process of decomposition of forest litter takes place and where nutrients are released for reuse.

Aquatic ecosystems such as lakes and oceans have strata determined by light penetration, temperature profile, and oxygen profiles (see Chapter 32). In the summer well-stratified lakes have a layer of freely circulating surface water, the *epilimnion;* a second layer, the *metalimnion,* which is characterized by a *thermocline* (a very steep and rapid decline in temperature); the *hypolimnion,* a deep, cold layer of dense water about 40°C, often low in oxygen; and a layer of bottom mud. In addition, two other structural layers are recognized, based on light penetration: an upper zone roughly corresponding to the epilimnion, which is dominated by plant plankton and is the site of photosynthesis, and a lower layer, in which decomposition is most active. The lower layer roughly corresponds to the hypolimnion and the bottom mud.

Ecosystems, terrestrial and aquatic, have similar biological structure. They possess an *autotrophic layer* concentrated where light is most available, which fixes the energy of the sun and manufactures food from organic substances. In forests this layer is concentrated in the canopy; in grasslands, in the herbaceous layer; and in lakes and seas, in the upper layer of water. Ecosystems also possess a *heterotrophic layer* that utilizes food stored by autotrophs, transfers energy, and circulates matter by means of herbivory, predation in the broadest sense, and decomposition.

The degree of vertical layering has a pronounced influence on the diversity of animal life in the community. A strong correlation exists between foliage height diversity and bird species diversity. Increased vertical stratification increases the availability of resources and living space, which favors a certain degree of specialization. Grasslands, with their two strata, hold six or seven species of birds, all ground nesters. An eastern deciduous forest may support 30 or more species occupying different strata. The scarlet tanager *(Piranga olivacea)* and wood pewee *(Contopus virens)* are canopy species, the Acadian flycatcher *(Empidonax virescens)* is an understory canopy species, and the hooded warbler *(Wilsonia citrina)* is a forest shrub species. Insects show similar stratification. Among the pine bark bettles inhabiting northeastern North America, the large red turpentine bettle *(Dendroctonus valens)* is restricted to the base of trees; the pine engraver beetle *(Ips pini)*, to the upper trunk and large branches. A third species, the small and abundant *Pityogenes hopkinsi*, lives on smaller branches in the crown.

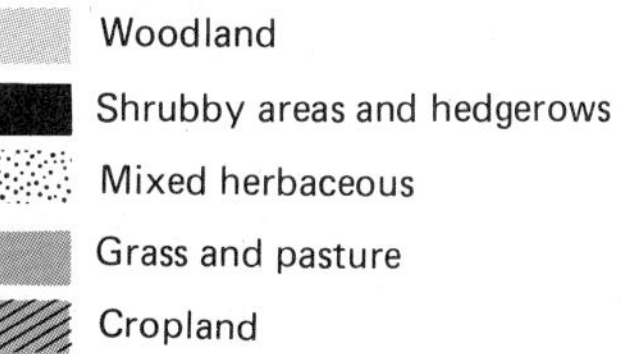

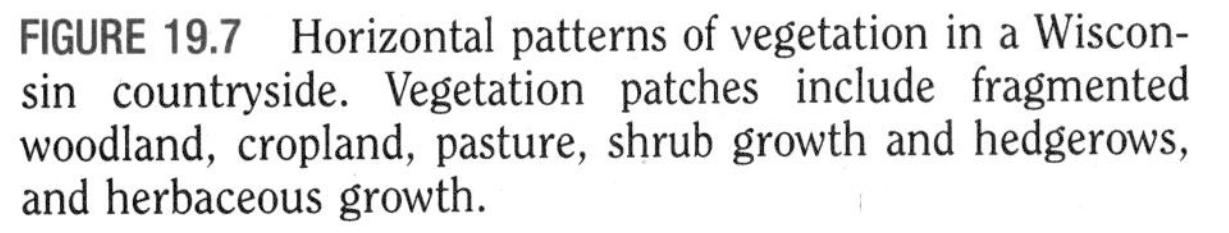

FIGURE 19.7 Horizontal patterns of vegetation in a Wisconsin countryside. Vegetation patches include fragmented woodland, cropland, pasture, shrub growth and hedgerows, and herbaceous growth.

HORIZONTAL PATTERNS

Walk across a typical old field. You move through patches of open grass, clumps of goldenrods, tangles of blackberry, and small thickets of sumac and other tall shrubs. Continue into an adjacent woodland and you may cross through open understory, patches of shade-tolerant undergrowth of laurel and viburnum, and come upon gaps in the canopy where dense thickets of new growth have claimed the sunlit openings. The vegetation patches form a quiltwork across the landscape (Figure 19.7). These patches, spatially separated, produce a horizontal pattern or patchiness that adds to the physical complexity of the environment.

This patchy distribution of plants shows both environmental and biological influences. In terrestrial communities, soil structure, soil fertility, moisture conditions, and aspect influence the microdistribution of plants. Patterns of light and shade shape the development of understory vegetation on the forest floor. Runoff and small variations in topography and microclimate produce well-defined patterns of plant growth. Grazing animals have subtle but important effects on the spatial patterning of vegetation, as do abiotic disturbances such as windthrow and fire. Plants with airborne seeds may be distributed widely, whereas plants with heavy seeds or with pronounced vegetative reproduction will be clumped near the parent plant. Plant toxins and shading suppress some plant species and encourage others. Pronounced horizontal patterning, *zonation,* is caused by differences in climate or soil that inhibit rooted vegetation. Such zonation is most conspicuous about bogs and ponds. Horizontal patchiness of plant life in turn influences the distribution of animal life across the landscape.

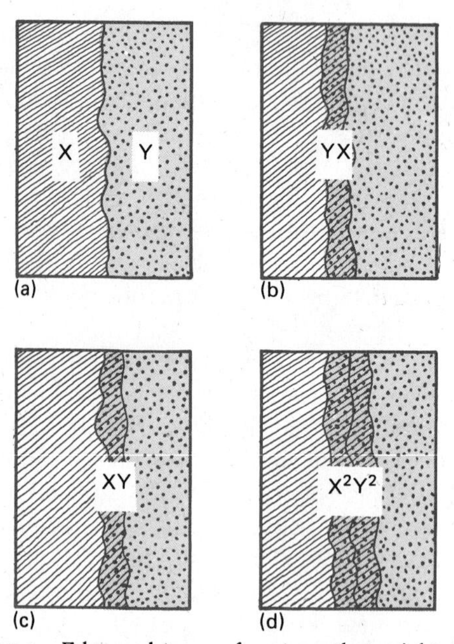

FIGURE 19.8 Edge and types of ecotone that might develop. (a) Abrupt, narrow edge with no development of an ecotone. (b) Narrow ecotone developed by advancement of community Y into community X. (c) Community X advances into community Y to produce ecotone XY. (d) Ideal ecotone development, in which plants from both communities invade each other to create a wide ecotone, X^2Y^2. This type of ecotone will serve the most edge species.

EDGE AND ECOTONE

Closely associated with horizontal patterns are edge and ecotone. Although the two terms are often used synonymously, they are different. An *edge* is where two or more different vegetational communities meet. An *ecotone* is where two or more communities not only meet but intergrade (Figure 19.8).

Edges may result from abrupt changes in soil type, topographic differences, geomorphic differences (such as rock outcrops), and microclimatic changes. Because the adjoining vegetation types are determined by long-term natural features, such edges are usually stable and permanent and are considered *inherent.* Other edges result from such natural disturbances as fire, storms, and floods or from such human-induced disturbances as grazing, timber harvesting, land clearing, and agriculture. The adjoining vegetational types are successional or developmental, and will change or disappear with time. Such edges are termed *induced.* They can be maintained only by periodic disturbance. Induced edges, too, may be abrupt, or they may be transitional, resulting in an ecotone.

Ecotones arise from the blending of two or more vegetational types. Plants competitively superior and adapted to environmental conditions in the edge advance as far into either community as their ability to maintain themselves will allow. Beyond this point interior plants of adjacent communities maintain themselves. As a result, the ecotone exhibits a shift in dominance.

Certain highly adaptable species tend to colonize such areas. Edge species of plants tend to be opportunistic (see Chapter 14), shade-intolerant, and tolerant of a relatively dry environment, including a high rate of evapotranspiration, reduced soil moisture, and fluctuating temperatures. Animal species of the edge are usually those that require two or more vegetational communities. For example, the ruffed grouse *(Bonasa umbellatus)* requires new forest openings with an abundance of herbaceous plants and low shrubs, dense sapling stands, pole timber for nesting cover, and mature forests for winter food and cover. Because the ruffed grouse spends its life in an area of 4 to 8 hectares, this amount of land must provide all of its seasonal requirements. Some species, such as the indigo bunting, are restricted exclusively to the edge situation (Figure 19.9).

The variety and density of life are often greatest in and about edges and ecotones. This phenomenon has been called the *edge effect.* Edge effect is influenced by

FIGURE 19.9 Map of territories of a true edge species, the indigo bunting (*Passerina cyanea*), which inhabits woodland edges, large gaps in forests creating edge conditions, hedgerows, and roadside thickets. The male requires tall, open song perches and the female a dense thicket in which to build a nest.

the amount of edge available—its length, width, and degree of contrast between adjoining vegetational communities. The greater the contrast between adjoining plant communities, the greater the species richness should be (Figure 19.10). An edge between forest and grassland should support more species than an edge between a young and a mature forest. The larger the adjoining communities, the more opportunity exists for flora and fauna of adjoining communities as well as species that favor edge situations to occupy the area. If patches of vegetation are too small to support their characteristic species the area becomes a homogeneous community dominated by edge species.

The edge effect comes about because environmental conditions differ from those of adjacent vegetational communities, especially adjoining forests. Increased solar radiation in the newly created edge, high temperature, and exposure to wind result in a high rate of evaporation. Plants place increased demands on soil moisture. Sudden exposure to sunlight subjects trees to stress from increased heat and light. Some mesic,

FIGURE 19.10 Contrast in edge is important in increasing species richness. A high-contrast edge (a) is more valuable to edge species than a low-contrast edge (b) because two quite different vegetation types adjoin. Low-contrast edges do not provide enough difference between vegetational communities to be of maximum value to edge species. Of greatest value is an advancing edge (c), such as woody vegetation invading an adjoining old field. An advancing edge not only provides variation in height but in effect creates two edges on the site.

shade-tolerant trees succumb. Others are injured by sun scald. Light-tolerant species respond by increasing crown growth and epicormic branching (new branching sprouting on the trunk). The effects of microclimatic changes are most pronounced on south-facing and west-facing edges, because they receive the greatest amount of solar radiation. Therefore an edge favors xeric, light-demanding species capable of competing successfully for available soil moisture.

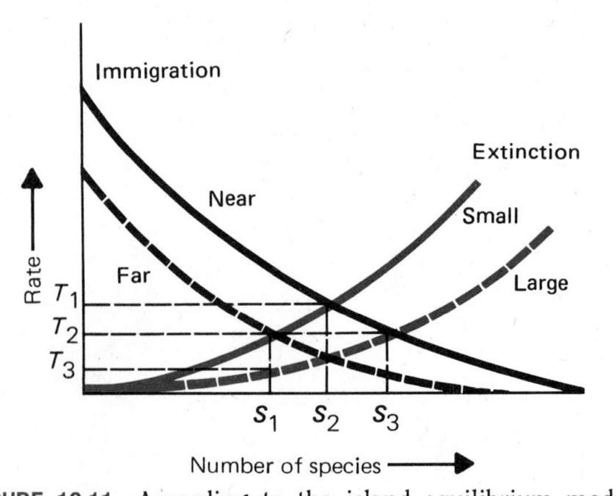

FIGURE 19.11 According to the island equilibrium model, immigration rates balance extinction rates. Immigration rates are distance-related. Islands near a mainland have a higher immigration rate than islands distant from a mainland. Extinction rates relate to area, being higher on small islands than on large ones. As the intersections of immigration and extinction curves indicate, the number of species at equilibrium is greater on larger islands than on small ones if both are the same distance from the mainland. The number of species expected on small islands close to the mainland is about the same as expected on large, distant islands. The turnover rate, immigration versus extinctions, is higher on small, near islands. They have small populations, which are more likely to become extinct, but at the same time they receive replacements easily from the nearby mainland. Large, far islands experience a low turnover rate. Their large populations reduce the chance of random extinction, and their distance from the mainland precludes a high immigration rate.

ISLAND ECOLOGY

As the size of an area increases, species richness also increases, up to some maximum point. Such a relationship between area and species richness is obvious on oceanic islands, a point noted by early naturalist explorers. Smaller oceanic islands held fewer species than larger islands, and remote oceanic islands, large and small, held the fewest species. The zoogeographer P. Darlington suggested a rule of thumb: A tenfold increase in area leads to a doubling of the number of species.

Such observed relationships were formally presented as a theory of island biogeography by Robert MacArthur and E. O. Wilson in 1963. The theory is simple. The number of species of a given taxon (a group of any given family or order) established on an island represents a dynamic equilibrium between the immigration of new colonizing species and the extinction of previously established ones (Figure 19.11). Immigration varies with the distance of the island from the mainland or pool of potentially colonizing species. Extinction will vary with the size of the island. The smaller the island the greater the probability of extinction.

A new or uninhabited island would be colonized rapidly by those species with the greatest dispersal ability. As the number of colonizing species increases, the number of immigrants arriving on the island decreases. Immigration declines for several reasons. As the number of colonizing species increases, fewer new immigrants are available from the source pool. Later immigrants may be unable to establish populations because habitats are filled or resources are already utilized. At the same time some island inhabitants go extinct. As more species arrive, the extinction rate increases because of interspecific competition and reduced population size, making the affected population vulnerable to random extinction. At equilibrium between immigration and extinction the number of species remains stable, although the composition of species may change. The rate at which one species is lost and a replacement gained is the *turnover rate.* The equilibrium level is influenced by the size of the island and the distance of the island from a pool of potential immigrants.

Although island biogeography theory was originally applied to oceanic islands, not all islands are oceanic. Mountain tops, bogs, ponds, dunes, fragmented woodlands and grasslands surrounded by a sea of agricultural land and urban developments, host plants and their insects, and animals and their parasites—all are essentially island habitats. However, considerable differences exist between habitat islands and oceanic islands. Oceanic islands are isolates. They are ecological units surrounded by an aquatic barrier to dispersal. They are inhabited by organisms of various taxa that arrived there by chance dispersal over a long period of time or represent remnant populations that existed on the area long before isolation. By contrast, habitat islands are samples of populations of a much larger area. These samples contain fewer species, fewer individuals within a species, and more species represented by only a few individuals. As a large area of habitat is fragmented, the total habitat area is reduced, and what is left is distributed in disjointed fragments of varying size. These fragments are separated by other types of habitats, particularly urban and suburban developments and agricultural lands. As more land area is carved out of the original habitat, the distinctiveness of the habitat patch becomes accentuated. Although surrounding areas may represent barriers to dispersal for habitat island inhabitants, these barriers are also terrestrial habitats with their own sets of species, including domestic animals such as cats, dogs, sheep, cattle, and species of wildlife highly adaptable to human habitations. These species invade the edges and move into the interior; cats, weasels, raccoons, and crows, for example, increase predatory pressure on interior species. Exposure to wind and solar radiation causes mortality in plant species on the edge, further reducing the integrity of the fragment.

WEAKNESSES

The theory of island biogeography too often is accepted uncritically. Although its major premise seems to hold, the theory has some weaknesses. Both immigrations and extinctions are poorly defined without a time limit. If a species appears, breeds, and disappears, then reappears later, should it be counted as an immigrant twice and an extinction once? The theory makes no allowances for chance arrivals and extinctions of species on islands. It ignores the effects of turnover on the composition of species. Species richness may remain constant, but species composition may change. On any given island or habitat patch the balance between immigration and extinction is explained as a function of size and degree of isolation, but the degree of isolation is relative. What is a short distance for a bird may be an insurmountable distance for a mammal or a lizard. The assumption that extinctions relate to an island's area overlooks the fact that immigrations and extinction may not be independent. Extinction of a dwindling population of a species may be slowed or even halted by an influx of immigrants, the rescue effect. Extinctions are influenced by life-history traits and are not due only to isolation. The model assumes that an island's area determines the number of species, and overlooks the role of habitat diversity, which might override island size. Further, it treats all species in a taxon as equals, with the same probability of extinctions and immigrations; it ignores population dynamics and life-history requirements of the species involved. In spite of its shortcomings, island biogeography theory has stimulated new insights and research into the distribution, diversity, and conservation of species.

TESTS

Few studies have been designed to test the island hypothesis. Several experimental studies of depopulation of arthropods on very small islets and their recolonization have been inconclusive. Studies of turnovers under more natural situations, based on recent surveys of species compared with species lists from previous surveys, have attempted to show dynamic equilibrium of island fauna, extinction rates, and turnover rates in which the total number of species remains constant. The appearance of new species (immigrants) in recent lists and the disappearance of old ones (extinctions) were used to calculate the average turnover rate.

Of considerably more interest are long-term annual censuses on a given island or habitat. An example is the seasonal census of confirmed nesting birds over a 26-year period in Eastern Wood of Bookham Common, Surrey, England (Figure 19.12). A 16-ha oak *(Quercus robur)* woods, it was one parcel of the 112-ha woods that covers much of the common. Although small, Eastern Wood is close to a pool of potential immigrants.

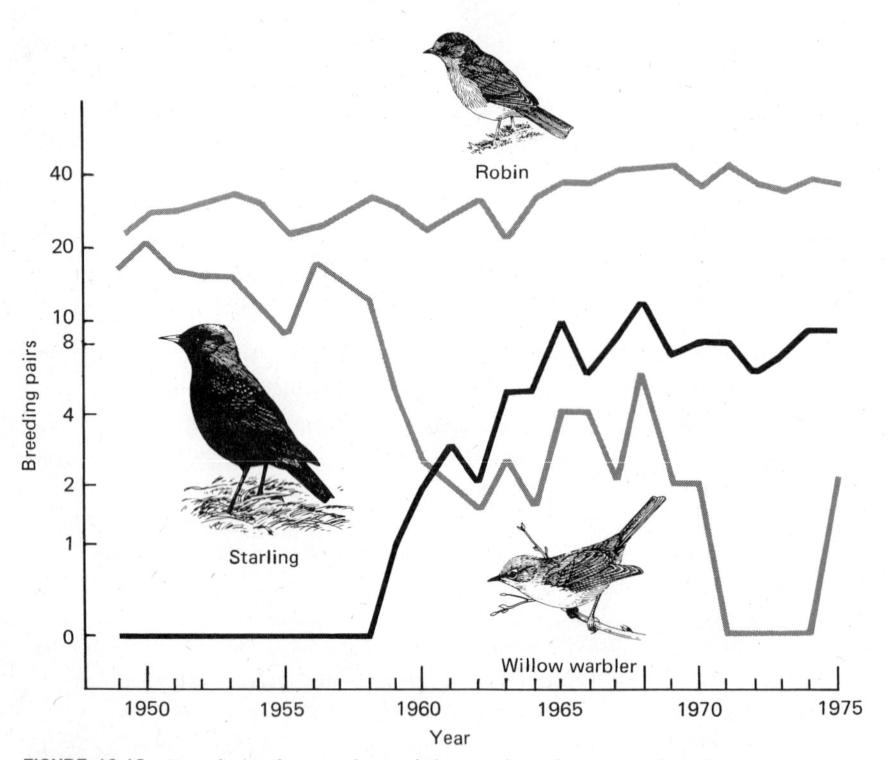

FIGURE 19.12 Trends in the numbers of three selected species of birds in Eastern Wood, a woodland island, Bookham Common, Surrey, England. The robin (*Erithacus rubecula*) is one of the most abundant breeding birds, obviously well adapted to small islands of woodlands. Its population fluctuated only slightly over the 27-year period. The willow warbler (*Phylloscopus trochilus*), abundant earlier, became extinct when logging on the tract stopped and woody growth encroached on the area. The willow warbler, a bird of open woods, is dependent on disturbance for maintenance of its habitat. This species points out the importance of some artificial disturbance in island situations to maintain variation in habitats. At the last census the willow warbler reappeared, thus experiencing both extinction and immigration. The starling (*Sturnus vulgaris*) immigrated into the area and apparently achieved equilibrium level in the oak wood. Its appearance may have been stimulated by improved nesting sites following cessation of logging.

The woods was under some form of management, including timber cutting, until 1952. After that, encroachment of woody growth changed the habitat. These changes resulted in the only apparent extinction—that of the willow warbler *(Phylloscopus trochilus),* associated with open woodland—and the appearance of one new immigrant, the starling *(Sturnus vulgaris).*

Some time during the 26 years, 44 species of birds appeared in the woods. Of these, 6 apparently did not nest, leaving 38 breeding species. Of these, 4 species had territories extending beyond the woods, and 9 species never had more than two pairs nesting in the woods itself. Eleven more species nested in fewer than 5 years during the 26 years. They had to be considered casual species, counted as immigrants and extinctions in calculating annual turnover. Only 14 species were regular breeders. Thus, much of Eastern Wood's avifauna was subsidized by nearby larger woodland. If the tract were suddenly transformed into a truly isolated island, those 14 species would make up—for a time, at least—the avifauna of the island. Ultimately, only those species with a minimum of 10 breeding pairs would persist. They are the great tit *(Parus major),* blue tit *(Parus caeruleus),* wren *(Troglodytes troglodytes),* robin *(Erithacus rubecula),* and blackbird *(Turdus merula).*

During the 26 years, Eastern Wood experienced considerable turnover of species, with an average of three immigrations and three extinctions a year. Species equilibrium, as determined by immigration and extinction curves, is 32 species, somewhat higher than the 27 species that on the average inhabited Eastern Woods over the years (Figure 19.13).

Data for Eastern Wood suggest that in general the theory of island biogeography is correct—that if the woods held 40 or more species, any new species that might breed in the woods would be a casual one, nesting infrequently at low density; that the addition of a species would be expected to lead to one additional extinction.

FRAGMENTATION OF HABITATS

Eastern Wood is an example of the effects of fragmentation—the breaking up of large tracts of habitat such as woodland and prairie into smaller units separated by seas of urbanization and agricultural crops. In the process of fragmentation, species requiring large parcels of habitat or food specialists, the interior species, disappear. Some maintain their populations if the population is supplemented by immigrants from a not too distant replacement pool, as in Eastern Wood. Other species, attracted by edge conditions, move in. Thus, some species are lost, some species remain, and some species are gained. The species composition of the habitat island shifts, usually toward edge or generalized species. The ability of interior species to maintain their presence depends upon the size of the fragment and its relationship to a pool of interior species.

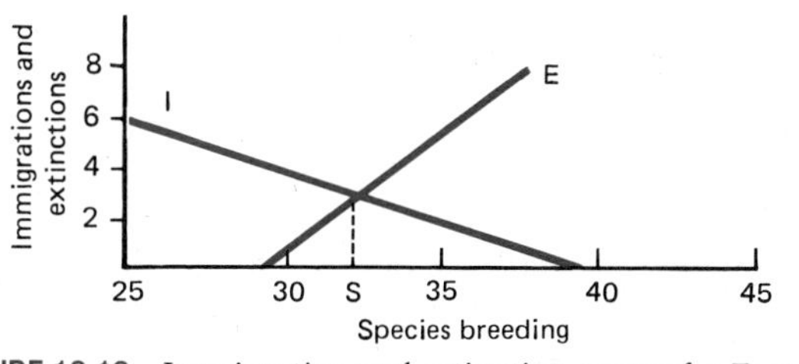

FIGURE 19.13 Immigration and extinction curves for Eastern Wood. The immigration curve cuts the x-axis at 39 species. The maximum number of species that bred at one time or another in Eastern Wood was actually 44. The extinction line is at 45°, indicating one extinction for every species present over 29. The equilibrium point (S), where immigration intersects extinction, is 32 species. The lines are straight rather than curved because data are discrete, not continuous.

What size island of remnant vegetation is needed to maintain regional populations and satisfy habitat requirements of the species concerned? At what island size does species richness reach its maximum? Such questions have stimulated a number of studies of the response of both plants and animals to habitat fragmentation.

There is a point in island size at which no interior species can exist (Table 19.5). The size of a fragmented woodland, for example, may be so reduced that the edge merges into the forest interior and the fragment, for all purposes, is forest edge (Figure 19.14). Mesic, shade-tolerant plants are replaced by shade-intolerant, xeric, opportunistic species over time, and animal species of the interior are replaced by edge species. As the size of the area increases, the ratio of edge to interior decreases, and the number of forest interior species may increase.

The minimum size of forest habitat needed to maintain interior species differs with plants and animals. For forest interior plants the minimum area depends upon the size at which moisture and light conditions become both mesic and shady enough to support shade-tolerant species. Size, however, depends in part on the nature of the edge about the stand—whether it is closed, cutting down on the penetration of light and wind—and on canopy closure. If the stand is too small and too open, the interior environment becomes so xeric that it prevents reproduction by mesic species, both herbaceous and woody. As a result, when mature residual mesic species—such as sugar maple and beech—die, they are replaced by xeric species such as oak.

TABLE 19.5 Area Requirements for Some Area-Sensitive Bird Species Based on Probability of Occurrence

Species	Area Required (ha)	
	Maximum Probability[a]	*50% of Maximum*[b]
Permanent residents		
Red-bellied woodpecker	85	0.3
Hairy woodpecker	200	6.8
Pileated woodpecker	3000	16.5
Tufted titmouse	52	0.5
Neotropical migrants		
Great crested flycatcher	72	0.3
Veery	250	20.0
Kentucky warbler	300	17.0
Ovenbird	450	6.0
Wood thrush	500	1.0
Red-eyed vireo	3000	2.5
Scarlet tanager	3000	12.0
Canada warbler	3000	400.0
Cerulean warbler	3000	700.0
Black-throated blue warbler	3000	1000.0

[a] Probability of occurrence increases with size.
[b] 50% maximum probability is suggested minimum area for breeding.

Several studies provide some insight into the impact of forest fragmentation on flora and fauna. Species richness of plants is greatest in edge situations where xeric species intermingle with some interior species. A study of Wisconsin woodlots showed that the total number of woody species increased with woodlot size up to approximately 2.3 ha. At that size vegetation achieved a maximum balance between edge and residual interior species. Beyond that size, species richness declined and finally leveled off at 9.4 ha (23 acres), as mesic conditions returned to the interior and shade-tolerant species persisted. Thus, there is a negative correlation between edge species and the size of forest islands and a positive correlation between interior species and size.

The nature of species replacement, turnover, and immigration in Wisconsin woodlands was influenced by distance of the forest island from seed sources. Seed dispersal to woodlands is aided by hedgerows and other narrow belts of vegetation linking one forest island with another. Species that are bird-dispersed are more successful immigrants than species dispersed by mammals and the wind. The latter two types tend to become local in distribution, leading to their extinction in many forest islands.

Among birds a similar pattern exists. Small forest islands of 5 ha or less are occupied by edge or ubiquitous species at home in any size forest tract. A New Jersey study showed that maximum bird diversity was achieved with woodlands 24 ha in size. However, these woodlands held no true forest interior species such as the worm-eating warbler *(Helmitheros vermivorus)* and the ovenbird *(Seiurus aurocapillus),* which are highly sensitive to forest fragmentation and require extensive areas of woods. The presence of forest interior species in smaller woodlands depends upon the nearness of those islands to a pool of replacement individuals. As with forest vegetation, species richness of forest interior species is positively correlated with island size.

Other studies of bird populations of large and small forest habitat islands (from 3 to 7620 ha) in agricultural regions in the United States and Canada showed that although two or more smaller forest habitats supported more species, long-distance migrants and interior species, typical of larger tracts, were missing or poorly

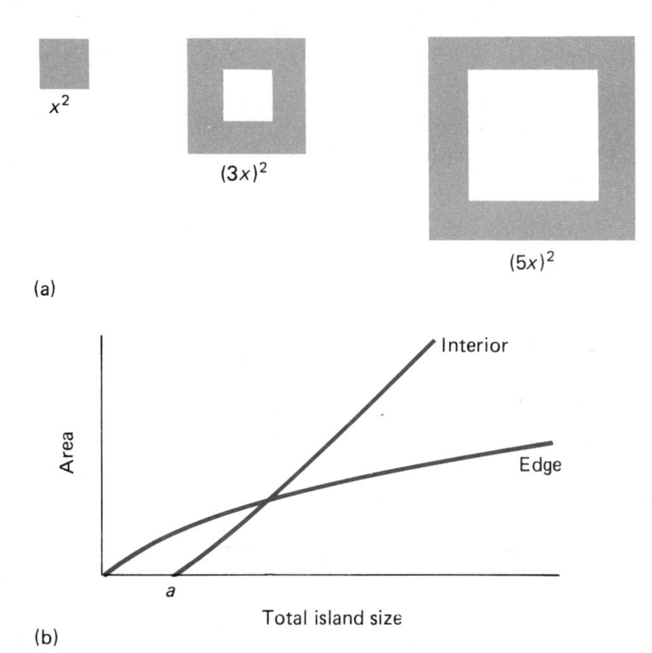

FIGURE 19.14 Relationship of island size to edge and interior conditions. All islands are surrounded by edge. (a) Assuming that the depth of the edge remains constant, the ratio of edge to interior decreases as island size increases. When the size of the island is large enough to maintain mesic conditions, an interior begins to develop. (b) The graph also shows this relationship. Below point *a*, the size at which interior species can exist, the island is all edge. As size increases, interior area increases and ratio of edge to interior decreases. This relationship of size to edge holds for square or circular islands. Long, narrow woodland islands whose width does not exceed the depth of the edge would be edge communities, even though their area might be the same as that of square or circular ones.

represented. Large forest tracts were important for increasing the number of forest interior neotropical migrants and certain resident species such as hairy woodpecker. Large forest tracts with a high degree of heterogeneity held the most species, supporting birds of both edge and interior.

SPECIES MANAGEMENT

It is easy to understand why island biogeography theory applies to the management of wild plants and animals. It is relevant to such problems as the size, shape, number, and distribution of parks and wildlife reserves, and the degree to which the increasing fragmentation of habitat is leading to species extinction.

Because of the lack of sufficient supporting data, there is vigorous debate over an important question: Which is more preferable, one large reserve or several smaller reserves that add up to the same area? Some argue that two or more smaller reserves will hold more species than one large reserve of the same area. Data on species richness seem to support that argument. Others argue that a large island is preferable because that area will not only support more rare species but also large populations of other species, making them less vulnerable to random extinction. Populations on small islands may not be large enough to be self-sustaining, and need to depend upon a subsidy of immigrants from distant sources, as exemplified by the bird populations of Eastern Wood.

There is more to consider. As large areas are fragmented into smaller ones, the populations they contain are going to decline to a much lower species equilibrium. (Recall the prediction that if Eastern Wood were completely isolated, the species equilibrium would collapse from 32 to 5 species.) Many species would suffer a sharp decline or crash, and the remnant population would be the progenitors of future generations. They would experience all the problems of small populations: a random sampling of gene frequency, genetic drift, erosion of genetic variation, increased inbreeding and accompanying loss of fitness, all increasing the probability of species extinction.

Another problem of small island size is the possible loss of *keystone species*—those upon which a number of other species depend in one way or another. Often keystone species are plants that provide nectar or fruit. Other species maintain certain conditions in the habitat upon which other species depend. For example, a sharp reduction in the rabbit population in southern England from myxomatosis resulted in thick growth of meadow grass in fields inhabited by the spectacular large blue butterfly *Maculinea arion.* Heavy grass resulted in the extinction of open-ground ant colonies, the nests of which were utilized by large blue caterpillars. As a result, the large blue is nearly extinct. The loss of one keystone species, the rabbit, a grazing herbivore, resulted in the extinction—locally, at least—of two other species.

Further, small islands may not be large enough to encompass the home range or territory of certain species. An island large enough to support a full complement of both edge species and interior species with small home ranges may not be large enough to support wide-ranging species. Concern should not be over large islands versus small islands but rather over the maintenance of an intact ecosystem. The size of an ecosystem might be dictated by the area needed to support a viable population of large predators or animals that require a large area to support them. Ecosystems large enough to support timber wolves, grizzly bears, lions, or elephants undoubtedly would encompass all species concerned. Management should be based not only on island theory and species richness but on a knowledge of species requirements, responses, and population genetics as well.

The answer, then, is not large or small reserves but a mix, with both heterogeneous and homogeneous habitats. An emphasis on diversity alone, so prevalent in species management, can lead to the decline and extinction of interior or area-sensitive species that need homogeneous habitats.

The adverse effect of fragmentation of habitats may be reduced for some species by corridors of hedgerows, riverine vegetation, and narrow strips of woodland or grassland that connect one habitat island with another. In some situations, corridors may be as small as culverts under highways that allow animals access to habitats bisected or isolated by road construction. Corridors are important for the dispersal and interchanges of animals, gene flow between habitats, and maintenance of viable populations.

Summary

A biotic community is a naturally occurring assemblage of plants and animals living in the same environment, mutually sustaining and interdependent, constantly fixing, utilizing, and dissipating energy.

All communities exhibit vertical layering and horizontal patterns, which largely reflect the life forms of plants and influence the nature and distribution of animal life in the community. The communities that are most highly stratified offer the richest variety of animal life, for they contain a greater assortment of microhabitats and available niches.

The makeup of any one community is determined in part by the species that happen to be distributed in the area and can grow under prevailing conditions. The exact species that settle in an area and the number that survive are rarely repeated in any two places at the same time, but there is a certain recurring pattern of more or less similar groups. Rarely can different groups of communities be sharply delimited, because they blend together to form a continuum along some environmental gradient.

The place where two different communities meet is an edge. The area where two communities blend is an ecotone. An edge may be inherent, produced by a sharp environmental change such as a topographical feature or soil type; or an edge may be induced, created by disturbance. Because it supports not only selected species of adjoining communities but also a group of opportunistic edge species, an edge has a high species richness.

Communities are organized about dominant species, especially in the temperate regions. The dominants may be the most numerous, possess the highest biomass, preempt the most space, or make the largest contribution to energy flow. The dominant species is not necessarily the most important in the community.

Communities may be characterized by species diversity. Species diversity involves two components: species richness, the number of species in a community, and evenness, how individuals are apportioned among the species. A species diversity index is useful only on a comparative basis, either within a single community over time or among communities. Species diversity within a community is called alpha diversity and that between communities, beta diversity.

A relationship exists between species diversity and area. In general, larger areas support more species than smaller areas. This species–area relationship is involved in the theory of island biogeography, which states that the number of species an island holds represents a balance between immigration and extinction. Immigration rates on an island are influenced by the distance of the island from the mainland or a pool of potential immigrants. Thus, islands the most distant from a mainland would receive the fewest immigrants and the ones closest to the mainland, the most. Extinction rates are influenced by the area of an island. Because small islands hold smaller populations and have less variation in habitat, they experience higher extinction rates than large islands.

The theory of island biogeography has practical applications. The fragmentation of natural habitats such as forests results in islands of remnant habitats in a sea of agricultural or urban lands. The effects of such fragmentation on species are predictable by island theory. It also serves as one guide to the establishment of nature reserves and the management of rare and endangered species.

In such situations, however, more than size and distance of islands are involved. Of critical importance are (1) interrelations of species, especially the role of keystone species in the maintenance of ecosystem integrity; (2) effective population size necessary to maintain genetic diversity and fitness; and (3) habitat heterogeneity, including early successional or disturbed areas required by some species.

Review and Study Questions

1. Contrast species dominance and species diversity.
2. Diversity is greatest in tropical rain forests. What does this fact tell you about dominance and species distribution there?
3. Contrast the stratification of an aquatic community and a terrestrial community.
4. What are edge, ecotone, and edge effect?
5. Discuss the utilization of a patchy environment by an edge species, such as the cottontail rabbit or the tiger swallowtail butterfly. What vegetational patches does it need? Why are such species found only in edge situations?
6. What is the theory of island biogeography?
7. Island biogeography theory, as originally formulated, considers immigration as a function only of distance from the source and extinction as a function only of population size. What else influences immigration and extinction?
8. How does the concept of island biogeography relate to the fragmentation of forests?
9. Contrast interior species with edge species. Are all small habitat island species edge species? Explain.

*10. How could an abrupt edge be modified to attract more edge species?

*11. Few reserves and parks in the world are large enough for the conservation of large animals. What problems, ecological, sociological, and others, might arise if a park were enlarged to enclose a whole ecosystem? (The journals *Conservation Biology* and *Biological Conservations* can help you answer.)

*12. Neotropical birds—those that spend most of the year in the tropics but nest in northern North America—are threatened by deforestation in the tropics. Might a greater threat be the fragmentation of nesting habitat, given that most of these birds are interior species? Discuss.

*13. Assess the impact of highways, housing developments, and agricultural practices in your area on habitat, relative to corridors. Conditions are best studied in winter when vegetational patterns are not masked by greenery. Are patches of woodlands and parks connected by hedgerows or greenbelts of vegetation? Are corridors for wildlife considered in land-use planning in your area?

Selected References

Burgess, R., and D. Sharpe, eds. 1981. *Forest island dynamics in man-dominated landscapes.* New York: Springer-Verlag.

Diamond, J., and T. J. Cade, eds. 1985. *Community ecology.* New York: Harper & Row. Theoretical discussions.

Harris, L. D. 1984. *The fragmented forest.* Chicago: University of Chicago Press. An important reference source.

MacArthur, R. H., and E. O. Wilson. 1967. *The theory of island biogeography.* Princeton, NJ: Princeton University Press. The original source.

Pielou, E. C. 1975. *Ecological diversity.* New York: Wiley Interscience.

Robbins, C. S., D. K. Dawson, and B. A. Dowell. 1989. Habitat area requirements of breeding forest birds of the middle Atlantic states. *Wildlife Monographs* 103.

Williamson, M. 1981. *Island populations.* Oxford, England: Oxford University Press. Good summary of island biogeography theory.

Wilson, E. O., ed. 1988. *Biodiversity.* Washington, DC: National Academy Press. A wide-ranging collection of papers on all aspects of biodiversity.

CHAPTER 20

Community Change: Temporal Patterns

Outline

Objectives

On completion of this chapter, you should be able to:

1. Define succession and describe its stages.
2. Discuss what mechanisms drive succession.
3. Evaluate the concept of the climax.
4. Discuss the role of time in succession.
5. Explain the relationship between plant and animal succession.

Abandoned cropland is a common sight in agricultural regions, particularly in areas once covered with forest. No longer tended, the lands grow up in grasses and nonwoody plants. If you watched long enough, you would observe that the fields are invaded by shrubby growth, followed by trees. Many years later, the abandoned croplands will be back in forest (Figure 20.1). Thus, over a period of years one assemblage of plant species succeeds another until a relatively stable forest occupies the ground. This replacement of one community by another over time is called *succession*. A more formal definition is the sequential change in relative abundances of dominant species (those with the highest biomass) in a community.

Succession on disturbed land is termed *secondary* because it begins on ground that supports or has supported life. Secondary succession can begin at any point along a gradient of disturbance, from open, bare ground to a clear-cut forest. Succession that begins on a site never before colonized by life, such as lava flows or newly exposed glacial till, is *primary succession*.

TERRESTRIAL SUCCESSION

PRIMARY SUCCESSION

Next time you visit the beach, look for primary succession on sand dunes. A product of pulverized rock, sand is deposited by wind and water. Along shores of lakes and oceans, sand particles may be piled up in long windward slopes to form dunes that move before the wind and often cover buildings and forests. With high surface temperatures by day and cold temperatures at night, dunes are rigorous environments for life to colonize. Grasses, particularly beach grass *Ammophila*, are the most successful pioneers and binding plants. When these and associated plants, such as beach pea *(Lathyrus)* and beach heath *(Hudsonia)*, have stabilized the dunes, at least partly, mat-forming shrubs such as beach plum *(Prunus maritima)* invade the area (Figure 20.2).

From this point vegetation may pass from pine to oak, or directly to oak. The low fertility of the dunes favors plants with low nutrient requirements. Because these plants are inefficient in cycling nutrients, especially calcium, soil fertility remains low. Low fertility and low moisture reserves in the sand inhibit the replacement of oak by more moisture- and nutrient-demanding trees. Such replacement happens only on favorable leeward slopes and depressions where the microclimate is more moderate and moisture can accumulate.

Newly deposited alluvial soil on a flood plain represents another barren primary site. Primary succession on a nutrient-rich alluvial floodplain in Alaska begins when seeds of all colonizers—willow, alder, balsam poplar *(Populus balsamifera)*, and white spruce—arrive more or less at the same time and give rise to seedlings. Willow, with its light, wind-dispersed seeds, is most abundant at first. Willow and alder grow rapidly, but the willow, naturally short-lived, is heavily browsed by snowshoe hare. Nitrogen-fixing alder then become dominant, eventually to be replaced by balsam poplar and long-lived white spruce. Thus primary succession may be influenced by random events and life-history traits of the colonists.

SECONDARY SUCCESSION

Secondary succession is most commonly encountered on abandoned farmland and in waste places such as fills, spoil banks, railroad grades, and roadsides, all artificially disturbed and frequently subject to erosion and settling.

Species most likely to colonize such places are the so-called weeds, species out of place from a human perspective. Although hard to define, weeds have two characteristics in common. They invade areas modified by human action; in fact, a few are confined to such habitats. Also they are exotics, not native to the region. Once native species move in, these plants disappear.

One of the classic examples of secondary succession is Catherine Keever's study of old-field succession in the Piedmont of North Carolina. The year a crop field is abandoned, the ground is claimed by annual crabgrass *(Digitaria sanguinalis)*, whose seeds, lying dormant in the soil, respond to light and moisture and germinate. But the crabgrass's claim to the ground is short-lived. In late summer the seeds of horseweed, a winter annual, ripen. Carried by the wind, they settle

FIGURE 20.1 Successional changes in an old field over 45 years. (a) The field as it appeared in 1942, when it was moderately grazed. (b) The same area in 1963. (c) A close view of the rail fence in the left background of (a). (d) Twenty years later the rail fence has rotted and white pine and aspen are growing. (e) In 1972 aspen has claimed much of the ground. (f) In 1987 the once open field is covered with a young forest dominated by quaking aspen and young maple.

FIGURE 20.2 Primary succession on a sand dune along the northeastern Atlantic coast. The dune is colonized by beach grass and beach plum, with an oak forest in the background.

on the old field, germinate, and by early winter produce rosettes. The following spring horseweed, off to a head start over crabgrass, quickly claims the field. During the summer the field is invaded by other plants—white aster *(Aster ericoides)* and ragweed *(Ambrosia artemissifolia)*. Competition from aster and inhibiting effects of decaying horseweed roots on horseweed itself allow aster to achieve dominance.

By the third summer broom sedge *(Andropogon virginicus)*, a perennial bunchgrass, invades the field. Abundant organic matter and the ability to exploit soil moisture efficiently permit broom sedge to dominate the field. About this time pine seedlings, finding room to grow in open places among the clumps of broom sedge, invade the field. Within five to ten years the pines are tall enough to shade the broom sedge. A layer of poorly decomposed pine needles (duff) that prevents most pine seeds from reaching mineral soil, dense shade, and competition for moisture among successfully germinating seedlings and shallow-rooted parent trees inhibit pines from regenerating on the site. Hardwoods grow up through the pines and, as the pines die (if they are not cut), take over the field. Further development of the hardwood forest continues as shade-tolerant trees and shrubs—dogwood, redbud, sourwood, hydrangea, and others—fill the understory. The sequence of communities has arrived at the mature or tolerant stage, in which only the dominant species of the crown can replace themselves in their own shade.

THE PROCESS

Although the impression is that as succession proceeds, one assemblage of species replaces another, the process takes place among individual plants of various species competing for available resources. The outcome is influenced by the timing of arrival, success at colonization, and the ability of individuals to exploit the available resources within the limitations imposed by the life-history characteristics common to the species.

The processes or mechanisms of succession have been described in terms of three models: facilitation, tolerance, and inhibition. The *facilitation model* regards succession as being driven by the organisms themselves. Early successional-stage species modify the environment and prepare the way for later-stage species, in effect facilitating their success.

The *tolerance model* involves the interaction of life-history traits, especially competition. It suggests that later successional species are neither inhibited nor aided by species of earlier stages. Later-stage species can invade a site, become established, and grow to maturity in the presence of those preceding them. They can do so because these later species have a greater tolerance to a lower level of resources than the earlier species. Such interactions lead to communities composed of those species most efficient in exploiting resources either by competing or by using resources unavailable to other species.

The *inhibition model* is purely competitive. No species is completely superior to another. A site belongs to those species that become established first and are able to hold their positions against all invaders. As long as they live, they maintain their positons, but the ultimate winners are the long-lived plants, even though early successional species may suppress them for a long time.

Succession on any site probably involves all three models, which together affect the efficiency of resource use and the outcome of competition among individuals of species of both early and late stages. Once a species has colonized an area, it has to compete with individuals of other species, as well as its own, for the available supply of limiting resources, chiefly light and soil nutrients. Succession comes about as the relative availability of resources, chiefly light and nutrients, and the ability of various plants to compete for those resources change through time. As plants grow they alter the environment, so the availability of resources changes, changing the rules for competitive success. No species can achieve maximum competitive ability under all circumstances. An inverse correlation develops between certain groups of traits. Species that are good competitors under one set of environmental conditions are poor competitors under another.

In the early stages of succession, soils typically are low in nutrients, especially nitrogen, and light is abundant. Plants that grow well under these conditions have the competitive advantage. As plants grow, ground cover and shading increase. Moisture and fertility improve as organic matter accumulates. Under these changing conditions, plants better able to exploit increased nutrients and reduced light gain the competitive advantage. Along this gradient of decreasing light and increasing nutrients, plant species and plant communities change (Figure 20.3a).

Under conditions typical of clear-cut forests and abandoned agricultural lands still high in nutrients, the resource gradient starts out relatively high in nutrients and high in light. Such conditions favor pioneer tree species, such as aspen, birches, yellow-poplar, and pines. Over time these species, being shade-intolerant, would yield to individuals of more shade-tolerant species, although gaps of high light would allow more long-lived individuals to retain their position (Figure 20.3b).

Some pioneer plant species, such as crabgrass, reduce their own competitive ability by producing chemicals that inhibit their own growth, an allelopathic effect that paves the way for invasion by grasses that are not affected by the toxins of weeds. Grasses in turn may inhibit nitrogen-fixing bacteria, thereby slowing succession to the next stage. Sassafras *(Sassafras albidum),* a pioneering tree species of later succession, maintains itself in relatively pure stands by releasing into the soil at different times of year plant toxins that inhibit the germination and growth of other plants.

COMMUNITY ATTRIBUTES

As succession proceeds, certain changes take place in the community. Early stages in succession are characterized by opportunistic species that respond quickly to disturbance and cope with a stressful environment. These plants generally are small and low-growing, have short life cycles, and produce large numbers of easily dispersed seeds. Their biomass is low and the source of nutrients is largely abiotic, coming from soil, air-blown dust, and rainfall. As succession proceeds from herbaceous to woody species—shrubs and trees—biomass accumulates and species diversity increases. These plants grow slowly and are long-lived. They produce few, heavy seeds, dispersed primarily by animals and gravity. Their seeds are large, providing an abundance of nutrients for the seedlings, but their vitality and longevity are low. These species are mostly specialists, adapted to a narrow range of environmental conditions. Much of their nutrient supply comes from rapid decomposition of organic matter and is recycled through the system. As the community matures, biomass accumulation in the form of new wood levels off, species diversity declines somewhat, and nutrients are tightly cycled from the soil and organic matter back through the plants. Only a small quantity of nutrients is lost from the undisturbed system.

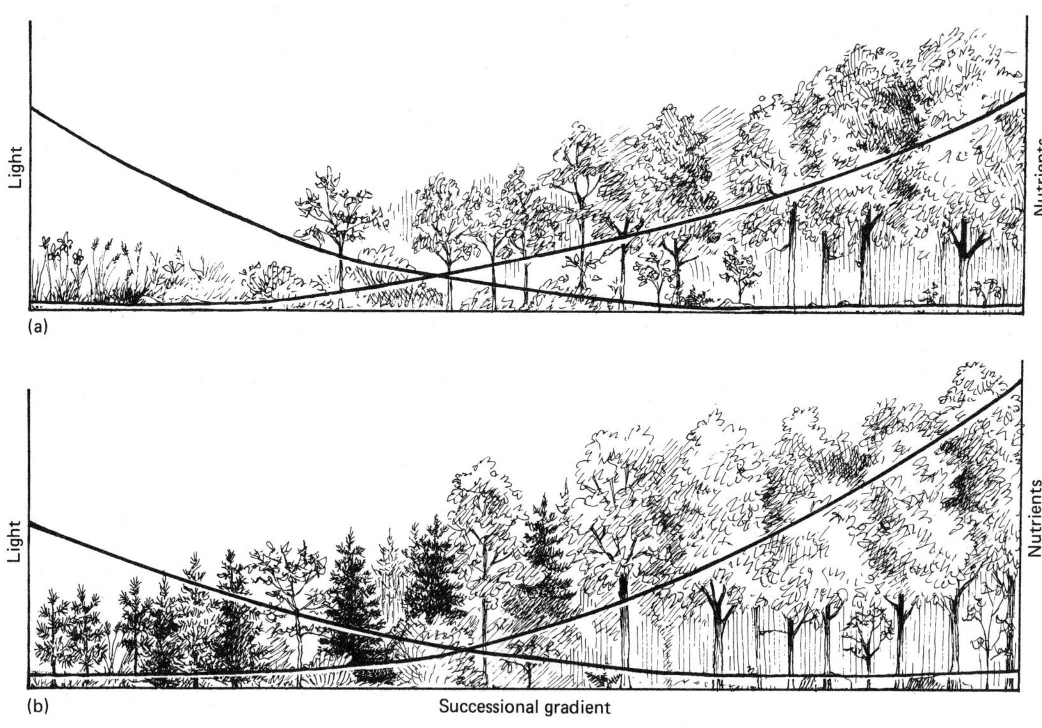

FIGURE 20.3 Competitive response of vegetation to a changing resource gradient as succession proceeds. (a) Succession on an open, barren area with an abundance of light and a low nutrient supply. (b) Succession on an open secondary with a high level of nutrients that supports pioneering woody vegetation.

AQUATIC SUCCESSION

The transition from pond to terrestrial community is a form of primary succession. It starts with open water and a bottom barren of life. The first forms of life to colonize the pond or lake are plankton, which may become so dense it clouds the water. If the plankton growth gets rich enough, the pond may support other forms of life—caddisflies, sunfish, perch, and bass.

At the same time the pond acts as a settling basin for sediment washed from the surrounding watershed. These sediments form an oozy layer that provides a substrate for rooted aquatics, such as the branching green algae *(Chara)* and pondweeds. These plants bind the loose mud and add materially to the accumulation of organic matter. Rapid addition of organic matter and sediments reduces water depth and increases the colonization of the basin by emergent vegetation, such as cattails and sedge, and submerged plants, such as pondweed. This activity enriches the water with nutrients and organic matter, further stimulating plankton and rooted plant production and sedimentation, and expands the surface area available for colonization by submerged, floating, and emergent plants. Eventually the substrate, supporting emergent vegetation such as sedges and cattails, develops into a *marsh*. As drainage improves and the land builds higher, emergents disappear, the soil rises above the water table, and organic matter, exposed to air, decomposes more rapidly. Meadow grasses invade to form a marsh meadow in forested regions and prairie in grassy country. Depending upon the surroundings, the area passes into grassland, forest, or peat bog.

In many instances aquatic succession reflects disturbance. Paleoecological investigations of the Indiana sand dunes suggest that the classic sequence of succession supposedly induced by plants is actually an effect of human disturbances to the surrounding land. The dune ponds remained more or less in the stages of floating and submerged vegetation until humans arrived; then changes were rapid. Increased sedimentation and changes in water chemistry brought about by nutrient-rich drainage from the surrounding watershed, pesticides, and runoff from roads and urban areas probably hastened the extinction of certain rooted plants and permitted the invasion of others. Thus rapid open water succession is a distortion of normal vegetational patterns.

HETEROTROPHIC SUCCESSION

Within each major community, and dependent on it for energy, are microcommunities. Dead trees, animal carcasses and droppings, and tree holes all furnish a substrate on which groups of plants and animals live, succeed each other, and eventually disappear, becoming incorporated into the soil. In these instances succession is characterized by early dominance of fungi and invertebrates that feed on dead organic matter, by maximum availability of energy and nutrients at the start, and by a steady decline of both as succession proceeds.

When a windstorm uproots or breaks a tree and sends it to the ground, the fallen tree becomes the stage for succession of plant and animal colonists that will stay with it until the log becomes part of the forest soil. The newly fallen tree, its bark and wood intact, is a ready source of shelter and nutrients. The first to exploit this resource are bark and wood-boring beetles that drill through the bark, excavate and feed on the inner bark and the cambium, reducing it to frass (droppings) and fragments, and tunnel galleries in which to lay eggs. Both adults and larvae drill more tunnels as they feed. Ambrosia beetles tunnel into the sapwood, creating galleries in which to grow fungi, on which both adults and larvae live. The tunnels provide a passageway and the frass and softened wood a substrate for bacteria and fungi, spores of which insects carry in. Loosened bark provides cover for predaceous insects soon to follow: centipedes, mites, pseudoscorpions, and beetles.

As decay proceeds, the softened wood holds more moisture, but the most accessible nutrients have been depleted, leaving behind more complex, decay-resistant compounds. The pioneering arthropods leave for other logs. Fungi possessing more sophisticated enzyme systems to break down cellulose and lignin move in to work on the sapwood. Moss and lichens find the softened wood an ideal habitat. Plant seedlings, too, take root on the softened logs, and their roots penetrate the heartwood, providing a pathway for fungal growth in the depths of the log.

Eventually the log is broken into light brown, soft, blocky pieces, and the bark and sapwood are gone. At this advanced stage of decay the log provides the greatest array of microhabitats and the highest species diversity. Invertebrates of many kinds find shelter in the openings and passages; salamanders and mice find shelter and dig tunnels in the rotten wood. Fungi and other microorganisms abound, and numerous species of mites feed on well-decomposed wood and fungi. At last the log crumbles into a red-brown mulchlike mound of lignin materials resistant to decay, its nutrients and energy largely depleted, and the tree is incorporated into the soil.

Such microcommunities illustrate one aspect of succession: The substrate can be changed by the organisms that depend upon it. When organisms exploit an environment, they may make the habitat unfavorable for their own survival and instead create a favorable environment for a different group of organisms. Our example began with specialists feeding on wood, and ended with generalists feeding on fungi and small invertebrates. Such succession, with its ultimate depletion of resources and reduction in species diversity, contrasts strongly with earlier examples, characterized by accumulation of biomass, energy, and nutrients, and increased feeding and habitat specialization as succession proceeds.

ANIMAL LIFE

As succession advances, animal life changes too (Figure 20.4). Each successional stage has its own distinctive groups of animals. Because animal life is influenced more by structural characteristics than by species composition, successional stages of animal life may not

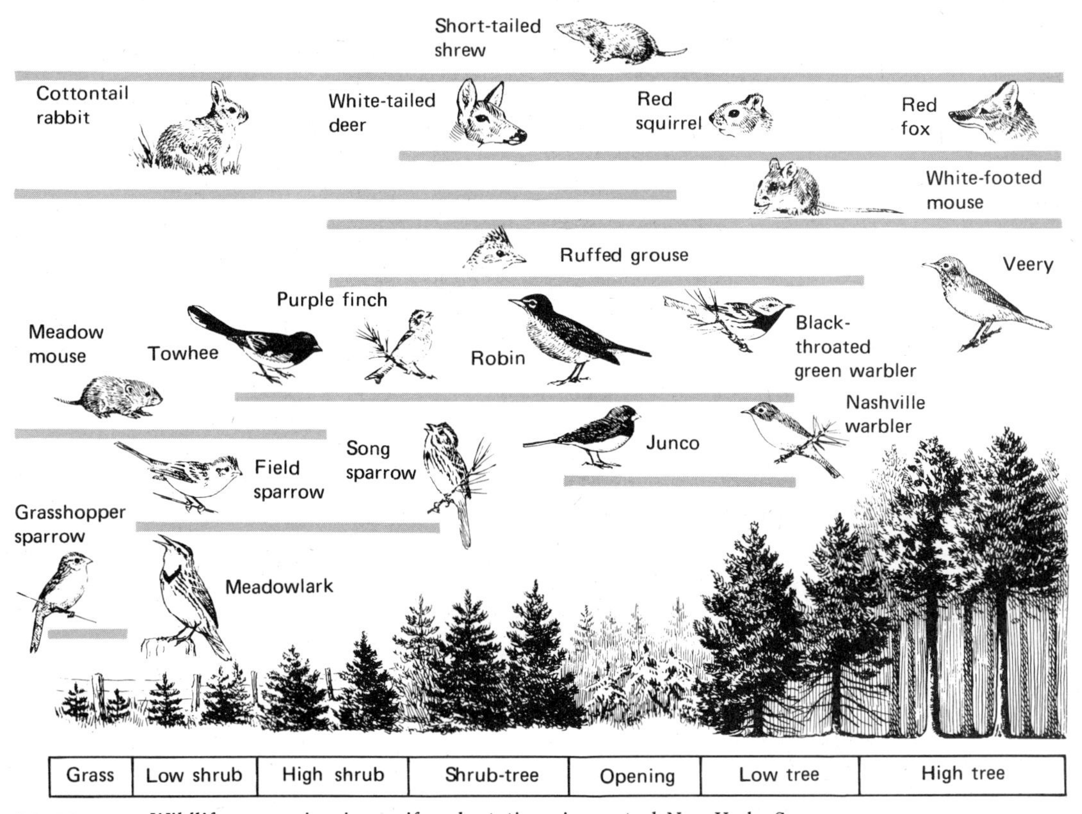

FIGURE 20.4 Wildlife succession in conifer plantations in central New York. Some species appear or disappear as vegetation density and height change. Other species are common to all stages.

correspond to the stages identified by plant ecologists. Animals, by their colonization, in effect classify a young stand of yellow-poplar or balsam fir under 6 m tall as a shrub stage of succession. A plant ecologist would consider the young yellow-poplar a shade-intolerant tree stage and the fir a shade-tolerant tree stage.

Animals can quickly lose their habitat by vegetation change. In eastern North America, where successional stages are most easily observed, early terrestrial stages of grasslands and old fields support meadowlarks, meadow mice, and grasshoppers. When woody plants, both young trees and shrubs, invade, a new structural element appears. Grassland animals disappear and shrubland animals take over. The thickets are claimed by towhees, catbirds, and goldfinches, and meadow mice give way to white-footed mice. When woody growth exceeds a height of 6 m, and the canopy closes, species of the shrubland decline, replaced by birds and insects of the forest canopy. As succession proceeds to mature forest, more structural elements are added. New species appear, such as tree squirrels, woodpeckers, and birds of the forest understory, like the hooded warbler and ovenbird.

The various stages of wetland succession support their own forms of life. Open water is inhabited by fish and used by waterfowl and herons as feeding areas. The floating aquatic stage supports hydras, frogs, diving beetles, gill-breathing snails, and insects colonizing the undersides of floating leaves. Emergent vegetation is occupied by nesting waterfowl and bitterns, redwinged blackbirds, muskrats, amphibians, flies and mosquitos, mayflies and dragonflies. Incoming woody vegetation such as alder and willow creates habitats for swamp sparrows, yellowthroats, woodcocks, and white-footed mice.

In general the diversity of animal life changes as the succession proceeds (Figure 20.5). Herbaceous and shrubland stages support the greatest diversity of animal species. These species depend upon the pioneering of plant succession for their habitat. As that stage passes away, so do the animals that occupy it. Therefore certain species of animal life are highly dependent on disturbances that restore or maintain early stages of succession. Among such animals are bobwhite quail, cottontail rabbit, prairie warbler, and woodcock. Because of their dense canopy and lack of understory vegetation, young forest stands hold the lowest diversity, which increases as the forest matures. Old-growth forests, the last stage in succession, support higher diversity than other forest stages because of more varied habitats, including dead and fallen trees and gaps in the canopy that stimulate new understory growth.

DIRECTION OF SUCCESSION

Classic successional theory holds that succession is directional and therefore predictable. The once dominant species or group of species will not become dominant again, unless a disturbance intervenes (Chapter 21).

We could predict with a high degree of probability that an old field in eastern North America will, barring further disturbance, return to forest. We would have much more difficulty predicting the kind of forest, even if we had knowledge of the previous vegetation. Each successional community is individualistic, a one-time product of the abiotic and biotic forces operating during its development. The exact interaction of these forces will not be repeated again. Any new successional community will be molded by current abiotic and biotic inputs. The exact original composition of species will not be duplicated. Thus we might predict the type of vegetation over a region, but not all the local communities. There are too many side roads succession can take (Figure 20.6).

CYCLIC REPLACEMENT

Successional stages that appear to be directional are often phases in a cycle of vegetational replacement. Such cycles come about when some periodic disturbance restarts succession. Such changes are part of community dynamics, usually occur on a small scale

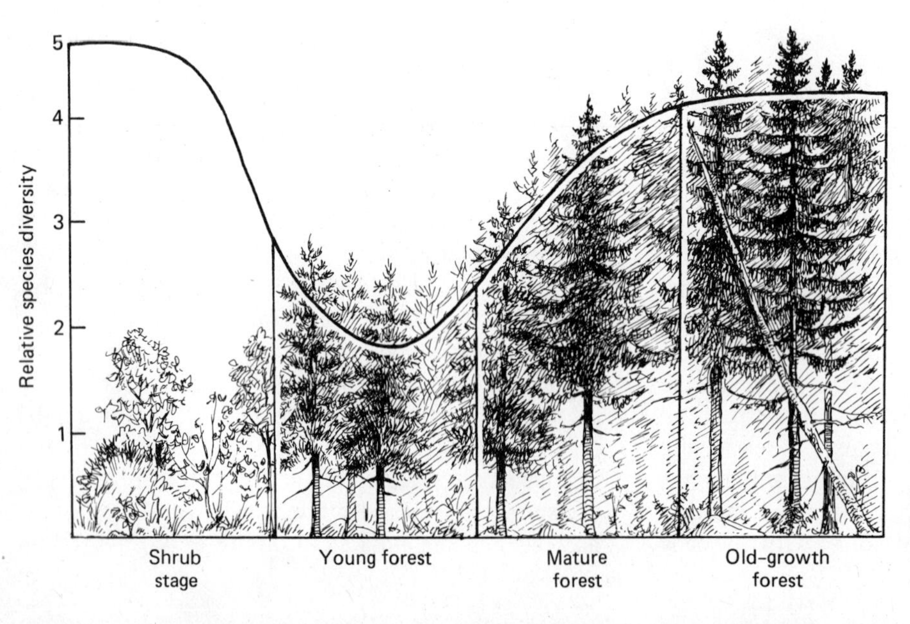

FIGURE 20.5 Relationship between successional stage in the Pacific Northwest forests and the number of mammal species present.

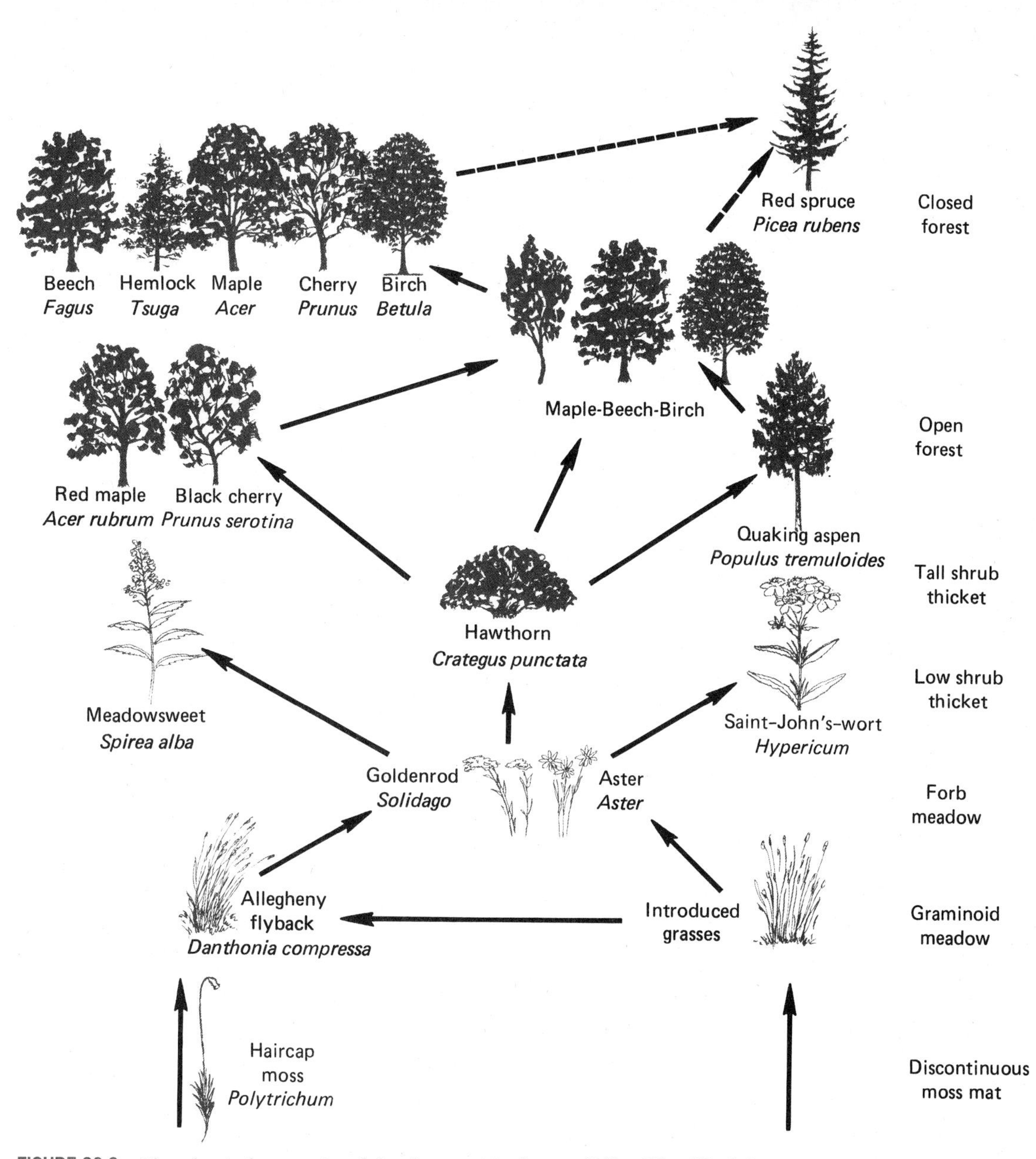

FIGURE 20.6 Flowchart of successional development in Canaan Valley, West Virginia, a high-altitude hanging valley surrounded by still higher mountains. Once covered with spruce forests and open glades, the valley was cut over, burned, and converted into farm and grazing land. Much of this marginal farmland has been abandoned and is reverting to natural vegetation (except where it is being converted into condominiums). Successional trends in this area vary with the nature of the soil and the depth of the water table. This diagram reflects successional development on a moderately well-drained, acidic soil. The starting point is either moss, a native grass (Allegheny flyback), or introduced grasses. Under certain conditions, succession may terminate in shrub communities of either Spirea or Saint-John's-wort, stands of which are so dense that vegetative growth beneath them is suppressed. In other instances, succession leads to a hemlock-beech-maple forest with only slight indication that succession might proceed to spruce.

within the community, and are repeated over the whole of the community. Each successive community or phase is related to the others by orderly changes. Such cyclic replacements contribute to community persistence.

The English ecologist Andrew Watt described such cycles in Scottish heaths. Scottish heather *(Calluna)* represents the peak of the upswing series. After the death of heather, a lichen *(Cladonia silvatica)* becomes dominant and covers the dead heather stems. Eventually the lichen disintegrates to expose bare soil, the last of the downswing communities. The bare soil is colonized by bearberry *(Arctostaphylos* spp.) to initiate another upswing. Heather then reclaims the area and dominates again. There are also shorter cycles.

Cyclic replacement is common and important in different ecosystems. Cyclic succession is frequently initiated by ants or ground squirrels in old-field communities in Michigan, where it involves lichens, mosses, Canada bluegrass, and dock. The overriding pattern of successional development on the coastal tundra of Alaska is cyclic, controlled primarily by changes in microrelief and drainage. Cyclic replacement also retains the long-term stability of pothole marshes in north-central North America (Figure 20.7). During periods of drought—about every 5 to 20 years—shallow marshes dry. Organic debris accumulated on the bottom decays rapidly, releasing nutrients for recycling and stimulating the germination of seeds. The upswing of the cycle begins with seed germination on exposed mud. That stage is followed by a newly flooded stage with sparse, often well-dispersed vegetation, dominated by annuals and immature perennials; a flooded, dense marsh dominated by perennials; and a deep, open marsh rimmed with emergents, fostered by the feeding activities of muskrats. The cycle begins anew when the marsh dries. Although these short-term cycles give the shallow marsh the appearance of an unstable ecosystem, such cyclic replacements ensure the long-term stability of the marsh ecosystem.

FLUCTUATIONS

Fluctuations are nonsuccessional or short-term reversible changes. Fluctuations differ from succession in that the floristic composition over time is stable: No new species invade the site and species may return to dominance. These changes result from such environmental stresses as soil-moisture fluctuations, wind, grazing, and the like.

Fluctuations in forest communities may involve an alternation of species in canopy gap replacements (see Chapter 21). In such forests each species tends to be replaced by its competitor. If one species becomes moderately abundant in the canopy, alternate species may be abundant beneath it. Thus over time, dominance in the canopy may shift in favor of the temporarily disadvantaged species. For example, in old northern hardwood stands, sugar maple tends to replace beech in small openings, and beech to replace sugar maple. In general the tendency is for the dominant tree to be replaced more than half the time by its competitor. Such alternation probably results for two reasons. First, the dominant tree usurps the site, concentrating the bulk of biomass at one particular place in a single tree. Its conspecifics are thinned out more severely than its competitors. Second, because of its influence on nutrient regeneration, light, and moisture, the canopy tree creates a somewhat species-specific microhabitat for seeds and seedlings beneath it. With an alternate species favored in the understory, the forest maintains species equilibrium.

Fluctuations may also involve replacement of one age class by another within a species. Such fluctuations are important in maintaining certain forest ecosystems, particularly coniferous forests. D. G. Sprugel described a wave regeneration pattern in balsam fir *(Abies balsamae)* forests in the northeastern United States. In this fluctuation trees die off continually at the edge of a "wave" and are replaced by vigorous stands of young balsam fir (Figure 20.8). The cycle is initiated when an opening occurs in the forest, exposing trees to the wind on one side of the opening. Desiccation of the canopy foliage by winter winds, the loss of branches and needles in winter from rime frost, and decreased primary production due to cooling of needles in summer cause the death of windward trees. Their death exposes the trees behind them to the same lethal conditions, and they die next. As this process continues, a wave of dying trees through the forest is followed by a wave of vigorous reproduction.

These regeneration waves follow each other at in-

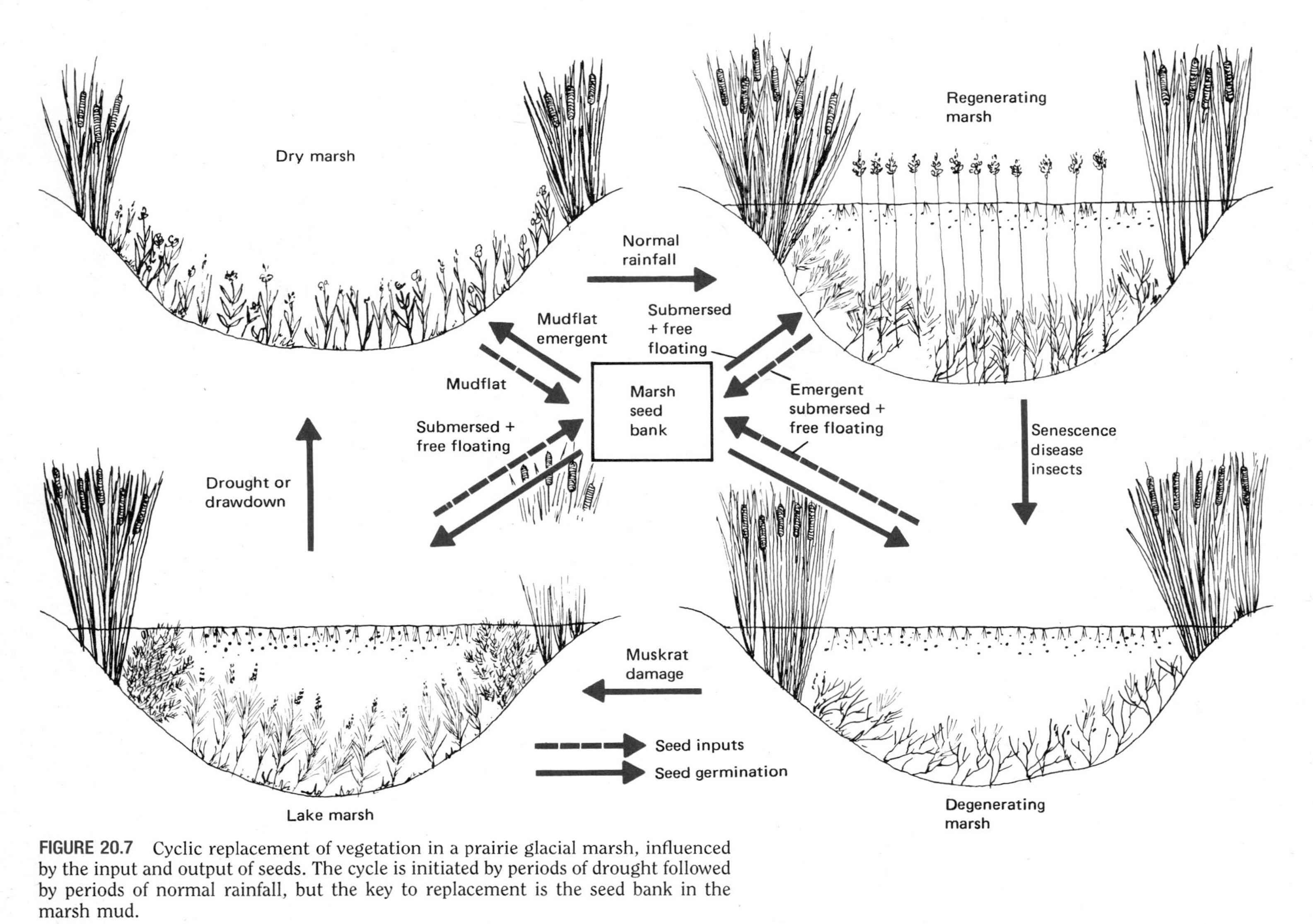

FIGURE 20.7 Cyclic replacement of vegetation in a prairie glacial marsh, influenced by the input and output of seeds. The cycle is initiated by periods of drought followed by periods of normal rainfall, but the key to replacement is the seed bank in the marsh mud.

FIGURE 20.8 A regeneration wave in a balsam fir forest. The wave is initiated at the location of the standing dead trees with mature trees beyond it and an area of vigorous reproduction below it. Where dead trees have fallen, a crop of young fir seedlings is developing. Beyond them is a dense stand of fir saplings, followed by a mature fir forest and then by a second wave of dying trees.

tervals of about 60 years. The process is so regular that all stages of degeneration and regeneration are found in the forest at all times, provided the stand is not cut. The cycle results in a steady state because the degenerative changes in one part of the forest are balanced by regenerative stages in another. The wave regeneration process ensures the stability of the forest and prevents its advancement into a hardwood stage.

THE CLIMAX

According to classic theory, succession stops when the community has arrived at an equilibrium or steady state with the environment. At this point the community is stable and self-replicating and, barring major disturbances, will persist indefinitely. This end point of succession is termed the *climax*. Many people misunderstand the concept and assume that any stand of large trees is a climax.

The climax, theoretically, takes on certain characteristics. The vegetation is tolerant of the environmental conditions it has imposed upon itself. It is characterized by an equilibrium between gross primary production and total respiration, between energy utilized from sunlight and energy released by decomposition, between the uptake of nutrients and the return of nutrients by litterfall. It has a wide diversity of species, a well-developed spatial structure, and complex food chains.

Every individual in the climax stage is supposedly replaced by another of the same kind; average species composition reaches an equilibrium. If offspring of the same species are favored over others, then a dead mature individual may be replaced by a plant of its own kind. If the offspring are concentrated about a mature parent, it may be replaced by its own progeny. This outcome is most likely if the replacement is essentially the same unit—a root or stump sprout from the dead individual. If conditions beneath the mature tree are less favorable for its own species than for other species, it will be replaced by another associated species. If conditions are neither more nor less favorable for the offspring than for other species, replacement individuals will be influenced by the relative abundance of seeds arriving on the site, suppressed individuals already present, and competitive interactions among them.

But self-destructive changes are continually taking place in the climax community. Trees grow old and die and more often than not are replaced by individuals of a different species. Changes are constantly occurring in patches across the community (see Chapter 21), slowly altering the average species composition. Although succession may slow down, it never ceases; and the stability and persistence of the climax reflect mostly the fact that the dominant species are long-lived in terms of a human time frame.

Even within a region the climax is not uniform vegetation. Climax vegetation is really a mosaic of vegetation types influenced by soil moisture, soil nutrients, topography, slope exposure, fire, animal activity, and availability and chance dispersal of seeds. The mosaic of climax vegetation changes as the environment changes. The climax community, then, represents a pattern of populations that corresponds to and changes with the pattern of environmental gradients to form ecoclines.

The idea of the climax has been weakened to the point of uselessness by widespread application to mature and old-growth vegetation that happens to be older than the lifetime of the visitor. Nevertheless it is useful in describing and mapping regional vegetation. This concept is important also because it affects management and preservation of old-growth forests, which depend upon the maintenance of certain environmental conditions. Old-growth sequoia, for example, even though 2000 to 3000 years old, is not climax vegetation. The potential climax is white fir *(Abies concolor)* and incense cedar *(Libocedrus decurrens).* Sequoia for thousands of years has depended upon low-intensity ground fires (see Chapter 21) to hold back invading species and to eliminate fungal pathogens. With fire suppression, a heavy understory of fir and cedar has developed, creating a potential fire ladder that could carry a devastating fire into the crowns of sequoia.

Vegetation is dynamic, and mortality is a feature of the climax. As old trees die, new individuals and species take their place. Thus species abundance and composition change, even though the forest's general appearance remains the same. Trees continue to grow, although slowly. The forest never achieves an equilibrium in growth and species composition. Rather, the standing crop of living biomass and total biomass (living + dead) fluctuates. The forest consists of patches in various stages of successional development, from ones with young growing trees to ones of senescence and downed timber.

Studies of old or mature ecosystems needed to test the climax theory are rare, especially in the eastern deciduous forests, where extremely few old untouched stands exist. One is the Dick Cove Natural Area, a mesic forest on the western slope of the Cumberland Plateau in Franklin County, Tennessee. It is a 40-ka tract surrounded by second-growth forest. The forest dominants are oaks (northern red oak, white oak, chestnut oak), then hickories, sugar maple, and yellow-poplar, associated with some 17 other species. Age is beginning to tell on the old trees. During an eight-year period from 1972 through 1981, 26 percent of the hickory and 18 percent of white oak and red oak over 43 cm dbh died from a combination of senescence, drought, insect damage, and blowdown. These dead trees ranged in age from 90 to 375 years. Dead red oak averaged 135 years and dead white oak, chestnut oak, and hickory averaged 210 years. The demise of the oaks after 350 years appears to be accelerating, bringing about a change in structure and species composition of the stand from the original apparent climax of oak and hickory to one of the slow-growing, shade-tolerant sugar maple and the fast-growing, shade-intolerant yellow-poplar with a strong component of hickory. There are few replacement trees of red oak, and yellow-poplar in all size classes is filling in the gaps (see Chapter 21). Even old-growth apparent climax forests are in flux over periods of several hundred years. What appears permanent to the observer is slowly transitory.

This point is even more applicable to the old-growth 450- to 1000-year-old stands of Douglas-fir of the Pacific Northwest. Technically, these old Douglas-fir forests are not climax, because Douglas-fir is a pioneer species that happens to be long-lived. The climax species that might replace Douglas-fir in another 500 years are western hemlock *(Tsuga heterophylla)* and Pacific silver fir *(Abies amabilis).* Only by human standards of permanency and by contrast to the deciduous forests of eastern North America are these old stands "climax." These old Douglas-fir stands are uneven-aged when, theoretically, they should be even-aged. Age classes range from 145 years and up within the stand, suggesting occasional disturbances and replacement, or perhaps failure of canopy closures over a period of time because of variable seed crops. Ratio of live biomass to dead biomass peaks at 300 to 400 years. Total biomass is its highest at about 750 years, and total dead biomass is greatest at 800 to 1000 years. The forest is a complex system of many species of fungi and epiphytic lichens (ones that grow up in the trees). Nutrient retention within the system involves complex pathways. Nitrogen is fixed by lichens in tree crowns and by microflora in decomposing logs. Structural diversity is high with a large range in the size of individual living trees and numerous large standing dead trees, large downed trees, and decomposing logs. When these old-growth stands are cut, an entire ecosystem is destroyed that cannot be replaced for hundreds of years, if ever. Like a species, once destroyed, an old growth forest with all its complexity and diversity can never be brought back.

SUCCESSION AND TIME

Time is an integral component of succession; but time is measured in terms of human experience. Climax vegetation is theoretically permanent, but what is permanent? Vegetation that remains the same over one or several human lifetimes? By that standard, old-field vegetation could be climax to ants, birds, or meadow mice.

Nevertheless, successional communities have their time spans, governed in part by the longevity of the plants that make up the stages. An old field with annual weeds may exist no more than 1 to 2 years. Pioneer lichens and moss on a granite outcrop may exist for hundreds of years. Grass stages may last 10 to 15 or even fewer years before being overtaken by woody growth. Woody growth in a shrub stage—whether true shrubs or incoming tree growth below a height of 6 m—may last an additional 10 to 15 years until pioneering trees take over or the canopy closes. If trees are pioneer, shade-intolerant species such as aspen, the stage may last 25 to 40 years before shade-tolerant

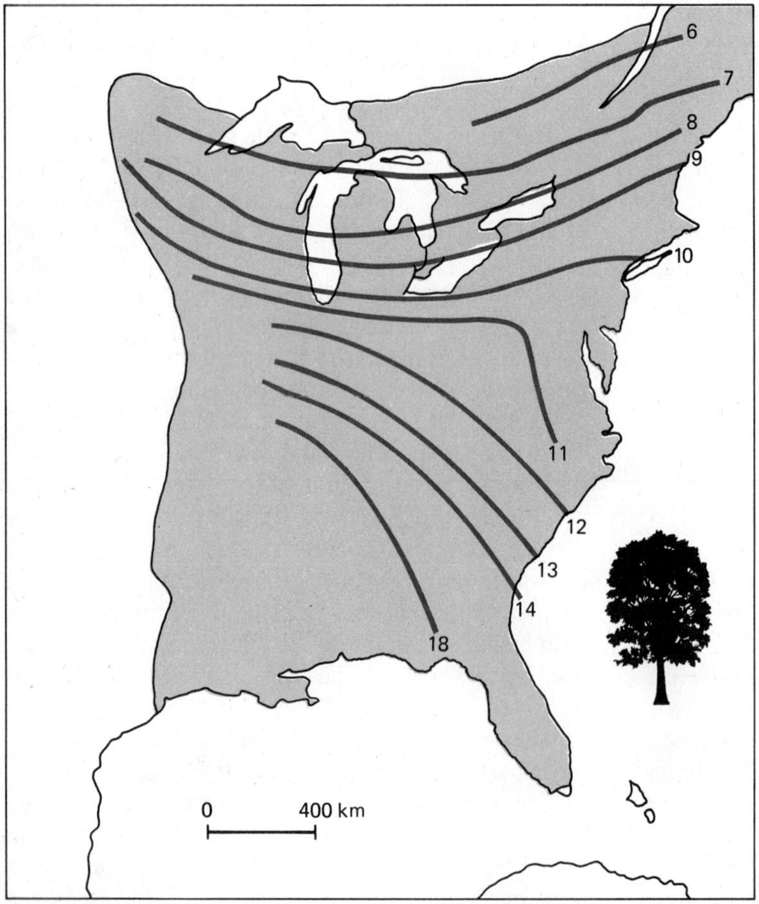

FIGURE 20.9 Following the melting of the last ice sheet, our familiar maples moved out of their refuge in the southern part of North America. The lines represent the leading edge of the northward-expanding population. The shaded area shows the modern range. The numbers tell age in thousands of years. Several species of maple reached their northern limits only about 6000 years ago.

species become dominant and hold the site 250 to 500 years.

This time line implies that shade-tolerant species replace shade-intolerant ones and that trees replace shrubs. Not always: A dense growth of shrubs such as meadow sweet, mountain laurel, or Saint-John's-wort may claim a site permanently—50, 60, 70 years without any indication of change. Shade-intolerant species can hold on to a site for many years. Yellow-poplar, white pine, and red pine, to mention several pioneer species, can colonize an abandoned field and remain dominant for well over 100 years. Douglas-fir, a pioneer species in western North America, may claim a site for over 1000 years before theoretically giving way to western hemlock, which may need another 500 years to become dominant.

How long a successional stage may last could be simply an academic question, except that time has implications in forestry and wildlife management. Certain early successional types of wildlife habitat are ephemeral. Their maintenance may be critical to the welfare of certain species, which require human interference with the successional process. Some commercially valuable timber trees are pioneering species that require periodic disturbance to ensure regeneration.

Long-term evolution and global climate swings bring long-term successional trends. For example, much of the present pattern of vegetational distribution has been influenced by the glacial events of the Pleistocene. During the last great ice sheet, when the climate was 5° C cooler than the present climate, spruce, fir, and various species of hardwood trees found refuge at varying distances south of the glacier. As the glacial ice retreated, temperatures gradually warmed to that of the present, and species of plants and animals dispersed northward, their speed of movement influenced by their rate of dispersal and by climatic conditions (Figure 20.9). Coniferous species displaced tundra, and temperate deciduous species replaced spruce and pine. If global temperature rises 2° C within the next 50 years, as some scientists predict, that would be an equivalent of a 200-mile northward shift in latitude. To remain in an equitable climate, both boreal and temperate plant species would have to disperse about 20 times faster than the rate following glacial retreat, a speed much too fast for species to adapt to a changing climate. As a result many species would disappear, associated animal species would become extinct, and new communities of adaptable species would develop. Today's familiar landscape would no longer exist.

Summary

With the passing of time natural communities change. Old fields of today return to forests tomorrow; weedy fields in prairie country revert to grasslands. This gradual sequential change in the relative abundance of dominant species in a community is succession. It is characterized by the replacement of opportunistic, early-stage species by late-stage species, by a progressive change in community structure, and by an increase in biomass and organic matter accumulation. Succession that begins on sites devoid of or unchanged by organisms is termed primary; and succession that begins when organisms already are present is called secondary.

The mechanisms of succession involve a pattern of species replacement driven by chance, differential longevity, and competition among individuals. Various combinations of life-history and physiological traits, characteristic of species but expressed by individuals, interacting with environmental conditions, produce the variety of community patterns.

Eventually communities arrive at some form of steady state with the environment. This stage, usually called the climax, is more or less self-sustaining. The climax usually involves a mosaic of regenerating patches in which new growth may not be the same species as the individuals being replaced.

Changes that are not truly successional include cyclic replacement and fluctuations. Periodic biotic or environmental disturbance can start regeneration again and again at some particular stage. Such cyclic succession aids in community persistence. Nonsuccessional, reversible changes or fluctuations in communities result from environmental stresses such as changes in soil moisture, wind, and grazing.

The most outstanding characteristic of natural communities is their dynamic nature. They are constantly changing through time—rapidly in early stages of development, more slowly in later stages. Even those communities that are seemingly the most stable slowly change through time. Successional changes in vegetation are accompanied by changes in animal life.

Some of the most pronounced changes occurred during the Pleistocene. At that time several slow advances and retreats of ice sheets dramatically altered vegetation over much of the Northern Hemisphere. Today global warming threatens to occur too fast for plants and animals to adapt.

Review and Study Questions

1. What is succession?
2. Distinguish between primary and secondary succession.
3. Discuss the role of cyclic replacement and fluctuations.
4. What part might allelopathy play in succession?
5. How does succession work in microcommunities?
6. What is the climax? Is it a valid concept?
7. What is the relationship of plant succession to animal habitats?
8. Relate r- and K-selected species (Chapter 14) to species of early- and late-stage succession. In these terms, compare a pine tree with crabgrass as an invader of old fields.

*9. Locate an area with which you were familiar long ago. What vegetational changes have taken place? What brought them about: land abandonment, logging, suburbanization? Old aerial photographs will provide insights.

Selected References

Birks, H. H., H. J. Birks, P. E. Kaland, and D. Moc. 1989. *The cultural landscape: Past, present, and future.* Cambridge, England: Cambridge University Press. Explores human interference with natural European vegetation and the results.

Bormann, F., and G. E. Likens. 1979. *Patterns and process in forested ecosystems.* New York: Springer-Verlag. A basic study of forest succession at Hubbard Brook, New Hampshire.

Davis, M. B. 1983. Holocene vegetational history of the eastern United States. In H. E. Wright, Jr., ed. *Late-quaternary environments of the United States. Vol II. The Holocene.* Minneapolis: University of Minnesota Press. Pp. 166–188.

Delcourt, P. A., and H. R. Delcourt. 1981. Vegetation maps for eastern North America, 40,000 yr BP to present. In R. Romans, ed. *Geobotany.* New York: Plenum Press. Pp. 123–166.

Golley, F., ed. 1978. *Ecological succession.* Benchmark Papers. Stroudsburg, PA: Dowden, Hutchinson, and Ross. Succession as viewed over time.

Huston, M., and T. Smith. 1987. Plant succession: Life history and competition. *American Naturalist* 130:168–198.

Keever, C. 1950. Causes of succession in old fields of the Piedmont, North Carolina. *Ecological Monographs* 20:229–250.

Knapp, R., ed. 1974. *Vegetation dynamics.* Vol. 8. *Handbook of vegetation science.* The Hague: W. Junk. Classical views of succession.

Leck, M. A., V. T. Parker, and R. L. Simpson. 1989. *The ecology of soil seed banks.* Orlando, FL: Academic Press. Relationship of seeds in soil to vegetation dynamics.

Maser, C., and J. M. Trappe, eds. 1984. *The seen and unseen world of the fallen tree.* USDA Forest Service Gen. Tech. Rept. PNW-164.

Olson, J. S. 1958. Rates of succession and soil changes in southern Lake Michigan sand dunes. *Botanical Gazette* 119:125–170.

Sprugle, D. G. 1976. Dynamic structure of wave-generated *Abies balsamea* forests in northeastern United States. *Journal of Ecology* 64:889–911.

Tilman, D. 1985. The resource-ratio hypothesis of plant succession. *American Naturalist* 116:362–369.

Walker, L. J., J. C. Zasada, and F. S. Chapin, III. 1986. The role of life history processes in primary succession on Alaskan flood plain. *Ecology* 67:1243–1253.

Watt, A. S. 1947. Pattern and process in the plant community. *Journal of Ecology* 35:1–22.

West, D. C., H. H. Shugart, and D. B. Botkin, eds. 1981. *Forest succession: Concepts and applications.* New York: Springer-Verlag. Excellent reference on patterns and processes of succession from temperate to tropical forests.

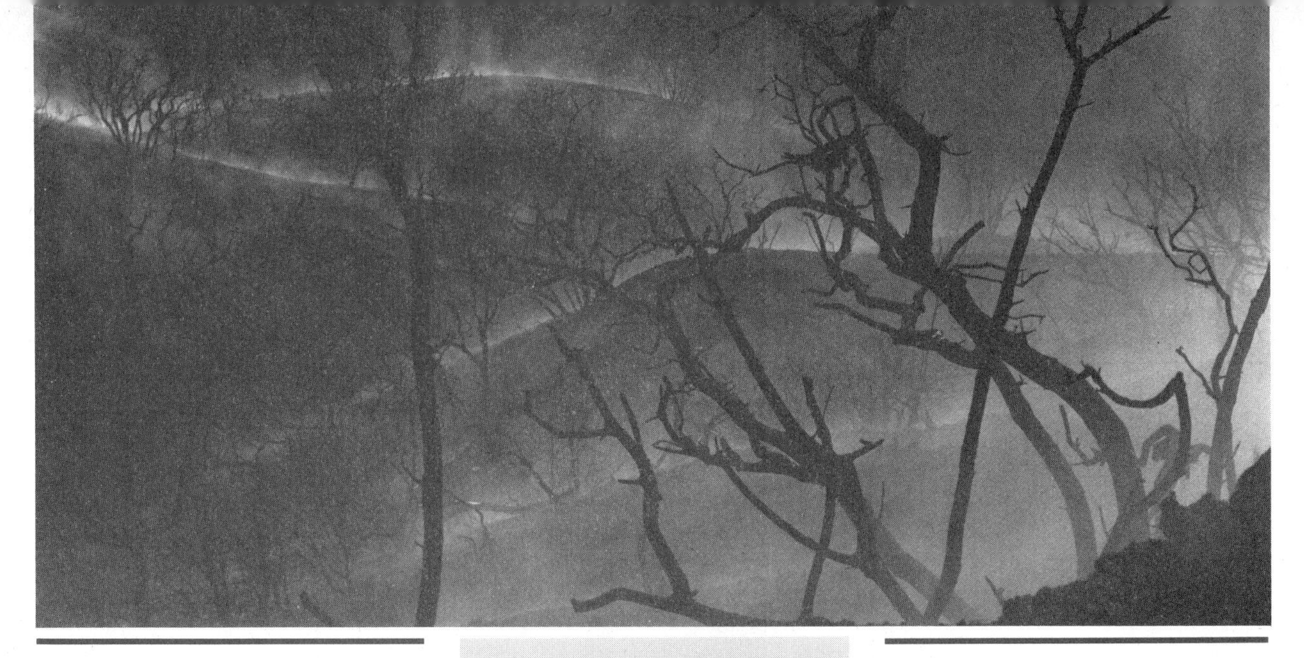

CHAPTER 21

Natural Disturbance and Human Impact

Outline

Objectives

On completion of this chapter, you should be able to:

1. Define disturbance, large-scale and small-scale.
2. Comment on the agents of disturbance, especially human ones.
3. Describe three types of forest fires and their effects.
4. Discuss the effects of fire on plant communities and on vertebrates, especially fire-dependent species.
5. Discuss the role of humans in the fire environment.
6. Comment on the relationship between disturbance and community stability.

Disturbance is a relatively discrete event in time that disrupts ecosystems, communities, or populations, changes substrates and resource availability, and creates opportunities for new individuals or colonies to become established. Disturbances have both spatial and temporal characteristics: size of the area disturbed; frequency, the mean number of events per unit time; turnover, the mean time between disturbances; intensity, the physical force of the event per area per time; and severity, the impact on the population or the community.

SCALE OF DISTURBANCE

Disturbance ranges from small, frequent events such as the death of a single tree in a forest to large-scale, rare events, in which extensive areas are swept by fire, buried under volcanic ash, torn by landslides, or denuded by human land-clearing schemes.

SMALL-SCALE DISTURBANCES

What constitutes a small-scale disturbance depends on the scale of the landscape in which it occurs. The loss of a group of trees in a very small woodland would have more impact than the loss of scattered individual trees or small groups of trees in a large forest.

In small-scale disturbances the death of individuals or groups opens the canopy or the substrate. Abrasive action of waves tears away mussels and algae from hard rocky tidal substrates. In grasslands, digging by badgers and groundhogs exposes patches of mineral soil for colonization by herbaceous plants. A tree dies or falls in a forest, opening up the canopy. In these instances the effect of the disturbance is to create a *gap,* a site for regeneration and new growth.

We tend to think of familiar natural landscapes as frozen in time, like an image in a photograph. To preserve an area as we see it, we protect it against fire, insect attack, and other events that seem harmful. In doing so we change it, because nature is not constant. Disturbance is the means by which landscape diversity is maintained.

Consider a scene in the Wind River Mountains of Wyoming. The area is largely grassland with patches of sagebrush, and the distant ridges and rock outcrops support scattered pockets of conifers, apparently swept by fire in the past (Figure 21.1). Protected from fire and other disturbances for over 90 years, the vista has changed. The slopes support lodgepole pine and limber pine. Sagebrush and bitterbrush, both fire-sensitive shrubs, have increased, as well as such cloning shrubs as willow and aspen, especially in the bottomlands. Protected from disturbance, the landscape has changed significantly over the years.

Gaps, large or small, in the forest are sites of increased availability of light, soil, warmth, and nutrients, and decreased soil moisture and relative humidity. Suppressed growth is quickly stimulated by this sudden abundance of resources. If gaps are small, such as those created by the death or loss of a tree, the response typically is reorganization of vegetation. The canopy expands to fill in above, while understory shrubs and seedlings and saplings of forest trees spring up below. Usually, small gaps favor the growth of tolerant species, so in a terminal or climax community, replacement species are likely to be those of the canopy.

If the gaps are large—the kind that result from some timber cutting, insect damage, windstorms, and ice storms—the response may involve both reorganization and invasion by opportunistic species. In deciduous forests, for example, the future composition of the gap will be determined in part by the competitive interactions of incoming growth. Intolerant species may outcompete tolerant species, which will remain in the understory, capable of filling in small gaps that appear later. Gap formation and gap replacement are important in maintaining diversity in tropical forests. In fact, they are essential for the regeneration of many species, both pioneer and primary.

Gaps result in patches of different stages of successional or compositional maturity. Chronic small-scale disturbances are important in the maintenance of species richness and structural diversity within a mature forest ecosystem.

LARGE-SCALE DISTURBANCES

Large-scale disturbances induced by fire, logging, landclearing, and other such events invite colonization by opportunistic species. What species colonize the dis-

FIGURE 21.1 (a) A northwest view of Chief Washakie's Shoshone encampment on Willow Creek, Wind River Mountains, Wyoming, August 1870. Shrubs in the foreground are sagebrush. Note the patches of conifers on the distant slopes. (b) The same area 97 years later. The distant slopes support lodgepole pine and limber pine. In the foreground are sagebrush, bitterbrush, and willow.

turbed area is influenced by seeds present on the site and nearby seedlings, saplings, soil conditions, amount of competition, and other ecological conditions. To take advantage of the disturbed conditions, opportunistic species should be nearby; their seeds should reach the site in sufficient numbers to ensure some seed survival; and seeds should settle on exposed mineral soil to favor rapid germination and seedling survival.

Some species colonize a disturbed area with dormant seeds or roots. Their response to the sudden removal of a forest canopy, as in clearcutting, is rapid. One example is pin cherry (*Prunus pensylvanicus*), whose seeds are carried to a forest by birds and small mammals or dropped by an earlier stand of cherry. Pin cherry seeds can remain dormant for up to 50 years. When the forest canopy is removed and moisture, temperature, and light conditions become favorable, pin cherry seeds germinate and young trees quickly dominate the site, crowding out the associated blackberry (*Rubus* spp.) that also colonizes the area. If the seedling growth is dense, a pin cherry canopy can close in four years, eliminating other species except highly shade-tolerant seedlings of sugar maple or beech (Figure 21.2). If seedling growth is moderately dense, species with wind-disseminated seeds, such as yellow birch and paper birch, will also occupy the site. Within 30 to 40 years pin cherry dies out, allowing birch, sugar maple, and beech to dominate the gap; but during its tenure pin cherry contributed numerous seeds to the forest floor, ready to reclaim the site when another disturbance provides the opportunity.

Although the response of woody vegetation to large-scale removal of timber is well understood, little attention has been given to the response of the shade-tolerant understory herbaceous plants. J. E. Ash and J. F. Barkham studied the response of the herbaceous understory layer of an English coppice forest after cutting. (A coppice forest is one in which the stump sprouts or root suckers are the main source of regeneration, with cutting rotations between 20 and 40 years.) Cutting coppice involves complete canopy removal, resulting in increased surface temperature on the forest floor and full exposure to light. Typically, open habitat or opportunistic species germinate and become established but are soon excluded by the developing canopy cover. In spite of the disturbance, characteristic woodland species persist throughout the cycle. Adapted to a high light regime in spring before

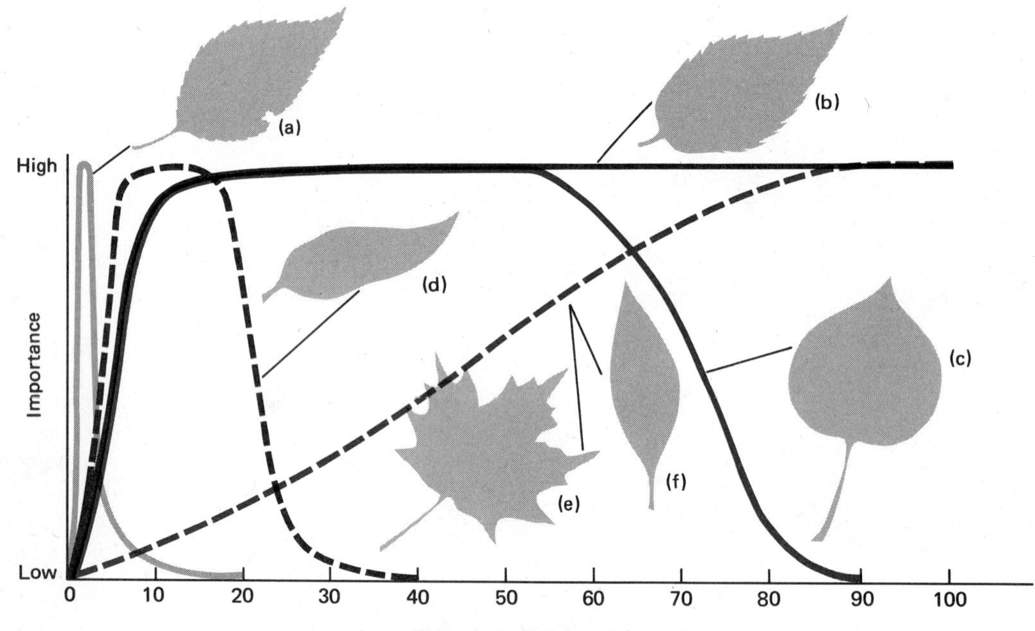

FIGURE 21.2 Species assume different degrees of importance along a gradient of time following the disturbance of a typical northern hardwoods forest. Immediately after disturbance, blackberries (a) dominate the site, but they quickly give way to yellow birch (b), quaking aspen (c), and pin cherry (d). Intolerant pin cherry assumes dominance early but within 30 years fades from the forest. Yellow birch, an intermediate species, assumes early dominance, which it retains into the mature stand. Quaking aspen, an intolerant species, begins to drop out after 50 years. Meanwhile sugar maple (e) and beech (f), highly tolerant species, slowly gain dominance through time. In about 100 years the forest is dominated by beech, maple, and birch.

the leaves are out, these plants have the ability to tolerate full exposure. At the same time they are able to coexist with annuals and open-habitat perennials that cast a shade on the ground like a tree cover. As opportunistic species disappear, woodland herbs again assume dominance, often developing into monospecific stands. Probably much the same response occurs in North American forests subject to similar large-scale disturbance.

NATURAL AGENTS OF DISTURBANCE

WIND

Wind is a subtle agent of disturbance to vegetation. It shapes the canopies of trees exposed to prevailing winds, affects their growth of wood, and uproots them from the ground. Especially vulnerable to windthrow are mature trees whose trunks lack the suppleness of youth and trees growing on shallow and poorly drained soils in which the roots, spreading along the ground, cannot get a firm grip in the soil (Figure 21.3). Trees weakened by fungal disease, insect damage, and lightning strikes and tropical forest trees carrying a heavy load of epiphytes in their crowns are also candidates for windthrow. The impact of wind is accentuated when strong winds accompany heavy snowfall that weighs down trees or heavy rains that soften the soil. Hurricanes and tornadoes have caused massive blowdowns to white pine stands in New England and to the pinelands of the southern United States. Hurricane Hugo, which hit southern United States in 1989, devastated much of the remaining large areas of old growth southern pines, habitat of the endangered red-cockaded woodpecker. This bird depends completely on old growth pines affected with a fungus-caused soft rot of the heartwood for nesting sites. The loss of these trees, highly vulnerable to storm damage, threatens the survival of the woodpecker.

The position trees occupy in the forest affects their vulnerability to wind damage. Trees bordering ragged forest gaps and trees growing along edges, roads, and power lines are more likely to blow down than trees in the forest interior.

MOVING WATER

Moving water is a powerful agent of disturbance. Storm floods scour stream bottoms, cut away banks, change the courses of streams and rivers, move and deposit sediments, and bury or carry away aquatic organisms. Strong waves on rocky intertidal and subtidal shores overturn boulders and dislodge sessile organisms. This

FIGURE 21.3 Wind can become a major disturbance, evident in the damage inflicted by Hurricane Hugo on this southern pine forest.

action clears patches of hard substrate for recolonization and maintains local diversity (see Chapter 35). High storm tides break down barrier dunes, allowing seawater to invade behind the dunes, and changes the shape of barrier islands.

DROUGHT

Prolonged drought can have a pronounced effect on vegetation composition and structure. On the grasslands of western Kansas during the drought years of the 1930s blue grama with its physiological ability to resist dry conditions became twice as dense as the drought-sensitive buffalo grass. When the rains came, buffalo grass quickly responded. Within two years it became five times as dense as blue grama. After ten years of favorable conditions the two species shared codominance. In temperate forests prolonged drought can result in heavy mortality of shallow-rooted tree species such as hemlock and yellow birch and understory trees and shrubs. Drought dries up wetlands, causing crisis conditions among waterfowl and other wetland wildlife, seriously reducing their populations. Drought in African savannas results in widespread deaths of elephants, antelopes, and other large herbivores from dehydration and starvation.

FIRE

Fire connotes warmth and cheer and also fear and destruction. In nature, fire has the same ambivalence. It is both a destroyer and a regenerator of life. It is also one of the major influences on community development.

Conditions and Causes

Three conditions are necessary for fire to assume ecological importance: (1) an accumulation of organic matter sufficient to burn, (2) dry weather conditions to render the material combustible, and (3) a source of ignition. The only two important sources of ignition are lightning and humans.

Certain regions possess climatic conditions conducive to the spread of fires set by lightning storms, which are not universally accompanied by precipitation. These conditions include a wet period during which fuel can build up and a dry period during which accumulated fuel can burn. Regions with such conditions include Africa, southern and western Australia, western North America, and the southern pinelands of the United States.

In the western United States, 70 percent of the forest fires are caused by dry lightning during the summer. Fires so caused are the most numerous during the growing season, from April through August. At that time fires generally are the least severe, but have the greatest impact as a selective force.

When humans appeared on the scene, fire became an even more powerful influence on vegetation, especially in grassland ecosystems, for they added a new dimension to it. Whereas lightning fires are random and periodic, human fires were often deliberately set to clear ground for agriculture, to improve forage for grazing, to open up the countryside, and to make travel easier. Other fires escaped from camps and debris burning. Whatever the reason, most fires set by humans burned in the nongrowing season, when fires were more intense and the damage more severe. As humans spread from fire-evolved grasslands and savannas to more humid forested regions, they introduced fire into vegetation types, such as hardwood forests, that are fire-resistant during the lightning season but highly flammable during a dry spring and fall.

The apparent destructiveness of fire has stimulated intensive suppression programs, especially in North America. Fire, once a relatively prevalent and selective force in shaping the nature and diversity of vegetation, was eliminated from the natural environment. Fire suppression allows debris and litter to accumulate, which once ignited support intense, forest-destroying fires. The effects of suppression are just as detrimental to vegetation as too frequent fires that do not allow the system to recover. Both alter the character of wildlands.

Types

Fires may be surface, crown, or ground (Figure 21.4), depending upon the kind and amount of fuel, moisture, wind, other meterological conditions, season, and vegetation. Surface fire, the most common type, feeds on the litter and debris on the ground. It kills herbaceous plants and woody seedlings, and scorches the bases and

FIGURE 21.4 Types of forest fires: (a) surface fire; (b) crown fire; (c) ground fire.

occasionally the crowns of trees. Surface fires may kill thin-barked trees by heating the inner bark and growing cambium layer. Thick-barked trees are better protected, but they can be scarred, allowing infection by fungi.

If the fuel load is high and the winds strong, surface fires may leap up into the canopy and cause a crown fire. Crown fires are most common in coniferous stands because their foliage is highly flammable. Crown fires, which can spread rapidly across the treetops, kill all aboveground vegetation; yet certain forest types, such as jack pine and lodgepole pine, require all-consuming crown fires to regenerate the stand.

Most destructive are ground fires that consume organic matter down to the mineral substrate. They are most prevalent in areas of deep, dried-out peat and extremely dry, light, organic matter. Such fires are flameless and extremely hot and persist until all available fuel is consumed. In spruce and pine forests with their heavy accumulation of fine litter, fire can burn down to rocks and mineral soil, eliminating any oppportunity for that type of vegetation to return (Figure 21.5).

FIGURE 21.5 Forest fires of great intensity have a profound influence. Intense fires burned over the Allegheny Plateau area in West Virginia known as Dolly Sods after the spruce-fir forest was cut. Fire consumed the peatlike ground layer to bedrock and mineral soil. The forest never recovered from the Civil War years' burn. The plateau is now a boulder-strewn landscape with scattered patches of blueberry, dwarf birches, mountain ash, and bracken fern.

Frequency

Fire is a semirandom, recurring event in ecosystems. How frequently a fire burns over an area, its return rate, is influenced by droughts, the accumulation and flammability of the fuel, the resulting intensity of the burn, and human interference. In grasslands of presettlement North America, fires occurred about every two to three years, time enough to allow dead stems, leaves, and mulch to accumulate. In forested ecosystems the frequency of fire varied, as indicated by the presence of fire scars in the growth rings of trees. Fires ranged from light surface ones with return intervals from 1 year to more than 25 years to crown fires with return intervals from 25 to 300 years. The types are not exclusive. For example, a red pine forest may experience a light to moderate surface fire every 25 to 30 years, followed by a crown fire once every 100 to 300 years.

Various forest ecosystems appear to burn and develop under certain fire frequencies, influenced by changes in climate. Frequent low-intensity surface fires every 5 to 20 years were typical in presettlement forests of ponderosa pine (*Pinus ponderosa*) in western North America. Such fires prevented a buildup of flammable ground litter, eliminated incoming shade-tolerant conifers, and thinned the stand. Infrequent crown and severe surface fires in combination burned across aspen (*Populus tremuloides*) and jack pine (*Pinus banksiana*) forests every 15 to 20 years with killing fires every 60 to 80 years. Such fires regenerated the stand.

Since the early 1900s, fire suppression in the United States has greatly reduced the frequency and increased the return intervals of fires to a point where they are measured in 1000+ years. Such exclusion has greatly altered the species composition and structure of fire-dependent ecosystems.

Effects on Soil

Fire affects not only vegetation but also soil. Fire raises the soil temperature, but variable amounts of moisture in the soil provide insulation. Soil temperature does not rise above 100°C until all moisture is evaporated.

Even under hot fires the temperature rarely exceeds 200°C at depths of 2.5 cm and cools off rapidly after the fire passes.

More important, fire reduces soil organic matter to ash, releasing it with CO_2 and nitrogen to the atmosphere. Most nitrogen and potassium are lost in a form unavailable to plants. Lost nitrogen is replaced by nitrogen-fixing legumes, whose growth is stimulated by fire and by increased activity of soil microorganisms.

As heat destroys organic matter, it breaks down soil aggregates. The larger pores that improve water infiltration and aeration are lost. Bulk density of the soil increases and permeability decreases. That reduces the infiltration of water into the soil, increases surface runoff, and promotes erosion on steep slopes.

Fire also affects wettability of certain soils. Decomposing plant material in ponderosa pine and white fir/sequoia forests and in chaparral releases water-repellent substances that stick to soil particles and accumulate in the surface and subsurface soil column between burns, creating a nonwettable layer. During a fire the temperature gradient through the soil stimulates the volatilization and downward diffusion of the water-repellent substances to lower soil layers. As a result, the nonwettable layer moves down into the soil (Figure 21.6). How far it penetrates depends upon the intensity of the fire. The nonwettable layer comes to rest above another wettable layer, where its downward movement is impeded. As the soil above the layer becomes saturated, it slips on the nonwettable layer and slides downslope.

Plant Response

Fire has other influences on the forest. It sets regeneration into motion by stimulating sprouting from roots and germination of seeds. Periodic surface fires thin coniferous stands such as ponderosa and longleaf pines. Importantly, fire acts as a sanitizer, terminating outbreaks of insects and such parasites as mistletoe.

Plants in fire-prone ecosystems respond to fire in three general ways. They can survive as mature plants with little or no damage. One defense is bark thick enough to insulate the cambium from the heat of surface fires. Because the heat of the fire is rarely uniform about the base of the tree, one part of the tree may be cool, while another part may burn to the cambium. A second defense is rapid growth, raising the crown high enough above the ground to escape damage from surface fire and to reduce the danger of a surface fire leaping up in the canopy.

For some plants the death of a mature stand from fire is a means of regenerating and perpetuating the forest. Such destruction results in even-aged stands that arise and grow old together, promoting severe but infrequent fires. Such a response works only if the time between fire is long enough to allow the plants to mature to produce seed. The seed may be stored either on the plant or in the soil, awaiting fire to release the seeds or stimulate germination. Jack pine and lodgepole pine, for example, retain unripened cones on the tree for many years. The seeds remain viable until a crown fire destroys the stand. The heat opens the cones, releases the seeds (*serotiny*), and prepares an open seedbed well fertilized with ash. Other species rely on stand destruction and fire-stimulated germination of seeds stored in the soil. The seed coat is impervious to water and other softening agents until the increased temperature cracks the hard seed coat or releases the seed from soil-stored chemical inhibitors imposed by overhead living vegetation. Examples of seeds opened by heat are such grassland legumes as beggar's-tick (*Trifolium* spp.).

A third widespread response to fire is resprouting. Although fire kills the tops and foliage, new growth appears as bud sprouts. Certain trees, particularly a number of eucalyptus species in Australia, possess buds protected beneath the thick bark of larger branches. These buds survive crown fires and break out to develop new foliage. Other plants sprout from buds on roots, rhizomes, root collars, and specialized structures call lignotubers—all protected from fire by the soil. Ferns have subterranean buds on rhizomes that respond to the loss of aboveground foliage. Shrubs such as blackberries and blueberries and trees such as aspen sprout vigorously from roots. Trees such as oaks and hickories sprout from buds that develop at the root collar just below the ground. Certain species of Mediterranean-type shrubs, including chamise (*Adenostoma fasciculatum*), as well as many Australian tree species, sprout from lignotubers.

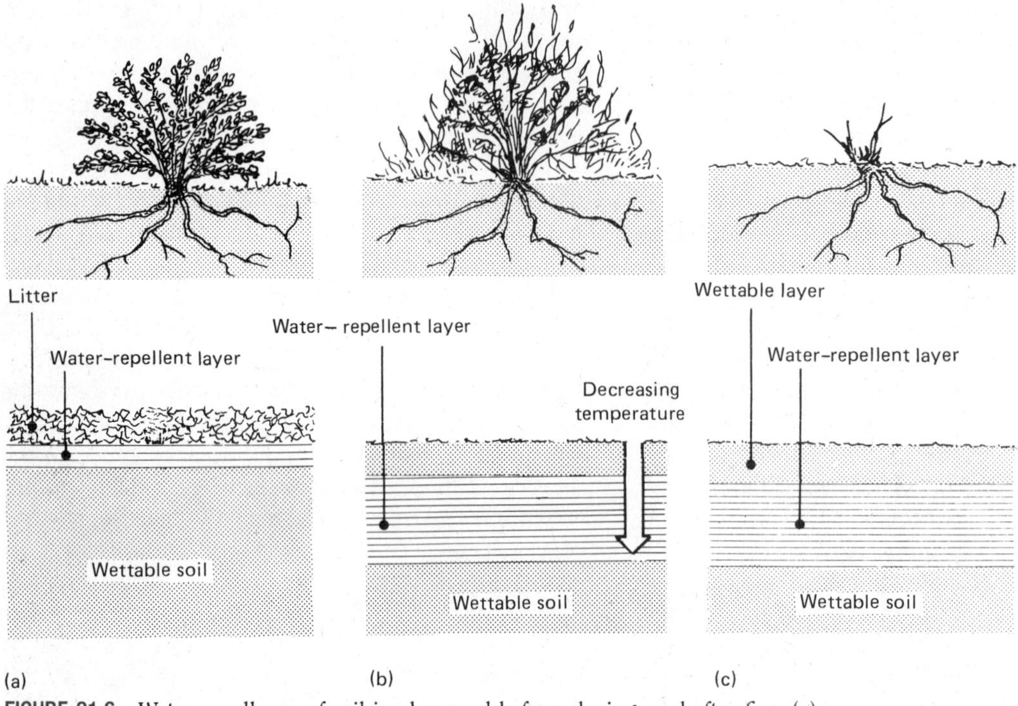

FIGURE 21.6 Water repellency of soil in chapparal before, during, and after fire. (a) Before fire, water-repelling substances accumulate in the litter and mineral soil immediately beneath it. (b) Fire burns vegetation and litter, causing the water-repelling substances to move downward along a temperature gradient. (c) After the fire a water-repellent layer is located below and parallel to the soil surface on the burned area. This layer can cause the soil to slide downslope.

Yellowstone

Some vegetation types, including Australian eucalyptus, California chaparral, the garrigue and matorral of the Mediterranean region, jack pine, and lodgepole pine, require the rejuvenating influence of periodic fires. Highly flammable at some point in their development, fire is an integral part of their life cycles.

An example of such fire-adapted vegetation is lodgepole pine, which dominated much of the Greater Yellowstone Area, the burning of which was a major event in 1988. Lodgepole pine (*Pinus contorta*), slow-growing, goes through four basic successional phases following fire: (1) the sapling stage, which does not burn easily, about 40 years in which the pines grow to canopy closure; (2) the immature forest stage from 40 to 150 years, which consists of densely clustered, even-aged trees susceptible to crown fires only in very windy conditions; (3) the mature forest, an even-aged closed canopy stand with a developing understory of spruce and fir lasting 150 to 300 years and increasing flammability; and (4) old-growth forest 200 to 350 years old, with pine, fir, and spruce of all ages, which act as a fire ladder reaching to the crown; this stage lasts until the next stand-replacing fire. One-third of the Yellowstone forests were 250 to 350 years old.

Stand-replacing fires in the 1700s had resulted in thousands of hectares of even-aged lodgepole pine in Yellowstone. Fire suppression from 1940 to 1970 had prevented spotty fires that would have created large patches of less flammable early successional stages within the forest. The result was a continuous landscape of old forest that would allow fires to burn without interference.

In 1988 both weather and vegetation conditions were ideal for a fire similar to that of the 1700s to sweep the area. The region was in the midst of a drought, fuel accumulation in the old-growth forest was high, relative humidity was low, and dry, gusty winds reached 100 km per hour. Many fires were lightning-set, but human fires were responsible for more than half the area burned, altering its pattern and extent.

Added to the effects of the fire itself were the scars from fire suppression activities. Camp and helicopter sites scattered through the park, bulldozed fire lanes, long corridors of tree removal, and turned-over soil will leave scars for years.

Fire did not sweep the Greater Yellowstone Area with equal intensity. Variations in topography, slope, wind, humidity, and fuel availability and condition produced a mosaic of burned, partially burned, and unburned areas that eventually will add diversity to the landscape. There will be larger meadows and young lodgepole pine and Engelmann spruce forests. Quaking aspen, a vigorous sprouter, will come up in thick stands, and herbaceous vegetation will be luxuriant for about a decade. In perhaps 30 years mature growth of sagebrush will return. The Yellowstone area is now as it was nearly 300 years ago; in 100 years it will appear much as it did before the fire; and according to the natural cycle of lodgepole pine, it will in due time burn again.

Animal Responses

A view of its immediate aftermath may give the impression that fire is a major agent of destruction to wildlife. On a short-term basis fire destroys or partially destroys habitat and causes some injury and death, either directly or indirectly by predators who take advantage of prey suddenly driven from cover. On the African savanna kites and other birds are grass-fire followers, hunting insects driven to flight by the advancing flames. Many flying insects, such as grasshoppers and moths, fly in front of the flames and are often engulfed by wind-driven gas clouds, but a surprising number go through the fire unscathed. Unless nesting, birds are rarely directly affected by fire other than short-term loss of habitat. Some species of birds decrease following fires, but ground-foraging birds seem to increase.

Many mammals, especially those that live in burrows, survive fires. Large mammals are adept at avoiding the front of the fire and working their way back through gaps and unburned patches to burned-over areas behind the flames. The Yellowstone fire of 1988 caused the deaths, mostly by smoke inhalation, of only one percent of the elk population. The major problem faced by these mammals is the short-term lack of food and cover. High populations of post-fire grazing herbivores feeding on newly regenerating plants may lead to overgrazing and elimination of palatable species.

Severe fires alter the habitat and eliminate for a time those species dependent on it. At the same time fire creates new habitat for a different set of species, especially those that favor open and shrubby land. Many animals favor both conditions and in fact are dependent on such fluctuations in habitat. Fire produces a mosaic of shrubs, timber, and open land, habitat essential for edge and shrubland species (see Figure 27.1).

A few species are wholly dependent on the disturbance of fire. One is the endangered Kirtland's warbler (*Dendroica kirtlandii*). Restricted to the jack pine forests of the lower peninsula of Michigan, the warbler requires large blocks (40+ ha) of even-aged stands of pine 1.5 to 4.5 m tall with branches close to the ground. Smaller or larger trees are unacceptable. Intervals of fire sufficient to maintain blocks of young jack pines are necessary to maintain the habitat of this warbler.

Another group of fire-dependent birds is the *Sylvia* warblers that occupy the Mediterranean shrublands of Sardinia. Of the five species, two are fire-dependent. One species, the Dartford warbler (*S. sarda*), can occupy a habitat patch that has been burned within six years. Another species, Marmora's warbler (*S. undata*), occupies only 18- to 20-year old, tall, shrubby growth. Two of the remaining species are fire-adapted, and one is fire-tolerant.

ANIMALS

Walk into a forest with a high population of deer or across an overgrazed grassland. It is easy to see the impact that herbivorous animals can have on communities. Overgrazing in native rangelands of the southwestern United States, for example, has reduced the organic mat and thus the incidence of fire. Because of infrequent fires, reduced competition from grasses, and dispersal of seed through cattle droppings, mesquite and other shrubs have invaded the grasslands. In many parts of eastern North America, large populations of white-tailed deer have eliminated certain trees such as white cedar and American yew from the forest, destroyed forest reproduction, and developed a browse line—the upper limits on a tree at which deer can reach foliage. In cutover areas of hardwood forests on the Allegheny Plateau in Pennsylvania, deer have greatly reduced pin cherry and blackberry, selectively reduced sugar maple, and favored the expansion of ferns and grass. Root-feeding pocket gophers (*Thomomys bottae*) apparently slow or inhibit the invasion of quaking aspen into deep-soil mountain meadows of Arizona, restricting the tree to rocky outcrops inaccessible to the rodents.

The African elephant influences the savanna ecosystem. When their numbers are in balance with the vegetation and their movements are not restricted, elephants have an important role in creating and maintaining the forest. When elephants exceed the capacity of their habitat to support them, their feeding habits combined with fires devastate flora, fauna, and soils. Elephant depredation on trees (Figure 21.7) acts as a catalyst to fires, which are the primary cause of converting forest to grassland.

Beaver create major disturbances in many forested regions of North America and Europe. By damming streams they alter the structure and dynamics of flowing water ecosystems (see Chapter 32). Pools behind dams become catchments for sediments and sites for organic decomposition. By flooding lowland areas, beaver convert forested stands into wetlands. By feeding on aspen, willow, and birch, beaver maintain stands of these trees, which otherwise would be replaced by later successional species.

Outbreaks of insects such as gypsy moth and spruce budworm defoliate large areas of forest, killing or stunting trees. Mortality ranges from 10 to 30 percent in hardwood forests infested by gypsy moth to 100 percent in spruce and fir stands. Outbreaks of bark beetles have much the same effect in pine forests. The impact of spruce budworm, bark beetles, and other major forest insects is most intense in large expanses of late-stage, homogeneous forest where natural fires

FIGURE 21.7 Grazing animals can have an important influence on succession and stability. The elephant, here uprooting a mopane tree (*Colophospermium*) at the end of the dry season, is important in the life cycle of mopane, which requires disturbance by elephants, and in the maintenance of the savanna. Too many elephants destroy it; too few elephants let the bush encroach.

have been suppressed for long periods, or where the stands have been stressed by air pollution and acid rain. In their own way these insects act to regenerate senescent or stagnated forests.

HUMAN IMPACTS

George Perkins Marsh back in 1864 wrote in his book *Man and Nature, or Physical Geography as Modified by Human Action* that the human being is a "... disturbing agent. Wherever he plants his foot the harmonies of nature are turned to discord." Nowhere is this observation more evident than in logging, cultivation, mining, and urbanization.

LOGGING

Disturbance by logging depends upon the methods employed: selection cutting or some form of even-aged management (clearcutting). In selection cutting, single trees or groups of trees are removed, based on their position in the stand and possibilities of future growth. Selection cutting produces only gaps in the forest canopy and favors reproduction of tolerant over intolerant trees. Forest composition essentially remains unchanged.

Even-aged management, more widely practiced, is removal of the forest and reversion to an early stage of succession. Unless followed by fire or badly disturbed by erosion caused by logging activities (Figure 21.8), the area fills in rapidly with herbs, shrubs, sprout growth, and seedlings of trees present in the understory. The area passes quickly through the shrub stage to an even-aged pole forest. Because many of the most valuable timber trees are intolerant to mid-tolerant species, they can be regenerated only by removal of mature trees, exposing the ground to sunlight.

There are three approaches to even-aged management. One is clearcutting 11- to 44-ha blocks of timber within large forest tracts. A second method is stripcutting, the removal of all salable timber and remaining trees in strips 15 to 30 m wide. Every third strip is removed, followed by removal of the remaining strips in two cuttings two to four years apart. A third method is shelterwood cutting, which leaves 10 to 70 percent of the stand on the first cut. When new growth is well under way, the remaining trees are removed. The first

FIGURE 21.8 Clearcutting can cause severe disturbance. The scars and erosion from this clearcut of Douglas-fir will remain for hundreds of years.

two methods of even-aged management favor regeneration of intolerant tree species. Shelterwood cutting retains some of the characteristics of the original forest yet permits intolerant species to regenerate.

A serious ecological problem with large block clearcutting in the coniferous forests of the Rocky Mountains, Pacific Northwest, and southeastern United States involves postharvest treatment of the site. Instead of relying on natural forest regeneration, which is slower, timber companies burn the slash, disturb the forest soil to eliminate remaining growth, and replant the cutover lands with nursery-grown seedlings or aerially reseed the area with the selected timber species. The result is a monocultural commercial forest that lacks the diversity of both the original and the naturally regenerated stand. Further short rotations preclude the return of stands of large trees, essential to many species of wildlife and forest flora.

Logging inevitably changes the composition and structure of the regenerating forest. Rarely is the composition of the incoming forest, particularly hardwoods, the same as that of the forest removed. Composition may be further affected by high-grading—the preferential removal of certain trees and species from the stand. Failure of certain species to regenerate under the conditions imposed adds to changes. In many oak forests of eastern North America, oaks fail to regenerate after the forest is cut. Foresters often further modify the regenerating forest by removing tree species not desired for timber or individuals of poor form. This removal may improve composition of the stand and the quality of the trees by economic standards, but not by ecological standards. Foresters may also eliminate dead trees and snags, essential habitat for cavity-dwelling wildlife. Short cutting rotation cycles of 30 to 50 years or less preclude the return of diverse old-growth forests (see Chapter 29) and wildlife species that depend on them.

The impact of logging disturbance depends on the species. It is clearly a case of "one man's meat being another man's poison." Small-scale disturbances, particularly treefall gaps, are helpful. The new growth filling in the gap provides low ground cover needed by species that depend on forest openings. Canopy gaps provide open areas for sallying flycatchers, yet the openings have no effect on canopy-dwelling species. Large clearcut areas create the expansive successional habitats needed by the opportunistic or ephemeral species that depend on short-lived early successional stages, such as woodcock (*Philohela minor*) and prairie warbler (*Dendroica discolor*). Such areas, on the other hand, eliminate habitat of canopy-dwelling species, vertebrate and invertebrate, that will not return for several decades, or may disappear. Most small ground-dwelling mammals are little affected by logging as long as the area is not converted to other land use. Cutting of old-growth forests, however, can permanently eliminate old-growth dependent species such as spotted owls and red-backed voles of the Pacific Northwest and the red cockaded woodpecker (*Picoides borealis*) of the southern pine forests of the United States.

CULTIVATION

When humans discovered the means of domesticating grains and animals, they emerged from a hunting-gathering to an agricultural way of life and radically transformed the vegetational patterns and related animal life on Earth. Humans cleared forests by both ax and fire to make room for agricultural fields. In doing so they destroyed much of the world's forests, developed new plant communities, eliminated many forms of animal life, influenced the spread of weedy plants, and affected the water cycle. Neolithic land clearance and expansion of stock raising in Europe converted the forests into pastures surrounded by scrubby woods of hazel, mixed oak, and birch. Where the soil was thin and poor, woody vegetation degenerated into woodless heaths that persist to this day. Excessive soil erosion from the destruction of forests and overgrazing by goats and sheep helped destroy the "grandeur of Greece" and the "glory of Rome." Buried beneath the sands of the ancient Fertile Crescent are the remnants of irrigation works. The remains of Mayan civilization, which apparently collapsed because of overexploitation of soil, leaving the unproductive soils of present-day Guatemala, are overgrown with tropical vegetation.

Such transformations continue at an accelerated pace. Little remains of the original North American prairie, and much of its original soil has blown away or has washed to the Gulf of Mexico. Wetlands have been drained and planted to wheat, corn, and soybeans.

Tropical forests of South America are being cleared at a phenomenal rate for grazing lands and settlement. Most of the rain forests of southeast Asia have been converted to plantations of rubber and palm nut.

In the course of such transformations certain weedy species of plants and insect pests have been carried around the world and native animal life sharply reduced or driven to extinction. A good deal of the most accessible and once fertile lands have been transformed into simple, highly artificial cultivated plant communities well adapted to grow on disturbed sites. Poorly managed, large acreages of formerly cultivated lands are abandoned, but they are so infertile that they can support only a depauperate vegetation, unlike the original plant cover. This situation is especially pronounced in tropical rain-forest country.

With cultivation and associated permanent settlement came methods of surveying and parceling land. The United States and Canada used the rectangular survey, apparently based on the Roman centuriation system of land division. The basic Roman unit was the *centuria,* which consisted of 100 square *heredia,* approximately 53 hectares (132 acres). The U.S. rectangular survey organizes landscape into six-by-six mile townships divided into 36 sections of one square mile (2.6 sq km) each. These east-west survey lines, unlike the Roman system, which had some regard for the lay of the land, ran straight lines across the country regardless of topography.

The rectangular survey delineated roads, county lines, township lines, farmstead boundaries, and more recently suburban lots. The survey lines are demarked by hedgerows and fencerows, square field boundaries, and blocks of woodland fragments in agricultural countryside. The rectangular survey has had an overwhelming permanent impact on the nature and evolution of the North American landscape, influencing the pattern of vegetational communities and the distribution and movement of animals.

URBANIZATION

Although cultivation has had a powerful impact upon the Earth, it cannot compare with urbanization. Urbanization is a developmental strategy much like ecological succession. Hunter-gatherers with a loose social structure became agrarian and established villages. As villages became centers of trade and crafts, they developed into cities. Over 10,000 years as agrarians, we humans have become urban. We are engaged in an ecological revolution the likes of which humans experienced only once before in history: the abandonment of a hunting way of life for an agrarian one.

In 1850 no society could be called urbanized. The world was still agrarian. In 1900 one country, Great Britain, was urbanized. By 1965 all industrial countries were urbanized and most of their people lived in cities. Like population growth, urbanization has increased rapidly. Urbanization from 1950 to 1960 equaled the rate from 1900 to 1950. Now in 1991 one-half of the world's population lives in cities of all sizes.

The urban revolution has obliterated the natural environment, its wildlife, and its wilderness. Amoeba-like, the city spreads out from its core. Following roads into the countryside, industrial sites, housing developments, and shopping malls engulf woods, fields, and croplands. As life in the inner city becomes less tolerable, many of us, still retaining some vestiges of our million years' attachment to the natural environment, move to a special twentieth-century semicountry institution, the suburbs. This movement aggravates rather than improves the problem. Suburban lots—the suburbanites' little peasant plots—eat up land at an enormous rate. Suburban development demands more highways and air-polluting automobiles to carry suburbanites back to the city and to work. Highways and beltways proliferate, burying more and more of the countryside beneath concrete and asphalt. The city moves outward to swallow up suburbs, creating large areas that are neither city nor countryside. These areas then become new core areas that, once firmly established, initiate a further spread of metropolitan areas. An extreme example is the large belt of coastal cities that extends from Boston, Massachusetts, to Miami, Florida. Trapped within this growth are pockets of natural vegetation and wildlife that may never survive.

The urbanized environment with its parks and suburbs is a different world ecologically than the one it displaced. The inner core is impoverished in species of both plants and animals, mostly exotic and able to endure urban conditions. Surrounding the core is an irregular ring of gradual change in vegetation and animal life. In a zone of transition or deterioration the plants are remnants of species such as maple, planted when the area was largely residential. On the edges of the city, plant life is dominated by small lawns with ornamental foundation plantings, largely conifers and scattered shade trees. In the residential zone or the semisuburbs, lawns, flower gardens, ornamental shrubs, scattered, often large, old shade trees, and some native species and some exotics dominate the plant life. In the outer suburbs, the commuter zone, remain some vestiges of predevelopment vegetation, including forest and shrubland. These forests, subject to landowner modifications, are a mix of native and exotic landscape species, and their overall structure has been simplified. Further, the region is permanently altered. Waterways are often channelized, soils have been disturbed and altered, plant life is subject to such stresses as air pollution and soil compaction, and native fauna is subject to competition and predation from domestic and exotic animals. From inner city to outer suburbs, urban plant communities are human creations, reflecting life-styles, values, and goals of the human inhabitants, not nature.

Adding to the physical disturbance is large-scale diversion of water from rivers and lakes for irrigation, seriously affecting the health and stability of the aquatic ecosystems exploited. Drainage from agricultural fields carries fertilizers and pesticides to groundwater, streams, and lakes, reducing water quality and impairing the health and reproduction of wildlife.

Modern agriculture has eliminated millions of acres of wildlife habitat once associated with agriculture. Farms prior to World War II were small by today's standards, grew a diversity of crops, fields of which were separated by hedgerows, windbreaks, and woodlots. This diversity created habitat for a number of native species considered rare at the time of settlement. These included bobwhite quail and cottontail rabbit in eastern United States, and the exotic ring-necked pheasant (which ecologically replaced the prairie chicken of the former grasslands) in the Midwest, along with many old-field and hedgerow species of birds and small mammals. Present-day agriculture, with its emphasis on large acreages of monocultural crops of corn, soybeans, and wheat, has eliminated this diversity of habitat, so even these species are declining or have disappeared from many areas. A similar loss of habitat has also taken place in Great Britain and Western Europe.

POLLUTION

Accompanying cultivation and urbanization is pollution of air, water, and soil, a topic considered in detail in Chapter 25. Use of fertilizers and pesticides in croplands, home gardens, and lawns, often in excessive amounts, contaminates the soil. Pesticides accumulated in the soil are taken up by plants, incorporated into their tissues, and transferred to humans in foods. Persistence of herbicides, insecticides, and fungicides in soil, together with toxic wastes in some areas, is a widespread problem. Leachates from croplands, septic tanks, and dumps find their way into the groundwater, streams, ponds, and lakes. Added to them is the heavy load of nitrogen, phosphorus, and organic matter from sewage and industrial effluents, all of which cause excessive nutrient enrichment of aquatic systems, producing significant chemical, biological, and ecological changes in streams, lakes, and estuaries.

SURFACE MINING

Demands for energy to support an urbanized society have intensified disturbance to the natural environment. Surface mining accounts for a high percentage of coal production and for most other extracted minerals, such as iron and gold ores. The magnitude of damage varies with the region and with the degree and success of reclamation efforts. The impact is most pronounced in mountainous regions, such as the Appalachians, where surface mining is on the contour or involves leveling mountaintops and filling in valleys (Figure 21.9). Whatever the method, surface mining does violence to the land. Deep, unweathered rock strata are broken and brought to the surface, where the material is subject to rapid weathering, releasing manganese, sulfate, iron, zinc, nickel, and other elements in toxic quantities. Carried away in high concentrations by water coming off mined sites, these elements reduce water quality downstream for both aquatic life and humans. Surface mining also alters the groundwater. Water tables, once deep in the underlying

FIGURE 21.9 In the mountainous Appalachian coal fields, surface mining permanently destroys the forest ecosystem and so alters the topography and geological structure of the region that it can never recover or even approach its predisturbance state.

rock strata, are exposed and flow freely to the newly created surface. Mining alters stream hydrology. It increases both peak stormwater flows and dry weather flows. Life in the streams is damaged by catastrophic stormflows and increased siltation. This altered hydrology reduces species richness, species diversity, and population densities of surviving species.

Reclamation efforts, however, successful in restoring contours and reducing erosion, rarely restore original vegetation. Hundreds of thousands of hectares of mined forested land in eastern United States have been altered into some type of grassland that will never revert to forest. Likewise, in the semiarid regions of the western United States, efforts to return the mined sites to the original shrub-steppe vegetation using direct-applied topsoil do not restore the species diversity of the original community.

Deep mining, too, disturbs the landscape. This is particularly true of longwall mining, a technique that involves the mechanical removal of the entire face of the coal seam. As the mining machinery advances through the coal seam, the supports over the mined area are pulled, allowing the overlying earth to drop into the mined-out area. An efficient method that allows virtually all of the coal to be removed, compared to old room-mining techniques that left a third or more of the coal in the ground, it nevertheless causes serious surface disturbances.

STABILITY AND RESILIENCE

Natural disturbance is part of the normal functioning of all communities, but they vary in their degree of stability and resilience. Both are important clues to how these systems will respond to human intervention and disturbance.

Stability is the tendency of an ecosystem to reach and maintain an equilibrium condition, either a steady state or a stable oscillation. If the system is highly stable, it resists departure from a steady state; and if disturbed, it recovers rapidly.

Stability may be local or global. Local stability is the tendency of a system to return to its original state when subjected to a small disturbance. Forest gaps filling in with similar tree species are examples of local stability. Global stability is the tendency of a community to return to its original condition when subjected to a large disturbance. Chaparral and eucalyptus forests returning quickly to their original condition and species composition after a fire represent global stability. Such systems exhibit low variability and strong resistance to change.

Communities most resistant to change characteristically have a large biotic structure such as trees and store nutrients and energy in standing biomass. A forest is relatively resistant to disturbance. It can withstand such environmental disturbances as sharp temperature changes, drought, and insect outbreaks. For example, a late spring frost may kill the new leaves of trees, but the forest draws on energy reserves to replace leafy growth. However, if the forest is highly disturbed by fire or logging, its return to original condition is slow. The system exhibits low resilience.

Resilience is the ability of a system to maintain its structure and pattern of behavior in the face of a disturbance. It is a measure of the ability of a system to absorb changes and still persist. Resilience does not imply high stability of individual populations. Populations within a system may fluctuate widely in response to environmental changes. The system may be highly resilient yet exhibit low stability.

For example, in the spruce-fir forests of northern North America the spruce budworm population under certain environmental conditions increases rapidly, escapes control of predators and parasites, and feeds heavily on balsam fir, killing many trees and leaving only the less susceptible spruce and birch. After the spruce budworm population collapses, having exhausted its food supply, young balsam fir grows back in thick stands with spruce and birch. Between budworm outbreaks balsam fir outcompetes spruce and birch, but during outbreaks spruce and birch are favored over balsam fir. These interactions among populations of budworm, balsam fir, spruce, and birch maintain system homeostasis. The spruce-fir forest is resilient, even though its population elements exhibit low stability.

Aquatic ecosystems, which lack long-term storage of energy and nutrients in biomass, exhibit little stability. An influx of sewage effluents disturbs the system by adding more nutrients than it can handle. However, because the system is limited in its capacity to retain and recycle nutrients, the system returns to its original condition soon after the disturbance is reduced or removed.

For example, Lake Washington, near Seattle, was used as a basin for sewage disposal. It received excessive nutrients, especially phosphorus. The input eliminated certain diatom and algal populations and encouraged others, especially filamentous algae, to increase, changing the clear-water lake ecosystem. Once the sewage input was diverted from the lake, phosphorus levels in the lake declined, algal populations shrank, and the lake returned to its clear-water condition. The lake had high resilience but low stability.

A very strong disturbance may carry a system into a different level of stability. The system may be so greatly disturbed that it is unable to return to its original state, and a different system with a different domain of stability takes its place. For example, when forests of Scotch pine were cut in northern Britain, they were replaced by moorlands; and when the spruce forests of the central Appalachians in North America were cut and burned, they were replaced by stands of blueberry and stunted birch. Extensive land clearing in the Amazon basin has transformed tropical forest areas into scrub savanna. Tropical forest trees have large seeds not easily dispersed and a short dormancy period, both of which prevent successful colonization of large disturbed areas.

Attempts by humans to interfere with natural disturbances and to reduce variablity in natural systems can backfire or create new, unforeseen problems. For example, spraying and other measures reduce budworm populations at first but allow buildup and persistence of foliage over large areas, providing the conditions for a huge outbreak of the pest. Accumulation of fuel that would have burned in small surface fires creates conditions for disastrous ground fires.

Summary

Disturbance is a relatively discrete event in time that disrupts communities and populations from outside, changes substrates and resource availability, influences species composition and system structure, initiates succession, and adds diversity to the landscape. Major natural agents of disturbance are wind, moving water, drought, fire, and animal activity, including insect outbreaks. Humans cause disturbance by logging, cultivation, urbanization, pollution, and mining.

Small-scale disturbances are typical of rocky intertidal shores and temperate and tropical forests. Wave action and moving water create gaps among sessile organisms on rocky substrates. Treefall and removal of individual trees by logging create small gaps in forests. The response to gap formation is canopy closure, invasion by opportunistic species, and growth of tolerant species, depending upon biotic conditions. Large-scale disturbances induced by major events such as fire, logging, or insect outbreaks invite colonization by opportunistic species. Such disturbances can modify the system by favoring certain species and eliminating others, or they can ensure regeneration of the system itself.

Of great ecological importance is the frequency and return interval of disturbances. Too frequent disturbances can eliminate certain species by destroying the plants before they have had time to mature and seed.

Too long a time between disturbances can reduce system diversity and set the stage for highly destructive disturbances.

Communities are subjected to many agents of disturbance. The most catastrophic are human activities like logging, urbanization, cultivation, and mining. Such disturbances destroy or simplify natural communities, eliminate or reduce the habitats and ranges of plants and animals, introduce exotic species to the detriment of native species, alter the direction of succession, and so change topography and landscape that they reflect culture and life-style rather than natural vegetation patterns.

Fire is the major natural large-scale disturbance to terrestrial ecosystems. It is both beneficial and adverse. It results in loss of soil nutrients but also makes nutrients available. It sets into motion regeneration of fire-adapted systems by stimulating root sprouting and germination of seeds. It can favor fire-resistant species and eliminate fire-sensitive ones, thereby influencing composition and structure of forest systems.

Response of animals to disturbance depends on the species. Short-term impacts are the loss of food and cover. Long-term effects may be the loss of habitat for some species and the gain of habitat for others. Some species depend on disturbance for the maintenance of their habitat, especially those associated with the more ephemeral stages of early succession. Other species depend on periodic fires to maintain their habitat and to provide a mosaic of vegetation types required in their life cycle. A few fire-dependent species would go extinct without periodic fires to maintain their habitat.

Communities vary in sensitivity to disturbance. Stability is the tendency of a system to return to and maintain an equilibrium condition after perturbation. If the system is highly stable, it resists change or returns rapidly to equilibrium conditions. Stable systems are highly resistant to disturbance but not necessarily resilient. Resilience is the ability of a system to absorb changes and still persist.

Review and Study Questions

1. What is meant by a disturbance?
2. What are the differences in ecological responses to small-scale and large-scale disturbances?
3. Discuss the effects of major types of human disturbances, including agriculture, logging, urbanization, and mining, on plant communities and their associated animals.
4. Distinguish among surface fire, crown fire, and ground fire.
5. Should we suppress forest fires? Explain the pros and cons.
6. How might the following responses be adaptations to fire: (1) resprouting from roots; (2) serotinous cones; (3) thick bark; (4) dormant buds under bark?
7. How can disturbance be an important aspect of community stability?

*8. Why is it ecological folly to build houses on chaparral-covered hillsides or in pine forests? (You might look up news on recent fires and mudslides in southern California, southern Florida, and southern France.)

*9. Find out how the following are adapted to fire: (1) longleaf pine, (2) giant sequoia, (3) jack pine, (4) Douglas-fir, (5) chamise, (6) grassland.

*10. Investigate in what ways humans have imposed their values and changed the landscape in your area. If possible consult old photos and aerial photos. What has happened to farmland, forest land? Discuss how these changes have affected the distribution of plants, wildlife habitat, and streams.

*11. What are the major pollution problems in your area, and what are their effects on soil and water quality?

Selected References

Johnson, H. B. 1976. *Order upon the land.* New York: Oxford University Press. An interesting study of the impact of the U.S. rectangular survey on landscape patterns in North America.

Komarek, E. V., Sr., ed. 1963–1976. *Proceedings: Tall timbers fire ecology conferences.* Vols. 1–15. Tallahassee, FL: Tall Timbers Research Station. The major reference on fire ecology.

Mooney, H. A. , et al., eds. 1981. *Proceedings: Symposia on fire regimes and ecosystem properties.* General Technical Report WO-26. Washington, DC: USDA Forest Service. A major reference on the subject.

Pickett, S. T. A., and P. White, eds. 1984. *Natural disturbance: The patch dynamics perspective.* Orlando, FL: Academic Press. Role of disturbance in succession and community diversity.

Pyne, S. J. 1982. *Fire in America: A cultural history of wildland and rural fire.* Princeton, NJ: Princeton University Press. History of attitudes toward fire in North America from the Indians to modern times.

Pyne, S. J. 1984. *Introduction to wildland fire: Fire management in the United States.* New York: Wiley Interscience. Covers all aspects of fire from causes to management and suppression.

Thomas, W. L., ed. 1956. *Man's role in changing the face of the earth.* Chicago: University of Chicago Press. Still an important book on human impact on the biosphere through the ages.

Watts, M. T. 1971. *Reading the landscape of Europe.* New York: Harper & Row. A companion volume to the above.

Watts, M. T. 1976. *Reading the American landscape.* New York: Macmillan. A delightful introduction to landscape ecology of the United States as influenced by humans. An update of the 1968 edition.

(See also references at the end of Chapter 20.)

PART VI
Ecosystem Dynamics

INTRODUCTION

The Ecosystem Concept

The idea of the ecosystem is at the heart of ecology today. The term is widely used but not widely understood. *Ecosystem* has been defined as a partially or completely self-contained mass of organisms, as all organisms present in an area together with their physical environment, and as all the energetic interactions and material cycling that link organisms in a community with one another and with their environment. The last definition comes closest to the real meaning.

As a biological term, *ecosystem* has a recent origin. It was coined by the English ecologist A. G. Tansley in 1935 in an article in the journal *Ecology:*

> The fundamental concept appropriate to the biome considered together with all the effective inorganic factors of its environment is the *ecosystem,* which is a particular category among the physical systems that make up the universe. In an ecosystem the organisms and the inorganic factors alike are *components* which are in relatively stable dynamic equilibrium. . . .
>
> It is the systems so formed which, from the point of view of the ecologist, are the basic units of nature on the face of the earth. Our natural human prejudices force us to consider the organisms (in the sense of the biologist) as the most important parts of these systems, but certainly the inorganic factors are also parts—there could be no systems without them, and there is constant interchange of the most various kinds within each system, not only between the organisms but between the organic and the inorganic.

Thus, the prefix *eco-* indicates environment, and *-system* refers to a complex of coordinated units.

An ecosystem is basically an energy-processing system whose components have evolved together over a long period of time. The boundaries of the system are determined by the environment—that is, by what forms of life can be sustained by the environmental conditions of a particular region. Plant and animal populations within the system are the objects through which the system functions.

Inputs into the system are both biotic and abiotic. The abiotic inputs are energy and inorganic matter. Radiant energy, both heat and light, imposes restraints on the system: (1) by influencing temperature and moisture and (2) by affecting the productive capability of the system. Inorganic matter consists of all nutrients, water, carbon dioxide, oxygen, and so forth that affect the growth, reproduction, and replacement of biotic material and the maintenance of energy flow. Some of the materials in this chemical environment are necessary for the maintenance of the system, while others may be toxic or detrimental to its functioning.

The biotic inputs include other organisms that move into the ecosystem as well as influences imposed by other ecosystems in the landscape. No ecosystem stands alone. One is influenced by others. A stream ecosystem, for example, is strongly influenced by the terrestrial ecosystem through which it flows.

In simplest terms, all ecosystems, aquatic and terrestrial, consist of three basic components—the producers, the consumers, and abiotic matter. Traditionally, the abiotic inputs—CO_2, O_2, H_2O, nutrients derived from weathering of materials and from precipitation, and so on—are considered as abiotic components of the system. Here they are considered as inputs to the system rather than as a part of the ecosystem (Figure VI.1).

The *producers,* or *autotrophs,* the energy-capturing base of the system, are largely green plants. They fix the energy of the sun and manufacture food from simple organic and inorganic substances. Autotrophic metabolism is greatest in the upper layers of the ecosystem—the canopy of the forest and the surface water of lakes and oceans.

The *consumers,* or *heterotrophs,* utilize the food stored by autotrophs, rearrange it, and finally decompose the complex materials into simple inorganic compounds. In this role their function is to regulate the rate of energy flow and nutrient cycling and to stabilize the system. The heterotrophic component is often subdivided into two subsystems, consumers and decomposers. The consumers feed largely on living tissue, and

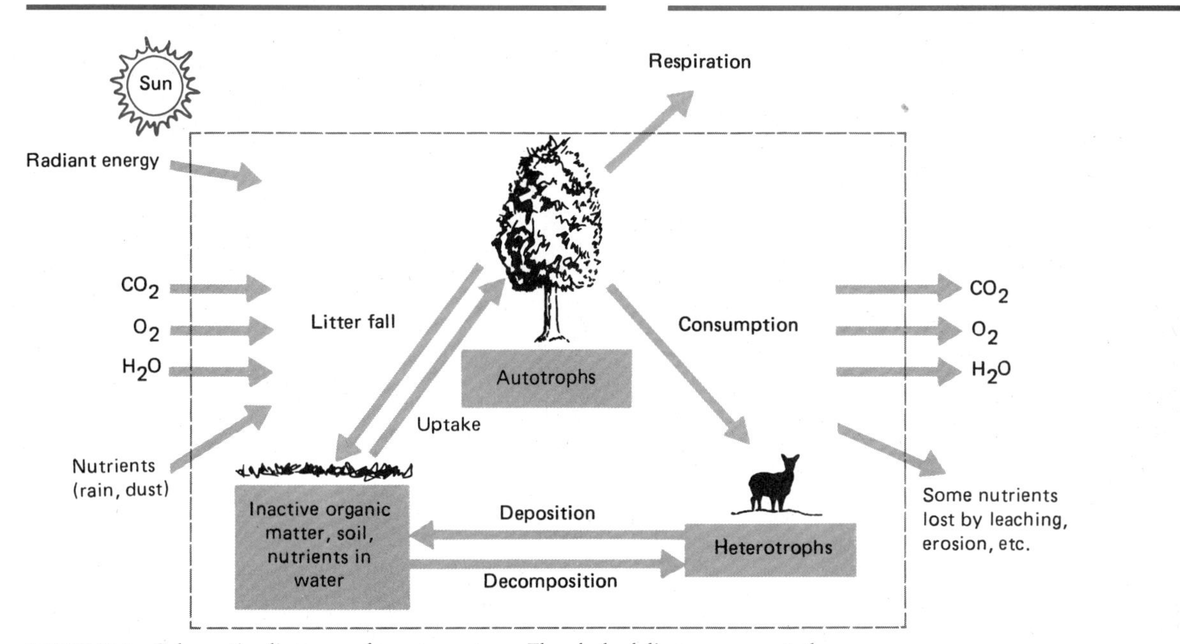

FIGURE VI.1 Schematic diagram of an ecosystem. The dashed lines represent the boundary of the system. The three major components are the producers, the consumers, and the abiotic elements—inactive organic matter, the soil matrix, nutrients in solution in aquatic systems, sediments, and so on. The lines indicate interactions within the system and with the environment.

the decomposers break down dead matter into inorganic substances. No matter how they may be classified, all heterotrophic organisms are consumers, and all in some way act as decomposers. Heterotrophic activity in the ecosystem is most intense where organic matter accumulates—in the upper layer of the soil and the litter of terrestrial ecosystems and in the sediments of aquatic ecosystems.

The third, or *abiotic,* component consists of the soil matrix, sediments, particulate matter, dissolved organic matter in aquatic ecosystems, and litter in terrestrial ecosystems. All the dead or inactive organic matter is derived from plant and consumer remains and is acted upon by the decomposer subsystem of the heterotrophs. Such organic matter is critical to the internal cycling of nutrients in the ecosystem.

The driving force of the system is the energy of the sun, which circulates all other inputs through the system. The various outputs—or more correctly, outflows—from one subsystem become inflows to another. While energy is utilized and dissipated as heat of respiration, the chemical elements from the environment are recycled by organisms within the system. How fast nutrients turn over in the system is influenced by these consumers.

CHAPTER 22

Production in Ecosystems

Outline

Objectives

On completion of this chapter, you should be able to:

1. Explain briefly the laws of thermodynamics and how they relate to ecology.
2. Define various types of ecological production.
3. Discuss how plants allocate net primary production.
4. Compare net primary production in various world ecosystems and explain why such variation occurs.
5. Describe secondary production and its allocation.
6. Compare assimilation efficiencies of poikilotherms and homeotherms and explain the difference.

Production in ecosystems involves the fixation and transfer of energy, the ultimate source of which is the sun. Fixation of solar energy is accomplished by plants through photosynthesis (Chapter 7). Fixed energy or photosynthate in the form of carbohydrates accumulates as plant biomass and becomes available to nonphotosynthetic organisms, which convert it to heterotrophic biomass. This fixation and transfer of energy are governed by the nature and laws of energy.

THE NATURE OF ENERGY

The energy of the sun comes as thermal energy (Chapter 5) and as light energy (Chapter 7). Thermal energy heats Earth; light energy is utilized in photosynthesis. In both instances radiation is the movement of particles from one point to another. Energy can be said to flow; but it can flow only if there is an energy source and an energy sink. For Earth the energy source is the sun. The planet absorbs part of the energy and gives up energy to a sink, outer space.

Thermal energy spreads rapidly among molecules in the system without any chemical reaction. It sets the molecules into a state of random motion and vibration. The hotter the object, the more thermal energy is being absorbed and the more vibrational and rotational change is taking place. These motions tend to spread from a hot body to a cooler one, transferring energy from one to the other. The energy of light waves, on the other hand, especially the blue and red wavelengths, excites electrons. The energy of light or photons sends one of a pair of electrons to a higher state or orbit. Uncoupled from its partner, it is free to be involved in photochemical reactions.

Energy exists in two forms, potential and kinetic. Potential energy is energy at rest—that is, capable of and available for performing work. Kinetic energy is energy of motion. It performs work at the expense of potential energy. Work is of at least two kinds: the storage of energy and the arranging or ordering of matter.

LAWS OF THERMODYNAMICS

The expenditure and storage of energy are described by two laws of thermodynamics. The first law of thermodynamics states that energy is neither created nor destroyed. It may change forms, pass from one place to another, or act upon matter in various ways, but regardless of what transfers and transformations take place, no gain or loss in total energy occurs. Energy is simply transferred from one form or place to another. When wood is burned, the potential energy present in the molecules of wood equals the kinetic energy released, and heat escapes. This reaction is *exothermic.*

On the other hand, energy from the surroundings may be paid into a reaction. Here, too, the first law holds true. In photosynthesis, for example, the molecules of the products store more energy than the reactants. The extra energy is acquired from the sunlight. Again, there is no gain or loss in total energy. When energy from outside surroundings is put into a system to raise it to a higher energy state, the reaction is *endothermic.*

Although the total amount of energy involved in any reaction, such as burning wood, does not increase or decrease, much of the potential energy involved is degraded into a form incapable of doing any further work. It ends up as heat, disorganized or randomly dispersed molecules in motion, useless for further transfer. The measure of this relative disorder is called *entropy.*

Transfer of energy involves the second law of thermodynamics. It states that when energy is transferred or transformed, part of the energy assumes a form that cannot be passed on any further. When coal is burned in a boiler to produce steam, some of the energy creates steam, and part is dispersed as heat to the surrounding air. The same thing happens to energy in the ecosystem. As energy is transferred from one organism to another in the form of food, a large part of that energy is degraded as heat—no longer transferable—and the remainder is stored as living tissue.

Biological systems seemingly do not conform to the second law, for the tendency of life is to produce order out of disorder, to decrease rather than increase entropy. The second law theoretically applies to the isolated, closed system in which no energy or matter is exchanged between the system and its surroundings. An isolated system approaches thermodynamic equilibrium—that is, a point at which all the energy has assumed a form that cannot do work. A closed, isolated system tends toward a state of minimum free energy (energy available to do work) and maximum entropy. An open system maintains a state of higher free energy and lower entropy. In other words, the closed system tends to run down; the open one does not. As long as there is a constant input of free energy to the system and a constant outflow of energy in the form of heat and waste, the system maintains a steady state. Thus, life is an open system maintained in a steady state.

PRIMARY PRODUCTION

The flow of energy through the ecosystem starts with the fixation of sunlight by plants, a process that in itself demands the expenditure of energy. A plant gets its start by living on the food stored in the seed until its production machinery is working. Once started, the green plant begins to accumulate energy. Energy accumulated by plants is called *production* or, more specifically, *primary production,* because it is the first and basic form of energy storage. The rate at which energy accumulates is known as *primary productivity.* All of the sun's energy that is assimilated—that is, total photosynthesis—is *gross primary production* (Box 22.1).

Because plants, like other organisms, must overcome the tendency of energy to disperse, free energy (that available to do work) must be expended for production as well as for reproduction and maintenance. This energy is provided by the reverse of photosynthesis: metabolic respiration, which results in the production of CO_2 and H_2O and the liberation of energy. Most plants have to contend with photorespiration (see Chapter 18), associated with photosynthesis but not with metabolism. That, too, acts as a drain on photosynthesis. Energy remaining after respiration and stored as organic matter is *net primary production,* or plant growth. Net primary production can be described by the following equation:

$$\text{Net primary production (NPP)} = \text{gross primary production (GPP)} - \text{autotrophic respiration (R)}$$

Production is usually expressed as kilocalories per square meter (kcal/m^2). However, production may also be expressed as dry organic matter in grams per square meter (g/m^2). If either of these two measures is employed to estimate efficiencies and other ratios, the same unit must be used for both the numerator and the denominator of the ratio. Only calories can be compared with calories, dry weight with dry weight.

BIOMASS

Net primary production accumulates over time as plant biomass. Part of this accumulation is turned over seasonally through decomposition. Part is retained over a longer period as living material. The amount of this accumulated organic matter found on a given area at a given time is the *standing crop biomass.* Biomass is usually expressed as grams of organic matter per square meter (g/m^2), calories per square meter (cal/m^2), or some other appropriate measure per unit of area. Thus, biomass differs from production, which is the rate at which organic matter is created by photosynthesis. The biomass present at any given time is not the same as productivity.

Biomass Allocation

Plants budget their production of photosynthate, distributing it in a systematic way to leaves, twigs, stems, bark, roots, flowers, and seeds. What each part receives depends upon its demands for energy, growth, and maintenance. How much is allocated to each component is difficult to determine because of the time-consuming chore of cutting trees and clipping herbs, extracting roots, separating each of the components, weighing them, and determining both their energy and their nutrient contents.

The pattern of allocation varies with the plant. Single-celled planktonic algae are well supplied, indeed immersed, in nutrients and light. They accumulate large supplies, growing rapidly and dividing into new individuals. The photosynthate increases population size, which is a measure of their growth.

Annuals begin their life cycles in the spring with the germination of overwintering seeds. With only one growing season in which to complete its life cycle, an annual has to allocate its photosynthate first to leaves,

BOX 22.1

Production Efficiencies

Production, both primary and secondary, can be expressed in a number of ways. The following are the important ones, including those used in the text.

Terms

GPP	Gross primary production
NPP	Net primary production
R	Respiration
P	Secondary production: tissue growth, reproduction, exoskeleton growth, biomass change
C	Consumption: plant material ingested
F	Egestion: feces, urine, gas, other products
A	Assimilation: food or energy absorbed

Equations

1. Photosynthetic efficiency = GPP/solar radiation
2. Assimilation efficiency, plants = GPP/light absorbed
3. Effective primary production = NPP/GPP
4. Assimilation efficiency, animals = A/C
5. Ecological growth efficiency = P/C
6. Growth efficiency = P/A

which, in turn, become involved in photosynthesis, which increases vegetation biomass. At the time of flowering, the plant decreases the amount of energy and nutrients allocated to leaves. They are supplied only with enough to maintain themselves; in fact, lower leaves may die. Most of the photosynthate is diverted to reproduction. In the sunflower, for example, the biomass of leaves declines from approximately 60 percent of the plant during growth to 10 to 20 percent by the time the seeds are ripe. When in bloom, the sunflower distributes 90 percent of its photosynthate to the flower head and the rest to leaves, stem, and roots. If the annual grows on a dry or nutrient-poor site, it has to allocate more energy and nutrients to roots and less to leaves and flowers, lowering its competitive ability.

Perennials maintain a vegetative structure over a period of several to many years. They begin their life cycles like an annual, but once established they allocate their photosynthate in a very different manner. Before perennials expend any energy on reproduction, they divert excess photosynthate to the roots. In some species, like the skunk cabbage, the roots develop into massive storage organs. Energy and nutrients stored in the roots make up a reserve upon which the plants draw when they begin early growth. When they are ready to flower, perennials divert energy going into storage to the production of flowers and fruit. As the flowers fade and the fruit ripens, the plants once more send photosynthate to the roots to build up the reserves they will need the next spring.

Consider the grass blue grama (*Bouteloua gracilis*), a C_4 species. Eighty-five percent of the net carbon fixed in photosynthesis is translocated. Of that, approximately 24 percent goes to shoots; 24 percent to the crown, from which new growth develops; and 52 percent to roots. Under environmental stress, perennials, including grasses, typically allocate more of their photosynthate to roots than to leaves and shoots.

Trees and woody shrubs live a long time, which greatly influences the manner in which they distribute photosynthate. Early in life leaves may make up more than one-half of their biomass (dry weight), but as trees age, they put down more woody growth. Trunks and stems become thicker and heavier, and the ratio of leaves to woody tissue changes. Eventually, leaves account for only 1 to 5 percent of the total mass of the tree. Thus, the production system that supplies the energy and nutrients is considerably less than the biomass it supports. Much of the photosynthate goes into permanent capital, inaccessible to the plant and ultimately available only to decomposers. Thus, much of the energy of woody plants goes into support and maintenance, which increases as the plants age.

When deciduous trees leaf out in spring, they expend one-third of their reserve energy on expansion and growth of leaves. This expenditure, of course, is repaid as the leaves carry out photosynthesis during the spring and summer. After leaves, trees give preference to flowers; then cambium, new buds, and deposits of starch in roots and bark; and finally, new flower buds. Reproduction and vegetative growth compete for energy allotments. If photosynthate is limited, vegetative growth gets first claim. Because energy demand by reproduction is high—up to 15 percent in pines, 20 percent in beech, and 35 percent or more in fruit trees—trees can afford an abundance of fruit only periodically, once every two or three years in deciduous trees and two to six years in conifers. As the growing season ends, photosynthates are withdrawn from leaves and the excess sent to roots, woody tissue, and buds.

Evergreen trees have a somewhat different situation. Because their photosynthetic machinery can function year round when temperature and moisture conditions permit, they do not need to draw on root reserves for new growth in spring. They can afford to wait until later in the growing season to send out new shoots. Then evergreens can draw on new photosynthate built up earlier in the spring. For the same reason, new growth develops rapidly and matures in a few weeks.

The allocation of biomass has been determined for some forest stands. For example, in a young oak-pine forest on Long Island, 25 percent of the net primary production was allocated to stem wood and bark, 40 percent to roots, 33 percent to twigs and leaves, and 2 percent to flowers and seeds. Among the shrubs, 54 percent of the net primary production was allocated to roots, 21 percent to stems, and 23 percent to leaves.

The proportionate allocation of net production to aboveground and belowground biomass, to shoot and root, tells much about different ecosystems and about different components within ecosystems. A low shoot–root ratio indicates that most net production goes into a plant's supportive functions. Plants with a large root biomass are more effective competitors for water and nutrients and can survive more successfully in harsh environments because most of their active biomass is below ground. Plants with a high shoot–root ratio have most of their biomass above ground and assimilate more light energy, resulting in higher productivity. Tundra communities in an environment with long, cold winters and a short growing season have shoot–root ratios that reflect a harsh environment. Wet grass and sedge communities have a shoot–root ratio of 1:2.1; dwarf shrub communities, 1:3.1; low shrub communities, 1:2.0. Even forested tundra has a relatively high ratio for a forest—1:0.8. Further south, midwest prairie grasses have a shoot–root ratio of 1:3, indicative, perhaps, of cold winters and a limited moisture supply. In forest ecosystems, with their high aboveground biomass, the shoot–root ratio is low. For the Hubbard Brook Forest in New Hampshire, the shoot–root ratio for trees is 1:0.213; for shrubs, 1:0.5; and for herbs, 1:1. As we could predict, the shoot–root ratio increases through the vertical strata from the canopy to the floor.

Vertical Distribution

Aboveground biomass is distributed vertically in the ecosystem. Vertical distribution of leaf biomass in terrestrial communities and the concentration of plankton and floating and submerged vegetation influence the penetration of light, which, in turn, influences the distribution of production in the ecosystem. The region of maximum productivity in the aquatic ecosystem is not the upper sunlit surface (strong sunlight inhibits photosynthesis) but some depth below, depending upon the clarity of the water and the density of plankton or vegetative growth (Figure 22.1). As depth increases, light intensity decreases until it reaches a point at which the light received by the vegetation is just sufficient to meet respiratory needs and production equals respiration (Figure 22.2). This level is known as the *compensation level.* In the forest ecosystem a similar situation exists. The greatest amount of photosynthetic biomass (Figure 22.3) as well as the highest net photosynthesis is not at the top but at some point below maximum light intensity. In spite of wide differences in plant species and in types of ecosystems, the vertical profiles of biomass of the various ecosystems appear to be similar.

Within the vertical profile, biomass varies seasonally and even daily. In grasslands and old-field ecosystems, much of the net production is turned over every year. The standing crop of living material in an old field in Michigan amounted to about 4×10^3 kg/ha in late summer, compared to 80 kg/ha in late spring. By then the standing crop in dead matter was nearly 3×10^3 kg/ha. The aboveground biomass of a tall-grass prairie that included both living and dead material was about twice that of the standing crop, the living material

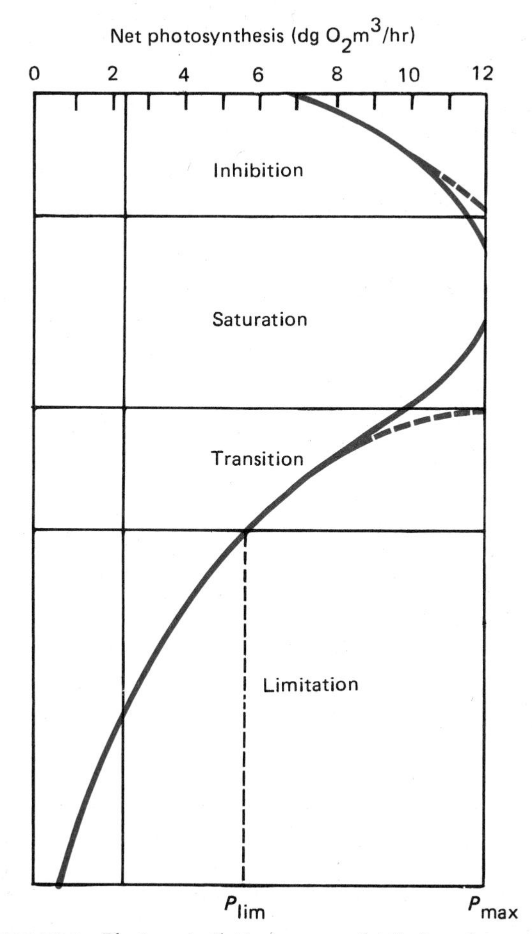

FIGURE 22.1 Photoassimilation versus depth in a homogeneous population of phytoplankton. Note that photosynthesis is inhibited at and near the surface because of high light intensity (irradiance). Maximum photosynthesis takes place somewhat below the surface where light is at saturation. Photosynthesis declines rapidly with depth because of reduction in irradiance and a change in the spectral composition of light. Under actual conditions, the photosynthetic profile would be irregular because of changes in biomass with depth under different environmental regimes.

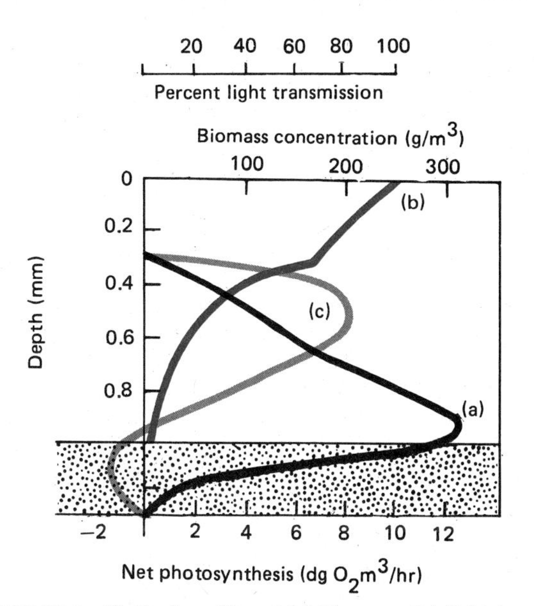

FIGURE 22.2 Vertical profiles of (a) biomass, (b) light transmission, and (c) net photosynthesis of a stand of a submerged macrophyte, *Vallisneria denseserrulata.* The biomass profile has little influence on the profile of net photosynthesis, which is influenced largely by irradiance and photosynthetic capacity. Most of the photosynthetically active biomass is just below the water's surface. Photosynthetic capacity of the vegetation, which decreases within the deeper parts of the stand, is independent of plant biomass because of morphological differences in the plants, decreased irradiance, and shading. Maximum photosynthesis is concentrated near the surface of the stand, whereas maximum biomass is near the base of the growth.

added during the growing season. The aboveground biomass had a turnover rate of approximately two years and the belowground biomass of roots, four years. In a forest ecosystem a considerably greater portion of net production is tied up in wood, as exemplified by an oak-pine forest on Long Island. Leaves, fruit, flowers, deadwood, and bark contributed 4342 g/m^2/year, for a total of 653 g/m^2, or about 58 percent of net primary production.

DIFFERENCES IN PRODUCTIVITY

Little of the energy assimilated by plants goes into organic production. Most of the light absorbed by plants is converted to heat and lost through convection and radiation. What fraction of light energy is used in photosynthesis goes into gross production, and what is left over after respiration goes into net production. Production efficiency (see Box 22.1), the ratio of net primary production to gross primary production, is on the average rather high. Algae and corn have an effi-

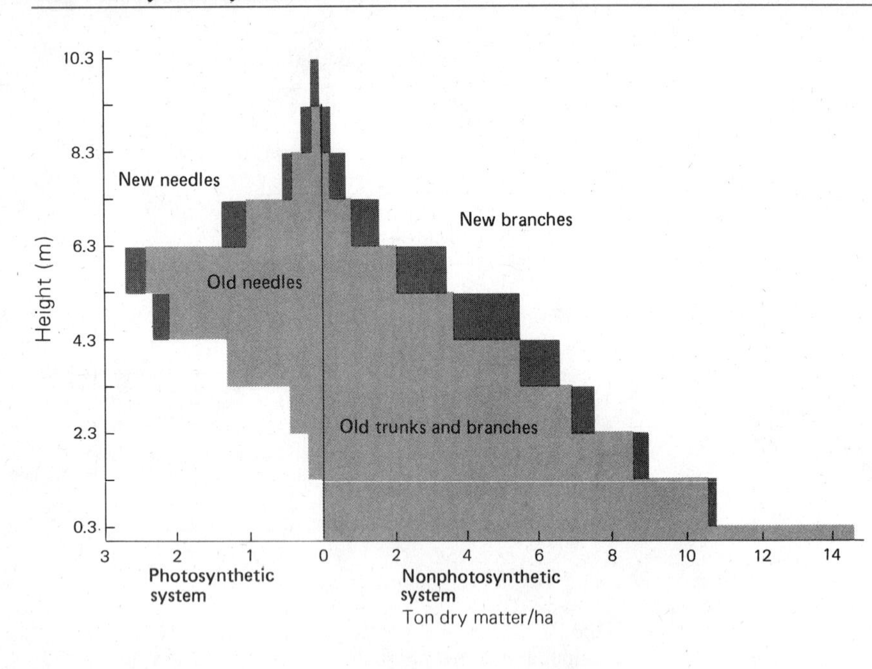

FIGURE 22.3 Structure and productive systems of a pine spruce-fir forest in Japan. The dark gray areas represent new needles and new branches. Note that most of the active photosynthetic area lies below the uppermost parts of the canopy, a pattern similar to that of the phytoplankton community.

ciency of 80 to 85 percent; eastern deciduous forests, 42 percent; prairie grasses, 66 percent; tundra vegetation, 50 percent; and submerged aquatic plants, 65 to 75 percent. The most efficient plants are those such as grasses that do not maintain a high supporting biomass as woody tissue (see Table 22.1).

Productivity of ecosystems is influenced by temperature and precipitation (Figure 22.4). Such relationships show up in variations in productivity of different global ecosystems (Figure 22.5). The most productive ecosystems are tropical rain forests and coral reefs (Figure 22.6). Deserts, tundra, and the open sea are the least productive.

Some ecosystems have consistently high production. Such high productivity usually results from an additional energy subsidy to the system. This subsidy may be a warmer temperature, greater rainfall, or circulating or moving water that carries food or additional nutrients into the community. In agriculture such subsidies include the use of fossil fuel for cultivation, application of fertilizer, and the control of pests. Swamps and marshes, ecosystems at the interface of land and water, may have a net productivity of 3300 g/m^2/year. Estuaries, because of inputs of nutrients from rivers and tides, and coral reefs, because of inputs from changing tides, may have a net productivity between 1000 and 2500 g/m^2/year. Among agricultural ecosystems, sugarcane has a net productivity of 1700 to 1800 g/m^2/year; hybrid corn, 1000 g/m^2/year; and some tropical crops, 3000 g/m^2/year.

CHANGES OVER TIME

Annual net production changes with time and age. For example, a Scots pine (*Pinus sylvestris*) plantation in England achieved a maximum annual net production of 22 $\times$ 10^3 kg/ha at the age of 20 years. It declined to 12 $\times$ 10^3 kg/ha at 30 years of age. Woodlands apparently achieve their maximum annual net production in the pole stage, when the dominance of trees is greatest and the understory is minimal (Figure 22.7, p. 370). The understory makes its greatest contribution to annual net production during the early and mature stages of the forest. As the age of a forest stand increases, more and more of the production is needed for maintenance, and very little gross production is left for growth (Figure 22.8, p. 370). The pattern is well illustrated by Douglas-fir in the Pacific Northwest. Seventy percent of the net production of a 20- to 40-year-old stand accumulates as stored biomass. In a 450-

TABLE 22.1 Estimates of Productivity and Respiration for Several North American Ecosystem Types (Carbon, $kg/m^2/yr$)

	Old-Growth Douglas-Fir Forest	*Eastern Deciduous Forest*	*Eastern Oak-Pine Forest*	*Prairie*	*Tundra*	*Potato Field*	*Rye Field*
Gross primary production (GPP)	7.72	1.62	1.32	0.64	0.24	1.29	1.00
Autotrophic respiration (R_A)	7.20	0.94	0.68	0.22	0.12	0.43	0.34
Net primary production (NPP)	0.52	0.68	0.60	0.42	0.12	0.85	0.66
Heterotrophic respiration (R_H)	0.36	0.52	0.37	0.27	0.11	0.50	0.31
Net ecosystem production (GPP − R_E)	0.16	0.16	0.27	0.15	0.01	0.36	0.35
Ecosystem respiration ($R_E = R_H + R_A$)	7.56	1.47	1.05	0.49	0.23	0.93	0.65
Production efficiency (R_A/GPP)	0.93	0.57	0.52	0.34	0.50	0.34	0.34
Effective production (NPP/GPP)	0.07	0.42	0.45	0.66	0.50	0.66	0.66
Maintenance efficiency (R_A/NPP)	13.80	1.38	1.13	0.51	1.00	0.52	0.52
Respiration allocation (R_H/R_A)	0.05	0.55	0.54	1.26	0.90	0.86	0.91
Ecosystem productivity (NEP/GPP)	0.02	0.10	0.20	0.23	0.05	0.28	0.35

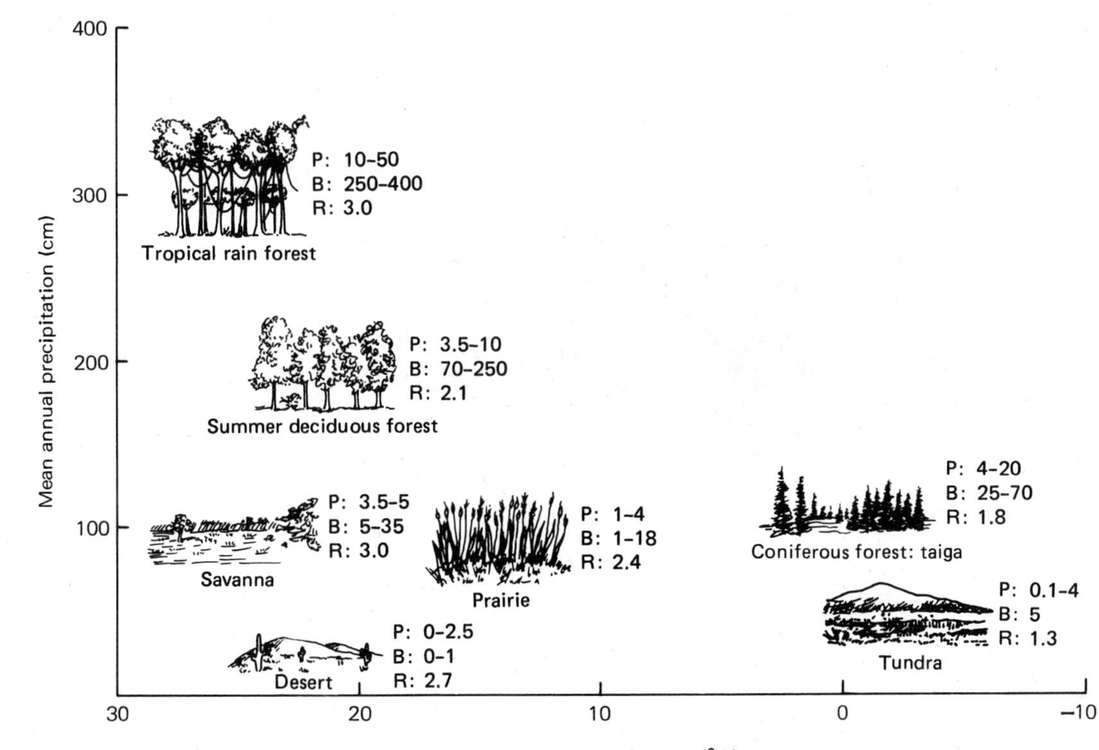

FIGURE 22.4 Distribution of primary production, biomass, and radiation input relative to temperature and rainfall. P = primary production (ton/ha); B = biomass (ton/ha); R = solar radiant input ($kcal/m^2/yr$, 0.3–3.0 microns).

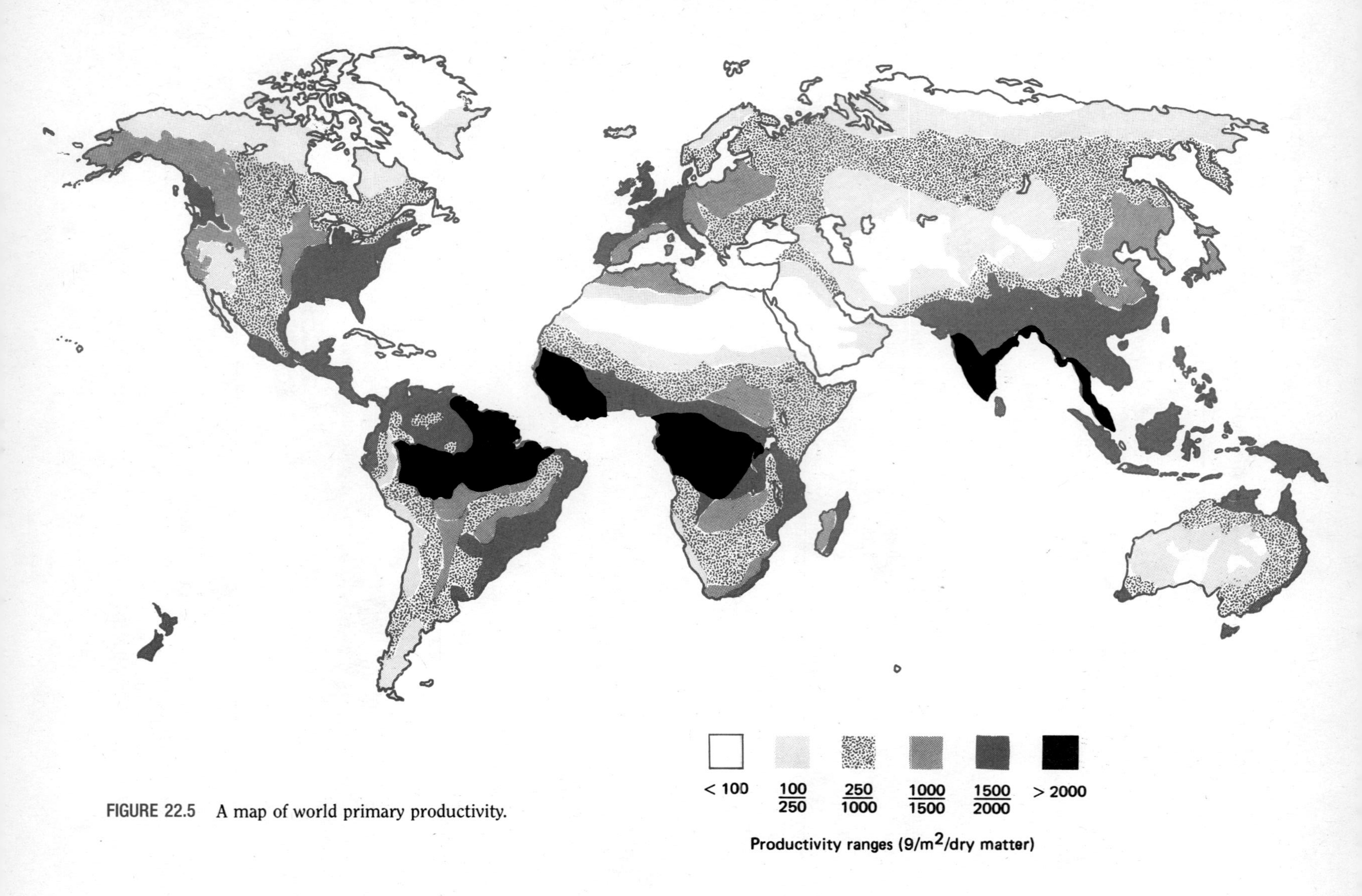

FIGURE 22.5 A map of world primary productivity.

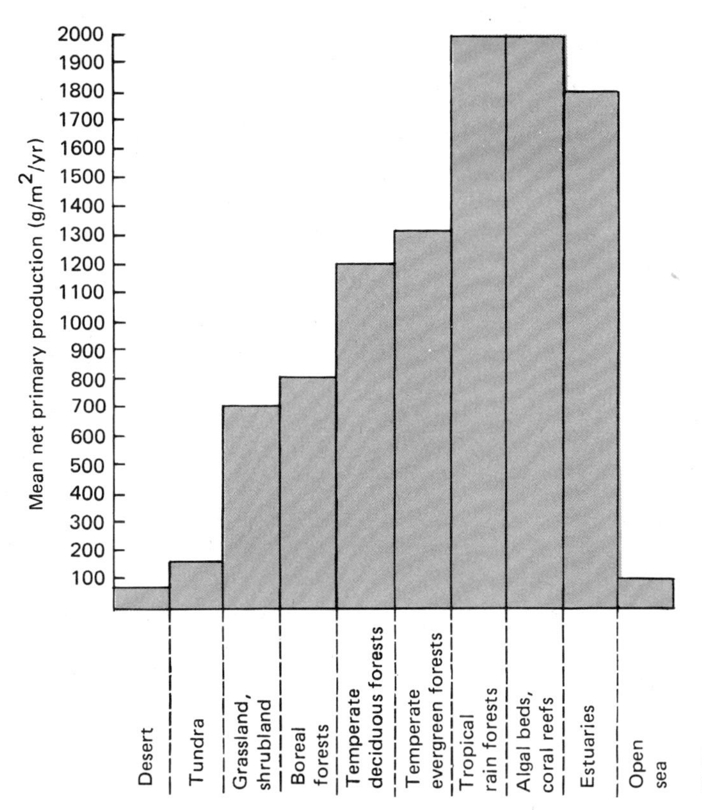

FIGURE 22.6 Comparative productivity of world ecosystems.

year-old stand, only 6 to 7 percent of gross photosynthesis is available for net production, and most of that is converted to detritus.

Production varies not only among different types of ecosystems but also among similar systems and within one system from year to year. Production is influenced by such factors as nutrient availability, precipitation, temperature, length of growing season, animal utilization, and fire. For example, herbage yield of a grassland may vary by a factor of 8 between wet and dry years. Overgrazing of grasslands by cattle and sheep or defoliation of forests by such insects as the gypsy moth can seriously reduce net production. Fire in grasslands may result in increased productivity if moisture is normal, but reduced productivity if precipitation is low. An insufficient supply of nutrients, especially nitrogen and phosphorus, can limit net productivity, as can mechanical injury of plants, atmospheric pollution, and the like.

SECONDARY PRODUCTION

Net production is the energy available to the heterotrophic components of the ecosystem. Theoretically, at least, all of it is available to the grazers or to the decomposers; but rarely is it all utilized in this manner. The net production of any given ecosystem may be dispersed to another food chain outside of the ecosystem through removal by humans or other agents such as wind or water currents. For example, about 45 percent of the net production of the salt marsh is lost to estuarine water. Much of the living material is physically unavailable to the grazers—they cannot reach the

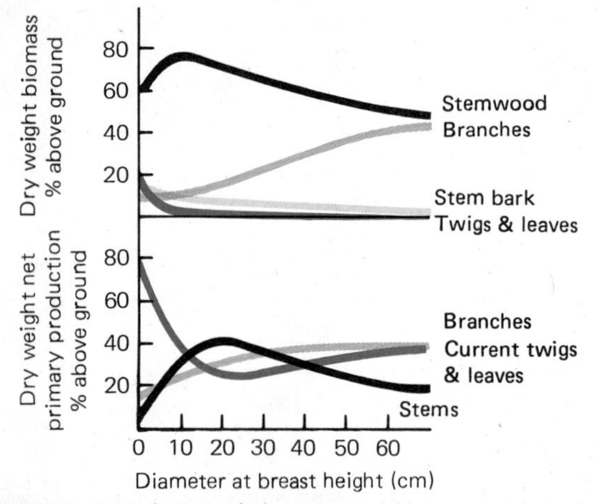

FIGURE 22.7 Relation of aboveground biomass and production to tree size in the Hubbard Brook Forest. Trends involve 63 sample trees of three major species: sugar maple, yellow birch, and beech. Note that as the trees increase in size, the ratio of branches to stems increases.

plants. Living organic matter, as long as it is alive, is unavailable to decomposers and detritus-feeders, and dead materials may not be relished by grazers. The amount of net production available to herbivores may vary from year to year and from place to place. The quantity consumed will vary with the type of herbivore and the density of the population.

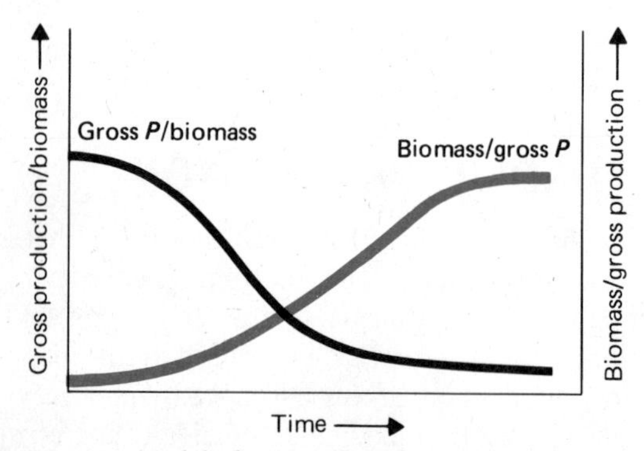

FIGURE 22.8 Model showing the change through time in ratio of gross community primary production to biomass (production efficiency) and ratio of biomass to gross community primary production (maintenance efficiency). Note the early high production efficiency and the later accumulation of biomass.

Once consumed, a considerable portion of the plant material, again depending on the kind of plant involved and the digestive efficiency of the herbivore, may pass through the animal's body undigested. A grasshopper assimilates only about 30 percent of the grass it consumes, leaving 70 percent as wastes. Mice, on the other hand, assimilate about 85 to 90 percent of what they consume.

Energy, once consumed, either is diverted to maintenance, growth, and reproduction or is passed from the body as feces and urine (see Figure 22.9). The energy lost through urine can be variable and often high. Another portion is lost as fermentation gases. Of the energy left after losses through feces, urine, and gases, part is utilized as "heat increment," which is the heat needed for metabolism above that required for basal or resting metabolism. The remainder of the energy is available for maintenance and production. It includes energy involved in capturing or harvesting food, muscular work expended in the animal's daily

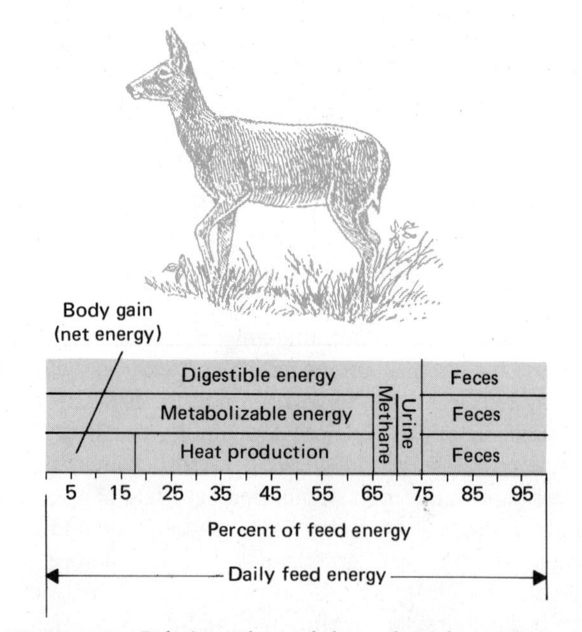

FIGURE 22.9 Relative values of the end products of energy metabolism in the white-tailed deer. Note the small amount of net energy gained (body gain) in relation to that lost as heat, gas, urine, and feces. The deer is a herbivore, a first-level consumer.

routine, and energy needed to keep up with the wear and tear on the animal's body. The energy used for maintenance is lost as heat.

Maintenance costs—highest in the small, active, warm-blooded animals—are fixed or irreducible. In small invertebrates energy costs vary with the temperature: A positive energy balance exists only within a fairly narrow range of temperatures. Below 5° C spiders become sluggish and cease feeding, and have to utilize stored energy to meet their metabolic needs. At approximately 5° C assimilated energy approaches energy lost through respiration. From 5° to 20.5° C spiders assimilate more energy than they respire. Above 25° C the ability of the spider to maintain a positive energy balance declines rapidly.

The energy left over from maintenance and respiration goes into production, including new tissue, fat tissue, growth, and new individuals. This net energy of production is *secondary production.* Within secondary production there is no portion known as gross production. That which is analogous to gross production is actually assimilation. Secondary production is greatest when the birthrate of the population and the growth rates of the individuals are highest. This usually coincides, for obvious reasons, with the time when net production is also highest.

This scheme is summarized in Figure 22.10. It is applicable to any consumer organism, herbivore or carnivore. The herbivore represents the energy source of the carnivore; and as in the case of the plant food of the herbivore, not all of the energy contained in the body of the herbivore is utilized by the carnivore. Part (such as hide, bones, and internal organs) is unconsumed, and the same metabolic losses can be accounted for. At each transfer considerably less energy is available for the next consumer level.

Just as net primary production is limited by a number of variables, so is secondary production. The quantity, quality (including the nutrient status and digestibility), and availability of net production are three limitations. So is the degree to which primary and available secondary production are utilized.

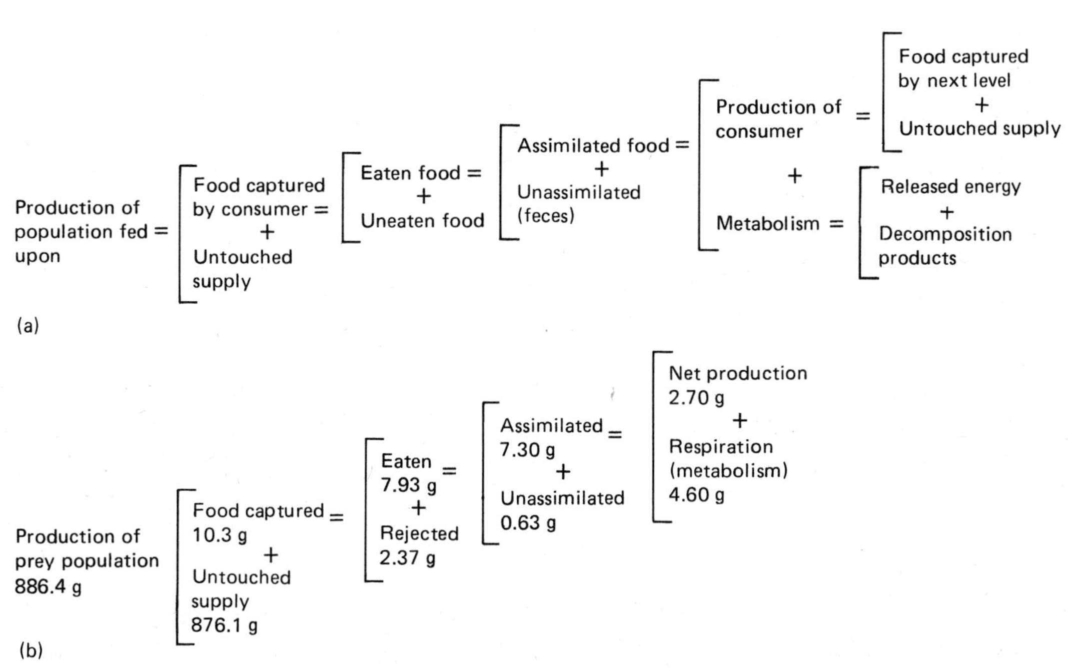

FIGURE 22.10 Components of energy metabolism in secondary production. (a) General model. (b) A field example, a ground-dwelling spider population in the spring.

TABLE 22.2 Secondary Production of Selected Consumers (kcal/m²/yr)

Species	*Ingestion (I)*	*Assimilation (A)*	*Respiration (R)*	*Production (P)*	*A/I*	*P/I*	*R/A*
Harvester ant (h)	34.50	31.00	30.90	0.10	0.89	0.0002	0.99
Plant hopper (h)	41.30	27.50	20.50	7.00	0.67	0.169	0.75
Salt marsh grasshopper (h)	3.71	1.37	0.86	0.51	0.37	0.137	0.63
Spider, small < 1 mg (c)	12.60	11.90	10.00	0.91	0.94	0.151	0.84
Spider, large > 10 mg (c)	7.40	7.00	7.30	−3.00	0.95	—	1.04
Savanna sparrow (o)	4.00	3.60	3.60	0.00	0.90	0.00	1.00
Old-field mouse (h)	7.40	6.70	6.60	0.1	0.91	0.014	0.98
Ground squirrel (h)	5.60	3.80	3.69	0.11	0.68	0.019	0.97
Meadow mouse (h)	21.29	17.50	17.00	—	0.82	—	0.97
African elephant (h)	71.60	32.00	32.00	8.00	0.44	—	1.00
Weasel (c)	5.80	5.50	—	—	0.95	—	—

NOTE: h = herbivore; o = omnivore; c = carnivore.

The latter can be examined using two different ratios. One is the ratio of assimilation (the energy extracted from the food) to ingestion (the food actually eaten), or A/C. This ratio measures the efficiency with which the consumer extracts energy from the food it consumes. The other is the ratio of productivity to assimilation, or P/A. This ratio measures the efficiency with which the consumer incorporates assimilated energy into new tissue, or secondary production.

The ability of the consumer population to convert the energy it ingests varies with the species and the type of consumer. Insects that feed on plant tissues, such as grasshoppers, are more efficient producers than insects that feed on plant juices, such as aphids. Larval stages of insects are more efficient producers than the adult stage. Homeotherms have a high assimilation, but because they utilize about 98 percent of that energy in metabolism, they have poor production efficiency (Table 22.2). Poikilotherms utilize about 79 percent of their total assimilation in metabolism. They convert a greater portion of their assimilated energy into biomass than do homeotherms. The difference, however, is balanced by assimilation efficiency. Poikilotherms have an efficiency of around 30 pecent in digesting food, whereas homeotherms have an efficiency of around 70 percent. The poikilotherm has to consume more calories to obtain sufficient energy to meet the needs of maintenance, growth, and reproduction.

Summary

A major function of ecosystems is energy flow, which supports life on Earth. Energy is governed by the laws of thermodynamics. The second law—which states that as energy is transferred or transformed from one state to another, a portion is no longer usable—is particularly applicable to energy flow in ecosystems. As energy moves through an ecosystem from sun to plants to consumers, much of it is lost as heat of respiration. Energy is degraded from a more organized to a less organized state, or entropy. However, a continuous flux of energy from the sun into ecosystems prevents them from running down.

To carry on photosynthesis and fix energy, plants must utilize part of the energy they fix. The total amount of energy fixed by plants is gross primary production. The amount of energy left after plants have met their respiratory needs is net primary production, which shows up as plant biomass. The amount of accumulated biomass on any given area at a particular time is standing crop biomass. Energy fixed by plants is allocated to different parts of the plant and to reproduction. Plants allocate energy first to leaves, then to flowers. Excess production goes to roots and other supporting tissue, where some of the reserves are available for growth the following year.

Efficiency of production varies among plants and

among ecosystems. Most efficient are those plants such as grasses and annuals that do not maintain a high supporting biomass. The least productive are old trees, which expend most of their gross production in maintenance. Productivity of ecosystems is influenced by temperature and precipitation. The most productive are topical rain forests, coral reefs, and estuaries; the least productive are tundras and warm deserts.

Net production is available to consumers either directly as plant tissue or indirectly through animal tissue. Once consumed and assimilated, energy is diverted to maintenance, growth, and reproduction and to feces, urine, and gas. Change in biomass, measured as weight change and reproduction, represents secondary production. Efficiency varies. Homeotherms have a high assimilation efficiency but low production efficiency because they have to expend so much energy in maintenance. Poikilotherms have low assimilation efficiency but high production efficiency. Much of the energy goes into growth rather than maintenance.

Review and Study Questions

1. What conditions are necessary for energy flow?
2. How do the first and second laws of thermodynamics relate to ecology?
3. Define primary production, primary productivity, gross primary production, net primary production, respiration, and secondary production.
4. What is the significance of the shoot–root ratio to production in terrestrial ecosystems?
5. How do plants allocate biomass and energy to growth, maintenance, and reproduction?
6. How does net production relate to the age of an ecosystem? What else influences productivity?
7. What world ecosystems have high net productivity? Low? Why?
8. How can we estimate primary and secondary production? (To answer consult the Selected References.)
9. What is secondary production? Aside from involving animals, how does it differ from primary production?
10. What is the difference in energy allocation and energy efficiency between homeotherms and poikilotherms?

Selected References

Bliss, L. C., O. W. Heal, and J. J. Moore, eds. 1981. Chapters 6, 7, 10, 11, and 12 in *Tundra ecosystems: A comparative analysis.* New York: Cambridge University Press.

Breymeyer, A. I., and G. M. Van Dyne, eds. 1980. Chapters 2, 3, 4, 5, and 6 in *Grasslands, systems analysis, and man.* New York: Cambridge University Press.

Cooper, J. P. 1975. *Photosynthesis and productivity in different environments.* New York: Cambridge University Press.

Edmonds, R. L., ed. 1982. Chapters 5 and 6 in *Analysis of coniferous forest ecosystems in the western United States.* US/IBP Synthesis, Series no. 14. Stroudsburg, PA: Dowden, Hutchinson & Ross Publishing Company.

Gates, D. M. 1985. *Energy and ecology.* Sunderland, MA: Sinauer Associates. Energy as it relates to human affairs.

LeCren, E. D., and R. H. Lowe-McConnell, eds. 1980. Chapters 5, 6, and 9 in *The functioning of freshwater ecosystems.* New York: Cambridge University Press.

Leith, H., and R. H. Whittaker, eds. 1975. *Primary productivity in the biosphere.* New York: Springer-Verlag.

National Academy of Science. 1975. *Productivity of world ecosystems.* Washington, DC: National Academy of Science. Good summary of world primary production.

Petrusewicz, K., ed. 1967. *Secondary productivity of terrestrial ecosystems.* 2 vols. Warsaw, Poland: Pantsworve Wydawnictwo Naukowe.

Phillipson, J. J. 1966. *Ecological energetics.* New York: St. Martin's Press. Although dated, this is still an excellent introduction to energy in ecosystems.

Reichle, D. E., ed. 1981. Chapters 7 and 8 in *Dynamic properties of forest ecosystems.* New York: Cambridge University Press.

Wiegert, R. G., ed. 1976. *Ecological energetics.* Benchmark Papers®. Stroudsburg, PA: Dowden, Hutchinson, & Ross. (Dist. by Academic Press, New York.) Collection of papers surveying the development of the concept.

The following are manuals on measuring productivity:

Edmondson, W. T., and G. G. Winberg. 1971. *A manual on methods for the assessment of secondary production in fresh waters.* IBP Handbook no. 17. Oxford, England: Blackwell Scientific Publications.

Milner, C., and R. E. Hughes. 1968. *Methods for the measurement of the primary production of grassland.* IBP Handbook no. 6. Oxford, England: Blackwell Scientific Publications.

Newbould, P. J. 1967. *Methods in estimating primary production of forests.* IBP Handbook no. 2. Oxford, England: Blackwell Scientific Publications.

Petrusewicz, K., and A. Macfadyen. 1970. *Productivity of terrestrial animals.* IBP Handbook no. 13. Oxford, England: Blackwell Scientific Publications.

Vollenweider, R. A. 1969. *A manual on methods for measuring primary production in aquatic environments.* IBP Handbook no. 12. Oxford, England: Blackwell Scientific Publications.

CHAPTER 23

Trophic Structure

Outline

Objectives

On completion of this chapter, you should be able to:

1. Describe a food chain and a food web.
2. Distinguish between grazing and detrital food chains.
3. Discuss the role of microorganisms and macroorganisms in the decomposition process.
4. Define trophic levels and describe ecological pyramids.
5. Discuss energy flow through the food chain.
6. Discuss the structure and dynamics of food webs.

FOOD CHAINS

The energy stored by plants is passed along through the ecosystem in a series of steps of eating and being eaten, the *food chain* (Figure 23.1). Food chains are descriptive. When worked out diagrammatically, they consist of a series of arrows, each pointing from one species to another for which it is a source of food. In Figure 23.1, for example, grass plants are consumed by grasshoppers, grasshoppers become food for clay-colored sparrows, and the sparrows are preyed upon by marsh hawks or harriers. This relationship can be written as follows:

Grass → grasshopper → sparrow → harrier

As the diagram indicates, no relationship is wholly linear. Resources are shared, especially at the beginning of the chain. The same plant is eaten by a variety of mammals and insects, and the same animal is consumed by several predators. Thus, food chains become interlinked to form a *food web*, the complexity of which varies within and between ecosystems.

COMPONENTS

Herbivores

Feeding on plant tissues is a host of plant consumers, the *herbivores*. They are capable of converting energy stored in plant tissue into animal tissue. Their role is essential in the community, for without them the higher trophic levels could not exist. The English ecologist Charles Elton, in his classic book *Animal Ecology* (1927), suggested that the term *key industry* be used to denote animals that feed on plants and are so abundant that many other animals depend on them for food.

Only herbivores are adapted to live on a high-cellulose diet. Modification in the structure of the teeth, complicated stomachs, long intestines, a well-developed caecum, and symbiotic flora and fauna enable these animals to use plant tissues. For example, ruminants, such as deer, have a four-compartment stomach. As they graze, these animals chew their food hurriedly. The material consumed descends to the first and second stomachs (the rumen and reticulum), where it is softened to a pulp by the addition of water, kneaded by muscular action, and fermented by bacteria. The bacteria convert part of the celluloses, starches, and sugars to short-chain volatile fatty acids. These acids are rapidly absorbed into the bloodstream and oxidized to provide the mammal's chief form of energy. At leisure, ruminants regurgitate the undigested portion, chew it more thoroughly, and swallow it again.

The lagomorphs—rabbits, hares, and pikas—have a simple stomach and a large caecum. In the formation of fecal pellets, part of the ingested material is attacked by microorganisms and is expelled into the large intestine as moist, soft pellets surrounded by a proteinaceous membrane. The soft pellets, much higher in protein and lower in crude fiber than the hard fecal pellets, are reingested (coprophagy). The amount of feces recycled by coprophagy may range from 50 to 80 percent. This reingestion is important, for it provides bacterially synthesized B vitamins and ensures more complete digestion of dry material and better utilization of protein.

Carnivores

Herbivores, in turn, are the energy source for *carnivores*, animals that feed on other animals. In a popular sense, carnivores are considered to be larger organisms that kill and eat smaller prey. In the broadest sense, any organism that feeds on another organism or the tissue of an organism is a carnivore, functionally speaking. Thus, carnivory could include *parasites*.

Organisms that feed directly on grazing herbivores are termed *first-level carnivores* or *second-level consumers*. First-level carnivores represent an energy source for second-level carnivores. The typical carnivore is well adapted for a diet of flesh. Hawks and owls have sharp talons for holding prey and hooked beaks for tearing flesh. Mammalian carnivores have canine teeth for biting and piercing. Cheek teeth are reduced, but many forms have sharp-crested shearing or carnassial teeth.

Other Feeding Groups

Not all consumers can be fitted neatly into a trophic level, for many do not confine their feeding to one level alone. The red fox feeds on berries, small rodents, and even dead animals. Thus, it occupies herbivorous and carnivorous levels, as well as acting as a scavenger.

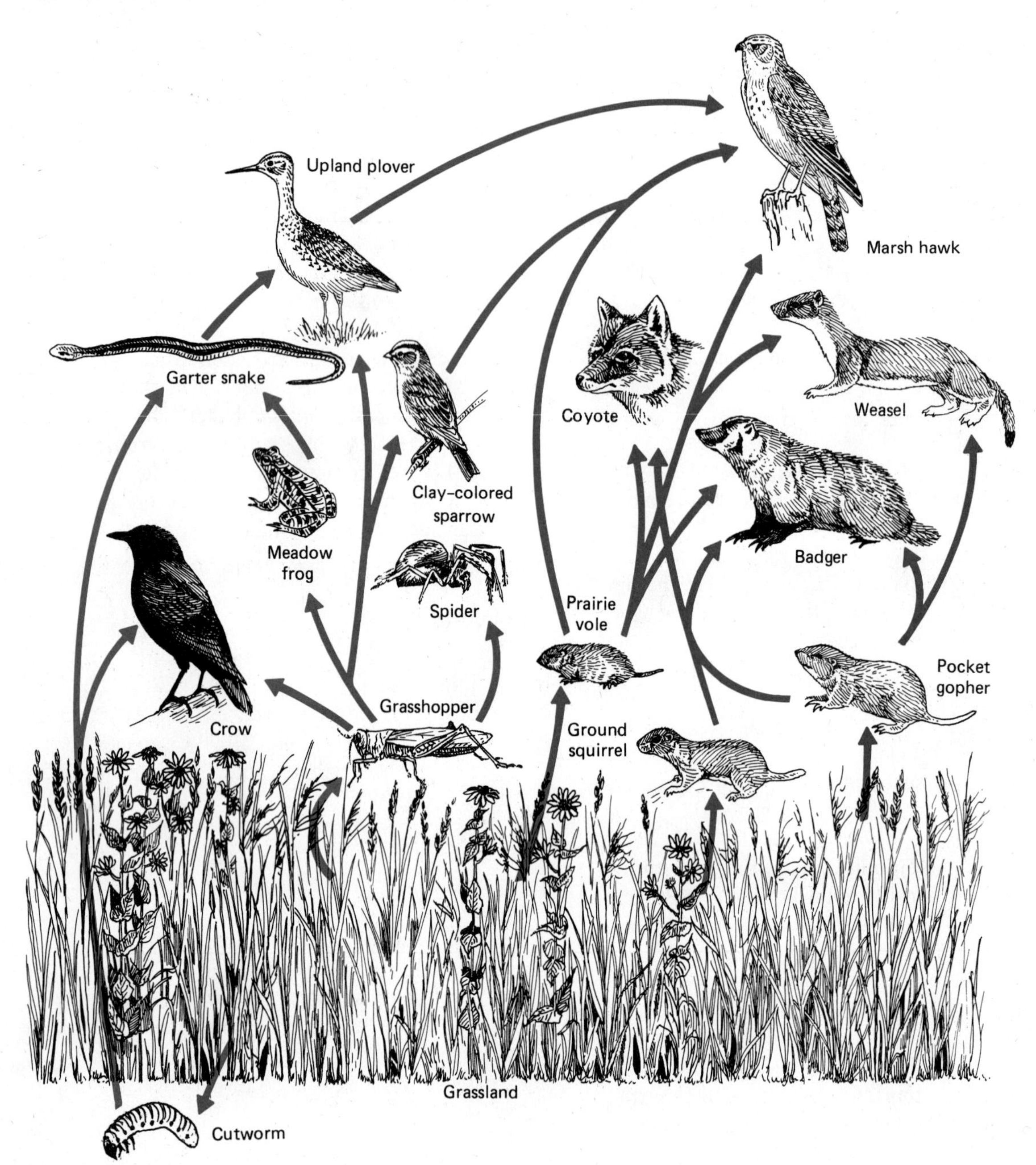

FIGURE 23.1 A food web within a prairie grassland community. Arrows flow from the eaten to the consumer.

Some fish feed on both plant and animal matter. The basically herbivorous white-footed mouse also feeds on insects, small birds, and bird eggs. The food habits of many animals vary with the seasons, with stages in the life cycle, and with the size and growth of organisms. Consumers that feed on both plant and animal matter are termed *omnivores* (Figure 23.2).

Scavengers are animals that eat dead plant and animal material. Among them are termites and various beetles that feed on dead and decaying wood, and crabs and other marine invertebrates that feed on plant particles in water. Bot flies, dermestid beetles, vultures, and gulls are only several of many animals that feed on animal remains. Scavengers may be considered either herbivores, carnivores, or saprophages.

Saprophytes are plant counterparts of scavengers. They draw their nourishment from dead plant and animal material, chiefly the former. Because they do not require sunlight as an energy source, they can live in deep shade or dark caves. Examples of saprophytes are fungi and Indian pipe. The majority are herbivores, but some do feed on animal matter.

Decomposers make up the so-called final feeding group. That is an oversimplified view of a complex functional group of organisms. All consumers to some degree function as decomposers. They either reduce enzymatically part of the material ingested or fragment it into smaller pieces, making it more accessible to other consumers, including bacteria and fungi.

Food chains involving the traditional decomposers, bacteria and fungi, usually reach up into the herbivore-carnivore food chains. In fact, decomposers—so frequently considered as some distant feeding group unclassifiable in the general scheme of food chains—actually function as herbivores and carnivores, depending upon the source of their food: dead plant or animal material. Perhaps only the bacteria and fungi that transform organic compounds into inorganic nutrients usable by photosynthetic plants should be considered outside of the general classification of herbivores and carnivores.

Biophages and Saprophages

Although descriptive, the terms *herbivore, carnivore,* and *omnivore* have probably outlived their usefulness as functional terms in energy transfer through food webs. Instead, heterotrophs should be considered either as biophages, those organisms utilizing living material, or saprophages, those utilizing nonliving matter. First-order biophages feed on living plants and are the traditional herbivores, whereas first-order saprophages feed on dead plant material as well as organic material egested by first-order biophages. First-order biophages, in turn, are utilized at death by second-order saprophages, which are really functional carnivores. In turn, first-order saprophages may be utilized by second-order biophages. Such an approach incorporates decomposers into general feeding groups at various levels in the food chain (Figure 23.3).

MAJOR FOOD CHAINS

Within any ecosystem there are two major food chains, the *grazing food chain* and the *detrital food chain* (see Figure 23.4). Because of the high standing crop and relatively low harvest of primary production, most terrestrial and shallow-water ecosystems are characterized by the detrital food chain. In deep-water aquatic systems—with their low biomass, rapid turnover of organisms, and high rate of harvest—the grazing food chain is the dominant one.

The amount of energy shunted down the two routes varies among communities. In an intertidal salt marsh, less than 10 percent of living plant material is consumed by herbivores, and 90 percent goes the way of the detritus-feeders and decomposers. In fact, most of the organisms of the intertidal salt marsh obtain the bulk of their energy from dead plant material. Fifty percent of the energy fixed annually in a Scots pine plantation is utilized by decomposers. The remainder is removed as yield or stored in tree trunks. In some communities, particularly undergrazed grasslands, unconsumed organic matter may accumulate and the materials remain out of circulation for some time, especially when conditions are not favorable for microbial action. The decomposer or detritus food chain receives additional materials from the waste products and dead bodies of both herbivores and carnivores.

Grazing Food Chain

The grazing food chain is the most obvious one (Figure 23.1). Cattle grazing on pastureland, deer browsing in

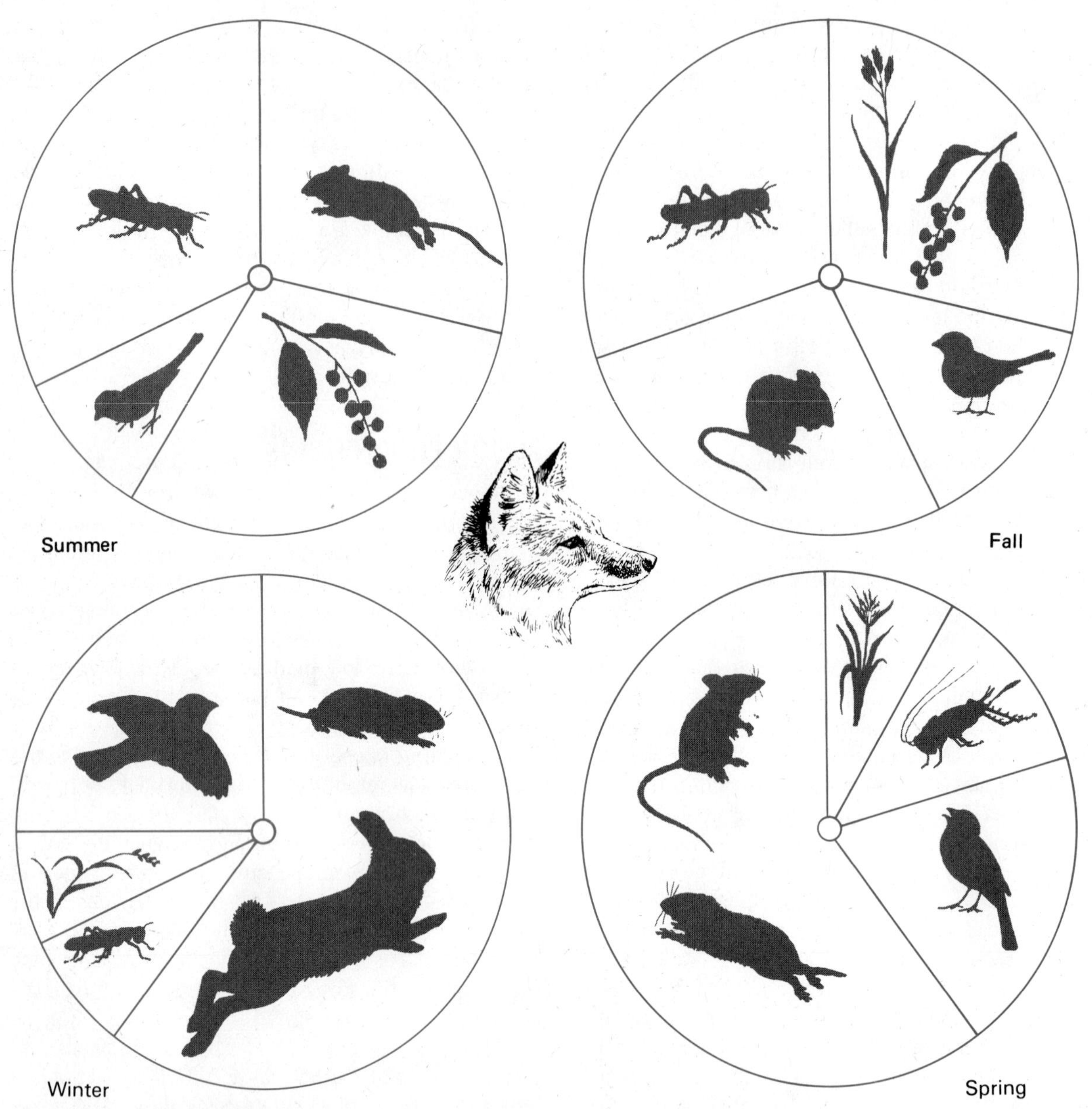

FIGURE 23.2 The red fox is an example of an omnivore, an animal that feeds on more than one trophic level. The food habits of the red fox are seasonal. The timing of flowering and the onset of breeding activities of animals influence the availability of food through the year. Note the prominence of fruits and insects in summer and of rodents in spring and fall.

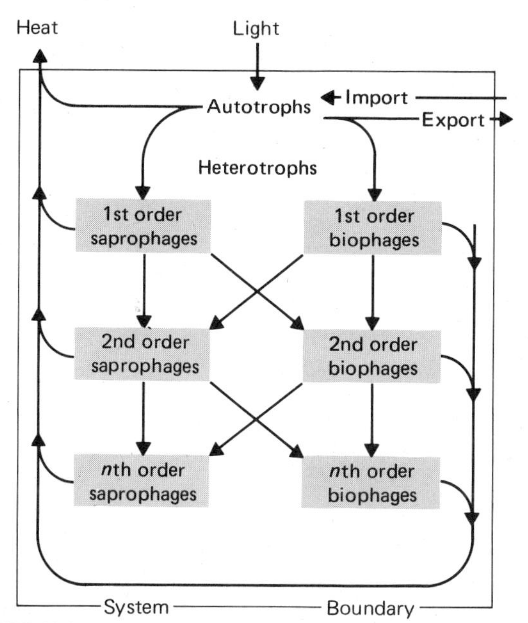

FIGURE 23.3 A two-channel model separates energy flow into two pathways, one utilized by organisms that feed on living matter, and the other by organisms that feed on nonliving matter.

the forest, rabbits feeding in old fields, and insect pests feeding on garden crops represent basic consumer groups of the grazing food chain. In spite of its conspicuousness, the grazing food chain is not the major one in terrestrial and many aquatic ecosystems. Only in some open-water aquatic ecosystems do the grazing herbivores play a dominant role in energy flow.

Although voluminous data exist on phytoplankton production, filtration rates by grazing zooplankton, and production efficiencies of zooplankton, relatively few data are available on the flow of energy, grazing rates, biomass turnover rates for phytoplankton, and turnover of zooplankton biomass within the same aquatic system. Some studies indicate that grazing protozoans feeding on certain planktonic algae consumed 99 percent of the population in 7 to 14 days. Turnover rates (production/biomass) in aquatic systems appear to be around 113 percent for phytoplankton, 7.1 percent for herbivorous zooplankton, and 1.2 percent for carnivorous zooplankton. The high turnover rate for phytoplankton reflects the high production rate and low biomass of unicellular algae.

In terrestrial systems a relatively small portion of primary production goes the way of the grazing food chain. Over a three-year period, only 2.6 percent of the net primary production of a yellow-poplar forest was utilized by grazing herbivores, although holes made in the growing leaves resulted in the loss of 7.2 percent of photosynthetic surface. Andrews and his associates studied energy flow through a short-grass plains ecosystem involving ungrazed, lightly grazed, and heavily grazed plots. Even on the heavily grazed grassland, cattle consumed only 30 to 50 percent of aboveground net primary production. About 40 to 50 percent of energy consumed by cattle is then returned to the ecosystem and the detrital food chain as feces.

Although the aboveground herbivores are the conspicuous feeders, belowground herbivores can have a pronounced impact on primary production and the grazing food chain. Andrews and his associates found that the belowground herbivores—consisting mainly of nematodes (Nematoda), scarab beetles (Scarabaeidae), and adult ground beetles (Carabidae)—accounted for 81.7 percent of total herbivore assimilation on the ungrazed short-grass plains, 49.5 percent on the lightly grazed grassland, and 29.1 percent on the heavily grazed one. Ninety percent of the invertebrate herbivore consumption took place below ground, and 50 percent of the total energy was processed by nematodes. On the lightly grazed grassland, cattle consumed 46 kcal/m^2 during the grazing season, and the belowground invertebrates consumed 43 kcal/m^2. When a nematicide was added to a midgrass prairie, aboveground net production increased 30 to 60 percent. Thus, belowground herbivorous consumption can impose a greater stress on a grassland ecosystem than aboveground herbivores.

Detrital Food Chain

The detrital food chain is common to all ecosystems, but in terrestrial and littoral ecosystems, it is the major pathway of energy flow, because so little of the net production is utilized by grazing herbivores. Of the total amount of energy fixed in a yellow-poplar (*Liriodendron*) forest, 50 percent of the gross production goes into maintenance and respiration, 13 percent is accumulated as new tissue, 2 percent is consumed by herbivores, and 35 percent goes to the detrital food chain.

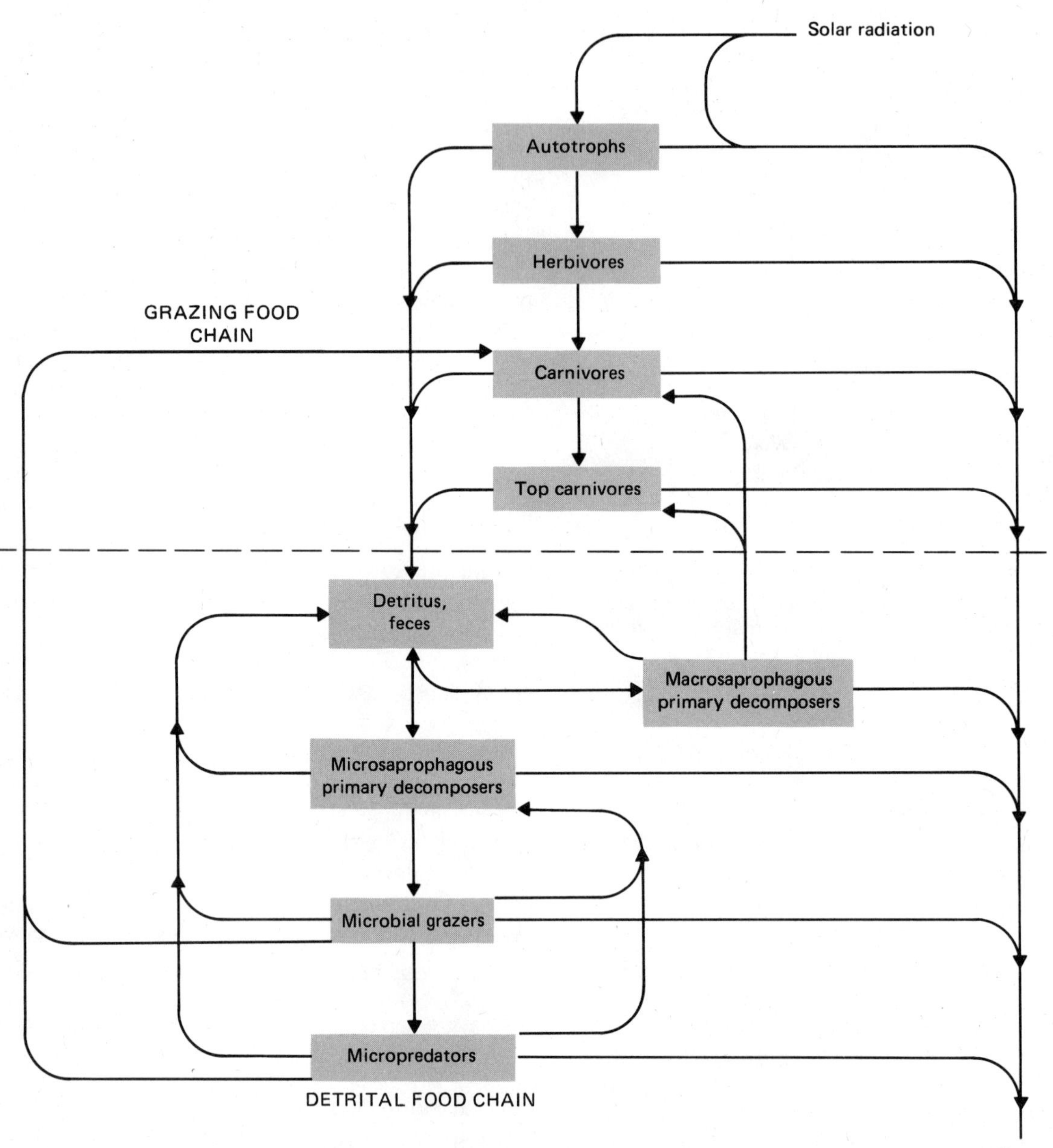

FIGURE 23.4 A model of detritals and grazing food chains. Two pathways lead from the autotrophs, one toward the grazing herbivores, the other toward the detritus feeders. Note the interrelationships between the two food chains.

Two-thirds to three-fourths of the energy stored in a grassland ecosystem undergrazed by domestic animals is returned to the soil as dead plant material, and less than one-fourth is consumed by herbivores. Of the quantity consumed by herbivores, about one-half is returned to the soil as feces. In the salt marsh ecosystem, the dominant grazing herbivore, the grasshopper, consumes just 2 percent of the net production available to it.

The forest litter, the habitat of a number of detrital-feeding invertebrates, is a good place to seek an example of a detrital food web. One such web (Figure 23.5) described involved five groups of litter feeders: millipedes (Diplopoda), orbatid mites (Cryptostigmata), springtails (Collembola), cave crickets (Orthoptera), and pulmonate snails (Pulmonata). Of these, the mites and springtails were the most important litter-feeders. These herbivores were preyed upon by small spiders (Araneidae) and predatory mites (Mesostigmata). The predatory mites fed on annelids, mollusks, insects, and other arthropods, and the spiders fed on predatory mites. Springtails, pulmonate snails, small spiders, and cave crickets were preyed upon by carabid beetles, while medium-sized spiders fed on cave crickets and other insects. The medium-sized spiders, in turn, became additional items in the diet of beetles. Beetles, spiders, and snails were consumed by birds and small mammals, members of a grazing food chain. In such a manner detrital food webs are linked, through predation, to grazing food chains at higher consumer levels.

Food chains involving *saprophages* may take two

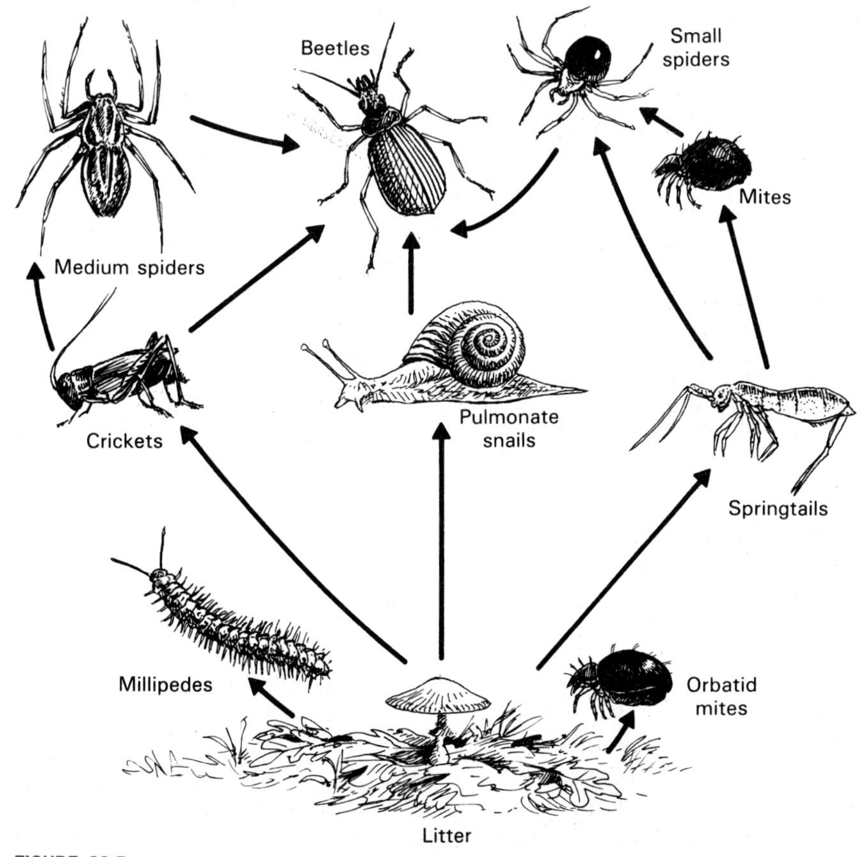

FIGURE 23.5 A detrital food chain involving litter-dwelling invertebrates.

directions: toward the carnivores or toward microorganisms. The role of these feeding groups in the final dissipation of energy has already been mentioned. They also serve as food for numerous other animals. Slugs eat the larvae of certain Diptera and Coleoptera, which live in the heads of fungi and feed on the soft material. Mammals, particularly red squirrels and chipmunks, eat woodland fungi. Dead plant remains are food sources for springtails and mites, which, in turn, are eaten by carnivorous insects and spiders. These, in turn, are energy sources for insectivorous birds and small mammals. Blowflies lay their eggs in dead animals, and within 24 hours the maggot larvae hatch. Unable to eat solid tissue, they reduce the flesh to a fetid mass by enzymatic action and feed on the proteinaceous material. These insects are food for other organisms.

DECOMPOSITION

Decomposition, the foundation of the detrital food chain, is the reduction of energy-rich organic matter by consumers (largely decomposers and detritivores) to CO_2, H_2O, and inorganic nutrients. Whereas photosynthesis involves the incorporation of solar energy and inorganic matter into biomass, decomposition involves the loss of heat energy and the conversion of organic nutrients into inorganic ones. Decomposition is not, as we are frequently led to believe, the final stop in the food web where plant and animal remains are processed by decomposer organisms. Actually, the greater part of the decomposition process is the conversion of organic matter into animal and microbial tissue (Figure 23.6), which moves into new food chains. Over varying periods of time, organic matter does end up inorganic, but by indirect routes.

Decomposition is a complex of many processes: the leaching of soluble compounds from dead organic matter, fragmentation, bacterial and fungal breakdown, consumption of bacterial and fungal organisms by animals, excretion of organic and inorganic compounds by organisms, and clustering of colloidal organic matter into larger particles. Organic matter acted upon by decomposers ranges in size from whole organisms to coarse particulate organic matter, fine particulate organic matter, and dissolved organic matter.

One pathway of decomposition begins when herbivores eat plants and carnivores eat prey. Not only does the animal extract minerals and nutrients from the food it eats, but it also deposits a substantial portion as partially decomposed material—feces. The amount and nature of fecal material deposited, especially by herbivores, depends partly on the digestibility of the plant and the ability of the animal's digestive system to handle plant material. Some herbivores select vegetation low in lignin and digest most of the easily soluble compounds. Others, notably the ruminants such as cattle and deer, rely upon bacteria and protozoans to break down the cellulose in the rumen and reticulum (the first two compartments of the four-compartment ruminant stomach) and produce volatile fatty acids, an important energy source for ruminants. Undigested and partially digested material passes through the gut and is available for microbial colonization.

Decomposition of leaves begins while they are on the plant. As the leaves approach senescence, the plants reabsorb much of their nutrients into roots and woody parts. During the growing season, plants produce varying quantities of exudates that support an abundance of surface microflora. These organisms feed on exudates and any cellular material that sloughs off. The same exudates account for nutrients leached from leaves during a rain.

While some microbes are using exudates of leaves, others are using organic material from the roots of living plants. The soil region immediately surrounding the roots supports a host of microbial feeders on root litter and on root exudates, which consist of simple sugars, fatty acids, and amino acids.

DECOMPOSER ORGANISMS

Dead organic matter, called *detritus*, is attacked by a variety of *saprophages*, organisms that live on dead matter. Saprophages can be divided into two groups: the microscopic and the macroscopic. Together there may be over a million of them in a square meter of the first 7 to 10 cm of a temperate hardwood forest soil. Approximately 40 percent of these organisms are bacteria; about 50 percent are microscopic fungi; 5 to 9 percent are protozoans; and 0.05 percent are true fungi

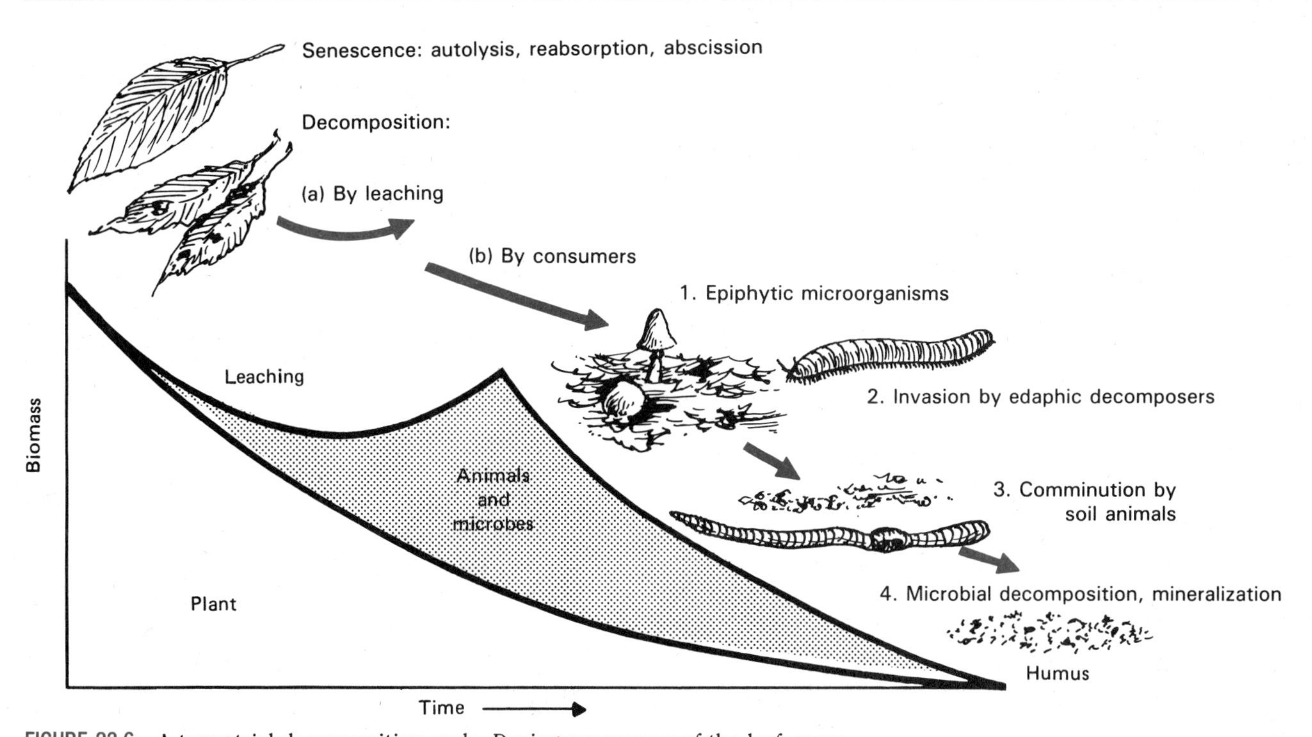

FIGURE 23.6 A terrestrial decomposition cycle. During senescence of the leaf, some nutrients are lost by leaching, autolysis, and reabsorption by the plant. Some nutrients in the leaf are lost to primary consumers such as caterpillars and aphids. The senescent leaf is also colonized by epiphytic saprophytes. Leaves fall to the ground to accumulate as leaf litter. Dead leaves are colonized by fungi and bacteria, which utilize soluble nutrients first. The leaves are then colonized by a succession of litter-feeders—reducer-decomposers—and by an array of microbes. Action of reducer-decomposers and microbes reduces litter to humic substances. Breakdown of organic matter continues, resulting in mineralization. Finally, the humus becomes mixed with mineral soil. Throughout the decomposition process, plant, animal, and microbial biomass has also been changing. Plant biomass, at its maximum at the time of initiation of decomposition, declines with time until it is incorporated into soil. Animal and microbial biomass increases as those organisms transform plant tissue into heterotroph tissue. Populations of these organisms reach their maximum at the leaf litter stage and then decline as less plant biomass is available to support them.

(mushrooms). Animals visible to the naked eye, the macrofauna, account for only 0.04 percent of the total population.

The dominant detrital-feeding organisms are heterotrophic bacteria and fungi. Bacteria may be *aerobic*, requiring oxygen as an electron acceptor; or they may be *anaerobic*, able to carry on their metabolic functions without oxygen by using some inorganic compound as the oxidant. Anaerobic bacteria commonly inhabit aquatic muds and sediments. As a group they are the major decomposers of animal matter. Major decomposers of plant material are the fungi, whose hyphae penetrate plant and animal matter. Both bacteria and fungi produce enzymes necessary to carry out specific chemical reactions. They secrete enzymes on plant and animal matter and absorb the resulting products as food. Once a group of bacteria and fungi has exploited the material as much as possible, another group moves in to continue the process. Thus a succession of microorganisms takes place in the detritus.

Macroorganisms include such small detritus-feeding animals as collembolas or springtails, mites, millipedes, earthworms, nematodes, and slugs in terrestrial ecosystems, and crabs, mollusks, mayfly, stonefly, and caddisfly larvae in aquatic ecosystems. Larger detritus feeders, such as earthworms and caddisfly larvae, break organic matter into smaller pieces by both mechanical action and digestion, mixing it with soil in the case of earthworms, excreting it, and even adding substances that stimulate microbial growth. These same organisms also consume bacteria and fungi associated with detritus, as well as small invertebrates and protozoans clinging to the material.

Other invertebrates, the microbivores, graze on the microbes. These invertebrates feed on detrital particles to remove bacteria and fungi and so prepare the surface for recolonization by other microbes. These grazers may reduce bacterial and fungal populations, inhibiting the effects of increased population density, accelerating the multiplication of soil microbes, and thus speeding up microbial activity.

A complementary relationship exists between macrodecomposers and microdecomposers. The macroorganisms fragment the detrital material, making it available to smaller detrital feeders and to bacteria and fungi. Ultimately, the material is reduced to a size at which even microbial activity cannot continue. At this point bacteria assimilate organic compounds, in effect concentrating them into larger particles that in turn are made even larger by bacterial aggregation. This material is again available to the macrodecomposers. They, in turn, may produce fecal pellets larger than the material digested, providing surfaces for microbial colonization. In such a manner, decomposing organic matter may be passed among feeding groups until it finally reaches an inorganic state.

THE FLOW OF NUTRIENTS

Organic detritus fallen on the ground, in streams, and at the edges of ponds is colonized by a growing population of mostly fungi and some bacteria. These microorganisms quickly extract the most soluble substances and soften the materials, making them available for detrital-feeding organisms. Among the first to invade the material are sugar-consuming fungi and bacteria. Once the glucose is utilized, the detritus is invaded by other bacteria and fungi feeding on cellulose.

As the microorganisms work on the plant debris, they assimilate nutrients and incorporate them in their tissues. As long as these nutrients are a part of the living microbial biomass, they are unavailable for recycling. That process is known as *nutrient immobilization*. The amount of mineral matter that can be tied up by microbes varies greatly. Some microbes exhibit luxury consumption, ingesting more than they need for maintenance and growth. Bacteria and fungi, however, are short-lived. They die or are consumed by litter invertebrates. Death and consumption, as well as the leaching of soluble nutrients from the decomposing substrate, release minerals contained in the microbial and detrital biomass. This process, known as *mineralization*, makes nutrients available for use by plants and microbes.

Thus, a cycle of immobilization and mineralization takes place within the soil. Nutrients are temporarily immobilized in microbial tissue. As microbes die, the nutrients are released or mineralized and become available for uptake again. Microbial uptake occurs simultaneously with mineralization. The amount of nutrients available for plants depends in part on the magnitude of uptake by microbial decomposers.

The process of decomposition is aided by the fragmentation of detritus by litter-feeding invertebrates. They consume parts of leaves, opening them up to microbial invaders. The action of such litter-feeders as millipedes and earthworms may increase exposed leaf area to 15 times its original size. Because the net assimilation of plant detritus by litter-feeders is less than 10 percent, a great deal of material passes through the gut of these organisms. They utilize only the easily digested proteins and carbohydrates. Mineral matter in fecal material is readily attacked by microbes. Some litter-feeders, such as earthworms, enrich the soil with vitamin B_{12}. In addition, they mix organic matter with soil, bringing the material into contact with other microbes.

Evidence suggests that in aquatic environments bacteria and, to a limited extent, fungi act more as converters than as regenerators of nutrients, whereas phytoplankton and zooplankton play a major part in the cycling of nutrients. Phytoplankton, macroalgae, and zooplankton furnish dissolved organic matter, with algae being the main contributor. Phytoplankton and other algae excrete quantities of organic matter at certain stages of their life cycle, particularly during rapid growth and reproduction. Twenty-five to 75 percent of the regeneration of nitrogen in the marine environment takes place by autolysis (enzymatic breakdown of plant and animal tissue upon death) of phytoplankton and zooplankton rather than by bacterial decomposition. In fact, 30 percent of the nitrogen contained in the bodies of zooplankton is lost by autolysis within 15 to 30 minutes after death, too rapidly for any significant bacterial action to occur.

Bacteria, phytoplankton, and zooplankton utilize inorganic nutrients as well as such organic nutrients as vitamin B_1 (necessary for the growth of both phytoplankton and zooplankton) and organic sources of nitrogen and phosphorus. They tend to concentrate these nutrients by incorporating them into their own biomass.

Important in the concentration of nutrients are the bacteria, which use dissolved organic matter as a substrate for growth. Both dissolved and colloidal matter condense on the surface of air bubbles in the water, forming organic particles on which bacteria flourish. Fragments of cellulose supply another substrate for bacteria. Bits of plant detritus, bacteria, and phytoplankton are consumed by both bacteria and planktonic animals.

Ciliates and zooplankton eat bacteria and excrete nutrients in the form of exudates and fecal pellets in the water. Zooplankton, too, in the presence of an abundance of food consumes more than it needs and can reduce microbial populations. Zooplankton excretes half or more of the ingested material as fecal pellets, which make up a significant fraction of suspended material. These pellets are attacked by bacteria that utilize the nutrients, growth substances, and energy they contain. Thus, the cycle starts over again (Figure 23.7).

Aquatic muds are largely anaerobic habitats. Fungi are absent, and the decomposer bacteria are largely facultative anaerobes. Incomplete decomposition often results in the accumulation of peat and organic muck. Nevertheless, particulate matter supports a rich bacterial population. For example, the snail *Hydrobia* feeds

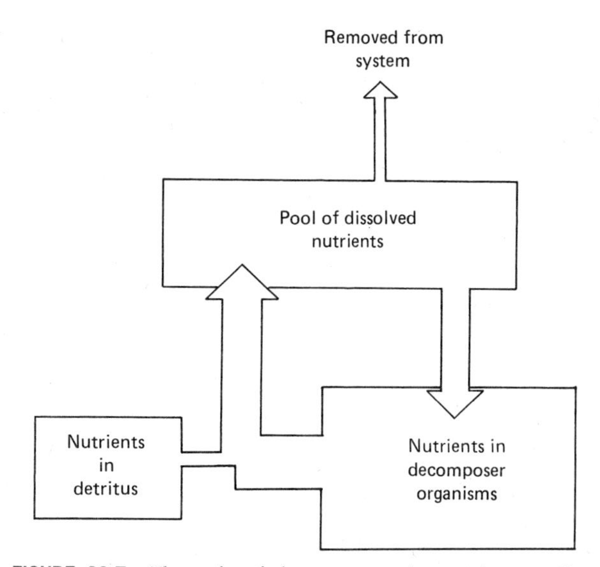

FIGURE 23.7 The role of decomposers in nutrient cycling and energy flow involves nutrient immobilization. Some of the nutrients they release move upward through the food chain.

on detritus found in mud flats. Its fecal pellets are devoid of nitrogen but rich in carbohydrates, suggesting that the snail cannot digest cellulose. If these pellets are held in filtered seawater, the nitrogen content quickly rises. The rise in nitrogen is accomplished by the growth of marine bacteria that colonize the fecal material and utilize the nitrogen dissolved in seawater. The fecal pellets, enriched with nitrogen in the form of bacterial protein, are reingested by the snail, a form of coprophagy. The snail digests the bacterial bodies, and the resultant fecal pellets are again devoid of nitrogen. Colonization of fecal pellets is repeated.

Thus, bacteria function primarily to concentrate nutrients rather than to release them to the environment by decomposition. That task is accomplished largely by algae, zooplankton, and detritus-feeding animals.

INFLUENCES

Decomposition is influenced by moisture, temperatures, exposure, type of microbial substrate, vegetation, and other variables. Temperature and moisture greatly influence microbial activity by affecting metabolic rates. Decomposing litter in wetlands undergoes conditions ranging from complete submergence to complete exposure to the atmosphere, from anaerobic to aerobic. Lower rates of decomposition take place under anaerobic conditions, whereas decomposition proceeds most rapidly under moist aerobic conditions. Alternate wetting and drying result in an increase in microbial growth and respiration during the rewetting period. In wetlands, decomposition proceeds most rapidly under alternate wetting and drying. Continuous dry spells and arid conditions reduce both activity and growth of microbial populations, inhibiting decomposition.

Slope exposure, especially as it relates to temperature and moisture, can speed or slow decomposition. Detritial material on cooler, moister north-facing slopes decomposes more rapidly than that on warmer, drier south-facing slopes.

Nutritional composition of leaves in litter has a powerful influence on decomposition. Easily decomposable and highly palatable leaves from such species as redbud, mulberry, and aspen support initially higher populations of decomposers than litter from oaks and pines, which is high in lignin. Earthworms have a pronounced preference for such species as aspen, ash, elm, and basswood; they take with less relish and do not entirely consume maple, eat sparingly of oak and beech, and do not eat pine or spruce needles. Thus decomposition of litter of certain species of plants proceeds more slowly than decomposition of litter from other species. On easily decomposable material, initially high populations of microbes decline as energy is depleted; but on more resistant pine and oak litter, initially low populations of microbes increase as decomposition proceeds.

How fast, then, does organic matter decompose? A number of methods have been devised to answer this question. One is to place litter of known weight in mesh bags on the top of litter or soil and record the loss of weight over time. Another is to label litter with radioactive carbon, ^{14}C, and follow the distribution of the carbon into the soil. A third is to measure the evolution and decline of CO_2 as decomposition proceeds.

In terrestrial and aquatic communities studies, initial decomposition is high because of the leaching of soluble compounds and consumption of highly palatable tissue by microbes. As decomposition proceeds, the rate slows because only less decomposable material remains. Leaves of deciduous trees placed in streams can lose 10 to 30 percent of their weight as dissolved organic matter in a few days. The remainder may take more than a year to decompose. In deciduous forests, easily decomposable material has a residence time of about 3 years. Slowly decomposable material has a residence time of about 59 years. Leaves and fine roots decompose most rapidly, logs most slowly. Depending upon size, logs may require 100 years or more to disappear. Bogs, being cold and wet, retain organic matter 100 times longer or more than tropical rain forests, in which decomposition is so rapid that organic material is lost in less than a year.

TROPHIC LEVELS

If all organisms that obtain their food in the same number of steps (that is, all those that feed wholly or in part on plants, wholly or in part on herbivores, and so on) are superimposed, the structure can be collapsed into a series of points representing the trophic or feeding levels of the ecosystem. Thus, each step in a food chain represents a trophic level. Animals feeding wholly on plants—for example, the grasshopper and vole—occupy a single trophic level. But most animals at higher levels, such as the red fox, participate simultaneously in several trophic levels because of variation in their diets. Their total food intake has to be apportioned among the several trophic levels involved. The first trophic level belongs to the producers, or plants; the second trophic level to the herbivores, or first-level consumers; the third level to the first-level carnivores, or second-level consumers; and so on.

Although trophic levels typically do not include the decomposers, logically they should. Decomposers should be considered herbivores or carnivores, depending on their food source. Decomposers feeding on dead plant material, as well as bacteria occupying the rumen of ungulate animals or the guts of termites, should be considered functional herbivores, or first-level consumers; decomposers feeding on the bodies of dead animals should be considered second-level consumers; and so on. In that manner, all the various steps in energy transfer in an ecosystem can be placed in some trophic level. That is the approach used in the construction of pyramids of biomass and energy, considered later.

ENERGY FLOW THROUGH A FOOD CHAIN

Energy flow through natural food chains is difficult to study. One energy flow pattern that has been worked out carefully involves old-field vegetation, the meadow mouse, and the least weasel (Figure 23.8). The mouse was almost exclusively herbivorous, and the weasel lived mainly on mice. The vegetation converted about 1 percent of the solar energy into net production, or plant tissue. The mice consumed about 2 percent of the plant food available to them and the weasels about 31 percent of the mice. Of the energy assimilated, the plants lost about 15 percent through respiration, the mice 68 percent, and the weasels 93 percent. The weasels used so much of their assimilated energy on maintenance that a carnivore preying on a weasel alone would not be able to survive.

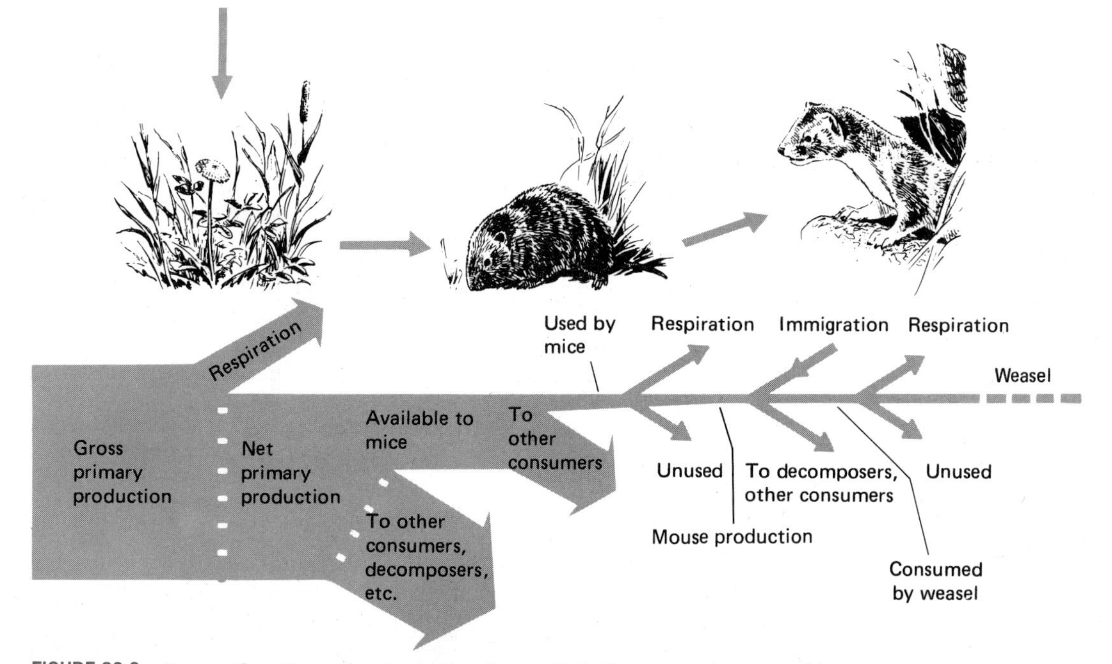

FIGURE 23.8 Energy flow through a food chain in an old field community in southern Michigan. The relative sizes of the arrows suggest the quantity of energy flowing through each channel.

An ecological rule of thumb allows a magnitude of 10 reduction in energy as it passes from one trophic level to another. Thus, if 1,000 kilocalories of plant energy were consumed by herbivores, about 100 kilocalories would be converted into herbivore tissue, 10 kilocalories to the first-level carnivore production, and 1 kilocalorie to the second-level carnivore. However, based on data available, a 90 percent loss of energy on the average from one trophic level to another may be high. Certainly, a wide range in the efficiency of conversion exists among various feeding groups (see Table 23.1). Production efficiency in plants (net production/solar radiation) is low, ranging from 0.34 percent in some phytoplankton to 0.8 to 0.9 percent in grassland vegetation. Plant production consumed by herbivores is utilized with varying efficiency. Herbivores consuming green plants are wasteful feeders, but not nearly as wasteful as those feeding on plant sap. More energy loss occurs in assimilation. Assimilation efficiencies vary widely among poikilotherms and homeotherms (Table 23.1). Homeotherms are much more efficient than poikilotherms. However, carnivorous animals, even poikilotherms, have high assimilation efficiency (see Box 23.1). Predaceous spiders feeding on invertebrates have assimilation efficiencies of over 90 percent. Because of high maintenance or respiratory costs, homoeotherms have low efficiency (production/assimilation) compared to poikilotherms (Table 23.1). Only about 2 to 10 percent of the energy consumed by herbivorous homeotherms goes into biomass production, less than the 10 percent average suggested by the rule of thumb. However, poikilotherms convert 17 percent of their consumption to herbivorous biomass. On midwestern grasslands average herbivore production efficiency, involving mostly poikilotherms, ranged from 5.3 to 16.5 percent (Table 23.2). Production efficiency on the secondary consumer, or carnivore, level ranged from 13 to 24 percent.

Transfer of energy from one trophic level to another tells the real story, but such data are hard to collect. The ratio of phytoplankton to secondary zooplankton production in open freshwater ecosystems is about 1:7.1, and the ratio of herbivore zooplankton production to carnivorous zooplankton production is 1:2.1. Efficiencies are lower in the benthic community: 2.2 for herbivores and 0.3 for carnivores.

Energy transfer efficiency (consumption at trophic level n/net production at level $n - 1$) among invertebrate consumers on a short-grass plain is about 9 percent for herbivores, 10 to 28 percent for saprophages, 38 percent for aboveground predators, and 56 percent for belowground predators. The trophic-level production efficiency (assimilation at trophic level n/net production at level $n - 1$) for soil and litter invertebrates in a deciduous forest ecosystem is approximately: saprophages, 0.11 to 0.17; phytophages, 0.02 to 0.07; and predators, 0.02.

ECOLOGICAL PYRAMIDS

If we sum all of the biomass or living tissue contained in each trophic level and all of the energy transferred between, we can construct pyramids of biomass and energy for the ecosystem (see Figure 23.9).

The pyramid of biomass indicates by weight or other means of measuring living material the total bulk of organisms or fixed energy present at any one time—the *standing crop*. Because some energy or material is lost in each successive link, the total mass supported at each level is limited by the rate at which energy is being stored below. In general, the biomass of the

TABLE 23.1 Assimilation and Production Efficiencies for Homeotherms and Poikilotherms

Efficiency	*All Homeotherms*	*Grazing Arthropods*	*Sap-feeding Herbivores*	*Lepidoptera*	*All Poikilotherms*
Assimilation					
A/C	77.5 ± 6.4	37.7 ± 3.5	48.9 ± 4.5	46.2 ± 4.0	41.9 ± 2.3
Production					
P/C	2.0 ± 0.46	16.6 ± 1.2	13.5 ± 1.8	22.8 ± 1.4	17.7 ± 1.0
P/A	2.46 ± 0.46	45.0 ± 1.9	29.2 ± 4.8	50.0 ± 3.9	44.6 ± 2.1

BOX 23.1

Ecological Efficiencies

$$\text{Assimilation efficiency (within a trophic level)} = \frac{\text{Assimilation}}{\text{Consumption}} \quad \frac{A}{C}$$

$$\text{Growth efficiency} = \frac{\text{Production}}{\text{Consumption}} \quad \frac{P}{C}$$

$$\text{Production efficiency} = \frac{\text{Production}}{\text{Assimilation}} \quad \frac{P}{A}$$

$$\text{Energy transfer efficiency} = \frac{\text{Consumption at trophic level } n}{\text{Production at trophic level } n-1} \quad \frac{C_n}{P_n - 1}$$

producers must be greater than that of the herbivores they support, and the biomass of the herbivores must be greater than that of the carnivores. That circumstance usually results in a gradually sloping pyramid for most communities, particularly the terrestrial and shallow-water ones, where the producers are large and characterized by an accumulation of organic matter, life cycles are long, and the harvesting rate is low.

This arrangement does not hold for all ecosystems. In such aquatic ecosystems as lakes and open seas, primary production is concentrated in the microscopic algae, characterized by a short life cycle, rapid multiplication of organisms, little accumulation of organic matter, and heavy grazing by herbivorous zooplankton. At any one point in time, the standing crop is low. As a result, the pyramid of biomass for these aquatic ecosystems is inverted; the base is much smaller than the structure it supports.

When production is considered in terms of energy, the pyramid indicates only the amount of energy flow at each level. The base on which the pyramid of energy is constructed is the quantity of organisms produced per unit of time or, stated differently, the rate at which food material passes through the food chain. Some organisms may have a small biomass, but the total energy they assimilate and pass on may be considerably greater than that of organisms with a much larger biomass. On a pyramid of biomass these organisms would appear much less important in the community than they really are. Energy pyramids are sloping because less energy is transferred from each level than was paid into it, in accordance with the second law of

TABLE 23.2 Consumer Efficiency (Secondary Production/Secondary Consumption)

		Producers		*Herbivores*		*Carnivores*	
Habitat	*Growing Season, Days*	*Production*	*% Efficiency*	*Production*	*% Efficiency*	*Production*	*% Efficiency*
Short-grass plains	206	3767	0.8	53	11.9	6	13.2
Midgrass prairie	200	3591	0.9	127	16.5	37	23.7
Tall-grass prairie	275	5022	0.9	162	5.3	15	13.9

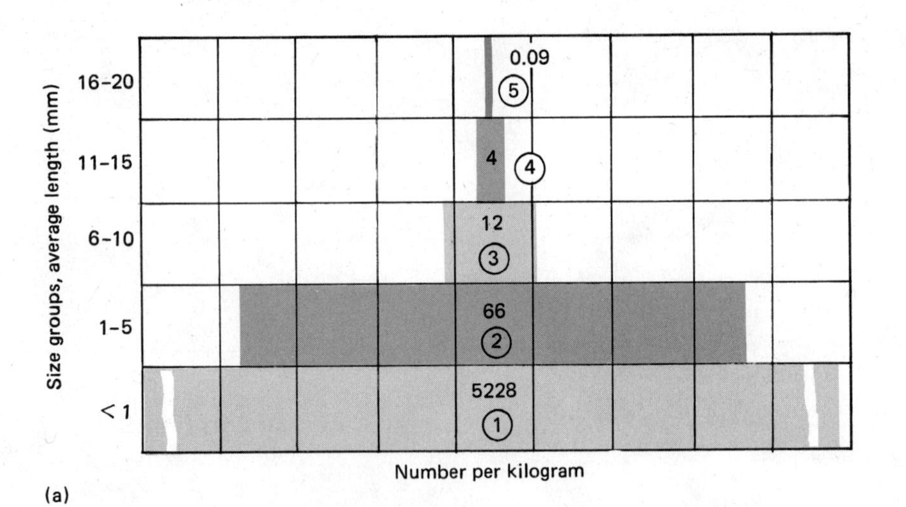

Size groups, average length (mm)
16-20
11-15
6-10
1-5
< 1
0.09
5
4
4
12
3
66
2
5228
1
Number per kilogram

(a)

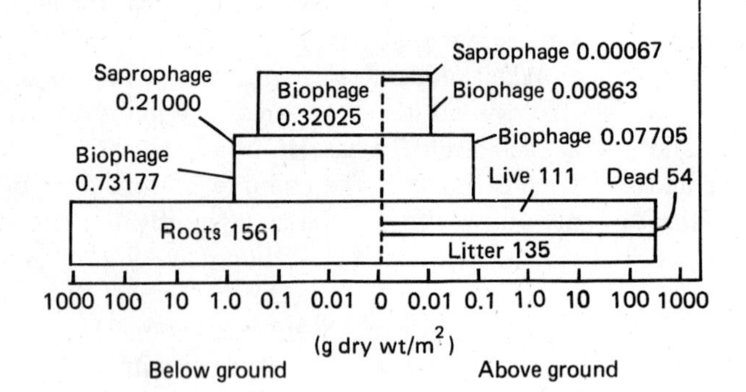

Saprophage 0.21000
Biophage 0.32025
Saprophage 0.00067
Biophage 0.00863
Biophage 0.07705
Biophage 0.73177
Live 111
Dead 54
Roots 1561
Litter 135
1000 100 10 1.0 0.1 0.01 0 0.01 0.1 1.0 10 100 1000
(g dry wt/m²)
Below ground
Above ground

(b)

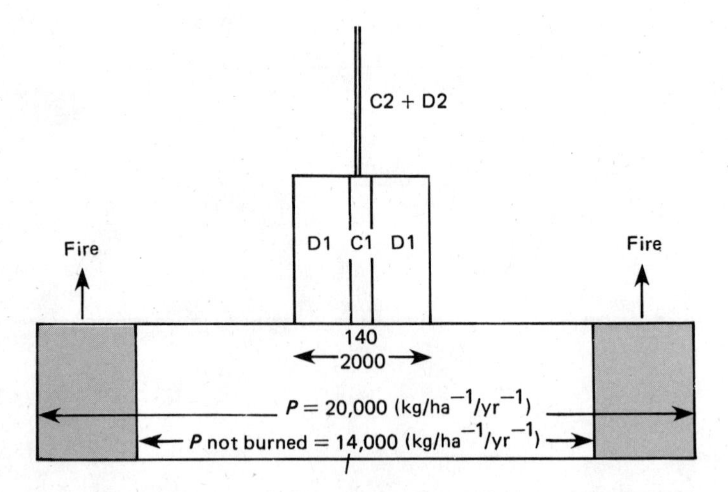

C2 + D2
D1 C1 D1
Fire
Fire
140
2000
P = 20,000 (kg/ha⁻¹/yr⁻¹)
P not burned = 14,000 (kg/ha⁻¹/yr⁻¹)

(c)

thermodynamics. In instances where the producers have less bulk than the consumers, particularly in open-water communities, the energy they store and pass on must be greater than that of the next level. Otherwise, the biomass that producers support could not be greater than that of the producers themselves. This high energy flow is maintained by a rapid turnover of individual plankton rather than by an increase in total mass.

Another pyramid commonly used in ecological literature is the pyramid of numbers. This pyramid was suggested by C. Elton, who pointed out the great difference in the number of organisms involved in each step of the food chain. Although the organisms on the lower trophic levels are the most abundant, the pyramid is occasionally inverted. A single tree, for example, would represent only one organism at the producer level; yet it supports thousands of consumer animals. Successive links of carnivores decrease rapidly in number until there are few carnivores at the top. The pyramid of numbers ignores the biomass of organisms. Although the numbers of a certain organism may be greater, the total weight, or biomass, of the organisms may not be equal to that of the larger organisms. Neither does the pyramid of numbers indicate the energy transferred nor the use of energy by the groups involved. Because the abundance of members varies so widely, it is difficult to show the whole community on the same numerical scale.

The pyramid of numbers is often confused with a similar one in which organisms are grouped into size categories and then arranged in order of abundance. Here the smaller organisms are again the most abundant; but such a pyramid does not indicate the relationship of one group to another.

FOOD WEBS

Trophic relationships in nature are not simple, straight-line food chains. Numerous food chains interlink into a complex food web, with all links leading from producers through an array of primary and secondary consumers. When food webs within a given habitat are unraveled by separating the numerous food chains, certain patterns emerge. All chains are interconnected. No matter how productive an ecosystem, each food chain rarely exceeds four links, because the length is limited by the inefficiency of energy transfer. Highly productive ecosystems may not support longer food chains, but they may support more species and thus more complex food webs.

Although omnivores at first glance may appear to be a significant component in food webs, they do not dominate. Omnivores typically feed on species in adjacent trophic levels, involving two different types of food. The ability to do that is limited by the morphological and physiological makeup of the omnivore. Birds with beaks adapted to tearing flesh are not well adapted to feed on fruits or seeds. Carnivorous mammals with teeth adapted for shearing meat and a short digestive system designed to digest animal protein cannot cope with a diet of grass. At the best they are limited to highly digestible fruits. For this reason om-

FIGURE 23.9 Examples of ecological pyramids. (a) A pyramid of numbers among the metazoans of the forest floor in a deciduous forest. (b) A pyramid of biomass for a northern short-grass prairie for July. The base of the pyramid represents biomass of producers; the second (middle) level, primary consumers; and the top, secondary consumers. Above-ground biomass (right) and below-ground biomass (left) are separated by a dashed vertical line. The trophic-level magnitudes are plotted on a horizontal logarithmic scale. The compartments are divided on a vertical linear scale according to live, standing dead, and litter biomass or biophagic and saprophagic consumer biomass. Note that unlike the conventional pyramids, this one recognizes the detrital as well as the grazing food chain components on the same trophic levels. (c) A pyramid of energy for the Lamto Savanna, Ivory Coast. *P* is primary production; C1, primary consumers; C2, secondary consumers; D1, decomposers of vegetable matter; D2, decomposers of animal material. Again, the detrital and grazing food chains have been collapsed into the same trophic levels.

nivory is not highly prevalent in food chains dominated by larger vertebrates and invertebrates. However, food webs dominated by insects and detritivores and their predators and parasitoids exhibit a more complex pattern of omnivory, which may involve feeding on nonadjacent trophic levels.

Predators may overlap in their exploitation of prey species. Foxes and kestrels, for example, feed on mice. Top predators feed on a number of species of primary and secondary consumers, or they may be more or less restricted to prey species on the adjacent trophic level below them. In general, the more species of prey an animal exploits, the fewer species of predators it faces. Consider the fox, which feeds on a wide range of prey species. Aside from humans, it has no major natural predator. By contrast, a screech owl, which feeds principally on mice, can succumb to predation by the great horned owl.

The study of food webs raises interesting questions. For example, what determines the size and complexity of food webs and the number of trophic levels? How do food webs acquire their structure? How are food webs affected by the successful invasion of new species and the elimination of others? Are complex food webs more stable than simple ones? What processes, including population dynamics and energy flow, go on in them?

Ecologists have both analyzed a number of actual food webs, terrestrial, marine, and freshwater, and simulated food webs to find some answers to these questions. Their analyses suggest that food webs in fluctuating environments—ones characterized by variations in temperature, salinity, pH, moisture, and other conditions—tend to have shorter food chains with fewer trophic links than those in more constant environments. Food chains in constant environments, such as pelagic regions of the ocean, are characterized by a greater species richness and more trophic links. Environmental variability alone, however, does not appear to constrain the average or maximum length of a food chain. Highly stratified environments such as a forest and a pelagic water column have longer food chains than poorly stratified habitats such as grassland, tundra, and stream bottoms. The widest food webs, those with the greatest number of herbivores, are the shortest. In contrast, narrow food webs have the greatest fraction of top carnivores.

How food chains might be assembled is another question. Are they a product of random assemblage or not? To find out, Peter Yodzis simulated the development of a food web based on natural parameters. Starting with a number of producer organisms, Yodzis added additional species, each of which had a certain ecological efficiency, making a fraction of its own consumption available to the next trophic level as a prey species. Each new species added had to obtain its energy from production available from other species. Each species had to choose a food source already used by another. Each new species introduced had a certain total production, a fraction of which had to be made available to a subsequently arriving species. Yodzis found an upper and lower limit to the number of species a predator could exploit. There arrived a point at which total production was too low to allow a new species to enter. Thus, the limit of energy transformation (the second law of thermodynamics) forced a pattern upon his food web. Yodzis found a degree of similarity between his simulated food webs and real-world food webs.

Generalist species most easily invade simple food webs. Specialists, capable of exploiting a restricted source of energy, are best able to invade complex webs. Removal of a generalist prey species, a generalist predator species, or any member of a simple, straight-line food chain has little effect on a food web. The removal of a key predator species can have a pronounced effect. Such a removal causes the greatest loss of species in the trophic level beneath when the predator has a controlling influence on the equilibrium density of the prey and the prey are generalists in their food habits. Removal has the least effect when the predator exerts a controlling influence on the equilibrium density of specialist prey species. In summary, field and computer simulation studies seem to support the hypotheses that complex food webs are less stable than simple ones, and that food webs are not random assemblages but rather are regulated by properties of existing food webs and the nature of invading species.

Two examples illustrate what can happen when a new top predator invades a food web and when one is removed. The first took place in 1967 when the Panamanian government and some businessmen introduced peacock bass *Cichla ocellaris*, a native of the Amazon River system, to Gatun Lake, Panama Canal. Their objective was to provide an outstanding sport fish and a highly edible market fish. A voracious fish, *Cichla*, quickly decimated the major planktivorous fish *Melaniris* and several other planktivores on which the native top predators—the tarpon, heron, and black tern—fed. These species consequently disappeared from most of the lake. *Cichla* became the top predator, and the planktivorous juvenile *Cichla* replaced the native planktivores. The juvenile *Cichla* and an algal feeding fish became the major prey species of the peacock bass. Thus the invasion of *Cichla* highly simplified the food web.

The second example involves the removal of the predaceous largemouth bass. When present, the predatory bass reduces the biomass of vertebrate planktivores, allowing invertebrate planktivores and large zooplankton herbivores to increase, and causing small herbivorous zooplankton and primary producers to decrease. When the bass was removed, vertebrate planktivores increased, invertebrate planktivores and large herbivorous zooplankton decreased, and small herbivorous zooplankton and primary producers increased.

Studies of actual food webs and computer simulations each have their shortcomings. Most studies set out to characterize certain environments, not to explore trophic relationships in detail. They are rather general at the lower trophic levels, especially at the producer level, and most detailed at the upper trophic levels. Often the food relations between eater and eaten are deduced from general observations rather than from detailed food habit studies of all species involved. Published studies mostly cover grazing food chains, ignoring the trophic levels and species in the detrital food web, just as important. Further, food webs vary seasonally in any ecosystem. Food webs are also subject to the foraging behavior of herbivores and carnivores involved and the heterogeneity and patchiness of the environment. Such spatial and temporal variations weaken any quantification of linkages and of other structural aspects of food chains.

Computer-simulated food webs attempt to mimic nature. Inputs are based on sets of assumptions and not on natural relationships. Although useful as tools to aid in understanding natural food webs, simulations cannot factor in all the subtleties involved, such as foraging patterns and prey numbers, behavior, and availability.

Summary

A basic function of the ecosystem is the flow of energy. The energy of sunlight fixed by the autotrophic component of the ecosystem is available to the heterotrophic component, of which the herbivores are the primary consumers. Herbivores, in turn, are a source of food for carnivores. At each step or transfer of energy in the food chain, a considerable amount of potential energy is lost as heat, until ultimately, the amount of available energy is so small that few organisms can be supported by that source alone. Animals further up on a food chain often utilize several sources of energy, including plants, and thus become omnivores.

Energy flow in the ecosystem takes two routes: one through the grazing food chain, the other through the detritus food chain. In the latter, the bulk of production is utilized as dead organic matter by saprophages.

Involved in the final dissipation of energy and return of nutrients to the ecosystem for recycling are a diverse group of decomposer organisms. The true organisms of decay, those responsible for the conversion of organic compounds to inorganic ions, are the heterotrophic bacteria and fungi. Biophagic consumers return part of the material ingested in a partially decomposed form to other decomposer organisms. Reducer-decomposers fragment detrital material into smaller particles, more accessible to bacteria and fungi. Another group, the microbivores, feeds on detrital particles, mainly for the bacteria and fungi growing on them. These organisms reduce microbial populations and thus influence microbial activity. Bacteria and fungi, as well as other decomposers, immobilize nutrients—remove them from circulation—by incorporating them in their body tissues. In terrestrial ecosystems, bacteria and fungi play the major role in decomposition. In aquatic ecosystems, bacteria and fungi act more as converters, while phytoplankton and zooplankton play a major role in the cycling of nutrients.

The loss of energy at each transfer limits the number of trophic levels, or steps, in the food chain to four or five. At each level, biomass usually declines; so if the total weight of individuals at each successive level is plotted, a sloping pyramid is formed. In certain aquatic situations, however, where there is a rapid turnover of small aquatic producers or phytoplankton, the pyramid of biomass may be inverted. In either case, energy decreases from one tropic level to another and is pyramidal.

The ratios of energy flow in or between trophic levels, in or between populations of organisms, or in or between individual organisms are called ecological efficiencies. Some of the most ratios useful are assimilation efficiency, consumption efficiency, production efficiency, and energy transfer efficiency.

Energy flow involves complex relationships known as food webs, about which food web theory has grown. This theory considers the size, organization, and structure of food webs, as influenced by environment, numbers of, invasions by, and loss of species, and the relationship of one trophic level to another.

Review and Study Questions

1. What is a food chain? A food web?
2. Distinguish between a biophage and a saprophage.
3. What are the two major food chains, and how are they related?
4. Why is decomposition more than just the traditional end point in the food chain?
5. Relate the following groups to terrestrial decomposition: saprophages, anaerobic bacteria, aerobic bacteria, fungi, microbial decomposers, macroorganisms, and microbial grazers.
6. What is nutrient immobilization? Mineralization?
7. What is the role of bacteria, phytoplankton, and zooplankton in the decomposition process in the aquatic environment?
8. How do decomposition rates vary among different plant parts?
9. How do temperature and moisture influence decomposition and organic accumulation in various ecosystems?
10. What is a trophic level? Relate the levels to ecological pyramids.
11. What is ecological efficiency?
12. What influences the length of food chains and pattern in food webs?
13. Are natural food webs free of the influence of humans? How might food webs be affected by the use of pesticides? By the overexploitation of fish, shrimp? By land disturbance? By the introduction of exotic species?

Selected References

Anderson, J. M., and A. MacFadyen, eds. 1976. *The role of terrestrial and aquatic organisms in the decomposition process.* Seventeenth Symposium, British Ecological Society. Oxford, England: Blackwell Scientific Publications.

Andrews, R. D., D. C. Coleman, J. E. Ellis, and J. S. Singh. 1975. Energy flow relationships in a short-grass prairie ecosystem. *Proc. 1st Inter. Cong. Ecol.*, 22–28. The Hague: W. Junk Publishers.

Briand, F. 1983. Environmental control of food web structure. *Ecology* 64:253–263.

Cohen, J. E. 1989. Food webs and community structure. In J. Roughgarden, R. May, and S. Levin, eds., *Perspectives in ecological theory*. Princeton, NJ: Princeton University Press, pp. 181–202.

Elton, C. 1927. *Animal ecology*. London: Sidgwick & Jackson. A classic, with original description of food chains and ecological pyramids.

Fletcher, M., G. R. Gray, and J. G. Jones. 1987. *Ecology of microbial communities*. New York: Cambridge University Press. A good overview of ecology at the decomposer level.

Mattson, W. J., ed. 1977. *The role of arthropods in forest ecosystems*. New York: Springer-Verlag. A look at the functioning of litter arthropods.

Pimm, S. L. 1982. *Food webs*. London: Chapman and Hall. A detailed discussion of aspects of food web theory.

Swift, M. J., O. W. Heal, and J. M. Anderson. 1979. *Decomposition in terrestrial ecosystems*. Oxford: Blackwell Scientific Publications. An excellent review of the decomposition process.

Yodzis, P. 1988. *An introduction to theoretical ecology*. New York: Harper & Row. Chapter 8 presents a theoretically interesting discussion of food webs.

Outline

CHAPTER 24

Cycles in Ecosystems

Objectives

On completion of this chapter, you should be able to:

1. Describe the water cycle and global water balance.
2. Define two types of biogeochemical cycles.
3. Describe the oxygen, carbon, and nitrogen cycles.
4. Describe the phosphorus cycle, terrestrial and aquatic, and the sulfur cycle.
5. Discuss nutrient flow and nutrient conservation in the ecosystem.

The existence of the living world depends upon the flow of energy and the circulation of materials through the ecosystem. Both influence the abundance of organisms, the metabolic rate at which they live, and the complexity of the ecosystem. Energy and materials flow through the ecosystem together as organic matter; one cannot be separated from the other. The continuous round trip of materials, paid for by the one-way trip of energy, keeps ecosystems functioning.

ESSENTIAL NUTRIENTS

Most of the nutrients required by life exist in mineral form in Earth's crust. They become available by weathering and chemical processes and enter biogeochemical cycles when plants take them up. Living organisms require at least 30 to 40 elements for growth, development, and reproduction (Table 24.1). The bulk of living matter consists of hydrogen, carbon, nitrogen, and sulfur. These *bulk elements* are concentrated in living tissue in grams per kilogram and are needed in gram amounts daily.

Another group of elements is needed in concentration lower than bulk elements. They include phosphorus, potassium, calcium, magnesium, and sodium. They are concentrated in grams per kilogram, but are required only in fractions of a gram per day. They are known as *macroelements* or *macronutrients*.

Plants and animals require numerous other elements in much lower concentrations, measured in milligrams or micrograms per kilogram of tissue. Because they were not easily quantified by analytical methods in the past, they came to be known as *trace elements*. Today we know them as *micronutrients*. Among them are iron, copper, zinc, iodine, boron, silicon, and nickel. All of the micronutrients, especially the heavy metals, can be toxic in quantities greater than needed.

BIOGEOCHEMICAL CYCLES

All the essential nutrients and many others besides, including a number of human-made materials such as chlorinated hydrocarbons, flow from the nonliving to the living and back to the nonliving parts of the ecosystem in a more or less circular path known as the *biogeochemical cycle* (*bio* for living; *geo* for water, rocks, and soil; and *chemical* for the processes involved). Some of the material is returned to the immediate environment almost as rapidly as it is removed; some is stored in short-term nutrient pools such as the tissues of plants and animals or the soil and sediment in lakes and ponds; and some is tied up chemically or buried deep in Earth in long-term nutrient storage pools before being released and made available to living organisms. Between the easily accessible and the relatively unavailable, there exists a slow but steady interchange.

The important roles in all nutrient cycles are played by green plants, which organize the nutrients into biologically useful compounds; by the organisms of decomposition, which return them to their simple elemental state; and by air and water, which transport nutrients between the abiotic and living components of the ecosystem. Without these factors no cyclic flow of nutrients would exist.

There are two basic types of biogeochemical cycles: *gaseous* and *sedimentary*. In gaseous cycles the main reservoir of nutrients is the atmosphere and ocean. In sedimentary cycles the main reservoir is the soil and the sedimentary rocks and other rocks of Earth's crust. Both involve biological and nonbiological agents, both are driven by the flow of energy, and both are tied to the water cycle.

THE WATER CYCLE

Water is the medium by which elements and other materials make their never-ending odyssey through the ecosystem. Without the cycling of water, biogeochemical cycles could not exist, ecosystems could not function, and life could not be maintained.

Solar energy is the driving force behind the water cycle (Figure 24.1). The heating of Earth's atmosphere (see Chapter 4) and its role in evaporation provide the basic mechanism of the cycle. Precipitation sets the cycle in motion. Water vapor in the atmosphere coalesces into droplets and ice crystals, which fall. Some of the water falls directly on the ground and bodies of

TABLE 24.1 Some Important Biological Elements

Element	*Use*
Carbon	Key atom of all forms of life. Along with oxygen and hydrogen makes up bulk of living matter.
Nitrogen	Component of all proteins, amino acids, and nucleic acids. Molecular nitrogen (N_2) makes up 78 percent of atmosphere. Plants can use it only in fixed form.
Oxygen	Makes up 21 percent of Earth's atmosphere. Necessary for oxidative processes.
Sulfur	Basic constitutent of protein. Supplied by precipitation and organic matter in soil.
Calcium	In animals, necessary for proper acid–base relationships and many physiological processes. Gives rigidity to skeleton of vertebrates, exoskeleton of arthropods, and shells of mussels. In plants, forms a cementing material between cells; allows growing root tips to develop normally.
Phosphorus	Component of proteins and many enzymes; major role in energy transfer at cellular level.
Magnesium	Integral part of chlorophyll; active in enzymes of plants and animals. Low intake causes grass tetany in ruminant animals.
Potassium	Formation of starches and sugars in plants. Synthesis of proteins, normal cell division, and carbohydrate metabolism in animals.
Sodium and chlorine	Indispensable to vertebrate animals; important in maintenance of acid–base balance, osmotic pressure of extracellular fluids, and formation and flow of gastric and intestinal secretions.
Iron	Active in nitrogen fixation and photosynthesis; transports electrons; part of complex proteins that activate and carry oxygen in blood.
Zinc	Formation of auxins in plant growth substances; component of enzyme systems in plants and animals.
Copper	Influences photosynthetic rate; involved in oxidation-reduction reactions; enzyme activator.
Molybdenum	Catalyst in conversion of gaseous N into usable forms by blue-green algae and N-fixing bacteria.
Iodine	Necessary for thyroid metabolism.
Boron	Essential for 15 functions in plants, including carbohydrate metabolism, water metabolism, and translocation of sugars.
Silicon	Needed by diatoms and grasses for support structures; conveys resistance to pathogenic bacteria and fungi.
Selenium	Needed by ruminants for Vitamin E activity.

water; some falls on vegetation, on litter on the ground, and on urban structures and streets. This water may be stored, may be hurried off, or, in time, may infiltrate the soil (see Chapter 9).

Because of interception, which can be considerable, various amounts of water never reach the ground but evaporate back to the atmosphere. In urban areas a great portion of rain falls on roofs, sidewalks, roads, and other paved areas, which are impervious to water. The water runs down gutters and drainage ditches to be hurried off to rivers.

The precipitation that reaches the soil moves into the ground by infiltration, the rate of which is influenced by soil, slope, type of vegetation, and the char-

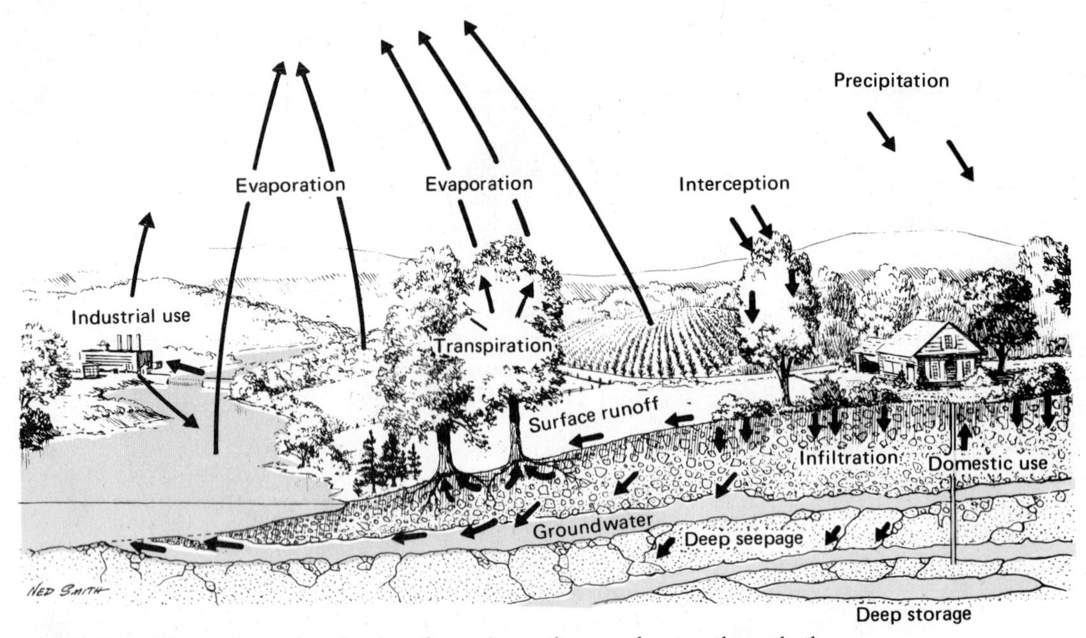

FIGURE 24.1 The water cycle, showing the major pathways of water through the ecosystem.

acteristics of the precipitation. In general, the more intense the rain, the greater the rate of infiltration, until the infiltration capacity of the soil, determined by soil porosity, is reached.

When the soil becomes saturated and when intense rainfall or rapid melting of snow exceed the infiltration capacity of the soil, water flows across the surface of the ground as overland flow or surface runoff. It concentrates in depressions and rills, where sheet flow changes to channelized flow. On city streets water moves in sheets across the pavement and becomes concentrated in streetside gutters.

In undisturbed forest soil, infiltration rates usually are greater than the intensity of rainfall and surface runoff does not occur. In urban areas infiltration rates may range from zero to a value exceeding the intensity of rainfall where soil is open and uncompacted. Because of low infiltration, runoff from urban areas might be as much as 85 percent of the precipitation.

Water entering the soil will percolate or seep down to an impervious layer of clay or rock to collect as groundwater. From here water finds its way into springs, streams, and eventually rivers and seas. A great portion of this water is utilized by humans for domestic and industrial purposes, after which it reenters the water cycle through discharge into streams and rivers or the atmosphere.

Water remaining on the surface of the ground and vegetation, as well as water in surface layers of streams, lakes, and oceans, *evaporates*, a process by which more molecules leave a surface than enter it. Plants take up water from the soil and lose it by transpiration through the leaves, where it evaporates. This moisture eventually condenses and falls once again.

THE GLOBAL PERSPECTIVE

The molecules of water that fall in a spring shower might well have been part of the Gulf Stream a few weeks before and in the Amazon tropical rain forest before that. The local storm is simply part of the mass movement and circulation of water about Earth, a movement suggested by the changing cloud patterns over the face of the planet. The atmosphere, oceans, and land masses form a single gigantic water system, which is driven by solar energy. The presence and movement of water in any one part of the system affects the presence and movement in all other parts.

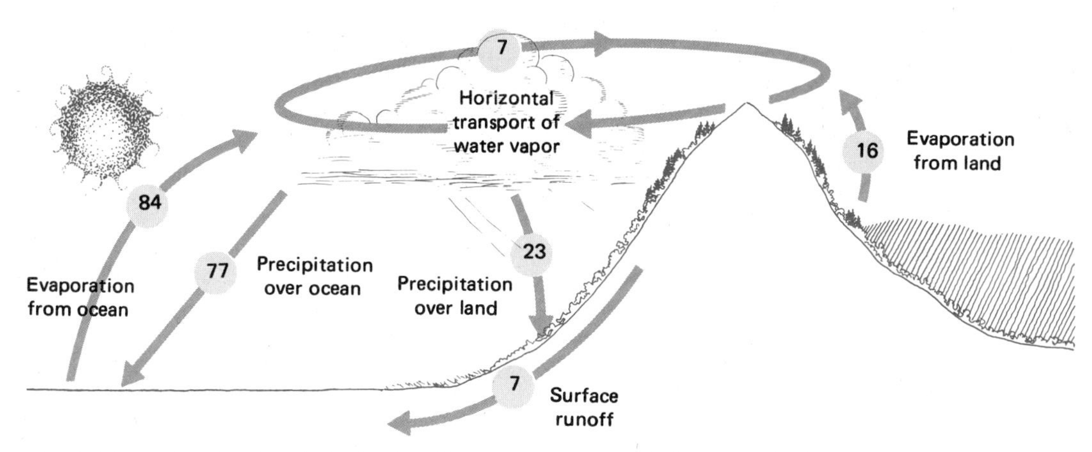

FIGURE 24.2 The global budget of water. The mean annual precipitation of 83.6 cm has been converted to 100 units.

The atmosphere is one key element in the world's water system. At any one time the atmosphere holds no more than a 10- to 11-day supply of rainfall in the form of vapor, clouds, and ice crystals. Thus, the turnover of water molecules is rapid. Because the source of water in the atmosphere is evaporation from land and sea, there are global differences in the amount of evaporation and the amount of moisture in the atmosphere at any given point. Evaporation at lower latitudes is considerably greater than evaporation at higher latitudes, reflecting the greater heat produced by the direct rays of the sun. Evaporation is greater over oceans than over land. Oceans account for 84 percent of evaporation—considerably more than they receive in return from precipitation (see Figure 24.2). Land areas contribute 16 percent of the annual evaporation; yet they intercept a greater amount.

Examined from a global point of view, the water cycle emphasizes the close interaction between the physical and geographic environments of the Earth. Thus, the water problem, often considered in local terms, is actually a global problem, and local water management schemes can affect the planet as a whole. Problems result not because an inadequate amount of water reaches Earth but because it is unevenly distributed to human population centers. Because humans have strongly interjected themselves into the water cycle, the natural usable water resources have decreased, and water quality has declined. The natural water cycle has not been able to compensate for the detrimental effects of humans on water resources.

THE DISTRIBUTION OF WATER

Although we view water as a local phenomenon, such as a stream or an autumn rain, it forms a single worldwide resource distributed in land, sea, and the atmosphere and unified by the hydrological cycle. It is influenced by solar energy, by the currents of the air and oceans, by heat budgets, and by the water balances of land and sea.

Oceans cover 71 percent of the Earth's surface (see Table 24.2). With a mean depth of 3.8 km (2.36 miles), they hold 93 to 97 percent of all the Earth's waters. Thus, fresh water usable by humans represents only 3 percent of the planet's water supply. Of the total fresh water on Earth, 75 percent is locked up in glaciers and ice sheets—enough to maintain all the rivers of the world at their present rate of flow for the next 900 years. If oceans contain 97 percent of the world's water, then nearly 2 percent of the remainder is tied up in ice. This leaves less than 1 percent of the world's water available and fresh. Freshwater lakes contain 0.3 percent of the freshwater supply, and at any one time rivers and streams contain only 0.005 percent of that supply. Soil moisture accounts for approximately 0.3 percent, and another very small portion of the Earth's water is tied up in living material.

TABLE 24.2 World Water Resources

Resource	*Volume (W), Thousands of km³*	*Annual Rate of Removal (Q) and Process*		*Renewal Period (T) (T = W/Q)*
Total water on Earth	1,460,000	520	Evaporation	2,800 years
Total water in the oceans	1,370,000	449	Evaporation	3,100 years
		37	Difference between precipitation and evaporation	37,000 years
Free gravitational waters in the Earth's crust (to a depth of 5 km)	60,000	13	Underground runoff	4,600 years
Amount of which is present in the zone of active water exchange	4,000	13	Underground runoff	300 years
Lakes	750		—	
Glaciers and permanent snow	29,000	1.8	Runoff	16,000 years
Soil and subsoil moisture	65	85	Evaporation and underground runoff	280 days
Atmospheric moisture	14	520	Precipitation	9 days
River waters	1.2	36.3[a]	Runoff	12(20) days[b]

[a] Not counting the melting of Antarctic and Arctic glaciers.
[b] 12 days for small river systems. 20 days for major rivers draining into the sea.
Note: Average error is probably 10–15 percent.

More stable is the groundwater supply, which accounts for 25 percent of our fresh water. Groundwater fills the pores and hollows within Earth just as water fills pockets and depressions on the surface. Estimates—necessarily rough and inaccurate—place renewable and cyclic groundwater at 7×10^6 km^3, or approximately 11 percent of the freshwater supply. Some of the groundwater is "inherited," as in aquifers in desert regions, where the water is thousands of years old. Because inherited water is not rechargeable, heavy use of these aquifers for irrigation and other purposes is mining the supply. In the foreseeable future, the supply could be exhausted. A portion of the groundwater, approximately 14 percent, lies below 1000 m. Known as fossil water, it is often saline and does not participate in the hydrological cycle.

The atmosphere, for all its clouds and obvious close association with the water cycle, contains only 0.035 percent fresh water, although it is the atmosphere and its relation to land and oceans that keep the water circulating over Earth.

Moisture in the atmosphere moves with the general circulation of the air. Air currents, hundreds of kilometers wide, are in fact giant, unseen rivers moving in great swirls above the Earth, only a part of whose moisture falls as precipitation in any one place (see Chapter 4).

The excess of precipitation over evaporation is eventually carried to the sea by rivers. Rivers are the primary movers of water over the globe, carrying many more times the amount of water than their channels hold. By returning water to the sea, they tend to balance the evaporation deficit of the oceans. Sixteen major rivers discharge 13,600 km^3 annually, or 45 percent of all the water carried by rivers. Adding the next 50 largest rivers brings the total to 17,600 km^3, or 60 percent of all the water discharged to the sea.

Evaporation, precipitation, detention, and transportation maintain a stable water balance on Earth. Consider the amount of water that falls on Earth in terms of 100 units (Figure 24.2). On the average, 84 units are lost from the ocean by evaporation, while 77 units are gained from precipitation. Land areas lose 16 units by evaporation and gain 23 units from precipitation. Runoff from land to the ocean makes up 7 units, which balances the evaporative deficit of the ocean. The remaining 7 units are circulating as atmospheric moisture.

In its global circulation, water also influences the heat budgets of the Earth. The highest heat budgets are in the low latitudes, the lowest in the polar regions, and a balance between incoming and outgoing cold and heat is achieved at 38° to 39° latitude (see Chapter 4). Excessive cooling of higher latitudes is prevented by the north and south transfer of heat by the atmosphere in the form of sensible and latent heat in water vapor and by warm ocean currents.

GASEOUS CYCLES

Because gaseous cycles are closely linked to the atmosphere and oceans, they are pronouncedly global. Most important are nitrogen, oxygen, and carbon dioxide. These three gases in stable quantities of 78, 21, and 0.03 percent, respectively, make up Earth's atmosphere, along with other trace gases.

Earth's atmosphere is extremely different from that predicted for a nonliving Earth and from that of other planets in the solar system. Such atmospheres are dominated by carbon dioxide and possess only a trace of oxygen. There are two views of the formation of Earth's atmosphere. One is that physical forces interacted to form life-sustaining conditions and life evolved to adapt to those conditions. The other is that organisms evolved with the physical environment and from the very beginning were directly involved in the development and control of the geochemical environment favorable to themselves. For example, emergence of photosynthetic algae in the early oceans first released O_2 into the atmosphere. Ocean microflora still provide 70 percent of our atmosphere's oxygen.

THE GAIA THEORY

The constancy of Earth's atmosphere over 3.6 billion years, with its high O_2 and low CO_2 content and moderate temperatures, suggests some feedback system. It has prompted James Lovelock, physical scientist, inventor of instruments to measure the Martian environment, and engineer, and microbiologist Lynn Margulis to postulate the Gaia theory (Gaia is the Greek word for Earth goddess) of global biogeochemical homeostasis. The homeostasis involves Earth's biosphere, atmosphere, oceans, and soil. Together they make up a feedback system that maintains an optimal physical and chemical environment for life on Earth. This feedback system could not have developed nor be maintained without the critical buffering activity of early life forms and continued coordinated activity of plants and microbes. Together they dampen the fluctuations of physical environment that would occur in the absence of a well-organized living system.

No control mechanisms have been discovered, but microorganisms are the only life forms that can function like a chemostat control system, making Earth one large cybernetic system. For example, maintenance of 21 percent O_2 in the atmosphere, to maximize aerobic metabolism just below the level that would make Earth's vegetation inflammable, is possibly the outcome of microbial activity. The accumulation of oxygen above 21 percent in the atmosphere might be dampened by the microbial production of CH_4 from the small amount of carbonaceous material of living matter buried each year.

The Gaia theory has not been accepted by all ecologists and atmospheric scientists; but it does help us understand the behavior of ecosystems and the interactions of biogeochemical cycling. Evidence seems to indicate that organisms do play a dynamic role in determining the composition of many chemicals in soil, water, and atmosphere. We need to look no further than the tremendous impact we humans have had on the physical aspects of Earth.

THE OXYGEN CYCLE

Oxygen, the by-product of photosynthesis, is involved in the oxidation of carbohydrates, with the release of energy, carbon dioxide, and water. Its primary role in biological oxidation is that of a hydrogen acceptor. The breakdown and decomposition of organic molecules proceeds primarily by dehydrogenation. Hydrogen is removed by enzymatic action from organic molecules in a series of reactions and is finally accepted by the oxygen, forming water.

Oxygen is very active chemically. It can combine with a wide range of chemicals in the Earth's crust and is able to react spontaneously with organic compounds and reduced substances.

The major supply of free oxygen that supports life is in the atmosphere. There are two significant sources of atmospheric oxygen. One is the photodisassociation of water vapor, in which most of the hydrogen released escapes into outer space. If the hydrogen did not escape, it would oxidize and recombine with the oxygen. The other source is photosynthesis, active only since life began on Earth. Because photosynthesis and respiration are cyclic, involving the alternate release and utilization of oxygen, one would seem to balance the other, so no significant quantity of oxygen would accumulate in the atmosphere. At some time in the Earth's history, the amount of oxygen introduced into the atmosphere had to exceed the amount used in the decay of organic matter and that tied up in the oxidation of sedimentary rocks. Part of the atmospheric oxygen represents the portion remaining from the unoxidized reserves of photosynthesis—coal, oil, gas, and organic carbon in sedimentary rocks. The amount of stored carbon in the Earth suggests that 150×10^{20} g of oxygen has been available to the atmosphere, over ten times as much as now present, 10×10^{20} g.

The main nonliving oxygen pool consists of molecular oxygen, water, and carbon dioxide, all intimately linked to each other in photosynthesis and other oxidation-reduction reactions and all exchanging oxygen with each other. Oxygen is also biologically exchangeable in such compounds as nitrates and sulfates, which are utilized by organisms that reduce them to ammonia and hydrogen sulfide.

On the surface the oxygen cycle might appear to be quite simple, but because oxygen is so reactive, its cycling is quite complex. As a constituent of carbon dioxide, it circulates freely throughout the biosphere. Some carbon dioxide combines with calcium to form carbonates. Oxygen combines with nitrogen compounds to form nitrates, with iron to form ferric oxides, and with many other minerals to form various other oxides. In these states oxygen is temporarily withdrawn from circulation. In photosynthesis the oxygen freed is split from the water molecule. This oxygen is then reconstituted into water during plant and animal respiration. Part of the atmospheric oxygen that reaches the higher levels of the troposphere is reduced to ozone (O_3) by high-energy ultraviolet radiation.

THE CARBON CYCLE

Because it is a basic constituent of all organic compounds and a major element in the fixation of energy by photosynthesis, carbon is so closely tied to energy flow that the two are inseparable. In fact, the measurement of productivity is commonly expressed in terms of grams of carbon fixed per square meter per year.

The source of all the fixed carbon in both living organisms and fossil deposits is carbon dioxide, CO_2, found in the atmosphere and dissolved in the waters of the Earth. To trace its cycling to the ecosystem is to redescribe photosynthesis and energy flow (see Chapter 7).

The carbon contained in animal wastes and in the protoplasm of plants and animals is eventually released by assorted decomposer organisms. The rate of release depends on environmental conditions such as soil moisture, temperature, and precipitation. In tropical forests most of the carbon in plant remains is quickly recycled, for there is little accumulation in the soil. The turnover rate of atmospheric carbon over a tropical forest is about 0.8 year. In drier regions such as grasslands, considerable quantities of carbon are stored as humus. In swamps and marshes, where dead material falls into the water, organic carbon is not completely mineralized and is stored as raw humus or peat and circulated only slowly. The turnover rate of atmospheric carbon over peat bogs is on the order of 3 to 5 years.

Similar cycling takes place in the freshwater and marine environments. Phytoplankton utilizes the carbon dioxide that has been diffused into the upper layers of water or is present as carbonates and converts it into carbohydrates. The carbohydrates so produced pass through the aquatic food chains. The carbon dioxide produced by respiration is reutilized by the phytoplankton in the production of more carbohydrates. Under proper conditions a portion is reintroduced into the atmosphere. Significant portions of carbon bound as carbonates in the bodies of shells, snails, and foraminifers become buried in the bottom mud at varying depths when the organisms die. Isolated from biotic activity, that carbon is removed from cycling and becomes incorporated into bottom sediments, which through geological time may appear on the surface as limestone rocks or coral reefs. Other organic carbon is slowly deposited as gas, petroleum, and coal at an estimated global rate of 10 to 13 g/m^2/year.

The cycling of carbon as carbon dioxide involves its assimilation and respiration by plants, its consumption in the form of plant and animal tissue by animals, its release through their respiration, the mineralization of litter and wood, soil respiration, accumulation of carbon in a standing crop, and its withdrawal into longer-term reserves such as humus and peat fossil deposits (Figure 24.3).

Diurnal and Seasonal Patterns

If you were to measure the concentration of carbon dioxide in the atmosphere above and within a forest on a summer day, you would discover that it fluctuates throughout the day (Figure 24.4). At daylight, when photosynthesis begins, plants start to withdraw carbon dioxide from the air, and the concentration declines sharply. By afternoon, when the temperature is increasing and the humidity is decreasing, the respiration rate of plants increases, the assimilation rate of carbon dioxide declines, and the concentration of carbon dioxide in the atmosphere increases. By sunset the light phase of photosynthesis ceases, carbon dioxide is no longer being withdrawn from the atmosphere, and its concentration in the atmosphere increases sharply. A similar diurnal fluctuation takes place in aquatic ecosystems.

Likewise, there is a seasonal course in the production and utilization of carbon dioxide that relates both to temperature and to the dormant and growing seasons (Figure 24.5). In the spring, when land is greening and phytoplankton is actively growing, the daily production of carbon dioxide is high. As measured by nocturnal accumulation in spring and summer, the rate of carbon dioxide production may be two to three times higher than winter rates at the same temperature. The transition from lower to higher rates increases dramatically at about the time of the opening of buds and falls off just as rapidly at about the time when the leaves of deciduous trees start dropping in the fall. Such fluctuations are most pronounced in the Northern Hemisphere with its much larger land area and greater plant cover.

The Global Carbon Cycle

The carbon budget of Earth is closely linked to the three major reservoirs—atmosphere, land, and ocean—and to the mass movements of air around the planet (Figure 24.6). The total global carbon pool is an estimated 44,748 gigatons (a gigaton is equal to 1 billion or 10^9 metric tons) (Figure 24.7). An estimated 4000 gt are in geological stores of recoverable fossil fuels.

The atmosphere is the smallest of the three active reservoirs, but its importance as a conduit between the other two stores overshadows its relatively small size. Carbon dioxide makes up only 0.03 percent of the atmosphere, 315 microliters of CO_2 per liter of air. The amount of CO_2 in the atmosphere, however, has been increasing exponentially since the Industrial Revolution (Figure 24.8), potentially altering Earth's climate (see Chapter 4).

Analysis of air trapped in air bubbles in polar ice shows that the amount of CO_2 in the atmosphere over the past 160,000 years has varied from 200 microliters per liter at the height of the last glaciation to between 260 and 300 microliters during the interglacial periods (Figure 24.8). From 1750 to 1800, the advent of the Industrial Revolution, which introduced mass burning of coal, CO_2 in the atmosphere increased to 279 microliters; by 1988 it had risen to 351 microliters or 748 gigatons, mostly from the burning of fossil fuels (coal, petroleum, and natural gas). Annual injection of carbon into the atmosphere amounts to 0.8 percent of the carbon content of the atmosphere. All of this input comes from Earth's geological reservoir, which otherwise would not enter the global carbon cycle. About 60 percent of this input remains in the atmosphere; the rest is scrubbed by terrestrial vegetation and the oceans. Oceans have taken up about 26 to 34 percent of the fossil fuel carbon put into the atmosphere between 1958 and 1980.

The major active reservoir of carbon is the ocean; it is also the major sink. The ocean holds an estimated

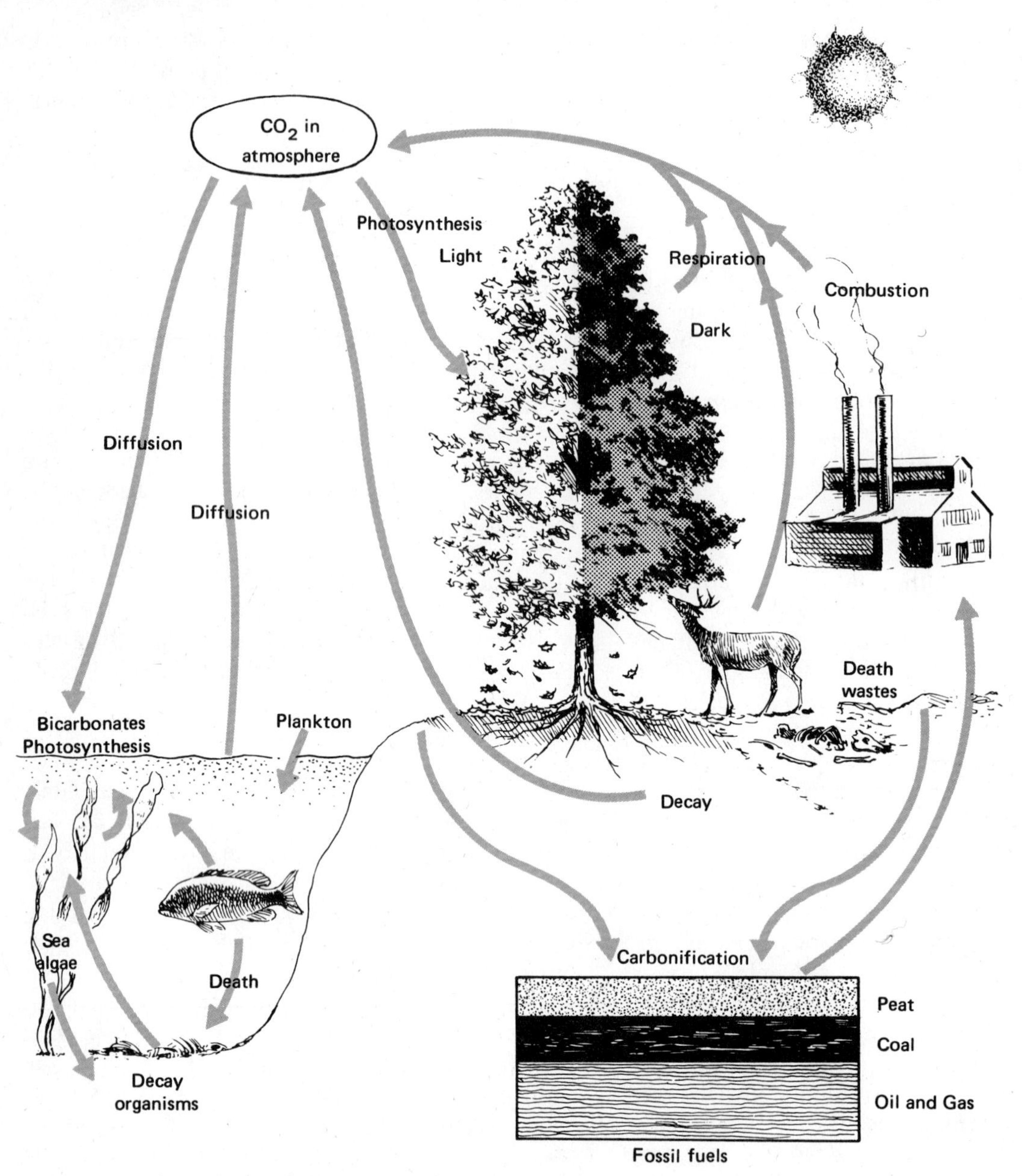

FIGURE 24.3 The carbon cycle in aquatic and terrestrial ecosystems.

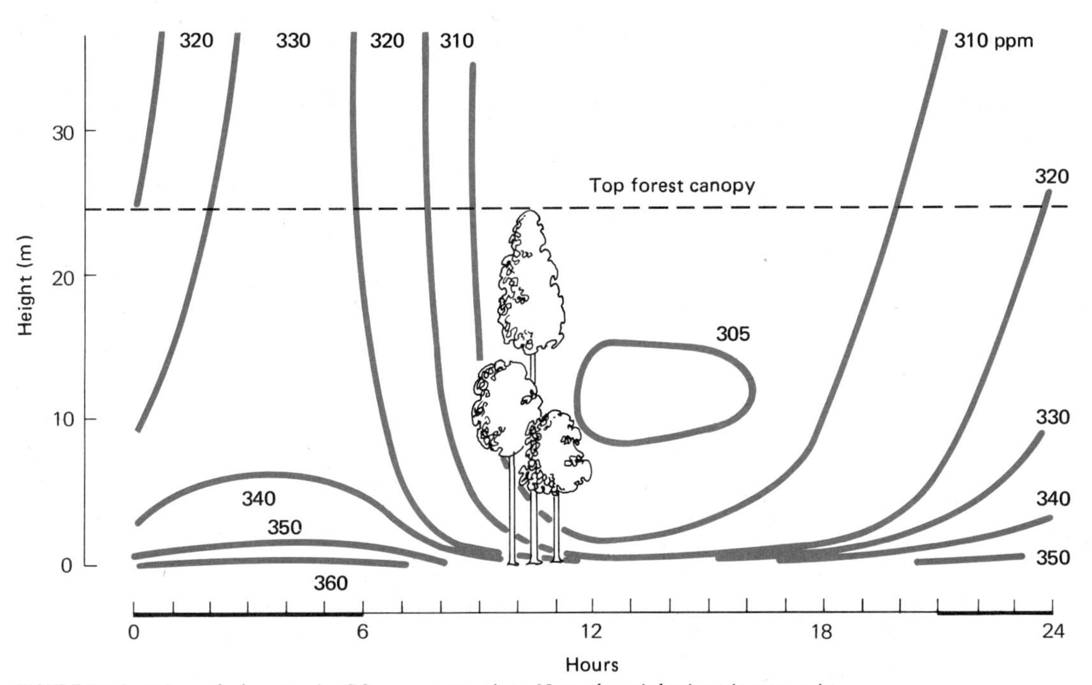

FIGURE 24.4 Diurnal changes in CO_2 concentration. Note the nighttime increase in CO_2 above and within the forest, especially above the forest floor. This increase results from the respiration of plants and soil microorganisms. During the day plants actively take in CO_2 in photosynthesis, reducing its concentration in the atmosphere above and within the forest. The concentration of CO_2 begins to increase in late afternoon and evening when photosynthesis winds down.

37,000 gigatons of inorganic carbon, mostly dissolved CO_2, and bicarbonate (HCO_3^-) and carbonate (CO_3^{-2}) ions. There is another 1000 gigatons of dissolved organic carbon and 30 gigatons of particulate carbon.

The oceans have a strong influence on the concentration of CO_2 in the atmosphere. Carbon is exchanged rapidly between the atmosphere and the surface waters. This exchange includes the physical processes of mixing and circulating sea water, chemical processes involving inorganic carbon compounds and ions, and biological processes of primary production and decomposition.

Little uptake of CO_2 could take place without the transfer of carbon to the deep, which reduces CO_2 in surface waters below equilibrium levels with the atmosphere. This transfer is accomplished by biological pumping. Phytoplankton through primary production take up dissolved inorganic carbon, passing much of it up through the food chain and converting considerable quantities into inorganic carbonate compounds such as shells of ocean invertebrates, reducing the level of inorganic CO_2 in surface water. Dead organic matter, dead bodies of marine animals, and feces sink to deep water, much richer in carbon than the surface. Some of this material undergoes mineralization and bacterial decomposition as it sinks; a portion is deposited on the ocean floor. Part of this carbon of the deep finds its way back up to the surface through upwellings. Much of the carbon in the sediments, however, is trapped for about 10^8 years. Carbon atoms in the water have a residence time of about 10^5 years, which means that in 100,000 years carbon atoms are completely replaced.

The terrestrial reservoir is an estimated 2400 gigatons. Most of this carbon is in slowly exchanging or-

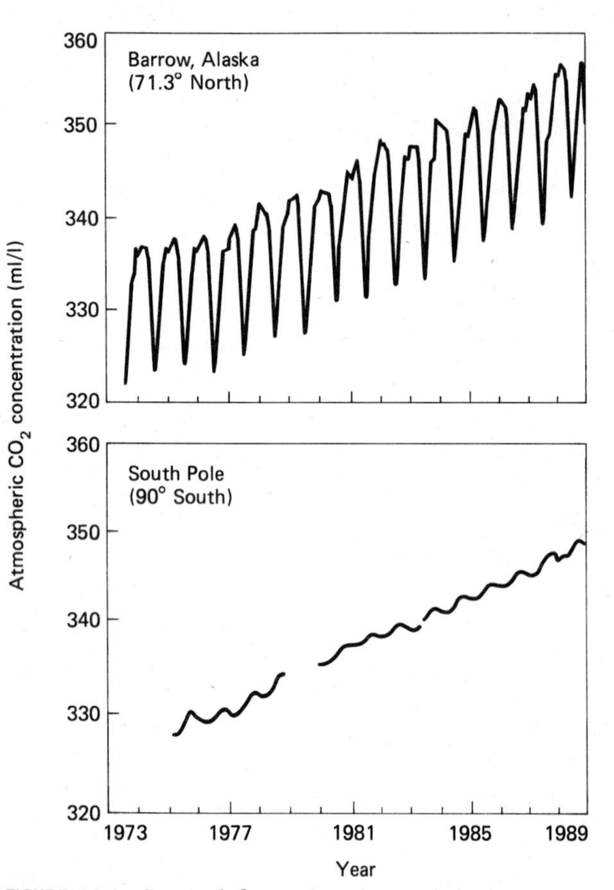

FIGURE 24.5 Seasonal fluctuations in atmospheric concentrations of CO_2 based on measurements by the National Oceanic and Atmospheric Administration in the Arctic and Antarctic. Note the strong fluctuations in the Northern Hemisphere compared to the Southern Hemisphere. The increasing amplitude reflects increased amounts of CO_2 in the atmosphere (see Figure 24.8).

ganic matter, about 420 to 830 gigatons in plants and 1200 to 1600 gigatons in soil and litter carbon. Exchanges between land mass and the atmosphere are nearly in equilibrium over the long term, although there may be short-term imbalances. About 62 gigatons are assimilated by photosynthesis and about 62 gigatons fall to the ground as litter. An estimated 42 gigatons of carbon in soil and litter are returned to the atmosphere through decomposition. Forests are the main consumers of carbon, fixing about 36 gigatons per year. Forests are also the major reservoir of the terrestrial active carbon pool, containing about 1485 gigatons. Much of this reserve is in woody parts with slow turnover. Nonwoody vegetation has a much faster turnover time. In tropical forests turnover time may be less than one year. Peatlands and wetlands store 0.1 to 0.3 gigatons per year, and dry desert soil stores about 0.01 gigaton of carbon per year as carbonates in the soil. Clearing of land and burning of tropical forests may return over 2.6 gigatons to the atmosphere, but the effects on the buildup of CO_2 in the atmosphere are debated. The continuing devastation of tropical forests may be reducing a major CO_2 sink. Feedback mechanisms in which CO_2 emissions lead to increased uptake by plants may balance the loss of production and storage.

THE NITROGEN CYCLE

Nitrogen is an essential constituent of protein, which is a building block of all living material. It is also the major constituent (79 percent) of the atmosphere. The paradox is that in its gaseous state, N_2, abundant though it is, is unavailable to most life. Before it can be utilized, nitrogen must be converted into some chemically usable form. Getting it into that form comprises a major part of the nitrogen cycle.

To be used, free molecular nitrogen has to be fixed. This *fixation* comes about in two ways. One is high-energy fixation. Cosmic radiation, meteorite trails, and lightning provide the high energy needed to combine nitrogen with the oxygen and hydrogen of water. The resulting ammonia and nitrates are carried to the Earth in rainwater. Estimates suggest that less than 8.9 kg N/ha is brought to the Earth annually in this manner. About two-thirds of this amount comes as ammonia and one-third as nitric acid, H_2NO_3.

The second method of fixation is biological. This method produces 100 to 200 kg N/ha, or roughly 90 percent of the fixed nitrogen contributed to the Earth each year. This fixation is accomplished by symbiotic bacteria living in association with leguminous and root-noduled nonleguminous plants, by free-living aerobic bacteria, and by blue-green algae. Fixation splits molecular nitrogen (N_2) into two atoms of N. The free N atoms then combine with hydrogen to form two molecules of ammonia, NH_3. This process requires

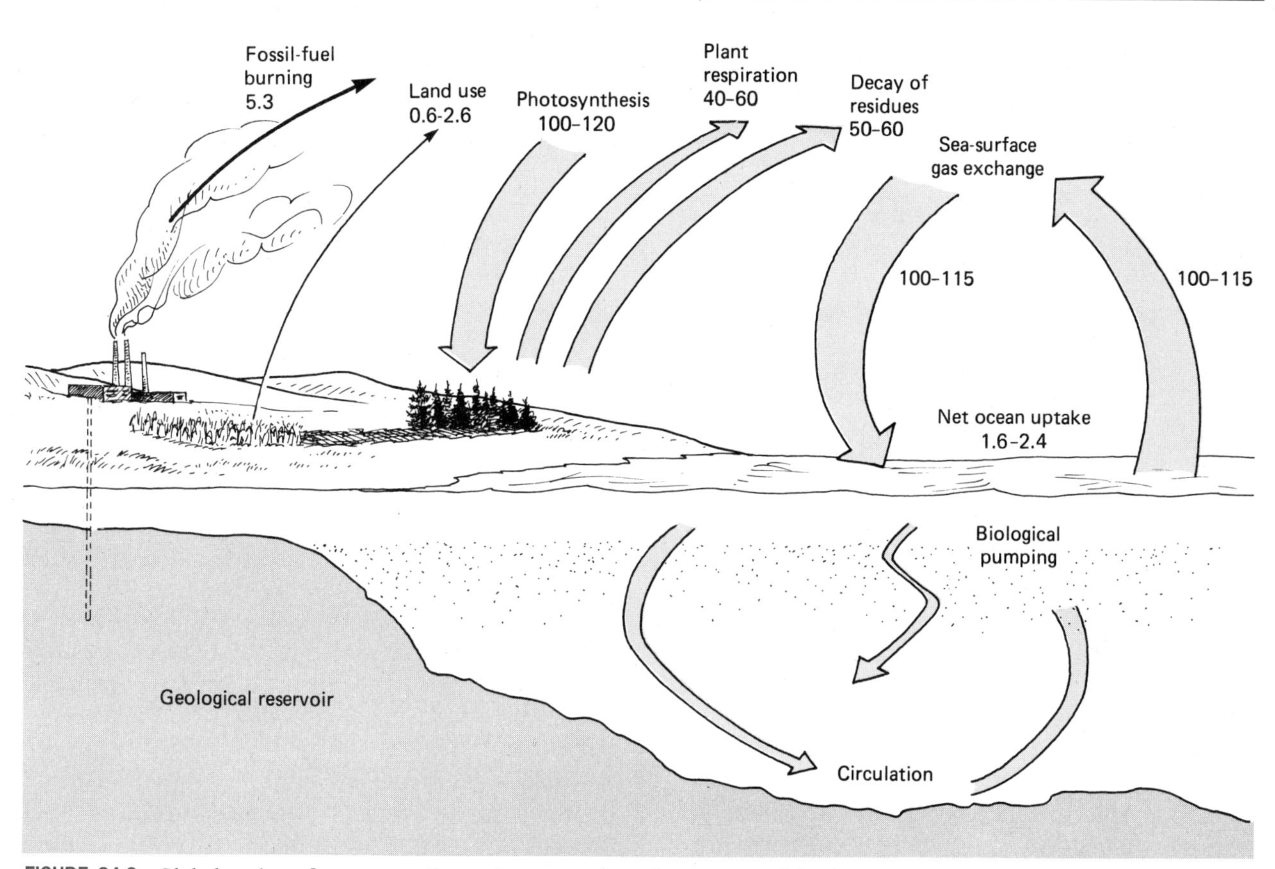

FIGURE 24.6 Global carbon flux among the major reservoirs: air, ocean, and land. All carbon fluxes are 1980 estimates in gigatons.

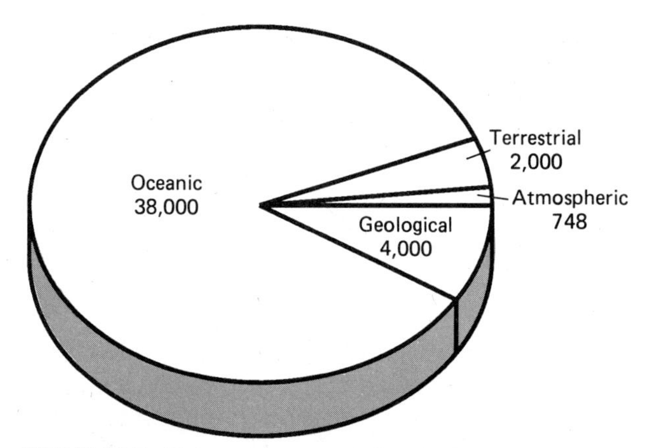

FIGURE 24.7 Major active reservoirs in the global carbon cycle. Reservoir sizes are in gigatons.

considerable energy. To fix 1 g of nitrogen, a nitrogen-fixing legume nodule bacteria must expend about 10 g of photosynthate or glucose.

In agricultural ecosystems the nodulated legumes of approximately 200 species are the preeminent nitrogen fixers. In nonagricultural systems some 12,000 species of plants, from free-living bacteria and blue-green algae to nodule-bearing plants, are responsible for nitrogen fixation. Also contributing to the fixation of nitrogen are free-living soil bacteria. The most prominent of the 15 known genera are the aerobic *Azotobacter* and the anaerobic *Clostridium*. Blue-green algae are another important group of largely nonsymbiotic nitrogen fixers. Of the some 40 known species, the most common are in the genera *Nostoc* and *Calothrix*,

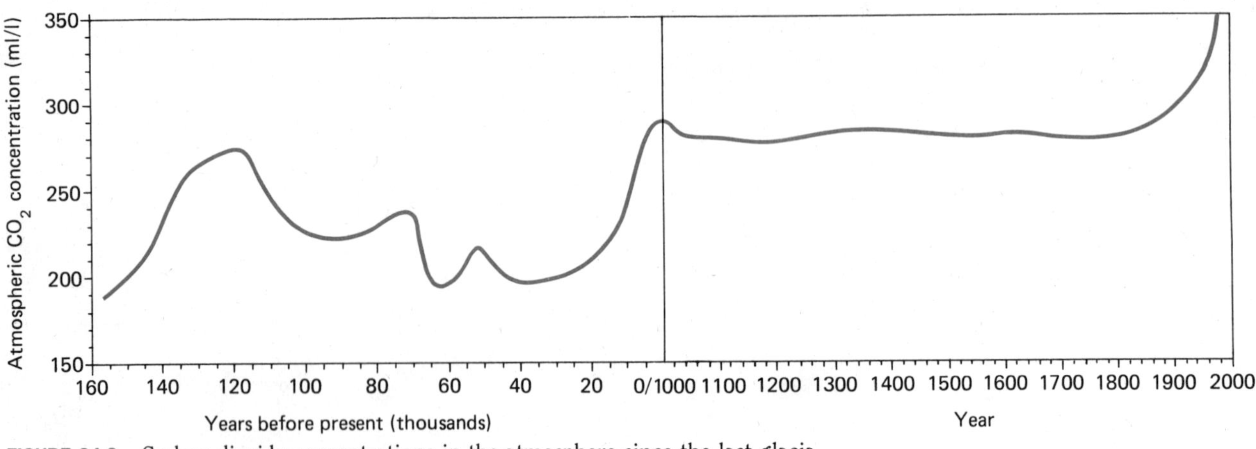

FIGURE 24.8 Carbon dioxide concentrations in the atmosphere since the last glaciation. Note the exponential increase in atmospheric carbon starting with the Industrial Revolution in the 1700s.

which are found both in soil and in aquatic habitats. Certain lichens (*Collema tunaeforme* and *Peltigera rufescens*) are also implicated in nitrogen fixation. Lichens with nitrogen-fixing ability possess nitrogen-fixing blue-green species as their algal component.

Another source of nitrogen is organic matter. The wastes of animals broken down by decomposition release nitrates and ammonia into the ecosystem. All of these nitrogenous products are involved in another phase of the nitrogen cycle: the processes of nitrification, denitrification, and ammonification.

In *ammonification* the amino acids are broken down by decomposer organisms to release energy. It is a one-way reaction. Ammonium, or the ammonia ion, is directly absorbed by plant roots and incorporated into amino acids, which are subsequently passed along through the food chain. Wastes and dead animal and plant tissues are broken down to amino acids by heterotrophic bacteria and fungi in soil and water.

Nitrification is a biological process in which ammonia is oxidized to nitrate and nitrite, yielding energy. Two groups of microorganisms are involved. *Nitrosomonas* bacteria utilize the ammonia in the soil as their sole source of energy. They promote its oxidation to nitrite ions and water. Nitrite ions can be oxidized further to nitrate ions in an energy-releasing reaction. The energy left in the nitrite ion is exploited by another group of bacteria, the *Nitrobacter*, which oxidize the nitrite ion to nitrate.

Nitrates are a necessary substrate for *denitrification*, in which nitrogen in the nitrate form is reduced to the gaseous form by the denitrifiers, represented by fungi and the bacteria *Pseudomonas*. Like nitrification, denitrification takes place under certain conditions: a sufficient supply of organic matter, a limited supply of molecular oxygen, a pH range of 6 to 7, and an optimum temperature of 60° C.

With the basic and necessary processes just described, the nitrogen cycle (Figure 24.9) can be followed briefly. The sources of nitrogen under natural conditions are the fixation of atmospheric nitrogen; additions of inorganic nitrogen in rain from such sources as lightning fixation and fixed "juvenile" nitrogen from volcanic activity; ammonia absorption from the atmosphere by plants and soil; and nitrogen accretion from windblown aerosols, which contain both organic and inorganic forms of nitrogen.

In terrestrial ecosystems, nitrogen, largely in the form of ammonia or nitrates, depending on a number of variable conditions, is taken up by plants, which

FIGURE 24.9 The nitrogen cycle in terrestrial and aquatic ecosystems.

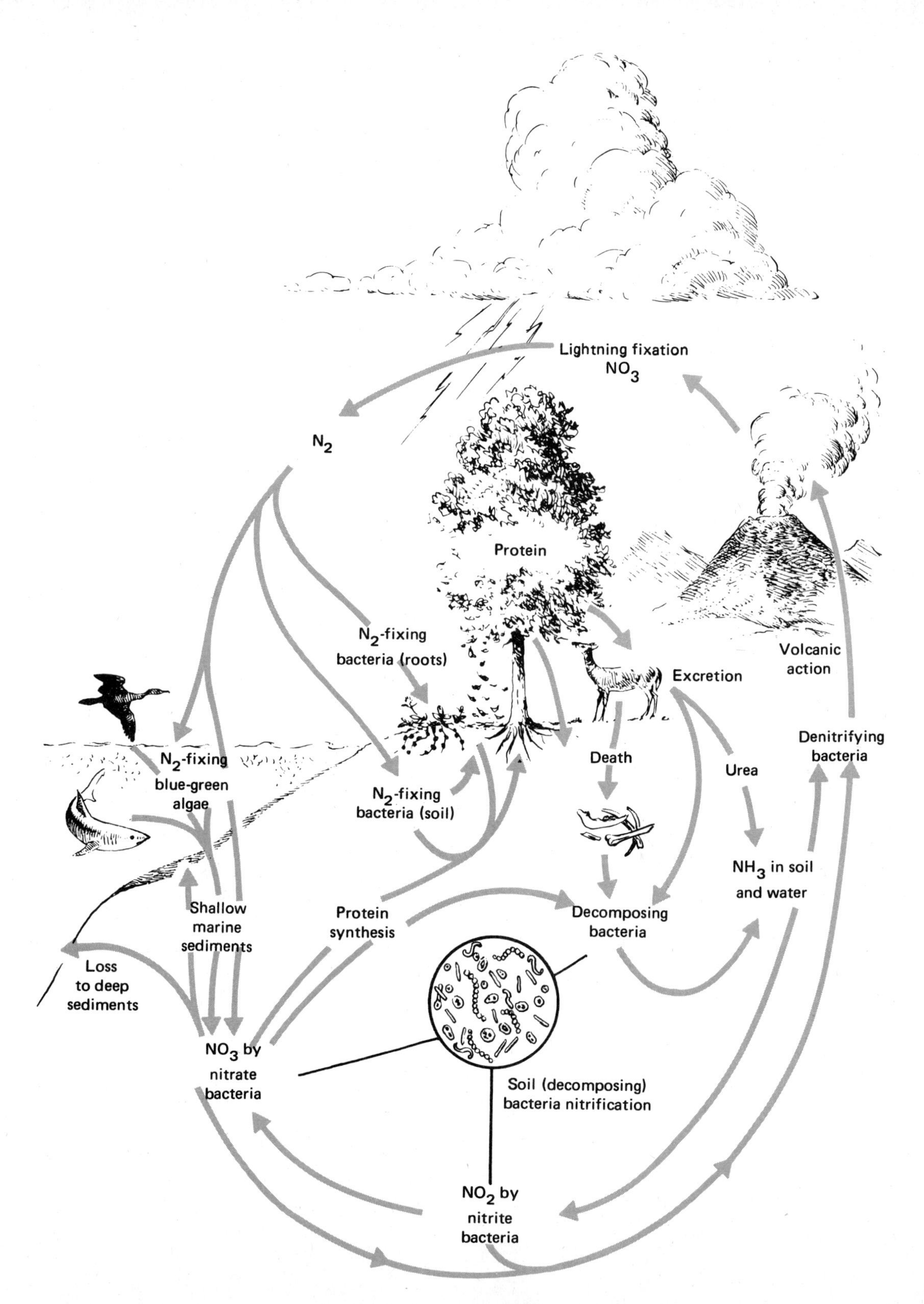
Lightning fixation
NO_3
N_2
Protein
N_2-fixing
bacteria (roots)
Volcanic
action
Excretion
N_2-fixing
blue-green
algae
N_2-fixing
bacteria (soil)
Death
Urea
Denitrifying
bacteria
NH_3 in soil
and water
Shallow
marine
sediments
Protein
synthesis
Decomposing
bacteria
Loss
to deep
sediments
NO_3 by
nitrate
bacteria
Soil (decomposing)
bacteria nitrification
NO_2 by
nitrite
bacteria

convert it into amino acids. The amino acids are transferred to consumers, which convert them into different types of amino acids. Eventually, their wastes (urea and excreta) and the decay of dead plant and animal tissue are broken down by bacteria and fungi into ammonia. Ammonia may be lost as gas to the atmosphere, acted upon by nitrifying bacteria, or taken up directly by plants. Nitrates may be utilized by plants, immobilized by microbes, stored in decomposing humus, or leached away. This material is carried to streams, lakes, and eventually the sea, where it is available for use in aquatic ecosystems.

In aquatic ecosystems, nitrogen is cycled in a similar manner, except that the large reservoirs contained in the soil humus are lacking. Life in the water contributes organic matter and dead organisms that undergo decomposition and the subsequent release of ammonia and, ultimately, nitrates.

Under natural conditions nitrogen lost from ecosystems by denitrification, volatilization, leaching, erosion, windblown aerosols, and transportation out of the system is balanced by biological fixation and other sources. Both chemically and biologically, terrestrial and aquatic ecosystems constitute a dynamic equilibrium system, in which a change in one phase affects the other.

SEDIMENTARY CYCLES

The mineral elements required by living organisms are obtained initially from inorganic sources. Available forms occur as salts dissolved in soil water or in lakes, streams, and seas. The mineral cycle varies from one element to another, but essentially it consists of two phases: the *salt solution phase* and the *rock phase*. Mineral salts come directly from the Earth's crust through weathering. The soluble salts then enter the water cycle. With water, they move through the soil to streams and lakes and eventually reach the seas, where they remain indefinitely. Other salts are returned to the Earth's crust through sedimentation. They become incorporated into salt beds, silts, and limestones. After weathering, they again enter the cycle.

Plants and many animals satisfy their mineral requirements from mineral solutions in their environments. Other animals acquire the bulk of their minerals from the plants and animals they consume. After the death of living organisms, the minerals are returned to the soil and water through the action of organisms and the processes of decay.

There are many different kinds of sedimentary cycles. Some, such as sulfur, are a hybrid between the gaseous and the sedimentary because they have reservoirs not only in the Earth's crust but also in the atmosphere. Others, such as phosphorus, are wholly sedimentary—the element is released from rock and deposited in both shallow and deep sediments of the sea.

THE PHOSPHORUS CYCLE

Because phosphorus is unknown in the atmosphere and none of its known compounds has an appreciable vapor pressure, the phosphorus cycle is closed. It can follow the hydrological cycle only part of the way, from land to sea (Figure 24.10). Under undisturbed natural conditions, phosphorus is in short supply. It is freely soluble only in acid solutions and under reducing conditions. In the soil, phosphorus is held in slightly soluble minerals that quickly establish an equilibrium between phosphorus in solution and phosphorus adsorbed on surfaces of soil particles. Much of it becomes immobilized as phosphates of either calcium or iron. Even superphosphate applied to cropland may be converted rapidly to unavailable inorganic compounds. Additional phosphorus occurs in stable but biologically inactive organic forms. Phosphorus's natural scarcity in aquatic ecosystems is emphasized by the explosive growth of algae in water receiving heavy discharges of phosphorus-rich wastes.

The main reservoirs of phosphorus are rock and natural phosphate deposits, from which the elements are released by weathering, leaching, erosion, and mining for agricultural use. Some of it passes through terrestrial and aquatic ecosystems by way of plants, grazers, predators, and parasites; and it is returned to the ecosystem by excretion, death, and decay. In terrestrial ecosystems organic phosphates are reduced by bacteria to inorganic phosphates. Some are recycled to plants, some become immobilized as unavailable chemical compounds, and some are immobilized by incor-

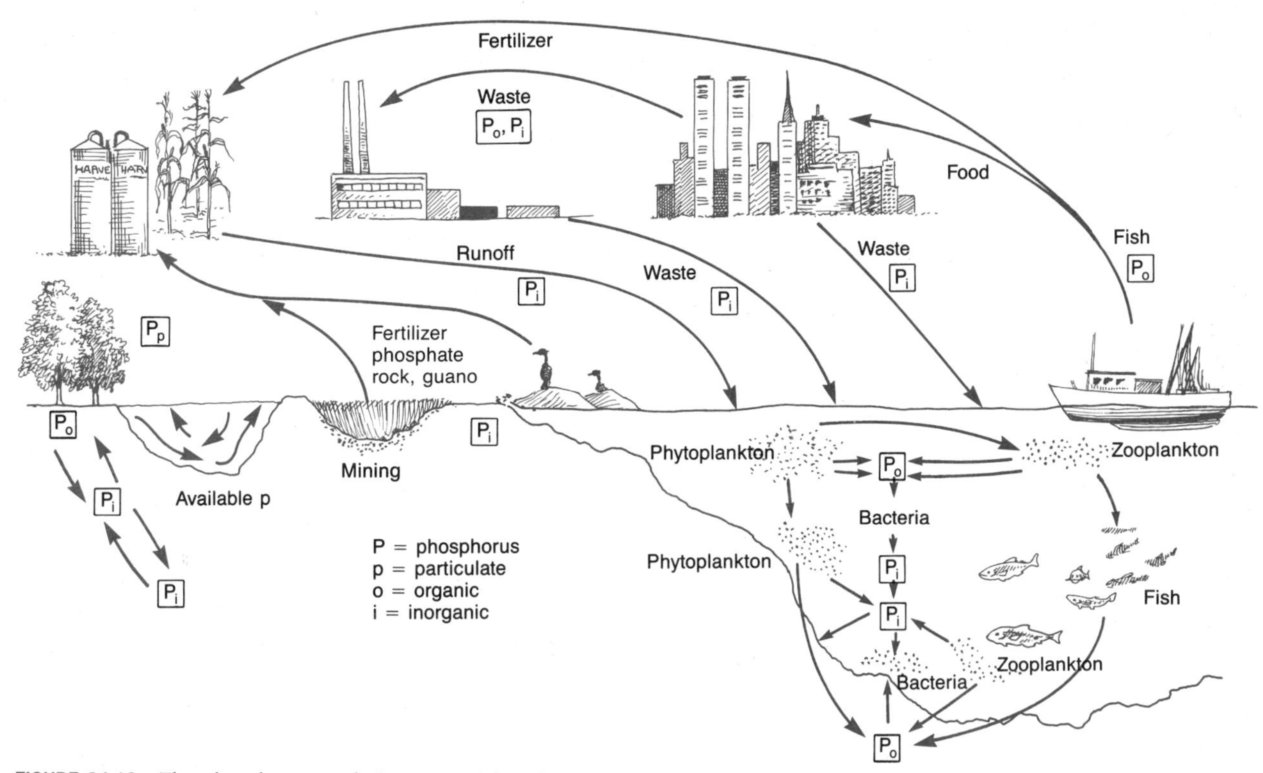

FIGURE 24.10 The phosphorus cycle in terrestrial and aquatic ecosystems.

poration into the bodies of microorganisms. Some of the phosphorus of terrestrial ecosystems escapes to lakes and seas, both as organic phosphates and as particulate organic matter.

In marine and freshwater ecosystems, the phosphorus cycle moves through three compartments: particulate organic phosphorus, dissolved organic phosphates, and inorganic phosphates. Inorganic phosphates are taken up rather rapidly by all forms of phytoplankton, which, in turn, may be ingested by zooplankton or detritus-feeding organisms. Zooplankton, in turn, may excrete as much phosphorus daily as is stored in its biomass, keeping the cycle going. More than half of the phosphorus zooplankton excretes is inorganic phosphate, which is taken up by phytoplankton. In some instances 80 percent of this excreted phosphorus is sufficient to meet the needs of the phytoplankton population. The remainder of the phosphorus in aquatic ecosystems is in organic compounds that may be utilized by bacteria, which fail to regenerate much dissolved inorganic phosphate. Bacteria are consumed by the microbial grazers, which then excrete the phosphate they ingest. Part of the phosphorus is deposited in shallow sediments and part in deep water. In the ocean some of the latter may be recirculated by upwelling, which brings the phosphates from the unlighted depths to the photosynthetic zones, where they are taken up by phytoplankton. Part of the phosphorus contained in the bodies of plants and animals is deposited in the shallow sediments and part in the deeper ones. As a result, the surface waters may become depleted of phosphorus and the deep waters saturated. Because phosphorus is precipitated largely as calcium compounds, much of it becomes immobilized for long periods in the bottom sediments. Upwelling returns some of it to the photosynthetic zones, where it is available to phytoplankton. The amount available is limited by the insolubility of calcium phosphate.

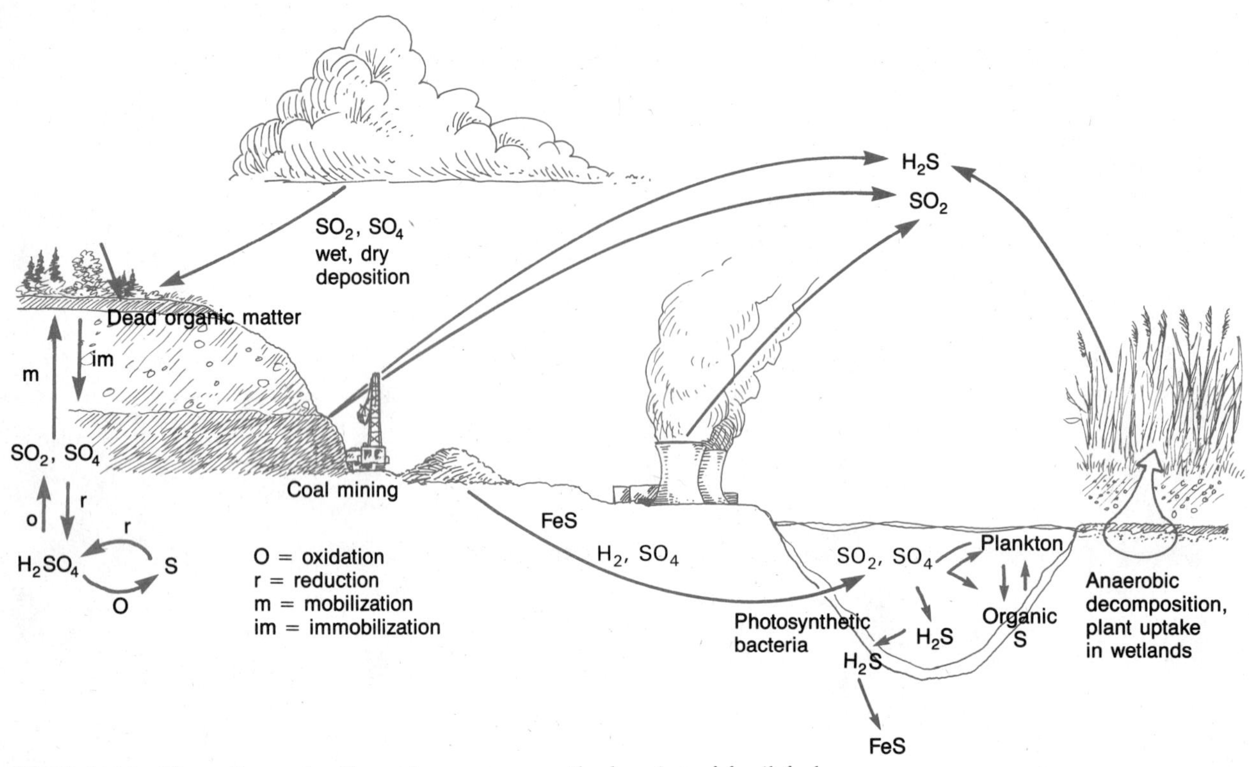

FIGURE 24.11 The sulfur cycle. The major sources are the burning of fossil fuels and acid drainage from coal mines.

THE SULFUR CYCLE

The sulfur cycle is both sedimentary and gaseous (Figure 24.11). It involves a long-term sedimentary phase in which sulfur is tied up in organic and inorganic deposits, is released by weathering and decomposition, and is carried to terrestrial and aquatic ecosystems in salt solution. The gaseous phase of the cycle permits the circulation of sulfur on a global scale.

Sulfur enters the atmosphere from several sources: the combustion of fossil fuels, volcanic eruptions, the surface of the oceans, and gases released by decomposition. It enters the atmosphere initially as hydrogen sulfide, H_2S, which quickly oxidizes into another volatile form, sulfur dioxide, SO_2. Atmospheric sulfur dioxide, soluble in water, is carried back to Earth in rainwater as weak sulfuric acid, H_2SO_4. Whatever the source, sulfur in a soluble form is taken up by plants and is incorporated through a series of metabolic processes, starting with photosynthesis, into such sulfur-bearing amino acids as cystine. From the producers, sulfur in amino acids is transferred to consumer groups.

Excretions and death carry sulfur in living material back to the soil and to the bottoms of ponds, lakes, and seas, where the organic material is acted upon by bacteria, releasing the sulfur as hydrogen sulfide or sulfate. One group, the colorless sulfur bacteria, both reduces hydrogen sulfide to elemental sulfur and oxidizes it to sulfuric acid. The green and purple bacteria, in the presence of light, utilize hydrogen sulfide as an oxygen acceptor in the photosynthetic reduction of carbon dioxide. Best known are the purple bacteria found in salt marshes and in the mud flats of estuaries. These organisms are able to carry the oxidation of hydrogen sulfide as far as sulfate, which may be recirculated and taken up by the producers or used by sulfate-reducing

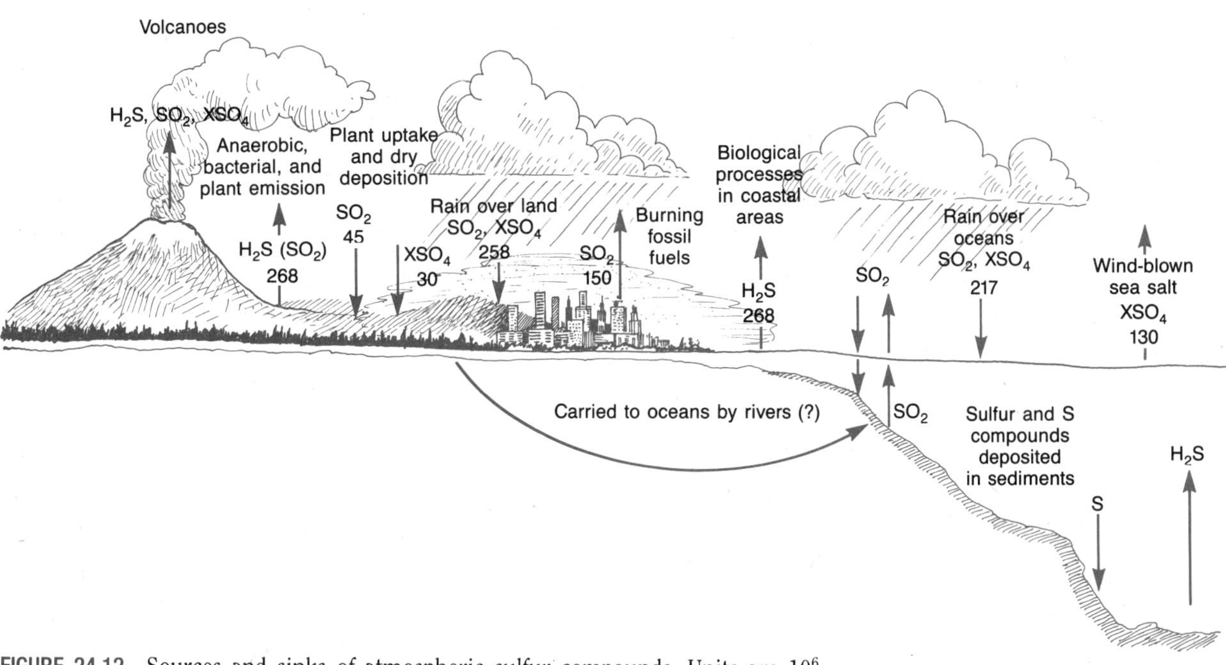

FIGURE 24.12 Sources and sinks of atmospheric sulfur compounds. Units are 10^6 tons per year, calculated as sulfate.

bacteria. The green sulfur bacteria can carry reduction of hydrogen sulfide to elemental sulfur.

Sulfur, in the presence of iron and under anaerobic conditions, will precipitate as ferrous sulfide, FeS_2. This compound is highly insoluble under neutral and alkaline conditions and is firmly held in mud and wet soil. Sedimentary rocks containing ferrous sulfide (called pyritic rocks) may overlie coal deposits. Exposed to air in deep and surface mining for coal, the ferrous sulfide oxidizes and in the presence of water produces ferrous sulfate ($FeSO_4$) and sulfuric acid. In other reactions ferric sulfate (Fe_2SO_4) and ferrous hydroxide ($FeOH_3$) are produced.

In this manner, sulfur in pyritic rocks, suddenly exposed to weathering by human activities, discharges heavy slugs of sulfuric acid, ferric sulfate, and ferrous hydroxide into aquatic ecosystems. These compounds destroy aquatic life and have converted hundreds of miles of streams in the eastern United States to highly acidic water.

As with nitrogen, oxygen, and other gaseous cycles, the biosphere plays an important role in the sulfur cycle. However, the sedimentary phase makes the total cycle more complex (Figure 24.12). Sources of sulfur include the weathering of rocks, especially pyrites, erosional runoff, industrial production, and decomposition of organic matter. The bulk of sulfur appears first as a volatile gas, hydrogen sulfide. In the hydrosphere, the soil, and the atmosphere, hydrogen sulfide is oxidized to sulfides and sulfates, the forms in which sulfur is most readily circulated. The atmosphere contains sulfate particles, sulfur dioxide, and hydrogen sulfide. The latter is most abundant over continents. The concentration of sulfur as hydrogen sulfide in the unpolluted atmosphere is estimated at 6 g/m^3; as sulfur dioxide, at 1 g/m^3. Part of the sulfur in the atmosphere is recirculated to land and sea by precipitation. The concentration of sulfur dioxide in rain falling over land has been estimated as 0.6 mg/liter and over sea as 0.2 mg/liter, excluding sea spray.

It is almost impossible to estimate the biological turnover of sulfur dioxide because of the complicated cycling within the biosphere. Net annual assimilation of sulfur by marine plants is on the order of 130 million tons. Added to the anaerobic oxidation of organic matter, that brings the total to an estimated 200 million tons. Both industrially emitted sulfur and fertilizer sulfur are eventually carried to the sea. These two sources probably account for the 50-million-ton increase annually of sulfur in the ocean.

INTERNAL NUTRIENT CYCLES

Nutrient cycling also takes place within the components of an ecosystem. A key element in such cycling is the vegetation, which takes up inorganic nutrients. The amount of nutrients moved through the system is influenced by nutrient availability to plants in soil and water. Availability depends in turn upon inputs to the detrital pool, the rate at which detritus is decomposed, the amount of nutrients that go into biomass storage and into the soil, humus, and detrital sediments, the release of nutrients from that pool, and removal by grazing herbivores.

How nutrients are utilized depends upon the size of the abiotic reserve, the proportion of nutrients stored in the abiotic and biotic components, the movement among them, and the rate of turnover in the recycling pool. Turnover depends upon the speed at which nutrients are released from the detrital pool to an available form, leaching, decomposition, and subsequent uptake.

Nutrients are constantly being removed or added by natural and artificial processes (see Figure 24.13). In woodland, shrub, and grassland ecosystems, nutrients are returned annually to the soil by leaves, litter, roots, animal excreta, and the bodies of the dead. Released to the soil by decomposition, these nutrients again are taken up, first by plants and then by animals. In freshwater and marine ecosystems, the remains of plants and animals drift to the bottom, where decomposition takes place. The nutrients again are recirculated to the upper layers by the annual overturns and by upwellings from the deep.

The cycle, however, is not a closed circuit within an ecosystem. Nutrients are continuously being imported into as well as carried out of any ecosystem. Appreciable quantities of plant nutrients are carried in by rain, snow, and airborne particles.

A small quantity of these nutrients is absorbed directly through the leaves, but it hardly offsets the quantity leached out. Rainwater dripping down from the canopy is richer in calcium, sodium, potassium, phosphorus, iron, manganese, and silica than rainwater collected in the open at the same time, although less rainwater reaches the forest floor. The nutrients leached from the foliage are taken up in time by the surface roots and translocated to the canopy. Such localized nutrient cycles may require only a few days for completion.

These little nutrient cycles can be followed by means of radioactive tracers. By inoculating white oak trees with 20 microcuries of ^{134}Cs (cesium 134), ecologists were able to follow the gains, losses, and transfers of this radioisotope. About 40 percent of the ^{134}Cs inoculated into the oaks in April moved into the leaves by early June (see Figure 24.14). When the first rains fell after inoculation, leaching of radiocesium from the leaves began. By September this loss amounted to 15 percent of the maximum concentration in the leaves. Seventy percent of this rainwater loss reached mineral soil; the remaining 30 percent found its way into the litter and understory. When the leaves fell in autumn, they carried with them twice as much radiocesium as was leached from the crown by rain. Over the winter, half of this amount was leached out to mineral soil. Of the radiocesium in the soil, 92 percent still remained in the upper 10 cm nearly 2 years after the inoculation. Eighty percent of the cesium was confined to an area within the crown perimeter, and 19 percent was located in a small area around the trunk. This finding suggests that cesium distribution in the soil was greatly influenced by leaching from rainfall and stemflow.

Movement of nutrients out of the system is slowed by internal mechanisms. One of the most prominent

FIGURE 24.13 Generalized nutrient budget of a forested ecosystem. Input of nutrients comes from precipitation, wind-blown dust, litterfall, weathering, and root decomposition. Outgo is through wood harvest, wildlife harvest, runoff, erosion, and leaching.

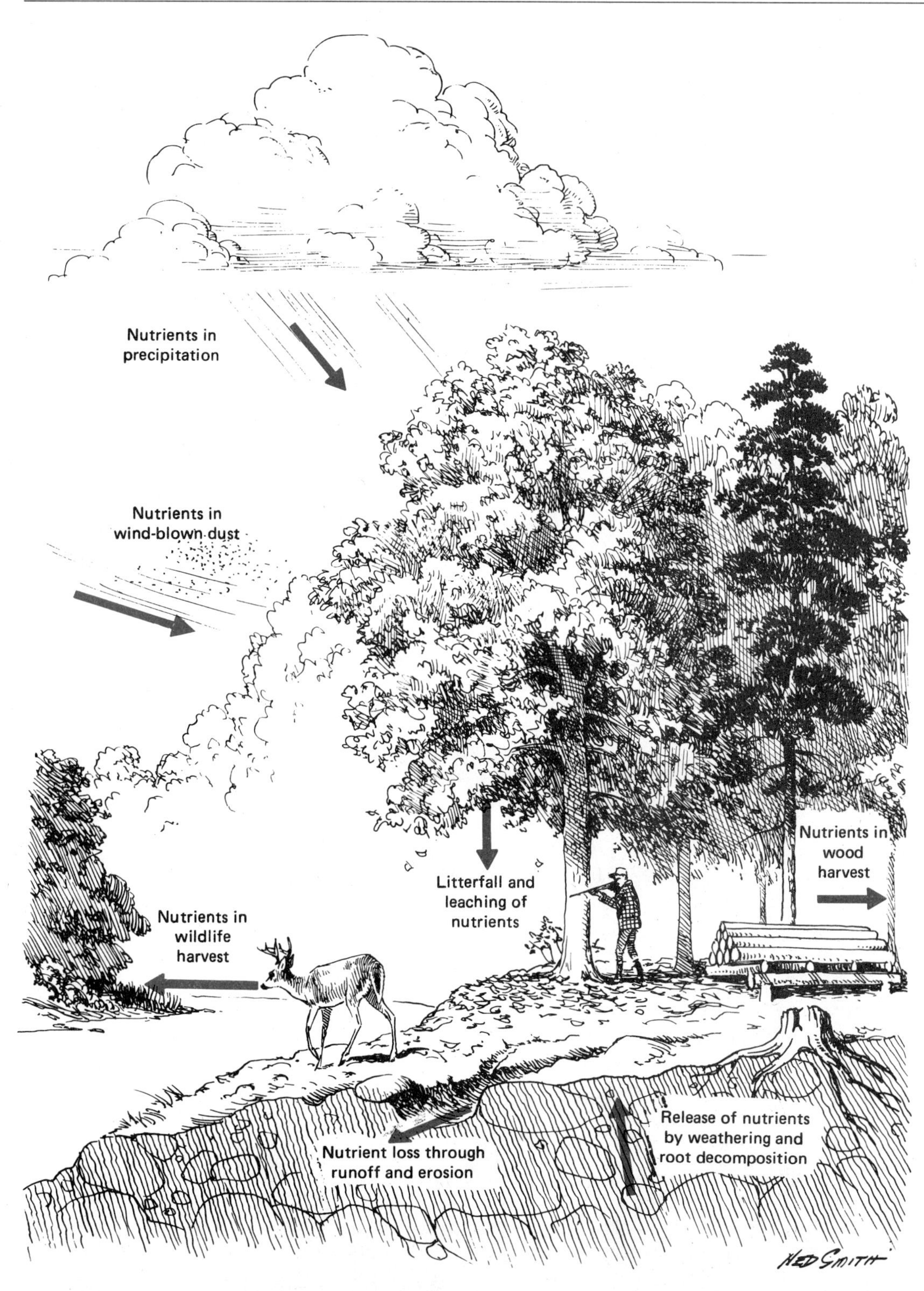
Nutrients in
precipitation
Nutrients in
wind-blown dust
Nutrients in
wood
harvest
Litterfall and
leaching of
nutrients
Nutrients in
wildlife
harvest
Release of nutrients
by weathering and
root decomposition
Nutrient loss through
runoff and erosion
NED SMITH

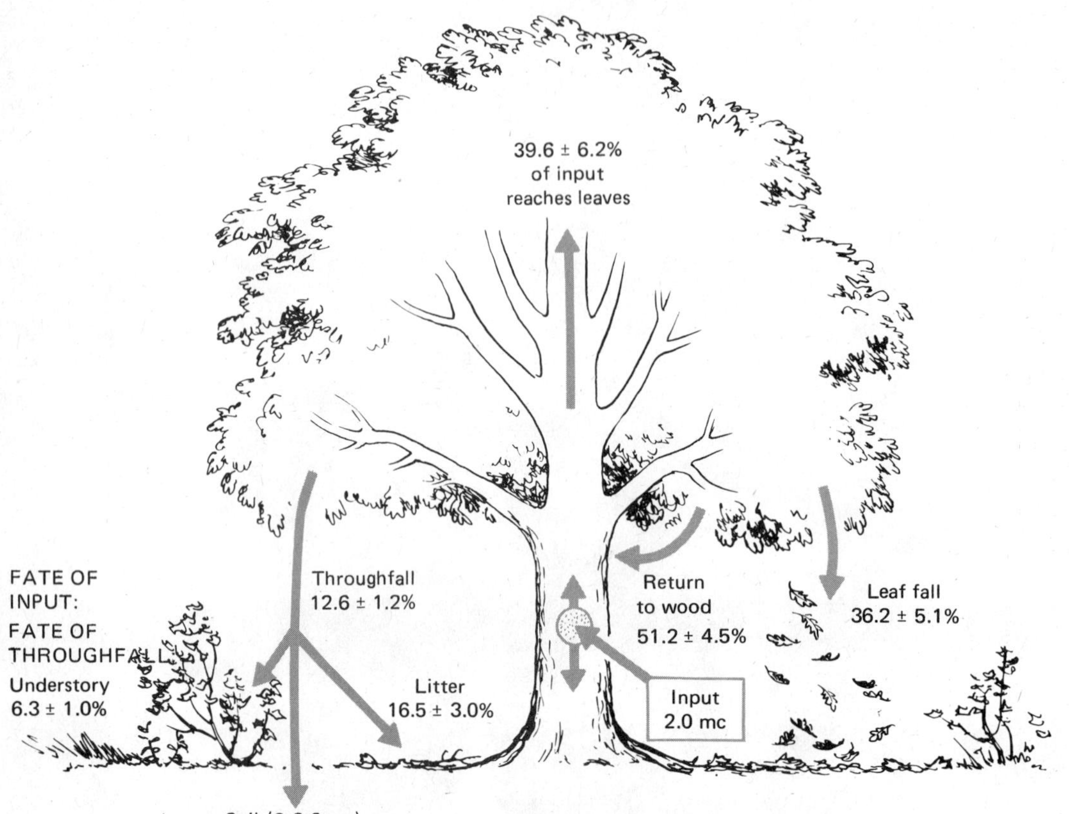

FIGURE 24.14 The cycle of ^{134}Cs in white oak, an example of nutrient cycling through plants. The figures are the average for 12 trees at the end of the 1960 growing season. Of the input reaching the leaves, 12.6 percent is leached by rainwater (throughfall), 36.2 percent is returned by leaf fall, and 51.2 percent returned to wood. Of the cesium leached from the canopy, 77.2 percent enters the soil, 16.5 percent is absorbed by the litter, and 6.3 percent is taken up by understory plants.

mechanisms to reduce nutrient losses is a mat of fine roots, humus, and mycorrhizal fungi (see Chapter 17) near the surface. Such mats are best developed in tropical rain forests and in coniferous forests of temperate regions. These mats physically absorb nutrients entering them almost immediately, with negligible leaching. Mycorrhizal fungi then take up nutrients from decomposing litter and transfer them to roots. Acidity and a high tannin content of the litter inhibit bacteria that could consume the limited nitrogen available.

Evergreenness is another nutrient-conserving mechanism. Thick cuticle, wax coating, and low water content make leaves more resistant to herbivores and parasites, reducing nutrient loss to consumers. The same features also enable evergreens to resist nutrient leaching by rainfall. Year-round rather than seasonal leaf fall allows trees to return nutrients slowly to the soil, in contrast to a rapid return by deciduous trees.

Other controls relate to energy flow. One mechanism is the diversion of energy production to plant populations with low biomass and with different responses to environmental conditions. Such plant populations reproduce rapidly when conditions are optimal and take up nutrients quickly. A succession of such plant populations through the seasons ensures continuous energy fixation and utilization of nutrients. A second mechanism puts energy conversion in individuals of great bulk and slow reproductive rates. Large individuals are able to survive unfavorable conditions and store large quantities of nutrients in biomass. Although the first mechanism is most typical of aquatic situations and the second of terrestrial ecosystems, especially forests, neither mechanism operates exclusively in only one type of ecosystem. Lakes have rooted aquatics with biomass accumulation, and forests have a seasonal parade of herbaceous understory plants.

Each mechanism has provisions for the conservation of nutrients. One provision is a pool of organic matter. Litter, soil organic matter, and biomass of autotrophs form such a pool in terrestrial ecosystems. In aquatic ecosystems the pool is particulate organic matter.

Turnover of the organic pool is relatively slow, roughly one magnitude lower than the turnover of the vegetative component. Organic matter has a key role in recycling nutrients because it prevents rapid losses from the system. Large quantities of nutrients are bound tightly in organic matter and are not readily available, but their release can be activated by decomposer organisms. If energy, and especially water, are limited, however, as in desert ecosystems, then recycling is minimal because the ecosystem cannot attain a standing crop large enough to deplete the nutrient supply.

Another mechanism is to partition the nutrient reserve between long-term and short-term nutrient pools. For example, of the structural components of individual plants—wood, bark, twigs, and leaves—leaves are recycled the fastest and wood the slowest. The leaves represent a short-term nutrient pool and wood a long-term reservoir. Thus, in a mature forest, nutrients stored in vegetation are recycled at various time intervals from 1 to 100 or more years. Such partitioning prevents excessive losses of nutrients and releases nutrients slowly to biogeochemical cycles.

Nutrient cycling within forest ecosystems in particular can be influenced by foliage-eating insects such as fall cankerworm (*Alsophila pometaria*) and saddled prominent (*Heterocampa guttivita*). Chronic defoliation by such insects results in an increased export of NO_3-nitrogen from the forest to stream ecosystems. Accompanying this export are a number of other changes in ecosystem function. Production of leaves increases and production of wood decreases as energy and nutrients are invested in new foliage. There are large inputs of nutrients to the forest floor through frass (fecal material of caterpillars) and increased litterfall. That results in increased microbial activity, including nitrifying bacteria and associated litter metabolism. The outcome is a significant increase in available nutrients (including mineral N) in soil and litter. Ecosystem responses involve a temporary shift from wood to leaf production, an increased rate of nutrient uptake, and an increased turnover rate of nutrients in litter and soil. In such a manner, forest insects also become a mechanism in the regulation of nutrient cycles.

Summary

Materials flow from the living to the nonliving parts of the ecosystem and back in a perpetual cycle. By means of these cycles, plants and animals obtain nutrients that are necessary for their well-being.

All of these biogeochemical cycles are closely tied to the water cycle. Most of Earth's water is in the oceans. Less than 1 percent is available as free fresh water. Water moves through the water cycle by precipitation, interception, infiltration, surface flow, and evaporation. The key to the water cycle is the atmosphere. Evaporation into and precipitation from the atmosphere, detention in oceans and land, and atmospheric transport maintain the global water balance. Oceans lose more water by evaporation and gain less from precipitation than land areas. The loss is made up by runoff from land.

There are two kinds of biogeochemical cycles: the gaseous (represented by oxygen, carbon dioxide, and nitrogen) and the sedimentary (represented by phosphorus). The sulfur cycle is a combination of the two.

The oxygen cycle is complex. The major sources of oxygen are the photodissociation of water vapor and photosynthesis. Oxygen also circulates freely as a constitutent of carbon dioxide. Active chemically, it combines with a wide range of inorganic chemicals, organic substances, and reduced compounds, and it is involved with the oxidation of carbohydrates, releasing energy, carbon dioxide, and water.

The carbon cycle is inseparable from energy flow. It involves the assimilation and respiration of carbon dioxide by plants, its consumption in the form of plant and animal tissue by heterotrophs, its release through respiration, the mineralization of litter and wood, soil respiration, the accumulation of carbon in standing crops, and its withdrawal into long-term reserves. The carbon cycle exhibits both diurnal and seasonal curves.

The main reservoirs in the global carbon cycle are the atmosphere, oceans, and land. The atmosphere, holding the smallest pool of active carbon, is the major conduit of carbon between the other two pools. Its carbon content has increased exponentially since the Industrial Revolution from 279 milliliters per liter of air to 352 milliliters in 1988. This increase results from the burning of fossil fuels, 60 percent of whose carbon remains in the atmosphere. About 25 percent of the input from fossil fuels is taken up by surface waters of the ocean, the major reservoir of global carbon. Carbon dioxide take-up by the surface waters is used by photosynthetic phytoplankton, passed along through the food chain, and eventually deposited in the carbon-rich deep water. This biological reduction of CO_2 in the surface waters upsets the equilibrium of carbon dioxide with air and surface water and allows more CO_2 to be removed from the atmosphere. The land reservoir of CO_2, most of which is in forest vegetation, soil, and litter, holds slowly exchanging short-term and long-term pools. Burning of forests and land clearing has increased input of CO_2 from the land; but a long-term equilibrium appears to exist between inputs to the land pool by photosynthesis or uptake by vegetation and outputs to the atmosphere by plant, animal, and soil respiration, and burning, with short-term imbalances.

Nitrogen is fixed both in the atmosphere and by bacteria found in symbiosis with plants, mostly legumes, and blue-green algae. Involved in the nitrogen cycle are the processes of ammonification, nitrification, and denitrification.

Sedimentary cycles involve two phases, salt solution and rock. Minerals become available through the weathering of Earth's crust, enter the water cycle as salt solutions, take diverse pathways through the ecosystem, and return to the sea or Earth's crust through sedimentation. The phosphorus cycle is wholly sedimentary, with reserves coming largely from phosphate rock. Mined phosphate either becomes immobilized in the soil or is carried back to the sea. Sulfur enters both gaseous and sedimentary phases. Major sources are the weathering of rocks, erosional runoff, and decomposition of organic matter. A significant portion of the sulfur released to the atmosphere is a by-product of the burning of fossil fuel. The atmosphere is the major mechanism of circulation of sulfur to land and sea.

Important roles in mineral cycling are played at one end by green plants, which take up nutrients, and at the other end by decomposers, which release nutrients for reuse, as well as by air and water, in which return trips are made. In ecosystems more or less in equilibrium, nutrient inputs equal nutrient outputs. Internal cycles through which nutrients move from soil to plant and back to soil assume great importance in the functioning of ecosystems.

Review and Study Questions

1. Explain the differences between micronutrients and macronutrients and their importance to organisms.
2. Describe the two types of biogeochemical cycles.
3. Describe the water cycle. Compare it over land and ocean.
4. What is the importance of the water cycle?
5. What are the features of the oxygen cycle?
6. What are the reservoirs in the global carbon cycle?
7. What is the impact of burning fossil fuels?
8. What is the role of the ocean in carbon cycling?
9. What biological and nonbiological mechanisms fix nitrogen?
10. What is the source of phosphorus? How does it circulate in aquatic ecosystems?
11. What are the sources and sinks of sulfur? Why is it both sedimentary and gaseous?
12. What influences nutrient utilization in the ecosystem?
13. What are the major external and internal inputs of nutrients in ecosystems?
14. What mechanisms conserve nutrients in an ecosystem?

Selected References

Alexander, M., ed. 1980. *Biological nitrogen fixation*. New York: Plenum. Ecology and physiology of nitrogen-fixing organisms.

Bolin, B., E. T. Degens, S. Kempe, and P. Ketner, eds. 1979. *The global carbon cycle*. SCOPE 13. New York: John Wiley. A comprehensive study of the carbon cycle.

Bormann, F. H., and G. E. Likens. 1979. *Pattern and process in a forested ecosystem*. New York: Springer-Verlag. A good discussion of nutrient cycling in the forest.

Hasler, A. D., ed. 1975. *Coupling of land water ecosystems*. New York: Springer-Verlag. Interrelations of nutrient cycles between land and freshwater ecosystems.

Pomeroy, L. R. 1974. *Cycles of essential elements*. Benchmark Papers in Ecology. Stroudsburg, PA: Dowden, Hutchinson, and Ross. Collection of important papers on mineral cycling.

Post, W. M., T. H. Peng, W. R. Emanuel, A. W. King, V. H. Dale, and D. L. DeAngelis. 1990. The global carbon cycle. *American Scientist* 78:310–326. An excellent updated review of the global carbon cycle—what we do and do not know about it.

Sprent, J. I. 1988. *The ecology of the nitrogen cycle*. New York: Cambridge University Press. Various processes and magnitudes of the nitrogen cycle.

CHAPTER 25

Human Intrusions upon Ecological Cycles

Outline

Objectives

On completion of this chapter, you should be able to:

1. Recognize human intervention in the water cycle.
2. Describe the role of nitrogen and sulfur as environmental pollutants.
3. Explain the role of ozone in the biosphere.
4. Describe the effects of acid precipitation on ecosystems.
5. Discuss the effects of excess deposition of phosphorus and heavy metals on ecosystems.
6. Describe how chlorinated hydrocarbons cycle through and impact ecosystems.

Humans have intruded on ecological cycles to such a degree that we cannot understand any cycle without taking human inputs into consideration. The discovery of fire and its use in modifying the environment marked the beginning of human intrusions. From then on these intrusions have increased exponentially since the Industrial Revolution. They range from modifying the water cycle by draining swamps, straightening streams, and polluting rivers to changing the atmosphere by pouring toxic substances from industrial processes into it and by burning fossil fuels.

Earth, too, has contributed a share of air pollution through volcanic eruptions with their sulfuric gases and particulate matter, lightning-set fires, and methane from large herds of grazing mammals and anaerobic decomposition in swamps and marshes. Earth's contributions, however, make up the background level of contaminations, amounts that the biosphere can handle. When humans added to that natural amount their own continuous inputs, air and water became truly polluted.

Pollution develops when gases, liquids, and particulate matter reside in the atmosphere or in water in quantities greater than normal amounts. (It may be difficult to determine what is normal in an already polluted world environment.) Particulate, gaseous, and liquid substances injected directly into the biosphere in large amounts are primary pollutants. Through photolytic and other reactions, primary pollutants give rise to more damaging secondary pollutants.

WATER

Globally, the water cycle remains fairly constant. Evaporation balances precipitation and rivers and streams carry runoff from land to sea. Humans interject themselves into the water cycle by withdrawing water for their own use from rivers, streams, lakes, groundwater, and aquifers, and by building dams and other catchments. Nevertheless, some 90 percent of this water withdrawn eventually finds its way back into the water cycle. The problem is the inequitable global and seasonal distribution of precipitation and storage. Because of the demand for water for home use in heavily populated regions and for irrigation in semiarid agricultural regions, available sources of water cannot meet the demand.

In the early days of soil and water conservation, the objective was to hold the raindrops where they fell. When rain and snow soaked into a receptive soil, water was held in place for plant growth, erosion-causing surface flow was checked, the groundwater table was recharged, and downstream flooding was reduced. Today, half a century later, we straighten and deepen stream channels to eliminate meanders and to hurry the water off from the uplands to the rivers.

Such a change reflects differences in land-use patterns. Forests, grasslands, and croplands are converted to roads, shopping malls, commercial centers, industrial sites, parking lots, airports, and urban and suburban developments. These areas intercept up to 100 percent of the rainfall, and lose it as surface flow channeled to storm drains and as evaporation. Residential lots, depending on their size, may intercept up to 80 percent of precipitation, and compacted by mowing and use, lawns and golf courses allow only a minimal amount of the water to infiltrate the soil. A suburban lawn may have an infiltration rate of only 0.021 cc per minute, compared to 1.27 cc/min for an undisturbed forest soil.

Major sources of water are unconfined and confined aquifers. Unconfined aquifers or groundwater storage areas are located in porous water-bearing layers of underground rocks, the upper level of which is known as the water table. Water percolating through the soil recharges the groundwater at a rate that can be high during rainy periods. However, much of the precipitation that normally would find its way to groundwater storage is intercepted by paved areas and rushed off the land. During periods of heavy rainfall this intercepted water intensifies downstream flooding.

Confined aquifers are imprisoned between layers of impenetrable rock. Often the water is under pressure and flows to the surface through cracks as artesian wells. Confined aquifers may be quite extensive. One of the largest is the Ogallala aquifer that extends under the farm belt in the United States from northern Nebraska to northwestern Texas. Confined aquifers recharge slowly, and the point of recharge may be hundreds of miles away.

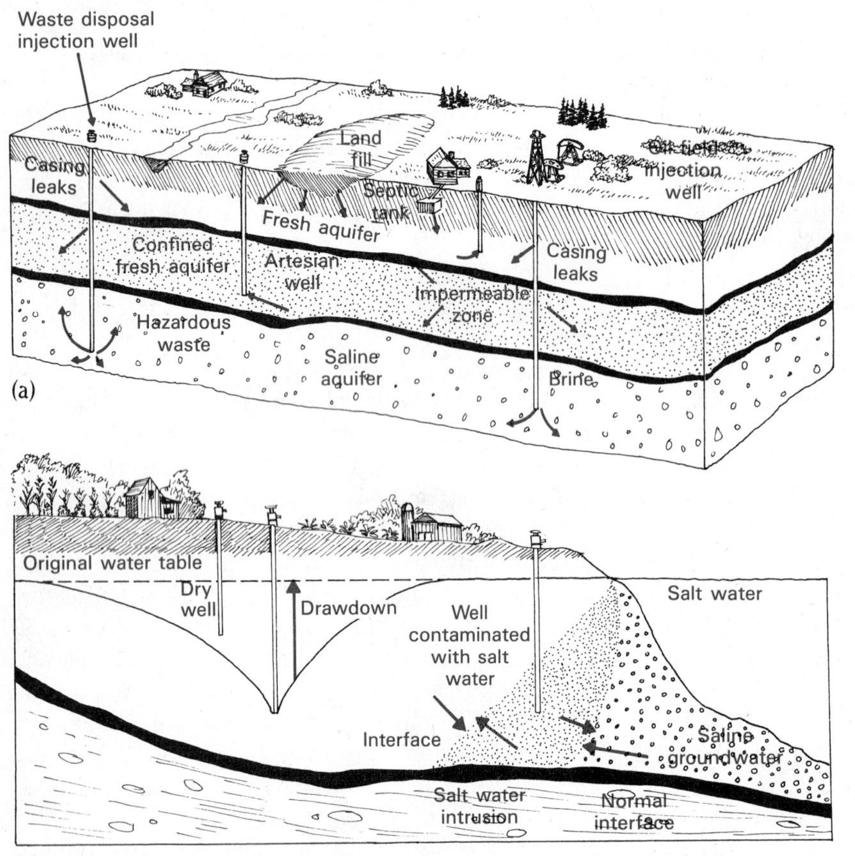

FIGURE 25.1 (a) Sources of groundwater pollution. (b) Effect of drawdown on aquifers, including cone of depression and salt-water intrusion.

In arid and semiarid regions, heavily populated areas and agricultural operations literally mine the aquifers and groundwater (Figure 25.1). When drawdown of groundwater exceeds recharge, the ground surface subsides, as has occurred in the southern United States and in the San Joaquin Valley of California, damaging houses, highways, and other structures. Along the Atlantic and Gulf Coast areas of the United States, and the coastal areas of Israel, Syria, and the Arabian Gulf States, depletion of both groundwater and aquifers breaks down the normal interface between salt water and fresh and allows the inland movement or intrusion of salt water. Excessive dependence on confined aquifers for urban and agricultural use draws down the aquifer, limiting its life. In the western United States one-fourth of the water withdrawn from aquifers, mostly for irrigation, is not replenished. Rapidly growing Tucson, Arizona, pumps five times as much water out of its aquifers as is replenished, and nearly 50 percent of that goes to keep lawns and golf courses green. Fifty percent of water removed for irrigation is lost to seepage and evaporation.

Use of aquifers is further restricted by groundwater pollution. Leachates from landfills, toxic waste sites, septic systems, feedlots, petroleum storage areas, pesticides, and agricultural chemicals find their way into aquifers, poisoning the water against further use. This problem is common and widespread, especially in the northeastern United States. Further polluting aquifers are toxic industrial wastes and brines from oil and gas pumped into the ground through deep injection wells (Figure 25.1) for disposal. Such pollutants work their way into aquifers where they are not easily diluted or dispersed.

To meet water demands, we further interfere with the normal flow of surface water through streams and rivers by building dams to store water. These dams reduce the downstream flow of water; and when we construct a series of dams the length of the river, we so reduce the downstream flow that little water is left in the lower reaches, adversely affecting the structure and function of the river ecosystem (see Chapter 33). Such is the situation on the Colorado River, where water use and allocations are tied up in regional competition and bureaucracy. Withdrawal of water from natural lakes for irrigation and urban uses in arid and semiarid regions causes major ecological problems. For example, at Mono Lake in California and the Aral Sea in Russia, drawdown has exposed great expanses of shoreline and disrupted the ecological integrity of the lake ecosystem.

Not only do we draw down lakes and reservoirs to meet water demand, we paradoxically also drain natural water storage areas, the wetlands. Wetlands help reduce floods by holding water and improve water quality by diluting and degrading toxic wastes. By holding water, wetlands increase infiltration into the ground and help recharge aquifers. In spite of their value, more than one-half of original coastal and inland wetlands of the United States have been drained, most for agriculture and the rest for development. In Iowa, for example, 95 percent of wetlands have been drained, and California has lost 90 percent of its wetlands.

SULFUR

Of all the atmospheric gases, sulfur dioxide (SO_2) is most strongly implicated in air pollution. Natural sources (see Chapter 24) include microbial activity, volcanoes, sea spray, and weathering processes; they produce about 60 percent of sulfur emissions to the atmosphere. Humans make the other 40 percent. Of the human-made emissions, 68 percent comes from the burning of fossil fuels, 40 percent from the burning of coal. Natural emissions are more widely distributed about the globe; human-made emissions are more concentrated regionally. Ninety percent of those inputs come from the urban and industrialized areas of Europe, North America, India, and the Far East. In 1990 alone, Europe and North America poured between 110 and 125 million metric tons of sulfur into the atmosphere. In these regions industrial input far exceeds natural input. Because of the imposition of emission controls, fluidized bed combustion, and changes in patterns of fuel consumption, sulfur dioxide emissions have been declining, but if coal consumption increases as predicted, sulfur dioxide levels could rise higher than ever.

Sulfur dioxide produces acute toxicity and major damage to vegetation in areas surrounding point sources. Examples are the destruction of forest cover in the vicinity of iron smelters at Sudbury and Wana, Ontario, and the copper smelters at Duckberry and Copperhill, Tennessee. Damage away from point sources is more longterm and insidious. The construction of tall stacks at power plants and industrial complexes has greatly reduced ground-level concentrations and point-source damage, but they eject into the atmosphere gases that are carried by upper-level winds to points far removed from the source. On the way some of the gases and particulate matter drop out as dry deposition relatively close to the source. In the moving plume, sulfur dioxide forms sulfur trioxide, which combines with atmospheric moisture to form sulfuric acid. Falling on land and water, this wet deposition is a significant part of acid rain:

$$SO_2 + O + M \rightarrow SO_3 + M$$
$$SO_3 + H_2O \rightarrow H_2SO_4$$

The concentration of SO_2 in the atmosphere is further increased by the removal of alkaline particulate matter, termed fly ash, from the stacks by scrubbers to eliminate local particulate fallout.

Sulfur dioxide in the atmosphere affects both humans and plants. In a few parts per million, it irritates the respiratory tract. In a fine mist or absorbed onto small particles, sulfur dioxide moves into the lungs and attacks sensitive tissues. High concentrations (over 1000 micromilligrams/m^3) have caused a number of air pollution disasters characterized by higher than expected death rates and increased incidents of bronchial asthma. Among such disasters are the Meuse Valley in

Belgium in 1930, Donora, Pennsylvania, in 1938, London in 1952, and New York and Toyko in the 1960s.

Sulfur dioxide injures exposed plants or kills them outright. Acidic aerosols present during periods of fog, light rain, and high relative humidity and moderate temperatures do most of the injury. The aerosols are absorbed by external surfaces of leaves. When dry, leaves and needles take up sulfur dioxide through the stomata. In the leaf it is rapidly oxidized to sulfates, which react on the surface of mesophyll cells as sulfuric acid. Symptoms of sulfur damage are a bleached look to deciduous leaves and red-brown needles on conifers, partial defoliation, and reduced growth.

NITROGEN

Although sulfur is the major atmospheric pollutant and the progenitor of acid rain, nitrogen in its various forms owns its share of the blame. Nitrogen is ambivalent. It makes up 78 percent of Earth's atmosphere, it is one of the most necessary nutrients for life, and it increases the fertility of soil and water. Simultaneously it acts as a photochemical oxidant, it is an important catalyst for the formation of ozone, it contributes to acid precipitation, and it induces public health risks.

When nitrogen became a pollutant, human intrusion into the natural cycle and balance of nitrogen was responsible. Atmospheric pollution by nitrogen involves various oxides of nitrogen; pollution of aquatic ecosystems involves nitrates leached from the soil. Natural leaching of nitrates and natural input of nitrogen oxides into the atmosphere have always occurred. Through their involvement in nitrification and denitrification processes, microorganisms in marine, freshwater aquatic, and terrestrial ecosystems release nitrous oxides to the atmosphere. Tropical forests and woodlands alone release three-fourths of the natural global flux of nitrous oxides (N_2O). However, nitrous oxides for the past 20 to 30 years have been increasing at the rate of 0.2 to 0.3 percent a year. Most of this increase comes from human sources.

Major human sources of nitrogen pollution are agriculture, industry, and automobiles. The first major intrusion probably came from agriculture, with its associated burning of forests and clearing of land for crops and pasture. Conversion of natural grasslands into grain fields has caused a steady decline in the nitrogen content of their soils. Breaking up and mixing the soil increase the rate of decomposition of deep organic manner, releasing oxides and nitrates of nitrogen. Removal of harvested crops and logging result in a heavy outflow of nitrogen from agricultural and forest ecosystems, not only in the material removed but in nitrate losses from the soil. Heavy application of artificial fertilizers to croplands disturbs the natural balance between denitrification and nitrogen fixation. A considerable portion of nitrogen fertilizers is lost as nitrates to groundwater, which eventually finds its way into aquatic ecosystems, reducing water quality and upsetting the natural balance of aquatic life. Excess nitrogen not leached out as nitrates is removed by microbial denitrification, increasing the atmospheric levels of nitrous oxide. About 10 percent of applied nitrogen fertilizers in time evaporates into the atmosphere. Added to these inputs are nitrogenous inputs from animal wastes at concentrated livestock feeding yards, from municipal sewage treatment plants, and from chemical fertilizer plants.

Automobile exhaust and industrial high temperature combustion add nitrous (N_2O) and nitric (NO) oxides, and nitrogen dioxide (NO_2). These oxides, relatively unreactive with a residence life of about 20 years, drift slowly up to the stratosphere. There ultraviolet light through a series of reactions reduces nitrous oxides to nitric oxide and atomic oxygen (O). Atomic oxygen reacts with molecular oxygen to form ozone (O_3). Ozone reacts with nitric oxides to form nitrogen and oxygen:

$$NO_2 + hv \rightarrow NO + O$$
$$O + O_2 + M \rightarrow O_3 + M$$
$$NO + O_3 \rightarrow N + 2O_2$$

where h is Planck's constant (cal/sec), v is the frequency of light, and M is any third-body molecule that aids in the reaction but is not changed by it.

Hydroxyl radicals (OH), photochemically generated, react with nitrogen dioxide to form nitric acid, which is carried to soil and water as a component of acid rain:

$$OH + NO_2 + M \rightarrow HONO_2 + M$$

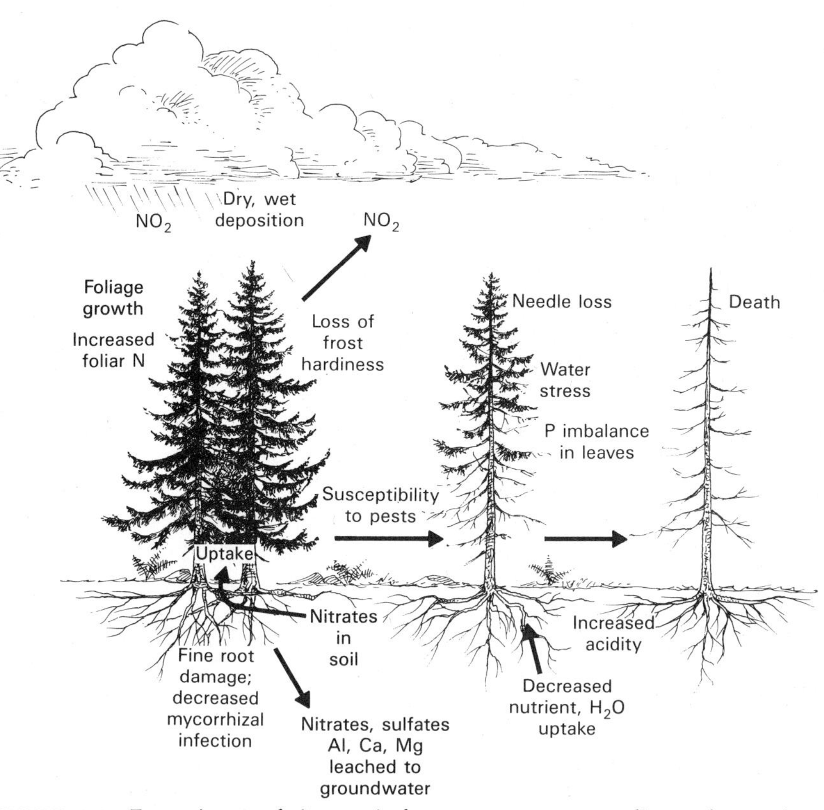

FIGURE 25.2 Excess inputs of nitrogen in forest ecosystems, according to the nutrient limitation hypothesis, affect many processes that result in the decline and death of forest trees and converts the ecosystem from a nitrogen sink to a nitrogen source.

Atomic oxygen from photochemical reduction of nitrogen oxides also reacts with a number of reactive hydrocarbons called radicals. These radicals take part in a series of reactions to form more radicals, which combine with oxygen, hydrocarbons, and NO_2. As a result nitrogen dioxide is regenerated, nitric oxide disappears, and ozone accumulates, along with a number of secondary pollutants, to produce photochemical smog.

One outcome of atmospheric pollution is an increased deposition of nitrogen, which initially benefits ecosystems that are traditionally nitrogen-limited, especially northern and high altitude forests. Such ecosystems, however, can suffer from too much of a good thing. Because nitrogen is limiting, these forests are efficient at retaining and recycling nitrogen from precipitation and organic matter. Only a few lose any significant amount of nitrates to streams. Now, many of these forests are receiving more nitrogen in the form of ammonium and nitrates than the trees and their associated microbial populations can handle and accumulate.

Evidence suggests that increased levels of nitrogen contribute to the decline and dieback of coniferous forests at high elevations (Figure 25.2). The first response to increased availability of nitrogen in a nitrogen-limited ecosystem is increased growth of both canopy and stems. If the increased growth in foliage continues into the summer, the late new growth may not have time to become frost-hardened and is killed during the winter. Overstimulated by nitrogen, tree growth exceeds the availability of other necessary nutrients in the soil, particularly phosphorus, and the tree begins to experience nutrient deficiencies. Experimental evidence suggests that the production of fine roots and ectomycorrhizae, which take up nutrients from the soil, is lower on sites rich in nutrients, especially nitrogen, than in nutrient-poor soils, and root turnover is higher. Trees on nutrient-poor soils have a longer-lived and higher-density root system and a lower turnover of root biomass, a condition that helps them to scavenge nutrients from poor soils. Thus as nitrogen levels increase, root biomass decreases, further inhibiting the uptake of nutrients other than nitrogen and impairing the ability of trees to pull water from the soil during periods of drought.

As nitrogen deposition increases, ammonium levels in the soil increase. Excess ammonium in the soil stimulates nitrification, denitrification, and mobility of nitrates in the soil, even at a low pH. As excess nitrates are leached from the soil, they add to increased anion movement through the soil, particularly aluminum, as has been observed in the spruce stands of the Great Smoky Mountains. This aluminum finds its ways to aquatic ecosystems, increasing their acidification. Increased nitrification and denitrification can also increase the emission of nitrous oxides to the atmosphere, adding to the greenhouse effect. Eventually nitrate losses approach nitrogen deposition and the ecosystem is no longer a nitrogen sink but a nitrogen source (Figure 25.2).

OZONE

Ozone is another of our ambivalent atmosphere gases. In the stratosphere, 10 to 40 kilometers above Earth, it shields the planet from biologically harmful ultraviolet radiation. Close to the ground, ozone is a damaging pollutant, cutting visibility, irritating eyes and respiratory systems, and killing and injuring plant life. In the stratosphere, ozone is diminished by its reaction with human-caused pollutants. In the troposphere, ozone is born from the union of nitrogen oxides with oxygen in the presence of sunlight.

A cyclic photolytic reaction maintains ozone in the stratosphere. Its production requires a breakage of the O—O bond in O_2, accomplished by solar radiation. Once free, the O atoms rapidly combine with O_2 to form O_3:

$$O_2 + h\nu \rightarrow O + O$$
$$O + O_2 + M \rightarrow O_3 + M$$

At the same time a reverse reaction consumes O_3:

$$O_3 + h\nu \rightarrow O + O_2$$
$$O + O_3 \rightarrow 2O_2$$

Under natural conditions in the stratosphere a balance exists between the rate of ozone formation and destruction. In recent times, however, a number of human-caused and some biologically derived catalysts injected into the stratosphere are reactive enough to reduce stratospheric ozone. Among them are chlorofluorocarbons (CCl_2F_2 and CCl_3F), methane (CH_4), both natural and human-caused, and nitrous oxide from denitrification and synthetic nitrogen fertilizer. Of particular concern is chlorine monoxide (ClO) derived from chlorofluorocarbons in aerosol spray propellants (banned in the United States), refrigerants, solvents, and other sources. Its chlorine can be tied up in harmless chlorine nitrates ($ClNO_3$) or it can be involved in the catalytic destruction of ozone:

$$Cl + O_3 \rightarrow ClO + O_2$$
$$O_3 + h\nu \rightarrow O + O_2$$
$$ClO + O \rightarrow Cl + O_2$$

In 1985 atmospheric scientists discovered a pronounced springtime thinning in the ozone layer over Antarctica. The question is whether the thinning results from chemical destruction by chlorofluorocarbons or from a natural effect of upper-level winds. Detection of chlorine monoxide in the hole hints that ClO is the problem. Models of the behavior of chlorofluorocarbon in the stratosphere suggest that the total ozone could be diminished as much as 16 percent in the lower stratosphere, where aerosols recycle the chlorine from inactive reservoirs. Reduction in the ozone layer can have adverse ecological effects on Earth, altering DNA and increasing skin cancer. In addition, changes in the ozone layer could increase temperature of the lower atmosphere, change air circulation patterns, and contribute to the greenhouse effect.

Down in the trophosphere, ozone and other photochemical oxidants including peroxacetyl nitrate (PAN), peroxyproprionyl nitrate (PPN), and hydrogen peroxide are created when nitrogen oxides and volatile hydrocarbons react with O_2 in the presence of sunlight. Automobile exhausts and industrial combustion add nitrogen dioxide (NO_2) to the atmosphere. There it is reduced by ultraviolet light to nitrogen monoxide (NO) and atomic oxygen (O) (see Nitrogen).

In the presence of sunlight, atomic oxygen from the photochemical reduction of NO_2 also reacts with a number of reactive hydrocarbons to form reactive intermediates called radicals. These radicals then enter a series of reactions to form still more radicals, which combine with oxygen, hydrocarbons, and NO_2. The outcome is a regeneration of nitrogen dioxide, elimination of nitrous oxide, accumulation of ozone, and the formation of a number of secondary pollutants, such as formaldehydes, aldehydes, and peroxyetyl nitrates, known as PAN. All of these substances collectively form photochemical smog.

Ozone, relatively insoluble in water, readily diffuses through stomatal cavities to the mesophyll cells, where it reacts rapidly on their surfaces. The degree of reaction varies among both crop plants and forest trees. Highly sensitive tobacco becomes flecked with white lesions; bean leaves show stippling and bleached areas; leaves of woody plants have reddish-brown lesions. Plants especially sensitive to ozone include white and ponderosa pines, red spruce, alfalfa, oats, spinach, and tomato. Foliar damage reduces photosynthetic capacity of plants, annual radial growth in trees, and nutrient retention in the foliage, predisposes trees to insect and fungal infections, and is associated with forest decline.

Other oxidants, especially PAN, are extremely toxic. By destroying some of the lower epidermal cells of leaves and by damaging chloroplasts, they interfere with the plants' metabolic processes and reduce growth. Damage to crop plants in the United States alone is an estimated $2 billion a year.

ACID DEPOSITION

Acid rain has been with us for years. In 1872 when the Industrial Revolution was in full swing, Robert Angus Smith introduced the term to describe the acidic rainfall around Manchester, England. In some areas acid rain was probably worse than now, but it was more localized. Only when we constructed large stacks in industrial and power plants, sending the emissions much higher into atmospheric circulation, did acid rain become a regional problem. The industrial midwestern United States and Ohio River Valley send acid pollutants to eastern Canada and the northeastern United States, eastern Canada sends acid rain to the northeastern United States, and central Europe and the United Kingdom send acid precipitation to Scandinavia.

Acid rain, which is only one component of atmospheric pollution, involves the mixing of sulfur dioxide, nitrogen oxides, hydrogen chloride, and other compounds with water vapor and oxygen to form dilute solutions of strong acids, notably nitric and sulfuric acid. They come to Earth in acid rain, snow, and fog, known as wet deposition. On the way down precipitation may wash out gaseous and particulate matter and carry them to the ground. Complementing wet deposition is dry deposition, which includes particulate matter and airborne gases (Figure 25.3).

Contrary to what many of us assume, rainwater is naturally acid. Unpolluted rainwater has a pH of 5.6, but rarely is rainwater pure water. Even in regions not subject to industrial pollution, atmospheric moisture is exposed to varying amounts of acids of natural origins, so that precipitation has a pH of about 5. In regions hundreds of kilometers about centers of human activity, however, the pH of precipitation is much lower, 3.5 to 4.5, or occasionally even less.

EFFECTS ON SOIL

Acid precipitation has its greatest impact on soils, especially in those regions whose soils, low in cations, are poorly buffered. Such soils, derived largely from granitic bedrock, are characteristic of the eastern and upper north central United States, southeastern United States, Canada, and northern Europe. Although not yet affected, parts of Asia, Africa, and South America are also acid-sensitive. In all these regions, terrestrial ecosystems are nutrient-poor and their soils acidic.

The effect of acid precipitation on soils varies with the ecosystem receiving it. Over short periods of time acid precipitation may have little effect. Many of the soils on which it falls are already acid and the vegetation is adapted to acidic conditions. In fact, the addition of sulfur and nitrogen to such soils may have a bene-

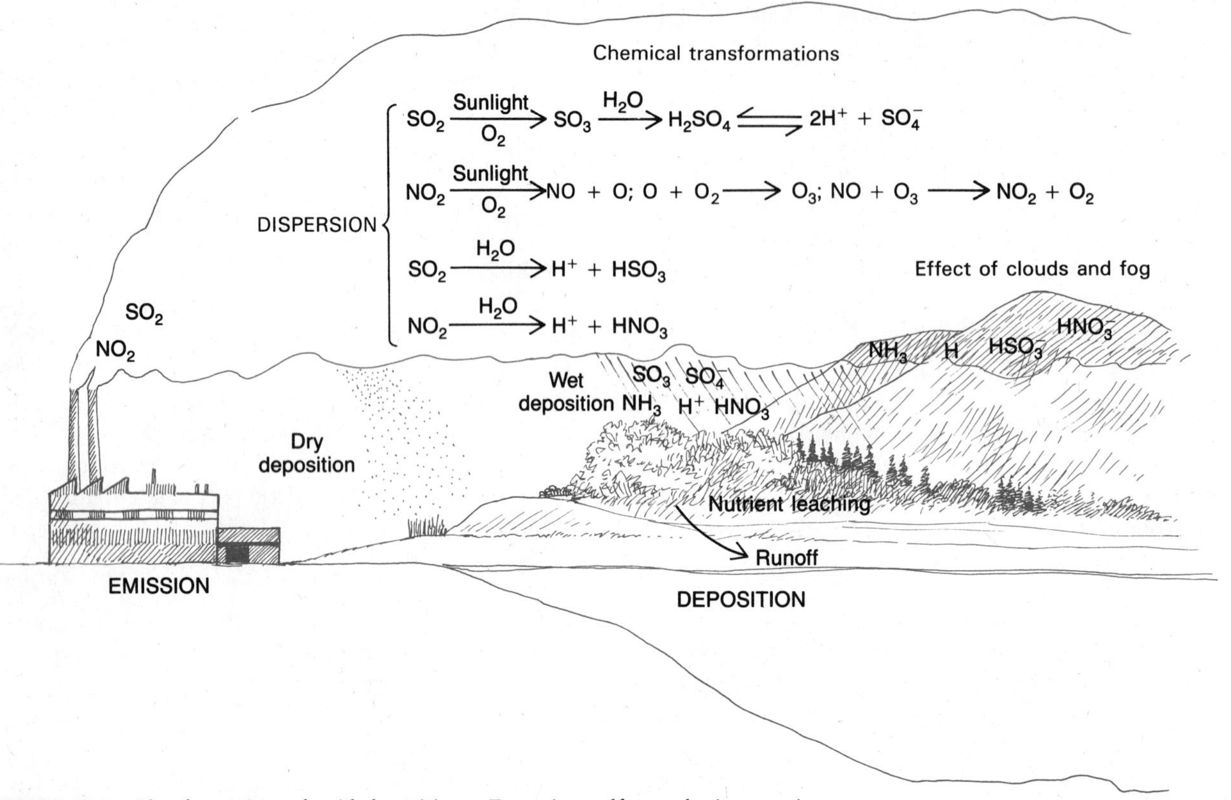

FIGURE 25.3 The formation of acid depositions. Excessive sulfur and nitrogen in several forms are being poured into the atmosphere. They are converted to sulfates, sulfides, and sulfuric and nitric acids and carried to Earth.

ficial fertilizing effect at first. However, over a longer period of time acid precipitation can have adverse effects. It increases the leaching of calcium, magnesium, and potassium from the receiving soil and replaces these cations with hydrogen ions, further increasing soil acidity. Acid precipitation can reduce the solubility and thus availability of phosphorus and the rate of nitrogen fixation. If the rate of leaching outstrips the replacement of these nutrients by weathering, acid precipitation upsets the nutrient balance of trees and other vegetation. Further, acid rainfall can inhibit the activity of fungi and bacteria in the soil, reducing the rate of humus production, mineralization, and the fixation of nutrients. All of these interactions result in nutrient-deficient soils.

The most serious effect, however, is the mobilization of toxic elements in the soil, particularly aluminum and manganese, as they are replaced by hydrogen ions on the soil particles (see Chapter 9). Aluminum affects the structure and function of fine roots and interferes with their uptake of calcium from the soil. It also suppresses the cambial growth in trees, which in turn reduces the formation of new sapwood. As sapwood growth declines, the ratio between living sapwood and dead heartwood declines. When sapwood forms less than 25 percent of the cross section of a tree, increased stress kills it.

EFFECTS ON VEGETATION

Little evidence exists to show that acid rain has a direct effect on most plants. Acid rain, intercepted by vegetation, leaches nutrients, particularly calcium, magnesium, and potassium, from the leaves and needles, but this, a normal process in nutrient cycling, has little effect on the trees' health, provided the trees can replace the nutrients lost by uptake from the soil. However, forests at high elevations are frequently enveloped by mists and fog. Cloud droplets are more acidic and hold higher concentrations of other pollutants than rainfall. When immersed in fog, needle-leafed conifers effectively comb moisture out of the air, and their wet surfaces permit the uptake of the pollutants it contains. As the water evaporates it leaves behind high concentrations of pollutants, some of which may be washed off during the next rain and deposited on the soil. Added to this wet deposition is a dusting of dry deposition that leaches nutrients from the leaves.

Over much of North America and Europe, forests, especially coniferous ones, are declining and dying. In Germany they call it *Waldschaden*, "forest damage" (replacing the older term *Waldsterben*, "forest death"). There 80 percent of the fir, 54 percent of both spruce and pine, and 60 percent of beech and of oak show symptoms of damage—defoliation, thinned crowns, yellowed and discolored foliage, premature leaf fall, and death. Over central Europe, Norway spruce, pine, beech, and oak suffer from chlorosis and defoliation. Fourteen percent of all Swiss forests, 22 percent of Austria's forests, and 29 percent of Holland's forests show symptoms of decline.

Pine forests in southern California northeast of Los Angeles and San Diego also suffer from chlorosis, decrease in radial growth, and death. In the high Appalachians from North Carolina to Vermont, red spruce has been declining in growth and dying since the 1960s. In the Great Smoky Mountains one-half of red spruce and Fraser fir are dead (Figure 25.4). In the southern United States, loblolly and slash pine are experiencing a 30 to 50 percent decline in radial growth and increasing mortality. Sugar maple in the northeastern United States and eastern Canada is experiencing crown dieback and bark peeling on larger branches. In the province of Quebec dieback affects 52 percent of the sugar maples. Throughout northeastern North America white pine shows needle discoloration and decreased growth in height, diameter, and needle length, all symptoms of ozone damage.

FIGURE 25.4 Forest decline at Richland Balsam, a nearly 1800-meter peak on the Blue Ridge Parkway in North Carolina near Waynesville. All of North Carolina's high peaks—Mount Mitchell, Clingman's Dome, Grassy Ridge on Roan Mountain, Grandfather Mountain, and others with dense forests of red spruce and Fraser fir—are experiencing a similar decline, which many credit largely to air pollution. Mount Mitchell, elevation 2,038 meters, has been recording 215 kg per ha (200 lb per acre) of sulfur dioxide deposition, over ten times the amount the ecosystem can accommodate.

Forest decline is not a new phenomenon. During the past 100 to 200 years our forests have experienced several declines with different species affected. What sets the current decline apart from all others is the variety of symptoms in the past and the similarity of symptoms among species today. Past declines could be attributed to natural stresses, such as drought and disease. What causes forest decline and dieback today is not established, but the widespread similarity suggests a common cause, air pollution. All of the affected forests are in the path of pollutants from industrial and urban sources.

Forests close to the point of origin of pollutants experience the most direct effects of air pollution, and their decline and death can be directly attributed to it.

Little evidence exists that acid precipitation alone is the cause of forest decline and death at distant points. The effects of acid deposition, however, in all its aspects can so weaken the trees that they succumb to other stresses such as drought and insect attack. The stressed stands of Fraser fir in the Great Smokies are succumbing to the attacks of the introduced balsam wooly adelgid. The once deep, fragrant stands of Fraser fir, especially on the windward side and peaks of the Great Smoky Mountains, are now stands of skeleton trees (Figure 25.4).

Aside from the economic and aesthetic loss of trees, air pollution and acid rain alter succession by changing the species composition of the affected forests. Just as the chestnut blight shifted dominance in the central hardwoods forest from chestnut to oaks, so air pollution is shifting dominance from pines and other conifers to deciduous trees more tolerant of air pollution.

Crops, too, are adversely affected by air pollution. Unlike the situation with forests, the impact of air pollution on crops, from vegetables to corn, soybeans, and wheat, is well documented. Visible effects include bleaching, yellowing of leaves, flecks, and other discoloration on the leaves. More insidious effects are the reductions in growth and yield. Economic losses range up to $7 billion a year. Researchers estimate that if the ambient level of ozone of 0.05 ppm, alone, were reduced by one-half to 0.025 ppm, we could increase the economic productivity of four crops—soybeans, corn, wheat, and peanuts—by $3 billion a year.

EFFECTS ON AQUATIC ECOSYSTEMS

Acid rain has its most pronounced effects on aquatic ecosystems in acid-sensitive regions. Acid precipitation has already caused acidification of many lakes and streams in the northeastern United States, Canada, Norway, Sweden, and the United Kingdom. In Norway and Sweden, over 69,000 lakes have lost their ability to buffer acidic inputs, and many have lost their fish life. Twenty-five percent of the lakes in the Adirondack Mountains of New York and over 1,600 lakes in the Province of Ontario, Canada, no longer support fish.

Acidification has eliminated many species of mayflies and stoneflies, and other invertebrates, the food of trout, from many streams. It has reduced reproduction of fish or wiped out the crustaceans and fish in affected lakes. Although adult fish may tolerate a pH below 5.5, juvenile fish and crustaceans cannot. As recruitment fails and food declines, fish life disappears.

Acidic inputs to aquatic ecosystems come directly from rainfall and snowmelt, and indirectly from the soils of the surrounding watershed. Acidic leaching of soils increases the nutrient levels of streams and lakes in a watershed, provided the soil has appreciable reserves of calcium. As acid rain percolates through the soil, it is neutralized while releasing basic ions and carrying them to streams and lakes. Such enrichment, however, is often canceled by snowmelt and spring rainwater flowing over the surface of the frozen ground, following old root channels and animal burrows, into receiving waters, discharging much of the winter precipitation in a slug of acidic waters. Such water then can become acid in spite of the buffering effects of the soil.

In the water, sulfate and nitrogen ions replace bicarbonate ions, and solutions once dominated by Ca^{++} and HCO^{-} become dominated by SO^{-}. The pH declines and the concentration of metallic ions increases. When the pH of the groundwater and surface water in the surrounding watershed is 5 or lower, high concentrations of aluminum ions are carried to lakes and streams. Aluminum then tends to precipitate the dark humics, increasing the transparency of the water. Increased light penetration into the water can stimulate phytoplankton production and the growth of benthic algae and bryophytes, but the number of species and biomass of zooplankton decrease.

Although adult fish and some other aquatic organisms can tolerate high acidity by itself, a combination of high acidity and a high level of aluminum, typical during snowmelt, can kill them. This sudden, heavy input can drop the pH levels of aquatic ecosystems quickly and release aluminum ions, which at the level of 0.1 to 0.3 mg/l retards growth and gonadal development of fish and increases their mortality.

Filling depressions and temporary ponds with snowmelt and surface runoff, even in regions outside of those experiencing acidification of lakes and streams, inhibits the reproduction of frogs and salamanders, whose eggs and larvae are sensitive to acidic water. This effect may account in part for the rapid decline of amphibians. Acid waters are toxic to invertebrates also, either killing them directly or interfering with their calcium metabolism, causing crustaceans to lose the ability to recalcify their shells after molting.

Can acidified lakes ever return to normal? Perhaps part way. Some Canadian lakes in the Sudbury region experienced a rapid increase in alkalinity and a decrease in aluminum and other trace metals when control of SO_2 and the closing of some smelters reduced acidic emissions. In response to the improved environment, some rotifer populations and phytoplankton populations recovered and several species of fish returned. Reintroduced trout survived but did not reproduce. Researchers working on the problem believe that the original pH values will not be restored for years because of the depletion of cations in the acidified soil. Unassisted biological recovery probably will never happen. Components of the lake community will have to be restocked. Even with that help, restoration of the original food chains and community structure will be difficult, if not impossible.

EFFECTS ON HUMAN STRUCTURES

Added to the damage to forest and crops and the acidification of aquatic ecosystems is the enormous cost of $75 to $80 billion per year attributed to the corrosion of materials, bridges, highways, buildings, and monuments exposed to the atmosphere. Iron and its alloys are rusted and weakened by chlorides and sulfides, aluminum by chlorides, copper and its alloys by sufides, and masonry and marble by sulfides and other atmospheric aerosols, whose corrosive effects appear to be enhanced by ozone and solar radiation. Our cultural heritage from antiquity, the monuments and buildings of ancient Greece and Rome, historical buildings dating back to the Middle Ages are being weathered away by air pollutants.

PHOSPHORUS

Human activities have altered the phosphorus cycle too. Because the cropping of vegetation depletes the natural supply of phosphorus in the soil, phosphate fertilizers must be added. The source of phosphate fertilizer is phosphate rock. Most of the phosphate applied as fertilizer, however, reacts with calcium, iron, and aluminum in the soil and becomes immobilized as insoluble salts.

Part of the phosphorus used as fertilizer is removed in harvested crops. Transported far from the point of fixation, this phosphorus in vegetables and grain is eventually released as waste when foodstuffs are processed or consumed. Concentration of phosphorus in the wastes of food-processing plants and feedlots adds excessive phosphates to natural waters. Greater quantities are released in urban areas, where phosphates are concentrated in sewage systems. Sewage treatment is only 30 percent effective in removing phosphorus, so 70 percent remains in the effluent and is added to the waterways. In aquatic ecosystems the phosphorus is taken up rapidly by the vegetation, resulting in a great increase in biomass. Unless new input of phosphorus continues, as it would through sewage effluents, the phosphorus is lost to the sediments. Eventually, all the phosphorus mobilized by humans becomes immobilized in the soil or in the bottom sediments of ponds, lakes, and seas.

Phosphorus requirements for agricultural production depend largely on the utilization of natural deposits of phosphate rock and, to a lesser extent, on the harvest of fish and guano deposits. Because so much of the phosphate applied to soil is eventually immobilized in deep sediments and because the activity of phytoplankton seems inadequate to keep phosphorus in circulation, more of the element is being lost to the depths of the sea than is being added to terrestrial and freshwater aquatic ecosystems.

HEAVY METALS

Mercury, cadmium, chromium, selenium, lead—these are some of the heavy metals, toxic to life in varying amounts, that enter biogeochemical cycles. Some, such as mercury and cadmium, are local, associated with industrial pollution, runoff from agricultural fields, toxic dumps, and landfills. Discharged into rivers or lakes or seeping into groundwater, these elements contaminate water supplies or build up in food chains. Although the serious problems they create are local in nature, that is, the materials do not enter global cycles, the effects of these metals are felt much more widely. Heavy metal contaminations associated with toxic waste dumps and discharges into lakes and streams have a severe impact on human health, including birth defects, associated cancers, and liver, kidney, and respiratory diseases. For example, industrial discharge of mercurial wastes in Minamata Bay and the Agano River in Japan contaminated the fish upon which bay residents depended. Consumption of fish resulted in an epidemic of mercurial poisoning with attendant deaths, birth defects, and impaired lives. So much toxic waste has been dumped or washed into the Great Lakes that in some parts fish are too contaminated to be eaten. Such contamination has serious economic effects on sport and commercial fisheries and serious ecological effects on fish-eating birds. Adding to the problem is the shipping of toxic wastes to points distant from their production, often foreign countries, thus spreading locally produced contaminants worldwide.

Some heavy metals, notably lead, join sulfur and nitrogen in atmospheric circulation, moving great distances from the point of origin. Automobiles burning leaded gasoline poured most of the lead into the air, until the use of unleaded fuel was mandated in the United States, Japan, Brazil, and European Common Market countries. Mining, smelting, and refining of lead, lead-consuming industries, coal combustion, burning of refuse and sewage sludge, and the burning and decay of lead-painted surfaces add additional quantities to the atmosphere.

Emitted into the air as small particles, less than 0.5 μm, lead is widely distributed to all parts of Earth. Regions near point sources of pollution, particularly roadsides, receive heavier depositions than remote areas. Urban areas, close to industry and heavy automobile traffic, may have a flux rate greater than 3000 g/ha/yr, whereas remote areas may experience a flux rate of less than 20 g/ha/yr.

Lead particles settle on the surface of soil and on vegetation. Forest canopies are particularly efficient at collecting lead from the atmosphere as dry deposition. Lead accumulates in the canopies during the summer and is carried to the ground by rain as throughfall and stemflow and by leaf fall in autumn. Between 1975 and 1984, the Hubbard Brook experimental forest in New Hampshire, remote from any major sources of lead contamination, experienced an input of 190 g/ha/yr and had an output of only 6 g/ha/yr, leaving the rest behind in the soil. Forests in the White Mountains of New Hampshire, exposed to air masses from industrial and urban areas, accumulated 200 g/ha/yr. Beech forests in Germany experienced a lead flux in precipitation collected beneath the canopy of 365 g/ha/yr, the more closed-canopy spruce forest 756 g/ha/yr, and adjacent open fields, 405 g/ha/yr. Some lead occurs naturally in the soil, but in uncontaminated areas this background lead amounts to only 10 to 20 μg per gram of dry soil.

In the forest soil, lead becomes bound to organic matter in the litter layer and reacts with sulfate, phosphate, and carbonate anions in the soil. In such an insoluble form, lead moves slowly, if at all, into the lower horizons. Its residence time in the upper soil layer is around 5000 years.

Once in the soil and on plants, lead enters the food chain. Plant roots take up lead from the soil and leaves pick it up from contaminated air or from particulate matter on the leaf. Lead is then taken up by herbivorous insects and grazing mammals, who pass it on to higher consumers. This uptake through the food chain is most pronounced along highway roadsides. Microbial systems also pick up lead and immobilize substantial quantities of it.

The long-term increase in concentrations of atmospheric lead in the industrialized areas of Earth has resulted in significant increases of lead in humans. The average body burden of lead among adults and children in the United States is 100 times greater than the natural burden, and existing rates of lead absorption are 30 times the level in preindustrial society. An intake of lead can cause palsy, mental retardation, partial paralysis, loss of hearing, and death.

CHLORINATED HYDROCARBONS

Of all human intrusions into biogeochemical cycles, none has attracted more attention than the chlorinated hydrocarbons, including the application of pesticides, the spraying of dioxin-containing herbicides, and the leakage of the industrial chemical PCB. The substance that first received attention is DDT. Because its behav-

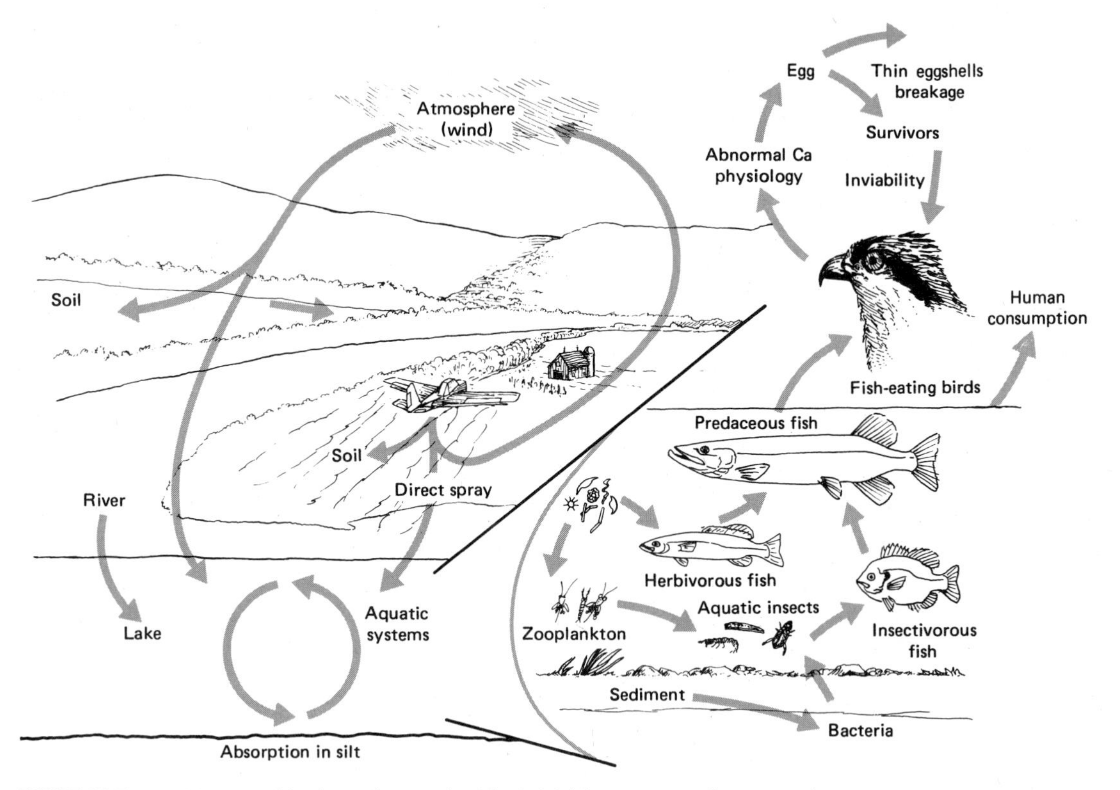

FIGURE 25.5 A chlorinated hydrocarbon cycle. The initial input comes from spraying on croplands and forest. A large portion fails to reach the ground and is carried on water droplets and particulate matter through the atmosphere.

ior in the environment has been so well studied, it serves as example of how pesticides circulate in the ecosystem.

As early as 1946 Clarence Cottam called attention to the damaging effects of DDT on ecosystems and nontarget species. But the impact of pesticides remained obscure until Rachel Carson exposed the dangers of hydrocarbons in her book *Silent Spring*. The detection of DDT in tissues of animals in the Antarctic, far removed from any applied source of the insecticide, emphasizes the fact that chlorinated hydrocarbons do indeed enter the global biogeochemical cycle and that they become dispersed around the Earth.

Chlorinated hydrocarbons have certain characteristics that make this global circulation possible (see Figure 25.5). They are highly soluble in lipids or fats and poorly soluble in water and therefore tend to accumulate in plants and animals. They are persistent and stable but undergo some degradation. DDT, for example, degrades to DDE, and has a half-life of approximately 20 years. It has a vapor pressure high enough to ensure direct losses from plants. It can become adsorbed to particles or remain as a vapor and, in either state, be transported by atmospheric circulation. It can return to land and sea on rainwater.

Insecticides are applied on a large scale by aerial spraying. One-half or more of the toxicant applied in this manner is dispersed to the atmosphere and never reaches the ground. If the vegetative cover is dense, only about 20 percent reaches the ground. For example, in a massive spraying of DDT over 66,000 acres of forest land in eastern Oregon in order to control an outbreak of Douglas-fir tussock moth, only about 26 percent of the intended DDT application reached the forest floor. In fact, as little as 0.02 percent of an aerially applied pesticide may actually hit target organisms.

On the ground or on the water's surface, the pesticide is subject to further dispersion. Apparently, little DDT moves from the surface soil to the subsoil. Pesticides reaching the soil are lost through volatilization, chemical degradation, bacterial decomposition, runoff, and the harvest of organic matter, which can amount to around 1 percent of the total DDT used on the crop.

DDT sprayed on forest and cropland enters streams and lakes, where it is subject to further distribution and dilution as it moves downstream. Insecticides released in oil solutions penetrate to the bottom and cause mortality of fish and aquatic invertebrates. Trapped in the bottom rubble and mud, the insecticide may continue to recirculate locally and kill for some days. In lakes and ponds, emulsifiable forms of DDT tend to disperse through the water, but not necessarily in a uniform way. DDT in oil solutions tends to float on the surface and is moved about by the wind. Eventually, pesticides reach the ocean, where they may concentrate in surface slicks. These slicks, which attract plankton, are carried across the seas by ocean currents. In the ocean, part of the DDT residue may circulate in the mixed layer. Some of it may be transferred to deep waters, and more may be lost through the sedimentation of organic matter.

Although considerable amounts of DDT and other pesticides are transported by water, the major movement of pesticide residue takes place in the atmosphere. Not only does the atmosphere receive the bulk of the pesticidal sprays, but it also picks up the fraction volatized from soil, water, and vegetation. If DDT remained as a vapor alone, the saturation capacity of the atmosphere would hold as much DDT as has been produced to date. But the capacity of the atmosphere to hold DDT is increased greatly by the adsorption of residues to particulate matter. Thus, the atmosphere becomes a large circulating reservoir of DDT and other chlorinated hydrocarbons. Residues are removed from the atmosphere by chemical degradation, diffusion across the air-sea interface, and (mostly) by rainfall and dry fallout.

Although the quantity of residues of DDT and other chlorinated hydrocarbons may be relatively small, amounting to one-thirtieth or less of the amount produced each year, the concentrations are still sufficient to have a deleterious impact on marine, terrestrial, and freshwater ecosystems. DDT and related compounds tend to concentrate in the lipids of living organisms, where they undergo little degradation.

The high solubility of DDT in lipids leads to its concentration through the food chain. While only a portion of the food is ingested by consumer organisms, most of the DDT contained in the food is retained in the fatty tissues of the consumer. Because it breaks down slowly, DDT accumulates to high and even toxic levels. DDT so concentrated is passed on to the next trophic level. The carnivores on the top level of the food chain receive massive amounts of pesticides.

High concentrations of DDT in their tissues often result in death or impaired reproduction and genetic constitution of organisms. A residue level of 5 parts per million in the lipid tissues of the ovaries of freshwater trout causes 100 percent dieoff of the fry, which pick up lethal doses as they utilize the yolk sac. High levels of DDT are correlated with the decline of such fish as sea trout and California mackerel. DDT and its degraded products interfere with calcium metabolism in birds. Chlorinated hydrocarbons block ion transport by inhibiting the enzyme ATPase, which makes available the needed energy. This reduces the transport of ionic calcium across membranes and can cause death. DDT also inhibits the enzyme carbonic anhydrase, which is essential for the deposition of calcium carbonate in the eggshell and the maintenance of a pH gradient across the membrane of the shell gland.

The prohibition of DDT in the United States and most Western nations in the early 1970s has resulted in the slow recovery of such raptors as bald eagles and ospreys and other fish-eating birds such as pelicans, now that these birds lay normal-shelled eggs again, but the ecological problems of pesticides have not lessened. Chlorinated hydrocarbons, including DDT, are used extensively in other parts of the world, notably Central and South America and Asia. There the problems associated with these pesticides still persist. They affect not only the native fauna, but also migratory birds from the Northern Hemisphere that come in contact with pesticides on their wintering grounds. Quantities of pesticides are sent back north to the United States,

Canada, and Europe on fruits and vegetables grown for the winter market.

More than 500 million kilograms of pesticides are used annually in the United States. Herbicides make up 60 percent of this total, insecticides 24 percent, and fungicides 16 percent. Of these pesticides 341 million kilograms are used on agricultural crops and pastures, 55 million by government and industry, 4 million on forests, and a surprising 55 million in and around urban and suburban homes. The most concentrated use is about the home grounds, not in agricultural fields. The dosage of pesticides about homes is 14 kg/ha compared to 3 kg on agricultural crops. As much as one-third of these household pesticides is never used but thrown into the trash and ultimately into the environment. These excess pesticides, together with losses from croplands and roadsides, expose pests to widespread selective pressures, increasing their resistance to pesticides.

Most studies have emphasized the effects of pesticides on target and nontarget species and environmental contamination. Largely unknown are the effects of pesticides on whole ecosystems: nutrient cycling, energy flow, food chains, genetics of populations, and species diversity. Pesticides that greatly reduce or eliminate a target species will reduce associated species as well, including predatory populations. With their predators reduced, some species held to a level of relative rarity by predation suddenly increase, becoming pests themselves. By destroying populations of certain insects or plants, pesticides and herbicides reduce species diversity. Ecosystems respond by replacing one individual species with another. The reduction of grazing zooplankton in certain aquatic ecosystems by pesticides resulted in an increase in phytoplankton populations. The use of herbicides on aquatic floating and emergent vegetation stimulated an increase in mats of blue-green algae. Reduction or loss of insect populations from pesticides also decreased bird and mammal populations dependent upon them. Spraying of endrin on crop fields can cause widespread death of passerine birds inhabiting them.

Herbicides and pesticides can interfere with the cycling of nutrients. Chemicals toxic to earthworms and other decomposers can decrease their populations and adversely affect decomposition and soil fertility. Pesticides can further influence nutrient cycling by altering the chemical composition of plants. Certain pesticides can increase the amount of nutrients such as N, P, Ca, and Mg in plants. These changes can influence insects feeding on them. For example, the application of the herbicide 2-4D to a cornfield increased the nitrogen content of the corn. Its improved nutritional quality resulted in a threefold increase in corn leaf aphid populations. Corn borer females were one-third larger than average and laid one-third more eggs. Thus the application of one type of pesticide can undo the effects of another.

Further, we must consider another ecological effect—that on humans. Humans are widely exposed to pesticides from fresh fruits and vegetables to use in homes and on lawns and gardens. Those who work directly with pesticides are particularly at risk. Annually in the United States some 45,000 humans are poisoned by pesticides. About 3,000 are admitted to hospitals and 200 die.

Summary

Humans have so intruded on ecological cycles that even in areas remote from human activity, cycles have been affected. Exploding populations and land-use changes have affected surface flows by paving enormous areas of land and compacting the soils of others, preventing the infiltration of water into the ground and the recharge of water tables. To rid the land of surface water, we have diverted those flows to rivers whose courses have been deepened and channelized. We have both polluted and overdrawn groundwater tables and aquifers, drained wetlands, the natural recharge points, and dammed rivers to divert their waters to human use.

The atmosphere has become a sink for excess sulfur dioxide, nitrogen oxides, and other gases and particulate matter. Normally in the atmosphere from natural sources, sulfur dioxide and nitrogen oxides, poured out from human sources, have become pollutants, material found in the biosphere in excessive and destructive amounts. When these primary pollutants interact with water and are involved in photochemical reactions, they transform into secondary pollutants.

Sulfur dioxide, emitted both by natural and human-related sources, especially the burning of fossil fuels, enters the atmosphere mostly as hydrogen sulfide. Hydrogen sulfide quickly oxidizes to sulfur dioxide, SO_2. Sulfur dioxide reacts with moisture in the atmosphere to form sulfuric acid, which is carried to Earth in precipitation. Sulfur dioxide is a major atmospheric pollutant, damaging and even killing plants, causing respiratory afflictions in humans and animals, and adding to acid deposition.

Humans also pour nitrogen oxides into the atmosphere and nitrates into aquatic systems. The major sources of nitrogen dioxide are automobiles and burning of fossil fuels. Nitrogen dioxide is reduced by ultraviolet light to nitrogen monoxide and oxygen. These substances react with hydrocarbons in the atmosphere to produce a number of pollutants, including ozone and PAN, which make up photochemical smog, a pollutant harmful to plants and animals. Excessive quantities of nitrates are added to aquatic ecosystems by excessive use of fertilizers on lawns and agricultural crops, by animal wastes, and by sewage effluents. More closely involved with pollution of aquatic systems is phosphorus from sewage effluents.

Nitrogen oxides are implicated in forest decline. According to the nutrient imbalance hypothesis, excessive nitrogen added as wet and dry deposition to forest ecosystems, especially coniferous ones, stimulates rapid growth in trees. This growth, in turn, increases the demand for other nutrients in the soil, especially phosphorus, in excess of their supply. This demand leads to nutrient depletion, which is intensified by an accelerated loss of nutrient cations through the acidification of soil. The end result is decline in growth and vigor of trees and their eventual death.

Ozone, produced by photochemical reactions in the atmosphere, and related to nitrogen oxide pollutions, presents its own set of problems. Its presence in the stratosphere is essential to reduce the influx of harmful ultraviolet radiation to Earth. However, human-produced chlorofluorocarbons rise to the stratosphere and become involved in the catalytic destruction of ozone. In the atmosphere ozone is created from nitrogen oxides by photochemical reactions. Further, such reactions produce related photochemical pollutants, all of which are highly toxic to vegetation.

Acid precipitation develops when sulfur dioxides and nitrogen oxides react with water and hydrogen in the atmosphere to produce sulfuric and nitric acids. These acids reach Earth as wet deposition in the form of rain, snow, and highly acidic mists and fog, and as dry deposition in the form of particulate matter and gases. Acid precipitation is implicated in the acidification of lakes and streams in northeastern North America, central Europe, and Scandinavia, where soils are poorly buffered. This acidification results in the loss of fish and their food sources, crustaceans and insects.

Acid precipitation promotes forest decline by increasing the acidity of poorly buffered soils, nutrient depletion, and aluminum toxicity in the soil, and by inhibiting the activity of soil fungi and bacteria. The effects of acid deposition can so weaken the trees that they succumb to other stresses such as drought and insect attacks.

Air pollution also adversely affects the quality and production of crops, with losses in the billions of dollars. Even greater damage is inflicted on human structures constructed of iron, aluminum, and masonry—roads, bridges, buildings, and monuments.

Of serious consequence globally is the insecticidal use of chlorinated hydrocarbons. These pesticides have contaminated global ecosystems and have entered food chains. Because they become concentrated at higher trophic levels, chlorinated hydrocarbons affect predaceous animals most. Fish-eating birds are endangered because chlorinated hydrocarbons interfere with their reproductive capability.

All of these forms of intrusions in ecological cycles pose serious problems for the future welfare of humans. As populations grow, adding more pollutants to the biosphere and drawing on more resources, especially water, the ability of Earth to sustain any quality of life, human and other, decreases sharply.

Review and Study Questions

1. Distinguish between primary and secondary pollutants.
2. Human intervention in surface runoff, infiltration, and river flow has had what effect?
3. What are confined and unconfined water tables, and how have humans jeopardized their integrity?

*4. The greatest population growth is occurring in the semiarid regions of the world. Discuss the effect of this population growth in relation to the depletion of aquifers. What are the social and political ramifications of water conservation in semiarid regions?
5. What are the major sources of nitrogen pollution?
6. Discuss the possible relationships between excessive nitrogen input to forest ecosystems and forest decline.
7. Show the relationship between nitrogen dioxide pollution and formation of ozone.
8. Explain the paradox of ozone: its beneficial role in the stratosphere and its harmful effects in the lower atmosphere.
9. What are the ecological effects of sulfur pollution?
10. What is acid deposition? What effects does it have on soils and on aquatic ecosystems?
11. If no strong evidence exists that acid precipitation has a direct harmful effect on vegetation, how might it be related indirectly to forest dieback and decline?
12. How are chlorinated hydrocarbons circulated in the biosphere?
13. What is the impact of chlorinated hydrocarbons in ecosystems?
*14. Most chlorinated hydrocarbons are either banned or highly restricted in the United States. However, many other pesticides are widely used, including organophosphates, carbamates, and pyrethroids. Although much less persistent in the ecosystem, many of these are more toxic to birds, mammals, and human beings than chlorinated hydrocarbons. What are some widely used pesticides, and what is their effect on nontarget species?

Selected References

Aber, J. D., K. N. Nadelhoffer, P. Steudler, and J. M. Melillo. 1989. Nitrogen saturation in northern forest ecosystems. *Bioscience* 39:378–386.

Anderson, T. L., ed. 1983. *Water rights: Scarce resource allocation, bureaucracy, and the environment*. San Francisco: Pacific Institute for Public Policy Research. Excellent analysis of politics involved in management of scarce water.

Binkley, D., C. T. Driscoll, H. L. Allen, P. Schoeneberger, and D. McAvoy. 1989. *Acidic deposition and forest soils: Context and case studies in southeastern United States*. New York: Springer-Verlag. Best information on effects of acidic depositions on soils and forests of the southeastern United States.

Dover, M. J., and B. A. Croft. Pesticide resistance and public policy. *Bioscience* 36:78–91. Problems created by pesticide resistance in insects.

Fradkin, P. L. 1981. *A river no more: The Colorado and the West*. New York: Knopf. Politics affecting water resources.

Leopold, L. B. 1974. *Water: A primer*. San Francisco: W. Freeman. Introduction to fundamentals of water resources.

MacKenzie, J. J., and M. T. El-Ashry. 1989. *Air pollution's toll on forests and crops*. New Haven, CT: Yale University Press. Excellent global overview of the problem.

Pimentel, D., and C. A. Edwards. 1982. Pesticides and ecosystems. *Bioscience* 32:595–600. A summary.

Pimentel, D., and L. Levitan. 1986. Pesticides: Amounts applied and amounts reaching pests. *Bioscience* 36:86–91. Revealing review.

Schulze, E.-D., O. L. Lange, and O. Oren. 1989. *Forest decline and air pollution: A study of spruce* (Picea abies) *on acid soils*. New York: Springer-Verlag. Detailed studies of forest decline in Germany, with emphasis on the nutrient decline hypothesis.

Smith, W. H. 1990. *Air pollution and forests: Interaction between air contaminants and forest ecosystems*, 2nd ed. New York: Springer-Verlag. A thorough and indispensable reference source.

Wellburn, A. 1988. *Air pollution and acid rain: The biological impact*. Essex, England: Longman Scientific and Technical. Copublished with Wiley, New York. A basic introduction to the topic.

PART VII
Diversity of Ecosystems

INTRODUCTION

Patterns of Terrestrial Ecosystems

A view from the window of a plane on a transcontinental flight from Boston to California is revealing to an ecology-minded passenger. Below, the pattern of vegetation changes from the mixed coniferous-hardwood forests of the northeast to the oak forests of the central Appalachians with patches of high elevation spruce forests. The forest cover, however fragmented, merges with midwestern croplands of corn, soybean, and wheat, land that once was the domain of tallgrass prairie. Wheat fields yield to high elevation shortgrass plains, and then the plains give way to the coniferous forest of the Rocky Mountains, capped by tundra and snowfields. Beyond the mountains to the southwest lie the tan-colored desert regions.

On a trip of less than eight hours the airborne ecologist can observe a wide range of vegetation that took months and years for early plant geographers and explorers to discover. Botanists, in particular, noted that the world could be divided into great blocks of vegetation—deserts, grasslands, and coniferous, temperate, and tropical forests. They called the divisions *formations.* In time plant geographers attempted to correlate vegetation formations with climatic differences and found that blocks of climate reflected blocks of vegetation with their own life form spectra.

Zoogeographers lagged behind plant geographers in their study of animal distribution. Complicating their studies were the great number of animal species and the lack of a clear-cut relationship between animal distribution and climate. Ultimately zoogeographers did accumulate basic information on the global distribution of animals. Alfred Wallace, also known for having developed the same general theory of evolution as Darwin, provided the major synthesis of animal distribution. Wallace's realms, with some modification, still stand today.

BIOGEOGRAPHICAL REALMS

There are six biogeographical *realms,* each more or less embracing a major continental land mass and separated by oceans, mountain ranges, or desert (see Figure VII.1). They are the Palearctic, the Nearctic, the Neotropical, the Ethiopian, the Oriental, and the Australian. Because some zoogeographers consider the Neotropical and the Australian realms to be so different from the rest of the world, these two are often considered as regions or realms equal to the other four combined. Then there are just three realms: Neogea (Neotropical), Notogea (Australian), and Metagea (the main part of the world). Each region possesses a certain distinction and uniformity in the taxonomic units it contains, and each to a greater or lesser degree shares some of the families of animals with other regions. Except for Australia, each has at some time in Earth's history had some land

FIGURE VII.1 The biogeographical realms of the world. Inset: No definite boundary exists between the Oriental and Australian region, where the islands of the Malay Archipelago stretch toward Australia. Two lines have been proposed to separate the two regions. One, Wallace's line, runs from the Philippines through Borneo and the Celebes. The other, Weber's line, lies east of Wallace's line. It separates the islands with the majority of Oriental animals from those with a majority of Australian animals. Because the islands between these two lines are sort of a transitional zone between the two regions, some zoogeographers call the area Wallacea.

BIOGEOGRAPHICAL OR FAUNAL REGIONS

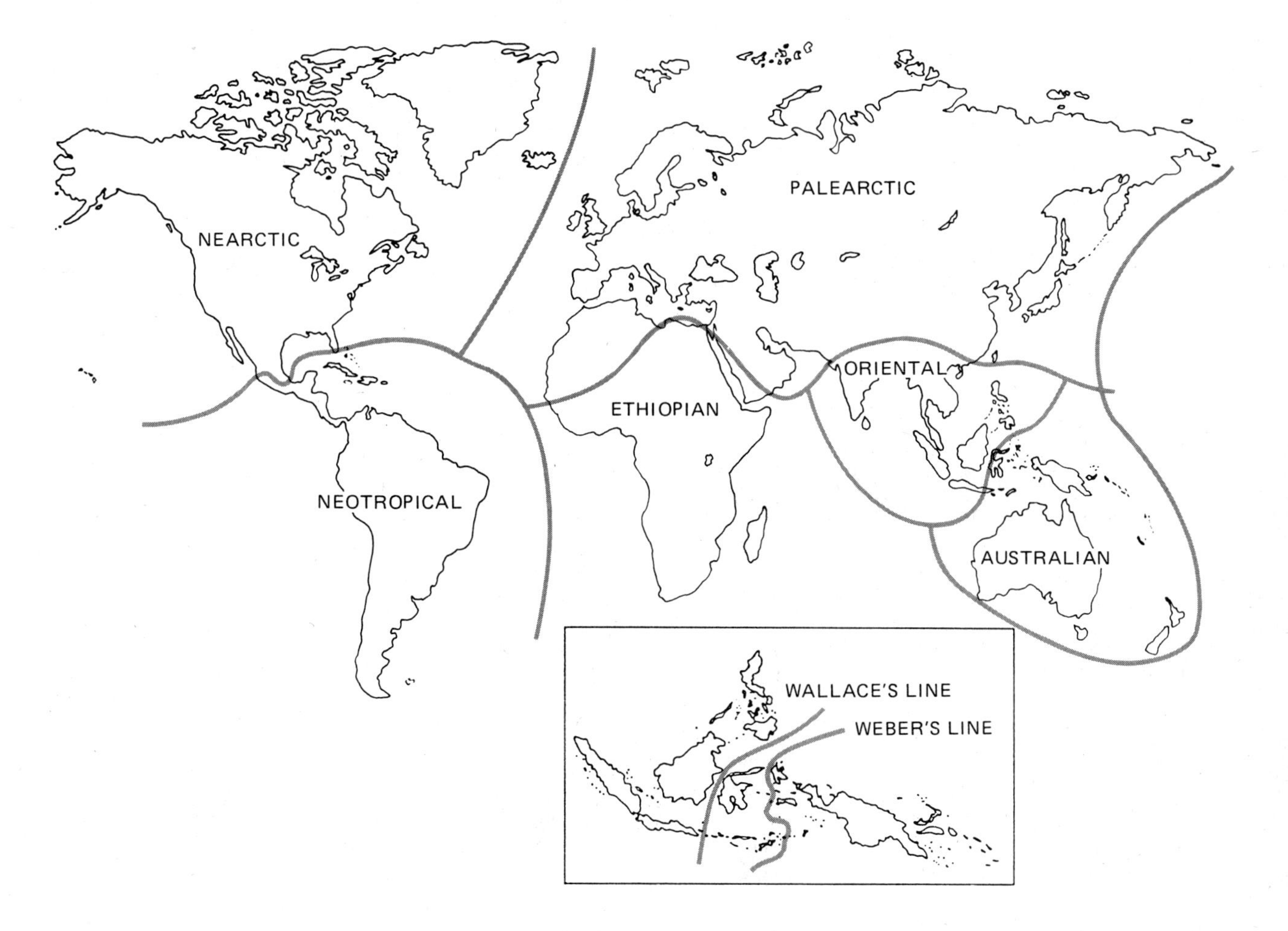

connection with another across which animals and plants could pass.

Two regions, the Palearctic and Nearctic, are quite closely related. In fact, the two are often considered as one, the Holarctic. Both are much alike in their faunal composition, and together they share, particularly in the north, such animals as the wolf, hare, moose (called elk in Europe), stag (called elk in North America), caribou, wolverine, and bison.

Below the coniferous forest belt, the two regions become more distinct. The Palearctic is not rich in vertebrate fauna, of which few are endemic. Palearctic reptiles are few and are usually related to those of the African and Oriental tropics. The Nearctic, in contrast, is the home of many reptiles and has more endemic families of vertebrates. The Nearctic fauna is a complex of New World tropical and Old World temperate families. The Palearctic is a complex of Old World tropical and New World temperate families.

Isolated until 15 million years ago, the fauna of the Neotropical is most distinctive and varied. In fact, about half of the South American mammals, such as the tapir and llama, are descendants of North American invaders, whereas the only South American mammals to survive in North America are the opossum and the porcupine. Lacking in the Neotropical is a well-developed ungulate fauna of the plains, so characteristic of North America and Africa. However, the Neotropical is rich in endemic families of vertebrates. Of the 32 families of mammals, excluding bats, 16 are restricted to the neotropical. In addition, 5 families of bats, including the vampire, are endemic.

The Ethiopian realm embraces tropical forests in central Africa and, in the mountains of east Africa, savanna, grasslands, and desert. During the Miocene and Pliocene, Africa, Arabia, and India shared a moist climate and a continuous land bridge, which allowed the animals to move freely among them. That accounts for some similarity in the fauna between the Ethiopian and Oriental regions. Of all the regions, the Ethiopian contains the most varied vertebrate fauna, and in endemic families it is second only to the Neotropical.

Of the tropical realms, the Oriental, once covered with lush forests, possesses the fewest endemic species and lacks a variety of widespread families. It is rich in primate species, including two families confined to the region, the tree shrews and tarsiers.

Perhaps the most interesting and the strangest region, and certainly the most impoverished in vertebrate species, is the Australian. Partly tropical and partly south temperate, this region is noted for its lack of a land connection with other regions; the poverty of freshwater fish, amphibians, and reptiles; the absence of placental mammals and the dominance of marsupials. Included are the monotremes, with two egg-laying families, the duckbilled platypus and the spiny anteaters. The marsupials have become diverse and have evolved ways of life similar to those of the placental animals of other regions.

LIFE ZONES

By the turn of the century, some biologists were attempting to combine plants and animals into one distributional scheme. C. Hart Merriam (1898), founder of the U.S. Bureau of Biological Survey (later to become the Fish and Wildlife Service), proposed the idea of *life zones.* These are transcontinental belts running east and west. The differences among them, expressed by the animals and plants living there, are supposedly controlled by temperature.

Merriam divided the North American continent into three primary transcontinental regions: the Boreal, the Austral, and the Tropical. Each of these regions Merriam further subdivided into life zones. For example, he subdivided the Boreal region into three zones: the Arctic-Alpine zone; the Hudsonian zone, or northern coniferous forest; and the Canadian zone, or southern coniferous forest.

Once widely accepted, life zones are rarely used today, although the terms still creep into the literature on vertebrates. The life zone concept failed because it is not a unit that can be recognized continentwide by a characteristic and uniform faunal or vegetational component. There are wide differences in the various zones between east and west.

BIOMES

Still another approach, pioneered by V. E. Shelford, was simply to accept plant formations as biotic units and to associate animals with plants. Such an approach works fairly well, because animal life does depend upon a plant base. These natural broad biotic units are called *biomes* (Figures VII.2, VII.3). Each biome consists of a

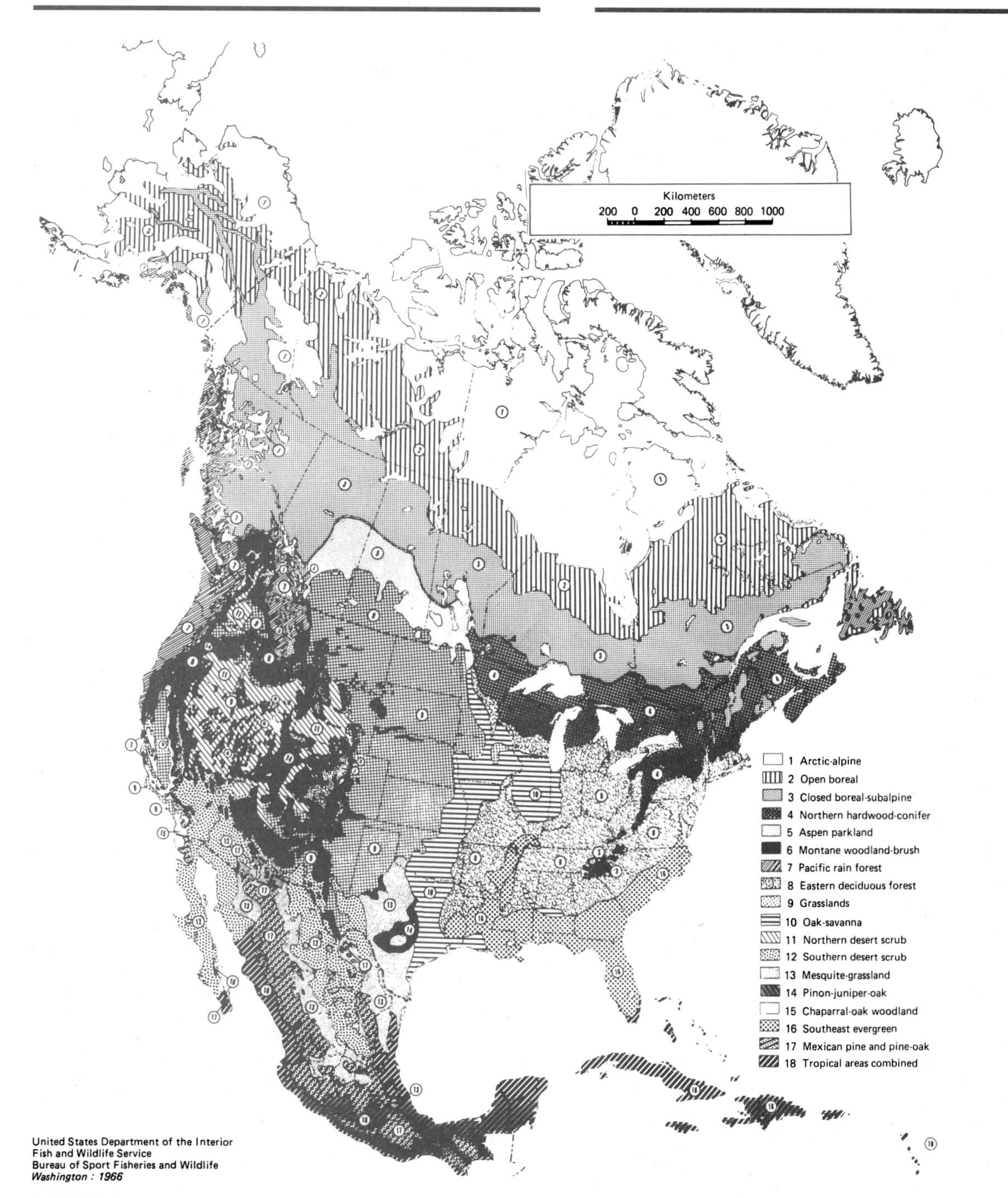

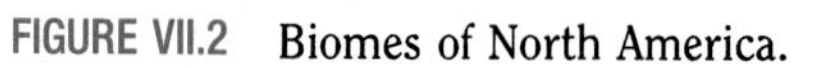

FIGURE VII.2 Biomes of North America.

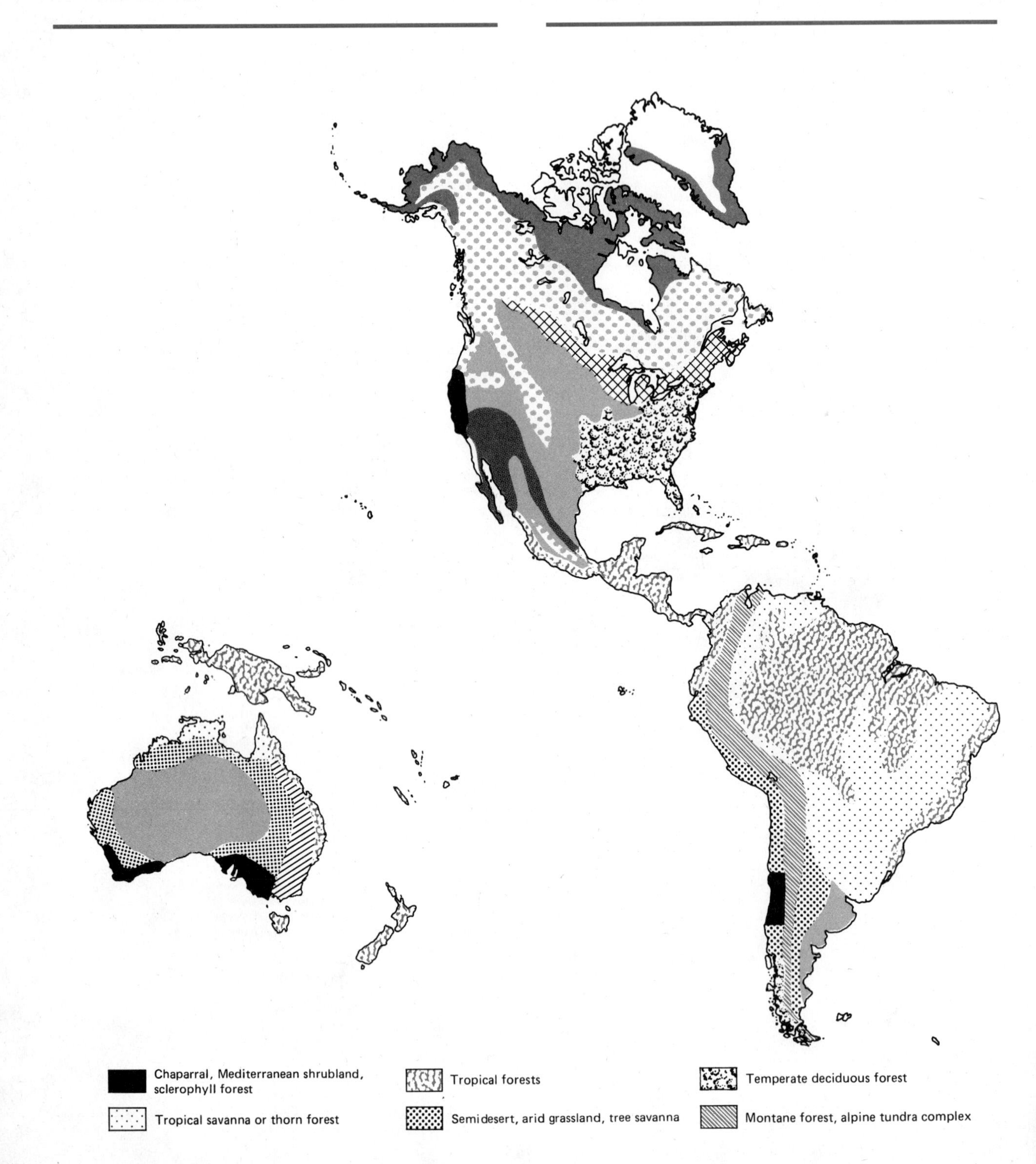
Chaparral, Mediterranean shrubland, sclerophyll forest
Tropical forests
Temperate deciduous forest
Tropical savanna or thorn forest
Semidesert, arid grassland, tree savanna
Montane forest, alpine tundra complex

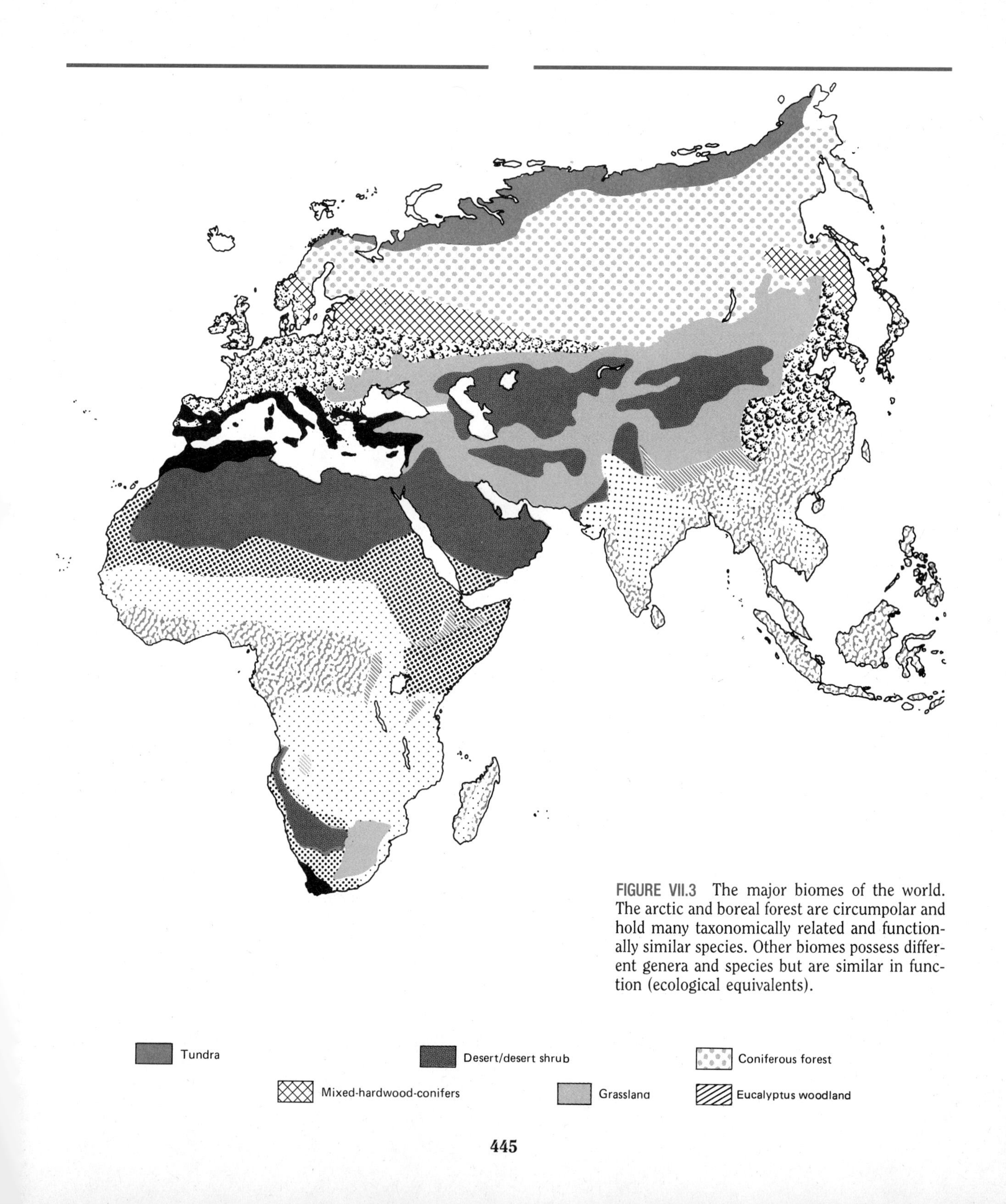

FIGURE VII.3 The major biomes of the world. The arctic and boreal forest are circumpolar and hold many taxonomically related and functionally similar species. Other biomes possess different genera and species but are similar in function (ecological equivalents).

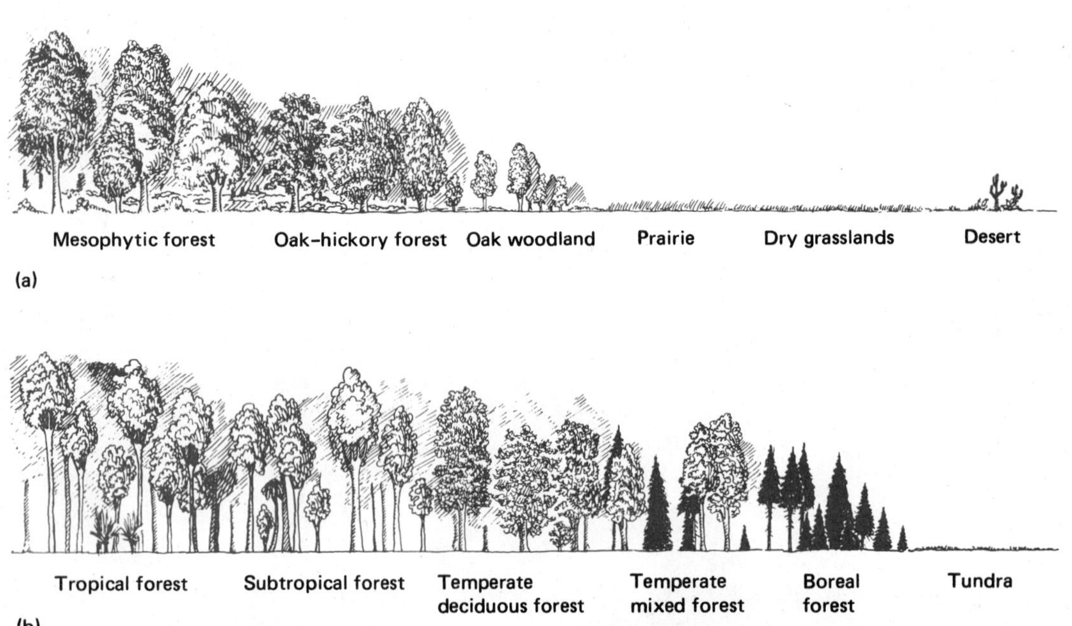

FIGURE VII.4 Gradients of vegetation in North America. (a) The east-west gradient runs from the mixed mesophytic forest of the Appalachians through the oak hickory forests of the central states, the ecotone of bur oak and grasslands, the prairie, the short-grass plains, and the desert. The transect does not cross the Rocky Mountains. This gradient reflects precipitation. (b) The north-south gradient reflects temperature. The transect cuts across the tundra, the boreal coniferous forest, the mixed northern hardwoods forest, the mixed mesophytic forests of the Appalachians, the subtropical forests of Florida, and the tropical forests of Mexico.

distinctive combination of plants and animals in the fully developed or climax community, and each is characterized by a uniform life form of vegetation, such as grass or coniferous trees. It also includes the developmental stages, which may be dominated by other life forms. Because the species that dominate the seral stages are more widely distributed than those of the climax, they are of little value in defining the limits of the biome.

On a local and regional scale, communities are considered as gradients, in which the combination of species varies as the individual species respond to environmental gradients. On a larger scale, we can consider the terrestrial and even some of the aquatic ecosystems as gradients of communities and environments on a continental scale. Such gradients of ecosystems are *ecoclines* (Figure VII.4).

In addition to gradual changes in vegetation, there are gradual changes in other ecosystem characteristics. From highly mesic situations and warm temperatures to xeric situations and cold temperatures, productivity, species diversity, and the amount of organic matter decrease. There is a corresponding decline in the complexity and organization of ecosystems, in the size of plants, and in the number of strata to vegetation. Growth form changes. The tropical rain forest is dominated by phanerophytes and epiphytes, the arctic tundra by hemicryptophytes, geophytes, and therophytes. Wherever similar environments exist on Earth, the same growth forms exist, even though species differences may be great. Thus, different continents tend to have communities of similar physiognomy.

There are six major terrestrial biomes: forest, grassland, woodland, shrubland, semidesert shrub, and desert. These six can be further divided into biome types, depending upon climatic conditions and elevation. These various biomes of the world fall into a distinctive pattern when plotted on a gradient of mean annual tem-

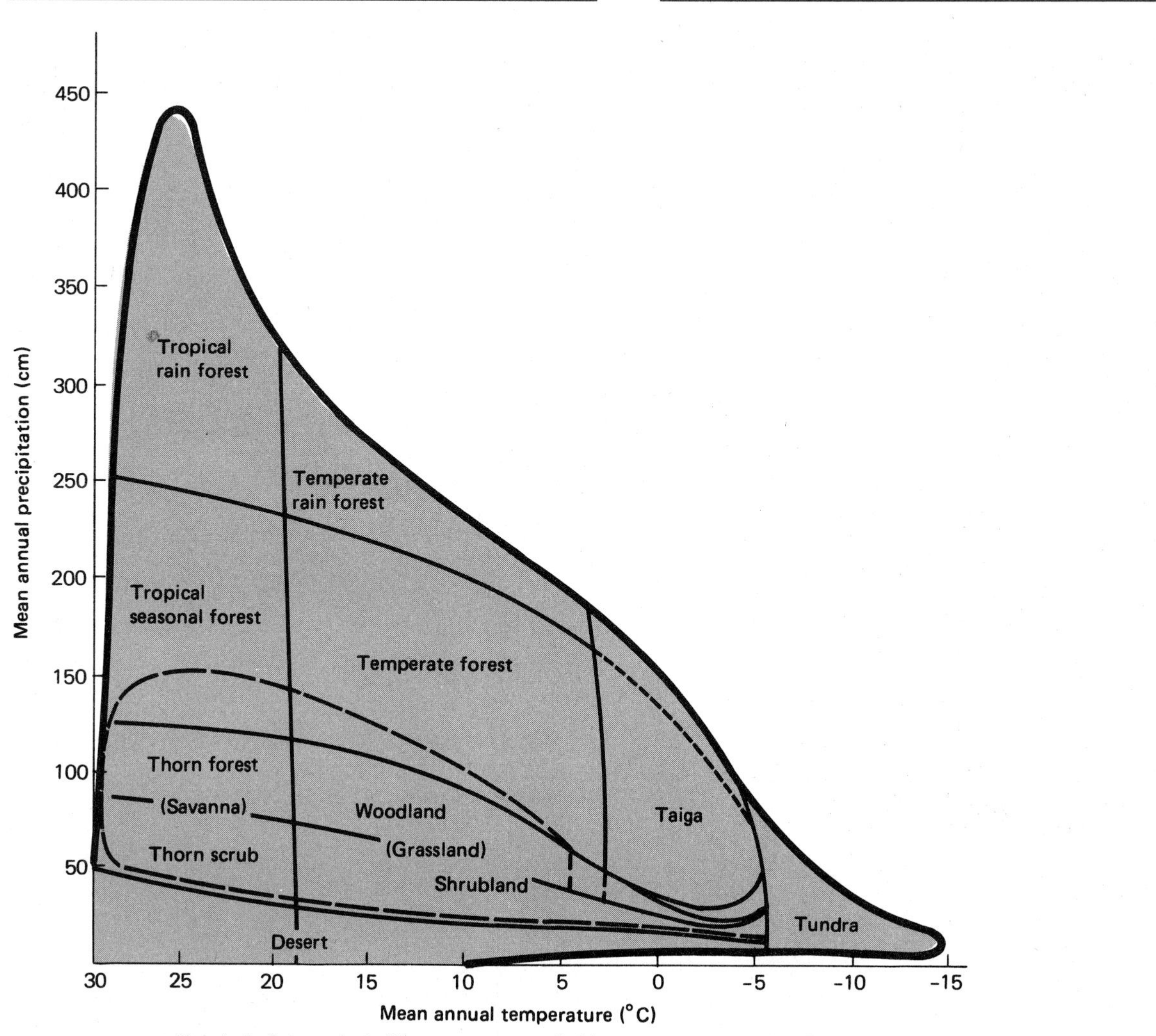

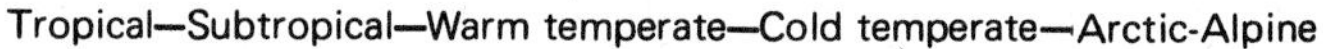

FIGURE VII.5 A pattern of world plant formation types in relation to climatic variables of temperature and moisture. In certain areas where the climate (maritime versus continental) varies, soil can shift the balance between such types as woodland, shrub, and grass. The dashed line encloses a wide range of environments in which either grassland or one of the types dominated by woody plants may form the prevailing vegetation in different areas.

perature and mean annual precipitation (Figure VII.5). The plots obviously are rough. Many types intergrade with one another, and adaptations of various growth forms may differ on several continents. Climate alone is not responsible for biome types. Soil and fire can also influence which one of several biomes will occupy a region. Structure of biomes is further influenced by the nature of the climate, whether marine or continental. The same amount of rain, for example, can support either shrubland or grassland.

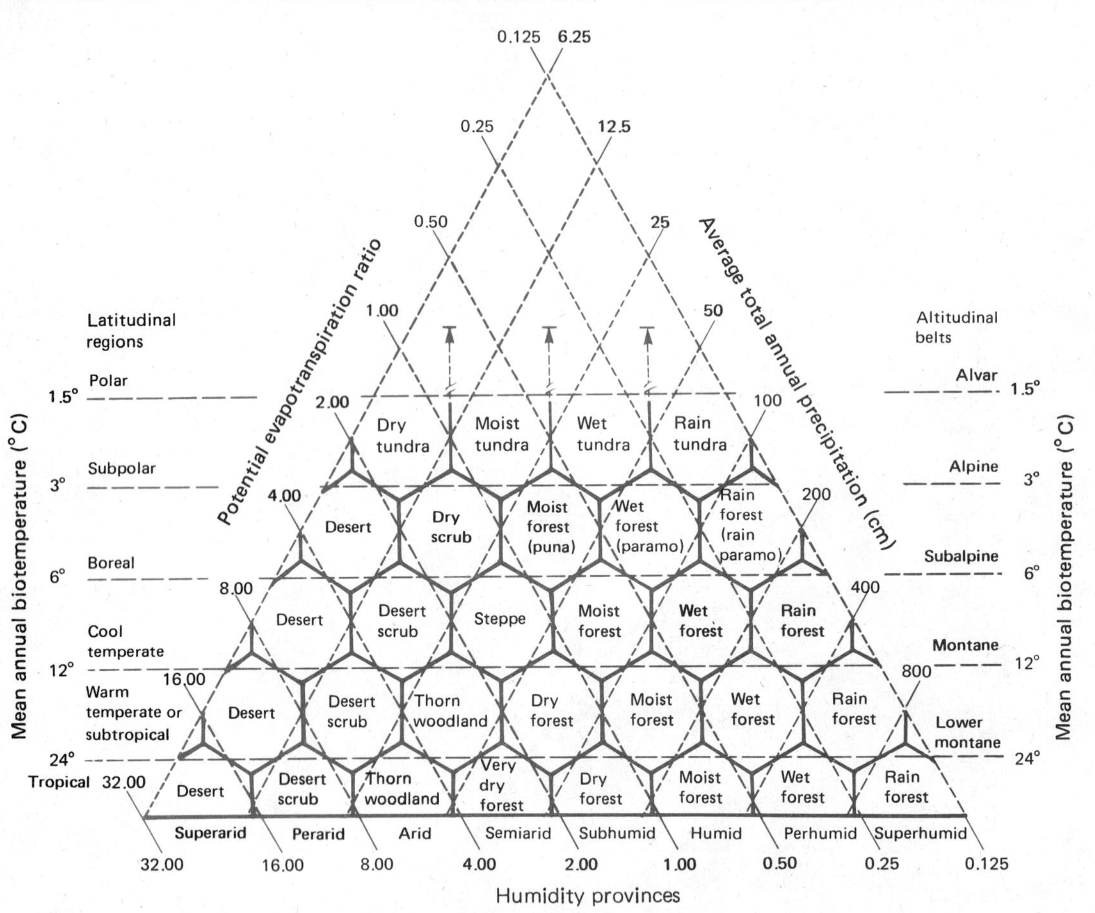

FIGURE VII.6 The Holdridge life zone system for classifying plant formation.

HOLDRIDGE LIFE ZONE SYSTEM

Another useful approach to the classification and study of ecosystems is the Holdridge *life zone* system. It is based on the assumptions that (1) mature, stable plant formations represent physiognomically discrete vegetation types recognizable throughout the world; and (2) geographical boundaries of vegetation correspond closely to boundaries between climatic zones. Vegetation is determined largely by an interaction of temperature and rainfall.

The Holdridge system divides the world into life zones arranged according to latitudinal regions, altitudinal belts, and humidity provinces (Figure VII.6). The boundaries of each zone are defined by mean annual precipitation and mean annual biotemperature. Altitudinal belts and latitudinal regions are defined in terms of mean biotemperatures and not in terms of latitudinal degrees or meters of elevation. The Holdridge classification is based on three levels: (1) climatically defined life zones; (2) associations, which are subdivisions of life zones based on local environmental conditions; and finally, (3) further local subdivisions based on actual cover or land use. The term *association,* as used by Holdridge, means a unique ecosystem, a distinctive habitat or physical environment and the naturally evolved community of plants and animals.

The subdivision into zones or associations is based on an interaction between environment and vegetation as defined by biotemperature, precipitation, moisture availability, potential evapotranspiration, and potential evapotranspiration ratio.

CHAPTER 26

Grasslands and Savannas

Outline

Objectives

On completion of this chapter, you should be able to:

1. Describe the types and general features of grasslands.
2. Describe the forms of animal life unique to grassland ecosystems.
3. Discuss energy flow and nutrient cycling in grassland ecosystems, contrasting belowground and aboveground production and consumption.
4. Describe the distribution and characteristics of savannas.
5. Point out the major structural and functional features of savanna ecosystems.
6. Discuss human impacts on grasslands and savannas.

GRASSLANDS

At one time grasslands covered about 42 percent of the land surface of Earth. In the northern hemisphere great expanses of grassland covered the midcontinent of North America and extended across the central part of Eurasia. In the southern hemisphere grasses covered much of the southern tip of South America and the high plateau of southern Africa. Today grasslands probably occupy less than 12 percent, most of them plowed under for cropland and degraded by overgrazing.

All grasslands have in common a climate characterized by rainfall between 250 and 800 mm (too light to support a heavy forest and too great to encourage a desert), a high rate of evaporation, periodic severe droughts, a rolling to flat terrain, and animal life dominated by grazing and burrowing species. Grasslands, however, are not exclusively a climatic formation, because most of them require periodic fires for maintenance, renewal, and elimination of encroaching woody growth.

Grasses have a mode of growth that adapts them to grazing and fire. The grass plant consists of leafy shoots called tillers. Each shoot has a leaflike blade or lamina, the base of which has a tubelike sheath. These tillers grow from short, stem-bearing root nodes, which grow upward only when the plant begins flowering. Tillers that group closely about a central stem and buds make up bunch or tussock grasses. Species spread by means of lateral buds on underground stems, producing a sod, are sod or turf grasses (Figure 26.1). Associated with grasses are a variety of legumes and composite plants.

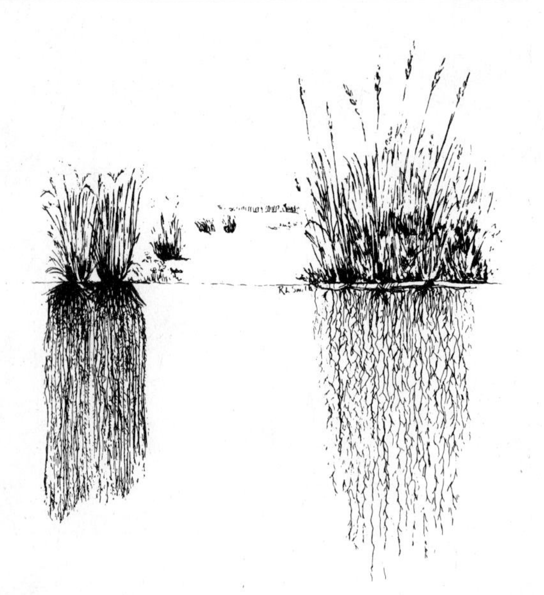

FIGURE 26.1 Growth forms of a sod grass, crested wheatgrass (right); and a bunchgrass, little bluestem (left). Also shown are root penetration (maximum depth, 2 m) and distribution in the soil.

TYPES OF GRASSLANDS

Anthropogenic Grasslands

The grasslands most familiar to a majority of us are hayfields and pasturelands, created and maintained by human efforts. Mostly they occupy former forested land, cleared for settlements and agriculture. Some grasslands, especially in Britain, Switzerland, and Scandinavia, have existed for centuries, becoming a sort of climax vegetation supporting its own distinctive vegetation. In other areas, such as eastern North America, abandoned agricultural grasslands have reverted to forest.

Tame grasslands are permanent, rotational (plowed every few years for other crops), or rough. The last are marginal, unimproved, semiwild lands used principally for grazing. Many seral or successional grasslands fit this category.

Ecologically, permanent hay and grazing lands differ from rotational hayfields. Permanent haylands and grazing lands consist of domesticated species adapted to periodic defoliation by mowing and grazing, but do support a diversity of herbaceous species. Fertilization reduces diversity by favoring more competitive grasses and forbs. Rotational hayfields are dominated by two or three cultivated species, usually two grasses and a legume. Such hayfields are denser and ranker in growth than are permanent grasslands. Their management consists of fertilization, mowing, and plowing at regular intervals for growing other crops in the rotation.

North American Grasslands

In North America, grasslands originally covered much of the interior between the Rocky Mountains and the eastern deciduous forest. These consisted of three main

FIGURE 26.2 Shortgrass plains give way abruptly to forested hills in the Black Hills of South Dakota.

types, distinguished by the height of the dominant species, influenced by climate and rainfall.

Tallgrass Prairie. The tallgrass prairie once occupied a narrow belt running north and south next to the deciduous forest of eastern North America. It was well developed in a region that could support forests. Oak-hickory forests did extend into the grassland along streams and rivers, on well-drained soils, sandy areas, and hills. Fires, often set by native Americans in the fall, stimulated a vigorous growth of grass and eliminated the encroaching forest.

Big bluestem (*Andropogon gerardi*), growing 1 m tall with flowering stalks 1 to 3½ m tall, was the dominant grass of moist soils and occupied the valleys of rivers and streams and lower slopes of hills. Associated with big bluestem were a number of forbs: goldenrods, compass plants, snakeroot, and bedstraw. The drier uplands were dominated by bunch-forming needlegrass (*Stipa*), side-oats grama (*Boutelou curtipendula*), and dropseed (*Sporobolus spp.*). The drier uplands grew such a diversity of forbs, especially composites, that they were nicknamed "daisy land."

Mixed-Grass Prairie. West of the tallgrass prairie was the mixed-grass prairie. Typical of the Great Plains, it embraced largely the needlegrass-grama grass community. Because the mixed-grass prairie is characterized by great annual extremes in precipitation, its aspect varies widely from year to year. In moist years midgrasses are prevalent, whereas in dry years short grasses and forbs are dominant. The grasses are largely bunch and cool-season species that begin their growth in early April, flower in June, and mature in late July and August.

Shortgrass Plains. South and west of the mixed prairie and grading into the desert are the shortgrass plains, one grassland that has remained somewhat intact (Figure 26.2). The shortgrass plains reflect a climate in which rainfall is infrequent and light (up to 400 mm in the west and 500 mm in the east), humidity low, winds high, and evaporation rapid. The shallow-rooted grasses utilize moisture in the upper soil layer, beneath which the roots do not penetrate. Sod-forming blue grama (*Bouteloua gracilis*) and buffalo grass (*Buchloe dactyloides*) dominate the shortgrass plains, accompanied by such midgrasses as wheatgrass (*Agropyron ssp.*), side-oats grama, and little bluestem (*Andropogon scoparius*). Because of the dense sod, few forbs grow on the plains, but prominent among them are lupines (*Lupinus ssp.*).

Desert Grasslands. From southeastern Texas to southern Arizona and south into Mexico lies the desert grassland, similar in many respects to the shortgrass plains, except that three-awn grass (*Aristida ssp.*) replaces buffalo grass. Composed largely of bunchgrasses, desert grasslands are widely interspersed with other vegetation types such as oak savanna and mesquite. The climate is hot and dry. Rain falls only during two seasons, summer (July and August) and winter (December to February), in amounts that vary from 300 mm to 400 mm in the west and 500 mm in the east; but evaporation is rapid, up to 2000 mm per year. Vegetation puts on most of its annual growth in August.

Annual Grasslands. Confined largely to the Central Valley of California is annual grassland. It is associated with Mediterranean-type climate, characterized by rainy winters and hot, dry summers. Growth occurs during early spring, and most plants are dormant in summer, turning the hills a dry tan color accented by the deep green foliage of scattered California oaks. The original vegetation was perennial grasses dominated by purple needlegrass (*Stipa pulchra*), but since settlement, grasses have been replaced by vigorous annual species well-adapted to a Mediterranean-type climate. Dominant species are wild oats (*Avena fatua*) and slender oats (*Avena barbata*) grass.

Eurasian Steppes

At one time the great grasslands of the Eurasian continent extended from eastern Europe to western Siberia south to Kazakhstan. Whereas the North American grasslands have an east-west zonation reflecting climatic changes especially in precipitation, the Eurasian steppes, treeless except for ribbons and patches of forest, are divided into four belts of latitude, from the mesic meadow steppes in the north to semiarid grasslands in the south. The meadow steppes occupy a region of chernozem soils in which the rainfall is 500–600 mm, extending south from the taiga. Dominated by bunch-forming fescues (*Festuca*) and feather grass (*Stipa*) along with many species of daisy (*Compositae*), the meadow steppes were once outstandingly beautiful in spring and early summer. Little remains of meadow steppes, turned under the plow for cereal grains. Further south where rainfall is 400 to 500 mm, tussock-forming species of *Stipa* dominate and flowering herbs are fewer. In the central Asian steppes with their cold dry spring no ephemeral plants exist, and grasses give way to woody and herbaceous species of drought-resistant *Artemensia*. About the Black Sea and in Kazakhstan, where the humidity is higher, steppe vegetation is dominated by large feather grasses and sheep's fescue (*Festuca ovina*) and by ephemeral spring plants such as tulips (*Tulipa*).

South American Pampas

In the southern hemisphere the major grasslands exist in southern Africa and in southern South America. Known as the pampas, the South American grasslands extend westward in a large semicircle from Buenos Aires to cover about 15 percent of Argentina. In the eastern part of the pampas rainfall exceeds 900 mm well distributed throughout the year. In this humid East the pampas are dominated by tallgrasses. South and west, where rainfall is about 450 mm, semidesert vegetation becomes prominent. South into Patagonia, where the rainfall averages about 250 mm, the pampas change to open steppe grasses dominated by *Stipa* and *Festuca* and xerophytic cushion plants. These pampas have been modified by the introduction of European forage grasses and alfalfa (*Medicago sativa*), and the eastern tallgrass pampas have been converted to wheat and corn.

South African Veld

The pampas of Argentina occupy the lowlands; by contrast the grasslands of southern Africa (not to be confused with the savanna) occupy the eastern part of a high plateau 1500 to 2000 m in the Transvaal and the Orange Free State. Most of the rainfall comes in the summer, brought in by the moist air masses from the Indian Ocean. The heaviest rainfall is in the east, the lowest in the west where the grasslands grade into the semiarid shrubland known as the Karoo. The nature of

FIGURE 26.3 A cut through a cultivated grassland, a hayfield, showing the several strata in grasslands.

the grasslands varies considerably, depending on soil parent material, climate, and grazing pressure. Typical grassveld, found at 1500 to 1700 m where rainfall is 650 to 750 mm, is dominated by red grass (*Themeda triandra*), growing on black turf soil. Areas of poor soils, called sourveld, support *Aristida* and *Erigristic*. At high elevations the dominant grasses are fescues and brome grass. Once inhabited by great herds of antelopes, these grasslands are now agricultural lands.

Australian Grasslands

Australia has four types of grasslands: arid tussock grassland in the northern part of the continent, where the rainfall averages between 200 and 500 mm, mostly in the summer; arid hummock grasslands dominated by *Triodia* and *Plectrachne* in areas with less than 200 mm rainfall; coastal grasslands dominated by *Sporobolus* in the tropical summer rainfall region; and subhumid grasslands dominated by such grasses as *Poa* and kangaroo grass (*Themeda*) along coastal areas where rainfall is between 500 and 1000 mm. Most of these grasslands have been changed by fertilization, introduced grasses and legumes, and sheep grazing.

STRUCTURE

Vegetation

The most obvious structural element of a grassland is the tall green growth so characteristic of summer fields. That layer is an ephemeral one, developing in spring and dying back in late autumn. It is but one of three strata that arise from the crowns, nodes, and rosettes of plants hugging the soil and making up the ground layer. The ground layer and the belowground root layer are the two major strata of the grassland.

The herbaceous layer, consisting of both grasses and forbs, has three or more sublayers, more or less variable in height, according to the grassland type (Figure 26.3). Low-growing and ground-hugging plants such as dandelion, strawberry, and mosses make up the first layer. As the growing season progresses, these plants become hidden beneath the middle and upper layers. The middle layer consists of shorter grasses and such forbs as wild mustard and daisy. The upper layer consists of leaves and flowering stems of tallgrasses and the leafy stalks and flowers of forbs.

The ground layer is most obvious in late winter and

early spring. Exposed to high light, the plants respond to early spring temperature and moisture. As the grasses and forbs grow taller and shade the ground, light intensity reaching the ground layer decreases. Temperature declines as solar insolation is intercepted by a blanket of vegetation, relative humidity increases, and wind flow decreases, creating a region of calm near the ground. Conditions on grazed lands are much different. Because the grass cover is closely cropped, the ground layer receives much higher solar radiation, higher temperatures, and greater wind velocity.

Grasslands that are unmowed, unburned, and ungrazed accumulate a thick layer of mulch (on your lawn it is called thatch). The oldest bottom layer, humic mulch, consists of decayed and fragmented remains of fresh mulch; the top layer consists of fresh herbage, leafy and largely undecayed, that is deposited throughout the growing season. As the mat increases in depth, it retains more moisture, creating favorable conditions for microbial activity. Three or four years must pass before natural grassland mulch decomposes completely.

Grazing reduces the mulch layer, as do fire and mowing. Light grazing tends to increase the weight of decayed humic mulch at the expense of fresh mulch; moderate grazing increases compaction, which favors microbial activity and a subsequent reduction in both fresh and humic mulch. Heavy grazing greatly reduces mulch accumulation. Burning reduces both fresh and humic mulch, but the mulch structure returns after a fire on lightly grazed and ungrazed lands. Mowing greatly reduces both fresh mulch and humic mulch. Haylands have a minimal amount of mulch on the ground.

The amount of mulch is ecologically important. Mulch increases soil moisture through its effects on infiltration and evaporation; it decreases runoff and erosion, stabilizes soil temperatures, and improves conditions for seed germination. How much mulch is necessary is the question. Where mulch can accumulate to the proper degree, grassland maintains itself. Heavy mulches can suppress growth of grasses and allow the invasion of forbs and woody vegetation. In areas of no accumulation, grassland regresses to weedy plants. Deep litter provides habitat for meadow mice and certain ground-nesting birds such as the bobolink, but inhibits the presence of others.

The root layer is more highly developed in grasslands than in any other major community. Half or more of the plant is hidden beneath the soil; in winter, roots represent almost the total grass plant. The bulk of the roots is fibrous and occupies rather uniformly the upper 15 cm or so of the soil profile; they decrease in abundance with depth. The depth to which roots extend is considerable. Little bluestem, for example, reaches 1 to 2 m and forms a dense mat as deep as 0.8 m. In addition many grasses possess underground stems or rhizomes that serve both to propagate the plant and to store food. Constantly dying, roots add finely divided organic matter to the mineral soil.

Roots develop in three or more zones. Some plants are shallow-rooted and seldom extend much below 0.5 m. Others go well below the shallow-rooted species but seldom more than 1.5 m. Deep-rooted plants extend even further into the soil and absorb relatively little moisture from the surface soil. Thus plant roots absorb nutrients from different depths in the soil at different times, depending on moisture.

Animal Life

Natural or tame, grasslands support similar forms of life, vertebrate and invertebrate. The invertebrate life includes an incredible number and variety of species and occupies all strata during some time of the year. During winter in temperate grasslands, insect life is confined largely to soil, litter, and grass crowns where they exist as eggs or pupae. In spring, soil occupants are chiefly earthworms and ants, the latter being the most prevalent if not the most conspicuous. The ground and mulch layers harbor scavenger carabid beetles and predaceous spiders, of which the majority are hunters rather than web builders. Life in the herbaceous layer varies as the strata become more pronounced from spring to fall. Here invertebrate life is most abundant and varied. Homoptera, Coleoptera, Diptera, Hymenoptera, and Hemiptera are all represented. Insect life reaches two highs during the year, a major peak in summer and a less defined one in the fall.

Large grazing ungulates and burrowing mammals

FIGURE 26.4 Bison, which once roamed the shortgrass plains in countless numbers, epitomize the North American grasslands.

are the most conspicuous vertebrates. All of the world's native grasslands support similar forms. The North American grasslands once were dominated by huge migratory herds of bison, numbering in the millions (Figure 26.4), and the forb-consuming pronghorned antelope (*Antilocarpa americana*). The most common burrowing rodent was the prairie dog (*Cynomys ssp.*), which along with gophers (*Thomomys* and *Geomys*) and the mound-building harvester ants appeared to be instrumental in the development, maintenance, and ecological structure of the shortgrass prairie.

The Eurasian steppes lack herds of large ungulates. The western steppes are home to the small migratory goat antelope, the saiga (*Saiga tartarica*), characterized by a large proboscis-like nose, which increases in size in the male during rut. Nearly extinct in 1917, it now numbers over one million animals. Further east lives the Mongolian gazelle (*Procarpa gutterosa*) and several species of rare wild horses. The dominant burrowing animals are the bobak marmot (*Marmotta bobak*), which looks much like an oversized prairie dog, the sousliks or ground squirrels (*Citellus*), and the common hamster (*Cricetus cricetus*).

The Argentina pampas also lack a large ungulate fauna. The two major large herbivores are the pampas deer (*Ozotoceras bezoarticus*) and further south the guanaco (*Lama guanaco*), small relatives of the camel, greatly reduced in number compared to historical times. Major burrowing rodents are the viscacha (*Lagostomus maximus*) and the Patagonian hare or mara (*Dolichotis patagonium*), a monogamous, cavylike rodent, with the long ears of a hare and the body and long legs of an antelope.

The African grassveld once supported great migratory herds of antelope and zebra along with their associated carnivores, the lion, leopard, and hyaena. Burrowing rodents include the kangaroo-like springhare (*Pedetes capensis*) and the gerbil (*Tatera brantsii*), and a most interesting carnivore, the meerkat (*Cynictis penicillata*), whose burrowing habits suggest those of the prairie dog. The rodents remain, but the great ungulate herds have been destroyed and replaced with sheep, cattle, and horses.

The Australian marsupial mammalian life evolved many forms that are the ecological equivalents of placental grassland mammals. The dominant grazing animals are a number of species of kangaroos, especially the red kangaroo (*Macropus rufus*) and the gray kangaroo (*M. giganteus*). The wombats (*Vombatus*) occupy the ecological niche of the viscachas of the pampas and the gophers of the prairies.

Three of the world's grasslands evolved unique unrelated birds with a poor ability to fly, large size, and high running speed. Australia has the emu (*Dromiceius casuarius*), the pampas the rhea (*Rhea americana*), and Africa the ostrich (*Struthio*). New Zealand, whose grasslands (not discussed) lacked any sort of herbivorous mammals, had flocks of the now extinct grass-consuming moa (*Dinorus*). Although the grasslands of the northern hemisphere lack such large birds, the European steppes do have the large great bustard (*Otis tarda*), weighing up to 16 kg. Its numbers have been reduced by loss of habitat to agriculture.

Vertebrate life in seral and tame grasslands is strongly affected by human management. Mowing of hayfields destroys habitat at a critical nesting time. Losses to birds, rabbits, and mice from mechanical injury and predation on exposed nests are often heavy, but most species will remain on the area to complete or reattempt nesting. Early mowing for grass ensilage at the very start of the nesting season eliminates nesting cover and forces the animals elsewhere. Early mowing is one of the reasons for the sharp decline in grassland birds. Pasturelands more often than not are so badly overgrazed they support little in the way of vertebrate life. The two most common inhabitants in eastern North America are the killdeer (*Charadrius vociferus*) and the horned lark (*Eremophila alpestris*).

FUNCTION

Grasslands are adapted to periods of drought and survive under low rainfall, but like all vegetation, grasses grow best under optimal conditions of moisture and temperature. Grasslands do the poorest where precipitation is the lowest and temperatures are high; they do best where the mean annual precipitation is greater than 800 mm and mean annual temperature is above 15° C. Production, however, is most directly related to precipitation (Figure 26.5). The greater the mean annual precipitation, the greater is aboveground production. This increased production comes about because increased moisture reduces water stress and enhances the uptake of nutrients. Some grass species are more efficient in using water than others. Grasses of arid regions use water inefficiently, because adaptations among desert grasses are not for the best use of water but for survival and persistence by other means. In humid, tallgrass country the vegetative canopy is dense and intercepts most of the solar radiation, so adaptations of plants are for efficient capture of light at the expense of efficient use of water; but in semiarid country, where water is limiting, plants have evolved adaptations to make the maximum use of water available.

We associate grassland production with the amount of aboveground growth of grass, but much of the net production is below ground. Except for tropical grasslands, which put most of their net production into aboveground biomass, seminatural and temperate grasslands send most of their production into the roots. Except for a short period of maximum aboveground biomass during the growing season, belowground biomass is two to three times that above ground. Seventy-five to 85 percent of grassland photosynthate is translocated to the roots for storage below ground.

Grasslands have evolved under grazing pressures of ungulates since the Cenozoic. Their structure and growth habits reflect this selective pressure. Critical growth tissues are at or below ground surface, protected from grazing and fire. As the grazers clip and eat the leaves, grasses respond by increasing the photosynthetic rate in remaining tissue, stimulating new growth, and reallocating nutrients and photosynthates from one part of the plant to another, especially from roots to stems. Moderate grazing actually improves the productivity of grasses by removing photosynthetically inefficient mature tissue to make room for young tissue and by increasing light intensity on the ground to stimulate new growth. Clipping also stimulates tillering and leaf growth, increasing density of ground cover. In addition, grazers recycle nutrients in grass through dung and urine.

Grasses, however, do not send all their production to rebuild aboveground biomass. They channel considerable amounts below ground to the roots (Figure 26.6). Grazed grasslands send about 63 percent of their total net primary production below ground and about 37 percent above ground. Ungrazed systems, perhaps

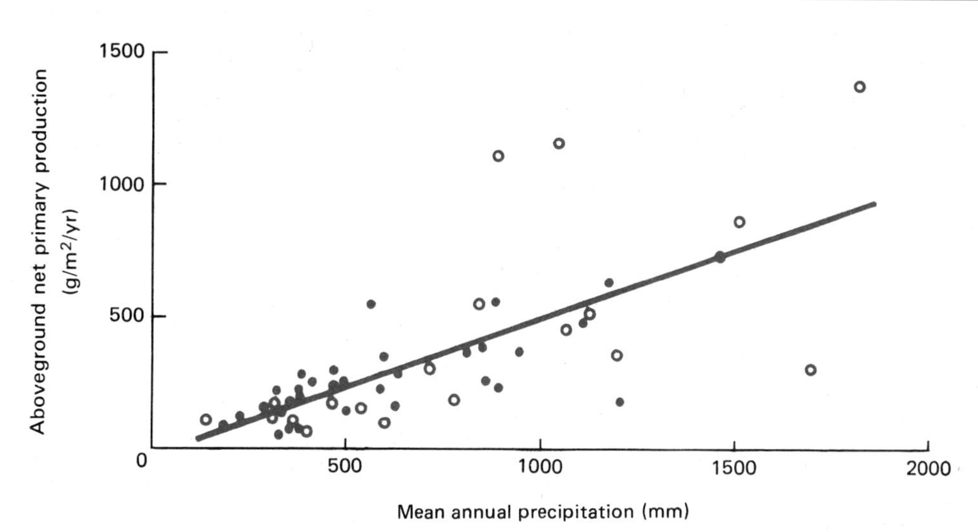

FIGURE 26.5 Relationship between aboveground primary production and mean annual precipitation for 52 grassland sites around the world. North American grasslands are indicated by solid dots.

not under the same pressure to build up belowground reserves, send about 52 percent of total net primary production below ground and about 42 percent above.

Grassland ecosystems respond to grazing in still another way by changing species composition. Some grasses and forbs tend to disappear, while other species increase. On desert grasslands of North America black grama is replaced by weedy species; on shortgrass plains, which are the most stable under grazing pressure, blue grama and prickly pear increase; on mixed-grass prairies midgrasses decrease and shortgrasses and sedges increase. On tallgrass sites tallgrasses disappear and little bluestem and tall dropseed increase; if grazing pressure is heavy the site may be invaded by the weedy Japanese chess. In tame pasturelands, heavy grazing increases the amount of unpalatable forbs, such as thistles and ironweed.

Because they are so conspicuous, we equate grazing pressure with aboveground herbivores such as cattle and rabbits and invertebrate grazers such as grasshoppers. In reality much more intense grazing takes place underground. The aboveground biomass of invertebrates—including plant consumers, saprovores, and predators—ranges from 1 to 50 g/m^2, and grazing mammals amount to about 2 to 5 g/m^2. Belowground invertebrates exceed 135 g/m^2, most of them nematodes. They account for 90, 95, and 93 percent of all belowground herbivory, carnivory, and saprophagous activity, respectively. Nematodes account for 46 to 67 percent of root and crown biomass consumed, 23 to 85 percent of fungi consumed, and 43 to 88 percent of belowground predation. Aboveground herbivores consume only 2 to 7 percent of primary production, or 3 to 10 percent if both the amounts consumed and wasted are considered. Total belowground consumption by all belowground herbivores, including wastage, amounts to 13 to 46 percent.

Not only is a large proportion of primary production consumed below ground, but also a greater proportion is utilized at each trophic level there (Figure 26.7). Some invertebrate aboveground consumers, particularly grasshoppers, are wasteful. The amount of aboveground vegetation they detach or otherwise kill about equals that consumed by vertebrate grazing herbivores.

Invertebrate consumers are also highly inefficient in assimilating ingested material and deposit much of their intake as highly soluble feces or frass. The nutrients they contain return rapidly to the system. Large grazing herbivores return a portion of their intake as dung, which is fed upon by a well-developed copropha-

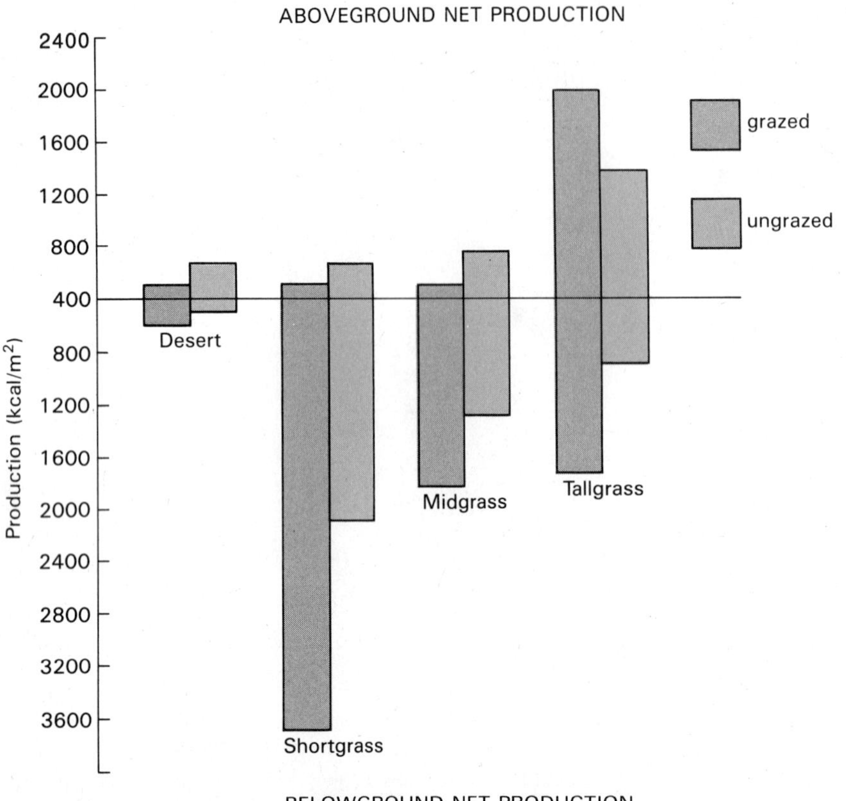

FIGURE 26.6 Aboveground and belowground net primary production for grazed and ungrazed North American grassland types.

gous fauna that speeds the decay of manure and accelerates the activity of bacteria in feces.

Most of the primary production, however, goes to the decomposers, dominated by fungi whose biomass is two to seven times that of bacteria. Overall the decomposer biomass exceeds that of invertebrates.

Central to the cycling of nutrients in grasslands is mulch or detritus, a large standing crop of which detritus can have detrimental effect on nitrogen cycling, particularly in tallgrass ecosystems. Detritus intercepts rainfall, from which microbes can assimilate inorganic nitrogen directly before it reaches plant roots, while the mulch itself inhibits nitrogen fixation by blue-green algae and free-living nitrogen-fixing microbes. By insulating the soil surface from solar radiation, mulch reduces root productivity and inhibits the activity of soil microbes and invertebrates. Periodic grassland fires clear away the mulch layer and release nutrients in detritus to the soil, but nitrogen, equal to about two years of nitrogen inputs to the system through rainfall, is lost to the atmosphere. Fires, however, stimulate the growth of nitrogen-fixing leguminous forbs and improve conditions for earthworms.

HUMAN IMPACT

Humans had their beginnings in the savanna grasslands of Africa and have inhabited and utilized grasslands throughout their evolutionary and cultural history. Crop agriculture and pastoralism began there and spread through the grasslands of the world, to the point where little natural grassland exists. Compared to forests, few grassland reserves exist in the world. We have broken up grasslands with the plow and converted the

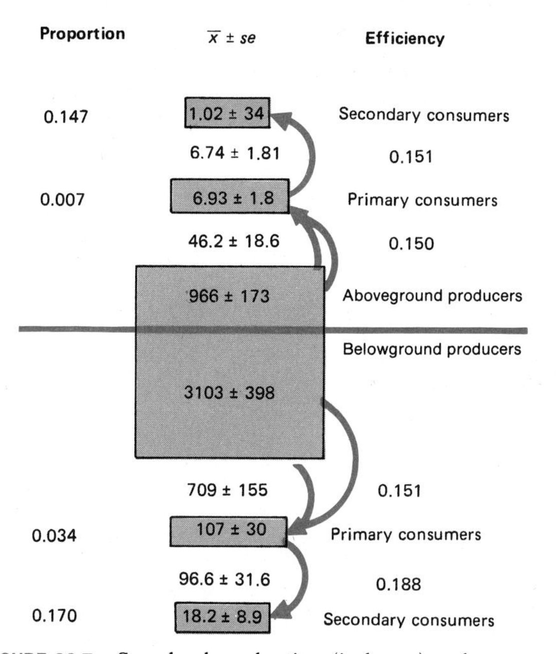

FIGURE 26.7 Grassland production (in boxes) and consumption (between boxes), representing means and standard errors for desert, tallgrass, shortgrass, and mixed-grass sites in North American grasslands. Production in one trophic level is represented as a proportion of production in the next trophic level and the efficiency as a ratio of production to consumption.

FIGURE 26.8 The effects of overgrazing by cattle are apparent in this shortgrass range. Prickly pear and mesquite have replaced grass. The ground is bare, most of the moisture from rains is lost as runoff, and erosion is serious.

most productive of them into the breadbaskets of the world, dominated by a monoculture of cereal grains. Conversion of the shortgrass plains of North America to wheat resulted in the Dust Bowl, when a seven-year drought hit the region in the 1930s.

On less productive grasslands we have replaced native plant species with highly productive forage plants accompanied by fertilization, pest control, irrigation, and other intensive practices. We have removed great herds of wild ungulates from natural grasslands and replaced them with domestic stock. Natural grasslands had experienced grazing pressure from free-ranging and often migratory populations of mammals. When we replaced these wild ungulates with domestic ones, we confined them with fences and overstocked the ranges, causing serious deterioration of grassland systems. Overgrazing desert grasslands in North America has increased the spread of mesquite because of lessened competition from grass and dispersal of seed by livestock (Figure 26.8). On other overgrazed grasslands mulch deteriorates and disappears because only a small amount of litter is added to the ground. Water flows over the surface, taking topsoil with it. Lacking moisture and nutrients, the original species cannot maintain themselves, and the vegetation cover continues to decrease until only an erosion pavement remains. In the semiarid regions of Africa, overgrazing has obliterated vegetation, converting those regions to deserts and dust bowls.

In forested regions, we have created new grasslands by clearing the forests and planting forage grasses. Such activity permitted the eastward spread of such grassland species as bobolinks, meadowlarks, and rabbits as their habitat in the plains was being destroyed. In Old World countries such as Britain and Scandinavia, human-created grasslands have existed for centuries, creating a seminatural climax vegetation. Like native grasslands, these, too, have been destroyed by plowing in recent times, threatening the extinction of such plants as bird's-eye primrose (*Primula farinosa*) and pasqueflower (*Amemone pulsatilla*) associated with undisturbed grasslands. Few areas of this type of grassland exist today; in effect they are the counterparts of the few relict areas of natural grassland in the Northern Hemisphere.

TROPICAL SAVANNAS

The one ecosystem that defies any general description is the tropical savanna. The problem is an old one, involving even its name. The word in its several origins, largely Spanish, referred to grasslands or plains; but over time the word was applied to an array of vegetation types representing a continuum of increasing cover of woody vegetation, from open grassland to widely spaced shrubs or trees to closed woodland (Figure 26.9). Moisture appears to control the density of woody vegetation (Figure 26.10), a function of both rainfall (amount and distribution) and soil—its texture, structure, and water-holding capacity.

Savannas cover much of central and southern Africa, western India, northern Australia, large areas of northwestern Brazil where they are known as cerrados, Colombia and Venezuela where they are called llanos, and to a more limited extent Malaysia. Some savannas are natural. Others are seminatural or anthropogenic, brought about in part and still maintained by centuries of human interference. In the African savannas, in particular, it is difficult to separate the effects of humans from the effects of climate. The savannas of central India, however, are the result of human degradation of original forest land.

Savannas, in spite of their vegetational differences, exhibit a certain set of characteristics. Savannas occur on land surfaces of little relief, often old alluvial plains. The soils are low in nutrients, due in part to infertile parent material and a long period of weathering. Savanna regions are associated with a warm continental type of climate with precipitation ranging between 500 mm and 2000 mm. Precipitation exhibits extreme seasonal fluctuations; in South American savannas, in particular, the soil–water regime may fluctuate from excessively wet to extremely dry, often below the permanent wilting point. Savannas are subject to recurrent fires, and the dominant vegetation is fire-adapted. Grass cover with or without woody vegetation is always present. When present the woody component is short-lived, with individuals seldom surviving for more than several decades (except for the African baobab trees). Detrital-processing termites are a conspicuous component of savanna animal life, especially in Africa.

STRUCTURE

Vegetation

The major and most essential stratum of the savanna ecosystem is grass, mostly bunch or tussock, with no vertical structure; its biomass decreases with height. A woody component adds one or two more vertical layers, ranging from about 50 to 80 cm when small woody shrubs are present to about 8 m in the tree savannas. Highly developed root systems make up the larger part of the living herbaceous biomass. The root system is concentrated in the upper 10 cm but extends down to about 30 cm. Savanna trees have extensive horizontal roots that go below the layer of grass roots. Competition may exist between grass and woody vegetation for soil moisture, but more intense competition takes place among trees, accounting for the spacing patterns of woody vegetation.

In contrast to the poorly developed vertical structure is a well-developed, although often unapparent, horizontal structure. The tussock grasses form an array of clumps set in a matrix of open ground, creating patches of low vegetation with frequent changes in microclimatic conditions. The addition of woody growth, the widely spaced shrubs and trees, increases horizontal structure extending to the soil. Trees add some organic matter and nutrients to the soil beneath them, reduce evapotranspiration, resulting in increased herbaceous and woody shrub growth, and provide patches of shade. On the African savanna in particular, large grazing herbivores rest in the shade during the heat of the day and concentrate nutrients from dung and urine beneath them.

Breaking up the monotony of the savannas are numerous marshy depressions that support wetland wildlife. Large ribbons of riverine or gallery forests weave through the savannas. The gallery forests support a

FIGURE 26.9 Savanna ecosystems in southern Africa: (a) grass savanna (note the termite mound); (b) shrub savanna; (c) tree savanna; (d) savanna woodland (dry season) dominated by *Combreteum*.

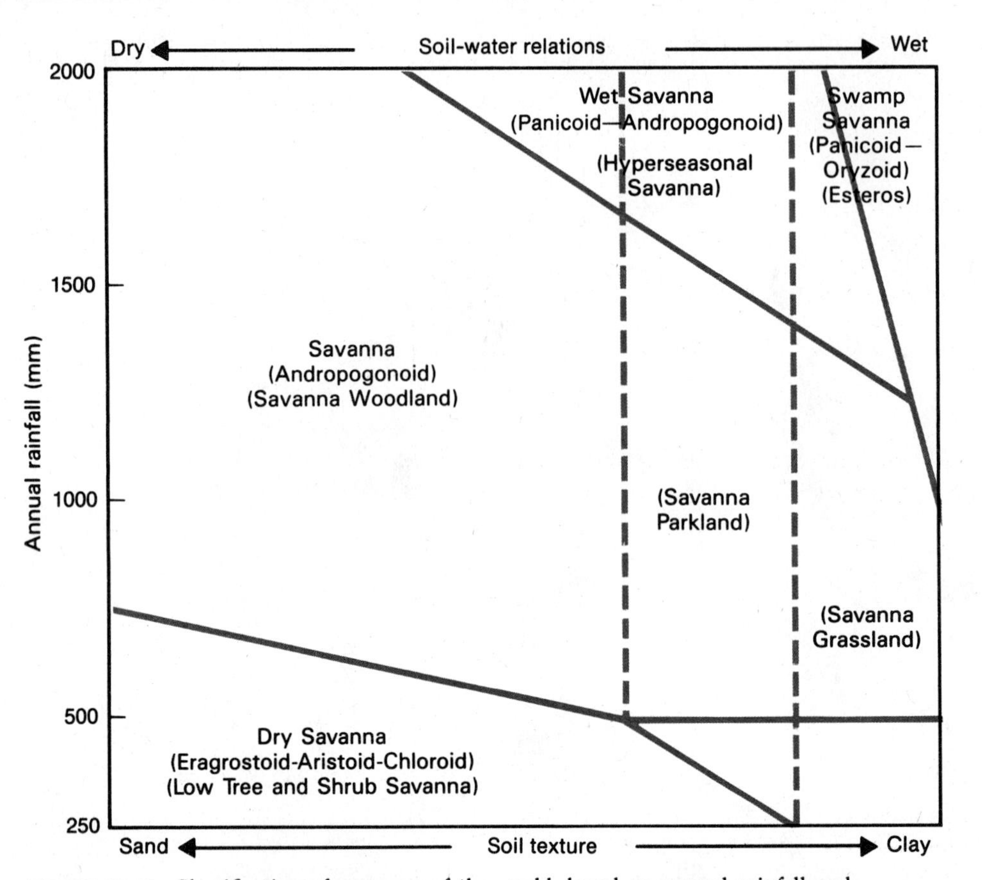

FIGURE 26.10 Classification of savannas of the world, based on annual rainfall and soil moisture as influenced by soil texture. Dominant grasses include the genera *Andropogon, Panicum, Oryza,* and *Eragrostis.*

diversity of wildlife and provide forage in the dry season for buffalo, waterbuck, and other large ungulates.

Animal Life

Savannas support or are capable of supporting a large and often varied assemblage of herbivores, invertebrate and vertebrate, grazing and browsing. Dominant herbivores are the invertebrates, including acrid mites, acridid grasshoppers, seed-eating ants, and detrital-feeding dung beetles and termites. Savanna vegetation supports an incredible number of insects: flies, grasshoppers, locusts, crickets, carabid beetles, and especially termites and ants, which dominate insect life. Insect abundance is seasonal and is strongly affected by burning, which can reduce populations by more than 60 percent. In South American savannas, where there is a strong element of grazing ungulates represented by pampas deer and the capybara (*Hydrochoerus hydrochaeris*), granivorous, insectivorous, and frugivorous birds become an important component of the consumer community.

The African savanna, visually at least, is dominated by a large and diverse ungulate fauna of at least 60 species that partition the vegetative resource among them. Some, such as the wildebeest and zebra, are migratory during the dry season; some, such as the impala, partially disperse during the dry season; others, such as the giraffe and Grant's gazelle, have little or no seasonal dispersal. Among the ungulates zebras and wildebeest are generalist grazers. Zebras, especially

FIGURE 26.11 Two major woody browsers of the African woodland savanna. The giraffe feeds on the tall woody growth; the endangered black rhino feeds on low shrubs.

during the migratory period, feed on upper grass leaves, low in protein. Wildebeest feed on the more nutritious grasses, and the small gazelles, being more refined feeders, live on the lower grasses left behind, especially the new short growth at the beginning of the rainy season. Other ungulates, such as giraffe, Thompson's gazelle, kudu, and black rhino (Figure 26.11), are woody browsers. A close interaction exists among the grazing herbivores, and intensive grazing pressure by one species can affect the populations of others. In spite of their visual dominance, large ungulates consume only about 10 percent of primary production.

Putting the level of consumption aside, herbivores have short-term and long-term impacts on the savanna. Over the short term, the grazing ungulates affect vegetation structure. Elephants can convert woodland to grassland (see Chapter 21), and large concentrations of grazers can turn grassland to eroded, bare ground. The species composition and structure of the African savanna vegetation would be different if it were not subject to heavy grazing, which alters competitive interactions among plants. Heavy grazing that reduces grass cover can competitively release woody growth, resulting in bush encroachment.

Over evolutionary time the selective pressures of grazing have resulted in the development of structural and chemical defenses against grazing, such as the concentration of silica in the leaves, and in the alteration of growth processes that respond favorably to grazing. Highly palatable plants retain a high carbohydrate concentration in their crowns and roots and respond quickly to defoliation. Some acacia trees respond to browsing by increasing growth. Others have their growth form altered and size reduced by browsing, especially by giraffes.

Living on the ungulate fauna are an array of carnivores, including the lion, leopard, cheetah, hyaena, and wild dog. Subsisting on leftover prey are a number of scavengers, including vultures and jackals.

FUNCTION

Because of the wide diversity of savanna types and limited studies, it is difficult to make any strong generalizations about primary production. Probably a wide range of production exists between grass savanna on one end of the gradient and tree savanna and woodlands on the other. It is initiated at the beginning of the wet season when moisture releases nutrients from materials accumulated in the dry season and stimulates nutrient translocation from the roots. This action is followed by a quick flush of growth into grass and woody plants. Nutrient movement between soil and vegetation is generally higher under the trees than in the open, because of greater organic matter accumulation and reduced evaporation, which keeps the soil moist. Savanna trees, especially the African acacias, exhibit tight internal cycling. Nitrogen concentration in the leaves, for example, decreases as the dry season approaches, with maximum withdrawal before leaf fall. The trees transfer some of the nitrogen into new woody growth, but much of it goes to the root reserve, where it is available to stimulate the flush of new season growth. A similar tight circulation exists in neotropical savannas. Most of the nitrogen in the dry aboveground biomass is lost to the atmosphere by volatilization if fire sweeps the savanna; otherwise a fraction will be transferred to the soil through leaching effects of rainwater.

Although the influence of large herbivores on the overall nutrient dynamics over the long term is a topic of current discussion, the impact of ants and especially termites cannot be questioned. Ants and termites consume and break down plant litter and modify the soil properties. Mound-building termites excavate and move tons of soil, mixing mineral soil with organic matter. Some species construct extensive subterranean galleries, and others accumulate organic matter. Comprising over 50 percent of soil biomass, termites have a considerable impact on the physical and chemical properties of savanna soil.

HUMAN IMPACT

Ever since their early evolution, humans have had such a close association with savanna vegetation, especially in Africa, that it is difficult to separate human influences from natural influences on the shaping of savanna ecosystems. In South America, savanna vegetation in many parts has been and still is relatively free of human impact. In Africa and Australia, however, fire, although always part of the environment, increased its importance with the arrival of early humans.

In modern times humans have had a severe and often adverse impact on the savanna ecosystem. Introduction of crops, grazing animals, and settlements has accentuated the dry season and increased the desiccation of the drier savannas. This desiccation has allowed the desert to encroach; such encroachment is most evident in the Sahelian zone south of the Sahara Desert in North Africa and in the Great Indian desert. Cutting and burning of trees for fuel wood, destruction by domestic grazing animals, and the loss of grass cover help to expose the soil to wind and water erosion. In some areas of Africa, savanna vegetation is being converted to pine and eucalyptus forests for wood and paper pulp. Widespread slaughter of the large grazing herbivores has also changed the character of the savanna vegetation. Many thousands of hectares have been converted to croplands of corn, pineapple, sisal, or improved pastures by addition of fertilizers and the introduction of exotic forage grasses and legumes. In Africa, savanna vegetation can support five times as much standing biomass of wild ungulates as it can domestic livestock, because wild ungulates can better utilize savanna forages, withstand heat stress, resist disease, gain weight, grow to marketable size, and yield more meat than domestic cattle. For this reason, an interest in game ranching is increasing, which would be a major step for the preservation of wild ungulates and the maintenance of savanna ecosystems.

Summary

Natural grasslands occupy regions where rainfall is between 250 mm and 800 mm a year, but they are not exclusively climatic. Many exist through the intervention of fire and human activity. Once covering extensive areas of the globe, grasslands have shrunk to a fraction of their original size because of conversion to cropland and grazing lands. Disappearing along with the native grasslands have been the large native grazing herbivores, replaced in part by domestic livestock. Conversion of forests into agricultural lands, the planting of hay and pasture fields, and the development of successional grasslands on disturbed sites extended the range of some grassland animals into once-forested regions. Successional and climax grasslands consist of sod formers, bunchgrasses, or both. Depending upon their fire history and degree of grazing, grasslands accumulate a layer of mulch that retains moisture, influences the character and composition of plant life, and provides shelter and nesting sites for some animals.

Productivity varies considerably, influenced greatly by precipitation. It ranges from 400 kcal/m^2/yr in semiarid grasslands to 30 times that much in subhumid, tame, and cultivated grasslands. The bulk of primary production goes underground to the roots. To a point grazing stimulates primary production. Although the most conspicuous grazers are the large herbivores, the major consumers are invertebrates. The heaviest consumption takes place below ground, where the dominant herbivores are nematodes. Most of the primary production goes to decomposers. Nutrients are recycled rapidly. A significant quantity goes to the roots to be moved above ground to next year's growth.

Savannas are grasslands with woody vegetation. They are characteristic of regions with alternating wet and dry seasons. Difficult to characterize precisely, savannas range from grass with an occasional tree to shrub and tree savannas. The latter grade into woodland and thornbush with an understory of grass. Much of the nutrient pool is tied up in plant and animal biomass, but nutrient turnover is high with little accumulation of organic matter. Savannas of Africa and Australia may have evolved under human influences of fire and the impact of grazing animals. All savannas in some manner or another have been altered by humans, including overgrazing by domestic animals, conversion to cropland, and elimination of great herds of grazing and browsing animals.

Review and Study Questions

1. What is the difference between a bunchgrass and a sod grass?
2. What characteristics do all grasslands have in common?
3. Why does the root system assume such importance in the grassland ecosystem?
4. What is the role of mulch in grassland ecosystems?
5. How have grasses adapted to grazing?
6. What characterizes grassland animal life?
7. Contrast the production and function of the aboveground and belowground components of grassland ecosystems. Why does so much energy flow (production) and nutrient cycling take place below ground?
8. What distinguishes savannas from grasslands in structure and function?
9. Where are the world's savannas located? Under what climatic conditions have they developed?
10. What is the relationship of humans to savannas?

*11. Speculate on why there are no grassland national parks and so few grassland reserves in North America. Should we have a Grassland National Park?

*12. Prior to 1930, Frederick Clements, the plant ecologist, considered the plowing of the shortgrass plains for wheat short-sighted and urged their preservation by restocking them with commercial herds of bison. Comment on his rejected suggestion. Does it have merit today? Also consider this question relative to game ranching in Africa.

Selected References

Grasslands

Breymeyer, A., and G. Van Dyne, eds. 1980. *Grasslands, systems analysis and man.* Cambridge, England: Cambridge University Press. Synthesis of International Biological Programme studies of grasslands; comprehensive.

Coupland, R. T., ed. 1979. *Grassland ecosystems of the world: Analysis of grasslands and their uses.* Cambridge, England: Cambridge University Press. Good reference.

Duffey, E. 1974. *Grassland ecology and wildlife management.* London: Chapman and Hall. Examines ecology of tame grasslands.

Falt, J. H. 1976. Energetics of a suburban lawn ecosystem. *Ecology* 57:141–150. Ecological concepts applied to the lawn. An interesting study.

French, N., ed. 1979. *Perspectives on grassland ecology.* New York: Springer-Verlag. A good summary of grassland ecology.

Reichman, O. J. 1987. *Konza prairie: A tallgrass natural history.* Lawrence: University of Kansas Press. An excellent overview of the tallgrass prairie as revealed by studies on a natural reserve.

Risser, P. G., et al. 1981. *The true prairie ecosystem.* Stroudsburg, PA: Dowden, Hutchinson, & Ross. Ecology of midgrass and tallgrass prairies.

Weaver, J. E. 1954. *North American prairie.* Lincoln, NE: Johnson. An old but important study of prairie vegetation written by an outstanding botanist who was familiar with the original condition of the prairies.

Savannas

Bourliere, F., ed. 1983. *Tropical savannas. Ecosystems of the world* 13. Amsterdam: Elsevier Scientific Publishers. A major reference work on savannas.

Sarimiento, G. 1984. *Ecology of neotropical savannas.* Cambridge, MA: Harvard University Press. A major study of South American savannas.

Sinclair, A. R. E., and M. Norton-Griffiths, eds. 1979. *Serengeti: Dynamics of an ecosystem.* A study of the African savanna ecosystem. Excellent.

Tothill, J. C., and J. J. Mott, eds. 1985. *Ecology and management of the world's ecosystem.* Canberra: Australian Academy of Science. Group of papers providing in-depth review of savannas and their management. An important reference.

CHAPTER 27

Shrublands and Deserts

Outline

Objectives

On completion of this chapter, you should be able to:

1. Describe the major characteristics of shrubs and shrublands.
2. Characterize the major types of world shrublands.
3. Discuss the relationship between growth forms of shrubs and nutrient cycling.
4. Describe the major features of deserts.
5. Discuss the unique aspects of nutrient cycling in desert ecosystems.
6. Discuss human impacts on both shrublands and deserts.

SHRUBLANDS

Covering large portions of the arid and semiarid world is climax shrubby vegetation. In addition, climax shrubland exists in parts of temperate regions because historical disturbances of landscapes have seriously affected their potential to support forest vegetation. Among such shrub-dominated human-induced climax communities are the *moors* of Scotland and the *macchia* of South America. Outside these regions, shrublands are seral, a stage in the land's progress back to forest. There they exist as second-class citizens of the plant world, given little attention by botanists, who tend to emphasize dominant plants. As a result, little work has been done on seral shrub communities.

CHARACTERISTICS

Shrubs are difficult to characterize. They have, as W. G. McGinnes points out, a "problem in establishing their identity." They constitute neither a taxonomic nor an evolutionary category. One definition is that a shrub is a plant with woody, persistent stems but no central trunk and a height of up to 4.5 or 6 m, but size does not set shrubs apart, because under severe environmental conditions many trees will not exceed that size. Some trees, particularly coppice stands, are multiple-stemmed, and some shrubs have large, single stems. Shrubs may have evolved either from trees or from herbs.

The success of shrubs depends largely on their ability to compete for nutrients, energy, and space. In certain environments shrubs have many advantages. They have a lower energetic and nutrient investment in aboveground parts than trees. Their structural modifications affect light interception, heat dissipation, and evaporative losses, depending on the species and environments involved. The more arid the site, the more common is drought deciduousness and the less common is evergreenness. The multistemmed forms influence interception and stemflow of moisture, increasing or decreasing infiltration into the soil. Because most shrubs can get their roots down quickly and form extensive root systems, they can utilize soil moisture deep in the profile. This feature gives them a competitive advantage over trees and grasses in regions where the soil moisture recharge comes during the nongrowing season. Because they do not have a high root–shoot ratio, shrubs draw less nutrient input into aboveground biomass and more into roots. Their perennial nature allows immobilization of limiting nutrients and slows the nutrient recycling process, favoring further shrub invasion of grasslands.

Subject to strong competition from herbs, some climax shrubs, such as chamise *(Adenostoma fasciculatum),* inhibit the growth of herbs by means of allelopathy. Only when fire destroys mature shrubs and degrades the toxins do herbs appear in great numbers. As the shrubs recover, herbs decline. The herb species affected have apparently evolved the ability to let their seeds lie dormant in the soil until they are released from suppression by fire.

TYPES OF SHRUBLAND

Mediterranean-Type Shrublands

In five regions of the world, lying for the most part between 32° and 40° north and south of the equator, are areas with a mediterranean climate: the semiarid regions of western North America, the regions bordering the Mediterranean Sea, central Chile, the Cape region of South Africa, and southwestern and southern Australia. The mediterranean climate is characterized by hot, dry summers with at least one month of protracted drought and cool, moist winters. About 65 percent of the annual precipitation falls during the winter months and for at least one month the temperature remains below 15° C.

All five areas support visually similar communities of xeric broadleaf evergreen shrubs and dwarf trees known as sclerophyll (*scleros,* "hard"; *phylll,* "leaf") vegetation with herbaceous understory. Sclerophyllous vegetation possesses such features as small leaves, thickened cuticles, glandular hairs, and sunken stomata. Although vegetation in all mediterranean-type ecosystems shares certain characteristics, exhibiting strong convergence in vegetation forms, each has evolved its own distinctive flora and fauna. In the Northern Hemisphere the vegetation evolved from tropical floras and developed in dry summer climates that did not exist until the Pleistocene.

In addition to similar forms, vegetation in each of the mediterranean systems also shows similar adaptations to fire and to low nutrient levels in the soil. In the mediterranean systems of Europe and North America, annuals make up 50 percent of the species and 10 percent of the plant genera; 40 percent of the species are endemics.

There are variations in the basic mediterranean-type ecosystems. In the Mediterranean region, shrub vegetation often results from forest degradation and falls into three major types. The *garigue,* resulting from degradation of pine forests, includes a number of different types of dwarf-shrub communities less than 0.5 m high, dominated by mostly aromatic evergreen shrublets on well-drained to dry, calcareous soil. The *maquis,* replacing cork forests, is a dense evergreen sclerophyllous shrub community growing in areas where the climate is comparatively moist. The *mattoral,* a Spanish term for all evergreen bush communities including garigue and maquis, appears to be equivalent to the North American chaparral. The Chilean mediterranean system, also called mattoral, varies across topographic positions from the coast to slopes of coastal ranges and the foothills of the Andean cordillera.

In North America the sclerophyllous shrub community is known as *chaparral,* a word of Spanish origin meaning a thicket of shrubby evergreen oaks. California chaparral is dominated by scrub oak *(Quercus dumosa)* and chamise (*Adenostoma fasciculatum*). Another shrub type, also designated as chaparral, is associated with the Rocky Mountain foothills in Arizona, New Mexico, and Nevada, and elsewhere. It differs from California chaparral in two ways. It is dominated by Gambel oak *(Q. gambelii)* and other species and lacks chamise; it is summer-active and winter-deciduous, whereas California chapparal is evergreen, winter-active, and summer-dormant.

In their original presettlement state, both the Mediterranean and California mediterranean plant communities were dominated by oaks, both shrub and tree. Stands of evergreen holm oak *(Quercus ilex)*, found in the warm, droughty regions, and the deciduous white oak *(Q. pubescens)*, occurring further inland and at higher altitudes where the growing season is shorter and the temperatures lower, were common. Cork oak *(Q. suber)* forests were widespread throughout the western Mediterranean. Natural oak forests remain only in scattered patches. In Spain, cork oak is a plantation tree, and many oak forests throughout the region have been converted to olive plantations. In California, four major oak forest communities supported a number of species including two endemics, the evergreen blue oak *(Q. douglassii)* and valley oak *(Q. lobata)*. The ranges of both have been greatly reduced by settlement.

Much of the South African mediterranean shrubland is heathland known as *fynbos,* discussed later. The more typical mediterranean-type shrubland, dominated by a broad-sclerophyll woody shrub, goes by the names of *strandveld, coastal renosterveld,* and *inland renosterveld.*

In southwest Australia the mediterranean shrub country, known as *mallee,* is dominated by low-growing *Eucalyptus,* 5 to 8 m high with broad sclerophyllous leaves. There are six types of mallee ecosystems, which intergrade. Three of them fall into mediterranean-type ecosystems, with a grassy and herbaceous understory (Figure 27.1). The other three types occur on nutrient-poor soils and fall under the category of heathland shrubs. Razed by fire at irregular intervals, the mallee ecosystem differs markedly from typical mediterranean-type ecosystems in its summer growth rhythm. Mediterranean-type shrublands typically initiate new growth during the spring; but the mallee retains its summer growth rhythm evolved from its origins in the subtropical Tertiary, which is out of phase with the mediterranean climate of the area. Growth takes place during the summer, the driest part of the year, using extensive root systems to draw on water conserved in the soil during the wet winter and spring showers.

For the most part, mediterranean-type shrublands lack an understory and ground litter, are highly inflammable, and are heavy seeders. Many species require the heat and scarring action of fire to induce germination. Others sprout vigorously after a fire.

For centuries periodic fires have roared through mediterranean-type vegetation, clearing away the old growth, making way for the new, and recycling nutrients through the ecosystem. When humans intruded on this type of vegetation, they changed the fire regime, either by attempting to exclude fire completely or by

FIGURE 27.1 Tall shrub mallee in Victoria, Australia, is an example of a mediterranean-type shrubland dominated by *Eucalyptus*. Note the canopy structure and the open understory of grass at the beginning of spring rains. This type of vegetation supports a rich diversity of birdlife.

FIGURE 27.2 The northern desert shrubland in Wyoming is dominated by sagebrush. Although classified as a cold desert, sagebrush forms one of the most important shrub types in North America.

overburning. In the absence of fire, chaparral grows tall and dense and yearly adds more leaves and twigs to those already on the ground. During the dry season the shrubs, even though alive, will nearly explode when ignited. Once set on fire by lightning or humans, an inferno follows.

After fire the land returns either to lush green sprouts coming up from buried root crowns or to grass, if a seed source is nearby. New grass and vigorous young sprouts are excellent food for deer, sheep, and cattle. As the sprout growth matures, chaparral becomes dense, the canopy closes, the litter accumulates, and the stage is set for another fire.

Northern Desert Scrub

In the Great Basin of North America, the northern, cool, arid region lying east of the Rocky Mountains, is the northern desert scrub. The climate is continental, with warm summers and prolonged cold winters. Although this region is perhaps more appropriately considered a desert, it is one of the most important shrublands in North America (Figure 27.2). Its physiognomy differs greatly from the southern hot desert (see Desert) and the dominant vegetation is shrub. The vegetation falls into two main associations: One is sagebrush, dominated by *Artemisia tridentata,* which often forms pure stands; the other is shadscale, *Atriplex confertifolia,* a C_4 species, and other chenopods, halophytes tolerant of saline soils. Inhabiting this shrubland are pocket and kangaroo mice, lizards, sage grouse, sage thrasher, sage sparrow, and Brewer's sparrow, four birds that depend on sagebrush.

A similar type of shrubland exists in the semiarid inland of southwestern Australia. Numerous chenopod species, particularly the saltbushes of the genera *Atriplex* and *Maireana,* form extensive low shrublands on low riverine plains (Figure 27.3).

FIGURE 27.3 Saltbrush shrubland in Victoria, Australia, is dominated by *Atriplex*. It is an ecological equivalent of the shrublands of the Great Basin in North America.

FIGURE 27.4 Heathlands, dominated by ericaceous shrubs, have a similar physiognomy around the world. Typical is this heathland in the Grampian Mountains of Australia.

Heathlands

Typically, heathlands have been associated with cool to cold temperate climatic regions of northwestern Europe. It was probably coincidental that the original name came to refer to land dominated by Ericaceae. The word *heath* comes from the German word *heide,* meaning "an uncultivated stretch of land," regardless of the vegetation. It just so happened that in parts of Germany uncultivated and waste land were dominated by Ericaceae.

Heathlands are found in all parts of the world, from the tropics to polar regions and from lowland to subalpine and alpine altitudes. Heathland flora probably evolved in the Mesozoic in the eastern to central portion of Gondwanaland. It survived the climatic changes of the Tertiary and Quaternary and retained most of its morphological and physiological characteristics. It expanded out of the fragments of Gondwanaland from Africa into western Europe and the northern part of Eurasia and North America, from India into southeastern Asia, and from Australasia into the Malay Archipelago.

Heathland vegetation is an assemblage of dense to mid-dense growth of ancient or primitive genera of angiosperms: evergreen sclerophyllous shrubs and subshrubs together with hemicryptophytes and therophytes, all adapted to fire (Figure 27.4). Heathland shrubs have leaves with thick cuticles, sunken stomata, thick-walled cells, and hard and waxy upper surfaces. Many species have leaves with small surface area—less than 25 mm^2—and others roll their edges in toward the midrib. Although mostly associated with heathlands, many heathland shrubs are usually present as shrubby understory in other ecosystems, such as the deciduous forest.

Heathlands invariably occur on nutrient-poor soils especially deficient in phosphorus and nitrogen. Although heathlands are most extensive in the arctic regions, they are also prominent in the mediterranean-type regions of South Africa, where they are known as *fynbos,* and in southeastern and western Australia. In subtropical to tropical climates true heathlands are confined to alpine areas and to lowland, poor soils subject to seasonal waterlogging. Some heathlands, such as the heather-dominated moors of Scotland, are human-induced and can be maintained only by periodic fires.

There are two distinct heathland ecosystems: dry heathlands and wet heathlands. Dry heathlands are on well-drained soils subject to seasonal drought, and wet heathlands are subject to seasonal waterlogging. In wet heathlands, grasses and sedges may become codominant with heathland shrubs; and in extreme wet heathlands the grass component is suppressed by *Sphagnum* moss (see Chapter 32). Because both foliage cover and height vary considerably with the habitat, heathlands are divided according to height of the uppermost stratum: shrubs taller than 2 m, scrub; shrubs 1–2 m, tall heathland; shrubs 25–100 cm, heathland; and shrubs less than 25 cm, dwarf heathland.

Successional Shrublands

On drier uplands, shrubs rarely exert complete dominance over herbs and grass. Instead the plants are scattered or clumped in grassy fields, the open areas between filled with the seedlings of forest trees, which in the sapling stage of growth occupy the same ecological position as tall shrubs (Figure 27.5). Typical are thicket-forming hazel (*Corylus* spp.), sumacs (*Rhus* spp.), and shrub dogwoods (*Cornus* spp.).

FIGURE 27.5 In eastern North America and in northern and western Europe, shrublands are usually successional communities, but if vegetation is dense, they may persist for a long time. This slope is claimed by Saint-John's-wort (*Hypericum virginicum*) and wild indigo (*Baptisia tinctoria*).

On wet ground the plant community often is dominated by tall shrubs and contains an understory intermediate between those of a meadow and a forest. In northern regions the common tall shrub communities found along streams and lake shores are thickets composed of alder or alder and a mixture of other species such as willow (*Salix* spp.) and red osier dogwood (*Cornus stolonifera*). Alder thickets are relatively stable and remain for some time before being replaced by forests. Outside of alder country, a shrub or *carr* community (carr is an English name for wet-ground shrub communities) occupies the low places. Dogwoods are some of the most important species in the carr. Growing with them are a number of willows, which as a group usually dominate the community.

Shrub thickets provide excellent food and cover for wildlife. Many shrubs, such as blackberry, hawthorn, greenbriar, and dogwoods, rank high as wildlife food. However, the overall value of different types of shrub cover, its composition, quality, and minimum amounts needed, have never been assessed. Seral shrub communities are ephemeral, lasting only about 15 to 20 years. There is some evidence, however, that even where forest is the normal end of succession, shrubs can form a stable community that will persist for many years. If incoming tree growth is removed either by selective herbicidal spraying or by cutting, shrubs eventually form a closed community resistant to further invasion by trees. This response has wide application, rarely used, in vegetation management of power line rights-of-way.

STRUCTURE

Shrub ecosystems, seral or climax, are characterized by woody structure, increased stratification over grasslands, dense branching on a fine scale, and low height, up to 8 m. Typically there are three layers—a broken upper canopy, an irregular low shrub canopy, and a grass/herbaceous layer—but the presence of these layers varies. Dense shrubland may have only a canopy layer, and stratification often decreases as the shrubs reach maximum height. This condition is particularly true in seral shrublands. Horizontal patterns vary with the vegetation type. Heathlands may exhibit little patchiness across the landscape. Mediterranean-type shrublands, notably the mattoral and the mallee, may be very patchy, with woody growth well interspersed with open areas of grass. Greatest patchiness probably occurs in seral shrublands, with scattered clumps of invading shrubs and trees.

Shrub communities have their own distinctive animal life. Seral shrub communities support not only species common to shrubby edges of forest and shrubby borders of fields but a number of species dependent on them, such as bobwhite quail, cottontail rabbit, prairie warbler *(Dendroica discolor),* and yellow-breasted chat *(Ictera virens).* In Great Britain some shrub communities, especially hedgerows, have been stable for centuries, and many forms of animal life, invertebrate and vertebrate, have become adapted to or dependent on them. Among these species are the whitethroat *(Sylvia communis),* linnet *(Acanthis cannabina),* blackbird *(Turdus merula),* and yellowhammer *(Emberiza citrinella).*

Climax shrub communities have a complex of animal life that varies with the region. Within the Mediterranean-type shrublands and heathlands, similarity in habitat structure and in the nature and number of niches has resulted in pronounced parallel and convergent evolution among bird species and some lizard species, especially between the Chilean mattoral and the California chaparral. In North America chaparral and sagebrush communities support mule deer *(Odocoileus hemionus),* coyotes *(Canis latrans),* a variety of rodents, jackrabbits *(Lepus* spp.), and sage grouse (*Centrocercus urophasianus*). The Australian mallee is rich in birdlife and the habitat of the endemic malleefowl (*Leipoa ocellata*), which incubates its eggs in a large mound. Among the mammalian life are the gray kangaroo and various species of wallaby. Rash clearance of mallee vegetation is endangering mallee wildlife as well as affecting millions of migrant, nectar-feeding birds supported by the mallee in spring.

FUNCTION

Precipitation, temperature, soil moisture, and nutrients are major influences on the function of mediterranean-type systems. Precipitation falls mostly in the cool winter months, and most of the plant growth and flowering is concentrated in spring, much of it at the end of the rainy season. Then the plants have to respond to the dry season, which imposes a great deal of environmental stress. How they respond is reflected in plant growth forms. Many plants of the mediterranean systems are shrubs, other are trees; some are deciduous, others are evergreen. Plants living under high environmental stress involving aridity, cold temperatures, short growing season, long periods of drought, and a low nutrient supply are mostly shrubs. Where drought periods are shorter and neither temperature nor nutrients strongly limit growth, trees grow. Whether the plant is deciduous or evergreen depends on the costs of maintaining leaves. If the physiological costs of growing new leaves each year are less than maintaining the same leaves, the plant acquires a deciduous habit. If the costs are greater, then the plants are evergreen. Evergreenness in mediterranean ecosystems confers some advantages. Present throughout the year, evergreen leaves can take advantage of favorable conditions for photosynthesis; and because the individual leaves remain on the plant for at least two years, they can store nutrients during the wet winters for later growth. Because of their structure, including thick leaves and sunken stomata, they possess a greater drought tolerance.

Soils of mediterranean-type ecosystems are low in nutrients and are especially deficient in nitrogen and phosphorus. During the dry period, nitrogen and other nutrients accumulate beneath the woody plants and remain fixed as the topsoil dries out. Wetting of the soil during the winter stimulates a flush of microbial activity involving decomposition of humus and mineralization of nitrogen and carbon. The concentration of nutrients made available stimulates a flush of growth. If heavy rains suddenly enter dry topsoil, quantities of nutrients may be lost by leaching and erosion.

Some plants of the mediterranean systems conserve nutrients by various means. *Ceanothus,* an early successional species in the California chaparral, is a nonleguminous nitrogen-fixer. In the Australian mallee *Atriplex vesticana,* a dominant plant, lowers the nitrogen content of the surrounding soil during the growing season and concentrates nitrogen directly beneath it through litterfall. Because the plant withdraws more nitrogen from the soil than it returns by litterfall and because its litter has lower nitrogen and phosphorus content than fresh leaves, *Atriplex* must transfer nitrogen and phosphorus from its leaves to its stems before leaf fall.

HUMAN IMPACT

Just as the various mediterranean ecosystems share many characteristics in common, so they also experience similar fates brought on by human activity. Many are in their present states because of past and current human disturbance or degradation of other vegetation types. The greatest impact has occurred in the lands about the Mediterranean Sea. Early colonization and long human occupancy dating back 10,000 years have had a profound influence on the vegetation. Cycles of logging for oak and pine to build ships and to construct settlements, wood cutting for fuel, clearing for agricultural lands, terracing hillsides for crops, vineyards, and olive groves, irrigation, land abandonment followed by destruction of terraced walls and decay of terraced hill land, overgrazing by cattle, sheep, and goats brought about by pastoral nomadism, followed by soil erosion, have degraded or destroyed vegetation. More recently, spreading urbanization, overcrowded settlements and recreational areas, and pollution threaten the future of these landscapes.

Mediterranean-type vegetation of California and Chile has undergone somewhat similar degradation, but over a much shorter period of time. Although native peoples exploited the plant communities for food, the major impact of humans began after Spanish colonization began in 1779. Many California oak communities have disappeared because of stock raising, charcoal burning, agricultural clearing, and rapidly spreading urbanization and suburbanization. What oak communities remain occur as scattered patches in inaccessible areas and in managed grazing land. Shrubby plant communities are extensive in rugged terrain, but they have been invaded by housing developments. In Chile the mattoral has been degraded by the grazing of sheep and goats, or cleared and replaced by cropland and tree plantations. In Australia grazing not only by sheep but by the introduced rabbit threatens the disappearance of numerous species of plants. Because of rough terrain, much of the South African fynbos remains in its natural state.

Seral shrub communities for the most part result from human disturbances such as land clearing and abandonment and logging. Once established, they are valuable as wildlife habitat and sources of food, such as fruits and nuts. At the same time shrublands are subject to destruction by humans. Too often unappreciated and regarded as worthless wasteland, shrublands are cleared for grazing land, converted into housing developments, and destroyed by land reclamation projects. Wetland shrub communities are drowned by flood control and hydroelectric projects or drained for other uses. Hedgerows, once a familiar part of the landscape in the United States and western Europe, have been torn out to enlarge agricultural fields to sizes required by mechanization. Because seral shrublands are ephemeral, lasting only 15 to 20 years before entering the forest stage of succession, they must be maintained by cutting, burning, or other techniques if they are to provide stable habitats for shrubland wildlife.

DESERTS

Geographers define deserts as land where evaporation exceeds rainfall. No specific amount of rainfall serves as a criterion; deserts range from extremely arid regions to those with sufficient moisture to support a variety of life. Deserts have been classified according to rainfall into *semideserts,* ones that have precipitation between 150 and 300 to 400 mm per year; *true deserts,* regions with rainfall below 120 mm per year; and *extreme deserts*, areas with rainfall below 70 mm per year. Deserts, which occupy about 26 percent of the continental area, occur in two distinct belts between 15° and 35° latitude in both the Northern and Southern Hemispheres—the Tropic of Cancer and the Tropic of Capricorn.

Deserts are the result of several forces. One force that leads to the formation of deserts and the broad climatic regions of Earth is the movement of air masses (see Chapter 4). High-pressure areas alter the course of rain. The high-pressure cell off the coast of California and Mexico deflects rainstorms moving south from Alaska to the east and prevents moisture from reaching the Southwest. In winter high-pressure areas move southward, allowing winter rains to reach southern California and parts of the North American desert. Winds blowing over cold waters become cold also. They carry little moisture and produce little rain. Thus the west coast of California and Baja California, the Namid Desert on coastal southwest Africa, and the coastal edge of the Atacama in Chile may be shrouded in mist, yet remain extremely dry.

Mountain ranges also play a role in desert formation by causing a rain shadow on their lee side. The high Sierras and Cascade Mountains intercept rain from the Pacific and help maintain the arid conditions of the North American desert. The low eastern highlands of Australia block the southeast trade winds from the interior. Other deserts, such as the Gobi and the interior of the Sahara, are so remote from the ocean that all of the water has been wrung from the winds by the time they reach those regions.

CHARACTERISTICS

All deserts have in common low rainfall, high evaporation (from 7 to 50 times as much as precipitation), and a wide daily range in temperature from hot by day to cool by night. Low humidity allows up to 90 percent of solar insolation to penetrate the atmosphere and heat the ground. At night the desert yields the accumulated heat of the day back to the atmosphere. Rain, when it falls, is often heavy and, unable to soak into the dry earth, rushes off in torrents to basins below.

The topography of the desert, unobscured by vegetation, is stark and, paradoxically, partially shaped by water. The unprotected soil erodes easily during violent storms and is further cut away by the wind. Alluvial fans stretch away from eroded, angular peaks of more resistant rocks. They join to form deep expanses of debris, the *bajadas*. Eventually, the slopes level off to low basins, or *playas,* which receive waters that rush down from the hills and water-cut canyons, or *arroyos.* These basins hold temporary lakes after the rains, but water soon evaporates and leaves behind a dry bed of glistening salt.

Deserts are not the same everywhere. Differences in moisture, temperature, soil drainage, topography, alkalinity, and salinity create variations in vegetation cover, dominant plants, and groups of associated species. There are hot deserts and cool deserts, extreme deserts and semideserts, ones with sufficient moisture to verge on being grasslands or shrublands, and gradations between those extremes within continental deserts. There is a certain degree of similarity among hot deserts and cold deserts of the world. The cold deserts, including the Great Basin of North America, the Gobi, Takla Makan, and Turkestan deserts of Asia, and high elevations of hot deserts are dominated by *Artemisia* and chenopod shrubs, and may be considered shrub steppes. The northern part of the North American Great Basin is dominated by nearly pure stands of big sagebrush (see Figure 27.2) and the southern part by shadscale and bud sage (*Artemisia*). The hot deserts range from no or scattered vegetation to ones with some combination of chenopods, dwarf shrubs, and succulents. The deserts of southwestern North America—the Mohave, the Sonoran, and the Chihuahuan—are dominated by creosote bush (*Larrea divaricata*) and bur sage (*Franseria* spp.). Areas of favorable moisture support tall growths of *Acacia* spp., saguaro (*Cereus giganteus*), palo verde (*Cercidium* spp.), and ocotillo (*Fouquieria* spp.).

STRUCTURE

Woody-stemmed and soft brittle-stemmed shrubs are characteristic desert plants (Figure 27.6). In a matrix of shrubs grows a wide assortment of other plants, the yucca, cacti, small trees, and ephemerals. In the Sonoran desert, in the Peru-Chilean, and South African Karoo and southern Namib deserts, large succulents rise above the shrub level and change the aspect of the desert far out of proportion to their numbers (Figure 27.7). The giant saguaro, the most massive of all cacti, grows on the bajadas of the Sonoran desert. Ironwood, smoketree, and palo verde grow best along the banks of intermittent streams, not so much because they require the moisture, but because their hard-coated seeds must be scraped and bruised by the grinding action of sand and gravel during flash floods before they can germinate.

Both plants and animals are adapted to the scarcity of water either by drought evasion or by drought resistance (see Chapter 6). Plant drought-evaders flower only in the presence of moisture. They persist as seeds

(a)

(b)

(c)

FIGURE 27.6 Three hot deserts dominated by woody, brittle-stemmed shrubs. (a) Edge of the Great Victorian desert in Australia. (b) Chihauhuan Desert in Nuevo Leon, Mexico. The substrate of this particular desert is sand-sized particles of gypsum. (c) The Arabian desert in the United Arab Emeriates. The sparseness and spacing of shrubs reflect extreme aridity.

FIGURE 27.7 Saguaro dominates the aspect of this portion of the Sonoran Desert in the southwestern United States. Note the water-shaped topography.

during drought periods, ready to sprout, flower, and produce seeds when moisture and temperature are favorable. There are two periods of flowering in the North American deserts: after winter rains come in from the Pacific Northwest, and after summer rains move up from the southwest out of the Gulf of Mexico. Some species flower only after winter rains, others only after summer rains, but a few bloom during both seasons. If no rains come, these ephemeral species do not bloom. Drought-evading animals, like their plant counterparts, adopt an annual lifestyle or go into estivation or some other stage of dormancy during the dry season. For example, the spadefoot toad (*Scaphiopus*) remains underground in a gelatinous-lined underground cell, making brief reproductive appearances during periods of winter and summer rains. If extreme drought develops during the breeding season, birds fail to nest and lizards do not reproduce.

Belowground biomass in the desert can be as patchy as the aboveground biomass. Desert plants may be deep-rooted woody shrubs, such as mesquite (*Prospis* spp.) and *Tamarix,* whose taproots reach the water table, rendering them independent of water supplied by rainfall. Some, such as *Larrea* and *Atriplex,* are deep-rooted perennials with superficial laterals that extend as far as 15 to 30 m from the stems. Other perennials, such as the various species of cactus, have shallow roots, often extending no more than a few centimeters below the surface. Ephemerals have shallow and poorly branched roots reaching a depth of about 30 cm, where they pick up moisture quickly from light rains.

The desert floor is stark, a raw mineral substrate of various types devoid of a continuous litter layer. Dead leaves, bud scales, and dead twigs, mostly associated with drought-resistant species that shed them to reduce transpiring surfaces, accumulate in wind-protected areas beneath the plants and in depressions in the soil.

FUNCTION

Primary production in the desert depends on the proportion of available water used and the efficiency of its use. Data from various deserts in the world suggest that annual primary production of aboveground vegetation varies from 30 to 200 g/m^2. Belowground production is also low but greater than aboveground production. It ranges from 100 to 400 g/m^2 in arid regions and from 250 to 1000 g/m^2 in semiarid regions.

The amount of biomass that accumulates and the ratio of annual production to biomass depend on the dominant type of vegetation. In those deserts in which trees, shrubs, and cactuslike plants dominate, annual production is about 10 to 20 percent of the total aboveground standing crop biomass. Annual or ephemeral communities have a 100 percent turnover of both roots and aboveground foliage, and their annual production is the same as peak biomass. In general, desert plants do not have a high root biomass relative to aboveground shoot biomass.

Adding to primary production in the desert are lichens and green and blue-green algae, abundant as soil crusts. Blue-green algal crusts have unusually high rates of nitrogen fixation, but less than one-half of their total nitrogen input becomes part of higher plants. Approximately 70 percent of the nitrogen is short-circuited back to the atmosphere as volatilized ammonia and as N_2 from denitrification, speeded by dry alkaline soils.

Nutrient cycling in arid ecosystems is tight. Two major nutrients, phosphorus and nitrogen, are in short supply; much of them is tied up in plant biomass, living and dead. Desert plants tend to retain in stems and roots certain elements, particularly nitrogen, phosphorus, and with some, potassium, before shedding any parts. The nutrients remaining in the shed parts collect and decompose beneath the plants, where microclimate conditions created by the plants favor biological activity. The soil is further enriched by animals attracted to the shade. The plants, in effect, create islands of fertility beneath themselves.

In spite of their aridity, desert ecosystems support a surprising diversity of animal life, including a wide assortment of beetles, ants, locusts, lizards, snakes, birds, and mammals. The mammals are notably herbivorous species (Figure 27.8). Grazing herbivores of the desert tend to be generalists and opportunists in their mode of feeding. They consume a wide range of species, plant types, and parts. Desert sheep feed on succulents and ephemerals when available and then switch to woody browse during the dry period. As a last resort, herbivores consume dead litter and lichens. Small herbivores—the desert rodents, particularly the family Heteromyidae, and ants—tend to be granivores, feeding largely on seeds, and are important in the dynamics of desert ecosystems.

Herbivores can have a pronounced impact on desert vegetation, especially if they are more abundant than the range's capacity to support them. Once grazers have eaten the annual production, they eat plant reserves, especially during long dry periods. Overbrowsing can so weaken the plant that the vegetation is destroyed or irreparably damaged. Areas protected from grazing, especially grazing by goats and sheep, have a higher biomass and a greater percentage of palatable species than grazed areas.

Native plant-eating herbivores in a shrubby desert, under most conditions, consume only a small part of the aboveground primary production, but seed-eating herbivores can eat most of the seed production, close

FIGURE 27.8 Typical herbivores of the desert. (a) A jerboa (*Alactaga*), a Middle-Eastern rodent whose long ears function as heat-radiating organs. (b) The desert oryx (*Oryx*) cope with desert environment by reducing unnecessary energy expenditure.

(a)

(b)

to 90 percent of it. This consumption can have a pronounced effect on plant composition and plant populations of the desert.

Carnivores, like the herbivores, are opportunistic feeders, with few specialists. Most desert carnivores, such as foxes and coyotes, have mixed diets that include leaves and fruits; even insectivorous birds and rodents eat some plant material. Omnivory, rather than carnivory and complex food webs, seems to be the rule in desert ecosystems.

The detrital food chain seems to be less important in the desert than in other ecosystems. Although most functional and taxonomic groups of soil microorganisms exist in the desert, fungi and actinomycetes are the most prominent. Microbial decomposition, like the blooming of ephemerals, is limited to short periods when moisture is available. For this reason dry litter tends to accumulate until the detrital biomass may be greater than the aboveground living biomass. Most of the ephemeral biomass disappears through grazing, weathering, and erosion. Decomposition proceeds mostly through detritus-feeding arthropods such as termites that ingest and break down woody tissue in their guts. In some deserts, considerable amounts of nutrients are locked up in termite structures, to be released when the structure is destroyed. Other important detritivores are acarids and various isopods.

HUMAN IMPACT

In spite of their aridity, deserts have not been spared impacts from humans. Only extremely arid deserts have escaped significant disturbance. In the past, human intrusions were limited to food-gathering and hunting forays by aborigines or to grazing by nomadic pastoralists. In recent times most aborigines have vanished, and many of the pastoralists have settled into agricultural communities. Most settlements and developments in these regions have taken place in the lowlands along major river systems, such as the Nile, the Tigris, and the Euphrates rivers. Desert regions, particularly in the Middle East, have been invaded by the oil industry, radically changing and polluting the desert environment. In parts of the Middle East and in North America, urban development has expanded into the desert, depleting the fossil water supply, which ultimately will limit the lifetime of such undertakings (Figure 27.9). Irrigation agriculture, dependent on the same water supplies, has turned some areas of desert green, but for how long? The deserts of the southwestern United States have experienced not only extensive urban development, replete with green lawns and swimming pools, but also massive degradation by unrestricted recreational use of all-terrain vehicles. Widespread collection of cacti of many species and sizes for the world plant trade in the deserts of United States, Mexico, Peru, Chile, and Brazil is destroying the integrity of desert ecosystems and threatening some species with extinction in the wild.

The greatest impact occurs on the semiarid edges of the natural deserts of the world, which support some agriculture and grazing. There mismanagement of land has created new deserts. Virgin lands, even in dry climates, are able to support some vegetation. The roots of trees, shrubs, and grasses tap the deeper water supply and bind the soil. However, expanding populations and the periods of adequate rainfall encourage encroachment on marginal lands. Overcultivated and overgrazed, the land is exposed to wind and water erosion. Because these regions are subject to unpredictable droughts, the human population finds itself faced with devastating famines and increasing land degradation. Eventually the destruction is total. Vegetation and topsoil are gone, dust storms become frequent, and sand dunes advance across the land. The land has reached a point of no return. The result is desertification—the creation of new deserts on the periphery of natural deserts in northern Africa, India, China, Argentina, Chile, Mexico, and the southwestern United States. Formation of new deserts has doubled over the past 100 years.

Supplied with water and managed well, many desert areas can be converted into productive agricultural land, but poor irrigation practices that allow the seepage of water from canals and overwatering of soil cause the water table to rise. As the moisture evaporates from the surface, it leaves behind a glistening surface layer of salt toxic to plants, and the land is abandoned to the wind. This salinization process has affected irrigated lands in India, Syria, Iraq, Soviet central Asia, California's San Joaquin Valley, and the Colorado River Basin. Because irrigation depends on "mined" fossil water

FIGURE 27.9 In spite of its aridity, humans encroach on deserts wherever they are. This settlement crowds the Oasis of Antoudi in the pre-Saharan desert in Morocco, which draws on the deep aquifers and spreads onto the surrounding desert.

beneath the desert and water drawn from rivers, irrigation severely affects the hydrology of deserts, further threatening the future of these regions.

Summary

Shrublands, which go by different names in various parts of the world, dominate regions with a mediterranean-type climate in which winters are mild and wet and summers are long, hot, and dry. Heathlands, dominated by ericaceous shrubs, are associated with cool to cold temperate climates of northern regions of the Northern Hemisphere and arctic, subalpine, and alpine regions of the world. Successional shrublands occupy land in transition from grassland to forest.

Shrublands characteristically have a densely branched woody structure and low height. The success of shrubs depends on their ability to compete for nutrients, energy, and space. In semiarid situations shrubs have numerous competitive advantages, including structural modifications that affect light interception, heat loss, and evaporation. Growth in mediterranean-type shrublands is concentrated at the end of the wet season, when nutrients in solution and a relative abundance of moisture produce a flush of vegetation. Nutrient cycling, especially of nitrogen and phosphorus, is tight. Many plants translocate nutrients from leaves to stems and roots before leaf fall; others concentrate nutrients in litterfall, which the plants take up again quickly in the wet season.

Deserts occupy about one-seventh of Earth's land surface and are largely confined to two worldwide belts, around the Tropic of Capricorn and the Tropic of Cancer. Deserts result largely from the climatic patterns of Earth, rain-blocking mountain ranges, and remoteness from oceanic moisture. Two broad types of deserts exist: cool deserts exemplified by the Great Basin of North America, and hot deserts, like the Sahara.

The desert is a harsh environment in which plants and animals have evolved ways of circumventing aridity and high temperature by becoming either drought-evaders or drought-resistors. Functionally, deserts are characterized by low net production, by opportunistic feeding patterns for herbivores and carnivores, and by a detrital food chain that is less important than in other ecosystems.

For centuries humans have occupied mediterranean-type shrublands, and lived on the periphery of deserts; some have made the desert their home. Overgrazing, wood cutting, and agricultural practices over the centuries have degraded most mediterranean-type shrublands and impoverished the soil. Poor land use about the desert's edge has caused the expansion of deserts, a process called desertification, which is continuing at an alarming rate. Settlement and agricultural development of the desert have been achieved by tapping deep but unreplenishable water reserves below the desert floor. The danger, however, already exists that human occupancy of the desert may so deplete the region's water resources and cause salinization of the soil that even more arid conditions will result.

Review and Study Questions

1. What characteristics do seral and climax shrublands share?
2. Describe mediterranean-type shrublands.
3. Argue that successional shrublands are not wastelands.
4. Contrast the benefits of evergreen and deciduous leaves in a mediterranean-type shrubland.
*5. Review the historical development of the Middle East and other Mediterranean lands relative to land degradation. What would those lands be like today had the people given some consideration to good land management? Were good land management techniques known and available in ancient times? Why are we making the same disastrous mistakes today?
6. What climatic forces lead to deserts? Human influences?
7. In what general ways are plants and animals of the desert adapted to aridity?
8. What is unique about nutrient cycling in the desert ecosystem?
*9. Human relationships to the desert are two-way. On the one hand, through the mismanagement of desert ecosystems, humans cause the spread of deserts in semiarid lands (desertification). On the other hand, humans attempt to bring the desert into greater productivity by irrigation. Discuss this two-way relationship and how both lead to the spread of deserts over time. (*Hint:* What is the relationship of desert reclamation to the mining of fossil water and increasing salinity of the soil?)
*10. Project future problems in the southwestern United States if development continues in the hot desert regions. What problems are already surfacing?
*11. Review the history of the Sahel in northern Africa as an example of how human errors, mismanagement, and poor understanding of desert ecology have led to massive economic and social problems across several African nations.

Selected References

Brown, G. W., Jr., ed. 1976–1977. *Desert biology.* 2 vol. New York: Academic Press. Basic information on biology and physical features of world deserts.

Castri, F. Di., and H. A. Mooney, eds. 1973. *Mediterranean-type ecosystems.* Ecological Studies No. 7. New York: Springer-Verlag. An excellent basic introduction to all aspects of mediterranean ecosystems.

Castri, F. Di, D. W. Goodall, and R. L. Specht, eds. 1981. *Mediterranean-type shrublands.* Ecosystems of the World No. 11. Amsterdam: Elsevier Scientific. A more comprehensive treatment of mediterranean ecosystems. A major reference.

Evenardi, M., I. Noy-Meir, and D. Goodall, eds. 1985–1986. *Hot deserts and arid shrublands of the world.* Ecosystems of the World 12A and 12B. Amsterdam: Elsevier Scientific. A major reference work on world deserts.

Groves, R. H. 1981. *Australian vegetation.* Cambridge, England:

Cambridge University Press. Concise descriptions of the diverse plant communities of Australia.
McKell, C. M., ed. 1983. *The biology and utilization of shrubs.* Orlando, FL: Academic Press. A major reference on shrubs worldwide.
Plumb, T. R., tech. coordinator. 1979. *Proceedings of the symposium on the ecology, management, and utilization of California oaks.* Gen Tech. Rept. PSW-44. Berkeley, CA: Pacific Southwest Forest and Range Experiment Station, Forest Service, U.S. Dept. of Agriculture. Thorough coverage of this important and degraded mediterranean ecosystem.
Polunin, O., and M. Walters. 1985. *A guide to the vegetation of Britain and Europe.* Oxford, England: Oxford University Press. An outstanding survey of the plant communities of Europe as far as the Russian border; well illustrated.
Specht, R. L., ed. 1979, 1981. *Heathlands and related shrublands.* Ecosystems of the World 9A and 9B. Amsterdam: Elsevier Scientific. A major reference on heathland communities worldwide.
Wagner, F. H. 1980. *Wildlife of the deserts.* New York: Harry W. Abrams. An excellent, well-illustrated introduction to the general characteristics of desert animals and their adaptation to arid environments.

CHAPTER 28

Tundra and Taiga

Outline

Objectives

On completion of this chapter, you should be able to:

1. Describe the characteristics of the tundra ecosystem.
2. Compare the arctic, alpine, and tropical alpine tundras.
3. Explain the role of permafrost in the arctic environment.
4. Describe the effects of human disturbances on arctic and alpine tundras.
5. Describe the characteristics of the taiga and its vegetation.
6. Explain the role of moss and lichens in the taiga.
7. Discuss the effects of human disturbance on the taiga.

TUNDRA

Encircling the top of the Northern Hemisphere is a frozen plain, clothed in sedges, heaths, and willows. Called the tundra, its name comes from the Finnish *tunturi,* meaning "a treeless plain." At lower latitudes similar landscapes, the alpine tundra, occur in the mountains of the world. Arctic or alpine, the tundra is characterized by low temperatures, a short growing season, and low precipitation (cold air carries very little water vapor).

The arctic tundra is a land dotted with lakes and crossed by streams. Where the ground is low and moist, extensive bogs exist. On high drier areas and places exposed to the wind, vegetation is scant and scattered, and the ground is bare and rock-covered. These regions are the fell-fields, an anglicization of the Danish *fjoeldmark,* or rock deserts. Lichen-covered, the fell-fields are most characteristic of highly exposed alpine tundra. The arctic tundra falls into two broad types: *tundra* with 100 percent cover and wet to moist soil, and *polar desert* with less than 5 percent cover and dry soil.

Conditions unique to the arctic tundra are a product of at least three interacting forces: permafrost, vegetation, and the transfer of heat. Permafrost is the perennially frozen subsurface that may be hundreds of meters deep. It develops where the ground temperatures remain below 0° C for many years. Its upper layers thaw in summer and refreeze in winter. Because the permafrost is impervious to water, it forces all water to remain and move above it. Thus the ground stays soggy on the tundra even though precipitation is low, enabling plants to exist in the driest parts of the Arctic.

Vegetation and its accumulated organic matter protect the permafrost by shading and insulation, which reduce and retard the warming and thawing of the soil in summer. Any natural or human disturbance, however slight, can cause the permafrost to melt. If the vegetation is removed, the depth of the thaw is 1.5 to 3 times that of the area still retaining vegetation. Thus vegetation and its organic debris impede the thawing of the permafrost and act to conserve it.

In turn, permafrost chills the soil, retarding the general growth of both aboveground and belowground parts of plants, limiting the activity of soil microorganisms, and impoverishing the aeration and nutrient content of the soil. The effect becomes more pronounced the closer the permafrost is to the surface of the soil, where it contributes to the formation of shallow root systems.

Alternate freezing and thawing of the upper layer of soil create the unique, symmetrically patterned landforms so typical of the tundra. This action of frost pushes stones and other material upward and outward from the mass to form a patterned surface. Frost hummocks, frost boils, and earth stripes, all typical nonsorted patterns, are associated with seasonally high water tables (Figure 28.1). Sorted patterns appear on better-drained sites. Best known are stone polygons, whose size is related to frost intensity and size of the material. On sloping ground, creep, frost thrusting, and downward flow of soil changes polygons into sorted stripes running downhill. Mass movement of supersaturated soil over the permafrost forms *solifluction* terraces, or "flowing soil." This gradual downward creep of soils and rocks eventually rounds off ridges and other irregularities in topography. This molding of the landscape by frost action, called *cryoplanation,* is far more important than erosion in wearing down the arctic landscape.

Alpine tundras have little permafrost, confined mostly to very high elevations; but frost-induced processes, such as small solifluction terraces and stone polygons, are present nevertheless. The lack of permafrost results in drier soils; only in alpine wet meadows and bogs do soil moisture conditions compare with those of the Arctic. Precipitation, especially snowfall and humidity, is higher in the alpine regions than in the arctic tundra, but steep topography induces a rapid runoff of water.

STRUCTURE

Vegetation

In spite of its distinctive climate and many endemic species, the tundra does not possess a vegetation type unique to itself. Structurally the vegetation of the tundra is simple. The number of species tends to be few; the growth is slow; and most of the biomass and functional activity are confined to relatively few groups. In the Arctic only those species able to withstand constant

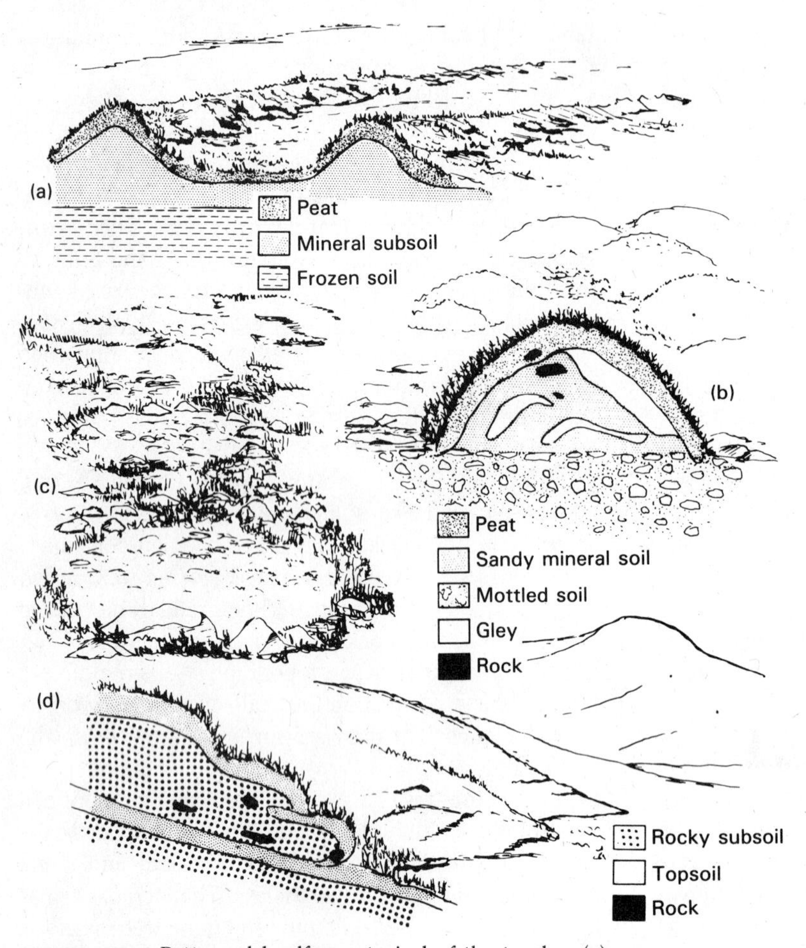

FIGURE 28.1 Patterned landforms typical of the tundra: (a) unsorted earth stripes; (b) frost hummocks; (c) sorted stone nets and polygons; and (d) a solifluction terrace.

disturbance of the soil, buffeting by the wind, and abrasion from wind-carried particles of soil and ice can survive. In the alpine tundra, the environment is even more severe for plants. It is a land of strong winds, snow, cold, and widely fluctuating temperatures. During the summer the temperature on the surface of the soil ranges from 40° to 0° C. The atmosphere is thin, so light intensity, especially ultraviolet, is high on clear days.

Although it appears homogeneous, the pattern of vegetation is patchy. A combination of microrelief, snowmelt, frost heaving, and aspect, among other conditions, produces an endless change in plant associations from spot to spot. In the arctic tundra (Figure 28.2), low ground is covered with a cotton grass, sedge, dwarf heath, sphagnum moss complex. Well-drained sites support heath shrubs, dwarf willows and birches, dryland sedges and rushes, herbs, mosses, and lichens. The driest and most exposed sites—the flat-topped domes, rolling hills, and low-lying terraces, all usually covered with coarse, rocky material and subject to extreme action by frost—support only sparse vegetation, often confined to small depressions. Plant cover consists of scattered heaths and mats of mountain avens,

FIGURE 28.2 The wide expanse of the arctic tundra. The photograph shows an area in the Arctic National Wildlife Refuge, 8 km from the Arctic Ocean. Note the frost polygons in the foreground and the caribou herd, a major arctic herbivore.

as well as crustose and foliose lichens growing on the rocks.

The alpine tundra is a land of rock-strewn slopes, bogs, alpine meadows, and shrubby thickets (Figure 28.3). Cushion and mat-forming plants, rare in the Arctic, are important in the alpine tundra. Low and ground-hugging, they are able to withstand the buffeting of the wind, and their cushionlike blanket traps heat. The interior of the cushion may be 20° C warmer than the surrounding air, a microclimate that is utilized by insects.

In spite of similar conditions, however, only about 20 percent of the plant species of the arctic and Rocky Mountain alpine tundra are the same, and they are of different ecotypes. Lacking in the Rocky Mountain alpine tundra are heaths and heavy growth of lichens and mosses between other plants; lichens are confined mostly to rocks and the ground is bare between plants. The alpine tundra of the Appalachian mountains, however, is dominated by heaths and sedge meadows, as is the alpine tundra of Europe, and mosses are common. The Australian alpine region supports a growth of heaths on rocky sites, and wet areas are covered with sphagnum bogs, cushion heaths, and sod tussock grasslands.

Alpine plant communities are not restricted to the northern and southern temperate regions of Earth. They also exist above the tree line of the high mountains in tropical regions: Central America, South America, Africa, Borneo, New Guinea, Java, Sumatra, and Hawaii. Tropical alpine vegetation and its environment contrast with those of temperate alpine regions. Tropical alpine regions undergo great seasonal variation in rainfall and cloud cover but little seasonal variation in mean daily temperature. Instead there is a strong daily fluctuation in temperature, from below freezing conditions at night to hot, summerlike temperatures during the day. Such a diurnal freeze–thaw cycle is unique to tropical alpine regions.

Tropical alpine tundras support tussock grasses, small-leaved shrubs, and heaths; but the one feature that sets tropical alpine vegetation apart from the temperate alpine is the presence of unbranched to little-branched, giant, treelike rosette plants (Figure 28.4), many of which belong to the family Compositae. Although the genera and species differ among the tropical alpine areas (for example, *Senecio* species in Africa and *Espeletia* species in the Andes), their growth forms and physiology are strikingly similar, suggesting convergent evolution. These species are the antithesis of the

(a)

(b)

(c)

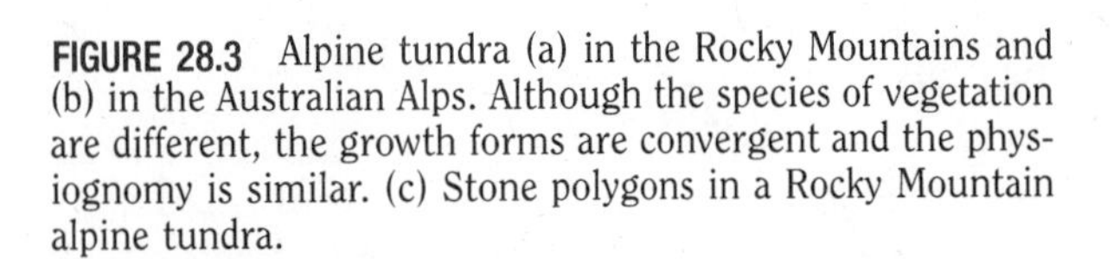

FIGURE 28.3 Alpine tundra (a) in the Rocky Mountains and (b) in the Australian Alps. Although the species of vegetation are different, the growth forms are convergent and the physiognomy is similar. (c) Stone polygons in a Rocky Mountain alpine tundra.

FIGURE 28.4 Tropical alpine tundra on Mt. Kenya, East Africa, with giant rosettes of *Lobelia*. Such alpine plants are able to resist freezing temperatures. Plant height tends to increase with increasing elevation.

usual low-growing tundra species. The higher the elevation, the taller these plants grow, reaching up to 6 meters. Many of these giant rosettes have well-developed water-sorting pith in the xylem; retain dead rosette leaves about the stem, which apparently acts as insulation against the cold; secrete a mucilaginous fluid about the bases of leaves that seems to function as a heat-storage device (water has a high heat-storing capacity); and possess dense pubescent hairs that reduce convective loss of heat.

Arctic plants propagate themselves almost entirely by vegetative means, although viable seeds many hundreds of years old exist in the soil. Alpine plants, including those of the tropics, propagate mostly by seeds. The short-lived adventitious roots of arctic plants are short and parallel to the rhizomes. Adventitious roots of alpine plants are long-lived, long, and, unimpeded by permafrost, can penetrate to considerable depths.

In both arctic and alpine tundras, topographic location and snow cover delimit a number of plant communities. On the arctic tundra, steep, south-facing slopes and river bottoms support the most luxurious and tallest shrubs, grasses, and legumes, whereas cotton grass dominates the gentle north-facing and south-facing slopes, reflecting higher air and soil temperatures and greater snow depth. Pockets of heavy snow create two types of plant habitat in both arctic and alpine tundra, the snow patch and the snow bed. Snow-patch communities occur where wind-driven snow collects in shallow depressions and protects the plants beneath. Snow beds are found where large masses of snow accumulate because of topographic peculiarities. Not only does the deep snow protect the plants beneath, but the meltwater from the slowly retreating snowbank provides a continuous supply of water throughout the growing season. Snow-bed plants have a short growing season, but are able to break into leaf and flower quickly because of the advanced stage of growth beneath the snow.

Microflora, the bacteria and fungi, live near the surface and are tolerant of cold temperatures. Bacteria are active at −7.5° C, and fungal growth continues at 0° C. Fungi, however, can break down plant structural carbohydrates at temperatures below 0° C, whereas aerobic bacteria at 0° C are restricted to nonstructural carbohydrates and products of fungal decomposition. Fungi and aerobic bacteria live mostly in the upper 7 to 10 cm of soil. Below that depth, only anaerobic bacteria exist.

Animal Life

The tundra world holds fascinating animal life even though the diversity of species is low. Invertebrate fauna are concentrated near the surface, where there are abundant populations of segmented whiteworms (Enchytraeidae), collembolas, and flies (Diptera), chiefly craneflies. Summer in the arctic tundra brings hordes of blackflies, deerflies, and mosquitoes. In alpine regions flies and mosquitoes are scarce, but collembolas, beetles, grasshoppers, and butterflies are common. Because of ever-present winds, butterflies keep close to the ground; other insects have short wings or no wings at all. Insect development is slow; some butterflies may take two years to mature, and grasshoppers three.

Dominant vertebrates on the arctic tundra are herbivores, including lemmings, arctic hare, caribou, and musk-ox. Although caribou may provide the greatest herbivore biomass, lemmings, which breed through the year, experience three- to four-year cycles. At their peak they may reach densities as great as 125 to 250 per hectare, consuming three to six times as much forage as caribou. Arctic hares that feed on willows disperse over the range in winter and congregate in more restricted areas in summer. Caribou are extensive grazers, spreading out over the tundra in summer to feed on sedges. Musk-ox are more intensive grazers, restricted to more localized areas where they feed on sedges, grasses, and dwarf willow. Herbivorous birds are few, dominated by ptarmigan and migratory geese.

The major arctic carnivore is the wolf (*Canis lupus*), which preys on musk-ox, caribou, and when they are abundant, lemmings. Medium-sized to small predators include the Arctic fox that preys on Arctic hare, and several species of weasel that prey on lemmings. Also feeding on lemmings are snowy owls (*Nyctea scandiaca*) and the hawklike jaegers (*Stencorarius* spp.). Sandpipers, plovers, longspurs, and waterfowl, which nest on the wide expanse of ponds and boggy ground, feed heavily on insects.

The alpine tundra, which extends upward like islands in mountain ranges, is small in area and contains few characteristic species. The alpine regions of western North America are inhabited by the hay-cutting pika, marmots, mountain woodchucks that hibernate over winter, mountain goats (not goats at all but related to the alpine-dwelling chamois of South America), mountain sheep, elk, voles, and pocket gophers. Eurasian alpine mammals include marmots and wild goats. The African alpine tundra is the home of the rock hydrax.

FUNCTION

Primary production on the tundra is low. Low temperature, a short growing season ranging from 50 to 60 days in the high arctic to 160 days in the low-latitude alpine tundra, and the low availability of nutrients help to keep it that way.

Plants are photosynthetically active on the arctic tundra about three months out of the year. As quickly as snow cover disappears, plants start photosynthetic activity, but it is limited initially because plant leaves are poorly developed. Alpine species, however, undergo a rapid burst of growth following snowmelt, at the expense of belowground root and rhizome carbohydrate reserves.

Arctic plants make maximum use of the growing season and light by carrying on photosynthesis during the 24-hour daylight period, even at midnight when light is one-tenth that of noon. They rarely become light-saturated, and possess a high leaf area index (0.5 to 1.0). The nearly erect leaves of some arctic plants permit the almost complete interception of the slanting rays of arctic sun.

Much of the photosynthate goes into the production of new growth, but about one month before the growing season ends, plants cease to allocate photosynthate to aboveground biomass. They withdraw nutrients from the leaves and move it to roots and belowground biomass, sequestering ten times the amount stored by temperate grasslands. In general, alpine tundras are more productive than arctic tundras.

Structurally, most of the tundra vegetation is underground. Root-to-shoot ratios of vascular plants range from 3:1 to 10:1. Roots are concentrated in the upper soil that thaws during the summer, and aboveground parts seldom grow taller than 30 cm. It is not surprising, then, that the net annual production of aboveground vegetation ranges from 40 to 110 g/m^2, whereas net annual belowground production ranges from 130 to 360 g/m^2. Efficiency of primary production ranges from 0.20 to 0.5 for the growing season. That rate of production is comparable to temperate ecosystems, but the growing season is so short—50 to 75 days—that overall annual production is greatly reduced.

Arctic and alpine tundras are short on nutrients because the short growing season, cold temperatures, and low precipitation slow weathering and restrict decomposition. Nutrient cycling has to be conservative and tight with minimal loss outside the system. Of all the terrestrial ecosystems, the tundra has the smallest proportion of its nutrient capital in live biomass. Dead organic matter functions as the nutrient pool, but most of it is not directly available to plants. To conserve nutrients, vascular plants retain and reincorporate nutrients, especially nitrogen, phosphorus, potassium, and calcium, in their tissues rather than release them to decomposers.

Because the tundra soil does not store available nutrients in any great quantity, plants depend on the release of nutrients from decomposition, the uptake of which is often aided by mycorrhizae. Decomposition is stimulated by rising temperature in spring. Because 60 percent of the active roots are in the upper 5 cm of soil, the root mass, once thawed, takes up most of the nutrients, except early in the season. Then plants have to compete with microbes and mosses, whose uptake of nutrients exceeds that of the plants. Once plants establish aboveground biomass, they have the competitive advantage over microbes.

Leaching or removal of nutrients is minimal, occurring mostly at the beginning of the growing season. Melting snow releases nutrients frozen over winter in the litter, excreta of animals, and microbes. Spring rains leach nutrients from dead plant material remaining from the previous growing season. In summer vascular plants leak nutrients, particularly phosphorus and potassium, from the cuticle of the leaves to the surface, where they are washed off by summer rains. Mosses often capture these nutrients before they reach the soil and release them slowly, thus functioning as a temporary nutrient sink.

A rapid upward movement of nutrients early in the season at the expense of belowground biomass supports fast shoot growth. Although more nutrients become available as the depth of thaw increases, they usually are in short supply. Tundra plants respond to the shortage, especially of nitrogen and phosphorus, by producing only a small biomass of leaves and stems that is well supplied with nutrients. Six weeks into the growing season, plants start to send nutrients belowground. As the cold approaches, the aboveground tissues die and their dead parts add to the accumulation of organic matter. Nutrients leached from the dead leaves are accumulated by mosses or are frozen into place until the following summer's snowmelt.

The two nutrients most limiting are nitrogen and phosphorus. Two major sources of nitrogen are precipitation and biological fixation. Anaerobic and free-living aerobic bacteria; blue-green algae in soil, water, and in the foliage of mosses where they live epiphytically; and lichens fix nitrogen biologically. Precipitation adds nearly as much. Phosphorus comes from decomposition of organic matter and animal defecation.

Animals contribute to the release of nutrients by either stimulating or short-circuiting decomposition. Soil invertebrates consume nearly all of the microbial populations near the surface of the soil. Grazing herbivores, especially lemmings, eat large amounts of aboveground production and distribute nutrients across the tundra through their droppings. During a cyclic high, lemmings may consume over 25 percent of the aboveground primary production or 10 percent of total plant production. They return about 70 percent as feces. Musk-ox, being selective grazers, restrict their feeding activity to certain areas, on which they may remove up to 85 percent of the herbage available. Overall this removal amounts to only about 15 percent of production. Decomposition of their dung is slow, taking 5 to 12 years to be recycled. Grazing herbivores also fell standing litter and live plant biomass, improving conditions for decomposers. Preying on the herbivores, mostly lemmings, are a number of carnivores. Efficient assimilators, the carnivores rarely pass off as feces more than 5 percent of total energy assimilated.

The activity of consumers stimulates a continuous overturn in both soluble and exchangeable pools. Nitrogen on the average is recycled 10 or more times and phosphorus 200 times during the growing season. Such efficient short-season cycling allows tundra ecosystems to make maximum use of a limited supply of nutrients.

HUMAN IMPACT

You might expect that the arctic tundra—remote, cold, and largely empty—would be relatively immune to serious disturbance. Human occupancy, once restricted mostly to coastal tundra, once consisted of native Eskimos who lived in close ecological harmony with the arctic environment. The arrival of Western culture has broken down that culture and weakened that strong ecological relationship. Winter igloos and summer tents have given way to permanent settlements with wooden houses, dog-drawn sleds have been replaced by snowmobiles, and rifles have replaced harpoons and spears as hunting instruments. All these social and cultural changes have affected the environment.

The discovery of oil has recently opened up the tundra for exploitation (Figure 28.5). Movement of heavy equipment, construction of all-season roads, airstrips, supply depots, oil pipelines, and oil spills have destroyed cover of mosses and grass, allowing the permafrost to melt and resulting in soil subsidence and gully erosion in the affected areas. Roads now obstruct wildlife movements and expose them to increased hunting pressure. Solid wastes and sewage, a disposal problem in the frozen landscape, pollute streams and surface waters, and toxic chemicals and heavy metals drain into arctic wetlands. Burning of vented gas from oil installations and sulfur dioxide originating from coal mining and smelting facilities, mostly in the USSR, pollute the arctic air. Because of the cold environment, tundra vegetation recovers slowly, if ever, from major disturbances.

The alpine tundra has not fared much better. Many alpine tundras, both temperate and tropical, have been overgrazed by domestic animals. Roads have opened the alpine tundra to recreational developments, ski trails, off-road vehicle tracks, and hiking trails. Such increased human activity damages thin soil and vegetation. Restricted in its distribution to mountaintop islands, tundra wildlife is vulnerable to human disturbance and faces a threatened future.

Because of the circumpolar nature of the arctic tundra, the worldwide threat of massive exploitation, and the international disturbance of wildlife, the maintenance of the arctic ecosystem is an international problem. Saving and managing alpine tundras depends on the action of individual countries.

KRUMMHOLZ

At high altitudes where the winds are too steady and strong for all but low ground-hugging plants, the forest is reduced to pockets of stunted, wind-shaped trees. This area where forest gives way to tundra is the Krummholz, or "crooked wood" (Figure 28.6). The Krummholz in the North American alpine region is best developed in the Appalachian and Adirondack mountains. On the high ridges, trees begin to show signs of stunting far below the timber line. As the trees climb upward, stunting increases until spruces and

FIGURE 28.5 A part of the Trans-Alaskan pipeline carrying oil from Prudoe Bay to Valdez across the arctic tundra.

(a)

(b)

FIGURE 28.6 The Krummholz. (a) In the Rocky Mountains the tree line is sharply defined. Note the narrow pockets of stunted trees. (b) The tree line in the Australian Alps in the Brindabella Range is marked by low, twisted snow gum *(Eucalyptus pauciflora)* in protected pockets that give way to low-growing heaths.

birches, deformed and semiprostrate, form carpets 0.6–1 m high, impossible to walk through but often dense enough to walk upon. Where strong winds come in from a constant direction, the trees are sheared until the tops resemble close-cropped heads, although the trees on the lee of the clumps grow taller than those on the windward side.

In the Rocky Mountains the Krummholz is much less marked, for there the timber line ends almost abruptly with little lessening of height. Most of the trees are flagged; that is, the branches remain only on the lee side. In the alpine regions of Europe, the Krummholz is characterized by dwarf mountain pine (*Pinus mugo*), which forms dense thickets 1–2 m high on calcareous soils. On acid soils it is replaced by dwarf juniper (*Juniperus communis* subsp. *nana*). In the Australian Brindabella Range, the Krummholz is marked by pockets of low, twisted snow gum (*Eucalyptus pauciflora*).

Though wind, cold, and winter desiccation are regarded as the cause of the dwarf and misshapen condition of the trees, the ability of some tree species to show a Krummholz effect is genetically determined, with species such as mountain (mugo) pine adapted to high alpine conditions. Eventually conditions become too severe even for the prostrate forms, and trees drop out entirely, except for those that have taken root behind the protection of high rocks. Tundra vegetation then takes over completely.

TAIGA

The largest vegetation formation on Earth is the boreal forest or taiga. This belt of coniferous forest encompassing the high latitudes of the Northern Hemisphere covers about 11 percent of Earth's terrestrial surface. Its northern limit is roughly along the July 13 isotherm, the southern extent of the Arctic front in summer, which also marks the beginning of the northward-stretching tundra. Its southern limit, much less abrupt, is more or less marked by the winter position of the Arctic front, roughly just north of 58° N latitude. In North America the boreal forest covers much of Alaska and Canada, and spills into Northern New England with a finger extending down the high Appalachians. In Eurasia the boreal forest begins in Scotland and Scandinavia and extends across the continent, covering much of Siberia, to northern Japan.

Four major vegetation zones make up the taiga: the forest-tundra ecotone with open stands of stunted spruce, lichens, and moss; the open boreal woodland of stands of lichens and black spruce; the main boreal forest with continuous stands of spruce and pines broken by poplar and birch on disturbed areas; and the boreal-mixed forest ecotone where the boreal forest grades into the mixed forest of southern Canada and the northern United States. Occupying, for the most part, glaciated land, the taiga is also a region of cold lakes, bogs, rivers, and alder thickets.

In Europe the forest is dominated by Norway spruce (*Picea abies*), Scots pine (*Pinus sylvestris*), and downy birch (*Betula pubescens*) (Figure 28.7); in Siberia by Siberian spruce (*P. obovata*), Siberian stone pine (*Pinus sibirica*), and larch (*Larix sibirica*); and in the Far East by Yeddo spruce (*P. jezoensis*). The North American taiga, richer in species, has four genera of conifers, *Picea, Abies, Pinus,* and *Larix,* and two genera of deciduous trees, *Populus* and *Betula* (Figure 28.8). Dominant tree species include black spruce (*Picea mariana*) and jack pine (*Pinus banksiana*).

A cold continental climate with strong seasonal variation dominates the taiga. The summers are short, cool, and moist, and the winters are prolonged, harsh, and dry, with a long-lasting snowfall. The driest winters and the most extreme seasonal fluctuations are in interior Alaska and central Siberia, which experience as much as 100° C seasonal temperature extremes. Like the tundra, much of the taiga is under the controlling influence of permafrost. As in the tundra, permafrost impedes infiltration and maintains high soil moisture.

FIGURE 28.7 Scots pine is a dominant tree in the Scandinavian taiga.

STRUCTURE

Vegetation

Like the tundra, the taiga is an endless sweep of sameness—a blanket of spire-shaped evergreens over the landscape. The appearance of the land can be deceptive, because variations in slope, aspect, topography, drainage, and presence of permafrost add variety to the vegetation. In the North American taiga, black spruce with its low nutrient requirements and ability to tolerate wet soils occupies cold, wet, north-facing slopes and bottomlands. White spruce and birch grow on permafrost-free south-facing slopes, and jack pine grows on the high, drier, warmer sites. Areas swept by fires come back in early successional hardwoods—quaking aspen, balsam poplar, white birch—and jack pine.

The boreal forest conifers fall into three growth forms: (1) the spire-shaped spruces and fir, with an open, narrow, upper canopy and a dense lower canopy that casts a deep shade on the forest floor; (2) the open, thin, light-penetrating upper canopy of pines; and (3)

FIGURE 28.8 Black spruce is a dominant conifer in the North American taiga.

the deciduous larch. Only a thick carpet of mosses grows in the dense shade of spruce, whereas under pine, light-tolerant lichens replace the shade-loving mosses.

The conifers are well suited to the cold taiga environment. The narrow, needle-like leaves with their thickened cuticles and sunken stomata reduce transpiration and assist in moisture conservation during periods of summer drought and winter freeze. Because they retain several years' growth of foliage at any one time, conifers can start photosynthesis quickly when environmental conditions are favorable.

Permafrost imposes its own authority on patterns of vegetation, which paradoxically encourages its formation. Permafrost impedes soil drainage, chills the soil, reduces its depth, slows decomposition, and reduces the availability of nutrients. Trees grow best where permafrost lies deep beneath the soil or is absent altogether; but the taiga trees worsen the situation for themselves by encouraging and maintaining the permafrost. Stands of spruce shade the ground and encourage a heavy growth of moss. Moss and an accumulation of the fine litter of undecomposed needles insulate the soil, immobilize nutrients, and increase the soil moisture. The colder the soil becomes, the closer to the surface moves the permafrost, and the more shallow becomes the soil. With little space in which to anchor their roots, the trees are subjected to frost heaving. During early periods of warm weather, the roots encased in frozen soil are unable to replace the moisture lost through the crowns. The result is winter kill.

In stands of pine, larch, and scattered spruce, conditions do not improve. Heavy mats of lichens replace mosses on the dry, nutrient-poor, highly acidic soil. Lichens retain soil moisture through the growing season, encouraging growth of trees on sites that otherwise would be too dry, but the thick lichen mat also insulates the soil, chilling it and inhibiting the decomposition of organic matter. In spite of these effects, lichens do appear to improve tree growth.

Fires are recurring events in the taiga. During periods of drought, fires can sweep over hundreds of thousands of hectares. All of the boreal species, both broadleaf trees and conifers, are well adapted to fire. Unless too severe, fire provides a seedbed for regeneration of trees. Light surface burns favor successional hardwoods. More severe fires eliminate hardwood competition and favor spruce and jack pine regeneration.

Animal Life

As on the tundra, caribou are the major herbivores. Inhabiting open spruce-lichen woodlands, caribou are wide-ranging and feed on grasses, sedges, and especially lichens. Joining the caribou is the moose, the largest of all deer. Called elk in Eurasia, the somewhat solitary moose is a lowland mammal feeding on aquatic and emergent vegetation as well as alder and willow. Competing with moose for browse is the cyclic snowshoe hare. The arboreal red squirrel inhabits the conifers and feeds on male flower buds and seeds of spruce and fir; and the quill-bearing porcupine feeds on leaves, twigs, and the inner bark of trees. Major mammalian predators are the wolf, which feeds on caribou and moose; the lynx, which preys on snowshoe hare and other small mammals; the pine marten, the major predator on red squirrels; and other species of weasels. Ruffed grouse and spruce grouse are conspicuous birds of the North American boreal forest, and capercaillie (*Tetrao* spp.) and hazel grouse (*Tetrastes bonasia*) in the Eurasian taiga. Crossbills and siskins extract conifer seeds from cones and occasionally move south in winter when the food supply fails. Importantly, the taiga is the nesting ground of many species of both neotropical and tropical migrant warblers. Major avian predators include numerous species of owls, such as the great gray owl (*Strix nebulosa*), and goshawks (*Accipiter* spp.).

Of great ecological and economic importance are major herbivorous insects, the larch sawfly (*Pristiphora erichsonii*), the pine sawfly (*Neodiprion sertifer*), and the spruce budworm (*Choristoneura fumiferana*). Although major food items for the insectivorous summer birds, these insects experience periodic outbreaks and defoliate and kill large expanses of forest.

FUNCTION

A cold environment, a short growing season, the presence of continuous or discontinuous permafrost, slow decomposition, low availability of nutrients, and infertile, acidic soils impede productivity in the taiga. Particularly important is the relationship between ground cover, aboveground biomass, and nutrient availability. Growing in the already cool dark shade of spruces, mosses lower soil temperature and increase soil moisture, both of which act to reduce decomposition and the release of nutrients. Increasing the nutrient shortage to trees is the sequestering of nutrients by mosses, which depend upon precipitation and tree wash for their mineral nutrition. Nutrients become available only when mosses, which have a slow turnover rate, die. Associated with drier, highly acidic soil, lichens, too, immobilize nutrients. Because lichens hold moisture, however, they do lessen the effects of summer drought.

Particularly acute is the shortage of nitrogen. In addition to being immobilized in mosses and lichens, nitrogen is tied up in poorly decomposed soil organic matter. The major source of nitrogen is precipitation and biological fixation by lichens, by alders, bog myrtle, soapberry, and other woody plants possessing root nodules, and by some free-living bacteria. Uptake of nitrogen by trees, however, exceeds the annual input by their litter. Such a slow turnover in nitrogen can have an adverse effect on the growth and vigor of trees and associated vegetation. Depending on its severity, fire releases various accumulated nutrients, but, more important, improves the conditions for decomposition by warming the soil and exposing organic matter to sunlight.

Because of the wide range of environmental conditions across the taiga, primary production is highly variable. In a general way, productivity there is much less than in deciduous forest, and biomass accumulation is slow. Primary production is influenced by herbivores not only through consumption but also by trampling and selective browsing on certain species, especially willow, birch, aspen, and alder. Moose can reduce young plant growth by as much as 50 percent. A study of plant–moose–wolf relationships on Isle Royale National Park showed that the moose ate about 3,000 tons of vegetation annually. In turn their major predator, the wolf, consumed about 45 tons of moose. The energy flow involved was plant, $11{,}000{,}000 \times 10^3$ kcal; moose, $46{,}000 \times 10^3$ kcal; and wolves, 780×10^3 kcal. Just as the nutrient and energy turnover among the plant components in the boreal forest is slow, so too is the turnover of biomass slow in the long-lived moose, which in turn controls the biomass of wolves.

HUMAN IMPACT

Exploitation and exploration of the taiga began in the late 1600s with the fur trade, over which wars were fought. Exploitation of the taiga's wildlife resources was followed by logging. The taiga is the world's richest source of softwood timber and pulpwood. One-half of the total exploitable reserve is in the USSR; the other half is in North America and Europe. Much of the logging is little more than timber mining with no overall effort at regeneration and restoration. In the USSR up to 3 million hectares are cut annually, and only about 35 percent is restored. Much of the boreal forest has a slow rate of regeneration; and in many areas removal of the forest results in the formation of bogs. Removal of trees for firewood and excessive grazing by reindeer along the taiga-tundra ecotone is causing a southward expansion of the tundra in the USSR at the expense of the taiga, a situation somewhat analogous to the spread of deserts into the African savanna.

The taiga is rich in metal ores, supplying a large portion of the world's minerals, including iron ore and gold, as well as in coal, gas, and oil. The extraction of these products began in the late nineteenth century with marked effects on the environment. Development gave rise to settlements, roads, and industrial complexes, including pulp and paper mills, mines, and ore smelting plants, which have contributed to massive air pollution in both Canada and the USSR. Huge freshwater systems of lakes and rivers have encouraged the development of massive hydroelectric schemes such as the ones at Angara in Siberia and on the Peace River in Canada, the installation of which lowered water in the Mackenzie Basin, damaging trapping grounds of northern native tribes and destroying waterfowl breeding habitat. Such hydroelectric developments drown hundreds of thousands of hectares of boreal forest, the breeding grounds of waterfowl and other northern wildlife, and destroy the cultural and social fabric of northern native American and Lapp communities.

Further ecological damage is caused by mining peat for fuel and moss litter and organic soils for the horticultural and gardening trade, and by draining for garden crops and pasture. Drainage of peatlands is extensive in Europe, especially the USSR.

Opening of the taiga by logging and industrial development makes the region accessible for expanded recreational use, including summer fishing camps and winter hunting lodges. What effects these and inevitable future developments hold for taiga wildlife and indeed the entire taiga ecosystem is a troublesome and unanswered question.

Summary

The arctic tundra that extends beyond the tree line of the far north and the alpine tundra of the high mountain ranges in the lower latitudes are at once similar and dissimilar. Both are characterized by low temperature, low precipitation, and a short growing season. Both possess a frost-molded landscape and plant species whose growth forms are low and whose growth rates are slow. The arctic tundra has a perpetually frozen subsurface, the permafrost; rarely does the alpine tundra. Arctic plants require longer periods of daylight than alpine plants and propagate mostly by vegetative means, whereas alpine plants propagate mostly by seeds. Over much of the Arctic, the dominant vegetation is cotton grass, sedge, and dwarf heaths. In the alpine tundra, cushion and mat-forming plants, able to withstand buffeting by the wind, dominate exposed sites. Net primary production is low, and most plant growth occurs underground. In spite of an assemblage of grazing ungulates and rodents, most of the production goes to decomposers. Decomposition, however, is slow, resulting in an accumulation of peat, which locks up nutrient supplies. Nutrient cycling, especially of N and P, is necessarily conservative and tight, operating on small pools of soluble and exchangeable nutrients. Most of the cycled nutrients are concentrated in the roots, translocated to growing shoots aboveground early in the season, and replaced by production and subsequent return of nutrients to the roots later in the growing season. Animals aid in the cycling of nutrients by consuming some of the primary production and distributing droppings and dung across the tundra.

Remote as it may seem, the tundra is subject to

strong human disturbance. Exploitation of its oil and mineral resources has caused considerable damage to permafrost and arctic vegetation through road building, construction, and settlement. Humans have introduced air, water, and soil pollution into a once pristine environment and have disturbed fish and wildlife, exposing them to potential dangers of increased exploitation. Huge hydroelectric projects have flooded vast areas of wetlands that are important as wildlife breeding grounds. Industrial intrusion has disrupted native culture, once in harmony with the arctic environment. The greatest threat to alpine tundras comes from recreational developments and use that damage the fragile alpine communities.

South of the tundra lies the circumpolar taiga or boreal forest, the largest vegetational formation on Earth. Characterized by a cold continental climate, the taiga consists of four major zones: the forest-tundra ecotone, open boreal woodland, main boreal forest, and boreal-mixed forest ecotone. The boreal forest is dominated by spruces and pine, with successional communities of birch and poplar. Ground cover below spruce is mostly moss; in open spruce stands and pine the cover is mostly lichens. Permafrost, the maintenance of which is influenced by tree and ground cover, has a strong influence on the pattern of vegetation, as do recurring fires. Because of cold, low nutrient availability, and a short growing season, productivity is low. Major herbivores include the caribou, moose (called elk in Europe), and snowshoe hare, preyed on by a colorful assortment of predators, including the wolf, lynx, and pine marten. The taiga is also the nesting ground of neotropical and tropical birds, and the habitat of such northern seed-eating birds as crossbills.

The taiga has a long history of human exploitation of its rich timber, mineral, and water resources. This exploitation will increase in the future with pronounced effects on the integrity of the taiga ecosystem.

Review and Study Questions

1. What physical and biological features characterize the tundra?
2. How does the alpine tundra differ from the arctic tundra?
3. What are the major contrasts between alpine tundras of temperate and tropical regions?
4. What is the relationship among permafrost, plant life, and nutrient cycling in the tundra?
5. Where does the bulk of net primary production of the arctic tundra accumulate and why?
6. What is the role of herbivores in the tundra ecosystem?
7. What major vegetation zones make up the boreal forest or taiga? What is the difference between them?
8. What is the relationship between tree cover and moss and lichen growth on the ground? How do they relate to permafrost?
9. How does the homogeneous forest cover relate to forest insect outbreaks and fire?

*10. Speculate why the taiga holds such a large number and diversity of important fur-bearing mammals. What effect did this abundance of valuable wildlife have on the history and settlement of the taiga?

*11. How does economic development clash with maintaining the integrity of the tundra and taiga ecosystems? Consider the ecological impacts of oil drilling in the Arctic National Wildlife Refuge, the James Bay hydroelectric development scheme in Quebec, and the proposed diversion of water southward from the taiga to the semiarid regions of the USSR.

Selected References

Tundra

Bliss, L. C., O. H. Heal, and J. J. Moore, eds. 1981. *Tundra ecosystems: A comparative analysis.* New York: Cambridge University Press. A major reference.

Brown, J., P. C. Miller, L. L. Tieszen, and F. L. Bunnell. 1980. *An arctic ecosystem: The coastal tundra at Barrow, Alaska.* Stroudsburg, PA: Dowden, Hutchinson & Ross.

Furley, P. A., and W. A. Newey. 1983. *Geography of the biosphere.* London: Butterworths. Chapter 10 gives a detailed discussion of the tundra biome.

Rosswall, T., and O. W. Heal, eds. 1975. *Structure and function of tundra ecosystems.* Ecological Bulletins 20. Stockholm: Swedish Natural Sciences Research Council.

Smith, A. P., and T. P. Young. 1987. Tropical alpine plant ecology. *Annual review of ecology and systematics,* 18:137–158. An informative review paper.

Sonesson, M., ed. 1980. *Ecology of a subarctic mire.* Ecological Bulletins 30. Stockholm: Swedish Natural Sciences Research Council. A detailed study of structure and function.

Wielgolaski, F. E., ed. 1975. *Fennoscandian tundra ecosystems.* Part I, *Plants and microorganisms;* Part II, *Animals and systems analysis.* New York: Springer-Verlag.

Zwinger, A. H., and B. E. Willard. 1972. *Land above the trees.* New York: Harper & Row.

Taiga

Bonan, G. B., and H. H. Shugart. 1989. Environmental factors and ecological processes in boreal forests. *Annual review of ecology and systematics,* 20:1–18. An informative review paper.

Knystautas, A. 1987. *The natural history of the USSR.* New York: McGraw-Hill. Contains an interesting, well-illustrated description of the Siberian taiga.

Larsen, J. A. 1980. *The boreal ecosystem.* New York: Academic Press. A summary of what we know about the ecology of the boreal ecosystem.

Larsen, J. A. 1989. *The northern forest border in Canada and Alaska.* Ecological Studies 70. New York: Springer-Verlag. An interesting description of the biotic communities and ecological relationships of the forest-tundra ecotone.

Polunin, O., and M. Walters. 1985. *A guide to the vegetation of Britain and Europe.* Oxford, England: Oxford University Press. Excellent description and guide to both tundra and taiga flora.

Outline

CHAPTER 29

Temperate Forests

Objectives

On completion of this chapter, you should be able to:

1. Describe the types and general features of coniferous and broadleaf forest ecosystems.
2. Describe the effects and role of environmental and biological stratification in forest ecosystems.
3. Compare the structure and function of coniferous and temperate deciduous forests.
4. Discuss the impact humans have had on forests ecosystems.

Temperate forests, in spite of their name, do not exist in a temperate environment. They occupy topographic positions that range from low-lying lands to mountaintops, and environmental conditions that range from warm and semiarid to cold and wet. They face great fluctuations in daily and seasonal temperatures, which stress the physiological activity of plants and animals. Deciduous trees are leafless during the winter and in northern regions remain so for the greater part of the year. They are exposed to droughts and in places to flooding. In spite of their intemperate environment, temperate forest ecosystems are able to maintain high productivity.

Temperate forests embrace coniferous, deciduous, or mixed stands. Regardless of type, all forests possess large aboveground biomass. This biomass creates several layers or strata of vegetation, which influence both the vertical structure and environmental conditions within the stand: light, moisture, temperature, wind, and carbon dioxide. These factors differ among forest types.

CONIFEROUS FORESTS

TYPES

Montane Forests

Many temperate coniferous forests are in the mountains. In Central Europe extensive coniferous forests, dominated by Norway spruce (*Picea abies*), cover the slopes up to the subalpine zone in the Carpathian Mountains and the Alps (Figure 29.1a). In North America several coniferous forest associations blanket the Rocky, Wasatch, Sierra Nevada, and Cascade mountains. In the southwestern United States such forests occur between 2500 and 4200 m elevation and in the northern United States and Canada between 1700 and 3500 m elevation. At high elevations in the Rocky Mountains, where winters are long and snowfall is heavy, grows a subalpine forest dominated by Engelmann spruce (*Picea engelmannii*) and subalpine fir (*Abies lasiocarpa*) (Figure 29.1b). Mid-elevations have stands of Douglas-fir, and lower elevations are dominated by open stands of ponderosa pine (*Pinus ponderosa*) (Figure 29.1c) and thick stands of the early successional pioneering conifer, lodgepole pine (*P. contorta*) (Figure 29.1d).

Similar forests grow in the Sierras and Cascades. There high-elevation forests consist largely of mountain hemlock, red fir (*Abies magnifica*), and lodgepole pine. We also find sugar pine (*Pinus lambertiana*), incense-cedar (*Libocedrus decurrens*), and the largest tree of all, the giant sequoia (*Sequoiadendron giganteum*), which grows only in scattered groves on the west slopes of the California Sierras.

A deciduous seral, but occasionally permanent, species common both to the montane and the boreal forest is trembling aspen (*Populus tremuloides*). It is the most widespread tree of North America (Figure 29.2).

Pine Forests

Pines form extensive stands in both Eurasia and North America. A major component of the Eurasian boreal forest (see Chapter 28), Scots pine is also widespread through central Europe, where it grows from the lowlands to the tree line in the mountains. It occurs largely as planted or seminatural stands in southern England and western France.

Unlike the Scots pine forests of Eurasia, the pine forests of the coastal plains of the South Atlantic and Gulf states are usually considered a seral stage of the temperate deciduous forest because without disturbance from fire and logging they give way to it. These pines maintain their presence by possessing a competitive advantage over hardwoods on nutrient-poor, dry, sandy soil and by their adaptation to a fire regime. At the northern end of the coastal pine forest in New Jersey, pitch pine (*Pinus rigida*) is the dominant species. Further south, loblolly (*P. taeda*), longleaf (*P. australis*), and slash (*P. caribaea*) pine are most abundant.

Temperate Rain Forests

South of Alaska the coniferous forest differs from the northern boreal forest, both floristically and ecologically. The reasons for the change are both climatic and topographic. Moisture-laden winds move in from the Pacific, meet the barrier of the Coast Range, and rise abruptly. Suddenly cooled by this upward thrust into the atmosphere, the moisture in the air is released as

FIGURE 29.1 Some coniferous forest types. (a) A Norway spruce forest in the Carpathian Mountains of central Europe. (b) Rocky Mountain subalpine forest dominated by subalpine fir (*Abies lasiocarpa*). This tree grows with Engelmann spruce and mountain hemlock. (c) A montane coniferous forest in the Rocky Mountains. The dry lower slopes support ponderosa pine; the upper slopes are cloaked with Douglas-fir. (d) Lodgepole pine forests cover extensive areas in pure stands in the Rocky Mountains. The tree is so called because western Indians use poles of this pine in erecting tepees.

FIGURE 29.2 Quaking aspen is the dominant deciduous tree in the western montane forest of North America. A successional to a long persisting species, aspen stands turn golden yellow in autumn.

FIGURE 29.3 Old-growth forest in the Pacific Northwest. Typical is this Douglas-fir stand with an abundant western hemlock understory. Such forests are the home of the spotted owl and murrelet.

rain and snow in amounts up to 635 cm/yr. During the summer, when winds shift to the northwest, the air is cooled over chilly northern seas. Although rainfall is low, cool air brings in heavy fog, which collects on the forest foliage and drips to the ground to add 127 cm or more of moisture. This land of superabundant moisture, high humidity, and warm temperatures supports the temperate rain forest, a community of luxuriant vegetation dominated by a variety of conifers well adapted to wet, mild winters, dry warm summers, and nutrient-poor soils. The forests are dominated by western hemlock (*Tsuga heterophylla*), mountain hemlock (*T. mertensiana*), Pacific silver fir (*Abies amabilis*), and Douglas-fir, all trees with high foliage and stem biomass (Figure 29.3). Further south, where precipitation still is high, grows the redwood (*Sequoia sempervirens*) forest, occupying a strip of land about 724 km wide.

STRUCTURE

Coniferous forests fall into three broad classes of growth form and growth behavior: (1) pines with straight, cylindrical trunks, whorled spreading branches, and a crown density that varies with the species from the dense crowns of red and white pine to the open, thin crowns of Virginia, jack, Scots, and lodgepole pine; (2) spire-shaped evergreens, including spruce, fir, Douglas-fir, and (with some exceptions) the cedars, with more or less tall pyramidal crowns, gradually tapering trunks, and whorled, horizontal branches; (3) deciduous conifers such as larch (*Larix* spp.) and bald cypress with pyramidal, open crowns that shed their needles annually. Growth form and behavior influence animal life and other aspects of coniferous ecosystems.

Vertical stratification in coniferous forests is not well developed. Because of a high crown density, the lower strata are poorly developed in spruce and fir forests, and the ground layer consists largely of ferns and mosses with few herbs. The maximum canopy development in spire-shaped conifers is about one-third down from the open crown, a profile different from that of pines. Pine forests with a well developed high canopy lack lower strata. The litter layer in coniferous forests is usually deep and poorly decomposed, resting on top of instead of mixing in with the mineral soil.

This poor vertical stratification influences the environmental stratification within the stand. When you walk into a spruce or fir forest, you are struck by the sharp diminution in light and deep shade. Light intensity is progressively reduced through the canopy to only a fraction of full sunlight. The upper crown of spruce and firs, a zone of widely spaced narrow spires, is open and well lighted, whereas the lower crown is dense and intercepts most of the solar radiation. Most pines form a dense upper canopy that excludes so much sunlight that lower strata cannot develop. Open-crowned pines allow more light to reach the forest floor, stimulating a grassy or shrubby understory. Because conifers retain their foliage through the year, light reaching the lower strata in coniferous forests is about the same throughout the year. Illumination is greater during midsummer when the sun's rays are most direct and lower in winter when the intensity of incident sunlight is the lowest.

The temperature profile of a coniferous forest also varies with the growth form. For example, in forests of sprucelike trees, temperatures tend to be coolest in the upper canopy, perhaps because of greater air circulation, and hottest in the lower canopy.

Animal life in the coniferous forest varies widely, depending upon the nature of the stand. Soil invertebrate litter fauna is dominated by mites. Earthworm species are few and their numbers low. Insect populations, although not diverse, are high in numbers and, encouraged by the homogeneity of the stands, are often destructive. Sawflies (*Neodriprion*), for example, attack a wide variety of pines, including pitch, Virginia, shortleaf, and loblolly, and the southern pine beetle (*Dendroctonus frontalis*) can reach outbreak proportions in southern pinelands.

A number of bird species are closely associated with coniferous forests. In North America they include chickadees, kinglets, pine siskins, crossbills, purple finches, and hermit thrushes. Related species, the tits and grosbeaks, are common to European coniferous forests.

Except for strictly boreal species, such as the pine marten and lynx, mammals have much less affinity for coniferous forests. Most are associated with both coniferous and deciduous forest; the white-tailed deer, moose, black bear, and mountain lion are examples. Their north-south distribution seems to be limited more by climate, especially temperature, than by vegetation. The red squirrel, commonly associated with coniferous forests, is quite common in deciduous woodlands in the southern part of its range.

FUNCTION

The great economic importance of forests and the increasing intensity of forest management practices demand considerable understanding of forest ecosystems. For these reasons forests have received the most intensive studies of ecosystem function. One coniferous forest ecosystem studied in detail is Douglas-fir, an important forest type in the western United States. Forest ecologists have researched nutrient budgets and cycling for both a 36-year-old young stand and a 450-year-old old-growth stand. Although, as we would expect, the old-growth stand had a considerably larger total biomass, both living and dead, than the younger stand, and both had similar foliar biomass, the younger forest had the higher percentage of its living biomass in foliage and roots—21 percent compared to 14 percent for the old-growth stand. Both age classes had most of their living biomass in branches and bole. Decaying logs and litter accounted for the most detrital matter (55 percent) in the old-growth stand; in the young stand, soil held most of the detrital organic matter (83 percent). Litter organic matter exceeded soil organic matter in the old-growth stand because of the accumulation and slow decompositon of needles and the long-term decomposition, over hundreds of years, of large fallen limbs and trunks.

Both old-growth and young stands cycle and store nutrients, particularly nitrogen, internally. The young stand accumulated most of its N in foliage and bole, whereas the old stand stored most of it in bole and root. The 450-year-old stand accumulated considerably more nitrogen, but invested a lower percentage of it in the soil. The young forest banked 98 percent of its detrital nitrogen in the soil.

Both systems had a low input of nitrogen, about 2 kg/ha/yr for each (Table 29.1). Of this input the 450-year-old stand returned somewhat more nitrogen to

TABLE 29.1 Nitrogen Transfers (kg/ha/yr) in Young and Old-Growth Douglas-Fir Ecosystems

Component	*Douglas-Fir 42-Year-Old*	*Douglas-Fir 450-Year-Old*
Input	1.67	2.0
Return to forest floor		
Throughfall	0.53	3.4
Litterfall	25.4	25.6
Total	25.93	29.0
Within vegetation		
Requirement	45.8	33.3
Redistribution	20.7	18.5
Uptake	25.1	14.5

the forest floor than the young stand, mostly because of greater throughfall. The young growing stand needed considerably more nitrogen, which was reflected in its budget. Uptake of nitrogen by the young stand amounted to 55 percent of its needs; for the old-growth forest, uptake meets 44 percent of the old-growth's requirement. To make up the deficiency, both stands recycled nitrogen within the biomass. This internal cycling is also typical for other nutrients with the exception of calcium and magnesium. Uptake of those nutrients far exceeds requirements, so internal cycling is not necessary.

Some coniferous ecosystems scavenge nutrients directly from rainfall through microcommunities of algae and lichens that colonize canopy leaves in balsam fir and western coniferous forests. Of particular interest is the complex community, including primary producers, consumers, and decomposers, supported by the canopy of old-growth Douglas-fir (Figure 29.4). Cyanophycophilous lichens fix atmospheric nitrogen. Organic nitrogen leached from these lichens combines with canopy moisture to form a dilute organic solution that in turn is taken up by microorganisms and other canopy epiphytes. Part of this microbial production is consumed by canopy arthropods. These nutrient cycles in the canopy tend to influence and even restrict the amount of nutrients that reach the forest floor by throughfall and stemflow.

Roots, fungi, and mycorrhizae (see Chapter 17) also have a role in nutrient cycling in coniferous stands. Fungal hyphae or rhizomorphs concentrate nutrients in their tissues, especially the fruiting body, and act as a living sink of nutrients. Resistant to leaching, the fungal rhizomorphs hold biologically important elements in the litter, which they may slowly release to the soil through exudates.

One component of forest ecosystems too often overlooked is dead wood. The mass of dead wood changes with tree mortality and disturbances. In general, wood litterfall increases through succession and becomes most conspicuous in old temperate forests, both deciduous and coniferous, where logs and large limbs decay slowly. Volume of woody debris is much greater in old-growth coniferous forest of the Pacific Northwest than in eastern deciduous forests.

Dead wood in the form of large standing dead trees or snags (Figure 29.5) or downed trunks and limbs makes up a unique and critical component of the forest ecosystem. Standing dead trees provide essential nesting and den sites for certain birds and mammals, food, and foraging areas. Fallen trees, which may make up 10 to 20 percent of the ground surface in forests, provide food, protection, and pathways for small mammals, and reproductive sites for certain woody plants. The elimination of standing and fallen dead trees greatly impoverishes animal life in the forest.

HUMAN IMPACT

Because of their early successional nature and dependence on disturbance for regeneration, temperate coniferous forest ecosystems can reoccupy a site that has been logged, although their return may follow a deciduous forest interlude. Historically, however, logging often has been followed by fire, which fed on the large amount of debris left behind. These human-caused fires

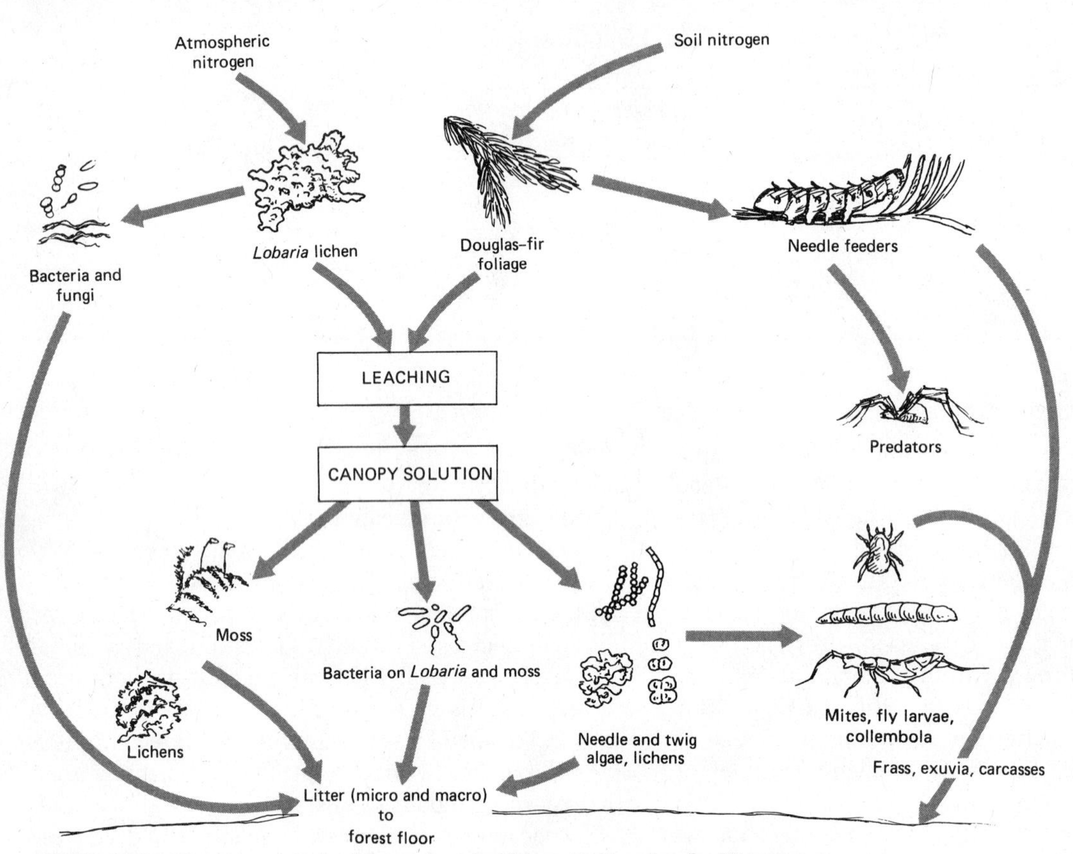

FIGURE 29.4 Nitrogen cycle in the canopy of old-growth Douglas-fir. Microecosystems exist in the high canopy of old Douglas-fir, and probably many other forests, including the tropical rain forest. The community consists of primary producers (lichens) and biophage and saprophage consumers. These microecosystems conserve and recycle nutrients such as nitrogen and influence nutrient return to the forest floor by leaching and throughfall.

have covered hundreds of thousands of hectares. The Cloquet-Moose Lake fire in Minnesota in fall of 1918 burned 4000 hectares in one day on logged-over white pine forest. The forest has never regenerated. In the spring of 1987 the Black Dragon fire, possibly the largest forest fire ever, burned over huge areas of large forest in China and the Soviet Union, which still fail to show strong signs of recovering. The devastation of such fires is much greater than of fires in uncut forest.

Montane coniferous forests and pinelands are a major source of lumber. Commercial foresters try to regenerate cutover forest for future cuts. One of the preferred methods is to clearcut blocks of forest, burn the debris on the ground to eliminate the fire hazard and to expose a seed bed. Then foresters allow the area to regenerate naturally, or they replant it to leapfrog an incoming deciduous forest stage. The latter practice is common on southern pinelands and in Douglas-fir and Engelmann spruce forests. Most European coniferous forests are artificially planted and maintained. An artificially regenerated stand, even of the same species as harvested, is not a forest (Figure 29.6). Although some foresters argue that this method is an improvement on nature, the result is a simplified, monocultural stand that lacks all the complexity of a natural forest.

Even a naturally regenerated forest will not replace the original old-growth forest. In the Pacific Northwest

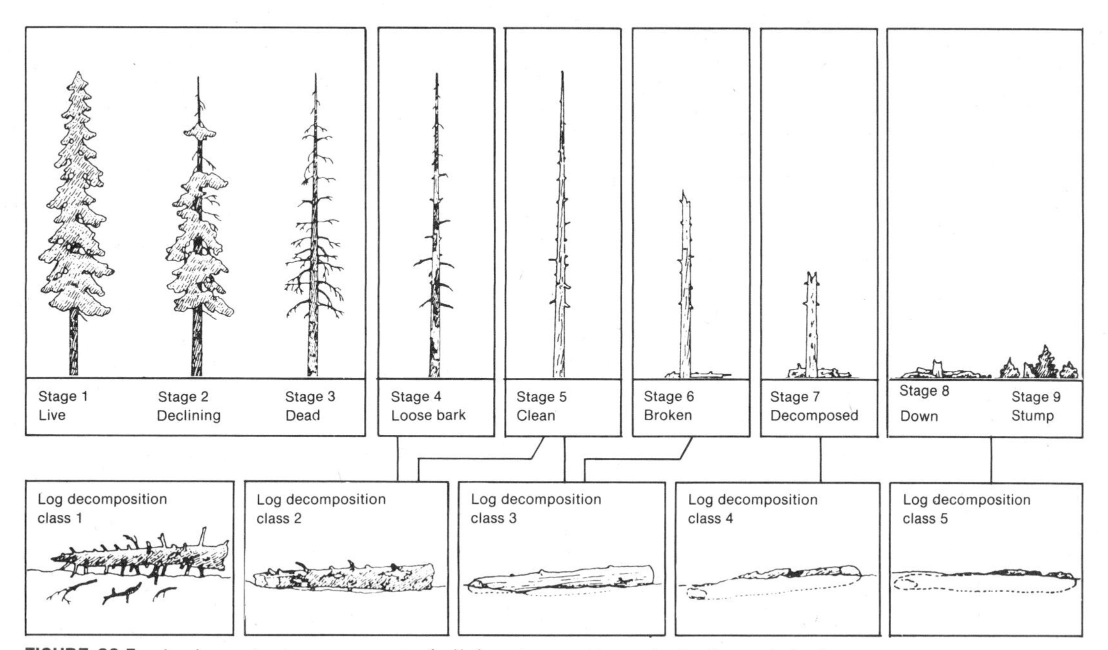

FIGURE 29.5 An important component of all forest ecosystems is dead wood, both standing and down. Snags or dead trees go through a successional process of decay, each stage of which supports its own group of animal life. When dead trees fall, they enter one of the four decomposition classes. Fallen logs provide habitats for an array of animal life and germination sites for forest tree seedlings.

FIGURE 29.6 A planting of trees does not make a forest. Ths monocultural stand of slash pine in Florida, one of the more important commercial species in the southern coastal plain, lacks the complexity of a natural slash pine forest, which includes loblolly pine, swamp tupelo, and other species. This stand has been thinned by the removal of every other row of trees.

the old-growth forest is an extremely complex ecosystem that evolved over hundreds of years. When such forests are cut, the system cannot be replaced, unless large tracts of old-growth forests are allowed to remain among younger stands, and the young stands are allowed to reach old age. Because old-growth forests are considered senescent, unproductive stands, the aim is to replace them with younger, faster-growing stands, which will not be allowed to reach an ancient age before they are cut. They will be harvested in cycles of 100 to 150 years, and the original ecosystem will have no time to return. The only way to maintain such old-growth ecosystems is to set aside sufficiently large tracts to sustain the ecosystem. Foresters regard retaining stands for that length of time as highly uneconomical and unproductive.

Serious damage is done to many montane coniferous forests by logging on steep slopes. Severe soil erosion degrades the site to a point where forest regeneration is impossible.

Montane coniferous forests are popular places for recreational use. Many forests have been carved up for ski slopes, trails, and access roads.These developments compact the soil, kill trees, and destroy natural values.

Coniferous forests have their share of diseases and insects. Humans often introduce a disease or insect or encourage them by logging and fire suppression. Examples are outbreaks of mountain pine beetle (*Dendroctonus monticolae*), which have wiped out thousands of hectares of lodgepole pine, and the western spruce budworm (*Choristoneura occidentalis*), which has become a problem because of the absence of light ground fires and selective cutting practices that encourage understory development in fir and spruce forests.

BROADLEAF FORESTS

TYPES

Deciduous Forests

The deciduous forest once covered large areas of Europe and China, parts of South America and the middle American highlands, and eastern North America. The deciduous forests of Europe and Asia have largely disappeared, cleared over the centuries for agriculture. What remains is seminatural, except for pockets in the more mountainous regions of central Europe. The two major forest types there are beech-oak-hornbeam and oak-hornbeam. Beech (*Fagus sylvatica*) is the most uniformly distributed tree throughout central Europe. Beech forests, which grow from lowlands into the mountains, are characterized by dense canopy and poorly developed understory. Occupying damper, more acid soils are oak-hornbeam forests dominated by pedunculate oak (*Quercus robur*) and sessile oak (*Q. petrata*). The Atlantic deciduous forest, originally dominated by beech, oaks, ash (*Fraxinus* spp.), and birch (*Betula* spp.), exists only in a seminatural state. Because of glacial history, the species diversity of the European deciduous forests does not compare with that of North America or China.

The Asiatic broadleaf forest, found in eastern China, Japan, Taiwan, and Korea, is similar to the North American deciduous forest and contains a number of plant species of the same genera as those found in North America and western Europe.

In North America the temperate deciduous forest reached its greatest development in the mixed mesophytic forest of the central Appalachians, where the number of temperate tree species is unsurpassed by any other area in the world (Figure 29.7). In eastern North America the deciduous forest consists of a number of associations, including the mixed mesophytic of the unglaciated Appalachian plateau; the beech-maple and northern hardwood forests (with pine and hemlock) in northern regions that eventually grade into the boreal forest; the maple-basswood forests of the Lake states; the oak-chestnut (now oak since the die-off of the American chestnut) or central hardwood forests, which cover most of the Appalachian mountains; the magnolia-oak forests of the Gulf Coast states; and the oak-history forests of the Ozarks. Fingers of forest extend along the rivers of the prairies, plains, and semiarid southwestern United States and Mexico. These forests, growing on the floodplains and banks of rivers and streams, are known as riparian woodlands. Growing on rich, moist, alluvial soil, riparian ecosystems are highly productive, add diversity to the landscape,

FIGURE 29.7 An Appalachian hardwood forest, dominated by oaks and yellow-poplar, in spring with an understory of flowering dogwood and redbud.

and provide habitat for a number of wildlife species disproportionate to their area.

Temperate Woodlands

In western parts of North America where the climate is too dry for montane coniferous forest, we find the temperate woodlands. These forests are characterized by open-growth small trees with a well-developed understory of grass or shrubs. They contain needle-leaved trees, deciduous broadleaf trees, sclerophylls, or any combination. An outstanding example is the piñon-juniper woodland (Figure 29.8). This ecosystem occurs from the Front Range of the Rocky Mountains to the eastern slopes of the Sierra Nevada foothills. In southern Arizona, New Mexico, and northern Mexico grow oak-juniper and oak woodlands, and in the Rocky Mountains, in particular, there are oak-sagebrush woodlands. In the Great Valley of California grows still another type—evergreen oak woodlands with grassy undergrowth.

FIGURE 29.8 Piñon (*Pinus edulis*) and juniper (*Juniperus osteosperma*) are two small, slow-growing trees characteristic of temperate woodlands in the southwestern United States. Seeds of piñon and fruits of juniper were staples in the diet of southwestern Indians.

Temperate Evergreen Forests

In several subtropical areas of the world are extensive mixed forests of both broadleaf evergreen and coniferous trees. Such forests include the eucalyptus in Australia (Figure 29.9), paramo forests and anacardia gallery forests of South America and New Caledonia, and false beech (*Nothofagus* spp.) forests in Patagonia. Representatives of temperate evergreen forests also occur in the Caribbean and on the North American continent along the Gulf Coast, in the hummocks of the Florida Everglades, and in the Florida Keys. Depending on location, these forests are characterized by oaks, magnolias, gumbo-lumbo (*Bursera simbaruba*), and royal and cabbage palms.

STRUCTURE

Highly developed, uneven-aged deciduous forests usually have four strata (Figure 29.10). The upper canopy consists of dominant and codominant trees, below which is the lower tree canopy and then the shrub layer. The ground layer has herbs, ferns, and mosses.

FIGURE 29.10 A view of a deciduous forest from a clearing shows the vertical stratification.

FIGURE 29.9 A eucalyptus forest in Victoria, eastern Australia.

Even-aged stands, the results of fire, clearcut logging, and other large-scale disturbances (see Chapter 21), often have poorly developed strata beneath the canopy because of dense shade. The low tree and shrub strata are thin and the ground layer is poorly developed, except in small, open areas.

The physical stratification of the forest influences the microclimate within the forest. The highest temperatures are in the upper canopy because this stratum intercepts solar radiation. Temperatures tend to decrease through the lower strata. The most rapid decline takes place from the leaf litter down through the soil.

Humidity in the forest interior is high in summer because of plant transpiration and poor air circulation. During the day, when the air warms and its water-holding capacity increases, relative humidity is lowest. At night, when temperature and moisture-holding capacities are low, relative humidity rises. The lowest humidity in the forest is a few feet above the canopy, where air circulation is best. The highest humidity is near the forest floor, kept that way by the evaporation of moisture from the ground and settling of cold air from the strata above.

Variation of humidity within the forest is influenced in part by the degree to which the lower strata are developed. Leaves add moisture to the immediate surrounding air; well-developed strata with more leaves have higher humidity. Thus layers of increasing and decreasing humidity may exist from the floor to the canopy.

Bathed in full sunlight, the uppermost layer of the canopy is the brightest part of the forest. Down through the forest strata, light intensity dims. In an oak forest only about 6 percent of the total midday sunlight reaches the forest floor; the forest floor is about 0.4 percent as bright as the upper canopy.

Light intensity within the forest varies seasonally (see Chapter 7). The forest floor receives its maximum illumination during early spring before the leaves appear; a second lower peak of maximum illumination during the growing season occurs in the fall. The darkest period is midsummer. Light intensity during summer is highly variable from point to point and time to time as sun shines through gaps in the canopy. Sun flecks can influence the distribution of herbaceous vegetation on the forest floor.

In general, the diversity of animal life is associated with stratification and the growth forms of plants (see Chapter 19). Some animals, particularly forest arthropods, are associated with or spend the major part of their lives in a single stratum; others range over two or more strata. The greatest concentration and diversity of life in the forest occurs on and just below the ground layer. Many animals, the soil and litter invertebrates in particular, remain in the subterranean stratum. Others, such as mice, shrews, ground squirrels, and forest salamanders, burrow into the soil or litter for shelter and food. Larger mammals live on the ground layer and feed on herbs, shrubs, and low trees. Birds move rather freely among several strata, but favor one layer over another. Some occupy the ground layer but move into the upper strata to feed, roost, or advertise territory.

Other species occupy the upper strata—the shrub, low tree, and canopy. The red-eyed vireo, the most abundant bird of the eastern deciduous forest of North America, inhabits the lower tree stratum and the wood pewee the lower canopy. The black-throated green warbler and scarlet tanager live in the upper canopy. Squirrels are mammalian inhabitants of the canopy, and woodpeckers, nuthatches, and creepers live amid the tree trunks between shrubs and the canopy.

FUNCTION

Compared to other ecosystems, an abundance of data exists on energy flow and nutrient cycling in deciduous forests. One of the most intensively studied was a mesic yellow-poplar (*Liriodendron tulipifera*) forest on Walker Branch, Oak Ridge, Tennessee, with an understory of redbud, dogwood, Virginia creeper, and Christmas fern.

Gross primary production amounted to 2162 g/C/m^2/yr, of which 66 percent was expended in plant respiration. This expenditure left a balance of 726 g C/m^2/yr as net primary production.

Total organic matter, both living and dead, amounted to 329,100 kg/ha, of which 42 percent was in living biomass. Ninety percent of the living biomass was in branches, bole, and root. Ninety-nine percent total detrital biomass, 190,500 kg/ha, was in the soil rooting zone. Only 1 percent was in the thin layer of forest litter. Thus in the forest, total detrital mass exceeded living organic biomass.

Nutrient cycling involves a balance between inputs to the biological system and outputs or losses through streamflow. The difference represents the amount of recharge to the system from the soil pool. In the yellow-poplar forest on Walker Branch, total input through precipitation for five elements—N, P, K, Ca, and Mg—amounted to 20.44 kg/ha, consisting mostly of nitrogen and calcium; losses amounted to 212.78 kg/ha, mostly calcium. Nitrogen and phosphorus were cycled more tightly than the other three elements. More of these two elements came into the system through precipitation than was lost through leaching. Calcium and magnesium sustained the greatest losses, but much of this loss probably resulted from the weathering of dolomitic limestone substrate.

Ecologists have drawn up annual nutrient balance sheets for several deciduous forest ecosystems. They

TABLE 29.2 Annual Element Balance of a 30- to 80-Year-Old Yellow-Poplar–Oak Forest, Oak Ridge, Tennessee (kg/ha/yr)

	N	*P*	*K*	*Ca*	*Mg*
Requirement	87.9	6.3	47.5	82.6	21.7
Uptake	58.1	3.4	40.0	87.6	12.4
Internal recycling	29.8	2.9	7.5	(−5.2)	9.3

Note: Requirement = annual increment of elements associated with bole and branch wood plus current foliage production.
Uptake = annual increment of elements associated with bole and branch wood plus annual loss through litterfall, leaf wash, and stem flow.
Internal recycling = Requirement − Uptake: deficiency in uptake made up by recycling elements within plant biomass.

found that the yellow-poplar forest at Oak Ridge (Table 29.2) could not meet its nutrient requirements by uptake from the soil, except for calcium, which it took up in excess of need (probably because of the limestone soil). Its uptake of nitrogen, for example, met only 66 percent of its requirements. The forest had to make up the deficiency by recycling the nutrient within its tree biomass. In the intensively studied northern hardwood forest ecosystem at Hubbard Brook, New Hampshire, on acidic granitic soil, outputs or losses exceeded the inputs of calcium, magnesium, and sodium, whereas potassium showed a gain (Table 29.3).

The litter layer is the most important short-term nutrient pool, because it quickly decomposes (average turnover time is four years) and recycles, although the bulk of the nutrient pool is in mineral soil. Nutrients stored in living biomass, especially in roots, are translocated and recycled through the living biomass, particularly the foliage. The foliage, in turn, translocates a considerable portion of its nutrients back to the roots before leaf fall. However, the forest ecosystem can maintain adequate mineral cycling only if nutrients are pumped from soil reserves, maintained in part by weathering of parent material.

The role of the various components of the forest ecosystem in nutrient cycling is illustrated by long-term studies of the nitrogen cycle in deciduous forests at Hubbard Brooks, New Hampshire, and Walker Branch, Oak Ridge, Tennessee. Both studies point out that natural forest ecosystems tend to accumulate and cycle large amounts of nitrogen. In both systems nearly 90 percent of the nitrogen was incorporated in the mineral soil. The remaining nitrogen was in vegetation and the forest floor. The most important mechanism for cycling nitrogen from vegetation to soil was death of small lateral roots, replaced by seasonal growth. The second most important mechanism was litterfall, 90 percent of which were leaves and reproductive parts. The third was leaching of nutrients from the leaves. Added to this input was nitrogen released through the decomposition of forest litter, input from precipitation, nitrogen fixation, and finally uptake from mineral soil.

Such studies of nitrogen cycling in the deciduous forest ecosystem emphasize important aspects of nutrient economics of the forest. Of the nitrogen added to long-term storage, about 54 percent is deposited in living matter and 46 percent is stored in organic matter in the forest soil. Of the amount of nitrogen coming into the system each year, the forest retains about 81 percent. Of the nitrogen used in plant growth, the forest withdraws about one-third from storage in the living plants. To replace that withdrawal, the forest

TABLE 29.3 Inputs and Losses of Nutrients, Hubbard Brook Forest (kg/ha)

	Ca	*Mg*	*Na*	*K*
Input	3.0 ± 0	0.7 ± 0	1.0 ± 0	2.5 ± 0
Output	8.0 ± 0.5	2.6 ± 0.06	5.9 ± 0.3	1.8 ± 0.1
Loss/Gain	−5.0 ± 0.5	−1.9 ± 0.06	−4.9 ± 0.3	+0.7 ± 0.1

takes a like amount from leaves before leaf fall and deposits it in the stems. So conservative are forest ecosystems in nitrogen cycling that only a small fraction of the nitrogen they add to the inorganic pool within the ecosystem is lost in streamflow. Thus an internal source of nitrogen normally not subject to loss from the system, annual uptake by living vegetation, and annual additions of nitrogen to woody biomass are important in promoting a tight cycling of nitrogen.

Deciduous forests appear to differ functionally from coniferous forests in the magnitude and nature of nutrient cycling. The differences are illustrated graphically in Figure 29.11, comparing nutrient cycling in a mixed oak forest and a spruce forest in Belgium, both of which have an annual production of 14.6 tn/ha/yr. From these data it is evident that considerably more nutrients are cycled through the deciduous forest than through the coniferous forest and that spruce retains relatively more nutrients in its biomass than oak. Spruce keeps more nitrogen, sulfur, and phosphorus than it returns through litterfall, throughfall, stemflow, and dead parts of the herbaceous layer. The oak forest, on the other hand, returns more of all elements than it retains. The deciduous forest is much more efficient at recycling calcium, largely through litterfall, whereas the spruce forest returns considerable quantities through stemflow and throughfall. Although the deciduous forest takes up more potassium than the spruce forest, it retains less. The spruce forest keeps nearly half its uptake of potassium. This difference confirms the idea that conifers tend to be nutrient accumulators.

HUMAN IMPACT

Although humans originally evolved in grassland and savanna ecosystems, we achieved our greatest cultural and economic development in temperate broadleaf deciduous forest regions. Destruction of the deciduous forest began early in Europe and was largely complete by the Middle Ages. Over most of Europe the broadleaf deciduous forests that do remain are seminatural, highly modified by humans. In North America great expanses of natural deciduous forest still exist in spite of mismanagement, logging, fires, and land clearing. Logging wiped out magnificent virgin stands in a wave of deforestation unmatched for speed in any other part of the world. Most of the eastern forest was cleared for agriculture so intensely that New England had a wood shortage in colonial days. As farming moved west into the prairie region, forest returned to abandoned land in New England and, later, elsewhere in the eastern United States. Much of the east now appears to be heavily forested. This appearance is illusionary, however. Much of the forest has been and is being fragmented by highways and industrial, urban, suburban, and recreational development (see Chapters 19 and 21). Large areas of forest land once held under one ownership are being divided into smaller and smaller tracts for suburban, recreational, and retirement housing, forever changing the nature and integrity of large forested tracts. This fragmentation of ownership has serious implications for both wildlife and wood production. Large reserves of strippable coal coupled with energy demands threaten the rich mesic forest of the Appalachians with its great diversity of tree species and endemic salamanders.

Many of the second, third, and fourth growth forests we see bear little resemblance in species to the original forest cover. A good proportion of these forests became reestablished by invading abandoned agricultural land, so the species composition was influenced by the available seed source and the response of colonizing species to the site conditions. The nature of the incoming forest on cutover land is influenced by the effects and type of logging, amount of disturbance to the residual stand and the forest floor, availability of seeds, suitable seedbed, and environmental conditions at the time of disturbance. For example, foresters have a difficult time regenerating Appalachian oak forests after harvesting, and the outcome is a replacement of oaks by red maple and black cherry, which may mean the demise of the great oak forests.

Introduced diseases and insects have not helped. Chestnut blight accidentally introduced into the United States from China by way of Europe in 1904 eliminated the American chestnut as a major component of the eastern forest. It holds on precariously by means of persistent, short-lived root sprouts. The gypsy moth, which escaped from a silk-moth breeder in Massachusetts in 1869, threatens the oaks of the central hardwood forest. Dutch elm disease spread by a bark beetle has devastated the American elm in North America.

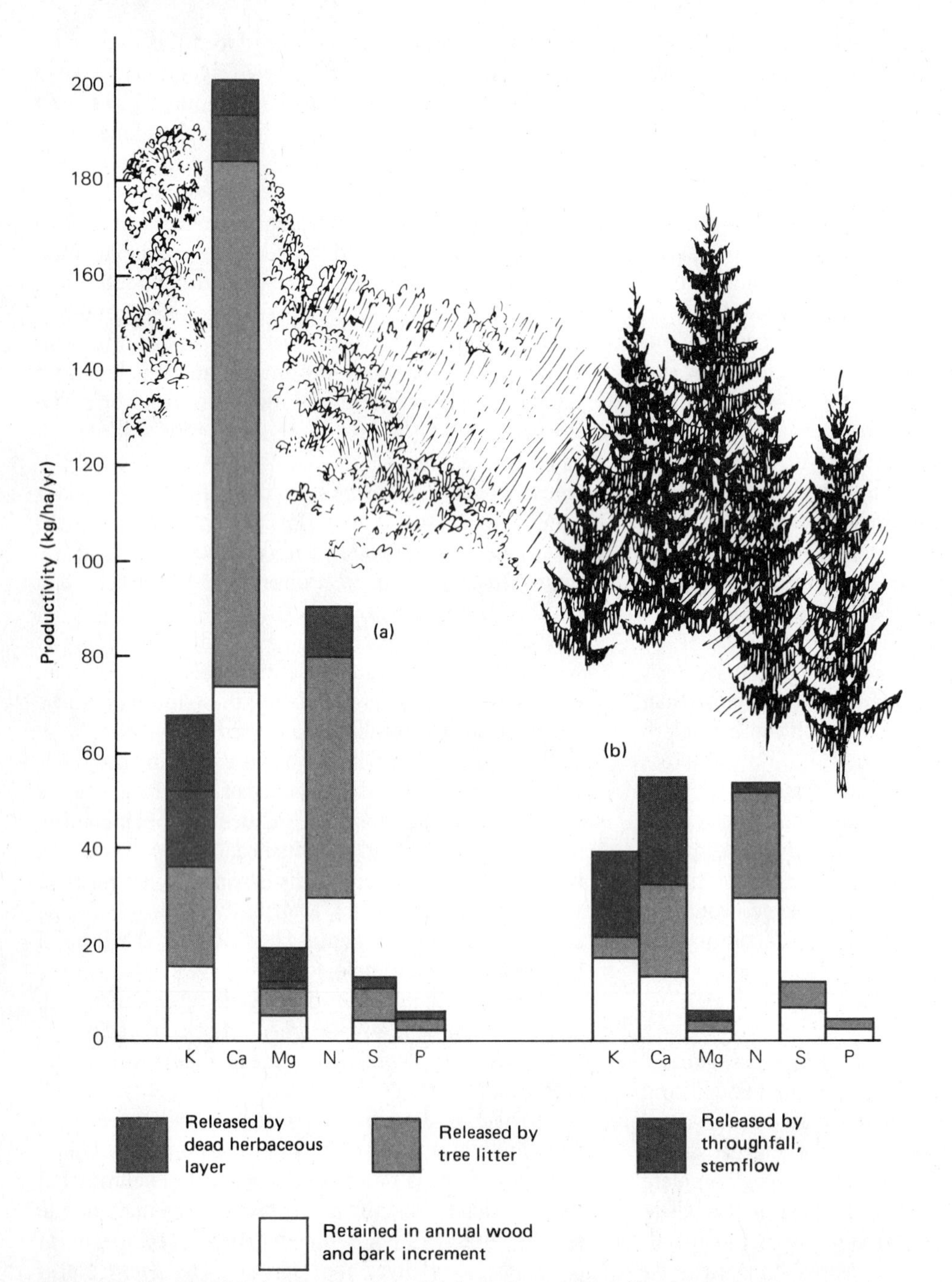

FIGURE 29.11 Simplified representation of biological nutrient cycling in two forests: (a) a mixed oak forest (*Quercus*) at Virelles, Belgium; and (b) a spruce forest (*Picea*) at Mitwart, Belgium. Productivity for both forests is 14.6 tn/ha/yr.

Especially impacted by human activity are the riparian forests of the prairie regions and particularly of the American southwest and Mexico. Occupying rich alluvial soils and drawing moisture from the rivers and streams, these areas appealed to settlers. They cut forests for wood, cleared land for urban development, crops, and grazing, and destroyed other forests by stream channelization and flood-control works. The loss of riparian woodlands has contributed to flooding and seriously reduced wildlife habitat, both nesting and wintering grounds and protective pathways for movement and migration.

As available forests dwindle, we see efforts to improve the situation. People have begun to study and protect the remaining riparian ecosystems in the prairie regions and the southwest. Forest ecologists have been studying nutrient cycling, biomass accumulation, successional trends, and forest development. Silviculturists have been investigating ways to improve harvesting and regeneration methods to reduce ecological damage to the forest. They have amassed a greater amount of information; but the problem is to get commercial interests to apply those findings to the forest.

Summary

Temperate coniferous and broadleaf deciduous forests are the dominant types of forest vegetation south of the boreal forest in the Northern Hemisphere. The coniferous forests include broadly the montane, pine, and temperate rain forests of the Pacific Northwest. Broadleaf deciduous forests include the hardwood forests of Europe and North America, the temperate woodlands and temperate broadleaf evergreen forests of southern North America, and the eucalyptus forest of southern Australia.

Both coniferous and deciduous forests are stratified into layers of vegetation. Accompanying this physical layering is a stratification of light, temperature, and moisture. The canopy receives the full impact of climate and intercepts light and rainfall. The forest floor is shaded through the year in most coniferous forests and in spring and summer in broadleaf deciduous forests.

The coniferous and deciduous forests hold different species of animal life, but animal adaptations are similar. The greatest concentration and diversity of life are on and just below the ground layer. Other animals live in various strata from low shrubs to the canopy. Whatever the forest, the trees that compose it create different environments that dictate the kinds of plants and animals that can live within it.

Mineral cycling in both coniferous and deciduous forests is tight. Nutrients accumulate in woody biomass to form a pool unavailable for short-term cycling. Although the bulk of nutrients is in the mineral soil, the most important pools in mineral cycling are root mortality, litterfall, and foliar leaching. Internal cycling of some nutrients, particular nitrogen, is important in nutrient conservation. In most undisturbed forest ecosystems only a small fraction of the nutrients is lost from the system through leaching and streamflow.

Coniferous forests exhibit short-term cycling between litterfall and uptake by trees. At the same time conifers appear to be accumulators, removing elements from the soil in quantities large enough to upset nutritional balance in ecosystems. Nutrient cycling in conifers appears to depend on mycorrhizae, fungi, and root activity. Fungi act as nutrient sinks, but a symbiotic relationship exists between mycorrhizae and roots in the uptake of nutrients.

Like grassland, desert, and shrubland ecosystems, coniferous and deciduous forests have been altered, degraded, or eliminated by human activity. The temperate forest regions of Eurasia and China have been cleared over the centuries for agriculture, so few forests remain. The forests that do exist are seminatural stands. In the United States the deciduous forest in the northeast and north central parts of the country also had been cleared for agriculture, but as agriculture declined in the northeast, forests returned. Now they are being fragmented by development. The species composition of most North American hardwood forests has been altered by logging and fire. Impacted the hardest are the narrow belts of riparian forest along streams and rivers in the plains and the American southwest, because in dry country they provided ideal sites for human occupancy and use. Coniferous forests, likewise, have been subject to logging. Fires that fed on the logging debris in many of them have destroyed their ability to return. Otherwise most coniferous forests are able to regenerate, but many of them are regenerated by artificial planting, developing into simple monocultural stands that are not true forest. The most complex of all coniferous forests are the old-growth temperate rain forests of the Pacific northwest, and their existence is threatened by logging.

Review and Study Questions

1. Discuss the various types of coniferous and deciduous forests.
2. What are the major reasons for stratification in forest ecosystems?
3. How does structural stratification affect environmental stratification?
4. What are the major life forms among conifers and what effect do they have on the structure of coniferous forests?
5. Contrast the structural stratification between coniferous and deciduous forests.
6. Contrast nutrient cycling of coniferous forests with that of deciduous forests.
7. What is the role of dead wood on the forest floor? Why should dead trees be allowed to stand?

*8. The effort to save the endangered spotted owl in the Pacific Northwest is not just an effort to save a species but to save an ecosystem. Why is it easier to gain support for an endangered species than an endangered ecosystem? Aside from the spotted owl, why should old-growth forests be saved?

*9. Trace human relationships to the forest through the centuries, especially in northern Europe, in North America, and in Australia. What are the similarities in all three continents, and why should those similarities exist?

Selected References

Bormann, F. H., and G. E. Likens. 1979. *Pattern and process in a forest ecosystem.* New York: Springer-Verlag. Excellent discussion of forest ecosystem functions, with special reference to Hubbard Brook.

Brinson, M. M., B. L. Swift, R. C. Plantico, and J. S. Barclay. 1981. *Riparian ecosystems: Their ecology and status.* FWS/OBS-81/17. Kearneysville, WV: Eastern Energy and Land Use Team, U.S. Fish and Wildlife Service. General information on riparian ecosystems is difficult to find. This is an excellent introduction.

Curtis, J. T. 1959. *The vegetation of Wisconsin.* Madison: University of Wisconsin Press. A classic study of forest vegetation of the Lake states.

Duvigneaud, P. (ed.). 1971. *Productivity of forest ecosystems.* Paris: UNESCO. Somewhat dated, but an excellent reference on the subject.

Edmonds, R. L. (ed.). 1981. *Analysis of coniferous forest ecosystems in the western United States.* Stroudsbury: Dowden, Hutchinson & Ross. Strong on function.

Franklin, J. F., and C. T. Dyrness. 1973. *Natural vegetation of Oregon and Washington.* USDA Forest Service Gen. Tech. Rept. PNW 8. Corvallis, OR: USDA Forest Service. Excellent description of old-growth forests and other vegetation.

Irland, L. C. 1982. *Wildland and woodlots: The story of New England's forests.* Hanover, NH: University Press of New England. Reviews the changing character of New England's forests since white settlement and emerging problems.

Lang, G. E., W. A. Reiners, and L. H. Pike. 1980. Structure and biomass dynamics of epiphytic lichen communities of balsam fir forests in New Hampshire. *Ecology* 63:541–550.

Norse, E. A. 1990. *Ancient forests of the Pacific Northwest.* Washington, DC: Island Press. Excellent description of the forest and its problems.

Packham, J. R., and D. J. L. Harding. 1982. *Ecology of woodland processes.* London: Edward Arnold. A concise, pleasant introduction to woodland ecology. Although it considers North American examples, it is largely Europe oriented, which makes it an excellent introduction to European forest ecology.

Polunin, O., and M. Walters. 1985. *A guide to the vegetation of Britain and Europe.* New York: Oxford University Press. Outstanding well-illustrated guide to seminatural and natural vegetation, including detailed description of forests.

Reichle, D. E. (ed.). 1981. *Dynamic properties of forest ecosystems.* Cambridge, England: Cambridge University Press. The major reference source on functions of forest ecosystems throughout the world.

USDA Forest Service. 1980. *Environmental consequences of timber harvesting in Rocky Mountain coniferous forests.* USDA Forest Service Gen. Tech. Rept. INT-90. Ogden, UT: Intermt. For. and Range Exp. Stat. Technical, detailed assessment of timber harvesting in the mountains.

Whittaker, R. H., and G. M. Woodwell. 1969. Structure and function, production, and diversity of the oak-pine forest at Brookhaven, New York. *J. Ecology* 57:155–174.

Outline

CHAPTER 30

Tropical Forests

Objectives

On completion of this chapter, you should be able to:

1. Describe the types of tropical forests.
2. Explain the proximate reasons for the high diversity of life in tropical rain forests.
3. Describe the structure of tropical rain forests.
4. Discuss the role of mutualism in functioning of tropical forests.
5. Describe the ways in which humans have affected tropical forests.

The term "tropical forest" brings to mind images of lush green jungles of the Amazon. Tropical forests, however, embrace more than the rain forest. Forty percent of the tropical land mass is open or closed forest; but of this forest only 25 percent is wet or rain forest; 32 percent is moist, and 42 percent is dry.

TYPES OF TROPICAL FORESTS

Plant geographers divide the tropical forests into a number of types and subtypes. One type grades into another, so no sharp boundary can be drawn.

RAIN FORESTS

The prototype of tropical forests is the tropical rain forest, restricted mostly to the equatorial climatic zone between latitudes 10° N and 10° S. The rain forest occupies those regions of the world where the temperatures are warm through the year and rainfall, measured in meters, occurs almost daily. Although there is little annual variation in temperature and precipitation, daily variations in heat, rainfall, and humidity are great.

Tropical forests do not form a continuous belt around the terrestrial equatorial region. They are discontinuous, broken up by differences in precipitation (governed by direction of the winds) and in the land masses. Most tropical forests, for example, grow below 1000 meters, and they are absent from the eastern part of equatorial Africa, the northwest corner of South America, and the southern tip of India, all places that experience long annual droughts.

Tropical rain forests fall into three main groups. The largest and most continuous is in the Amazon basin of South America. The second is in the Indo-Malaysian area from the west coast of India and southeast China through Malaysia and Java to New Guinea. This forest has the greatest diversity of plant species. The third forest is in West Africa around the Gulf of Guinea and extending into the Congo Basin. Smaller rain forests occur on the eastern coast of Australia, the windward side of the Hawaiian Islands, the South Sea Islands, and the east coast of Madagascar. There the trees are mainly evergreen with little or no bud protection.

Within the rain forest are many subtypes. The most luxuriant is the multilayered lowland rain forest (Figure 30.1). At higher elevations the lowland forest grades into mountain forest with its abundant undergrowth, tree ferns, and small palms. The mountain forest gives way to the cloud forest with its understory thickets and trees burdened with epiphytes. Although it receives less rainfall, the cloud forest is continually wrapped in clouds and mists. Fingers of rain forest, called gallery forest, follow river courses into savannas. Swamp forests occupy perennially wet soils, and peat forests grow on nutrient-poor ones.

SEASONAL FORESTS

The rain forest flows into tropical and subtropical seasonal forests, also called semi-evergreen and semideciduous forest. Although they retain many characteristics of the rain forest, these forests are subject to droughts of two to four months. Some 30 percent of the trees of the upper canopy lose their leaves during the dry period, but the lower canopy trees and understory retain their leaves through the year. In the Indo-Malaysian forest new leaves emerge about a month after the coming of the monsoon rains, the emergence trig-

FIGURE 30.1 A primary dipterocarp tropical rain forest bordering the Sungai Tembeling River, Peninsular Malaysia.

gered by increasing temperature. In other regions leaves emerge about a month before the rainy season begins, often coinciding with flowering. Fruits develop at the beginning of the dry season. As with the rain forest, there are lowland and mountain types.

DRY FORESTS

Overlooked in our concern over tropical forests are the dry tropical forests (Figure 30.2). The largest proportion of dry tropical forest is in Africa and on tropical islands, where it comprises about 80 percent of the forested area. About 22 percent of South American and 55 percent of Central American forested areas are dry tropical forest. Much of the original forest is gone, especially in Central America and India. It has been converted to agricultural and grazing land, or it has regressed through disturbance to thorn woodland, savanna, and grassland. Dry tropical forests experience a dry period, the length of which is based on latitude. The more distant the forest is from the equator, the longer is the dry season, up to eight months (see Chapter 4). During the dry period the trees and shrubs drop their leaves. Before the start of the rainy season, which may be much wetter than the wettest time in the rain forest, the trees begin to leaf; during the rainy season the landscape becomes uniformly green.

FIGURE 30.2 A dry tropical forest in Africa composed largely of Buffalo thorn, *Ziziphus mucronata.*

TROPICAL RAIN FORESTS

Tropical rain forests are noted for their diversity of plant and animal life. Tree species number in the thousands. A 10-square kilometer area of tropical rain forest may contain 1500 species of flowering plants and up to 750 species of trees. The richest is the lowland tropical forest of Peninsular Malaysia, which contains some 7900 species. There one of the major groups, the Dipterocarpaceae, contains 9 genera and 155 species, of which 27 are endemic. (The Asian dipterocarps have 12 genera and 470 species.) Tropical rain forests also account for several million species of flora and fauna, one-half of all known plant and animal species, and 20 to 25 percent of all known arthropods.

STRUCTURE

Vegetation

The tropical rain forest can be divided into five general layers (Figure 30.3), most apparent in the undisturbed forest. Stratification, however, is often poorly defined, because the growth plan of many tree species is the same, differing only in size. The uppermost layer consists of emergent trees over 50 to 80 m high whose deep crowns billow above the rest of the forest to form a discontinuous canopy. The second layer, consisting of mop-crowned trees less than 50 m high, forms another, lower, discontinuous canopy. Not clearly separated from one another, these two layers form an almost complete canopy. The third layer, the lowest tree stratum, is made up of trees with conical crowns. It is continuous, often the deepest layer, and well defined. The fourth layer, usually poorly developed in deep shade, consists of shrubs, young trees, tall herbs, and ferns. Many of these plants have elongated, downward-curving leaf blades, called drip-tips. These blades apparently enable the leaves to rid themselves of excess water in their permanently wet environment, increase transpiration, and reduce nutrient leaching. The fifth stratum is the ground layer of tree seedlings and low herbaceous plants and ferns.

A conspicuous part of the rain forest is plant life dependent on trees for support. Such plants include epiphytes, climbers, and stranglers. Climbers, the *lianas,* are vines with fine stringlike to massive cable-

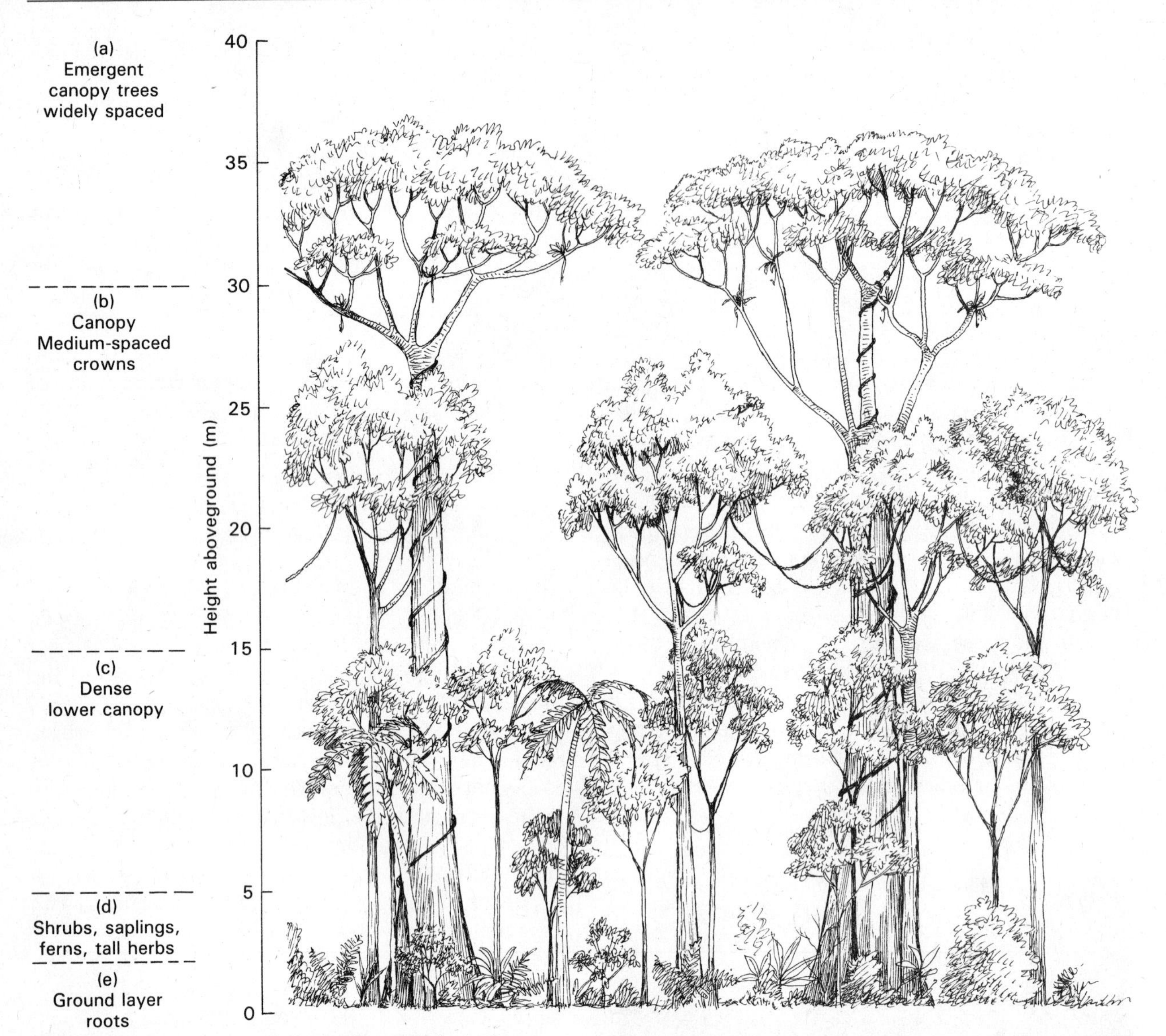

FIGURE 30.3 Vertical stratification of a tropical rain forest.

like stems that reach the tops of trees and expand into the form and size of a tree crown. They may loop to the ground and ascend again. Climbers grow prolifically in openings, giving rise to the image of the impenetrable jungle. This popular image applies more to secondary forest, second growth that develops where primary forest has been disburbed.

Stranglers and epiphytes share some characteristics. Stranglers start life as epiphytes. As they grow, they send roots to the ground and increase in number and girth until they eventually encompass the host tree and claim the crown and limbs as support for their own leafy growth. Epiphytes inhabit niches on the trunks, limbs, and branches, and even leaves of trees, shrubs, and climbers. Their roots are aerial. Epiphytes come in various types. One group, the microepiphytes, consists of mosses, lichens, and algae. Macroepiphytes, such as orchids and members of the Ericaceae, are vascular plants. Their roots never reach the ground. Hemiparasites, they attach themselves to a tree and take up water, nutrients, and some photosynthate. Some of the epiphytes are important in recycling minerals leached from the canopy.

The floor of the tropical rain forest is thickly laced with roots, both large and small, forming a dense mat on the ground. Except for a few that reach down to weathered parent rock, rain forest roots are shallowly concentrated in the upper 0.3 m of soil, where inorganic nutrients are available. Associated with the fine roots are mycorrhizae (see Chapter 17) that aid in the uptake of nutrients from the decaying organic matter.

Reaching upward as much as 5 m from the widespreading roots to the trunks of many species of large trees are thin, strong, planklike outgrowths called *buttresses* (Figure 30.4). These buttresses function as prop roots, providing support for trees rooted in soil that offers poor anchorage.

The mature tropical forest, like the mature temperate forest, is a mosaic of continually changing vegetation. Death of tall trees, brought about by senescence, lightning, wind storms, hurricanes, defoliation by caterpillars, and other causes, creates gaps (see Chapter 21), which shade-intolerant pioneer species quickly fill.

FIGURE 30.4 The planklike buttresses help to support tall rain forest trees.

These trees are replaced eventually by shade-tolerant late successional species, perhaps over a period of 100 years; but continuous random disturbances across the forest ensure persistence of the species in the mature forest. A high frequency of tree fall may account for the low density of large trees (1 m+ dbh) in mature rain forests. Enhancing this diversity are local changes in soil, topography, and drainage that support varying arrays of different species.

Most tropical rain forest trees reach full height when they have achieved only about one-third to one-half of their final bole diameter. Thus stratification or layering results when a group of species of similar mature height dominates a stand. Layering is also influenced by crown shape, which in turn is correlated with tree growth. Young trees still growing in height have a single stem and a tall narrow crown; they are *monopodial.* As many species mature, large limbs diverge from the upper stem or trunk; they become *sympodial.* This change happens when the bud of the main stem axis ceases to grow and the lateral buds take over their role. The process repeats itself, adding to crown growth and producing a pattern that suggests the spokes of an umbrella. Looking up into the canopy, the observer gains the impression that crowns of trees fit together like a jigsaw puzzle with the pieces about a meter apart (Figure 30.5). This growth pattern is called *crown shyness.* Among the dipterocarps of the Far East, this reiteration of the growth pattern within the crown to produce many dense subcrowns gives a cauliflowerlike appearance to the canopy. Within the crown there is minimal overlap among the leaves, each positioned to receive the maximum amount of light available.

Layering of vegetation influences the internal microclimate of the forest. The crowns of emergent trees experience conditions similar to open land. The level of CO_2 and amount of humidity increase through the canopy, and temperature and evaporation decrease. From the ground up to about 1 m the levels of CO_2 are high, and humidity at over 90 percent is oppressive. Temperature on the average is 6° C cooler inside than outside the forest, with a strong nocturnal inversion. Although light penetrates the upper canopy, the lianas, epiphytes, and the lower tree layers block out most of it. The amount of light that reaches the floor of a Malaysian rain forest is about 2 to 3 percent of incident

FIGURE 30.5 A view through the canopy of a secondary forest reveals crown shyness. The upper canopy has sympodial crown structure. Note the monopodial crown of an understory tree.

radiation, and half of that comes from sun flecks, shafts of light that pass through the leaves and change through the day; about 6 percent comes from breaks in the canopy, and 44 percent from reflected and transmitted light.

Animal Life

Stratification of animal life in the tropical rain forest is pronounced. J. L. Harrison in 1962 described six distinct feeding communities of birds and mammals in a lowland tropical forest of the Far East. (1) A group feeding above the canopy is made up mostly of insectivorous and some carnivorous birds and bats. (2) A top of the canopy group consists of a large variety of birds, fruit bats, and other species of mammals that eat leaves, fruit, and nectar. A few are insectivorous and mixed feeders. (3) Below the canopy, in a zone of tree trunks, is a world of flying animals—birds and insectivorous bats. (4) Also in the middle canopy are scansorial mammals, such as squirrels, that range up and down the trunks, entering the canopy and the ground zone to feed on the fruits of epiphytes growing on tree trunks, on insects, and on other animals. (5) The forest floor is occupied by large herbivores, such as the gaur, or seladang (Figure 30.6), tapir, and elephant, which feed on ground vegetation and low-hanging leaves, and their attendant carnivores, such as leopards and tigers, all of which range over a large area. (6) The final feeding stratum includes the small ground and undergrowth animals, birds and small mammals capable of some climbing, that search the ground litter and lower parts of tree trunks for food. This stratum includes insectivorous, herbivorous, carnivorous, and mixed feeders.

The enormous diversity, but low species population density, of animal life in the tropical rain forest mirrors the great diversity of microhabitats and niches. Insect fauna number in the millions, with many species still to be discovered. Not only is invertebrate life distributed vertically, but horizontally as well. Many species are restricted to certain forests, and within them inhabit only certain plants. Numerous species live in the epiphytes and small pools of water caught in epiphytic plants, where they may be joined by small canopy-dwelling frogs.

FIGURE 30.6 A gaur, or seladang (*Bos gaurus*), a large, black, endangered bovid of Southeast Asia, emerges from a rain forest to visit a salt lick.

Nearly 90 percent of all primates live in the tropical rain forests of the world. Sixty-four species of New World primates, small with prehensile tails, live in the trees. The Indo-Malaysian forests are inhabited by a number of primates, many of which are restricted to certain regions. The orangutan, an arboreal ape, is confined to the Island of Borneo. Peninsular Malaysia has seven species of primates, including three gibbons, two langurs, and two macaques. The long-tailed macaque is common to disturbed or secondary forests, and the pig-tailed macaque is a terrestrial species, adaptable to human settlements. The tropical forest of Africa is home to mountain gorilla and chimpanzees. The diminished rain forest of Madagascar holds 39 species of lemurs.

FUNCTION

Tropical forests are among the most productive ecosystems in the world. Because sites, soils, location, precipitation, and other conditions vary among tropical forests, so does their productivity (Figure 30.7). In general, tropical forests use about 70 to 80 percent of their assimilated energy in maintenance and 20 to 30 percent for net production. Mean annual net production is about 22 mtn/ha. This amount exceeds the net production of temperate forests (about 13 mtn/ha/yr) by a factor of 1.7 and that of boreal forests (an average of 8 mtn/ha/yr) by a factor of 2.7. The rate of wood production by tropical forests is about the same as that of temperate hardwood forests, but tropical forests turn more of their net production into foliage and fruit.

Because high year-round temperatures and abundant rainfall accelerate geological cycling, biological cycles function as retainers of nutrients in the living portion of the system. Tropical forests may store nutrients in the living biomass where they are protected from leaching, or they may reduce to a minimum the time nutrient elements remain in the soil. This mechanism of nutrient cycling contrasts with that of the temperate forest, in which the soil plays a significant role in nutrient cycling.

Tropical forests have a large standing-crop biomass, averaging about 300 tn/ha, two times that of the temperate hardwood forest. They tend to concentrate more calcium, silica, sulfur, iron, magnesium, and sodium, and less potassium and phosphorus in their biomass than do temperate forests. Because of the great variation within and among tropical rain forests, large differences in mineral storage and nutrient cycling exist among them.

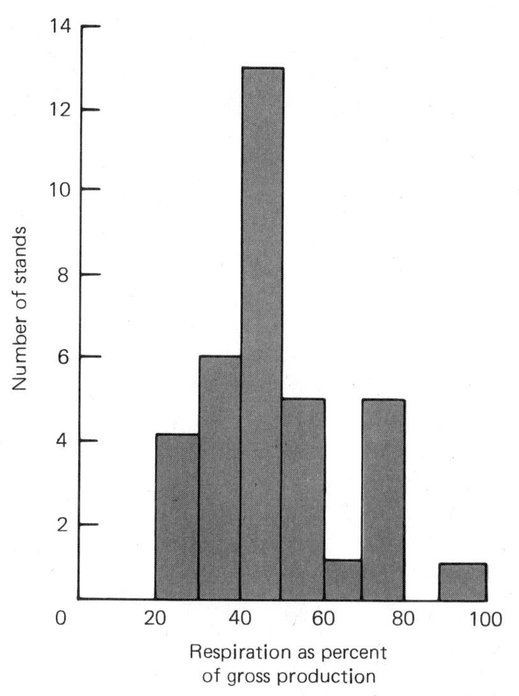

FIGURE 30.7 Variation in production in tropical rain forests, shown by autotrophic respiration as a precentage of gross primary production. All types of vegetation are represented. Variations among stands make generalization difficult.

Tropical rain forests maintain their nutrient balance by internal cycling (see Chapter 24). Nutrients leached from foliage and epiphytes by throughfall are rapidly taken up by the roots and returned to the vegetation. Nutrients not taken up are captured and held by fungal rhizomorphs and mycorrhizal fungi. From their studies, ecologists learned that a tropical forest on the Ivory Coast of Africa recycled more than 60 percent of its potassium and 15 to 56 percent of other nutrients such as calcium, magnesium, and nitrogen gained by throughfall. At the end of the dry season, when leaching is the greatest, the soil-root system picked up the nutrients leached from dead leaves.

The soil-root system is highly dependent on mycorrhizal fungi (see Chapter 17). The mycorrhizae transfer nutrients directly from dead organic matter to the living roots with a minimum of leakage to the soil, so minerals remain tied up in living and dead organic matter. This concentration and rapid recycling of nutrients in the uppermost layers of the soil explain why many tropical soils cleared of their forest vegetation are nutrient-poor and can support agricultural systems for only a few years.

Other important mutualistic relationships keep the tropical forest running. Ninety-eight to 99 percent of all flowering species of the tropical lowland rain forest are pollinated by animals. The major exception is the emergent dipterocarps, mainly the genus *Shorea,* whose pollen is carried by the wind. Animal pollinators range from the tiny fig wasps, only 1 to 2 mm in length, to flying fox bats with a wingspread of two meters. Bats, birds, bees, moths, beetles, large flies, and wasps comprise the major pollinators. Bats in both the neotropical and paleotropical rain forests are important pollinator in the high canopy. Flowers pollinated by bats open at dusk. They have a sour scent, a sticky nectar, and pale yellow or white flowers. Among the birds, hummingbirds are dominant pollinators in the neotropical forests, whereas in paleotropical forests sunbirds, white eyes, and honeyeaters are major bird pollinators. Bird-pollinated flowers are usually scentless with bright colors—reds, oranges, yellows, and greens—and with abundant, watery nectar. In all rain forests, bees are the major insect pollinators, aided by moths and beetles. Bee flowers are rich in nectar, weakly scented, and often colored yellow, blue-green, or blue. Beetle flowers are heavily scented with a rich sweet smell or a carrion odor, depending on the insects to be attracted. Pollination systems are specialized to the extent that a given plant species is pollinated by only one or a few species belonging to the same taxonomic order. One outstanding example is the one-to-one mutualistic relationship between the fig wasp and figs (*Ficus* spp.), which are pollinated only by that insect.

Tropical plants also use animals to disperse their seeds. Across all the tropical forests, 50 to 90 percent of trees and shrubs depend on animal dispersers. To attract them, the fruits are easily obtained and easily digested, and the seeds are discarded or pass through the digestive tracts (see Chapter 17). The problem for the fruit eaters is seasonal availability of fruits, which can be scarce at the end of the rainy season and the beginning of the dry season. The fruits may be spatially or temporally patchy, depending on the species, the sequence of fruiting, and distribution of plants through the forest. The wide-ranging fruit-eating bats generally eat the larger fruits of widely dispersed tropical trees. In the neotropical forest little dietary overlap exists among birds, bats, and primates. In the tropical forests of the Far East, however, which hold a lower diversity of fruit-eating species, considerable dietary overlap exists, especially among the primates.

HUMAN IMPACT

For at least 20 centuries, humans have been associated with tropical forests. They entered the rain forest first as hunter-gatherers. Nomadic, they lived off the forest, hunting its wildlife and gathering its abundance of fruits, roots, and tubers for food, and leaves, bark, and wood for clothing and shelter. Because they were a part of the forest ecosystem and their populations were small—about one person per 4 square kilometers—they had little impact on the forest. Other groups, less nomadic, the hunter-gardeners, established small semipermanent settlements and cultivated garden plots of native plants to supplement food obtained by hunting and gathering. Still others became shifting cultivators, developing the technique of slash-and-burn agriculture. These people cut down and burned the vegetation on small plots. The ashes fertilized the soil in which the cultivators interplanted the seeds of a variety of crops. For one to two years the plots remained productive; after that the people abandoned them to vigorous native weeds and cleared new ones. Meanwhile the old plots, now supporting pioneering tropical vegetation, remained fallow for 8 to 20 years, at which time they were cleared again for crops. As long as the populations were not large, the tropical forest was able to absorb shifting cultivation; but as populations grew and pressure on the land increased, long fallow periods were no longer possible. Land degradation and erosion ate away at the forest.

The real onslaught to the tropical forest came with the Europeans. They exploited the forests for hardwood timber (Figure 30.8a), cleared it for pasture and cropland, and in Southeast Asia converted it to rubber and

(a)

(b)

(c)

(d)

FIGURE 30.8 The tropical rain forest faces destruction. (a) A logging truck has delivered a load of *Shorea* logs to this large Malaysian sawmill. (b) A large land-clearing scheme doomed this Amazonian tropical rain forest. (c) A rubber plantation. When latex production falls at about 25 years of age, the trees are clearcut, the logs are used for wood products, and a new plantation is started on the site. (d) Hundreds of hectares of tropical rain forest were cleared to establish this new oil palm plantation.

palm plantations. Since World War II the rate of exploitation has accelerated. Worldwide demand for tropical hardwood timber has greatly reduced some species, such as various species of *Shorea,* commercially sold as lavan and Philippine mahogany, and driven others, including African and Central American mahogany (*Khaya* spp.) and Brazilian rosewood (*Dalbergia nigra*), to the brink of extinction.

Logging in itself is not the direct cause of loss of tropical forests. Secondary forest can grow back. The problems with logging are the damage done to residual vegetation when the large-crowned trees are felled and erosion caused by road building and heavy machinery. More important, roads open up inaccessible areas to settlers who swarm in and clear off the remaining forest for cultivation (Figure 30.8b). In a short time, the land grows to unproductive shrub and grassland with little hope of reverting to tropical forest.

In addition to logging, millions of hectares of lowland tropical rain forest have been and are being converted to other uses. The hillsides of Malaysia are green not because they are covered with tropical forests but because they are blanketed with rubber and palm plantations (Figure 30.8c, d). In the Amazon and Central

America, large areas of tropical forest have been cleared, burned, and converted to pastureland with devastating results. Response of pasture grasses, stimulated by the fertilizing effects of the ashes, is vigorous at first, but the fertility quickly declines because of leaching, binding of some essential nutrients to the soil, and removal in crops and cattle. Within five to eight years the pasture degenerates into weedy fields. Because the cleared areas are so large, the soil so infertile, and the distance from any primary forest so great, the tropical forest cannot regenerate as it did in smaller plots of slash-and-burn agriculture.

As logging and clearing eat deeper into the tropical forests, these activities come in direct conflict with native peoples of the forests. Contact with the western world has tragic effects: loss of home and livelihood because of a fragmented environment, outright murder, introduction of disease, and destruction of native culture. Unable to exist in the forest any longer, tribes are forced to move to settlements and take up a foreign way of life or resist removal, slowing the inevitable. For example, the Bateq in Peninsular Malaysia cling to their fragmented tropical forest home, too small to support them through the year (Figure 30.9). During the fruit-poor monsoon season, they either move to nearby villages and leave the forest by day to earn money for food or stay in settlement camps. At the end of the monsoon season, they disappear into the forest.

FIGURE 30.9 A semipermanent camp of a group of Bateq. A number of such camps, at which a group remains for only several days at a time, are located throughout the forest.

The greatest tragedy of the tropical deforestation is the loss of biological diversity. Although occupying only 7 percent of Earth's surface and disappearing at the rate of an estimated 56×10^6 ha per year, the tropical rain forest region holds 50 to 80 percent of the world's plant species; and most of the animal species are endemic, resident, and nonmigratory. As tropical rain forests are reduced and fragmented, species lose their habitat, and mutualistic relationships involving pollination and seed dispersal necessary for the survival of both plants and animals are broken. Lost, too, are major and potential sources of medicine and wild genotypes needed for improving agricultural crops and livestock. As native peoples disappear or lose their culture, we will lose their knowledge of the ecology and usefulness of the biotic resources of the tropical rain forest.

There is hope that the destruction may slow or be contained. The costs of developing pastures and rangeland in the Amazonian tropical forests do not balance short-term yields, and the investments in fertilizer and weed control to maintain productivity are prohibitive. Agriculturalists are developing tree crops and new agricultural methods for tropical soils to slow slash-and-burn agriculture. Interest in managing rain forests on a sustained yield basis, especially in the Far East, is gaining momentum. The Forestry Research Institute of Malaysia is developing methods and markets for utilizing trees that otherwise would be wasted. Tropical forest silviculturists have developed management plans to ensure continued production from forest reserves and are establishing plantations of fast-growing tropical species. Many species of wildlife, fortunately, do well in secondary forests, including gibbons, deer, gaur, and elephants. Because of burgeoning world populations and demand for wood, ceasing all cutting of tropical forests is unrealistic; but we must ensure that we use the tropical rain forests without destroying them.

Summary

Tropical forests embrace a number of types, from lowland tropical rain forests to montane and cloud forests, tropical and subtropical seasonal forests, and dry tropical forests. Tropical rain forests, noted for their enormous diversity of life, divide into five general layers: the emergent trees, high upper canopy, low-tree stratum, shrub understory, and a ground layer of herbs and ferns. Conspicuous parts of the rain forest are the lianas or climbing vines, epiphytes growing up in the trees, and stranglers that grow downward from the canopy to the ground. Many of the large trees develop buttresses for support. Horizontally, the rain forest is a mosaic of continually changing vegetation that adds to its diversity. Reflecting this vegetative stratification is the stratification of animal life into six pronounced feeding groups from the canopy down to the ground layer.

Tropical forests possess a large standing-crop biomass that ties up great quantities of nutrients. Much of the mineral cycling takes place between rapid decomposition of litter and the rapid uptake of the nutrients it contains. Roots are concentrated in the top layer and on the surface of the soil in close contact with the litter, where a symbiotic relationship between roots and mycorrhizae facilitates nutrient uptake. Functional relationships within the forest involve numerous mutualistic interactions for pollination and seed dispersal.

Humans have had a powerful impact on tropical forests, beginning with slash-and-burn cultivators who, as their population grew, cleared increasingly large amounts of forest. They were followed and accompanied by western agricultural developments that still involve massive clearing of forests for cattle ranches and rubber and palm plantations. Logging's worst effect is making remote areas accessible to settlers, who further clear the forests. The major impact of extensive tropical deforestation is the rapidly increasing loss of cultural and biological diversity.

Review and Study Questions

1. Name and characterize the types of tropical forests.
2. What are the major strata in the tropical forest?
3. What are emergents, lianas, and epiphytes? What are their positions in the rain forest?
4. What are buttresses?
5. What is crown shyness?
6. If tropical forest soils are so nutrient-poor, how can they support such a high plant biomass and diversity?
7. How are mutualistic interactions involved in the functioning of the tropical rain forest?
8. How do logging, shifting cultivation, cattle ranching, and other agricultural developments affect tropical rain forests?
9. What are the long-term effects of tropical deforestation?

*10. The media have focused on deforestation in the Amazon. Investigate the rate and consequences of deforestation in Central America, Sarawak, Borneo, or Thailand.

*11. Select a native people of South or Central America or the Ibans of Sarawak, and report on the impact of deforestation on them.

Selected References

Ashton, P. S. 1988. Dipterocarp biology as a window to understanding of tropical forest structure. *Ann. Rev. Ecol. Syst.* 19:347–370.

Bawa, K. S. 1990. Plant-pollinator interactions in tropical rain forests. *Ann. Rev. Ecol. Syst.* 21:399–422.

Buschbacker, R. J. 1986. Tropical deforestation and pasture development. *Bioscience* 36:22–28. Good review of the various aspects of the problem.

Colchester, M. 1989. *Pirates, squatters and poachers: The political ecology of the dispossession of the native peoples of Sarawak.* London: Survival International. Excellent review of the complex problems surrounding forestry, economic development, and welfare and rights of native peoples.

Collins, M., ed. 1990. *The last rain forest.* New York: Oxford University Press. The best general reference available; good atlas of tropical forest regions.

Denslow, J. S. 1987. Tropical rainforest gaps and tree species diversity. *Ann. Rev. Ecol. Syst.* 18:431–451.

Erwin, T. L. 1988. The tropical forest canopy: The heart of biotic diversity. In E. O. Wilson and F. M. Peters, eds., *Biodiversity.* Washington, DC: National Academy Press.

Fleming, T. H., R. Breitwisch, and G. H. Whitesides. 1987. Patterns of tropical vertebrate frugivore diversity. *Ann. Rev. Ecol. Syst.* 18:71–90. Concise review of tropical frugivory.

Furtado, J. I., and K. Ruddle. 1986. The future of tropical forests. In N. Polunin, ed., *Ecosystem theory and application.* New York: Wiley, pp. 145–171.

Golley, F. B., ed. 1983. *Tropical forest ecosystems: Structure and function.* Ecosystems of the World No. 14A. Amsterdam, Holland: Elsevier Scientific Publishing Company. An important reference.

Jansen, D. J. 1986. The future of tropical ecology. *Ann. Rev. Ecol. Syst.* 17:305–324. Discusses approaches needed to maintain the integrity of tropical ecosystems.

Jordan, C. F., and J. R. Kline. 1972. Mineral cycling: Some basic concepts and their application in a tropical rain forest. *Ann. Rev. Ecol. Syst.* 3:33–50.

Lathwell, D. J., and T. L. Grove. 1986. Soil-plant relations in the tropics. *Ann. Rev. Ecol. Syst.* 17:1–16.

Leigh, E. G., Jr. 1975. Structure and climate in tropical rain forest. *Ann. Rev. Ecol. Syst.* 6:67–86. Concise review of effects of climate on lowland and montane rain forests.

Martin, C. 1991. *The rainforests of West Africa.* New York: Birkhauser. A comprehensive treatment of the ecology and conservation of African rain forests.

Murphy, P. G., and A. E. Lugo. 1986. Ecology of tropical dry forests. *Ann. Rev. Ecol. Syst.* 17:67–88.

Myers, N. 1983. *A wealth of wild species.* Boulder, CO: Westview Press. Excellent reference on the value of tropical plants to human welfare.

Richards, P. W. 1952. *The tropical rain forest: An ecological study.* Cambridge, England: Cambridge University Press. (Reprint, 1972.) A classic reference.

Simpson, B. B., and J. Haffer. 1978. Spatial patterns in the Amazonian rain forest. *Ann. Rev. Ecol. Syst.* 9:497–518.

Sutton, S. L., T. C. Whitmore, and A. C. Chadwick, eds. 1983. *Tropical rain forests: Ecology and management.* Oxford, England: Blackwell.

Tomlinson, P. B. 1987. Architecture of tropical plants. *Ann. Rev. Ecol. Syst.* 18:1–21.

Walter, H. 1971. *Ecology of tropical and subtropical vegetation.* Edinburgh: Oliver & Boyd.

Whitmore, T. C. 1984. *Tropical rainforests of the Far East.* Oxford: Clarendon Press. The authoritative reference on Southeast Asian rain forests.

Wilson, E. O., Jr., ed. 1988. *Biodiversity.* Washington, DC: National Academy Press. Contains information section on preserving biological diversity.

CHAPTER 31

Lakes and Ponds

Outline

Objectives

On completion of this chapter, you should be able to:

1. Describe the seasonal stratification of light, temperature, and oxygen in lakes and ponds.
2. Explain how this stratification comes about.
3. Discuss horizontal and vertical zonation of life in lakes and ponds.
4. Describe the mechanisms of energy flow and nutrient cycles in lakes.
5. Distinguish among eutrophic, oligotrophic, and dystrophic lakes.
6. Discuss the impact of humans on pond and lake ecosystems.

No feature in the landscape attracts more interest than ponds and lakes. We are drawn toward them as if pulled by a magnet. We seek them out for beauty and for recreation, and we like to live or camp along their shores. Yet how many of us give any thought to the life they hold and how it functions beneath the surface, or how lakes came about in the first place, or how our intrusion affects them?

Lakes and ponds are inland depressions containing standing water (Figure 31.1). They vary in depth from one meter to over 2000 meters. They range in size from small ponds of less than a hectare to large seas covering thousands of square kilometers. Small bodies of water so shallow that rooted plants can grow over much of the bottom are ponds. Some lakes are so large that they mimic marine environments. Most ponds and lakes have outlet streams; and both may be more or less temporary features on the landscape, geologically speaking.

Some lakes formed by glacial erosion and deposition. Glacial abrasion of slopes in high mountain valleys carved basins, which filled with water from rain and melting snow to form tarns. Retreating valley glaciers left behind crescent-shaped ridges of rock debris that dammed up water behind them. Numerous shallow kettle lakes and potholes were left behind by glaciers that covered much of northern North America and northern Eurasia.

Lakes also form when silt, driftwood, and other debris deposited in beds of slow-moving streams dam up water behind them. Loops of streams that meander over flat valleys and floodplains often become cut off, forming crescent-shaped oxbow lakes.

Shifts in Earth's crust, either by the uplifting of mountains or the displacement of rock strata, sometimes develop water-filled depressions. Craters of some extinct volcanoes have also developed into lakes. Landslides block off streams and valleys to form new lakes and ponds. In any given area all natural ponds and lakes have the same geological origin and similar characteristics; but because of varying depths at time of origin, they may represent several stages of development.

Many lakes and ponds are formed through nongeological activity. Beavers dam streams to make shallow but often extensive ponds. Humans create huge lakes by damming rivers and streams for power, irrigation, or water storage (see Chapter 33) and construct smaller pounds and marshes for water, fishing, and wildlife. Quarries and strip mines form other ponds.

PHYSICAL CHARACTERISTICS

Unlike most terrestrial ecosystems, lakes and ponds have well-defined boundaries—the shoreline, sides of the basin, surface of the water, and bottom sediments. Within these boundaries, environmental conditions vary from one pond or lake to another. However, all still-water ecosystems share certain characteristics. For one, life in still-water ecosystems depends on light. The amount of light penetrating the water is influenced not only by its natural attenuation in clear water (see Chapter 7), but also by turbidity from silt and other material carried into the lake and from the growth of phytoplankton. Temperatures vary seasonally and with depth. Oxygen can be limiting, especially in summer, because only a small proportion of the total volume of water is in direct contact with oxygen-rejuvenating air and the process of decomposition on the bottom consumes it. These variations in oxygen, temperature, and light strongly influence the distribution and adaptations of life in lakes and ponds.

Life in lakes and larger ponds experiences seasonal shifts, prompted by changes in temperature and oxygen throughout the basin brought about by the heating and cooling of surface waters. In late spring and early summer both increasingly direct solar radiation and warming air temperatures heat the surface water faster than the deep water. Because water reaches its maximum density at 4° C (see Chapter 6), the surface water becomes lighter as its temperature increases. Soon a layer of lighter, warm water, called the *epilimnion,* rests on top of a heavier mixed layer of cooler water (Figure 31.2). This layer, known as the metalimnion, becomes cooler with depth. For approximately every meter downward, the temperature declines 1° C, a drop called the *thermocline.* (If you have dived into deep water you must have become suddenly aware of the thermocline.) When the temperature of the water reaches 4° C and its greatest density, it lies as a layer of cold water on the bottom called the *hypolimnion.* The ther-

(a)

(b)

(c)

FIGURE 31.1 Lakes and ponds fill basins or depressions in the land. (a) A glacial-formed lake in a North Dakota prairie. (b) A beaver-constructed pond in a wooded landscape. (c) A human-constructed old New England mill pond. Note the floating vegetation.

mocline acts as a barrier between the epilimnion and hypolimnion. The lake basin is much like a sandwich, with the epilimnion and hypolimnion forming the top and bottom of the roll and the thermocline the filling in between. The filling is thick enough to prevent any contact between the top and bottom water, and little circulation takes place.

Oxygen produced by phytoplankton and the action of wind keep the upper layer of water aerated. The waters below, however, may be deficient in oxygen,

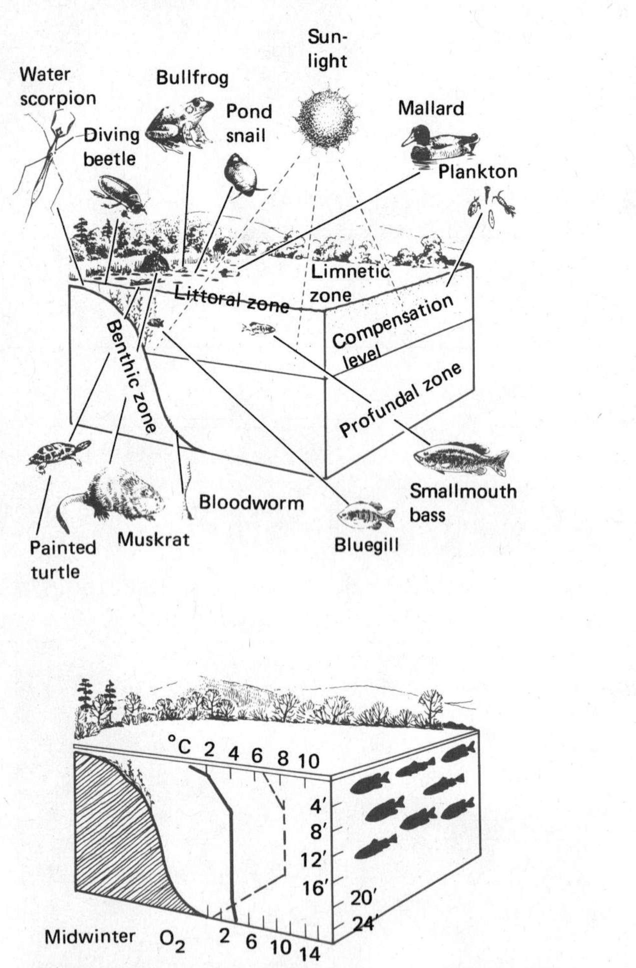

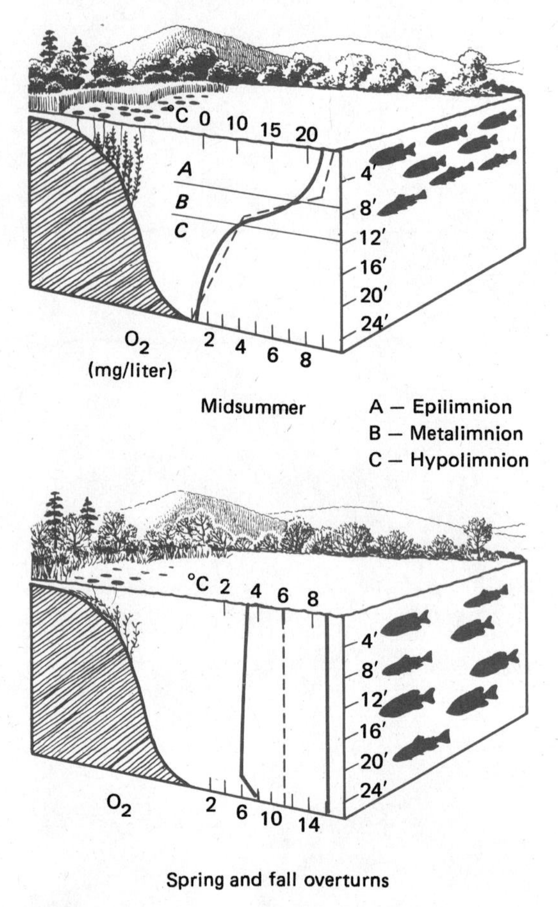

FIGURE 31.2 Seasonal variations in the stratification of oxygen and temperature and the distribution of aquatic life in lake ecosystems. The diagram on the upper left shows the major zones—littoral, limnetic, profundal, and benthic. The compensation level is the depth at which light becomes too low for photosynthesis. Surrounding the lake is a variety of organisms typical of a lake community. The other diagrams show how distribution of oxygen (dashed line) and temperature (solid line) in a lake during the different seasons affects the distribution of fish life. The narrow fish silhouettes represent trout, or cold water species. The wider silhouettes are bass, or warm water species. Note the pronounced horizontal stratification in midsummer and the nearly vertical oxygen and temperature curves during the spring and fall overturns.

consumed by decomposers (Figure 31.3). Because sediments accumulate on the bottom, the deep waters are relatively high in nutrients, but those nutrients are not available to phytoplankton in the upper layer. Deprived of the riches below, phytoplankton may experience nutrient depletion late in summer.

By fall, conditions begin to change, and a turnabout takes place. Air temperatures and solar insolation decrease, and the surface water starts to cool. As it does, the water becomes more dense and sinks, lifting the warmer water below to the surface, where it cools in turn. This cooling continues until the temperature is

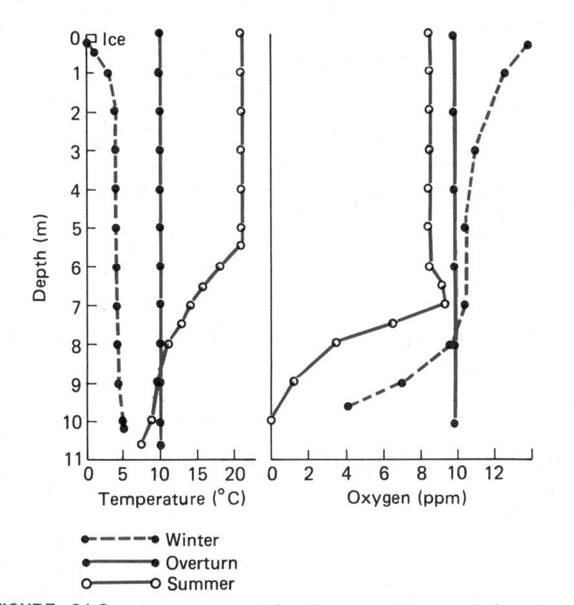

FIGURE 31.3 Oxygen stratification in Mirror Lake, New Hampshire in winter (January), summer (August), and late fall (November). The late fall overturn results in both a uniform temperature and uniform distribution of oxygen throughout the lake basin. In summer a pronounced stratification of both temperature and oxygen exists. Oxygen declines sharply in the thermocline and is nonexistent on the bottom, because of decomposition taking place in the sediments. In winter oxygen is also stratified, but it is present at a low concentration in deep water.

uniform throughout the basin (Figure 31.3). Now pond and lake water circulate throughout the basin and the nutrients denied the phytoplankton in late summer are at the surface. This circulation, which recharges oxygen and nutrients through the basin, is called the fall overturn. Through the stirring action of the wind, the overturn may last until ice forms.

Then comes winter, and the surface water cools to below 4° C. It becomes lighter again and remains on the surface. (Remember, water becomes lighter above and below that temperature.) If the winter is cold enough, surface water freezes; otherwise it remains close to 0°. Now the warmest place in the pond or lake is on the bottom. A slight inversion temperature stratification develops, in which the water becomes warmer, up to 4° C with depth. The water beneath the ice may be warmed by solar radiation through the ice. Because that increases its density, this water flows to the bottom, where it mixes with water warmed by heat conducted from the bottom mud. The result is a higher temperature on the bottom, although the overall stability of the water is undisturbed.

With the spring breakup of ice and the heating of the surface water up to 4° C another overturn, the spring one, occurs. Again oxygen and nutrients are recharged throughout the basin. The surface waters are now both nutrient and oxygen rich, ready for the spring growth of phytoplankton. As the season wears on, the lake water again become stratified into the three familiar layers.

Not all lakes experience such seasonal changes in stratification, and you should not consider this phenomenon as characteristic of all deep bodies of water. In shallow lakes and ponds, temporary stratification of short duration may occur; in others stratification may exist, but no thermocline develops. In some very deep lakes, the thermocline may simply descend during periods of overturn and not disappear at all. In such lakes the bottom water never becomes mixed with the top layer. However, some form of thermal stratification occurs in all very deep lakes, including those of the tropics.

STRUCTURE

Ponds and lakes may be divided into both vertical and horizontal strata based on penetration of light and photosynthetic activity. The horizontal zones are obvious to the eye; the vertical one, influenced by depth of light penetration, is not (Figure 31.1). Surrounding most lakes and ponds and engulfing some ponds completely is the *littoral zone* or shallow-water zone in which light reaches the bottom, stimulating the growth of rooted plants. Beyond the littoral is open water, the *limnetic zone,* which extends to the depth of light penetration. It is inhabited by plant and animal plankton and *nekton,* free-swimming organisms such as fish that can move about freely. Beyond the depth of effective light penetration is the *profundal zone.* Its beginning is marked by the *compensation level* of light, the point at which respiration balances photosynthesis. The profundal zone depends on a rain of organic material from the limnetic as its energy source. Common to both the littoral and profundal zones is the third vertical stratum, the *benthic zone* or bottom region, which is the place of decomposition. Although these zones are named and often described separately, all are closely dependent on one another in the dynamics of lake ecosystems.

LITTORAL ZONE

Aquatic life is richest and most abundant in the shallow water about the edges and in other places within lakes and ponds where sediments have accumulated on the bottom, descreasing water depth. Dominating these areas is emergent vegetation, plants whose roots are anchored in the bottom mud, whose lower stems are immersed in water, and whose upper stems and leaves stand above water (Figure 31.4). The distribution and variety of plants vary with water depth and fluctuation of water levels. Very shallow depths support spike rushes and small sedges; deeper water is occupied by plants with narrow tubular or linear leaves, such as bulrushes, reeds, and cattails. With them are associated such broadleaf emergents as pickerelweed (*Pontederia* spp.) and arrowhead (*Sagittaria* spp.). Beyond the emergents and occupying even deeper water is a zone of floating plants such as pondweed *(Potamogeton)* and pond lily (*Nuphar* spp.). Many of these floating plants have poorly developed root systems but highly developed aerating systems. In depths too great for floating plants live submerged plants, such as certain species of pondweed. Lacking cuticles, these plants absorb nutrients and gases directly from the water through thin and finely dissected or ribbonlike leaves. Associated with the emergents and floating plants is a rich community of organisms, among them hydras, snails, protozoans, and sponges. Insects include dragonflies and diving insects such as water boatmen and diving beetles that carry a bubble of air with them when they go underwater in search of prey. Fish such as pickerel and sunfish find shelter, food, and protection among the emergent and floating plants. Fish of lakes and ponds lack strong lateral muscles characteristic of fish living in swift water, and some, such as sunfish, have compressed bodies that permit them to move with ease through the masses of aquatic plants. The littoral zone contributes heavily to the large input of organic matter into the system.

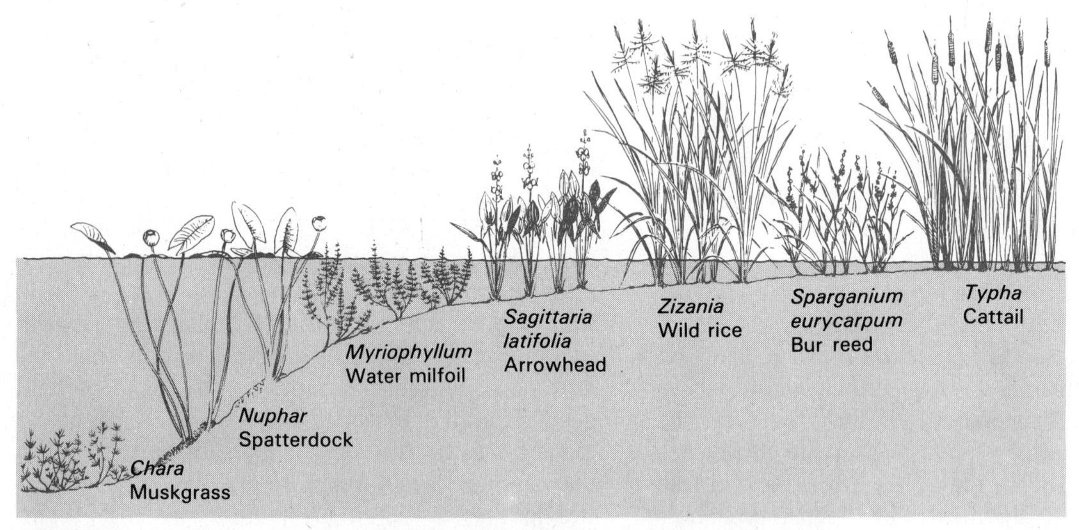

FIGURE 31.4 Zonation of emergent, floating, and submerged vegetation at the edge of a lake or pond. Note the changes in vegetation with water depth. Such zonation does not necessarily reflect successional stages, as is often inferred, but rather water depth.

LIMNETIC ZONE

People usually associate the open water of a lake with fish; but the main forms of life in the limnetic zone are not the large obvious creatures, but minute plant and animal organisms—phytoplankton and zooplankton. Because the tiny plants that make up the phytoplankton, including desmids, diatoms, and filamentous algae, carry on photosynthesis in open water, they are the base on which the rest of limnetic depends. Their presence is most obvious during bloom, the season when the populations of these plants are most abundant. Suspended with the phytoplankton are small animals, mostly tiny crustaceans, that graze on minute plants. These animals form an important link in energy flow in the limnetic zone.

Light sets the lower limit at which phytoplankton can exist, so populations of these plants are concentrated in the epilimnion. Because the zooplankton feeds on these minute plants, it too is concentrated in the limnetic zone. By its own growth phytoplankton limits light penetration into the water and through the summer reduces the depth at which it can live. As the zone becomes more shallow, phytoplankton can absorb more light, increasing organic production.

Within the limits of light penetration, the depth at which various species of phytoplankton can live is influenced by optimum conditions for their development. Some phytoplankton species live just below the surface; others are more abundant a few feet beneath; and those requiring colder temperatures live deeper still. Cold-water plankton, in fact, is restricted to lakes in which phytoplankton growth is scarce in the epilimnion and in which the oxygen content of the deep water is not depleted by decomposition of organic matter.

Animal plankton may be seasonally stratified because it is capable of independent movement. In winter some animal plankton species distribute themselves evenly to considerable depths; in summer they concentrate in layers most favorable to them and to their stages of development. At that season zooplankton undertake a vertical migration during some part of the 24-hour period. Depending on the species, they spend the night or day in the deep water or on the bottom and move up to the surface during the alternate period to feed on phytoplankton (Figure 31.5).

During the spring and fall overturns, plankton is carried downward, but at the same time nutrients released by decomposition on the bottom are carried upward to the impoverished surface layers. In spring when surface waters warm and stratification develops, phytoplankton has access to both nutrients and light. A spring bloom develops, followed by a rapid depletion of nutrients and a reduction in planktonic populations, especially in shallow water.

Fish make up most of the nekton in the limnetic zone. Their distribution is influenced mostly by food supply, oxygen, and temperature. During the summer largemouth bass, pike, and muskellunge inhabit the warmer epilimnion waters, where food is abundant. In winter they retreat to deeper water. Lake trout, on the other hand, move to greater depths as summer advances. During the spring and fall overturn, when oxygen and temperature are fairly uniform throughout, both warm-water and cold-water species occupy all levels.

PROFUNDAL ZONE

The profundal zone depends not only on the supply of energy and nutrients from the limnetic zone above but also on the temperature and availability of oxygen. In highly productive waters, oxygen may be very limiting because the decomposer organisms so deplete it that little aerobic life can survive. The profundal zone of a deep lake is much larger in proportion to total volume, so production of the epilimnion is relatively low, and decomposition does not deplete the oxygen. In these lakes the profundal zone supports some life, particularly fish, some plankton, and such organisms as certain cladocerans that live in the bottom ooze. Some zooplankton may occupy this zone during some part of the day, but migrate up to the surface to feed. Only during spring and fall overturns, when organisms from the upper layers enter this zone, is life abundant in profundal waters.

Easily decomposed substances drifting down through the profundal zone are partly mineralized while sinking. The remaining organic debris—dead bodies of plants and animals of the open water, and decomposing plant matter from shallow-water areas—

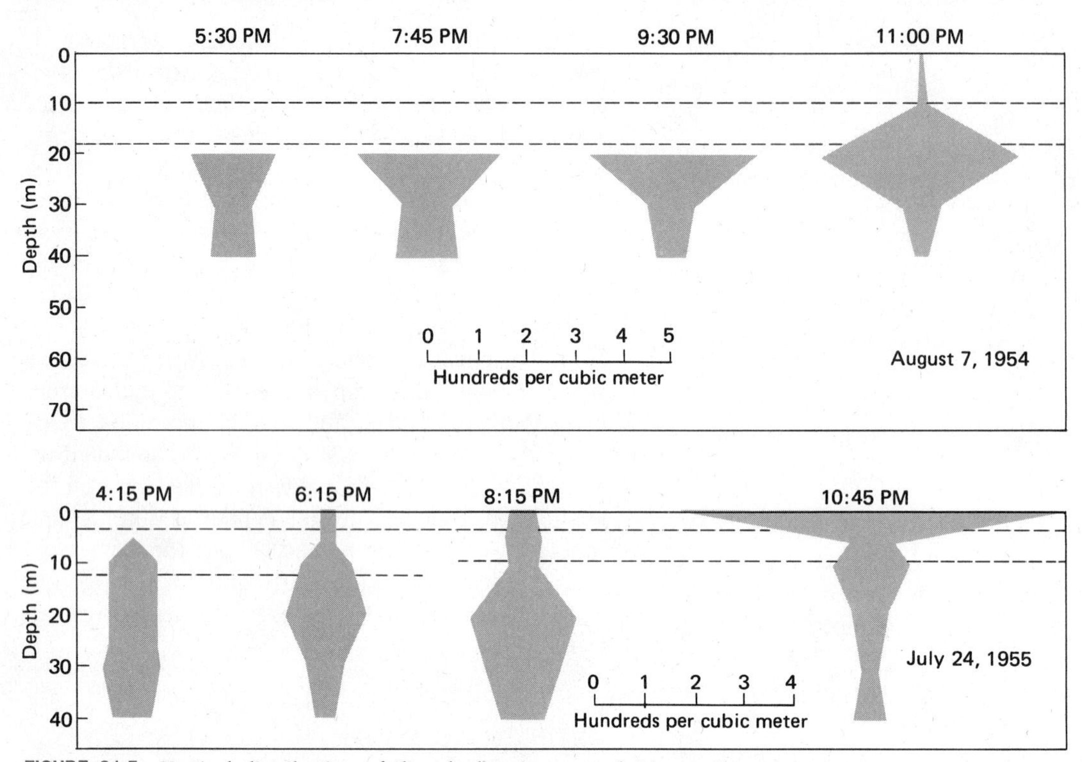

FIGURE 31.5 Vertical distribution of the planktonic copepod *Limnocalanus marcurus* on two midsummer days. The maximum number reach the surface 1.5 to 4 hours before sunset. This organism inhabits deeper water. The dashed lines represent the metalimnion.

settles on the bottom. Together with quantities of material washed in by inflowing water, they make up the bottom sediments, the habitat of benthic organisms.

BENTHIC ZONE

The bottom ooze is a region of great biological activity, so great in fact that the oxygen curves for lakes and ponds show a sharp drop in the profundal water just above the bottom (Figure 31.3). Because the organic muck is so low in oxygen, the dominant organisms there are anaerobic bacteria. Under anaerobic conditions, however, decomposition cannot proceed to inorganic end products. When the amounts of organic matter reaching the bottom are greater than can be utilized by bottom fauna, odoriferous muck rich in hydrogen sulfide and methane results. Thus lakes and ponds with highly productive limnetic and littoral zones have an impoverished fauna on the profundal bottom. Life in the bottom ooze is most abundant in the lakes with a deep hypolimnion in which some oxygen is still available.

As the water becomes more shallow, the benthos changes. The action of water, plant growth, drift materials, and recent organic deposits modify the bottom material—stones, rubble, gravel, marl, and clay. Increased oxygen, light, and food encourage a richness of life not found on the profundal bottom.

Closely associated with the benthic community are organisms collectively called *periphyton* or *aufwuchs*. They are attached to or move on a submerged substrate, but do not penetrate it. Small aufwuchs communities colonize the leaves of submerged aquatics, sticks, rocks, and other debris. Periphyton, mostly algae and diatoms, living on plants are fast growing and lightly attached. Because the substrate is so short-lived, the associated periphyton rarely lives for more than one summer. Aufwuchs on stones, wood, and debris form a more crustlike growth of bluegreen algae, diatoms, water moss, and sponges.

FUNCTION

Lakes and ponds appear to be self-contained ecosystems, lying in basins surrounded by totally different terrestrial ecosystems. Nevertheless they are strongly influenced by inputs of materials from surrounding terrestrial ecosystems and other sources outside the basin (Figure 31.6). Nutrients and other substances move across the boundaries along biological, geological, meteorological, and hydrological pathways.

Wind-borne particulate matter, dissolved substances in rain and snow, and atmospheric gases make up meteorological inputs. Outputs along the same pathway are small, mainly spray aerosols and gases such as carbon dioxide and methane. Geological inputs include nutrients dissolved in groundwater and inflowing streams and particulate matter flowing into the basin from the surrounding watershed. Geological outputs include dissolved and particulate matter carried out of the lake by outflowing waters and nutrients buried in

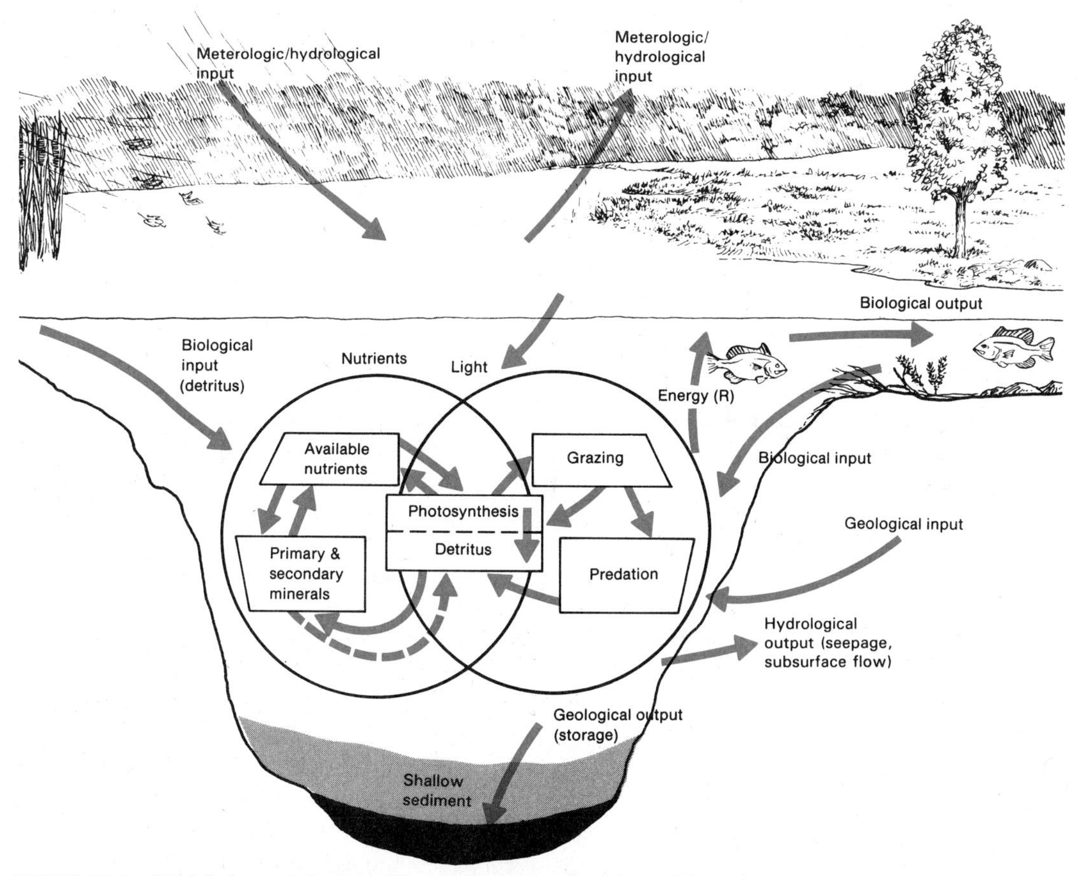

FIGURE 31.6 Model for nutrient cycling and energy flow in a lake ecosystem. Meteorological, geological, and biological inputs enter the lentic system from the watershed that contains it. The nutrients and energy entering and generated within the system move through a number of pathways. Part of the nutrients and energy fixed accumulate in bottom sediments.

deep sediments that are removed from circulation for a long time. Biological inputs and outputs are relatively small, mostly animals such as fish that move in and out of the lake. Hydrological inputs involve precipitation and drainage of surface waters. Outputs include seepage through the walls of the lake basin, subsurface flows, and evaporation. Nutrients and energy move through lakes and ponds by the way of grazing and detrital food chains.

Although studies of lake metabolism have emphasized the phytoplankton-zooplankton grazing food chain, in reality lakes, like terrestrial communities, are dominated by the detrital food chain. Most of this detritus (which is all dead organic carbon) comes from the littoral zone. In total, however, detritus includes particulate and dissolved organic carbon (POC and DOC) that comes from external sources and cycles within the system, and organic matter lost to a particular trophic level through egestion, excretion, and secretion.

Lake ecosystems function mostly within a framework of organic carbon transfer. The central pool, which comes from both internal and external sources, represents the major flow through the system. Particulate organic carbon comes from three sources: (1) imports into the system from the outside; (2) the littoral zone; (3) the lentic zone. Most detrital metabolism takes place in the open-water zone during sedimentation and in the benthic zone, where particulate matter is decomposed.

Phytoplankton contributes the primary production in the limnetic zone, and macrophytes provide the same in the littoral zone. The contribution each makes varies among lentic systems. Availability of nutrients in the water influences phytoplankton production. If nutrients are not limiting and the only losses are respiratory, the rate of net photosynthesis and biomass accumulation is high. In fact, a linear relationship exists between phytoplankton production and phytoplankton biomass. However, as phytoplankton biomass increases, shading also increases, which reduces net photosynthesis and increases respiration. As a result production declines. When nutrients are low, respiration and mortality increase, reducing net photosynthesis and thus biomass. However, if zooplankton grazing and bacterial decomposition are high, nutrients recycle rapidly, resulting in a high rate of net photosynthesis even though the concentration of nutrients and biomass accumulation is low.

Macrophytes also contribute heavily to lake production. The ratio of macrophytic production in the littoral to microphytic production in the limnetic is influenced by the fertility of the lake. Highly fertile lakes support a heavy growth of phytoplankton that shades out macrophytes and reduces their contribution. In less fertile lakes where phytoplankton production is low, light penetrates much deeper into the water and rooted aquatics grow. Macrophytes are little affected by nutrient availability in the open water because they draw their nutrients from the bottom sediments.

Nutrient transfers within lentic ecosystems take place largely between the water column and sediments. They involve uptake by phytoplankton, zooplankton, bacteria, and other consumers as well as sedimentation in both the water column and benthic muds (Figure 31.7). In spring when phytoplankton bloom is at its height, nitrogen and phosphorus become depleted in the limnetic zone, because of the high rates of photosynthesis, sinking of dead phytoplankton, and sedimentation. At the same time decomposition decreases particulate N and P. This action increases dissolved P, but the dissolved N is lost through denitrification.

In summer conditions change. Because of a decline in phytoplankton in the limnetic zone and a slower sinking rate, as much N and P enter solution as are taken up by phytoplankton in photosynthesis. N and P increase in the dissolved and particulate pools and in bottom sediments. Phosphorus, in particular, becomes trapped in the hypolimnion, unavailable to phytoplankton until the fall overturn.

Macrophytes, however, can change this situation by moving phosphorus from sediments to the water column and on to phytoplankton, at the same time building up more bottom sediments. Steven Carpenter found that in Lake Wingra, Wisconsin, macrophytes significantly increased the amount of phosphorus available to phytoplankton that it otherwise would have to obtain from direct release from the sediment. Macrophytes obtained 73 percent of the phosphorus incorporated in shoot tissue from sediments. Eventually some 550 kg of P from this source became available to phytoplankton, compared to 470 kg available from sed-

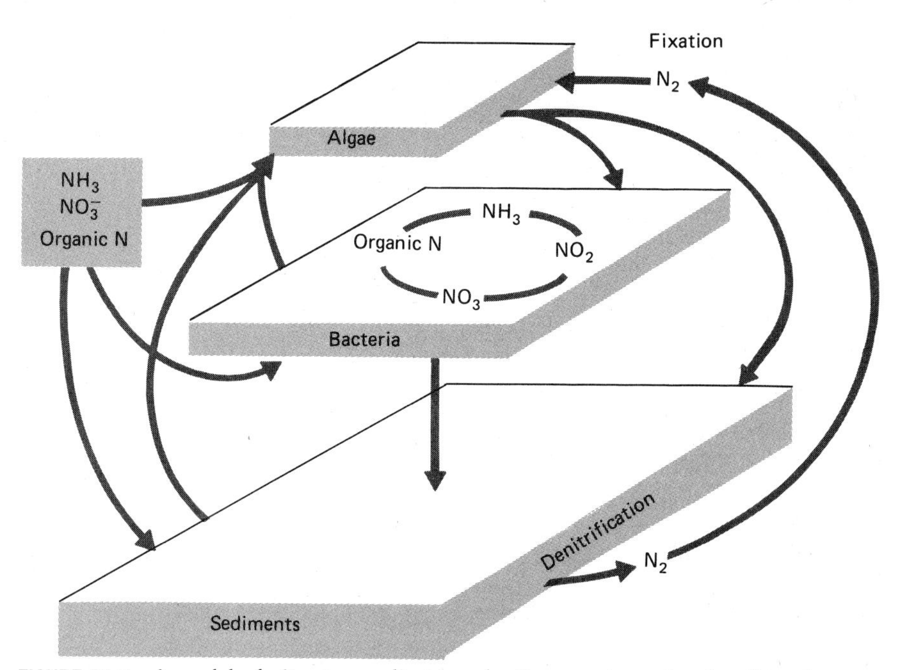

FIGURE 31.7 A model of nitrogen cycling in a lentic ecosystem, showing the relationship between water column and sediments. Sediments are both a storehouse and a source of nutrients in the lentic system.

iments through overturns and disturbance and resuspension of sediments. This uptake stimulates the production of both macrophytic and phytoplankton biomass, adding to sediment accumulation on the bottom. This buildup of sediment creates new areas for colonization and provides additional phosphorus for phytoplankton. Thus rooted aquatic plants enhance the recycling of phosphorus by mobilizing it from the sediments. Such activities accelerate the nutrient enrichment of lakes.

Although the productivity of a lake is related to the nutrient richness of its waters, other internal forces influence it. Any two lakes with a similar nutrient load may differ in productivity. Various species of phytoplankton differ in their rates of metabolic activity and nutrient recycling, both size-dependent. Small phytoplankters, for example, have higher maximum growth rates overall, higher maximum growth at low levels of nutrients, and a lower sinking rate than large phytoplankton species.

Feeding on phytoplankters are the zooplankton, essential to the recycling of nutrients, particularly N and P (see Chapter 24). Various species and sizes of zooplankton graze on different sizes of phytoplankton. Depending on the size relationship of the dominant zooplankters, they influence the species composition and size structure of the phytoplankton community. Zooplankters, in turn, are eaten by invertebrate planktivores (insect larvae and crustaceans) and vertebrate planktivores (minnows and small spiny fish). These predators, too, are size-dependent in their food selection (see Chapter 16). The vertebrate planktivores may eat their associated invertebrate plantivores as well. The vertebrate planktivores, in turn become prey for fish-eating predators (piscivores).

Interactions among these feeding groups flow down through the food web, influencing productivity at each trophic level. A rise in the biomass of predatory fish (bass, pike, trout) can result in changes in density, species composition, and behavior of zooplanktivorous fish. That relationship, in turn, can affect invertebrate planktivores. Vertebrate planktivores, taking the largest available prey, reduce the density of large zooplankters, forcing invertebrate planktivores to select smaller species. Any changes in the relative densities of these two groups of planktivores influence the structure and density of zooplankton, thus affecting grazing intensity and rates of nutrient recycling. With fewer herbivorous grazers, phytoplankton increases. Each change in biomass at one trophic level has the opposite response at the next trophic level. Throughout the food web, however, maximum production is achieved at intermediate levels of density. Such trophic interactions influence and regulate productivity of lake ecosystems.

NUTRIENT STATUS

A close relationship exists between land and water ecosystems. Primarily through the hydrological cycle one feeds on the other. The water that falls on land runs from the surface or moves through the soil to enter streams, springs, and eventually lakes. The water carries with it silt and nutrients in solution, all of which enrich aquatic ecosystems. Human activities, including road construction, logging, mining, construction, and agriculture, add an additional heavy load of silt and nutrients, especially nitrogen, phosphorus, and organic matter. The outcome of these inputs is nutrient enrichment of aquatic systems. This enrichment is termed *eutrophication.*

The term *eutrophy* (from the Greek *eutrophos,* "well nourished") means a condition of being nutrient-rich. The opposite of eutrophy is *oligotrophy,* the condition of being nutrient-poor. The terms were first introduced by the German limnologist C. A. Weber in 1907 when he applied the terms to the development of peat bogs. E. Naumann later associated the terms with phytoplankton production in lakes: Eutrophic lakes found in fertile lowland regions hold high populations of phytoplankton; oligotrophic lakes, common to regions of primary rocks, contain little plankton. This concept of oligotrophy and eutrophy ignores the input of of highly productive littoral zones.

EUTROPHIC SYSTEMS

A typical eutrophic lake (Figure 31.8) has a high surface-to-volume ratio; that is, the surface area is large relative to depth. It has an abundance of nutrients, especially nitrogen and phosphorus, that stimulate a heavy growth of algae and other aquatic plants. Increased photosynthetic production leads to an increased regeneration of nutrients and organic compounds, stimulating even further growth. Phytoplankton becomes concentrated in the upper layer of the water, giving it a murky green cast. The turbidity reduces light penetration and restricts biological productivity to a narrow layer of surface water. Algae, inflowing organic debris and sediment, and remains of rooted plants drift to the bottom, adding to the highly organic sediments. Bacteria partially convert this dead organic matter into inorganic substances. The activities of these decomposers deplete the oxygen supply of the bottom sediments and deep water to the point where this region of the lake cannot support aerobic life. The number of species declines, although the biomass and numbers of organisms remain high. As the basin continues to fill, the volume decreases and the resulting shallowness speeds the cycling of available nutrients and further increases plant production. Positive feedback carries the lake or pond to extinction—filling in the basin and developing a marsh or swamp, and ultimately a terrestrial community.

OLIGOTROPHIC SYSTEMS

Oligotrophic lakes have a low surface-to-volume ratio, water that is clear and appears blue to blue-green in the sunlight, bottom sediments that are largely inorganic, and a high oxygen concentration throughout the hypolimnion. Nutrient content of the water, however, is low; and although nitrogen may be abundant, phosphorus is highly limited. Low input of nutrients from surrounding terrestrial ecosystems and other external sources are mostly responsible for this condition. Low availability of nutrients causes a low production of organic matter, particularly phytoplankton. Low organic matter production leaves little for decomposers, so oxygen concentration remains high in the hypolimnion. These oxidizing conditions are responsible for the low release of nutrients from the sediment. The lack of decomposable organic matter means low bacterial populations and slow rates of microbial metabolism. Although the numbers of organisms in oligotrophic lakes and ponds may be low, species diversity

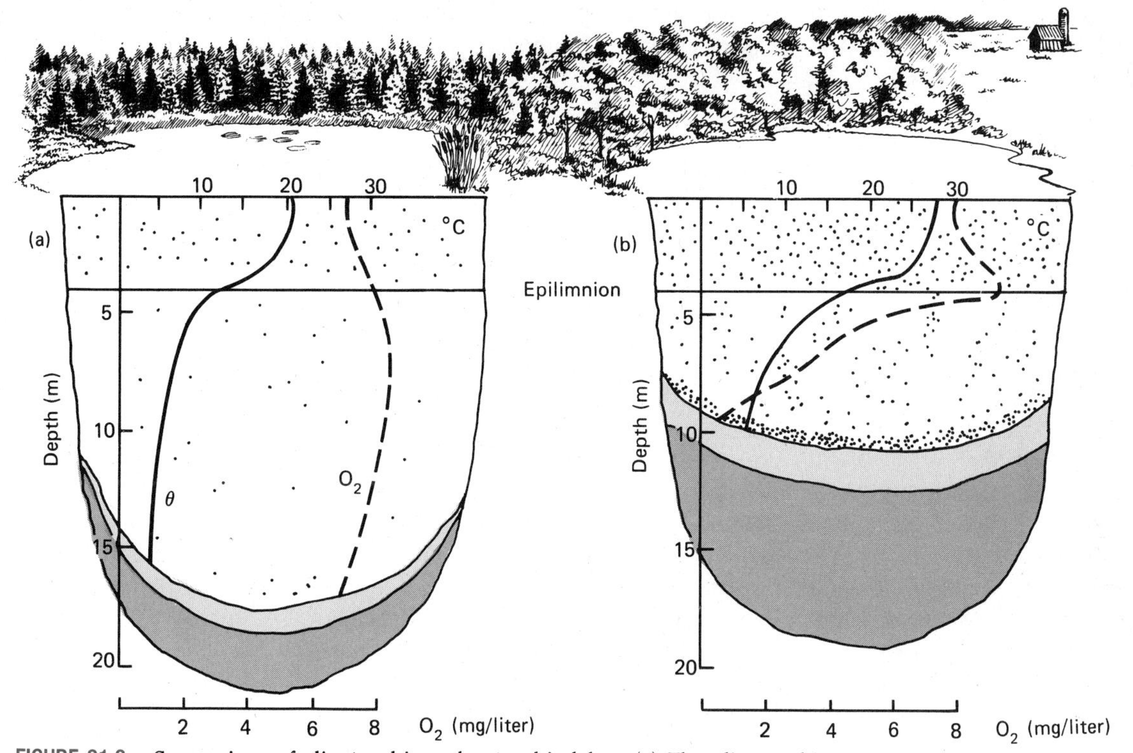

FIGURE 31.8 Comparison of oligotrophic and eutrophic lakes. (a) The oligotrophic lake is deep and has relatively cool water in the epilimnion. The hypolimnion is well supplied with oxygen. Organic matter that drifts to the bottom falls through a relatively large volume of water. The watershed surrounding the lake is largely oligotrophic, dominated by coniferous forests on thin and acid soil. (b) The eutrophic lake is shallow and warm, and oxygen is nearly depleted in the deeper water. The amount of organic detritus is large in relation to the volume of water. The watershed surrounding the lake is eutrophic, consisting of a nutrient-rich deciduous forest and farmland.

is often high. Fish life is dominated by members of the salmon family.

DYSTROPHIC SYSTEMS

Lakes that receive large amounts of organic matter from surrounding land, particularly in the form of humic materials that stain the water brown, are called *dystrophic*. Although the productivity of dystrophic lakes is considered low, they are low only in planktonic production. Dystrophic lakes generally have highly productive littoral zones, particular those that develop bog flora. This littoral vegetation dominates the metabolism of the lake, providing a source of both dissolved and particulate organic matter.

HUMAN IMPACT

A pristine lake is one of the ecosystems highly vulnerable to human-induced changes. Its end begins when the first house appears on its shores or when it is opened to recreational use, however limited. Any activity and settlement begins to unravel the fabric of the lake ecosystem. As more people are attracted to the lake, developers move in. The shore is parceled into lots, cottages and residences spring up, boats stir up the otherwise placid waters, and nutrient-rich seepages from septic tanks and pesticides find their way into the lake. Before long the entire structure of life changes by degrees.

Consider an oligotrophic lake whose shoreline is newly developed. Drainage from the development moves into the lake. When nutrients in moderate amounts are added to this oligotrophic lake, they are taken up rapidly and circulated. It so happens that tissues of aquatic algae growing there in low populations contain phosphorus, nitrogen, and carbon in the ratios of 1 P : 7 N : 40 C per 500 g wet weight. If nitrogen and carbon are in excess and phosphorus is limiting, the addition of phosphorus will stimulate algal growth; if nitrogen is limiting, the addition of that nutrient will do the same. In most oligotrophic lakes, phosphorus rather than nitrogen is limiting. Because of its low ratio, 1 P per 500 g, an addition of even a moderate amount of phosphorus generates considerable growth of algae. As increasing quantities of nutrients are added to a lake or pond, it begins to change from oligotrophic to *metatrophic* (having a moderate amount of nutrients) to eutrophic. This change has been happening to clear oligotrophic lakes around the world at an increasing rate.

In fact, this galloping eutrophication has been changing naturally eutrophic lakes into *hypertrophic* ones. An excessive nutrient content results from a heavy influx of wastes, raw sewage, drainage from agricultural lands, river basin development, runoff from urban areas, and burning of fossil fuels. This accelerated enrichment, which results in chemical and environmental changes and causes major shifts in plant and animal life, has been called *cultural eutrophication.*

Destroying floating and emergent vegetation around lakes is a proliferation of permanent and summer homes and marinas along the shores. Wakes created by motorboating disturb littoral vegetation and birds that nest within it, notably loons on northern lakes. Motorboats discharge an oily mixture with gas exhausts beneath the surface of the water, where it escapes immediate detection. One gallon of oil per million of water imparts an odor to lake water; eight gallons per million taint fish. These oily discharges can lower oxygen levels and adversely affect the growth and longevity of fish.

The problems are more than just excessive quantities of nutrients flowing into the lake. Added to them are life-threatening, organism-deforming pesticides feeding in from surrounding farmland, suburban lawns, and golf courses. Toxic wastes enter from lakeshore industries. Some areas of the Great Lakes have become so contaminated with these wastes that fish taken from them are unfit for human consumption. Silt from construction, road building, logging, and other sources fills in the lake shores and destroys littoral vegetation. Many lakes are suffering from acid deposition (see Chapter 25). Already over 10 percent of northern lakes are so acidified they cannot support fish life and their associated invertebrate fauna.

Pollution, overfishing, and introduction of exotic species, accidentally or on purpose, into lakes and larger ponds have upset the original, natural assemblage of species, changed the nature of the animal community, and the stucture of food webs. So badly have many lakes been modified that even with the best of pollution controls and other restoration efforts, they cannot be brought back to their original condition, although they can be greatly improved.

Because of the great amount of water they contain and the volume of water flowing into them, some lakes have been exploited for urban consumption and irrigation. Because water from the rivers flowing into the Aral Sea in the southern USSR has been diverted for irrigation, the Aral Sea has dropped 9 meters and is expected to drop another 8 to 10 m, which would reduce its volume by one-half. The surrounding shoreline and exposed lake bottom are nearly desert, and a thriving fishery industry has been destroyed. On a smaller scale, diversion of water from Mono Lake in California has shrunk its surface area by one-third, increased its salinity, and is threatening the future of resident and migratory wildlife that depend on it.

Summary

Lakes and ponds comprise lentic ecosystems, standing bodies of water that fill a depression in the landscape. Geologically speaking, lakes and ponds are ephemeral features. In time they fill in, grow smaller, and may finally be replaced by a terrestrial community.

A nearly self-contained ecosystem, a lake exhibits gradients that are seasonally stratified in light, temperature, and dissolved gases. In summer the lake has a surface layer of warm, circulating water, the epilimnion; a middle zone, the thermocline or metalimnion, in which the temperature rapidly drops; and a bottom stratum, the hypolimnion, a layer of denser water approximately 4° C, often low in oxygen and high in carbon dioxide. When the surface waters cool in the fall, the temperature becomes uniform throughout the basin and the water circulates throughout the lake. A similar mixing of water takes place in the spring, when the lake warms. These seasonal overturns are important in recirculating nutrients and mixing the bottom water with the top. The area where light penetrates to the bottom of the lake, a zone called the littoral, is occupied by rooted plants. Beyond this zone is the open-water or limnetic zone, inhabited by plant and animal plankton and fish. Below the depth of effective light penetration is the profundal region, where the diversity of life varies with temperature and oxygen supply. The bottom or benthic zone is a place of intense biological activity, for here decomposition of organic matter takes place. Anaerobic bacteria are dominant on the bottom beneath the profundal water, whereas the benthic zone of the littoral is rich in decomposer organisms and detritus feeders. Although lake ecosystems are often considered autotrophic systems dominated by phytoplankton and the grazing food web, lakes are strongly dependent on the detrital food web. Much of that detrital input comes from the littoral zone.

Lakes may be classified as eutrophic or nutrient-rich, oligotrophic or nutrient-poor, or dystrophic, acidic and rich in humic material. Most lakes are subject to cultural eutrophication, which is the rapid addition of nutrients, especially nitrogen and phosphorus, from sewage and industrial wastes. Cultural eutrophication has produced significant biological changes, mostly detrimental, in many lakes.

Adding to this intrusion is residential, recreational, and industrial development of the shore, all of which eliminate or impoverish the littoral vegetation, contribute silt and pesticides, and add other contaminants to the water. These pollutants together with overfishing and introduction of exotic fish and other aquatic organisms change the lake community and feeding relationships. Destruction of littoral vegetation and disturbance from human activity eliminate or greatly reduce wildlife use of the lake.

Review and Study Questions

1. Explain why seasonal stratification of temperature and oxygen takes place in lakes and deep ponds.
2. What characterizes the epilimnion, the hypolimnion, and the thermocline (also called the metalimnion)?
3. What distinguishes the littoral zone from the limnetic, and the limnetic from the profundal?
4. What conditions distinguish the benthic zone from the other strata, and what is its role in the lake ecosystem?
5. What are the main sources of nutrients and energy in the lake ecosystem?
6. What is the relationship between nutrient availability and phytoplankton production?
7. Describe the mechanisms of nutrient transfers in the lake ecosystem.
8. What are the major lentic feeding groups, and how do they interact to structure the lake ecosystem?
9. Distinguish among oligotrophy, eutrophy, and dystrophy.
10. What is cultural eutrophication?
11. How has human intrusion in lake ecosystems and their terrestrial surroundings affected their function and structure?

*12. Consider a residential town built on the shores of a small natural lake. The town has no central sewage system and relies on residential septic tanks. The lake, once known for its recreation and fishing, is now badly polluted. Fish life has declined, algal growth has increased, and in summer bacterial counts become so high that swimming is restricted. The town's solution to the problem is to dump chlorine about the swimming areas in summer. The town is not high above the level of the lake, and the soils are sandy. What is the reason for the problem in the lake, what ecological events are taking place, and what is the long-term solution to the problem?

*13. Investigate the ecological health of a pond or lake in your area. How have the water quality and life in the lake changed over the years? What current human pressures impinge on it?

Selected References

Brock, T. D. 1985. *A eutrophic lake: Lake Mendota, Wisconsin.* New York: Springer-Verlag. A study of the process of eutrophication.

Carpenter, S. A. 1980. Enrichment of Lake Wingra, Wisconsin, by submerged macrophyte decay. *Ecology,* 61:1145–1155.

Carpenter, S. A., and J. F. Kitchell. 1984. Plankton community structure and limnetic primary production. *Am. Nat.,* 124:159–172.

Carpenter, S. A., J. F. Kitchell, and J. Hodgson. 1985. Cascading trophic interactions and lake productivity. *Bioscience,* 35:634–639. Carpenter's three papers provide an excellent introduction to the functioning of lake ecosystems.

Hutchinson, G. E. 1957–1967. *A treatise on limnology.* Vol. 1. *Geography, physics, and chemistry.* Vol. 2. *Introduction to lake biology and limnoplankton.* New York: Wiley. A classic reference.

Likens, G. E., ed. 1985. *An ecosystem approach to aquatic ecology: Mirror Lake and environment.* New York: Springer-Verlag. An outstanding detailed study of a lake ecosystem and its interactions with the surrounding terrestrial communities.

Macan, T. T. 1970. *Biological studies of English lakes.* New York: Elsevier.

Macan T. T. 1973. *Ponds and lakes.* New York: Crane, Russak. Macan's two books are classic studies of English lakes.

Rich, P. H., and R. G. Wetzel. 1978. Detritus in the lake ecosystem. *Am. Nat.* 112:57–71. Details the role of detritus in the economy of a lake ecosystem.

CHAPTER 32

Freshwater Wetlands

Outline

Objectives

On completion of this chapter, you should be able to:

1. Define a wetland.
2. Describe the various types of wetlands.
3. Explain the role of hydrology and hydroperiod in wetlands.
4. Compare production and nutrient cycling among wetlands.
5. Explain the ecologic and economic value of wetlands.
6. Discuss effects of human attitudes and activities on wetlands.

What is a wetland? This question seems to require only a simple answer: an area covered with water and supporting aquatic plants. Although this answer is not wrong, it is not correct either, not in this day when the life or death of a wetland rests on a precise definition.

Some wetlands are easy to distinguish. A water area supporting submerged plants such as pondweed, floating plants such as pond lily, and emergents such as cattails and sedges is unquestionably a wetland. But what about a piece of ground where the soil is more or less permanently wet and supports some ferns and such trees as maple that also grow on the uplands? Where do you draw the line between wetlands and uplands on this gradient of soil wetness?

It is obvious that vegetation alone does not define a wetland. First we must consider the hydrological conditions; then we may use vegetation as an indicator. *Wetlands* are areas that range along a gradient from permanently flooded to periodically saturated soil and support hydrophytic (water-loving) vegetation at some time during the growing season (Figure 32.1). Hydrophytic plants are those adapted to grow in water or on soil that is periodically anaerobic (deficient in oxygen) because of excess water (see Chapter 6). Hydrophytic plants include several groups: (1) obligate wetland plants, such as the submerged pondweeds, floating pond lily, and emergent cattails and bulrushes, and trees such as bald cypress *(Taxodium distichum);* (2) facultative wetland or amphibious plants that can grow in standing water or saturated soil and rarely grow elsewhere, such as certain sedges and alders; (3) facultative species such as red maple *(Acer rubrum),* which have about a 50:50 probability of growing in either wetland or nonwetland situations; and (4) facultative upland species such as beech *(Fagus grandifolia),* which have a 1 to 30 percent probability of growing in a wetland. It is the last group of plants that is critical

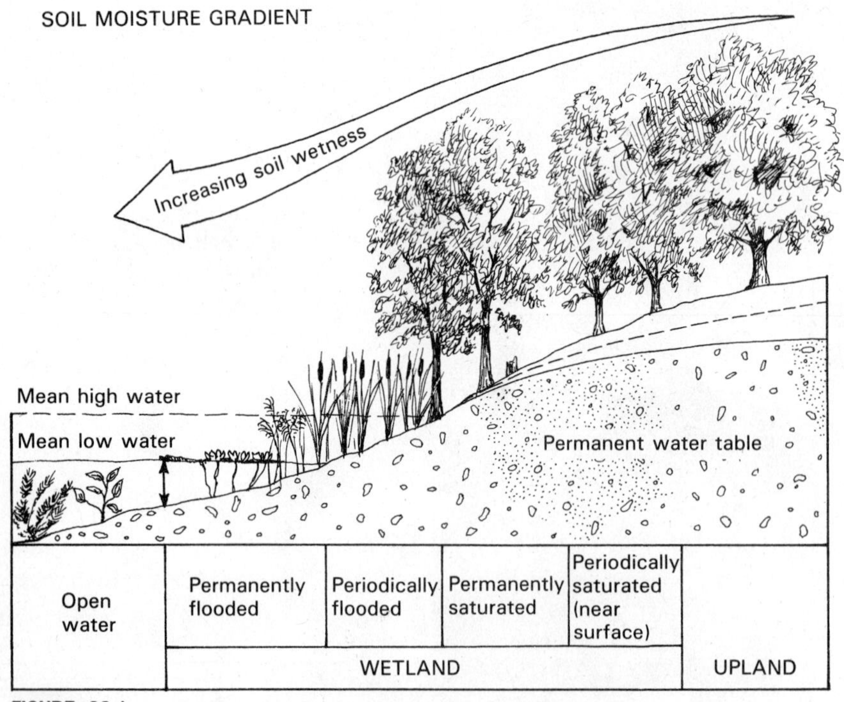

FIGURE 32.1 Location of wetlands along a soil moisture gradient.

in determining the upper limit of a wetland on the soil moisture gradient, when species designation alone is insufficient.

For example, some species of trees usually associated with uplands adapt and grow quite well in wetland environments. An example from eastern North America is red maple. It thrives in the drier uplands, but is also a conspicuous species in forested wetlands. In the uplands red maple has a deep taproot; in wetland situations the tree has a shallow root system that enables it to avoid anaerobic stress. The red maple is a facultative species that has evolved ecotypes adapted to different soil moisture conditions. Black gum *(Nyssa sylvatica),* too, grows in both upland and wetland situations. Pitch pine *(Pinus rigida),* associated with the drier ridge tops of the southern Appalachians, has ecotypes that grow in poorly drained soils and muck of swamps. In fact, pitch pine is a dominant species in the wetlands of the extensive New Jersey pine barrens. Hemlock *(Tsuga canadensis),* a shallow-rooted species, is at home in wetland situations.

The point is that species of vegetation alone do not define a wetland. They are important indicators, especially in the wettest situations, but ecotypes of upland species confuse the situation. It is essential to consider the hydrologic conditions and soil properties along with the vegetation.

TYPES OF WETLANDS

A wide variety of wetlands exists, and classifying them for management and conservation has presented problems. An old, short, but still useful classification appears in Table 32.1. A much more comprehensive classification is *Classification of Wetlands and Deepwater Habitats of the United States.*

Wetlands most commonly occur in three topographic situations (Figure 32.2). Many develop in shallow basins, ranging from upland depressions to filled-in lakes and ponds. They are basin wetlands. Other wetlands develop along shallow and periodically flooded banks of rivers and streams. They are riverine wetlands. A third type occurs along the coastal areas of large lakes and seas and is known as fringe wetlands. Some of the best-developed fringe wetlands are mangrove swamps associated with a marine environment, considered in Chapter 36.

What separates the three types is the direction of water flow (Figure 32.2). Water flow in the basin wetlands is vertical, involving precipitation and capillary flow. In riverine wetlands water flow is unidirectional. In fringe wetlands flow is in both directions, because it involves rising lake levels or tidal action. The flows may bring in and carry away nutrients and sediments. They may stress systems by exporting or importing too much.

Wetlands dominated by emergent herbaceous vegetation are *marshes* (Figure 32.3). Growing to reeds, sedges, grasses, and cattails, marshes are essentially wet prairies. Forested wetlands are commonly called *swamps.* They may be deep-water swamps dominated by cypress, tupelo, and swamp oaks (Figure 32.4); or they may be shrub swamps dominated by alder and willows. Along many large river systems are extensive tracts of *riparian woodlands* (Figure 32.5), which are occasionally or seasonally flooded by river waters but are dry for most of the growing season.

Wetlands in which considerable amounts of water are retained by an accumulation of partially decayed organic matter are *peatlands* or *mires* (Figure 32.6). Mires fed by water moving through mineral soil, from which they obtain most of their nutrients, and dominated by sedges are known as *fens.* Mires dependent largely on precipitation for their water supply and nutrients and dominated by *Sphagnum* moss are *bogs.* Mires that develop on upland situations where decomposed, compressed peat forms a barrier to the downward movement of water, resulting in a perched water table above mineral soil, are *blanket mires* and *raised bogs* (Figure 32.2). Blanket bogs are popularly known as *moors.* Because bogs depend on precipitation for nutrient inputs, they are highly deficient in mineral salts and low in pH. Bogs also develop when a lake basin fills with sediments and organic matter carried by inflowing water. These sediments divert water around the lake basin and raise the surface of the mire above the influence of groundwater. Other bogs form when a lake basin fills in from above rather than from below (Figure 32.7), creating a floating mat of peat over open water. Such bogs are often termed *quaking.*

TABLE 32.1 Types of Wetlands

Type	*Site Characteristics*	*Plants and Animals*
Inland Fresh Areas		
Seasonally flooded basins or flats	Soil covered with water or waterlogged during variable periods, but well drained during much of the growing season; in upland depressions and bottomlands	Bottomland hardwoods to herbaceous growth
Fresh meadows	Without standing water during growing season; waterlogged to within a few inches of surface	Grasses, sedges, rushes, broadleaf plants
Shallow fresh marshes	Soil waterlogged during growing season; often covered with 15 cm or more of water	Grasses, bulrushes, spike rushes, cattails, arrowhead, smartweed, pickerelweed; a major waterfowl production area
Deep fresh marshes	Soil covered with 15 cm to 1 m of water	Cattails, reeds, bulrushes, spike rushes, wild rice; principal duck-breeding area
Open fresh water	Water less than 3 m deep	Bordered by emergent vegetation such as pondweed, naiads, wild celery, water lily; brooding, feeding, nesting area for ducks
Shrub swamps	Soil waterlogged; often covered with 15 cm or more of water	Alder, willow, buttonbush, dogwoods; nesting and feeding area for ducks to limited extent
Wooded swamps	Soil waterlogged; often covered with 0.3 m of water; along sluggish streams, flat uplands, shallow lake basins	North: tamarack, arborvitae, spruce, red maple, silver maple; south: water oak, overcup oak, tupelo, swamp black gum, cypress
Bogs	Soil waterlogged; spongy covering of mosses	Heath shrubs, *Sphagnum*, sedges
Coastal Fresh Areas		
Shallow fresh marshes	Soil waterlogged during growing season; at high tide as much as 15 cm of water; on landward side, deep marshes along tidal rivers, sounds, deltas	Grasses and sedges; important waterfowl areas
Deep fresh marshes	At high tide covered with 15 cm to 1 m of water; along tidal rivers and bays	Cattails, wild rice, giant cutgrass
Open fresh water	Shallow portions of open water along fresh tidal rivers and sounds	Vegetation scarce or absent; important waterfowl areas
Inland Saline Areas		
Saline flats	Flooded after periods of heavy precipitation; waterlogged within few cm of surface during the growing season	Sea blite, salt grass, saltbush; fall waterfowl-feeding areas

TABLE 32.1 *(Continued)*

Type	*Site Characteristics*	*Plants and Animals*
Saline marshes	Soil waterlogged during growing season; often covered with 0.61 to 1 m of water; shallow lake basins	Alkali hard-stemmed bulrush, wigeon grass, sago pondweed; valuable waterfowl areas
Open saline water	Permanent areas of shallow saline water; depth variable	Sago pondweed, muskgrasses; important waterfowl-feeding areas
Coastal Saline Areas		
Salt flats	Soil waterlogged during growing season; sites occasionally to fairly regularly covered by high tide; landward sides or islands within salt meadows and marshes	Salt grass, sea blite, saltwort
Salt meadows	Soil waterlogged during growing season; rarely covered with tide water; landward side of salt marshes	Cordgrass, salt grass, black rush, waterfowl-feeding areas
Irregularly flooded salt marshes	Covered by wind tides at irregular intervals during the growing season; along shores of nearly enclosed bays, sounds, etc.	Needlerush, waterfowl cover area
Regularly flooded salt marshes	Covered at average high tide with 15 cm or more of water; along open ocean and along sounds	Atlantic: salt-marsh cordgrass; Pacific: alkali bulrush, glassworts; feeding area for ducks and geese
Sounds and bays	Portions of saltwater sounds and bays shallow enough to be diked and filled; all water landward from average low-tide line	Wintering areas for waterfowl
Mangrove swamps	Soil covered at average high tide with 15 cm to 1 m of water; along coast of southern Flordia	Red and black mangroves

STRUCTURE

The structure of a wetland is influenced by the phenomenon that creates it—its hydrology. Hydrology has two components. One is the physical aspects of water and its movement: precipitation, surface and subsurface flow, direction and kinetic energy of water, and the chemistry of the water. The other is *hydroperiod,* which includes the duration, frequency, depth, and season of flooding. Length of the hydroperiod varies among types of wetlands. Basin wetlands have a longer hydroperiod and usually experience flooding during periods of high rainfall and drawdown during dry periods. Both phenomena appear to be essential to the long-term existence of wetlands (see Chapter 26). Riverine wetlands have a short period of flooding associated with peak stream flow. The hydroperiod of fringe wetlands, influenced by wind and lake waves, may be short and regular, and does not undergo the seasonal fluctuation characteristic of many basin marshes.

Hydroperiod influences plant composition, for it affects germination, survival, and mortality at various stages of the plants' life cycles. The effect of hydroperiod is most pronounced in basin wetlands, especially those of the prairie regions of North America. In basins (called potholes in the prairie region) deep enough to have standing water throughout periods of drought, the dominant plants will be submergents. If the wetland goes dry annually or during a period of drought, tall or midheight emergent species such as cattails will

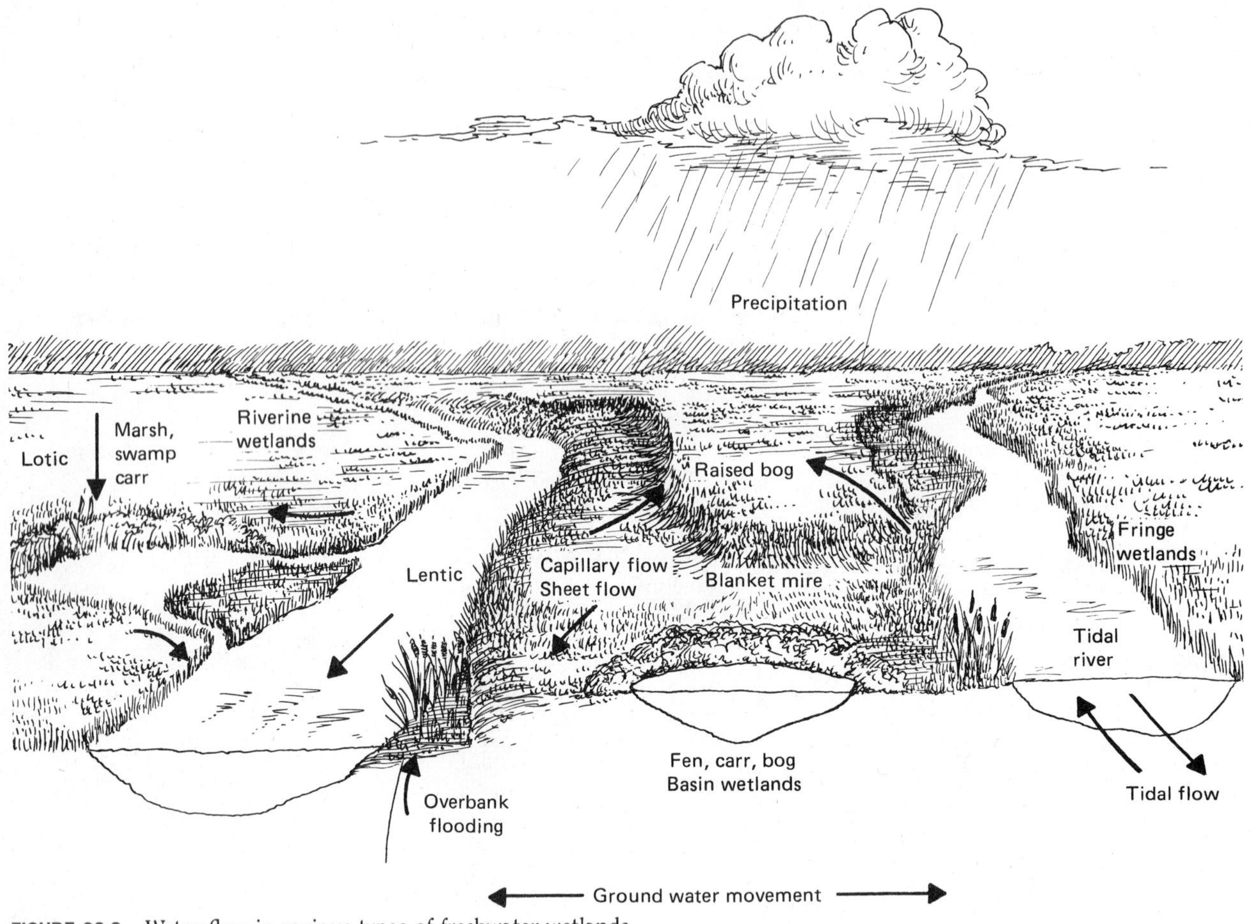

FIGURE 32.2 Water flow in various types of freshwater wetlands.

dominate the marsh. If the pothole is shallow and flooded only briefly in the spring, then grasses, sedges, and forbs will make up a wet-meadow community.

If the basin is sufficiently deep toward its center and large enough, then zones of vegetation may develop (see Chapter 31), ranging from submerged plants to deep-water emergents such as cattails and bulrushes, shallow-water emergents, and wet-ground species such as spikerush. Zonation reflects the response of plants to hydroperiod. Those areas of wetland subjected to a long hydroperiod will support submerged and deep-water emergents; those with a short hydroperiod and shallow water are occupied by shallow-water emergents and wet-ground plants.

Periods of drought and wetness can induce vegetation cycles associated with changes in water levels. Periods of above-normal precipitation can raise the water level and drown the emergents to create a lake marsh dominated by submergents. During a drought the marsh bottom is exposed by receding water, stimulating the germination of seeds of emergents and mudflat annuals. When water levels rise again the mudflat species drown, and the emergents survive and spread vegetatively.

Peatlands differ from other freshwater wetlands in that their rate of organic production exceeds the rate of decomposition, and much of the production accumulates as peat. In northern regions acid-forming,

FIGURE 32.3 A northern wetland with well-developed emergent vegetation and patches of open water, an ideal environment for wetland wildlife.

FIGURE 32.4 A forested wetland of bald cypress in the southern United States.

water-holding sphagnum mosses add new growth on top of the accumulating remains of past moss generations; and their spongelike ability to hold water increases water retention on the site. As the peat blanket thickens, the water-saturated mat of moss and associated vegetation is raised above and insulated from mineral soil. The peat mat then becomes its own reservoir of water, creating a perched water table.

Peat bogs and mires generally form under oligotrophic and dystrophic conditions (see Chapter 31). Although usually associated with and most abundant in boreal regions of the Northern Hemisphere, peatlands also exist in tropical and subtropical regions. They develop in mountainous regions or in lowland or estuarine regions where hydrological situations encourage an accumulation of partly decayed organic matter. Examples are the Everglades in Florida and the pocosins of the southeastern United States coastal plains.

Biologically, wetlands are among the richest and most interesting of ecosystems. They support a diverse community of benthic, limnetic, and littoral invertebrates, especially crustaceans and insects. These inver-

(a) (b)

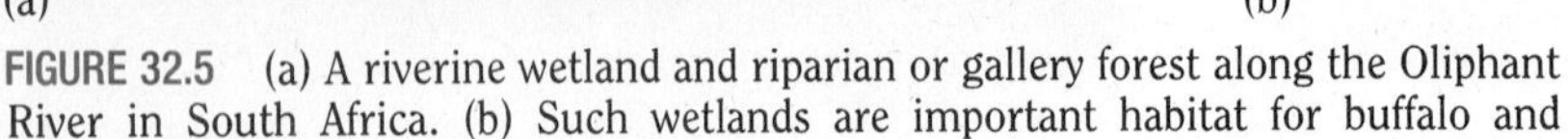

FIGURE 32.5 (a) A riverine wetland and riparian or gallery forest along the Oliphant River in South Africa. (b) Such wetlands are important habitat for buffalo and hippopotamus during the dry season.

FIGURE 32.6 A bog. This is the "muskeg furthest south," Cranberry Glades, Pocahontas County, West Virginia. Tamarack is absent, as is Labrador tea, but this is the southernmost point of the bog rosemary. Note the hummock effect of moss and lichens.

tebrates, along with small fishes, provide a food base for waterfowl, herons, gulls, and other birds, and supply the fat-rich nutrients needed by ducks for egg production and the growth of young. Wetlands support a diversity of amphibians and reptiles, notably frogs, toads, and turtles.

Herbivores make up a conspicuous component of animal life. Microcrustaceans filter algae from the water column. Snails eat algae growing on the leaves and litter; geese graze on new emergent growth; coots and mallards and other surface-feeding ducks feed on algal mats. The dominant herbivore in the prairie marshes is the muskrat *(Ondatra zibethicus)*. During population highs, muskrats can eliminate emergent vegetation, creating "eat-outs" and transforming an emergent-dominated marsh into an open-water one. Introduced into Eurasia, the muskrat has become the major herbivore in many marshes on that continent. Muskrats are the major prey for mink, the dominant carnivore on the marshes. Other predators include raccoon, fox, weasel, and skunk, which can seriously reduce the reproductive success of waterfowl on small marshes surrounded by agricultural land.

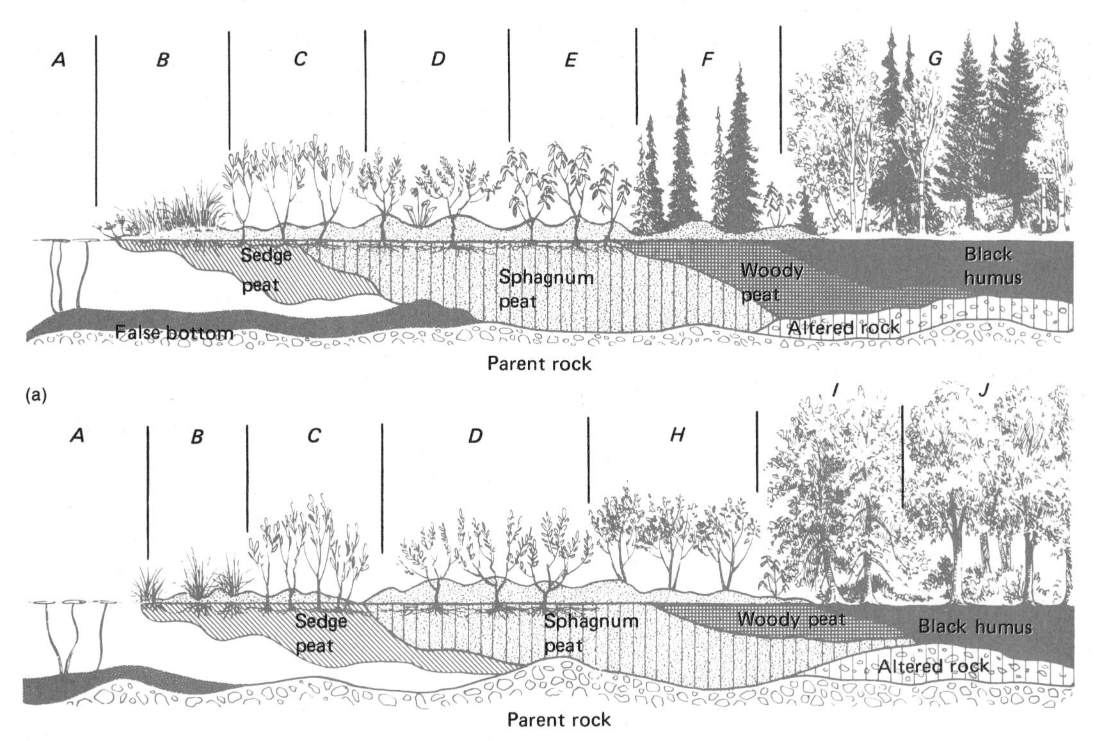

FIGURE 32.7 (a) Transect through a quaking bog, showing zones of vegetation, sphagnum mounds, and floating mats. A, pond lily in open water; B, buckbean and sedge zone; C, sweet gale zone; D, leatherleaf; E, Labrador tea; F, black spruce; G, birch-black spruce-balsam fir forest. (b) Alternative vegetational sequence: H, alder; I, aspen, red maple; J, mixed deciduous forest.

FUNCTION

Freshwater wetlands are highly productive ecosystems, but their complexity and differences make generalizations about their functions difficult. In fact, we know much less about wetland functions than about forests and grasslands.

Wetlands are sedimentary or detrital systems. They accumulate carbon, nitrogen, phosphorus, and other materials and exchange them among the wetland, the atmosphere, and the landscape. Productivity of freshwater marshes is influenced by hydrological regimes: groundwater, surface runoff, precipitation, drought cycles, flooding in riverine wetlands, and the like. Wetlands, whatever their type, are closely associated with the total landscape in which they reside, its size, soils, land use, nutrient availability, and types of vegetation. Basin wetlands accumulate muck and peat. Riverine wetlands experience a throughflow of water, importing and exporting materials. All of these conditions influence the nature of each wetland's vegetation. In turn, the life-history pattern of the species involved further influences the productivity of the wetlands.

Aboveground biomass varies with the proportionate abundance of annual and perennial species, whose dominance changes through the growing season. Annual emergents increase their biomass through the growing season, reaching a maximum in late summer. Perennials increase their biomass during the first part of the growing season, then see it decline or level off as they become senescent (Figure 32.8). In general, however, the average maximum standing crop of biomass matches annual aboveground productivity.

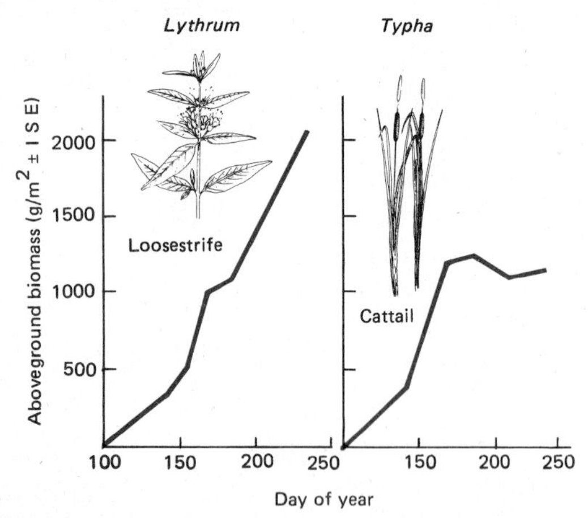

FIGURE 32.8 Pattern of aboveground biomass accumulation through the growing season for a freshwater annual, loosestrife (*Lythrum*), and a perennial, cattail (*Typha*). Note the linear increase in biomass in the annual and the sigmoid growth curve of the perennial.

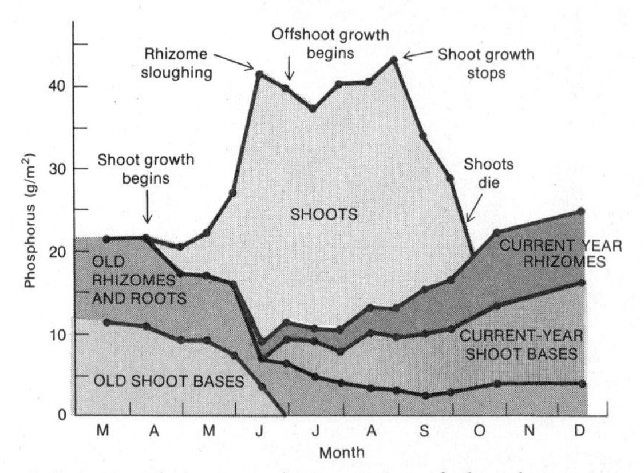

FIGURE 32.9 The seasonal economics of phosphorus in cattails (*Typha latifolia*) in a marsh at Lake Mendota, Wisconsin, with deposits and withdrawals in the various parts of the plant.

Belowground production, much more difficult to estimate, appears for some species to be highest in summer, at the same time the peak aboveground biomass is achieved. Others, such as cattails *(Typha)* and sedges *(Scirpus),* reach peak production in the fall, when nutrients are stored in the roots. Such species may have minimal root biomass in the summer because of nutrient transfer to aboveground biomass.

The cycling of phosphorus in a cattail marsh on Lake Mendota, Wisconsin, outlined in Figure 32.9, provides an example. In spring the new shoots of cattail draw on the phosphorus reserves in the rootstalks for their initial growth before mobilizing P from the soil. By June cattails are accumulating P at a rate higher than they are accumulating biomass. By midsummer *Typha* has accumulated about 40 g P/m^2 in the shoots, 78 percent of total P in biomass. At the same time the belowground pool is minimal. As the season progresses P accumulates in the rhizomes until December, but at a rate slower than accumulation of belowground biomass. In fall large amounts of P begin to disappear. Most of it is lost through leaching and death of the shoots. Only about 28 percent of the summer accumulation is returned to the rhizomes, which will be rapidly depleted again by spring growth.

To balance losses, cattails and other emergent plants must draw on the P supply in the soil, which is derived from decomposition of the litter of previous years. By doing so, the plants act as a nutrient pump, drawing nutrients from the soil, translocating them into the shoots, and then releasing them to the surface soil by leaching and death of shoots during the growing and postgrowing season. In this way marsh plants make nutrients sequestered in the soil available for growth.

Nitrogen, possessing a gaseous form, experiences a considerable exchange between the wetland and atmosphere, involving nitrogen fixation, volatilization of NH_3, denitrification, and possibly nitrification. Denitrification may be the major source of loss of nitrogen from wetlands. Wetland plants mobilize nitrogen from the soil, much as they do phosphorus, and concentrate it in their tissues. Under eutrophic conditions, the accumulation of nitrogen may be high. Nevertheless, wetland plants undertake considerable internal cycling, through which they meet 40 percent of their requirements for both N and P.

Wetlands contribute a great amount of litter or detrital material to the system. This material at first decomposes rapidly, as leaching in a watery environment removes soluble compounds as dissolved organic matter (Figure 32.10). After initial leaching, decomposition proceeds more slowly. Permanently submerged leaves decompose more rapidly than those on the marsh surface, because they are more accessible to aquatic detritivores such as crayfish, and because the constantly moist environment is more favorable for

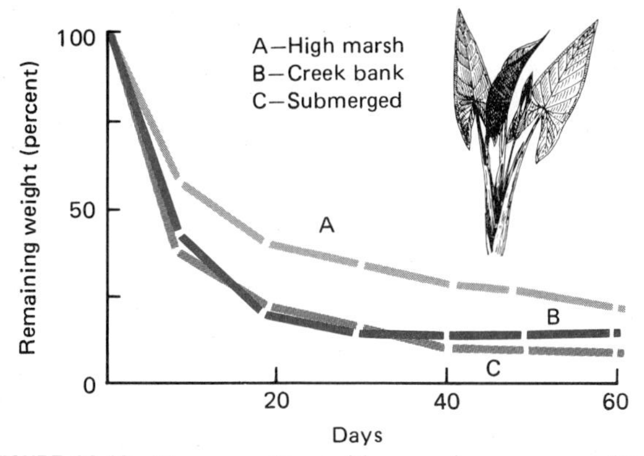

FIGURE 32.10 Decomposition of leaves of arrow arum (*Peltandra virginica*) as measured by the percentage of ash-free dry weight remaining in litter bags under three conditions: (A) irregularly flooded high marsh exposed to alternate wetting; (B) creek bed flooded two times daily; (C) permanently submerged. Note that the detrital material consistently wet showed the highest rate of decomposition, although the overall pattern of decomposition is similar.

microbial decomposition. Microbes that use the litter as a carbon source add to it additional nitrogen and phosphorus, which they obtain from the surrounding water and sediments. This accumulation of N and P in the decomposed material helps to retain N and P in the wetland, where it is available to emergent and other plants. How long these nutrients remain in the depository depends on how rapidly the litter decomposes. Much of the finer material becomes incorporated into anaerobic bottom muds where decomposition results in the production of methane and hydrogen sulfide.

Peatlands present a different situation. Because bog vegetation is not in contact with mineral soil and because inflowing groundwater is blocked, bogs depend mostly on precipitation for their nutrients. In addition, both blue-green algae living in close association with bog mosses and bog myrtle fix nitrogen; and carnivorous plants, such as sundews, extract nitrogen from captured and digested insects. Bogs, however, face a scarcity of nutrients, a shortage compounded by the plants themselves. Most of the nutrients they fix in their tissues remains in the accumulating peat.

However, bog plants do possess some means of conserving nutrients. Consider how the trailing plant cloudberry *(Rubus chamaemorus)* manages its phosphorus budget. This plant increases its uptake of phosphorus through the roots prior to budbreak. After budbreak, the cloudberry increases the amount of phosphorus in stem, leaf, and root. In summer after the plant has completed its shoot growth, it moves phosphorus from its shoots to developing fruits and to roots and rhizomes. As senescence sets in, cloudberry moves most of the phosphorus remaining in its shoots to its winter buds, where it accumulates and is available for the next year's growth.

Energy flow in peatlands differs from that in other wetlands because the detrital food chain is impaired. In most ecosystems, material that enters the detrital food web is eventually recycled, and the energy that enters the system is liberated or stored in living material. In bogs, material resulting from primary production accumulates in a partially decomposed state, and energy is locked up in peat until environmental conditions change to favor decomposition or until the material burns.

Because of low temperatures, acidity, and nutrient immobilization, primary production in peatlands is low, as little as 300 g/m^2/year in sphagnum bogs. Likewise, decomposition of that primary production is slow. Shrub litter decomposes slowly and sphagnum decomposes hardly at all. Ecologists estimated that in English mires, turnover of 95 percent of the organic matter for the system as a whole takes 3000 years and for the top 20 cm of material, 70 years. This slow rate of decomposition accounts for the accumulation of peat.

VALUE OF WETLANDS

Just as we like to dam rivers, so are we motivated to drain wetlands and convert them into dry land. The Romans drained the great marshes about the Tiber to make room for the city of Rome. In spite of the enormous amount of vacant dry land about him, George Washington proposed draining the Great Dismal Swamp (most of which has been done since his time). Many of us consider wetlands as wastelands, areas to be drained for more productive uses by human standards: agricultural land, solid waste dumps, housing, industrial developments, and roads. We also look on wetlands as forbidding mysterious places, sources of pestilence, the home of dangerous and pestiferous insects, abode of slimy sinister creatures that rise out of swamp waters. We are blind to the ecological, hydrological, and economic values of wetlands.

Wetlands assume an importance, ecologically and economically, out of proportion to their size. Their major contribution is to the hydrology of a region. Basin wetlands, in particular, are groundwater recharge points. They hold rainwater, snowmelt, and surface runoff in their basins and discharge the water slowly into the aquifers. These same basins also function as natural flood-control reservoirs. As little as 5 percent of the watershed or catchment area in wetlands can potentially reduce flood flows by as much as 50 percent (Figure 32.11).

Wetlands act as water-filtration systems. Wetland vegetation takes up excessive nitrogen, phosphorus, sulfates, copper, iron, and other heavy metals brought by surface runoff and inflow, incorporates them into plant biomass, and deposits much of them in anaerobic bottom muds. Because of their ability to filter out heavy metals and to reduce pH, we are beginning to treat urban wastewater and drainage from surface mines by diverting these flows into natural or especially created wetlands.

Wetlands contribute to the human economy in other ways. They provide places of recreation, sources of horticultural peat and timber—notably bald cypress and bottomland hardwoods in southern United States—and sites for growing cranberries in the northeastern United States. These uses, however, tend to interfere with the natural function and integrity of the wetland ecosystems.

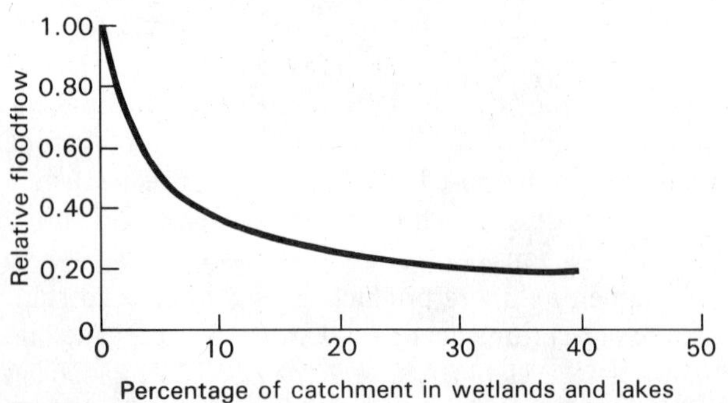

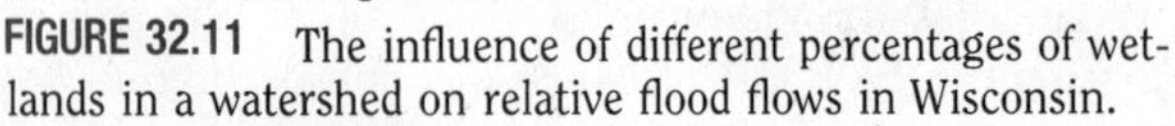
FIGURE 32.11 The influence of different percentages of wetlands in a watershed on relative flood flows in Wisconsin.

Wetlands are vitally important as wildlife nesting and wintering habitat. Many species of wildlife, some of them endangered, are dependent on wetlands. Worldwide, wetlands are home to numerous species of amphibians and reptiles, including alligators and crocodiles. Many species of fish of the Amazon and other tropical rivers depend on seasonal flooding of riverine swamps and floodplains as places to forage for terrestrial foods and to spawn. Waterfowl, wading birds, gulls and terns, herons and storks depend on marshes and wooded swamps for nesting and foraging. In fact, the prairie pothole region of north central United States and Canada is used by two-thirds of the continent's 10 to 12 million waterfowl as a nesting area. Waterfowl in concentrated numbers use southern marshes and swamps as wintering habitat. Moose, hippopotamous, waterbuck, otters, and muskrat are mammalian inhabitants of wetlands. In addition to wetland dwellers, wetlands support animal life in other ecosystems. A mosaic of wetlands in an upland terrestrial environment increases the abundance and diversity of wildlife populations.

HUMAN IMPACT

That we have little regard for wetlands and their values is underscored by the destruction we have imposed on them. Wetlands, both forested and nonforested, once made up about 3 percent of Earth's surface, but much of that area, especially in the Northern Hemisphere, has been converted to other land uses. In colonial times the area embraced by the 50 United States contained some 392 million acres of wetlands. Of these, 221 million acres were in the lower 48 states, 170 million acres in Alaska, and 59,000 acres in Hawaii. Now, 200 years later, Alaska has lost a fraction under 1 percent, Hawaii 12 percent, and the lower 48 states have lost well over 50 percent of their wetlands. Among the states California has lost 91 percent of its wetlands. Wetlands, which once made up 5 percent of that state's total land area, have shrunk to one-half of 1 percent. Over the continental United States the 392 million acres of wetland have decreased to 274 million acres (Figure 32.12), and many of these remnants are degraded.

The rationales for drainage are many. The most per-

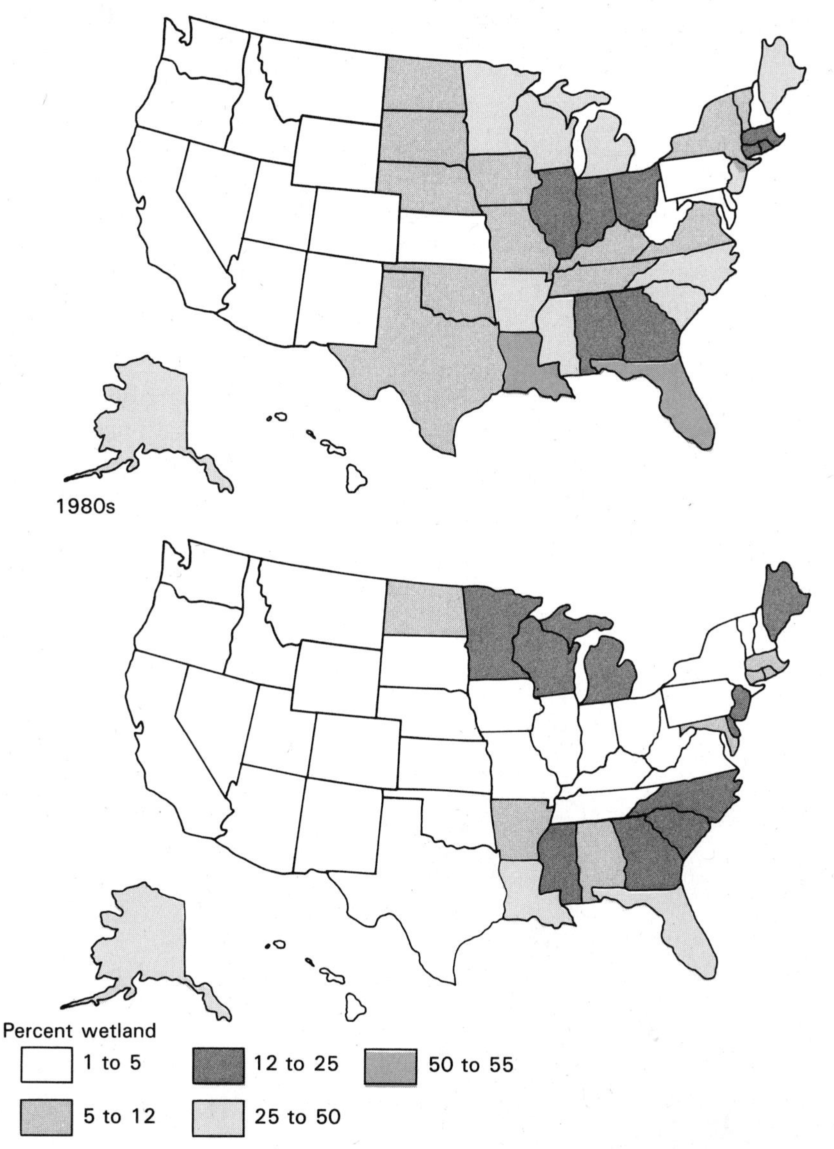

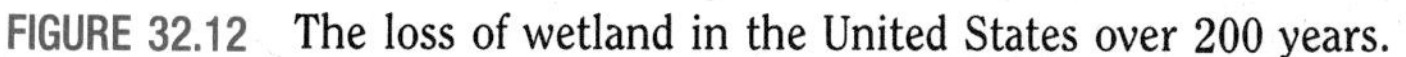

FIGURE 32.12 The loss of wetland in the United States over 200 years.

suasive relates to agriculture. Drainage of wetlands opens many hectares of rich organic soil for crop production. In the prairie country the innumerable potholes are viewed as a nuisance to efficient agriculture. Draining them tidies up fields and allows unhindered use of large agricultural machinery. There are other reasons, too. Wetlands are viewed as an economic liability by landowners and by local governments. They produce no economic return and they provide little tax revenue. Many regard the wildlife wetlands support as threats to grain crops. Elsewhere wetlands are considered valueless lands, at best filled in and used for development. Some major wetlands have been in the way of dams. For example, the large Pymatuning Lake in the states of Pennsylvania and Ohio covers a 4200 ha sphagnum-tamarack bog. Peat bogs in the northern United States, Canada, Ireland, and northern Europe are excavated for fuel, horticultural peat, and organic soil. In some areas such exploitation threatens the existence of peatland ecosystems.

Many remaining wetlands, especially in the north central and southwestern United States, are contaminated and degraded by pesticides and heavy metals carried into them by surface and subsurface drainage and sediments from surrounding croplands (Figure 32.13). Although inputs of N and P increase the productivity of wetlands, the concentration of herbicides, pesticides, and heavy metals poisons the water, destroys invertebrate life, and has debilitating effects on wildlife, including deformities, lowered reproduction, and death. Waterfowl in wetlands scattered throughout agricultural lands are also more subject to predation, and without access to natural upland vegetation they breed less successfully.

Wetlands can suffer even from the best intentions. Management attempts to maintain some water stability reduce or eliminate the fluctuations in water level so necessary to the heatlh of the wetland. Lack of fluctuations increases anaerobic decomposition, tying up nutrients in the bottom muck, and changes plant composition, eliminating those species that require a cycle of drawdown for germination and flooding for growth.

The loss of wetlands has reached a point where both environmental and socioeconomic values—including waterfowl habitat, groundwater supply and quality, floodwater storage, and sediment trapping—are in jeopardy. Although we have made some progress in the United States toward preserving the remaining wetlands through legislative action and land purchase, the future of freshwater wetlands is not secure. Apathy,

FIGURE 32.13 Wetlands at the Stillwater National Wildlife Refuge in Nevada are so contaminated with pesticides, heavy metals, and saline water draining in from surrounding irrigated agricultural land that the refuge is nearly worthless as a wildlife habitat.

hostility toward wetland preservation, political maneuvering, court decisions, and arguments over what constitutes a wetland allow the continued destruction of wetlands at a rate of over 200,000 ha a year.

Summary

Wetlands are best defined as a community of hydrophytic plants occupying a gradient of soil wetness from permanently flooded to periodically saturated during the growing season. Hydrophytic plants are those adapted to grow in water or on soil periodically deficient in oxygen. Wetlands dominated by grasses and herbaceous hydrophytes are marshes; those dominated by wooded vegetation are forested wetlands or swamps. Wetlands characterized by accumulation of peat are mires. Mires fed by water moving through the mineral soil and dominated by sedges are fens; those dominated by sphagnum moss and dependent largely on precipitation for moisture and nutrients are bogs. Bogs are characterized by blocked drainage, an accumulation of peat, and low productivity. In bogs, organic production, however, exceeds the rate of decomposition, so most of the nutrients fixed in plants are removed from circulation and stored in slowly decomposing peat. In freshwater marshes and swamps nutrients are taken up by plants. A portion is recycled within the plants; some is recycled through decomposition; and the rest is stored in benthic muds and litter for varying periods of time.

Structure and function of wetlands are strongly influenced by their hydrology—both the physical movement of water and the hydroperiod. Hydroperiod is the depth, frequency, and duration of flooding. Hydroperiod's influence on vegetation is most evident in basin wetlands that exhibit zonation from deep-water submerged vegetation to wet-ground emergents.

Wetlands have an ecologic and economic value out of proportion to their extent. They function as recharge points for groundwater aquifers, water-storage basins that reduce intensity of flooding, water-filtration systems for pollutants, sources of wood products, and habitat for a rich diversity of wetland wildlife.

In spite of their importance, over 50 percent of the original wetlands of the United States (excluding Alaska and Hawaii) have been drained for other land uses, mostly agriculture and development. Many remaining wetlands are contaminated by pesticides, herbicides, heavy metals, and excessive inputs of nitrogen and phosphorus from surrounding watersheds. We are making an effort to save remaining wetlands, but progress is slow because of human resistance to save what appears to many as worthless land.

Review and Study Questions

1. What is a wetland? A hydrophyte?
2. How does the definition of a wetland relate to the gradient of soil wetness?
3. Hydrologically, what are the three major types of wetlands?
4. Characterize the various types of wetlands based on vegetation.
5. What is the hydroperiod, and how does it relate to the structure of wetlands?
6. How do wetlands relate to the landscapes in which they reside?
7. Contrast nutrient cycling in marshes and peatland.
8. Discuss the value of wetlands, emphasizing a rationale for their preservation.
9. What major impacts have humans made on wetlands, and why?
10. What is the paradox of draining wetlands, then building large flood-control dams?

*11. What has been the impact of drainage of prairie potholes, swamps, and bottomland hardwood forests on waterfowl populations?

*12. What has been the fate of wetlands in your region? Are figures available for losses? To what use has the drained land been put?

*13. Trace the history of the attitudes toward wetlands from Roman times to the present. Why are we so fearful of them?

*14. How could you argue for preservation of wetlands on an economic basis to developers and local governments? Is it difficult to put monetary values on wetlands?

Selected References

Cowardin, L. M., V. Carter, and E. C. Golet. 1979. *Classification of wetlands and deepwater habitats of the United States.* U.S. Department of Interior, Fish and Wildlife Service FWS/OBS-79/31. A revised classification of wetlands.

Dahl, T. E. 1990. *Wetland losses in the United States, 1780's to 1980's.* U.S. Department of Interior, Fish and Wildlife Service.

Ewel, K. C. 1990. Multiple demands on wetlands. *Bioscience* 40:660–666. Societal benefits of wetlands, with cypress swamps serving as a case study.

Ewel, K. C., and H. T. Odum, eds. 1986. *Cypress swamps.* Gainesville: University Presses of Florida. In-depth studies of the structure, function, and management of cypress swamps in southern United States.

Good, R. E., D. F. Whigham, and R. L. Simpson, eds. 1978. *Freshwater wetlands: Ecological processes and management potential.* New York: Academic. A review of functional aspects of wetlands and their management implications.

Gore, A. P. J., ed. 1983. *Mires, swamp, bog, fen, and moor.* Ecosystems of the world 4A and 4B. Amsterdam: Elsevier. A review of the world of freshwater wetlands.

Greesen, P. S., J. R. Clark, and J. E. Clark, eds. 1979. *Wetland functions and values: The state of our understanding.* Minneapolis: American Water Resources Asssociation. An earlier work still of value.

Lugo, A. E. 1990. *The forested wetlands.* Amsterdam: Elsevier. Excellent overview and discussion of structure and function of the world's forested wetlands.

Mitsch, W. J., and J. G. Gosselink. 1986. *Wetlands.* New York: Van Nostrand Reinhold. A pioneering text and a major reference.

Moore, P. D., and D. J. Bellemany. 1974. *Peatlands.* New York: Springer-Verlag. An outstanding introduction to the ecology and development of peatlands.

National Audubon Society. 1990. The last wetlands. *Audubon* 92(4). A highly informative issue devoted entirely to wetlands, their management and preservation.

Tiner, R. W. 1991. The concept of a hydrophyte for wetland identification. *Bioscience* 41:236–247. Reviews problems and means of identifying wetlands.

Van der Valk, A., ed. 1989. *Northern prairie wetlands.* Ames: Iowa State University Press. Detailed studies of major wetlands rapidly disappearing.

Weller, M. W. 1981. *Freshwater wetlands: Ecology and wildlife management.* Minneapolis: University of Minnesota Press. An excellent, nontechnical introduction.

CHAPTER 33

Flowing-Water Ecosystems

Outline

Objectives

On completion of this chapter, you shoud be able to:

1. Describe the physical characteristics of flowing-water ecosystems.
2. Compare fast streams with slow streams and rivers.
3. Describe nutrient cycling in flowing-water ecosystems.
4. Discuss the role of various feeding groups in streams and rivers.
5. Explain the role of detritus in lotic ecosystems.
6. Point out the effects of pollution on streams and rivers.
7. Discuss the problems channelization causes.
8. Explain the meaning of regulated rivers and discuss the impact of dams on flowing-water ecosystems.

Even the largest of rivers begin somewhere back in the hinterlands as springs or seepage areas, becoming headwater brooks and streams; or they arise as outlets of ponds or lakes. A very few emerge full-blown from glaciers. As the brook drains away from its source it flows in a direction and manner dictated by the lay of the land and underlying rock formations. Its course may be determined by the original slope; or water, seeking the least resistant route to lower land, may follow joints and fissures in bedrock near the surface and shallow depressions in the ground. Whatever its direction, water concentrates in rills that erode small furrows, which soon grow into gullies. Moving downstream, especially where the gradient is steep, the moving water carries with it a load of debris collected from its surroundings that cuts the channel wider and deeper. Sooner or later, the stream deposits this material on its bed or along its banks. In mountainous areas, erosion continues to eat away at the head of the gully, cutting backward into the slope and increasing the drainage area. Joining the new stream are other small streams, spring seeps, and surface water.

Just below its source the stream may be small, straight, and swift, with waterfalls and rapids. Further downstream, where the gradient is less, velocity decreases, meanders become common, and the stream deposits its load of sediment as silt, sand, or mud. At flood time, a stream drops its load of sediment on surrounding level land, over which floodwaters spread to form floodplain deposits. These floodplains are a part of a stream or river channnel used at the time of high water—a fact few people recognize.

Where a stream flows into a lake or a river into the sea, the velocity of water is suddenly checked. The river then is forced to deposit its load of sediment in a fan-shaped area about its mouth to form a delta. Here its course is carved into a number of channels, which are blocked or opened with subsequent deposits. As a result, the delta becomes an area of small lakes, swamps, and marshy islands. Material the river fails to deposit in the delta is carried out to open water and deposited on the bottom.

Because streams become larger on their course to rivers and are joined along the way by many others, we can classify them according to order. A small headwater stream without any tributaries is a first-order stream. When two streams of the same order join, the stream becomes one of higher order. If two first-order streams unite, the resulting stream becomes a second-order one; and when two second-order streams unite, the stream becomes a third-order one. An order of a stream can increase only when a stream of the same order joins it. Its order cannot be increased with the entry of a lower-order stream. In general, headwater streams are orders 1 to 3; medium-sized streams, 4 to 6; and rivers, greater than 6.

The area of land a stream or river drains is its *watershed*. Each watershed is different, characterized by vegetative cover, geology, soils, topography, and land use. Streams and rivers provide the drainage pathways. Ponds, lakes, and wetlands act as the catch basins. Thus a watershed includes *lotic* or flowing-water systems and *lentic* or still-water systems.

STRUCTURE

PHYSICAL STRUCTURE

Velocity of the current molds the character and the structure of a stream. Shape and steepness of the stream channel, its width, depth, roughness of the bottom, and intensity of rainfall and rapidity of snowmelt affect velocity. Fast streams (Figure 33.1) are those whose velocity of flow is 50 cm per second or higher. At this velocity, the current will remove all particles less than 5 mm in diameter and will leave behind a stony bottom. High water increases the velocity; it moves bottom stones and rubble, scours the streambed, and cuts new banks and channels. As the gradient decreases and width, depth, and volume of water increase, silt and decaying organic matter accumulate on the bottom. The character of the stream changes from fast water to slow, with associated change in species composition (Figure 33.2).

Flowing-water ecosystems often alternate two different but interrelated habitats, the turbulent riffle and the quiet pool (Figure 33.3). The waters of the pool are influenced by processes occurring in the rapids above, and the waters of the rapids are influenced by events in the pool.

FIGURE 33.1 A fast mountain stream. The gradient is steep and the bottom is largely bedrock.

FIGURE 33.3 Two different but related habitats in a stream, the riffle (foreground) and the pool (background).

FIGURE 33.2 A slow stream is deeper and has a lower gradient and velocity.

Riffles are the sites of primary production in the stream. Here the periphyton, or *aufwuchs,* organisms that are attached to or move on submerged rocks and logs, assume dominance. Periphyton, which occupies a position of the same importance as phytoplankton of lakes and ponds, which consists chiefly of diatoms, blue-green and green algae, and water moss.

Above and below the riffles are the pools. Here the environment differs in chemistry, intensity of current, and depth. Just as the riffles are the sites of organic production, so the pools are the sites of decomposition. They are catch basins of organic materials, for here the velocity of the current is reduced enough to allow part of the load to settle. Pools are the major sites of CO_2 production during the summer and fall. That is necessary for the maintenance of a constant supply of bicarbonate in solution. Without pools, photosynthesis in the riffles would deplete the bicarbonates and result in smaller and smaller quantities of available carbon dioxide downstream.

Free carbon dioxide in rapid water is in equilibrium with that of the atmosphere. The amount of bound carbon dioxide is influenced by the nature of the surrounding terrain and decomposition taking place in pools of still water. Most of the carbon dioxide in flowing water occurs as carbonate and bicarbonate salts. Streams fed by groundwater from limestone springs receive the greatest amount of carbonates in solution.

The degree of acidity or alkalinity, or pH, of the water reflects the CO_2 content as well as the presence of organic acids and pollution. The higher the pH of stream water, the richer natural waters generally are in carbonates, bicarbonates, and associated salts. Such streams support more abundant aquatic life and larger fish populations than streams with acid waters, generally low in nutrients.

The constant churning and swirling of stream water over riffles and falls give greater contact with the atmosphere. Thus, oxygen content of the water is high and often near the saturation point for existing temperatures. Only in deep holes or in polluted waters does dissolved oxygen show any significant decline.

Temperature of a stream is variable. Small, shallow streams tend to follow, but lag behind, air temperatures, warming and cooling with the seasons, but rarely falling below freezing in winter. Streams with large areas exposed to sunlight are warmer than those shaded by trees, shrubs, and high banks. That fact is ecologically important because temperature affects the stream community, influencing the presence or absence of cool-water and warm-water organisms.

ADAPTATIONS

Living in a moving-water environment, inhabitants of streams and rivers have a major problem of remaining in place and not being swept downstream. They have evolved unique adaptations for dealing with life in the current (Figure 33.4). A streamlined form, which offers less resistance to water current, is typical of many animals of fast water, such as the dace and the brook trout. The larval forms of many species of insects cling to the undersurfaces of stones, where the current is very weak. They possess extremely flattened bodies and broad, flat limbs that allow the current to flow over them. Typical are many species of mayflies and stoneflies. Other forms, such as the blackfly (*Simulidae*) larvae, attach themselves in one way or another to the substrate, and they obtain food by straining particles carried to them by the current. The larvae of certain species of caddisflies construct protective cases of sand or small pebbles and cement them to the bottoms of stones. Larvae of net-spinning caddisflies (*Hydropsyche*) firmly attach to stones funnel-shaped, food-collecting nets whose open ends face upstream. Sticky undersurfaces aid snails and planarians to cling tightly and move about on stones and rubble in the current.

Among the plants water moss (*Fontinalis*) and heavily branched filamentous algae cling to rocks by strong holdfasts. Other algae grow in cushionlike colonies or closely appressed sheets that are covered with a slippery, gelatinous coating and follow the contours of stones and rocks.

All inhabitants of fast-water streams require high, near saturation concentrations of oxygen and moving water to keep the absorbing and respiratory surfaces of animals in continuous contact with oxygenated water. Otherwise a closely adhering film of liquid impoverished of oxygen forms a sort of cloak about their bodies.

In slow-flowing streams where current is at a minimum, streamlined forms of fish give way to species such as small-mouthed bass, shiners, and darters (Figure 33.4b). They trade strong lateral muscles needed in fast current for compressed bodies that enable them to move through beds of aquatic vegetation. Pulmonate snails and burrowing mayflies replace rubble-dwelling insect larvae. Bottom-feeding fish, such as catfish, feed on life in the silty bottom, and back swimmers and water striders inhabit sluggish stretches and still backwaters.

FUNCTION

The flowing-water or lotic system is open and largely heterotrophic (Figure 33.5). A major energy source is detrital material carried to it from the outside. Much of this organic matter input comes as coarse particulate organic matter (CPOM), leaves and woody debris dropped from streamside vegetation, particles larger than 1 mm in size. Another type of organic input is fine particulate organic matter (FPOM), material less than 1 mm in size, including leaf fragments, invertebrate feces, and precipitated dissolved organic matter. A third input is dissolved organic matter (DOM), material less than 0.5 micron in solution. One source of

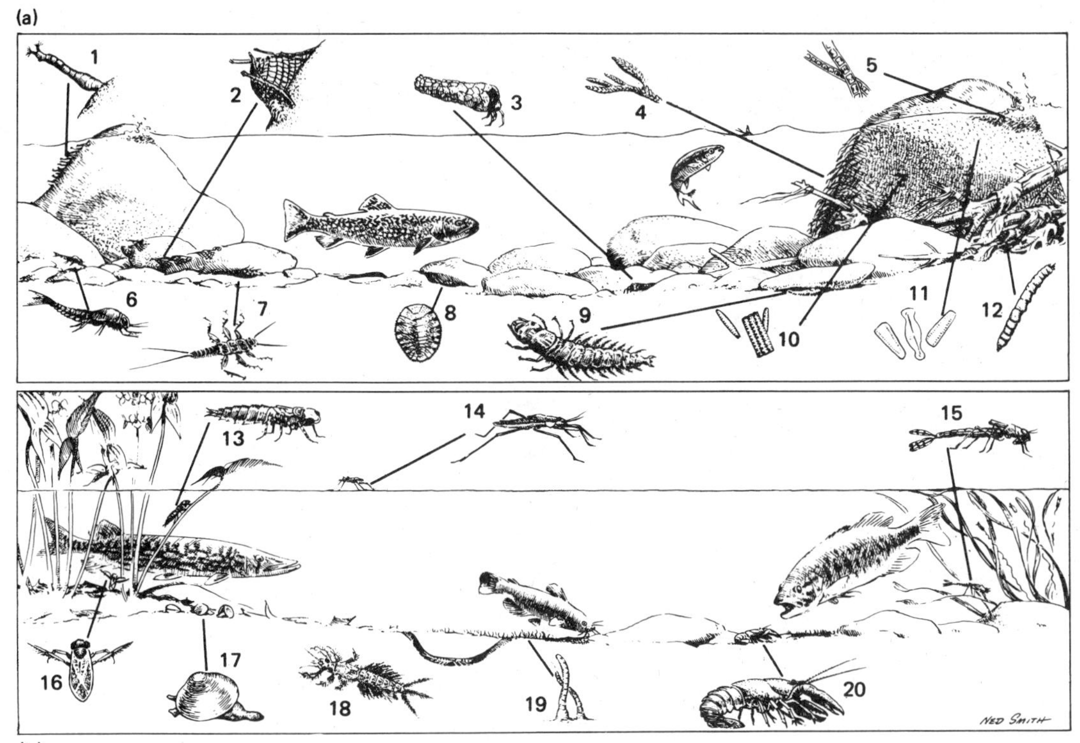

FIGURE 33.4 Comparison of life in a fast stream (a) and a slow stream (b). Fast stream: (1) blackfly larva (Simuliidae); (2) net-spinning caddisfly (*Hydropsyche* spp.); (3) stone case of caddisfly; (4) water moss (*Fontinalis*); (5) algae (*Ulothrix*); (6) mayfly nymph (*Isonychia*); (7) stonefly nymph (*Perla* spp.); (8) water penny (*Psephenus*); (9) hellgrammite (dobsonfly larva, *Corydalis cornuta*); (10) diatoms (Diatoma); (11) diatoms (*Gomphonema*); (12) cranefly larva (Tipulidae). The fish in the fast stream is a brook trout. Slow stream: (13) dragonfly nymph (Odonata, Anisoptera); (14) water strider (*Gerris*); (15) damselfly larva (Odonata, Zygoptera); (16) water boatman (Corixidae); (17) fingernail clam (*Sphaerium*); (18) burrowing mayfly nymph (*Hexegenia*); (19) bloodworm (Oligochaeta, *Tubifex* spp.); (20) crayfish (*Cambarus* spp.). The fish in the slow stream are left to right: northern pike, bullhead, and smallmouth bass.

DOM is rainwater dripping through overhanging leaves, dissolving the nutrient-rich exudates on them. Other DOM input comes by a geological pathway through subsurface seepage, which brings nutrients leached from adjoining forest, agricultural, and residential lands. Many streams receive inputs from mechanical pathways through the dumping of industrial and residential effluents. Supplementing this detrital input is autotrophic production in streams by diatomaceous algae growing on rocks and by rooted aquatics such as water moss. Energy is lost through two pathways: geological (through streamflow feeding downstream systems) and biological (from respiration).

FOOD WEBS

The processing of this organic matter involves both physical and biological mechanisms (Figure 33.6). In fall, leaves drift down from overhanging trees, settle on the water, float downstream, and lodge against banks, debris, and stones. Soaked with water, the leaves sink to the bottom, where they quickly lose 5 to 30

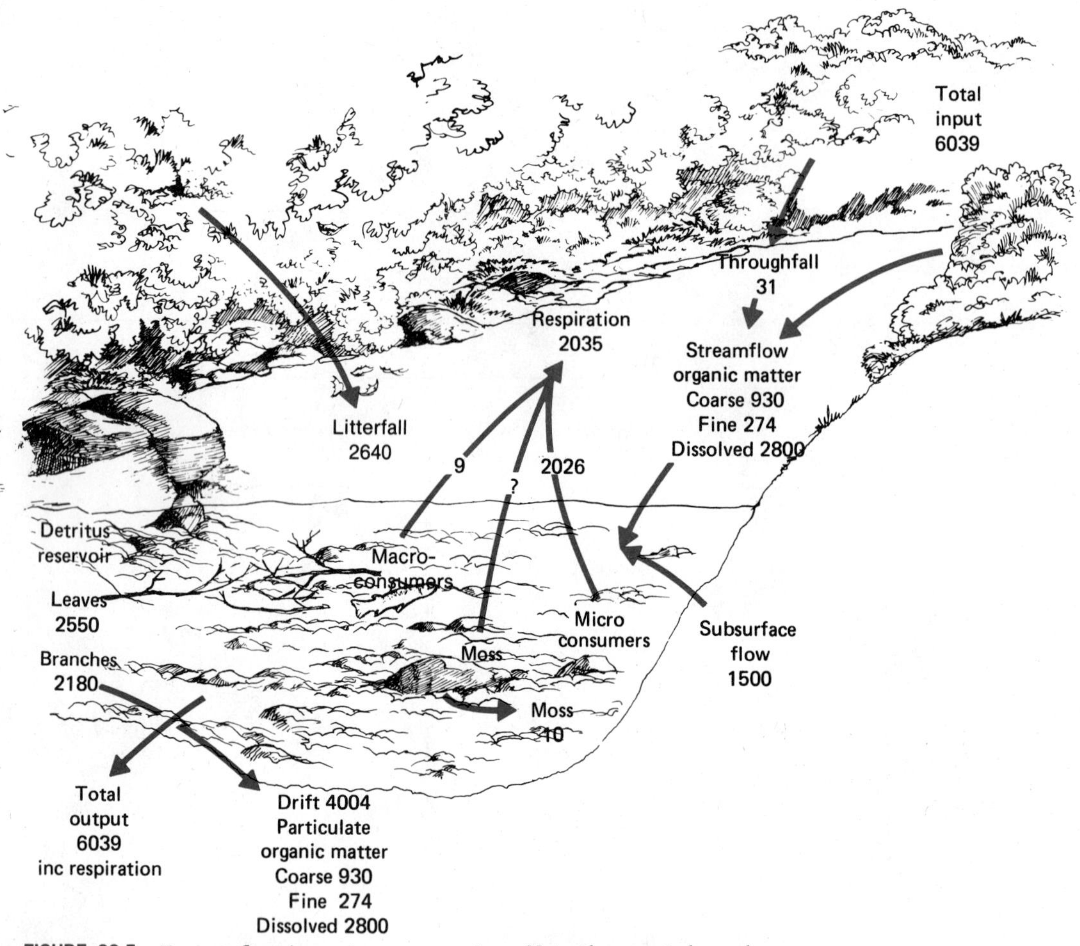

FIGURE 33.5 Energy flow in a stream ecosystem. Note the great dependence on materials from terrestrial sources and inflow from upstream and the role of coarse, fine, and dissolved organic matter. Primary production contributes little to energy flow. Energy values in kcal/m^2/yr, are based on Bear Brook, Hubbard Forest, New Hampshire.

percent of their dry matter as water leaches soluble organic matter from their tissues. Much of this DOM is either incorporated onto detrital particles or precipitated to become part of the FPOM. Another part is incorporated into microbial biomass.

Within a week or two, depending on the temperature, the surface of the leaves is colonized by bacteria and fungi, largely aquatic hyphomycetes. Fungi are more important on CPOM because large particles offer more surface for mycelial development. Bacteria are associated more with FPOM. Microorganisms degrade cellulose and metabolize lignin. Their populations form a layer on the surface of leaves and detrital particles that is much richer nutritionally than the detrital particles themselves. Leaves and other detrital particles are attacked by a major feeding group, the *shredders,* insect larvae that feed on coarse particulate organic matter. Among these shredders are craneflies (Tipulidae), caddisflies (Trichoptera), and stoneflies (Plecoptera). They break down the CPOM, feeding on the material not so much for the energy it contains but for the bacteria and fungi growing on it. Shredders assim-

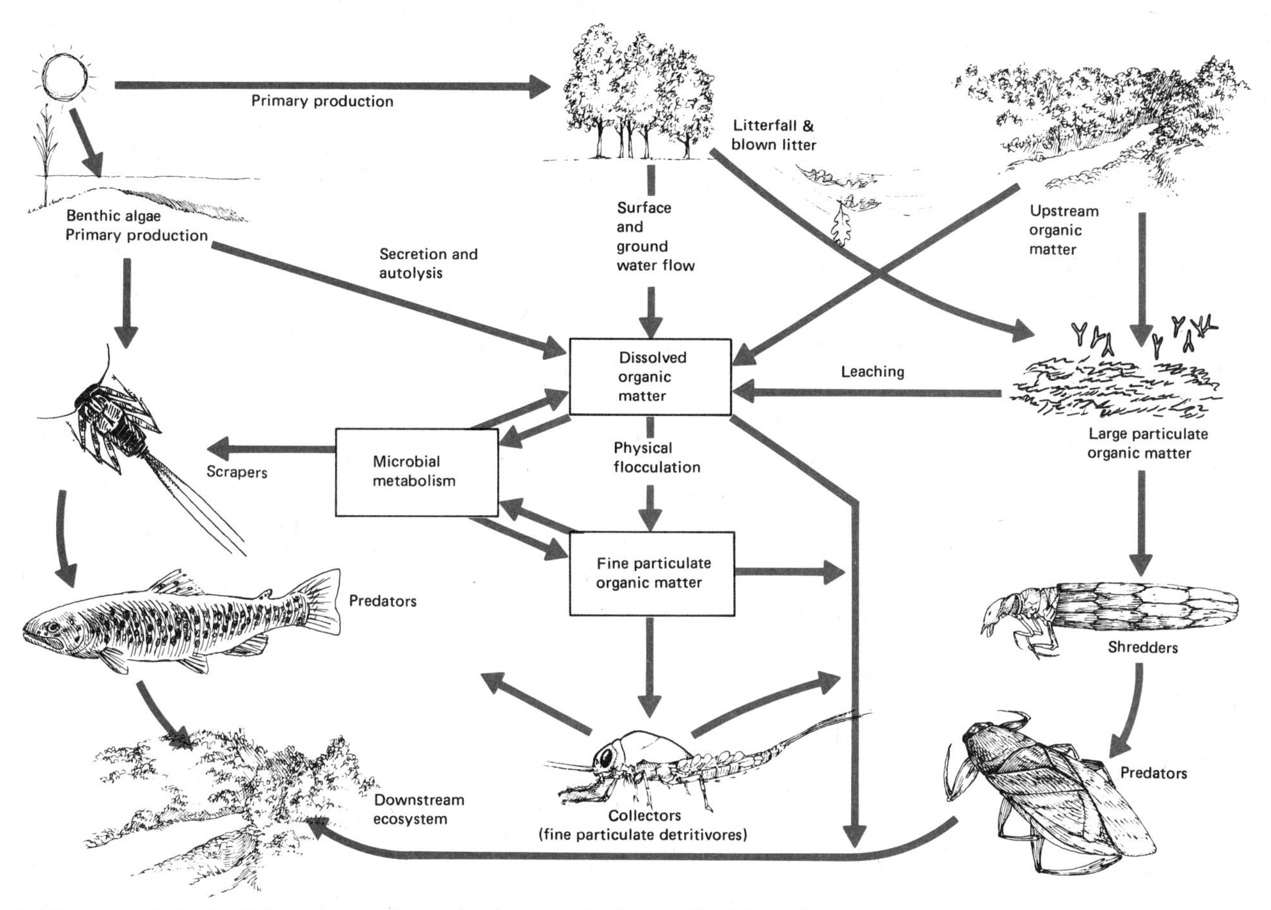

FIGURE 33.6 A food web in a stream. Processing leaves and other particulate matter and dissolved organic matter are bacteria and fungi, and functional groups of organisms, mostly invertebrates: the shredders, collectors, grazers, and predators.

ilate about 40 percent of the material they ingest and pass off 60 percent as feces.

Broken up by the shredders and partially decomposed by microbes, the leaves, along with invertebrate feces, become part of the FPOM, which also includes some precipitated DOM. Drifting downstream and settling on the stream bottom, FPOM is picked up by another feeding group of stream invertebrates, the *filtering* and *gathering collectors.* The filtering collectors include, among others, the larvae of blackflies (Simulidae), with filtering fans, and net-spinning caddisflies, including *Hydropsyche.* Gathering collectors, such as larvae of midges, pick up particles from stream bottom sediments. Collectors obtain much of their nutrition from bacteria associated with the fine detrital particles.

While shredders and collectors feed on detrital material, another group feeds on the algal coating of stones and rubble. These are the *grazers,* which include the beetle larvae, water penny (*Psephenus* spp.), and a number of mobile caddisfly larvae. Much of the material they scrape loose enters the drift as FPOM. Another group, associated with woody debris, are the *gougers,* invertebrates that burrow into water-logged limbs and trunks of fallen trees.

Feeding on the detrital feeders and grazers are predaceous insect larvae such as the powerful dobsonfly larvae (*Corydalus cornutus*) and fish such as the sculpin (*Cottus*) and trout. Even these predators do not depend solely on aquatic insects; they also feed heavily on terrestrial invertebrates that fall or are washed into the stream.

Because of current, quantities of CPOM, FPOM, and invertebrates tend to drift downstream to form a sort of traveling benthos. This is a normal process in streams, even in the absence of high water and abnormal currents. Drift is so characteristic of streams that a mean rate of drift can serve as an index of the production rate of a stream.

ENERGY FLOW AND NUTRIENT CYCLING

Energy flow in lotic ecosystems has been documented for only a few streams. One energy budget is for the well-studied, small, forested Bear Brook in Hubbard Forest of northern New Hampshire. That budget is summarized in Figure 31.5. Over 90 percent of the energy input came from the surrounding forested watershed or from upstream. Primary production by mosses accounted for less than 1 percent of the total energy supply. Algae were absent from the brook. Inputs from litter and throughfall accounted for 44 percent of the energy supply, and geological inputs from subsurface flows accounted for 56 percent. Energy was introduced in three forms: CPOM represented by leaves and other debris; FPOM represented by drift and small particles; and DOM. In Bear Brook 83 percent of input from surface and subsurface flow and 47 percent of the total energy input was in the form of DOM. Sixty-six percent of the organic input was exported downstream, leaving 34 percent to be utilized locally.

Although nutrient cycling is downhill in all ecosystems, the problem in flowing water is how to keep nutrients upstream and reduce losses to downstream. Nutrients in terrestrial and lentic systems are recycled more or less in place. An atom of nutrients passes from soil or water column to plants and consumers back to soil or water in the form of detrital material or exudates. Then it is recycled within the same segment of the system, although losses do occur. Cycling essentially involves time. Flowing water has an added element, spatial cycle. Nutrients in the form of DOM and POM are constantly being carried downstream. How quickly these materials are carried downstream depends on how fast the water moves and what physical and biological means hold nutrients in place. Physical retention involves storage in wood detritus such as logs and snags in the stream, accumulation of debris in pools formed behind logs and boulders, leaf sediments, and beds of macrophytes. Biological retention is uptake and storage in animal and plant tissue for later recycling.

The processes of recycling, retention, and downstream displacement may be pictured as a spiral lying longitudinally in a stream (Figure 33.7). *Spiraling* combines nutrient cycling and downstream transport. One cycle in the spiral involves the uptake of an atom or nutrient from DOM, its passage through the food chain, and its return to water, where it is available for reuse. Thus one cycle or loop begins with the water compartment and ends with return to the same. Spiraling is measured as the distance needed for completion of one cycle. The longer the distance required, the more open the spiral; the shorter the distance, the tighter the spiral. If leafy detritus can be physically held in place long enough to allow the biological component of the stream, especially the shredders and microbes, to process the organic matter, then the spiral may be tight. This retention is especially important in fast headwater streams, which can rapidly lose unprocessed particulate organic matter downstream.

At the same time, however, shredders contribute to downstream flow of nutrients by fragmenting CPOM and excreting fecal material as FPOM, which joins with invertebrate drift and algal growth torn from the stream bottom. By trapping some of this FPOM and consuming it in place, collector organisms, especially net-spinning filter feeders, tighten the spiral.

Ecologists at Oak Ridge, Tennessee, experimentally determined how quickly one nutrient, exchangeable phosphorus in the form of $^{32}PO_4$, moved downstream in a small woodland brook, Walker Branch. The tagged P moved downstream at the rate of 10.4 m a day and cycled once every 18.4 days. Thus the average downstream distance of one spiral was 190 m. In other words, one atom of P on the average completed one cycle from the water compartment and back again for every 190 m of downstream travel. The spiraling length was partitioned into an uptake length of 165 m, asso-

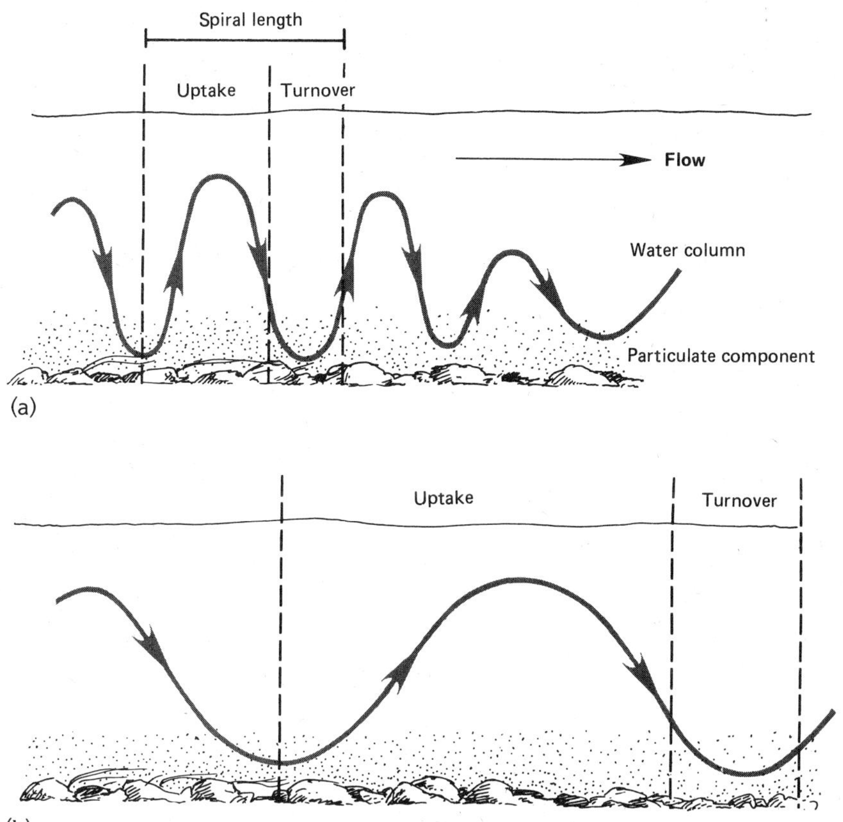

FIGURE 33.7 Nutrient spiraling between particulate organic matter, including microbes, and the water column in a lotic ecosystem. Uptake and turnover take place as nutrients flow downstream. (a) Tight spiraling. (b) Open spiraling. The tighter the spiraling, the longer the nutrients remain in place.

ciated with transport in the water column, mostly as DOM; a particulate turnover length of 25 m, associated with FPOM; and a consumer turnover length of 0.05 associated with consumer drift. CPOM accounted for 60 percent of the uptake, FPOM 35 percent, and aufwuchs 5 percent. Turnover time of P in CPOM ranged from 5.6 to 6.7 days, and in FPOM 99 days. These data indicated that phosphorus remained in a segment of the stream for that period of time. Only 2.8 percent of P uptake from particulate matter was transferred to consumers; most of the P was transferred back to water. About 30 percent of the consumer uptake was transferred to predators.

The bottom and width influence overall production of a stream. Pools with sandy bottoms are the least productive because they offer little substrate for the aufwuchs. Bedrock, although a solid substrate, is so exposed to currents that only the most tenacious organisms can maintain themselves. Gravel and rubble bottoms support the most abundant life because they provide the greatest surface area for aufwuchs, offer many crannies and protected places for insect larvae, and are the most stable. Food production decreases as the particles become larger or smaller than rubble. Bottom production in streams 6 m wide decreases by one-half from sides to center, and in streams 30 m wide, it decreases by one-third. Streams 2 m or less in width are four times as rich in bottom organisms as those 6 to 7 m wide. That is one reason why headwater streams make such excellent trout nurseries.

THE RIVER CONTINUUM

From its headwaters to its mouth the lotic ecosystem is a continuum of changing environmental conditions (Figure 33.8). Headwater streams (stream orders 1 to 3) are usually swift, cold, and in forested regions shaded. They are strongly heterotrophic, heavily dependent on the input of detritus from terrestrial streamside vegetation, which contributes more than 90 percent of the organic input. Even when headwater streams are exposed to sunlight and autotrophic production exceeds heterotrophic inputs, organic matter produced invariably enters detrital food chains. Dominant organisms are shredders, processing large-sized litter and feeding on CPOM, and collectors, processors of FPOM. Populations of grazers are minimal, reflecting the small amount of autotrophic production, and predators are mostly small fish—sculpins, darters, and trout. Headwater streams, then, are accumulators, processors, and transporters of particulate organic matter of terrestrial origin. As a result the ratio of gross primary production to community respiration is less than 1. As streams increase in width to medium-sized creeks and rivers (orders 4 to 6), the importance of riparian vegetation and its detrital input decreases. Exposed to the sun, water temperature increases; and as the gradient declines, the current slows. These changes bring about a shift from a dependence on a terrestrial input of particulate organic matter to primary production by algae and rooted aquatic plants. Gross primary production now exceeds community respiration. Because of the lack of CPOM, shredders disappear, and collectors, feeding on FPOM transported downstream, and grazers, feeding on autotrophic production, become the dominant consumers. Predators show little increase in biomass but shift from cold-water species to warm-water species, including bottom-feeding fish such as suckers and catfish.

As the stream order increases from 6 through 10 and higher, riverine conditions develop. The channel is wider and deeper. The volume of flow increases, and the current becomes slower. Sediments accumulate on the bottom. Both riparian and autotrophic production decrease, with a gradual shift back to heterotrophy. A basic energy source is FPOM, utilized by bottom-dwelling collectors, now the dominant consumers. However, slow, deep water and DOM support a minimal phytoplankton and associated zooplankton population.

Throughout the downstream continuum, the lotic community capitalizes on upstream feeding inefficiency. Downstream adjustments in production and the physical environment are reflected in changes in consumer groups (Figure 33.8). Through the continuum the lotic ecosystem achieves some balance between the forces of stability, such as natural obstructions in flow that aid in the retention of nutrients upstream, and the forces of instability, such as flooding, drought, and temperature fluctuations.

HUMAN IMPACT

POLLUTION

For centuries humans have used streams and rivers as depositories of human, industrial, and solid wastes with the idea that these materials would be diluted and carried downstream. So pervasive has that idea been that few larger streams and rivers (and not many smaller streams) have escaped pollution. The magnitude of ecological changes brought about depends on the type of pollutant and its quantity both in time and space.

Industrial pollution of large streams and rivers is the most serious because of its concentration and the chemical complexity of many materials involved. Water withdrawn from rivers for cooling in power plants and certain industrial processes and then returned heated raises river temperatures and lowers dissolved oxygen. Water used for flushing and chemical treatment and returned to the river imparts bad tastes and odors and introduces toxic substances that affect downstream use. Discharges from chemical plants and sulfurous wastes

FIGURE 33.8 The lotic system is a continuum from headwater to the river's mouth. The headwater stream is strongly heterotrophic, dependent on terrestrial input of detritus. As stream size increases, the input of organic matter shifts from particulate organic matter to primary production by algae and rooted vascular plants. This shift is influenced by shading. As the stream grows into a river, the lotic system shifts back to heterotrophy.

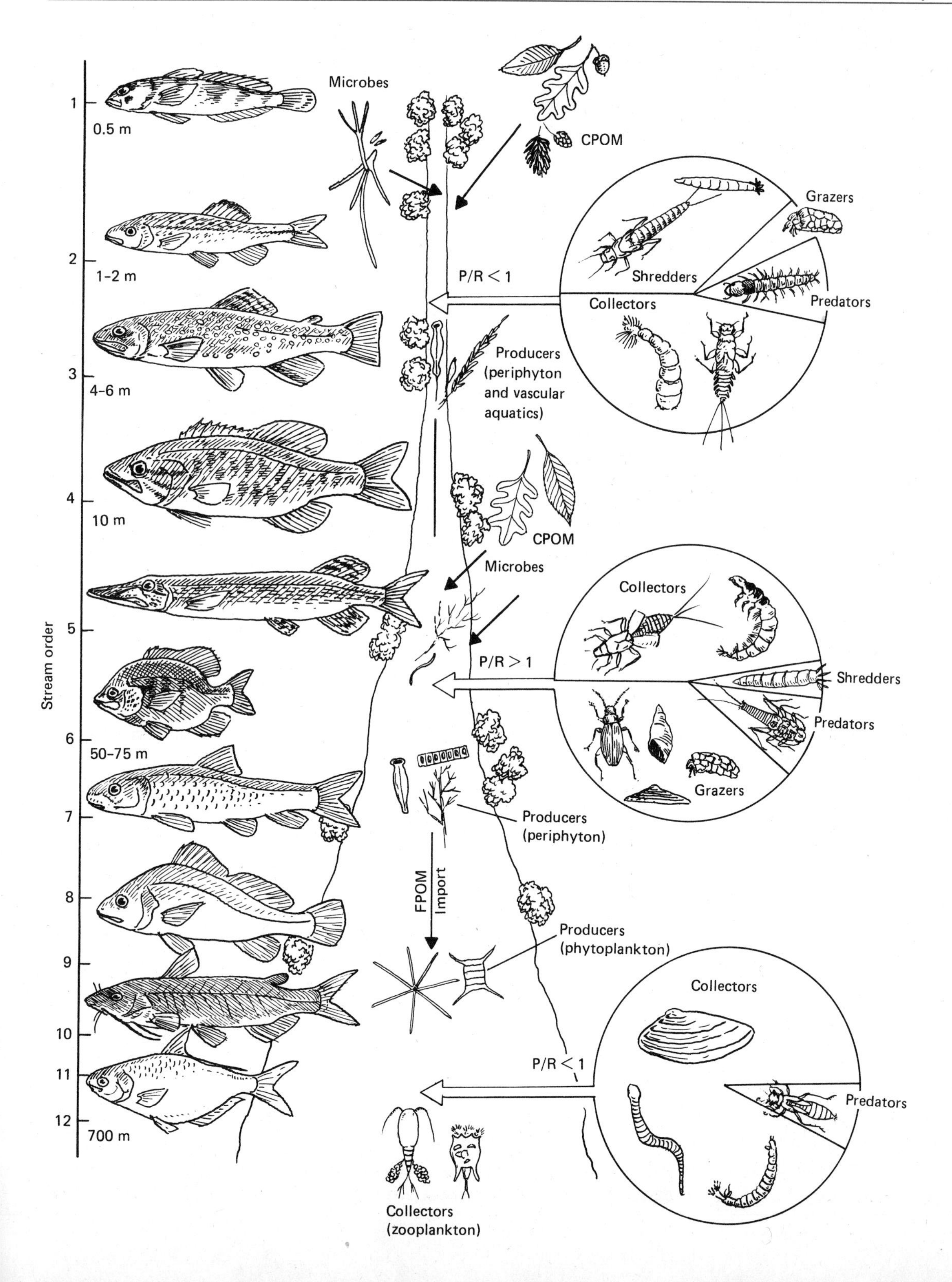
Stream order
1
2
3
4
5
6
7
8
9
10
11
12
0.5 m
1-2 m
4-6 m
10 m
50-75 m
700 m
Microbes
CPOM
Grazers
Shredders
P/R < 1
Collectors
Predators
Producers
(periphyton
and vascular
aquatics)
CPOM
Microbes
Collectors
P/R > 1
Shredders
Predators
Grazers
Producers
(periphyton)
FPOM
Import
Producers
(phytoplankton)
Collectors
P/R < 1
Predators
Collectors
(zooplankton)

from pulp and paper mills are highly poisonous to aquatic life. Many chemical wastes, perhaps harmless alone, react with other chemicals to produce highly toxic conditions. Sudden influxes of such chemicals have frequently caused spectacular and tragic kills of fish and other aquatic life. Acid water from both deep and surface coal mines has destroyed stream life in coal country and has so reduced bacterial activity that biological purification of sewage and other organic wastes in water is impeded.

Radioactive wastes from uranium mills and nuclear power plants that find their way into streams do not break down naturally but do become increasingly diluted as they move downstream from the source. Some radioactive materials are deposited on the bottom, while others are taken up by aquatic organisms, both plant and animal. Even minute amounts of radioactive substances may become concentrated many thousands of times in living issue and be passed along in the food chain (Chapter 25).

Raw and inadequately treated sewage poured into streams and rivers sharply changes biological conditions. Even effluents from sewage treatment plants can upset ecological stability. As sewage enters a stream, it is dispersed and the solids settle to the bottom, where they are attacked by aerobic bacteria. This bacterial activity depletes oxygen, but the loss is offset by the absorption of more oxygen from the air into the stream. Streams can purify themselves by natural, bacterial breakdown of organic matter. The time required depends on the degree of pollution and the character of the stream. A fast-flowing stream constantly saturated with oxygen can purify itself much faster than a slow stream, which does not have the luxury of rapid oxygenation.

The carbon dioxide and hydrogen content of the water is high at the point of discharge. This condition eliminates normal stream life, particularly vertebrates and mollusks, and replaces them with a new group of dominant organisms, including protozoans, mosquito larvae, and tubifex worms. Below this zone of active decomposition, flowing waters dilute the pollutants. Although conditions improve, the stream still is far from normal. Green algae are present, but reduced in numbers; bacteria are abundant and oxygen low. Downstream the pollutants are diluted further, dissolved oxygen is higher, and organisms tolerant of such conditions, such as carp, catfish, chironomid larvae, and protozoans, inhabit the area. Eventually the water becomes clean and fresh again, and normal populations of fish and invertebrates reappear.

In far too many streams, however, conditions become worse downstream. No sooner has a river somewhat recovered from its polluted condition than another town, city, or industry dumps its sewage. As a result the river carries a load greater than it can handle and aerobic conditions no longer exist. Aerobic bacteria are replaced by anaerobic ones, and normal stream life is destroyed. Putrefying bacteria alone remain, and the stream becomes a foul-smelling, open sewer. Thus pollution by upstream industry and communities severely affects downstream communities. It decreases the quality of water for commercial and domestic use and increases the costs of water filtration and purification.

Siltation, caused by the erosion of farmlands, road construction, surface mining, logging, and other forms of soil disturbance (Figure 33.9), is the most insidious form of pollution, for it is widespread, it often goes unnoticed, and the damage it does is often permanent. Clay soils suspended in water block out light and prevent the growth of aquatic plants. Silt settles on the stream bottom, covering substrate for insect larvae and smothering larvae, mussels, and other bottom organisms. It blankets sewage and other organic material and retains them in place, reducing the oxygen supply. Silt clogs the opercular cavities and gill filaments of fish and the mantles and gills of mollusks, killing them both. Silty water flowing through the gravel nests of trout and salmon causes heavy mortality of eggs. Thousands of miles of trout and salmon streams have been destroyed by siltation, which more than any other cause limits the natural reproduction of these fish.

REGULATED RIVERS

The effects of damming are still worse. As children we enjoy damming up small drainage ditches filled after a rain. As adults we support small to massive schemes to straighten and throw dams across streams and rivers for various reasons. We can trace such damming activities as far back as 5000 years, but the greatest outburst of dam construction began after World War II. It has

FIGURE 33.9 Banks of pasture streams broken down by livestock are an important, but often overlooked, source of siltation in smaller streams.

continued until over 60 percent of the world's streams and rivers have been dammed. Dams' flows are regulated (Figure 33.10), which profoundly affects the lotic systems, their hydrology, ecology, and biology. Life in lotic environments is adapted to natural flow regimes, with their periods of high, low, and average flow. Dams change the environment in which lotic organisms live, more often than not to their detriment.

Under normal conditions free-flowing streams and rivers experience seasonal fluctuation in flow. Snowmelt and early spring rains bring scouring high water; summer brings low water levels that expose some of the stream bed and speed decomposition of organic matter along the edges. Life in the lotic environment has adapted to these seasonal changes. Damming a river or stream interrupts both nutrient spiraling from upstream and the lotic continuum. Downstream flow is greatly reduced as a pool of water fills behind the dam, developing characteristics similar to those of a natural lake, yet retaining some features of the lotic system, such as a constant inflow of water. Heavily fertilized by the decaying detrital material on the newly flooded land, the lake develops a heavy bloom of phytoplankton and in tropical regions dense growths of floating plants. Species of fish, often introduced exotics, adapted to lakelike conditions replace fish of flowing water.

The type of pool allowed to develop depends on the purpose of the dam and has a strong effect on downstream conditions. Some are single-purpose dams for flood control or water storage; others are multiple-purpose dams, providing hydroelectric power, irrigation water, and recreation, among other uses. Flood-control dams have a minimum pool; the dam fills only during a flood, at which time inflow exceeds outflow. Engineers release the water slowly to minimize downstream flooding. In time water behind the dam recedes to original pool depth. During flood and postflood

FIGURE 33.10 A series of locks and dam, such as this one, regulates the depth and flow of rivers to convert unnavigable rivers into navigable ones. Such dams can serve a dual function as small hydroelectric dams.

periods the river below carries a strong flow for some time, scouring the riverbed. During normal times, flow below the dam is stabilized. If the dam is for water storage, the reservoir holds its maximum pool; but during periods of water shortage and drought, drawdown of the pool can be considerable, exposing large expanses of shoreline for a long time and stressing or killing littoral life. Only a minimal quantity of water is released downstream, usually an amount required by law, if such exists. Hydroelectric and multiple-purpose dams hold a variable amount of water, determined by consumer needs. During periods of power production pulsed releases are strong enough to wipe out or dislodge benthic life downstream, which under the best of conditions has a difficult time becoming established.

Reservoirs with a large pool of water become stratified, with a well-developed epilimnion, metalimnion, and hypolimnion (see Chapter 31). If water is discharged from the upper layer of the reservoir, the effect of the flow downstream is similar to that of a natural lake. Warm, nutrient-rich, well-oxygenated water creates highly favorable conditions for some species of fish below the spillway and on downstream. If the discharge is from the cold hypolimnion, downstream receives cold, oxygen-poor water carrying an accumulation of iron and other minerals and a concentration of soluble organic materials. Such conditions inimical to stream life may persist for hundreds of kilometers downstream before the river reaches anything near normal conditions. Gated selective withdrawal structures or induced artificial circulation to increase oxygen concentration reduce such problems at some dams.

Impacts of dams on lotic systems are compounded when a number of multipurpose dams are built on a river. The amount of water released and moving downstream becomes less with each dam until eventually all available water is consumed and the river simply dries up. That is the situation on the Colorado River, the most regulated river in the world. The river is nearly dry by the time it reaches Mexico.

Effects of dams go beyond simply changing the nature of the lotic system. Large dams on such rivers as the Columbia in North America interfere with the migratory patterns of anadromous fish such as salmon. Dams obstruct the upstream movement of fish. Although fish ladders are of assistance, many local populations have been excluded from their traditional spawning streams. Even if spawning is successful, unnatural timing of high and low flows in the rivers may induce premature seaward migration of the young or lengthen their downstream passage, exposing them to high temperatures. Because of these environmental conditions and losses caused by passage through hydroelectric dams, 90 percent of juveniles perish on the journey from their home stream to the estuary.

Dams have an impact on the economic, social, and cultural patterns of human society. Although dams provide many economic benefits—at great ecological cost—such as power, water, industrial and agricultural developments, and some fisheries, they have exacted some great economic and human costs. Communities have been displaced from lands to be flooded, and historic sites and productive agricultural lands have been covered with water. Sediments that might have built up floodplains or seasonally fertilized croplands, as the annual flooding of the Nile once did, remain behind the dam, lowering agricultural production. Reduction of freshwater inflow with its rich supply of nutrients impoverishes estuaries and allows salt water to intrude upriver, destroying or greatly reducing the estuarine fishery resource. In tropical regions dams, with their hundreds of miles of shoreline, provide excellent conditions for vectors of such parasitic disease as malaria, arborviruses, and schistosomiasis. Schistosomiasis, the intermediate host of which is an aquatic snail, is picked up by humans, the definitive host, when they play, wash, bathe, walk, or work in infested waters. The problem is especially prevalent and widespread in tropical Africa. Because of the great ecological changes brought on by damming rivers and streams that affect human health and welfare, and because of the increasing rareness of free-flowing streams with their own valuable ecological and economic contributions, a moratorium on dam building may be a wise decision.

Associated with dam building is stream and river channelization, the dredging and straightening of streams and rivers for flood control, navigation, and agricultural development. Thousands of kilometers of meandering, productive, fish-filled streams and rivers have been channeled into sterile, unattractive drainage ditches. Such channelization has eliminated streamside vegetation, important as wildlife habitat, and destroyed associated wetlands (see Chapter 32). Newly cut channels support little bottom fauna, and fish lack food, shelter, and breeding sites. Paradoxically, channelization, designed to speed water from the uplands to the rivers and sea, actually intensifies downstream flooding because it increases the volume and rapidity of flow.

Summary

The characteristics that set flowing-water ecosystems apart from other aquatic systems are their dependence on detrital material from terrestrial sources and their currents. Current shapes the nature of life in streams and rivers and carries nutrients and other materials downstream. Lotic systems exhibit a continuum of physical and ecological conditions from the source to the mouth. There is a longitudinal gradient in temperature, depth and width of the channel, velocity of the current, and nature of the bottom. Headwater streams are strongly heterotrophic and dependent on inputs of detritus. They are inhabited by organisms that are well adapted to life in the current. They may be streamlined in shape, flattened to conceal themselves in crevices and underneath rocks, or attached to rocks and other substrate. Larger streams, open to sunlight, shift from a heterotrophic to an autotrophic condition. Primary production from algae and rooted aquatics becomes an important energy source. Large rivers return to a heterotrophic condition. They are dependent on fine particulate organic matter and dissolved organic matter as sources of nutrients and energy. Downstream systems in effect depend on the inefficiencies of energy and nutrient processing upstream. This inefficiency develops because the current keeps moving nutrients downstream.

Energy comes in three detrital fractions: coarse particulate organic matter (CPOM), fine particulate organic matter (FPOM), and dissolved organic matter (DOM), as well as from autotrophs, algae on stones, and rooted aquatics. Processing this organic matter are fungi and bacteria feeding on CPOM; shredders feeding on CPOM and its associated bacteria and fungi; collectors feeding on bacteria and FPOM carried by the current; scrapers working on algae; and piercers utilizing plant juices of rooted aquatics. The ratio of these feeding groups changes along the lotic continuum. A major problem in flowing-water ecosystems is the retention of nutrients in any segment of the system. Nutrients cycle among particulate matter, the water column, and consumers as they move downstream. This concurrence of nutrient cycling and downstream transport is called spiraling.

The integrity of flowing-water ecosystems has been damaged by pollution from human, industrial, and toxic wastes, by stream channelization to speed up the flow of water, and by dams that impound water and interrupt the lotic continuum. The impact of dams on downstream ecosystems and their biota has been pronounced, completely changing the character of affected streams and rivers. Dams hold back sediments from floodplains and riparian habitats, increase and decrease the intensity and volume of flow, and change fish and invertebrate populations, eliminating some species and benefiting others. Impacts vary, depending on the purpose and management of the dams.

Review and Study Questions

1. What physical characteristics are unique to flowing-water ecosystems?
2. In what way are stream organisms adapted to living in flowing water? How do adaptations change as fast streams become slow?
3. What is the basic energy source of headwater streams?
4. Characterize the major functional groups of stream invertebrates and describe their role in the food web of streams.
5. What is spiraling, and how does it function in nutrient cycling in streams? What compartments and consumer groups are involved?
6. How do downstream lotic systems relate to upstream systems? What is the continuum concept?
7. How does stream channelization—the straightening and deepening of a streambed—affect the structure and function of a lotic system?

*8. Determine the extent of stream channelization in your area. What was the reason for channelization? Use an old aerial photograph to assess the amount of change. What was lost? What steps could be taken to minimize the effects of stream channelization? (See, for example, Gorr and Petts 1989.)

*9. What dams exist in your area? What are their types? What have been the ecological and economic benefits? What were the ecological costs of their construction? What are downstream conditions?

*10. Each major dam in the world has its own characteristics and impacts. Select one major dam—for example, the dams on the Columbia River, the Glen Canyon Dam, the Aswan High Dam in Egypt, the Ord River Dam of tropical Australia, the Kariba Dam on the Zambezi River, or the large dams on the River Volga in the USSR—for a detailed report. How is it harmful? How is it helpful?

Selected References

Cummins, K. W. 1974. Structure and function of stream ecosystems. *Bioscience,* 24:631–641.

Cummins, K. W. 1979. Feeding ecology of stream invertebrates. *Ann. Rev. Ecol. Syst.,* 10:147–172. A good review of functional aspects of stream organisms.

Gore, J. A., and G. E. Petts. 1989. *Alternatives in regulated river management.* Boca Raton, FL: CRC Press. Focuses attention on ways to reduce ecological effects of river regulation.

Hynes, H. B. N. 1970. *The ecology of running water.* Toronto: University of Toronto Press. Dated, but a classic and valuable work; a major reference.

Meyer, J. L. 1990. A blackwater perspective on riverine ecosystems. *Bioscience,* 40:643-651. A more detailed look at stream ecosystem function, especially food webs, than can be presented in this chapter.

Petts, G. E. 1984. *Impounded rivers; Perspectives for ecological management.* New York: Wiley. Detailed analysis of the effects of dams on the world's rivers, especially downstream.

Stanford, J. A., and A. P. Covich, eds. 1988. Community structure and function in temperate and tropical streams. *J. North Amer. Benthol. Soc.,* 7:261–529. A valuable special issue of the journal.

Vannote, R. L., G. W. Minshall, K. W. Cummins, J. R. Sedell, and C. E. Cushing. 1980. The river continuum concept. *Can. J. Fish. Aquat. Sci.,* 37:130–137.

Ward, J. V. 1979. *The ecology of regulated streams.* New York: Plenum.

Whitten, B. A., ed. 1975. *River ecology.* Berkeley: University of California Press.

CHAPTER 34

Oceans

Outline

Objectives

On completion of this chapter, you should be able to:

1. Discuss the salinity of the oceans.
2. Describe the formation and types of waves and currents in the ocean.
3. Explain the cause of tides.
4. Describe the major zones in the sea and their relationship to temperature stratification.
5. Discuss the role of phytoplankton, zooplankton, and nekton in nutrient cycling in the open sea.
6. Discuss the effect of human activities on the structure and function of the open sea.

Freshwater rivers eventually empty into the oceans, and terrestrial ecosystems end abruptly at the edge of the sea. For some distance there is a region of transition. Rivers enter the saline waters of the ocean, creating a gradient of salinity. That gradient provides a habitat for organisms uniquely adapted to the half-world between salt water and fresh. The coastal regions exposed to the open sea are inhabited by other organisms able to live in the often severe environments dominated by tides. Beyond lies the open ocean—shallow seas overlying continental shelves and the deep oceans.

FEATURES OF THE MARINE ENVIRONMENT

The marine environment is marked by a number of differences from the freshwater world. It is large, occupying 70 percent of Earth's surface, and it is deep, in places nearly 7 kilometers. The surface area lighted by the sun is small compared to the total volume of water. This small volume of sunlit water and the dilute solution of nutrients limit primary production. All of the seas are interconnected by currents, dominated by waves, influenced by tides, and characterized by salinity, restricting life (see Chapter 6).

SALINITY

The salinity of the open sea is fairly constant, averaging about 35 parts per thousand (‰). Two elements, sodium and chlorine, make up some 86 percent of sea salt. These, along with other major elements such as sulfur, magnesium, potassium, and calcium, whose relative proportions vary little, comprise 99 percent of sea salts. Determination of the most abundant element, chlorine (see Table 34.1), is used as an index of salinity of a given volume of sea water. Salinity is expressed in ‰ as the amount of chlorine in grams in a kilogram of seawater.

The salinity of parts of the ocean is variable because of physical processes. Salinity is affected by evaporation and precipitation, most pronounced at the interface of sea and air; by the movement of water masses; by the mixing of water masses of different salinities, especially near coastal areas; by the formation of insoluble precipitates that sink to the ocean floor; and by the diffusion of one water mass to another.

TABLE 34.1 Composition of Seawater of 35‰ Salinity,* Major Elements

Elements	*Grams/kg*	*Millimoles/kg*	*Milliequivalents/kg*
Cations			
Sodium	10.752	467.56	467.56
Potassium	0.395	10.10	10.10
Magnesium	1.295	53.25	106.50
Calcium	0.416	10.38	20.76
Strontium	0.008	0.09	0.18
			605.10
Anions			
Chlorine	19.345	545.59	545.59
Bromine	0.066	0.83	0.83
Fluorine	0.0013	0.07	0.07
Sulphate	2.701	28.12	56.23
Bicarbonate	0.145	2.38	—
Boric acid	0.027	0.44	—
			602.72

* Chlorinity can be converted to salinity, the total amount of solid matter in grams per kilogram of seawater. The relationship of salinity to chlorinity is expressed as follows:

$$S(‰) = 1.80655 \times \text{chlorinity}$$

Note: Surplus of cations over strong anions (alkalinity):2.38.

The elements most affected by these physical processes are the conservative ones not involved in biological processes. The most variable elements in the sea are the nonconservative ones, such as phosphorus and nitrogen, because their concentrations are related to biological activity. Taken up by organisms, these elements are usually depleted near the surface and enriched at lower depths. In parts of the ocean, some of these nutrients are returned by upwelling.

TEMPERATURE AND PRESSURE

What has already been written about temperature in fresh water (Chapter 31) also applies to the sea. The range of temperature is far less than that on land, although it is considerable—from −9° C in arctic waters to 27° C in tropical waters. In general, seawater is never more than 2° to 3° below the freezing point of fresh water nor warmer than 27° C. At any given place the temperature of deep water is almost constant and cold, below the freezing point of fresh water. Seawater has no definite freezing point, although there is a temperature for seawater of any given salinity at which ice crystals form. Thus, pure water freezes out, leaving even more saline water behind. Eventually, it becomes a frozen block of mixed ice and salt crystals. With rising temperatures the process is reversed.

Unlike fresh water, seawater (with a salinity of 24.7‰ or higher) becomes heavier as it cools and does not reach its greatest density at 4° C. Thus, the limitation of 4° C as the temperature of bottom water does not apply to the sea. The temperature of the sea bottom generally averages around 2° C even in the tropics if the water is deep enough. The temperature of the ocean floor over 1 km deep is 3° C.

Another aspect of the marine environment is pressure. Pressure in the ocean varies from 1 atmosphere at the surface to 1000 atmospheres at the greatest depth. Pressure changes are many times greater in the sea than in terrestrial environments, and pressure has a pronounced effect on the distribution of life. Certain organisms are restricted to surface waters, where the pressure is not so great, while others are adapted to life at great depths. Some marine organisms, such as the sperm whale and certain seals, can dive to great depths and return to the surface without difficulty.

WAVES AND CURRENTS

Waves are generated by wind on the open sea. The frictional drag of the wind on the surface of smooth water ripples the water. As the wind continues to blow, it applies more pressure to the steep side of the ripple, and wave size begins to grow. As the wind becomes stronger, short, choppy waves of all sizes appear; and as they absorb more energy, they continue to grow. When the waves reach a point at which the energy supplied by the wind is equal to the energy lost by the breaking waves, they become whitecaps. Up to a certain point, the stronger the wind, the higher the waves.

The waves that break on a beach are not composed of water driven in from distant seas. Each particle of water remains largely in the same place and follows an elliptical orbit with the passage of the wave form. As a wave moves forward it loses energy to the waves behind and disappears, its place taken by another. Thus, the swells that break on a beach are distant descendants of waves generated far out at sea.

As the waves approach land, they advance into increasingly shallow water. The height of each wave rises until the wave front grows too steep and topples over. As the waves break on shore, they dissipate their energy, pounding rocky shores or tearing away sandy beaches at one point and building up new beaches elsewhere.

Surface waves are the most obvious ones, but in the ocean there are also internal waves. Similar to surface waves, internal waves appear at the interface of layers of waters of different densities. In addition, there are stationary waves or seiches.

Just as there are internal waves, so there are internal currents in the sea. Surface currents are produced by wind, heat budgets, salinity, and the rotation of Earth (see Chapter 4). Water moving in surface currents must be replaced by a corresponding inflow from elsewhere. Because the surface waters are cooled and salinity changes, high-density water formed on the surface, largely at high latitudes, sinks and flows toward low latitudes. These currents are subject to the Coriolis effect (see Chapter 4) and are deflected or obstructed by submarine ridges and modified by the presence of other water masses. The result is three main systems of subsurface water movements: the bottom, the deep, and the intermediate ocean currents, each of which runs counter to the others.

In coastal regions, winds blowing parallel to the coast cause surface waters to be blown offshore. This water is replaced by water moving upward from the deep, a process known as *upwelling.* Although cold and containing less oxygen, upwelling water is rich in nutrients that support an abundant growth of phytoplankton. For this reason, regions of upwellings are highly productive, teeming with fish and bird life.

TIDES

The gravitational pulls on Earth of the sun and the moon each cause two bulges in the waters of the oceans. The two caused by the moon occur at the same time on opposite sides of Earth on an imaginary line extending from the moon through the center of Earth. The tidal bulge on the moon side is due to gravitational attraction; the bulge on the opposite side occurs because the gravitational force there is less than at the center of Earth. As Earth rotates eastward on its axis, the tides advance westward. Thus, any given place on Earth will in the course of one daily rotation pass through two of the lunar tidal bulges, or high tides, and two of the lows, or low tides, at right angles to the high tides. Since the moon revolves in a 29½-day orbit around Earth, the average period between successive high tides is approximately 12 hours and 25 minutes.

The sun also causes two tides on opposite sides of Earth, and these tides have a relation to the sun like that of the lunar tides to the moon. Because the gravitational pull of the sun is less than that of the moon, solar tides are partially masked by lunar tides except for two times during the month—when the moon is full and when it is new. At these times, Earth, moon, and sun are nearly in line, and the gravitational pulls of the sun and the moon are additive. This combination causes the high tides of those periods to be exceptionally large, with maximum rise and fall. These are the fortnightly *spring tides,* a name derived from the Saxon *sprungen,* which refers to the brimming fullness and active movement of the water. When the moon is at either quarter, its pull is at right angles to the pull of the sun, and the two forces interfere with each other. At this time the differences between high and low tide are exceptionally small. These are the *neap tides,* from an old Scandinavian word meaning "barely enough."

Tides are not entirely regular, nor are they the same all over Earth. They vary from day to day in the same place, following the waxing and waning of the moon. They may act differently in several localities within the same general area. In the Atlantic, semidaily tides are the rule. In the Gulf of Mexico, the alternate highs and lows more or less efface each other, and flood and ebb follow one another at about 24-hour intervals to produce one daily tide. Mixed tides are common in the Pacific and Indian oceans. These tides are combinations of daily and semidaily tides in which one partially cancels out the other. Local tides around the world are inconsistent for many reasons. These include variations in the gravitational pull of the moon and the sun due to the elliptical orbit of Earth, the angle of the moon in relation to the axis of Earth, onshore and offshore winds, depth of water, contour of the shore, and internal waves.

ZONATION AND STRATIFICATION

Just as lakes exhibit stratification and zonation, so do the seas. The ocean itself has two main divisions: the *pelagic,* or whole body of water, and the *benthic,* or bottom region (Figure 34.1). The pelagic is further divided into two provinces: the *neritic,* water that overlies the continental shelf, and the *oceanic.* Because conditions change with depth, the pelagic is divided into three vertical layers or zones. From the surface to about 200 m is the *photic* zone, in which there are sharp gradients in illumination, temperature, and salinity. From 200 to 1000 m is the *mesopelagic* zone, where little light penetrates and the temperature gradient is more even and gradual, without much seasonal variation. It contains an oxygen-minimum layer and often the maximum concentration of nitrate and phosphate. Below the mesopelagic is the *bathypelagic* zone, where darkness is virtually complete, except for bioluminescence, temperature is low, and the pressure is great.

The upper layers of ocean water are thermally stratified. Depths below 200 m are usually thermally stable. In high and low latitudes, temperatures remain fairly constant throughout the year. Polar seas, covered with ice most of the year, exhibit no thermocline in winter, spring, and fall. The waters are well mixed and nutrients are not limiting. Slight stratification takes place in the polar summer (July and August). At that time the ice melts, the water warms enough, and light is sufficient to support a bloom of phytoplankton. In tropical seas, the upper waters are well lighted and the continuous input of energy maintains a high temperature throughout the year. Light and temperature are optimum for phytoplankton production, but the waters are permanently stratified. That prevents mixing and upward circulation of nutrients. The result is low productivity. In temperate seas, thermal structure changes seasonally, reflecting the amount of light and solar thermal energy entering the water. Water in the summer is thermally stratified with no mixing. In spring

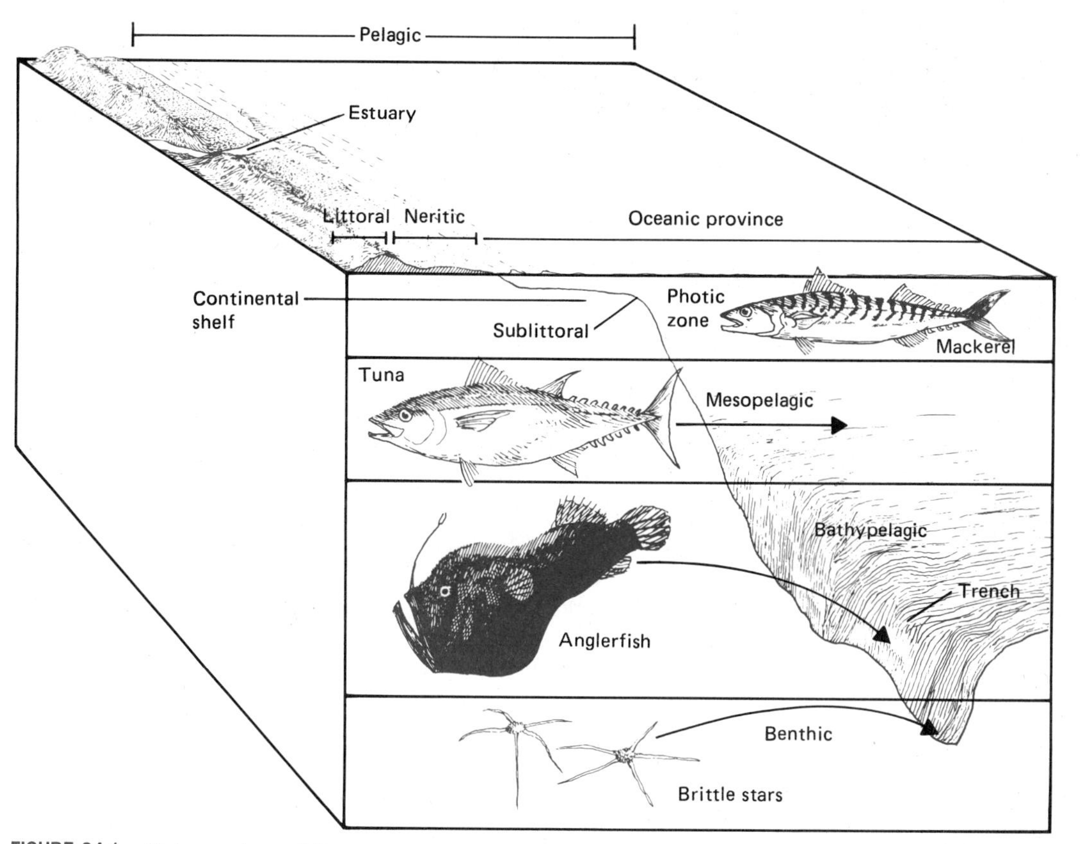

FIGURE 34.1 Major regions of the ocean.

and fall, when the surface water warms and cools, respectively, thermal stratification decreases, and the waters become mixed to varying degrees, recharging nutrients in the surface waters.

THE OPEN SEA

Viewed from the deck of a ship or from an airplane, the open sea appears to be monotonously the same (Figure 34.2). Nowhere can you detect any strong pattern of life or well-defined communities as you can over land. The reason is that pelagic ecosystems lack the supporting structures and framework of large dominant plant life, and their major herbivores are not large, conspicuous mammals like elephants and deer, but tiny zooplankton.

There is a reason for the smallness of sea plants. Surrounded by a chemical medium that contains in varying quantities the nutrients necessary for life, they absorb their food directly from the water. The smaller the organism, the greater the surface area exposed for the absorption of nutrients and solar energy. Seawater is so dense that there is little need for supporting structures.

Nevertheless, differences based on physical characteristics and life forms do allow a division of the sea into ecological regions. The Arctic Ocean lies north of the land masses in the Northern Hemisphere and is open only to the Atlantic Ocean. It holds its unique forms of life, as does the Southern or Antarctic Ocean, which lies about the continent of Antarctica and is open to three oceans, the Atlantic, Pacific, and Indian. Warm oceanic waters making up the Atlantic and Pacific have some of their own distinctive communities.

FIGURE 34.2 The open sea, birthplace of waves that crash against rocky and sandy shores thousands of miles away.

The deep-sea benthic ecosystems are quite different from the lighted waters. Other marine ecosystems include the coral reefs (Chapter 35) and upwelling systems off the coasts of California, Peru, Northwest Africa, Southwest Africa, India, and Pakistan. Important and distinctive are the shelf-sea ecosystems, such as the Georges and Grand Banks of the North Atlantic. Shallow, productive, and nutrient-rich, they support a diversity of fish and invertebrate life.

STRUCTURE

Because the global regions vary from oligotrophic to eutrophic and from cold to tropical, a brief discussion of structure must be general and emphasize the major structural groups.

Phytoplankton

Requiring light, phytoplankton is restricted to the upper surface waters, which, determined by the light penetration, vary from tens to hundreds of meters. Because of seasonal, annual, and geographic variations in light, temperature, and nutrients, as well as grazing by zooplankton, the distribution and species composition of phytoplankton vary from ocean to ocean and place to place within them.

Each ocean or region within an ocean appears to have its own dominant forms. Littoral and neritic waters and regions of upwelling are richer in plankton than mid-oceans. In regions of downwelling, the dinoflagellates, a large, diverse group characterized by two whiplike flagellae, concentrate near the surface in areas of low turbulence. They attain their greatest abundance in warmer waters. In summer they may concentrate in the surface waters in such numbers that they color it red or brown. Often toxic to other marine life, such concentrations of dinoflagellates are responsible for red tides. In regions of upwelling, the dominant forms of phytoplankton are diatoms. Enclosed in a silica case, diatoms are particularly abundant in arctic waters. Smaller than diatoms are the Coccolithophoridae, so small that they pass through plankton nets (and so are classified as *nanoplankton*). Their minute bodies are protected by calcareous plates or spicules embedded in a gelatinous sheath. Universally distributed in all waters except the polar seas, the Coccolithophoridae have the ability to swim. Droplets of oil aid in buoyancy and serve as a means of storing food. In equatorial currents in shallow seas, the concentration of phytoplankton is variable. Where both lateral and vertical circulation of water is rapid, the composition reflects in part the ability of the species to grow, reproduce, and survive under local conditions.

Zooplankton

Grazing on the phytoplankton is the herbivorous zooplankton, largely copepods—planktonic arthropods (the most numerous animals of the sea) and the shrimplike euphausiids, commonly known as krill. Other planktonic forms are the larval stages of such organisms as gastropods, oysters, and cephalopods. Feeding on the herbivorous zooplankton is the carnivorous zooplankton, which includes such organisms as the larval forms of comb jellies (Ctenophora) and arrowworms (Chaetognatha).

Like phytoplankton, the composition of zooplankton varies. In general, zooplankton falls into two main groups, the larger forms characteristic of shallow coastal waters and, generally, the smaller forms characteristic of the deeper open ocean. Zooplankton of the continental shelf contains a large portion of the larvae of fish and benthic organisms. It includes a greater diversity of species, reflecting a greater diversity of environmental and chemical conditions. The open ocean, being more homogeneous and nutrient-poor, supports a less diverse zooplankton. Zooplanktonic species of polar waters, having spent the winter in a dormant state in the deep water, rise to the surface during short periods of diatom blooms to reproduce. In temperate regions, distribution and abundance depend on the temperature of the water. In tropical regions, where temperature is nearly uniform, zooplankton is not so restricted, and reproduction occurs throughout the year.

Also like phytoplankton, zooplankton lives mainly at the mercy of the currents; but possessing sufficient swimming power, many forms of zooplankton exercise some control. Most species migrate vertically each day to arrive at a preferred level of light intensity. As darkness falls, zooplankton rapidly rises to the surface to feed on phytoplankton. At dawn, it moves back down to preferred depths.

Nekton

Feeding on zooplankton and passing energy along to higher trophic levels is the nekton, swimming organisms that can move at will in the water column. They range in size from small fish to large predatory sharks and whales, seals, and marine birds such as penguins. Some of the predatory fish, such as herring and tuna, are more or less restricted to the photic zone. Others are found in the deeper mesopelagic and bathypelagic zones, or move between them as the sperm whale does. Although the ratio in size of predator to prey falls within certain limitations, some of the largest nekton organisms in the sea, the baleen whales, feed on disproportionately small prey, euphausiids. By contrast, the sperm whale attacks very large prey, the giant squid.

Living in a world that lacks any refuge against predation or site for ambush, inhabitants of the pelagic zone have evolved various means of defense and of securing prey. Among them are the stinging cells of the jellyfish, streamlined shapes that allow speed both for escape and for pursuit, unusual coloration, advanced sonar, a highly developed sense of smell, and social organization involving schools or packs. Some animals, such as the baleen whale, have specialized structures that permit them to strain krill and other plankton from the water. Others, such as the sperm whale and certain seals, have the ability to dive to great depths to secure food. Phytoplankton lights up darkened seas, and fish take advantage of that bioluminescence to detect their prey.

Residents of the deep also have special adaptations for securing food. Some, like the zooplankton, swim to the upper surface to feed by night; others remain in the dimly lit or dark waters. Darkly pigmented and weak-bodied, many of the deep-sea fish depend on luminescent lures, mimicry of prey, extensible jaws, and expandable abdomens (which enable them to consume large items of food). Although most of the fish are small (usually 15 cm or less in length), the region is inhabited by rarely seen large species such as the giant squid. In the bathypelagic region bioluminescence reaches its greatest development—two-thirds of the species produce light. Bioluminescence is not restricted to fish. Squid and euphausiids possess searchlightlike structures complete with lens and iris; and squid and shrimp discharge luminous clouds to escape predators. Fish have rows of luminous organs along their sides and lighted lures that enable them to bait prey and recognize other individuals of the same species.

The Benthos

The term *benthal* refers to the floor of the sea, and *benthos* refers to plants and animals that live there. There is a gradual transition of life from the benthos on the rocky and sandy shores to that in the ocean's depths. From the tide line to the abyss, organisms that colonize the bottom are influenced by the nature of the substrate. Where the bottom is rocky or hard, the populations consist largely of organisms that live on the surface of the substrate, the *epifauna* and the *epiflora*. Where the bottom is largely covered with sediment, most of the inhabitants, chiefly animals, live within the deposits and are known collectively as the *infauna*. The kind of organism that burrows into the substrate is influenced by the particle size, since the mode of burrowing is often specialized for a certain type of substrate.

The substrate varies with the depth of the ocean and with the relationship of the benthic region to land areas and continental shelves. Near the coast, bottom sediments are derived from the weathering and erosion of land areas along with organic matter from marine life. The sediments of deep water are characterized by fine-textured material that varies with depth and with the types of organisms in overlying waters. Although these sediments are termed *organic,* they contain little decomposable carbon, consisting largely of skeletal fragments of planktonic organisms. In general, with regional variations, organic deposits down to 4000 m are rich in calcareous matter. Below 4000 m, hydrostatic pressure causes some forms of calcium carbonate to dissolve. At 6000 m and lower, sediments contain even less organic matter and consist largely of red clays rich in aluminum oxides and silica.

Within the sediments are layers that relate to oxidation reduction reactions. The surface, or oxidized layer, yellowish in color, is relatively rich in oxygen, ferric oxides, nitrates, and nitrites. It supports the bulk of benthic animals, such as polychaete worms, bivalves, and copepods, and a rich growth of aerobic bacteria. Below this surface is a grayish transition zone to the black layer, characterized by a lack of oxygen, iron in the ferrous state, nitrogen in the form of ammonia, and hydrogen sulfide. This layer is inhabited by anaerobic bacteria, chiefly reducers of sulfates and methane.

In a world of darkness, no photosynthesis takes place, so the bottom community is strictly heterotrophic, depending entirely on what organic matter finally reaches the bottom as a source of energy. In spite of the darkness and depth, the benthic communities support a high diversity of species. In the shallow benthic regions the recorded number of polychaete worms is over 250 species and of pericarid crustaceans (shrimplike mysidaceans, cumaceans, the small tanaidaceans, and isopods) well over 100. But the deep sea benthos supports a surprisingly higher diversity. The number of species collected in over 500 samples of which the total surface area sampled was only 50 m^2 was 707 species of polychaetes and 426 species of pericarid crustaceans. Most of the species are small, but among the deep-sea amphipods, isopods, and copepods, a few are large by crustacean standards, reaching 15 to 42 cm in length.

There are several hypothesized reasons for this deep-sea diversity. One is the lack of widespread disturbance or environmental extremes. The temperature is nearly constant and the bottom is not stirred by storms, as the shallow bottoms are. Small local disturbances are created as the crustaceans and other bottom dwellers move, creating mounds and pits on the surface, interrupting the smoothness and adding diversity. Increasing the variety of living conditions is the patchy distribution of food. The benthos depends on the rain of organic matter drifting to the bottom, which is small and scattered. Patches of dead phytoplankton, the bodies of dead whales, seals, birds, fish, and invertebrates, all provide a diversity of foods for different feeding groups and species. The low input and patchy distribution of food probably result in less competition by direct interference among species for available food.

Bottom organisms have four feeding strategies. They may filter suspended material from the water, as the stalked coelenterates do; they may collect food particles that settle on the surface of the sediment, as sea cucumbers do; they may be selective or unselective deposit feeders, as the polychaete worms are; or they may be predatory, like the brittle stars and the spiderlike pycnogonids.

Important in the benthic food chain are the bacteria of the sediments. Common where large quantities of organic matter are present, bacteria may reach several tenths of a gram per square meter in the topmost layer of silt. Bacteria synthesize protein from dissolved nutrients, and in turn become a source of protein, fat, and oils for deposit feeders.

Hydrothermal Vents

In 1977 oceanographers first discovered along volcanic ridges in the ocean floor of the Pacific near the Galápagos Islands high-temperature deep-sea springs. These springs vent jets of hydrothermal fluids that heat the surrounding water to 8° to 16° C, considerably higher than the 2° ambient water. Since then oceanographers have discovered similar vents on volcanic ridges along

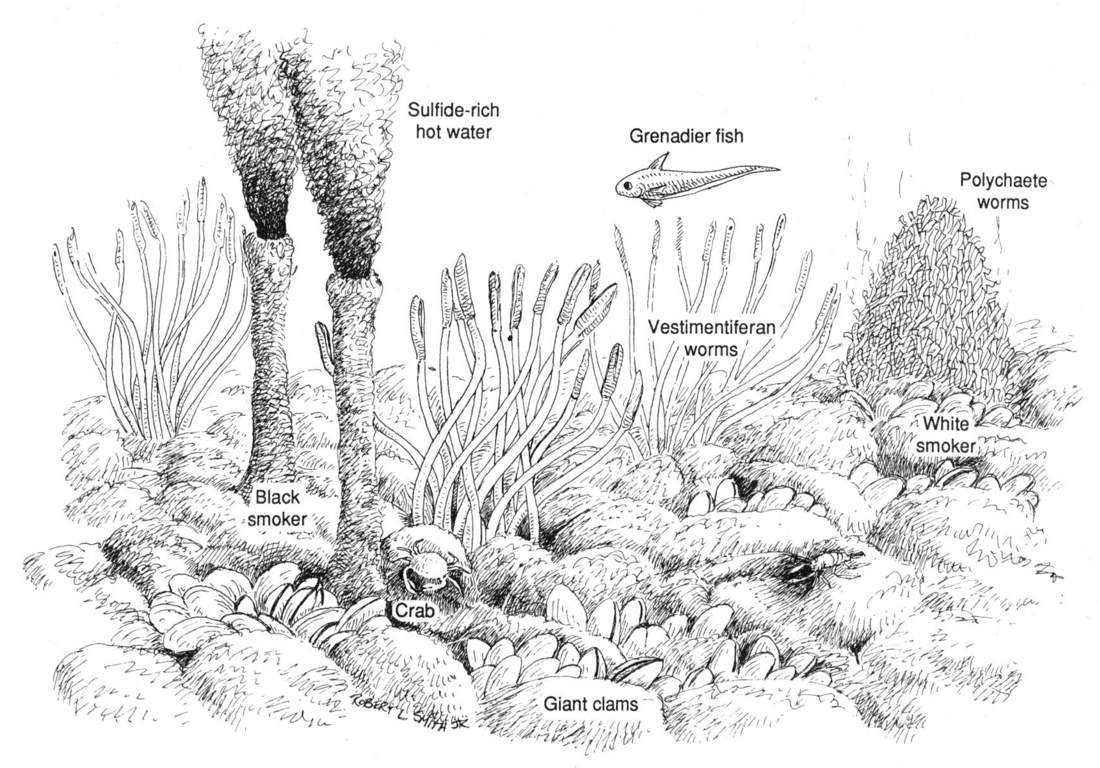

FIGURE 34.3 A typical hydrothermal vent mound resting on flows of black basaltic lava.

fast-spreading centers of ocean floor, particularly in the eastern Pacific.

Vents form when cold seawater flows down through fissures and cracks in the basaltic lava floor deep into the underlying crust. The waters react chemically with the hot basalt, giving up some minerals but becoming enriched with others such as copper, iron, sulfur, and zinc. Heated to a high temperature, the water re-emerges through mineralized chimneys rising up to 13 meters above the sea floor. Among the chimneys are white smokers and black smokers (Figure 34.3). White-smoker chimneys rich in zinc sulfides issue a milky fluid under 300° C. Black smokers, narrower chimneys rich in copper sulfides, issue jets of clear water from 300° to over 450° C that are soon blackened by precipitation of fine-grained sulfur-mineral particles.

Associated with these vents is a rich diversity of newly discovered forms of deep-sea life confined within a few meters of the vent system. They include giant clams, mussels, polychaete worms that encrust the white smokers, crabs, and vestimentifera worms lacking a digestive system. The primary producers are chemosynthetic bacteria that oxidize the reduced sulfur compounds, such as H_2S, to release energy, which they use to form organic matter from carbon dioxide. Primary consumers, the clams, mussels, and worms, filter bacteria from water and graze on bacterial film on rocks. The giant clam *Calyptogena magnifica* and the large vestimentifera worm *Riftia pachyptila* contain symbiotic chemosynthetic bacteria in their coelomic tissues. These bacteria need a reduced sulfide source, which is carried to them by the blood of these animals. *Riftia* has in its blood a sulfide-binding protein that concentrates sulfide from the environment and transports it to the bacteria. Such concentrations would poison other animals, but the sulfide-bearing protein of the worm and apparently the clam has a high affinity for free sulfides, preventing them from accumulating in the blood and entering the cells.

FUNCTION

Although the oceans dominate Earth's surface, they contribute much less to Earth's primary production than do terrestrial ecosystems. Oceans are less productive because only a superficial illuminated area up to 100 m deep can support plant life; and most of the open sea is nutrient-poor, with an almost nonexistent nutrient reserve. Phytoplankton, zooplankton, and other organisms' remains sink below the lit zone into the dark benthic water. While this sinking supplies nutrients to the deep, it robs the upper layers.

This depletion of nutrients is most pronounced in tropical waters. There, a permanent thermal stratification—with its layer of warmer, less dense water lying on top of a colder, denser layer of deep water—prevents an exchange of nutrients between the surface and the deep. Thus, in spite of high light intensity and warm temperatures, tropical seas are the lowest in productivity, which amounts to an estimated 18–50 g C/m^2/yr.

The temperate oceans are more productive, largely because a permanent thermocline does not exist. During the spring and to a limited extent during the fall, temperate seas, like temperate lakes, experience a nutrient overturn (see Chapter 31). This recirculation of phosphorus and nitrogen from the deep stimulates a surge of spring phytoplankton growth. As spring wears on, the temperature of the water becomes stratified and a thermocline develops, preventing a nutrient exchange. The phytoplankton growth depletes the nutrients, and the phytoplankton population suddenly declines. In the fall a similar overturn takes place, but the rise in phytoplankton production is slight because of decreasing light intensity and low winter temperatures. Reduced production in winter holds down the annual productivity of temperate seas to a level a little above that of tropical seas, 70–110 g C/m^2/yr.

Most productive are coastal waters and regions of upwelling, whose annual productivity may amount to 1000 g C/m^2/yr. Major areas of upwelling are largely on the western sides of continents: off the southern California coast, Peru, northern and southwestern Africa, and the Antarctic. Upwellings result from the differential heating of polar and equatorial regions that produces the equatorial currents (see Chapter 4) and the winds. For example, as the water is pushed northward to the equator by winds blowing out of the south, it is deflected from the coast by the Coriolis force. As the deflected surface waters move away, they are replaced by an upwelling of colder, deeper water that brings a supply of nutrients into the warm, sunlit portions of the sea. As a result, regions of upwelling are highly productive, supporting an abundance of life. Because of their high productivity, upwellings support important commercial fisheries such as the tuna fishery off the California coast, the anchovy fishery off Peru, and the sardine fishery off Portugal. Other zones of high production are coastal waters and estuaries, where productivity may run as high as 380 g C/m^2/yr. Turbid, nutrient-rich waters are major areas of fish production.

Thus, between upwellings and coastal waters, the most productive areas of the sea are the fringes of water bordering the continental land masses. A great deal of the measured productivity of the coastal fringes comes from the benthic as well as the surface waters, since the seas are shallow. Benthic production, largely unavailable, is not considered in the productivity of the open sea. An estimate of the total production for marine plankton is 50 Gt of dry matter per year; if benthic production is considered, total production may be 55 Gt of dry matter per year.

Carbohydrate production by phytoplankton, largely diatoms, is the base on which the life of the seas exists (Figure 34.4). Conversion of primary production into animal tissue is accomplished by zooplankton, the most important of which are the copepods. To feed on the minute phytoplankton, most of the grazing herbivores must also be small, measuring between 0.5 and 5.0 millimeters. In the oceans most of the grazing herbivores are members of the genera *Calanus, Acartia, Temora,* and *Metridia,* probably the most abundant animals in the world. The single most abundant copepod is *Calanus finmarchicus,* with its close relative *C. helgolandicus*. In the Antarctic the shrimplike euphausiids, or krill, fed on by the baleen whales and penguins, are the dominant herbivores. The herbivorous copepods, then, become the link in the food chain

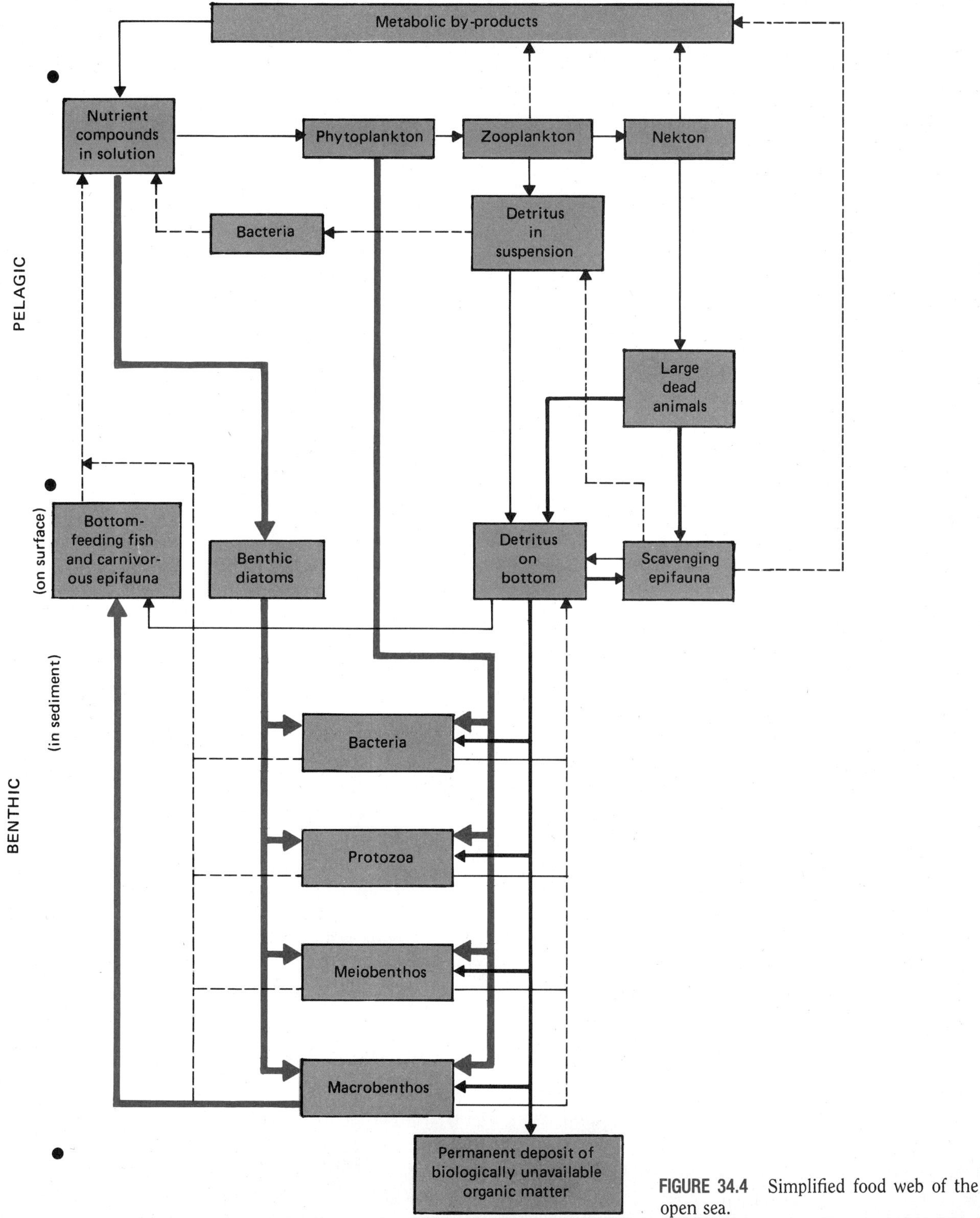

FIGURE 34.4 Simplified food web of the open sea.

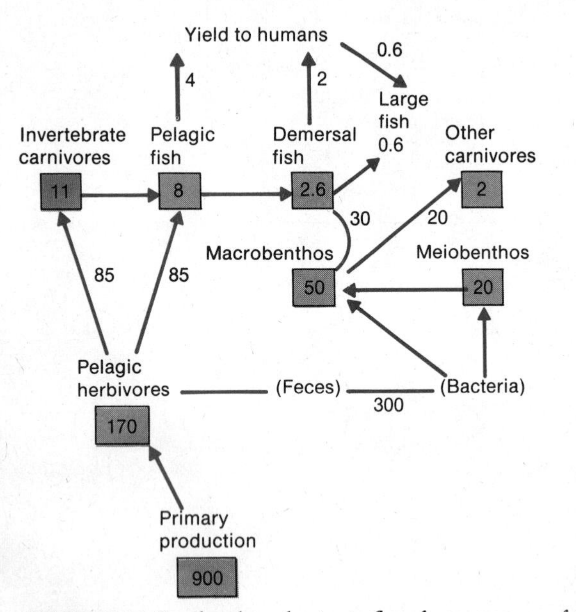

FIGURE 34.5 Food web and energy flow from an area of the North Sea. The values are in kcal/m^2/yr. Note the low-energy yield to humans, 6.6 kcal/m^2/yr from a primary production of 900 kcal/m^2/yr.

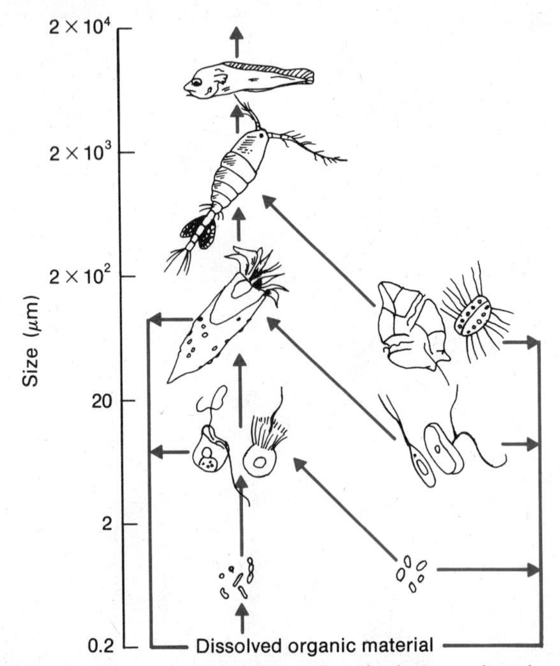

FIGURE 34.6 A microbial loop that feeds into the classic marine food web. On the right are representative photoautotrophs, nanoflagellates, and cyanobacteria, and on the left are associated heterotrophs, bacteria, nanoflagellates, and larval fish.

between the phytoplankton and the second-level consumers, as illustrated in the North Sea food web (Figure 34.5).

However, part of the food chain begins not with the phytoplankton, but with organisms even smaller. Recent investigations show that bacteria and protists, both heterotrophic and photosynthetic, make up one-half of the biomass of the sea, and are responsible for the largest part of energy flow in pelagic systems. Photosynthetic nanoflagellates (2–20 μm) and cyanobacteria (1–2 μm) are responsible for a large part of photosynthesis in the sea. These cells excrete a substantial fraction of their photosynthate in the form of dissolved organic material that is utilized by heterotrophic bacteria. Populations of such bacteria are dense, around 1 million cells per milliliter of seawater. These heterotrophic bacteria account for about 20 percent of primary production. Bacterial growth efficiency does not exceed 50 percent, so one-half of phytoplankton primary production in the form of dissolved organic material is consumed by bacteria. Bacterial numbers in the sea remain relatively stable, suggesting predation, but filter-feeding zooplankton cannot retain particles of bacterial size, so the consumption of bacteria by heterotrophic nanoflagellates, experimentally demonstrated, accounts for the disappearance of bacterial production. This uptake by heterotrophic bacteria of dissolved organic matter produced by the plankton and their subsequent consumption of bacteria nanoplankton introduces a feeding loop, termed the *microbial loop* (Figure 34.6), and adds several more trophic levels in the plankton food chain.

HUMAN IMPACT

The ocean, Thor Heyerdahl reported after completing his trip across the southern Atlantic in 1968 on the papyrus raft *Ra II,* is polluted. "For weeks along the route," Norman Baker, navigator on the *Ra II,* wrote,

"We saw no land, no ship, no light, no man; all we saw was his garbage, and we saw that all the time." Among this garbage were floating gobs of oil. ". . . As far as I could see," Baker continued, "below as well as to the sides, the entire visible layer of the ocean was infested with hanging bits of hardened oil." Conditions have not improved over the past quarter century. They have only gotten worse.

Despite natural leakages from deposits on the ocean floor, only since oil became the major energy source, exploited, transported, and used about the world, has it become a major pollutant of the seas. Leakage from rusting oceanic pipelines, errors in handling oil cargo, accidents involving tankers and barges, sinking of ships during wars, illegal washing of tanker bilges, spills and leakage from off-shore drilling rigs, all contribute to a growing oil pollution.

Oil is a collection of hundreds of substances that react with the environment. When released in water it spreads in a film over the surface. The lighter fractions evaporate or are absorbed by particulate matter and sink to the bottom. Some becomes dissolved in seawater. Much is oxidized by bacteria, yeast, and molds that attack different fractions of crude oil. Bacteria can digest straight-chain hydrocarbons. They have more problems with branch-chained hydrocarbons, which persist for a long time as tarry chunks bobbing on the surface and lying on the bottom of the sea.

The aspect of oil pollution that captures our attention is the damage to coastal ecosystems (Chapter 35). We deplore massive oil spills from tanker wrecks along the coast, or the disastrous releases of oil from the Kuwait oil fields. Much less obvious but in the long run much more important are the insidious effects of oil and other pollutants on the world's oceans. We know little about what effect oil has on the benthos, where it comes in direct contact with life in the deep. We do know that oil reaching the bottom of bays and harbors kills bottom life, and that certain fractions of oil soluble in water are highly toxic to many forms of marine life.

The problem is aggravated by the presence of many other toxic materials in the seas. The oceans have become the great dumping grounds for the world. Toxic substances of all sorts—heavy metals such as mercury, cadmium, lead, zinc, copper; pesticides, herbicides—enter the oceans from many sources and add to the dissolved hydrocarbons from oil. These pollutants, although in low concentrations, are foreign to the marine ecosystem, and some, such as mercury, become magnified in the food web. Particularly sensitive to this chronic pollution are the phytoplankton and many forms of zooplankton and crustaceans. Such pollution inhibits photosynthesis, growth, and cell division of marine phytoplankton. It affects growth and development of microzooplankton filter feeders, and the early development stages of other forms of life. The impact of marine pollution has barely been studied, but the fact that pollution in the North Atlantic has decreased the number of species of phytoplankton and zooplankton warns us of future problems. The long-term effects will be a decrease in primary production in the ocean accompanied by a disruption of trophic interactions beginning with phytoplankton and zooplankton and working its way up the food web.

The vastness and depth of the ocean, far removed from direct human contact, make it a seemingly ideal place for disposing of the wastes of human activity: sewage, sewage sludge, industrial wastes, garbage, and radioactive wastes. Discharge of sewage from southern California cities out to the mainland shelf has contaminated a 3640-km^2 area of ocean bottom. This pollution has degraded benthic invertebrates, killed beds of kelp, and caused disease in fish. The bottom of a 105-km^2 area of the New York Bight is covered with black toxic sludge. The ocean has long been a favorite place to dump solid wastes, including urban garbage, industrial and hazardous wastes, construction materials, and junk, from cars to military shells and old ships. Although some of the metallic junk has provided sheltering reefs for fish, other material entangles or poisons marine birds and mammals.

Already the recipient of low-level radioactive wastes, the deep benthic regions, because of the relative stability of their sediments, are viewed as possible safe places for the disposal of accumulating highly radioactive wastes from nuclear power plants. Given our general ignorance of the structure and function of the deep-sea benthos and deep-sea food webs, we have no idea of the long-term effects and dangers of such activity.

Summary

The marine environment is characterized by salinity, waves, tides, depth, and vastness. Salinity is due largely to sodium and chlorine, which make up 86 percent of sea salt. Although sea salt has a constant composition, salinity varies throughout the oceans. It is affected by evaporation, precipitation, movement, and mixing of water masses of different salinities. Because of its salinity, seawater does not reach its greatest density at 4° C but becomes heavier as it cools.

Like lakes, the marine environment experiences both zonation of life and stratification of temperature. The open sea can be divided into three vertical zones. The bathypelagic is the deepest, void of sunlight and inhabited by darkly pigmented, weak-bodied animals characterized by bioluminescence. Above it lies the dimly lit mesopelagic zone, inhabited by characteristic species such as certain sharks and squid. Both the bathypelagic and mesopelagic zones depend on a rain of detrital material from the upper lighted zone, the photic, for their energy source. The sea bottom or benthic region is inhabited by its own unique fauna adapted to life in total darkness and high pressure. The species diversity of the deep-sea benthos is surprisingly high. The detrital feeding and predaceous fauna are dominated by polychaete worms and pericarid crustaceans. Along the volcanic oceanic ridges, especially in the mid-Pacific, are hydrothermal vents inhabited by unique and newly discovered forms of life, including clams, worms, and crabs. The source of primary production for these hydrothermal vent communities are chemosynthetic bacteria that use sulfates as an energy source.

Because of the impoverished nutrient status of ocean water, productivity is low. Nutrient reserves in the upper layer of water are low; phytoplankton and other life sink to the deep water; and a thermocline, permanent in deep water, prevents the circulation of deep water to the upper layer. Most productive are shallow coastal waters and areas of upwelling, where nutrient-rich deep water comes to the surface. Copepods are the major herbivores, the critical link in the food chain between producers and second-level consumers. However, recent studies show that bacteria and protists, both heterotrophic and photosynthetic, make up one-half of the biomass of the sea and are responsible for the largest part of energy flow in the pelagic system.

The oceans have been chronically polluted from human activity and from the use of the sea as a dumping ground. Major chronic pollutants are oil released from tanker accidents, seepage from off-shore drilling and other sources, and toxic materials such as pesticides and heavy metals from industrial, urban, and agricultural sources. The long-term effects of these pollutants, as well as the effects of the potential use of the deep sea as a dump for radioactive wastes, are unknown.

Review and Study Questions

1. Why is the ocean salty?
2. What causes tides? What are spring tides? Neap tides?
3. What type of currents other than surface currents exists in oceans? What is their significance?
4. What is upwelling, and what is its ecological importance?
5. Characterize the major regions of the ocean, both vertical and horizontal.
6. How does temperature stratification in the tropical seas differ from that in the polar seas? What sea develops the most pronounced thermocline, and why?
7. Distinguish between epifauna and infauna.
8. What makes up the nekton, and what is its significance in the sea?
9. What might account for the high diversity of the deep-sea benthos? How are these organisms adapted to darkness?
10. What are hydrothermal vents? What makes life about them unique?
11. What is the role of phytoplankton, zooplankton, and the nekton in the marine food web?
12. What is the significance of the microbial loop in the marine food web?
13. Why is primary production in the seas so low? What effect does that have on the overall productivity of the oceans?
14. What characteristics of oil make it a chronic pollutant of the open ocean?

15. What are some other major sources of ocean pollution and their potential effects?
*16. Find out the extent of the use of oceans for sewage and solid waste disposal. Does any of your garbage end up there?
*17. Report on the cause and effects of a major oil spill.
*18. How can interior agriculture regions pollute the ocean?
*19. Debate the pros and cons of using the deep-sea bottom for disposal of radioactive wastes.

Selected References

Boucher, G. 1985. Long-term monitoring of meiofauna densities after the *Amoco Cadiz* oil spill. *Mar. Pollution Bull.* 16:328–333.

Carson, R. 1961. *The sea around us.* New York: Oxford University Press. A classic book on the sea.

Fenchel, T. 1987. Marine plankton food chains. *Ann. Rev. Ecol. Syst.* 19:19–38.

Geyer, R. A., ed. *Marine environmental pollution.* New York: Elsevier.

Grassle, J. F. 1985. Hydrothermal vent animals: Distribution and biology. *Science* 229:713–717.

Grassle, J. F. 1989. Species diversity in deep-sea communities. *Trends. Ecol. Evol.* 4:12–15.

Grassle, J. F. 1991. Deep-sea benthic diversity. *Bioscience* 41:464–469.

Gross, M. G. 1982. *Oceanography: A view of Earth,* 3rd ed. Englewood Cliffs, NJ: Prentice-Hall. An excellent reference on the physical aspects of the sea.

Hardy, A. 1971. *The open sea: Its natural history.* Boston: Houghton Mifflin. A classic introduction.

Hayman, R. M., and R. C. McDonald. 1985. The ecology of deep sea hot springs. *Am. Sci.,* 73:441–449.

Kinne, O., ed. 1978. *Marine ecology.* 5 vol. A major and often technical reference source.

Marshall, N. B. 1980. *Deep-sea biology: Developments and perspectives.* New York: Garland STMP Press.

National Academy of Sciences. 1985. *Oil in the sea, inputs, fates, and effects.* Washington, DC: National Academy Press.

Nelson-Smith, A. 1972. *Oil pollution and marine ecology.* London: Elek Science. Good, concise basic reference.

Nybakken, J. W. 1988. *Marine biology: An ecological approach,* 2nd ed. New York: Harper & Row. A solid, up-to-date reference on marine life and ecosystems.

Powell, M. A., and C. N. Somero. 1983. Blood components prevent sulfide poisoning of respiration of the hydrothermal vent tubeworm *Riftia pachyptila. Science* 219:297–299.

Rex, M. A. 1981. Community structure in the deep-sea benthos. *Ann. Rev. Ecol. Syst.* 12:331–353. A good review of deep-sea species diversity.

Steele, J. 1974. *The structure of a marine ecosystem.* Cambridge, MA: Harvard University Press.

CHAPTER 35

Intertidal Zones and Coral Reefs

Outline

Objectives

On completion of this chapter, you should be able to:

1. Describe the major features and zonation of rocky, sandy, and muddy shores and coral reefs.
2. Discuss the adaptations of life to an intertidal environment.
3. Explain the role of disturbance in intertidal and coral-reef ecosystems.
4. Describe the formation of coral reefs.
5. Explain the symbiotic relationship of coral anthozoans and algae and its ecological significance.
6. Discuss the impact of human activity on the intertidal zones and coral reefs.

Where the edge of the land meets the edge of the sea we find the fascinating and complex world of the seashore. Rocky, sandy, muddy, protected, pounded by incoming swells, all shores have one feature in common: They are alternately exposed and submerged by the tides. Roughly, the region of the seashore is bounded on one side by the height of the extreme high tides and on the other by the height of the extreme low tides. Within these confines conditions change from hour to hour with the ebb and flow of the tides. At flood tide the seashore is a water world; at ebb tide it belongs to the terrestrial environment, with its extremes in temperature, moisture, and solar radiation. In spite of all this change, the seashore inhabitants are essentially marine, adapted to withstand some degree of exposure to the air for varying periods of time.

FIGURE 35.1 The broad zones of life exposed at low tide on the rocky shore of the Bay of Fundy. Note the heavy growth of *Fucus* on the lower portion and the white zone of barnacles above. Nestled in a large depression in the rocks is a tidal pool.

ROCKY SHORES

STRUCTURE

As the sea recedes at ebb tide, rocks, glistening and dripping with water, begin to appear. Life hidden by tidal water emerges into the open air layer by layer. The uppermost layers of life are exposed to air, wide temperature fluctuations, intense solar radiation, and desiccation for a considerable period, while the lowest fringes on the intertidal shore may be exposed only briefly before the flood tide submerges them again. These varying conditions result in one of the most striking features of the rocky shore, the zonation of life (Figure 35.1). Although this zonation may differ from place to place as a result of local variations in aspect, substrate, wave action, light intensity, shore profile, exposure to prevailing winds, climatic differences, and the like, the same general features are always present. All rocky shores have three basic zones, characterized by dominant organisms (Figure 35.2).

Where the land ends and the seashore begins is hard to determine. The approach to a rocky shore from the landward side is marked by a gradual transition from lichens and other land plants to marine life dependent at least partly on the tidal waters (Figure 35.3). The first major change from land shows up on the *supralittoral fringe,* where salt water comes only every fortnight on the spring tides. It is marked by the black zone, a patchlike or beltlike encrustation of lichens of the Verrucaria-type and Myxophyceae algae such as *Calothrix* and *Entrophsalis.* Capable of existing under conditions so difficult that few other plants could survive, these blue-green algae, enclosed in slimy, gelatinous sheaths, and their associated lichens represent an essentially nonmarine community, on which graze the periwinkles, basically marine animals. Common to this black zone is the rough periwinkle, which grazes on the wet algae covering the rocks. On European shores lives a similarly adapted species, the rock periwinkle, the most highly resistant to desiccation of all the shore animals.

Below the black zone lies the *littoral zone,* covered and uncovered daily by the tides. The littoral tends to be divided into subzones. In the upper reaches barnacles are most abundant. The oyster, the blue mussel, and the limpets appear in the middle and lower portions of the littoral, as does the common periwinkle.

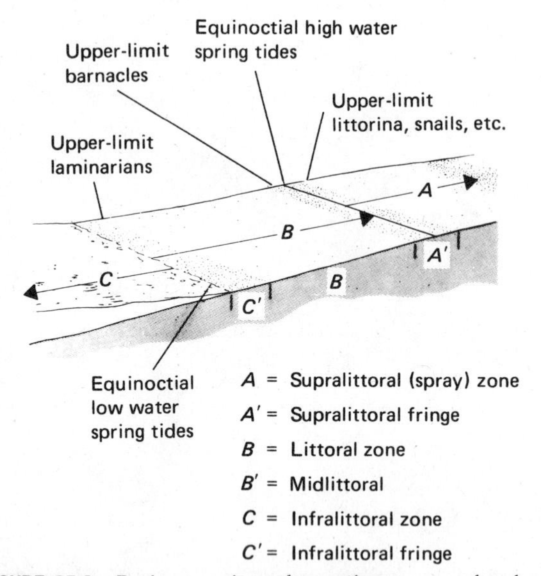

FIGURE 35.2 Basic or universal zonation on a rocky shore. Use this diagram as a guide when studying the subsequent drawings of zonation and reading the discussion in the text. *A*, supralittoral (spray) zone; *A'*, supralittoral fringe; *B*, littoral zone; *B'*, midlittoral zone; *C*, infralittoral zone; *C'*, infralittoral fringe.

Occupying the lower half of the littoral zone (midlittoral) of colder climates and in places overlying the barnacles is an ancient group of plants, the brown algae, commonly known as rockweeds (*Fucus* spp.) and wrack (*Ascophyllum nodosum*). Rockweeds attain their finest growth on protected shores, where they may grow 2 m long; on wave-whipped shores they are considerably shorter.

The lower reaches of the littoral zone may be occupied by blue mussels instead of rockweeds, particularly on shores where hard surfaces have been covered in part by sand and mud. No other shore animals grow in such abundance; the blue-black shells packed closely together may blanket the area.

Near the lower reaches of the littoral zone, mussels may grow in association with red algae, *Gigartina,* a low-growing, carpetlike plant. Algae and mussels together often form a tight mat over the rocks. Here, well protected in the dense growth from waves, live infant starfish, sea urchins, brittle stars, and bryozoan sea mats or sea lace (*Membranipora*).

The lowest part of the littoral zone, uncovered only at the spring tides and not even then if wave action is strong, is the *infralittoral fringe.* This zone, exposed for short periods of time, consists of forests of the large brown algae, *Laminaria,* one of the kelps, with a rich undergrowth of smaller plants and animals among the holdfasts.

Beyond the infralittoral fringe is the *sublittoral zone,* the open sea. This zone is principally neritic and benthic and contains a wide variety of fauna, depending on the substrate, the presence of protruding rocks, gradients in turbulence, oxygen tensions, light, and temperature.

The pattern of life on rocky shores is heavily influenced by biotic interactions of grazing, predation, competition, and larval settlement. Where wave action is heavy on New England intertidal rocky shores, periwinkles are rare, permitting a more vigorous growth of algae. Lack of grazing favors ephemeral algal species such as *Ulva* and *Enteromorpha,* whereas grazing allows the perennial *Fucus* to become established. On the New England coast the mussel *Mytilus edulis* outcompetes barnacles and algae; but predation by the starfish *Asterias* spp. and the snail *Nucella lapillus* prevents dominance by mussels except on the most wave-beaten areas. A similar situation exists on the Pacific coast. Barnacles of several species tend to outcompete and displace algal species, but in turn the dominant mussel *M. californianus* destroys the barnacles by overgrowing them. However, where present the predatory starfish *Pisaster ochraceus* prevents the mussel from completely overgrowing barnacles. The end result of such interactions of the physical and the biotic is a patchy distribution of life across the rocky intertidal shore. The ebbing tide leaves behind pools of water in rock crevices, in rocky basins, and in depressions (see Figure 35.1). They represent distinct habitats, which differ considerably from exposed rock and the open sea, and even differ among themselves. At low tide all pools are subject to wide and sudden fluctuations in temperature and salinity, changes most marked in shallow pools. Under the summer sun the temperature may rise above the maximum many organisms can tolerate. As water evaporates, especially in the shallower pools, salt crystals may appear around the edges. When rain or land drainage brings fresh water to the

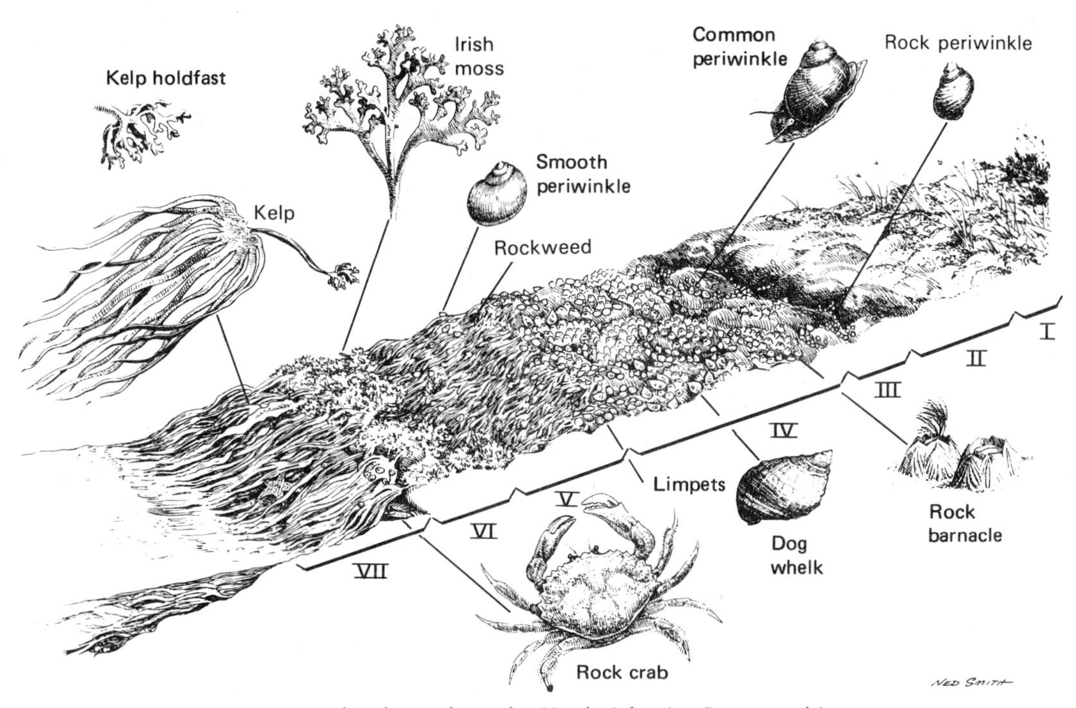

FIGURE 35.3 Zonation on a rocky shore along the North Atlantic. Compare this with the diagram in Figure 35.2. I, land: lichens, herbs, grasses; II, bare rock; III, black algae and rock periwinkle (*Littorina*) zone; IV, barnacle (*Balanus*) zone: barnacles, dog whelks, common periwinkles, mussels, limpets; V, fucoid zone: rockweed (*Fucus*) and smooth periwinkles; VI, Irish moss (*Chondrus*) zone; VII, kelp (*Laminaria*) zone.

pool, salinity may decrease. In deep pools such fresh water tends to form a layer on top, developing a strong salinity stratification in which the bottom layer and its inhabitants are little affected. If algal growth is considerable, oxygen will be high during the daylight hours, but low at night, a situation that rarely occurs at sea. The rise of CO_2 at night lowers the pH.

Pools near low tide are influenced most briefly by the rise and fall of tides; those that lie near and above the high tide line are exposed the longest and undergo the widest fluctuations. Some may be recharged with seawater only by the splash of breaking waves or occasional high spring tides. Regardless of their position on the shore, most pools suddenly return to sea conditions on the rising tide and experience drastic and instantaneous changes in temperature, salinity, and pH. Life in the tidal pools must be able to withstand those wide and rapid fluctuations in the environment.

FUNCTION

Life on the rocky shore is both autotrophic and heterotrophic. Many organisms, such as barnacles, depend on tides to bring them food; others, such as periwinkles, graze on algal growth on the rocks. In fact, the functioning of the rocky seashore ecosystem involves complex interactions between physical and biotic aspects of the ecosystem. Like a salt marsh that receives a daily energy subsidy from tidal flooding, rocky shores receive an energy subsidy, although indirectly, from the waves (Figure 35.4). Wave-generated energy impinging on a rocky shore is roughly twice that of solar

FIGURE 35.4 Waves pounding on rocky shores provide an energy subsidy to intertidal life.

FIGURE 35.5 A long stretch of sandy beach washed by waves. Although the beach appears barren, life is abundant beneath the sand.

radiation, amounting to 0.045 watts/cm^2 compared to 0.017 to 0.025 watts/cm^2 of solar radiation. Energy input is reflected in productivity, which in the kelp beds of the Pacific Northwest exceeds that of the tropical rain forest. Part of this productivity relates to the much higher leaf surface per square meter—about 20 m^2 of growing surface compared to about 8 m^2 for a tropical forest. The highest standing crops of kelp, mussels, and other organisms are in areas receiving the heaviest wave action.

The intertidal organisms do not use the wave energy directly. Rather, the force of the waves favorably affects conditions for productivity. For one, heavy wave action reduces the activity of such predators of sessile intertidal invertebrates as starfish and sea urchins. Waves bring in a steady supply of nutrients and carry away products of metabolism. They keep in constant motion the fronds of seaweeds, moving them in and out of shadow and sunlight, allowing more even distribution of incident light and thus more efficient photosynthesis. By dislodging organisms, both plants and invertebrates, from the rocky substrate, waves open up space for colonization by algae and invertebrates and reduce strong interspecific competition. In effect disturbance (see Chapter 21), which influences community structure, is the root of intertidal productivity.

SANDY AND MUDDY SHORES

Sandy and muddy shores appear barren of life at low tide, in sharp contrast to the life-studded rocky shore; but the sand and black mud are not as dead as they seem, for beneath them life lurks, waiting for the next high tide.

The sandy shore is a harsh environment; indeed, the matrix of this seaside environment is a product of the harsh and relentless weathering of rock, both inland and along the shore. Through eons the products of rock weathering are carried away by rivers and waves to be deposited as sand along the edge of the sea. The size of the sand particles deposited influences the nature of the sandy beach, water retention during low tide, and the ability of animals to burrow through it. Beaches with steep slopes are usually made up of larger sand grains and are subject to more wave action. Beaches exposed to high waves are generally flattened, for much of the material is carried away from the beach to deeper water and fine sand is left behind (Figure 35.5). Sand grains of all sizes, especially the finer particles in which the capillary action is the greatest, are more or less cushioned by a film of water, reducing further wearing action. The retention of water at low tide by sand is one of the outstanding environmental features of the sandy shore.

In sheltered areas of the coast, the slope of the beach may be so gradual the surface appears to be flat. Because of the flatness the outgoing tidal currents are slow, leaving behind a residue of organic material

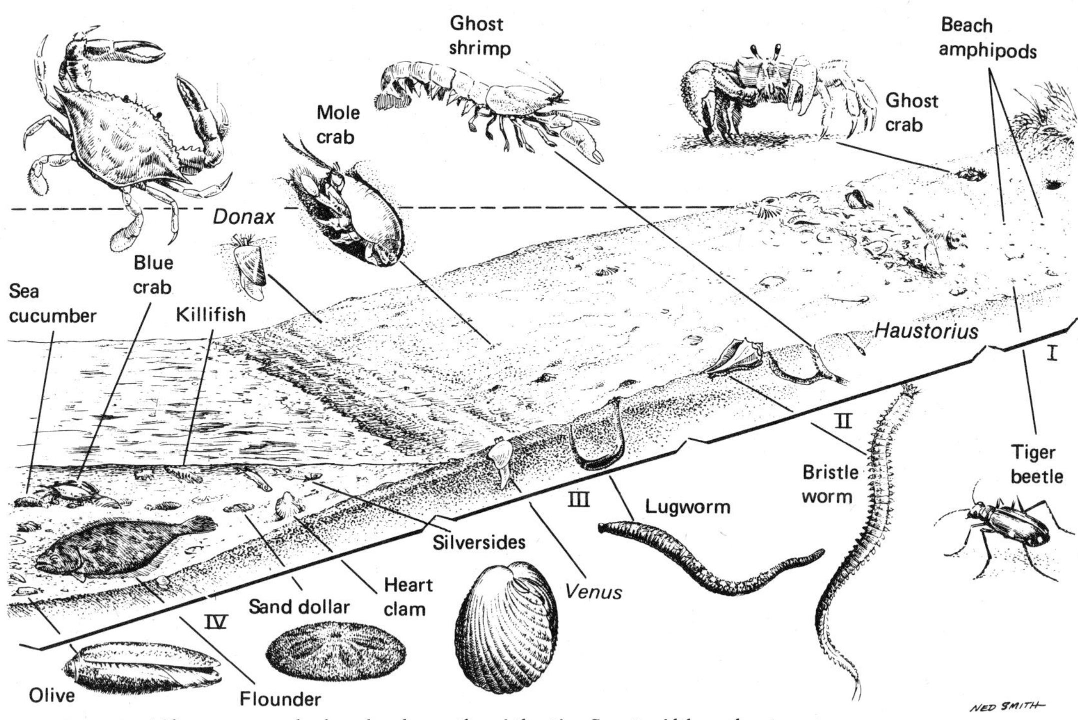

FIGURE 35.6 Life on a sandy beach along the Atlantic Coast. Although strong zonation is absent, organisms still change on a gradient from land to sea. I, supratidal zone: ghost crabs and sand fleas; II, supralittoral zone: ghost shrimp, bristle worms, clams; III, littoral zone: clams, lugworms, mole crabs; IV, infralittoral zone: sand dollar, blue crab. The dashed line indicates high tide.

settled from the water. In these situations mudflats develop.

STRUCTURE

Life on the sand is almost impossible. Sand provides no surface for attachment of seaweeds and their associated fauna; and the crabs, worms, and snails characteristic of rocky crevices find no protection here. Life, then, is forced to live beneath the sand.

Life on sandy and muddy beaches consists of epifauna and infauna. Most infauna either occupy permanent or semipermanent tubes within the sand or mud or are able to burrow rapidly into the substrate. Multicellular infauna obtain oxygen either by gaseous exchange with the water through their outer covering or by breathing through gills and elaborate respiratory siphons.

Within the sand and mud live vast numbers of *meiofauna* with a size range between 0.5 mm and 62 μm, including copepods, ostracods, nematodes, and gastrotrichs. Interstitial fauna are generally elongated forms with setae, spines, or tubercles greatly reduced. The great majority do not have pelagic larval stages. These animals feed mostly on algae, bacteria, and detritus. Interstitial life, best developed on the more sheltered beaches, shows seasonal variations, reaching maximum development in summer months.

Sandy beaches also exhibit zonation related to tidal influences (Figure 35.6), but it must be discovered by digging. Sandy and muddy shores can be divided roughly into supralittoral, littoral, and infralittoral zones, based on animal organisms, but a universal pattern similar to that of the rocky shore is lacking. Pale, sand-colored ghost crabs and beach hoppers occupy the upper beach, the supralittoral. The intertidal beach, the littoral, is a zone where true marine life appears. Although sandy shores lack the variety found on rocky shores, the populations of individual species of largely burrowing animals often are enormous. An array of animals, among them starfish and the related sand dollar, can be found above the low-tide line and in the infralittoral.

Organisms living within the sand and mud do not experience the same violent fluctuations in temperature as those on rocky shores. Although the surface temperature of the sand at midday may be 10° C or more higher than the returning seawater, the temperature a few inches below remains almost constant throughout the year. Nor is there a great fluctuation in salinity, even when fresh water runs over the surface of the sand. Below 25 cm, salinity is little affected.

Associated with these essentially herbivorous animals are the predators, always present whether the tide is in or out. Near and below the low-tide line live predatory gastropods, which prey on bivalves beneath the sand. In the same area lurk predatory portunid crabs such as the blue crab and green crab, which feed on mole crabs, clams, and other organisms. They move back and forth with the tides. The incoming tides also bring other predators, such as killifish and silversides. As the tide recedes, gulls and shorebirds scurry across the sand and mudflats to hunt for food.

FUNCTION

For life to exist on the sandy shore, some organic matter has to accumulate. Most sandy beaches contain a certain amount of detritus from seaweeds, dead animals, feces, and material blown in from shore. This organic matter accumulates within the sand, especially in sheltered areas. In fact, an inverse relationship exists between the turbulence of the water and the amount of organic matter on the beach, with accumulation reaching its maximum on the mudflats. Organic matter clogs the spaces between the grains of sand and binds them together. As water moves down through the sand, it loses oxygen from both the respiration of bacteria and the oxidation of chemical substances, especially ferrous compounds. The point within the mud or sand at which water loses all its oxygen is a region of stagnation and oxygen deficiency. It is marked by the formation of a layer of dark iron sulfides of variable depth. On mudflats such conditions exist almost to the surface.

The energy base for sandy beach and mudflat fauna is organic matter. Much of it becomes available by bacterial decomposition, which goes on at the greatest rate at low tide. The bacteria are concentrated around organic matter in the sand, where they escape the diluting effects of water. Each high tide dissolves and washes away into the sea the products of decomposition and brings in more organic matter for decomposition. Thus sandy beaches and mudflats are important sites for biogeochemical cycling, supplying offshore waters with phosphates, nitrogen, and other nutrients.

At this point energy flow in sandy beaches and mudflats differs from that in terrestrial and aquatic ecosystems, for the basic consumers are bacteria. In more usual energy-flow systems, bacteria act largely as reducers responsible for the conversion of dead organic matter into a nutrient form that can be used by producer organisms. In sandy beaches and mudflats, bacteria not only feed on detrital material and break down organic matter but also are a major source of food for higher level consumers.

A number of deposit-feeding organisms ingest organic matter largely as a means of obtaining bacteria. Prominent among them are numerous nematodes and copepods (Harpacticoida), the polychaete worm *Nereis,* and the gastropod mollusks. Deposit feeders on sandy beaches obtain their food by actively burrowing through the sand and ingesting the substrate to feed on the organic matter it contains. Most common among them is the lugworm *Arenicola,* which is responsible for the conspicuous coiled and cone-shaped casts on the beach.

Other sandy-beach animals are filter feeders, obtaining their food by sorting particles of organic matter from tidal water. Two of these "surf fishers" that advance and retreat up and down the beach with flow and ebb of tide are the mole crab *Emerita* and the coquina clam *Donax.*

Because of their dependence on imported organic

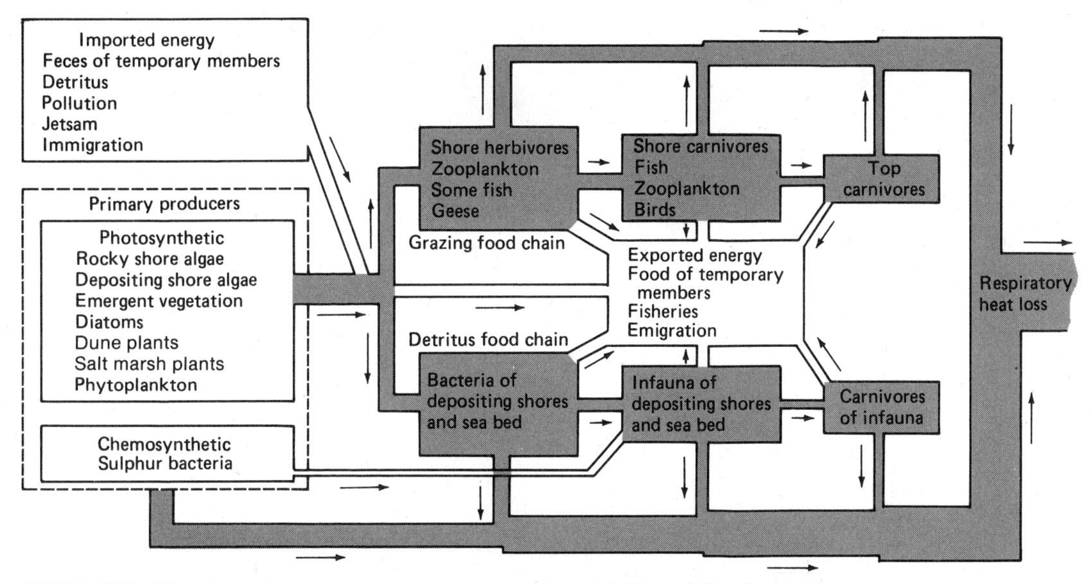

FIGURE 35.7 The coastal ecosystem, a supraecosystem consisting of the shore, the fringing terrestrial regions, and the sublittoral zones. It involves two food webs: that of the rocky shore with its algae, herbivores, and zooplankton; and the detrital food web, involving the bacteria of the depositing shore and sublittoral muds and dependent detritivores and carnivores. Coastal ecosystems are extremely productive; because energy imports exceed exports, the system is continuously gaining energy.

matter and their essentially heterotrophic nature, the sandy beaches and mudflats perhaps cannot be considered ecosystems apart from the whole coastal environment (Figure 35.7). Except in the cleanest of sands, some primary production does take place in the intertidal zone. The major primary producers are the diatoms, confined mainly to fine-grained deposits of sand containing a high portion of organic matter. Productivity is low; one estimate places productivity of moderately exposed sandy beaches at 5 g C/m^2/yr. Production may be increased temporarily by phytoplankton carried in on the high tide and stranded on the surface. Again, the majority are diatoms. More important as producers are the sulfur bacteria in the black sulfide or reducing layer. These chemosynthetic bacteria use the energy released when ferrous oxide is reduced. Some mudflats may be covered with the algae *Enteromorpha* and *Ulva,* whose productivity can be substantial.

In effect, the shore and the mudflats are part of a larger coastal ecosystem involving the salt marsh, the estuary, and coastal waters. They act as sinks for energy and nutrients, because the energy they utilize comes not from primary production but from organic matter that originates outside of the area. And many of the nutrient cycles are only partially contained within the borders of the shore.

CORAL REEFS

Lying in the warm, shallow waters about tropical islands and continental land masses are colorful rich oases in nutrient-poor seas, the coral reefs (Figure 35.8). What makes them unique is the fact that they are an accumulation of dead skeletal material built up by carbonate-secreting organisms, living coral (Cnidaria, Anthozoa), coralline red algae (Rhodaphyta, Corallinaceae), foraminifera, and mollusks. Built only underwater at shallow depths, coral reefs need a stable foundation upon which to grow. Such foundations are provided by shallow continental shelves and submerged volcanos.

Coral reefs, which build up to sea level, are of three

FIGURE 35.8 The coral reef supports a rich diversity of coral species, algae, and colorful fish that occupy the great variety of its habitats, as evidenced by this scene in a Caribbean reef.

types with many gradations among them. (1) *Fringing reefs* grow seaward from the rocky shores of islands and continents. (2) *Barrier reefs* parallel shorelines of continents and islands and are separated from land by shallow lagoons. (3) *Atolls* are horseshoe-shaped rings of coral reefs and islands surrounding a lagoon, formed when a volanic mountain subsided. Such lagoons are about 40 m deep, are usually connected to the open sea by breaks in the reef, and may hold small islands of patch reefs. Reefs build up to sea level.

STRUCTURE

Coral reefs are complex ecosystems that begin with the complexity of the corals themselves. Corals are modular animals, anemone-like cylindrical polyps, with prey-capturing tentacles surrounding the opening or mouth. Most corals form sessile colonies supported on the tops of dead colonies and cease growth when they reach the surface of the water. There are close relationships between coral animals and algae. In the tissues of the gastrodermal layer live *zooxanthellae*, symbiotic, photosynthetically active, endozoic dinoflagellate algae upon which coral depend for most efficient growth. On the calcareous skeletons live still other kinds of algae, both the encrusting red and green coralline species and filamentous species, including turf algae. Associated with coral growth are mollusks, such as giant clams (*Tridacna, Hippopus*), echinoderms, crustaceans, polychaete worms, sponges, and a diverse array of fishes, both herbivorous and predatory.

Zonation and diversity of coral species are influenced by an interaction of depth, light, grazing, competition, and disturbance. The basic diversity gradient is established by light, which influences the depth at which the zooxanthellae can survive. Diversity is lowest at the crest near the surface, where only species such as massive pillar-shaped corals tolerant of intense or frequent disturbances by waves can survive. Diversity increases with depth to a maximum of about 20 m, a region occupied by brain, crustose, and delicate branched and fan corals, and then decreases as light becomes attenuated, eliminating shade-intolerant species. Imposed upon this gradient are other abiotic and biotic disturbances that vary in intensity and decrease with depth. Growth rates of photosynthetic coral are highest in shallow depths; and a few species, especially the branching corals, can easily dominate the reef by overgrowing and shading the crustose corals and algae. Disturbances by wave action, storms, and grazing reduce the rate of competitive displacement among corals. Heavy grazing of overgrowing algae by sea urchins and fish, such as parrotfish, increases encrusting coraline algae. Light grazing allows rapidly growing filamentous and foliose algal species to eliminate crustose algae.

Intense disturbances or loss of species can have pronounced, even disastrous effects on coral reefs. In a recent population outbreak a major coral predator, the crown-of-thorns starfish *Acanthaster planci,* ate nearly all of the corals on certain reefs in the western Pacific, destroying the reefs. Another disaster involved the most important predator of coral and algae in the Caribbean, the black sea urchin *Diadema antillarium.* Although it preyed mostly on newly settled young coral, the urchin did create algal-free settling areas for the larvae. An epizootic disease destroyed 95 percent of the *Diadema* population in 1983–1984, shortly after Hurricane Allen in 1980 had scattered two dominant elkhorn coral species. The absence of black sea urchin allowed dense growths of microalgae to smother newly developing colonies of the corals. The result was the population collapse of these once dominant species along the north Jamaican coast.

FUNCTION

Coral are partially photosynthetic organisms and partially heterotrophic. During the day zooxanthellae carry on photosynthesis and directly transfer organic material to coral tissue. At night coral polyps feed on zooplankton, securing phosphates and nitrates and other nutrients needed by the anthozoans and their symbiotic algae. Thus nutrients are recycled in place between the anthozoans and the algae. In addition, carbon dioxide concentrations in animal tissue enable the coral to extract the calcium carbonate needed to build the coral skeletons.

Adding to the productivity of the coral are crustose coralline algae, turf algae, macroalgae, seagrass, sponges, phytoplankton, and a large bacterial population. Coral reefs are among the most highly productive ecosystems on Earth. Net productivity ranges from 1500 to 5000 g C/m^2/yr compared to 15 to 50 g C/m^2/yr for the surrounding ocean. Because of the ability of the coralline community to retain nutrients within the system and to act as a nutrient trap, coral reefs are oases of productivity within a nutrient-poor sea.

This high productivity and the diversity of habitats within the reef support a high diversity of life—thousands of kinds of exotic invertebrates, some of which, such as sea urchins, feed on coral animals and algae, hundreds of kinds of herbivorous fish that graze on algae, and many predatory species. Some of these predators, such as the puffers and filefish, are corallivores, feeding on coral polyps. Others lie in ambush for prey in coralline caverns. In addition there is a wide array of symbionts, such as cleaning fish and crustaceans, that pick parasites and detritus from larger fish and invertebrates.

HUMAN IMPACT

Humans are strongly attracted to the seashore. Travel brochures show scenic empty sandy beaches backed up by pristine shore vegetation. Rarely do they show the real picture: crowded beaches backed by strands of hotels, boardwalks, and shops. Recreational and commercial development of the seashores, along with intensive seasonal human use, has had a long-term impact on the intertidal ecosystems, the severity of which will increase as human populations grow.

This use has had serious effects on intertidal wildlife, especially that of sandy shores. Beach-nesting birds such as the piping plover (*Charadrius melodus*) and least tern (*Sterna antillarum*) are so disturbed by bathers and dune buggies that both species are in danger of extinction. Other terns and shore birds are subjected to both competition for nest sites and egg predation by rapidly growing populations of species of large gulls that are highly tolerant of humans and thrive on human garbage. Sea turtles and the horseshoe crab (*Limulus polyphemus*), dependent on sandy beaches for nesting sites, find themselves evicted and are declining rapidly for that reason.

Habitat destruction is only one aspect of human impact on intertidal ecosystems. Another is increased pollution. Seashore cottages use septic tanks that drain into sandy soil; commercial developments, intentionally or unintentionally, drain all sorts of wastes into the ground; and coastal cities and small towns pour raw sewage into shallow waters off the coast.

Each incoming tide brings into the beaches fecal-contaminated water that, added to septic tank inputs, makes beaches unsafe for humans and contaminates invertebrate life. Tides also carry in old fish lines, plastic debris and other wastes, and blobs of oil, all hazardous to humans and wildlife.

Oil is probably the most dramatic intertidal pollutant. Always subject to some degree of oil pollution from natural sources, intertidal life is affected most adversely when oil spills, large or small, off the coast reach the shores (see Chapter 34). The damage to intertidal life depends on the extent and intensity of the spill and the type of oil involved: heavy crude oil (involved in the Persian Gulf and Prince William Sound spills), diesel oil, or split oils.

Controlling oil pollution on shores is difficult because incoming tides wash more oil up onto the shore. Millions of sea birds such as cormorants and diving ducks and thousands of marine mammals have been victims of oil pollution. A heavy coating of oil mats feathers, impairing the ability of birds to fly and swim, reduces the feathers' water repellency, and destroys the insulating effects of down. During cold weather a spot of oil as small as a button over a bird's vital organ can induce death from hypothermia. As they preen to clean their feathers, birds ingest fatal amounts of oil. Marine mammals, especially seals and sea otters, fare no better. Oil destroys the insulating effects of their fur, clogs their ears and nostrils, and irritates their eyes. Many such mammals succumb to hypothermia and dehydration.

Less visible to humans is the damage done to invertebrate and plant life of the shores. Particularly vulnerable to smothering oil pollution on sandy shores are interstitial life and sand crabs. Even after the sand appears to be scrubbed by tides, a layer of oil may still reside several meters below the surface. Some intertidal invertebrates, particularly mollusks, appear to be resistant to oil pollution, and their flesh becomes tainted, a taste that is passed along in the food chain. On rocky shores, barnacles resist oil pollution, but the grazing periwinkles, limpets, and whelks are particularly vulnerable. Oil works its way beneath their shells, causing these gastropods to lose their hold, and they are carried away by the tides. As oil eliminates these grazing invertebrates, the barren patches are colonized by algae and seaweeds, particularly the red algae. They too become encrusted with oil and are torn away by the waves. Aside from damage to particular species, the ultimate outcomes of oil pollution on the intertidal zone include a great reduction in species diversity, a simplification of the food web, and an increase in the populations of resistant species.

Summary

Sandy shore and rocky coast are places where sea meets the land. The drift line marks the furthest advance of tides on the sandy shore. On the rocky shore the tide line is marked by a zone of black algal growth. The most striking feature of the rocky shore, zonation of life, results from alternate exposure and submergence by the tides. The black zone marks the supralittoral, the upper part of which is flooded only every two weeks by spring tides. Submerged daily by tides is the littoral, characterized by barnacles, periwinkles, mussels, and fucoid seaweeds. Uncovered only at spring tides is the infralittoral, which is dominated by large brown laminarian seaweeds, Irish moss, and starfish. Distribution and diversity of life across the rocky shore are also influenced by wave action, competition, herbivory, and predation. Left behind by outgoing tides are tidal pools. These are distinct habitats subject over a 24-hour period to wide fluctuation in temperature and salinity and inhabited by varying numbers of organisms, depending on the amount of emergence and exposure.

By contrast, sandy and muddy shores appear barren of life at low tide, but beneath the sand and mud conditions are more amenable to life than on the rocky shore. Zonation of life is hidden beneath the surface. The energy base for sandy and muddy shores is organic matter made available by bacterial decomposition. They are important sites for biogeochemical cycling, supplying nutrients for offshore waters. The basic consumers are bacteria, which in turn are a major source of food for both deposit-feeding and filter-feeding organisms. Sandy shores and mudflats are a part of the larger coastal ecosystem, including the salt marsh, estuary, and coastal waters.

Nutrient-rich oases in nutrient-poor tropical seas are coral reefs. Coral reefs are complex ecosystems based on anthozoan corals and their symbiotic endozoan dinoflagellate algae and coralline algae. Recycling nutrients within the system and functioning as nutrient sinks, coral reefs are among the most productive ecosystems in the world. Their productive and varied habitats support a high diversity of colorful invertebrate and vertebrate life.

Wedged between human communities on the land side and the open sea on the other and subjected to heavy seasonal recreational use, the intertidal communities are especially vulnerable to intense human disturbance. Occupancy and recreational use have destroyed important wildlife habitats and are bringing some dependent species close to extinction. Sewage and solid waste carried in by the tide periodically pollute the shores, endangering human health and wildlife. Chronic oil pollution and especially major oil spills that occur too frequently not only cause massive economic damage; they also decimate vertebrate and invertebrate life of the intertidal communities, affect natural relationships between predator and prey, reduce species diversity, and affect community structure.

Review and Study Questions

1. What are the three major zones of the rocky shore? What are the distinctive features of each?
2. What adaptations enable inhabitants of rocky shore to avoid desiccation, maintain position, and survive flooding?
3. What role do predation and competition play in the organization of the rocky shore community?
4. C. M. Yonge, the British marine ecologist, called rock or tide pools "microcosms of the sea," but with at least two significant qualifications relative to environmental conditions. What are these two differences?
5. What are the major zones of life on sandy shores? How do they differ from those of the rocky shore?
6. In what way does life on a sandy shore survive in the harsh environment?
7. Contrast the energy source of a sandy or muddy shore with that of the rocky shore. What groups of organisms are the basic consumers of sandy shores?
8. What are coral reefs and how are they formed?
9. Explain how the symbiotic relationship between algae and the anthozoans influences the vertical distribution and functioning of corals.
10. Why are coral reefs so productive?
11. In what ways does human activity adversely affect sandy and rocky shores?

*12. Make a list of endangered and threatened shore birds. How does their status relate to human use of the seashore?

*13. Explain how oil affects life on sandy and rocky shores. What is the difference?

*14. Report on the ecological and economic effects of some major oil spills, such as the *Amoco Cadiz* off the Brittany coast of France, the *Exxon Valdez* in Prince William Sound, Alaska, and the release of oil into the Persian Gulf during the Gulf War.

Selected References

Carson, R. 1955. *The edge of the sea*. Boston: Houghton Mifflin. A classic introduction.

Dayton, P. 1971. Competition, disturbance, and community organization: The provision and subsequent utilization of space in a rocky intertidal community. *Ecol. Monogr*. 45:137–159.

Eltringham, S. K. 1971. *Life in mud and sand*. New York: Crane, Russak. A concise, informative introduction.

Hiatt, R. W., and D. W. Strasburg. 1960. Ecological relationships of the fish fauna on coral reefs of the Marshall Islands. *Ecol. Monogr*. 30:66–120. An important, well-illustrated reference on coral reef fish.

Huston, M. 1985. Patterns of species diversity on coral reefs. *Ann. Rev. Ecol. Syst*. 16:149–177. An important reference on the roles of light and disturbance.

Jackson, J. B. C. 1991. Adaptation and diversity of reef corals. *Bioscience* 41:475–482. Reviews patterns of species distribution relative to life history and disturbance.

Jones, O. A., and R. Endean, eds. 1973, 1976. *Biology and geology of coral reefs*. Vols. II, III. New York: Academic Press.

Leigh, E. G., Jr. 1987. Wave energy and intertidal productivity. *Proc. Natl. Acad. Sci. USA* 84:1314.

Lessios, H. A. 1988. Mass mortality of *Diadema antillarum* in the Caribbean: What have we learned? *Ann. Rev. Ecol. Syst.* 19:371–393. Detailed review of the ecological impact of the die-off.

Lubchenco, J. 1978. Algal zonation in the New England rocky intertidal community: An experimental analysis. *Ecology* 61:333–344. An outstanding study and informative reference.

Newell, R. C. 1970. *Biology of intertidal animals*. New York: Elsevier. Adaptations of animals to the intertidal environment.

Nybakken, J. W. 1982. *Marine biology: An ecological approach*. New York: Harper & Row. Informative chapters on rocky, sandy, muddy shores and coral reefs.

Paine, R. T. 1969. The *Piaster-Tegula* interaction: Prey patches, predator food preference, and intertidal community structure. *Ecology* 59:150–961. An outstanding paper on the role of predation in the rocky shore community.

Pomeroy, L. R., and E. J. Kuenzler. 1969. Phosphorus turnover by coral reef animals. In D. J. Nelson and F. E. Evans, eds., *Symposium on radioecology conf.* 670503. Springfield, VA: National Technical Information Services. Pp. 478–483.

Reaka, M. J., ed. 1985. *Ecology of coral reefs*. Symposia series for undersea research, NOAA Underseas Research Program 3. Washington DC: US Department of Commerce.

Sale, P. F. 1980. The ecology of fishes on coral reefs. *Oceanogr. Mar. Biol. Ann. Rev.* 18:367–421. Sweeping review of the coral reef fish ecology.

Stephenson, T. A., and A. Stephenson. 1973. *Life between the tidemarks on rocky shores*. San Francisco: Freeman. A detailed description of the structure of intertidal life around the world.

Underwood, A. J., E. J. Denley, and M. J. Moran. 1983. Experimental analyses of the structure and dynamics of mid-shore rocky intertidal communities in New South Wales. *Oecologica* 56:202–219.

Wellington, G. W. 1982. Depth zonation of corals in the Gulf of Panama: Control and facilitation by resident reef fishes. *Ecol. Monogr*. 52:223–241.

Wilson, R., and J. Q. Wilson. 1985. *Watching fishes: Life and behavior on coral reefs*. New York: Harper & Row. Although written for a general audience, this well-illustrated book is one of the most accessible and informative references on the subject.

Yonge, C. M. 1949. *The seashore*. London: Collins. An old well-illustrated classic.

CHAPTER 36

Estuaries, Salt Marshes, and Mangrove Forests

Outline

Objectives

On completion of this chapter, you should be able to:

1. Define an estuary and describe its characteristics.
2. Describe how freshwater and tidal inputs relate to the functioning of estuaries.
3. Describe the major features of a salt marsh.
4. Discuss the relationship of the salt marsh to the estuarine ecosystem.
5. Discuss the unique features of the mangrove forest.
6. Explain the importance of the mangrove forest to the tropical coastal ecosystem.
7. Discuss the impact humans have had on the estuary, salt marsh, and mangrove forest.

ESTUARIES

Waters of all streams and rivers eventually drain into the sea; and the place where this fresh water joins the salt water is called an *estuary* (Figure 36.1). Estuaries are semienclosed parts of the coastal ocean where the seawater is diluted and partially mixed with water coming from the land. Estuaries differ in size, shape, and volume of water flow, all influenced by the geology of the region in which they occur. As the river reaches the encroaching sea, the stream-carried sediments are dropped in the quiet water. They accumulate to form deltas in the upper reaches of the mouth and shorten the estuary. When silt and mud accumulations become high enough to be exposed at low tide, tidal flats develop. These flats divide and braid the original channel of the estuary. At the same time, ocean currents and tides erode the coastline and deposit material on the seaward side of the estuary, also shortening the mouth. If more material is deposited than is carried away, barrier beaches, islands, and brackish lagoons appear.

STRUCTURE

Salinity, Temperature, and Nutrient Traps

The one-way flow of streams and rivers into the estuary meets the inflowing and outflowing tides. This meeting sets up a complex of currents that varies with the season, amount of rainfall, tidal oscillations, and winds. The interaction of fresh water and salt influences the salinity of the estuarine environment.

Salinity varies vertically and horizontally, often within one tidal cycle (Figure 36.2). Vertical salinity may be the same from top to bottom, or it may be completely stratified, with a layer of fresh water on top

FIGURE 36.1 An unspoiled wild estuary at Coos Bay on Oregon's Pacific Coast preserved as a sanctuary.

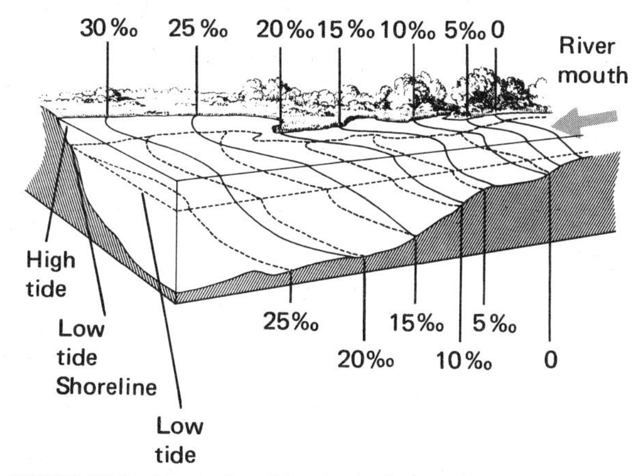

FIGURE 36.2 Vertical and horizontal stratification of salinity from the river mouth to the estuary at both low and high tides. At high tide the incoming seawater increases the salinity toward the river mouth; at low tide salinity is reduced. Note also how salinity increases with depth, because lighter fresh water flows over denser salt water.

and a layer of dense salty water on the bottom. Salinity is homogeneous when currents, particularly eddy currents, are strong enough to mix the water from top to bottom. The salinity in some estuaries is homogeneous at low tide, but at flood tide a surface wedge of seawater moves upstream more rapidly than the bottom water. Salinity is then unstable, and density is inverted. The seawater on the surface tends to sink as lighter fresh water rises, and mixing takes place from the surface to the bottom. This phenomenon is known as *tidal overmixing*. Strong winds, too, tend to mix salt water with fresh water in some estuaries, but when the winds are still, the river water flows seaward on a shallow surface over an upstream movement of seawater, more gradually mixing with the salt.

Horizontally, the least saline waters are at the river entrance and the most saline at the mouth of the estuary (see Figure 36.2). The configuration of the horizontal zonation is determined mainly by the deflection caused by the incoming and outgoing currents. In all estuaries of the Northern Hemisphere, outward-flowing fresh water and inward-flowing seawater are deflected to the right because of Earth's rotation. As a result, salinity is higher on the left side.

The salinity of seawater is about 35‰; that of fresh water ranges from 0.065 to 0.30‰. Because the concentration of metallic ions carried by rivers will vary from drainage to drainage, the salinity and chemistry of estuaries differ. The portion of dissolved salts in the estuarine waters remains about the same as that of seawater, but the concentration varies in a gradient from fresh water to sea.

Exceptions to these conditions exist in regions where evaporation from the estuary may exceed the inflow of fresh water from river discharge and rainfall (a negative estuary). This condition causes the salinity to increase in the upper end of the estuary, and horizontal stratification is reversed.

Temperatures in estuaries fluctuate considerably diurnally and seasonally. Waters are heated by the sun and inflowing and tidal currents. High tide on the mudflats may heat or cool the water, depending on the season. The upper layer of estuarine water may be cooler in winter and warmer in summer than the bottom, a condition that, as in a lake, will result in spring and autumn overturns.

Mixing waters of different salinities and temperatures acts as a nutrient trap (Figure 36.3). Inflowing river waters more often than not impoverish rather than fertilize the estuary, except for phosphorus. Instead, nutrients and oxygen are carried into the estuary by the tides. If vertical mixing takes place, these nutrients are not swept back out to sea but circulate up and down among organisms, water, and bottom sediments (Figure 36.4).

Estuarine Organisms

Organisms inhabiting the estuary are faced with two problems—maintenance of position and adjustment to changing salinity. Most estuarine organisms are benthic. They are securely attached to the bottom, are buried in the mud, or occupy crevices and crannies about sessile organisms. Mobile inhabitants are mostly crustaceans and fish, largely young of species that spawn offshore in high-salinity water. Planktonic organisms are wholly at the mercy of the currents. Because the seaward movement of streamflow and ebb tide transport plankton out to sea, the rate of circulation or flushing time determines the nature of the plankton population. If the circulation is too vigorous, the plankton population may be small. Phytoplankton in summer is mostly near the surface and in low-

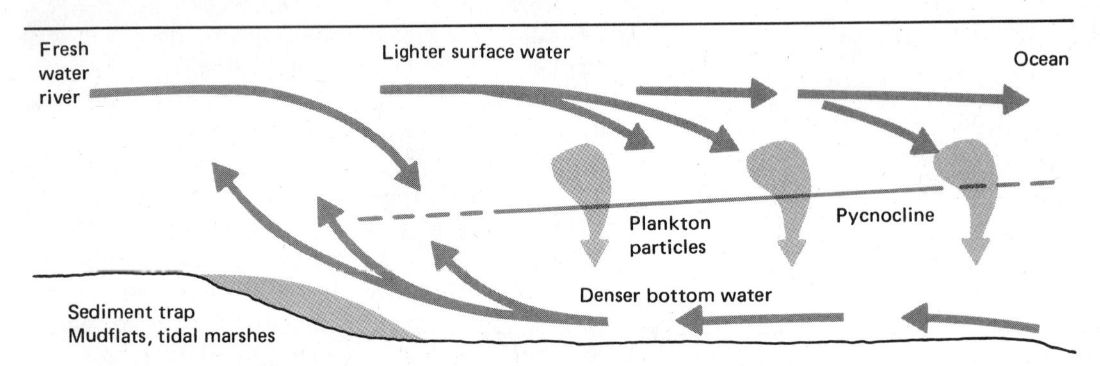

FIGURE 36.3 Circulation of fresh and salt water in an estuary creates a nutrient trap. A salt wedge of intruding seawater on the bottom produces a surface flow of lighter fresh water and a counterflow of heavier brackish water. This countercurrent serves to trap nutrients, recirculating them toward the tidal marsh. The same countercurrent also sends phytoplankton up the estuary, repopulating the water. When nutrients are high in the upper estuary, they are taken up rapidly by tidal marshes and mudflats. These areas tend to trap particulate nitrogen and phosphorus, convert them to soluble forms, and export them back to open waters of the estuary. Plants on the tidal marshes and mudflats act as nutrient pumps between bottom sediments and surface water.

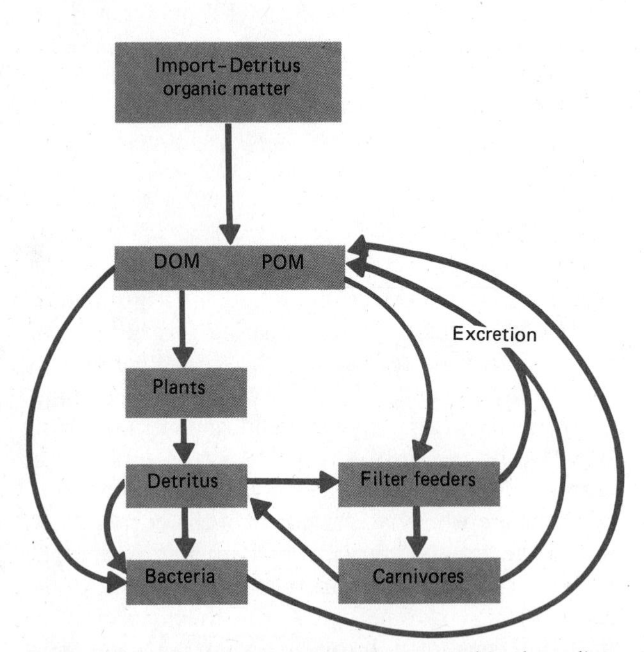

FIGURE 36.4 The estuary is mostly heterotrophic, depending on organic material from sea and land. This diagram shows how organic matter is cycled in the estuary.

salinity water. In winter phytolankton is more uniformly distributed. For any planktonic growth to become endemic in an estuary, reproduction and recruitment must balance loss from the physical processes that disperse the population.

Salinity dictates the distribution of life in the estuary. Essentially, the organisms of the estuary are marine, able to withstand full seawater. Except for anadromous fish, no freshwater organisms live there. Some estuarine inhabitants cannot withstand lowered salinities, and these species decline along a salinity gradient. Sessile and slightly motile organisms have an optimum salinity range within which they grow best. When salinities vary on either side of this range, populations decline.

Size influences the range of motile species within the estuarine waters, particularly fish. Some, such as the striped bass, spawn near the interface of fresh and low-salinity water (Figure 36.5). The larvae and young fish move downstream to more saline waters as they mature. Thus, for the striped bass the estuary serves as both a nursery and as a feeding ground for the young. Anadromous species such as the shad (*Alosa*) spawn in fresh water, but the young fish spend their first summer in the estuary, then move out to the open

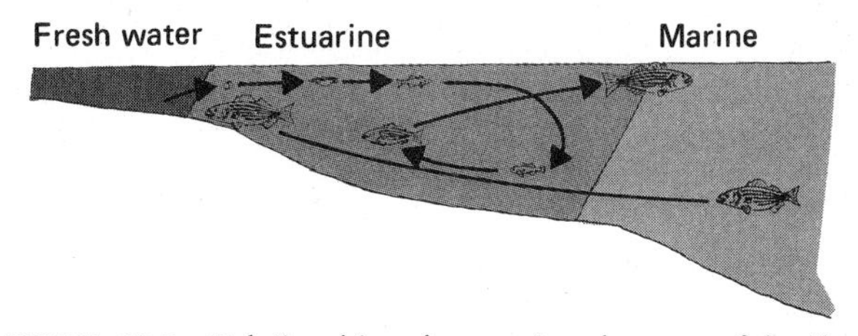

FIGURE 36.5 Relationship of a semianadromous fish, the striped bass, to the estuary. Adults live in the sea, but young fish grow up in the estuary.

sea. Other species such as the croaker (Sciaenidae) spawn at the mouth of the estuary, but the larvae are transported upstream to feed in the plankton-rich low-salinity areas. Still others such as the bluefish (*Pomatomus saltatrix*) move into the estuary to feed. In general, marine species drop out toward fresh water and are not replaced by freshwater forms. In fact, the mean number of species decreases progressively from the mouth of the estuary to upstream stations.

Salinity changes often affect larval forms more severely than adults. Larval veligers of the oyster drill *Thais* succumb to low salinity more easily than the adults. A sudden influx of fresh water, especially after hurricanes or a heavy rainfall, sharply lowers the salinity and causes a high mortality of oysters and their associates.

The oyster bed and oyster reef are the outstanding communities of the estuary. The oyster is the dominant organism about which life revolves. Oysters may be attached to every hard object in the intertidal zone or they may form reefs, areas where clusters of living organisms grow cemented to the almost buried shells of past generations. Oyster reefs usually lie at right angles to tidal currents, which bring planktonic food, carry away wastes, and sweep the oysters clean of sediment and debris. Closely associated with oysters are encrusting organisms such as sponges, barnacles, and bryozoans, which attach themselves to oyster shells and depend on the oyster or algae for food.

FUNCTION

Estuarine systems function on both plankton-based and detrital-based food webs. The producer component, particularly in the middle and lower estuary, consists of dinoflagellates and diatoms. The latter convert some of the carbon intake to higher caloric fats and lipids rather than low-energy carbohydrates typical of most green plants. This fat provides a high-energy food base for higher trophic levels.

Although inflowing water from rivers and coastal marshes carries nutrients into the estuary, phytoplankton production is regulated more by internal nutrient cycling than by external sources. This internal cycling involves excretion of mineralized nutrients by herbivorous zooplankton (see Chapter 22) and release of nutrients remineralized by invertebrates of the bottom sediments, by the roiling of sediments, and by steady-state exchanges between nutrients present in the particulate and dissolved phases.

Nutrients accumulated over winter in temperate estuaries stimulate a winter–spring bloom. As the nutrients become depleted and the phytoplankton experiences intensive predation by zooplankton, the bloom collapses and the phytoplankton falls to the bottom, where it is fed upon by bivalves and other filter-feeding invertebrates. In well-mixed estuaries, nutrients remineralized in the benthos are returned to the water column, stimulating a summer bloom.

In shallow estuarine waters, rooted aquatics such as widgeon grass and eelgrass (*Zostera marina*) assume major importance. These aquatic plants are complex systems supporting a large number of epiphytic and epizoic organisms. Such communities are important to certain vertebrate grazers such as brant, Canada geese, the black swan in Australia, and sea turtles, and provide a nursery ground for shrimp and bay scallops.

HUMAN IMPACT

For centuries estuaries have carried the burden of human disturbance. Because they are semienclosed, provide natural harbors, and are located at the juncture of navigable rivers and the sea, estuaries have long been sites for cities and industry. As a result of this concentration of human activity, estuaries have become the dumping grounds for untreated sewage, industrial effluents and wastes, and chronic oil pollution from ships and industry. Because rivers draining vast inland areas flow into them, estuaries receive a load of pollutants from toxic chemicals, pesticides, and sewage to silt from mining, agriculture, construction, and lumbering. The commercial and recreational values of estuaries stimulate bulkheading, channel dredging, and filling to create waterfront industrial, recreational, and residential sites. Thermal power plants pour heated effluents into the water, raising the temperature near the shore. This warm water and the erosion of bottom sediments by its discharge reduce oxygen content of the estuarine water. All of these activities alter the flow of tidal currents, increase the anoxic conditions of the estuarine bottom, and in other ways change the nature of the estuary.

These intrusions have a pronounced ecological and economic effect on the estuary. Decreased oxygen on the bottom favors only a few benthic organisms, such as polychaete worms, and eliminates others, such as oysters. Excessive inputs of phosphorus stimulate the explosive growth of a few algae at the expense of others, as well as the spread of nonendemic plants. This reduction in species diversity simplifies the natural food web, affecting all life in the estuary.

Estuaries, a rich source of food and a safe haven from predators, are major nurseries for commercially and recreationally important finfish and shellfish species such as mullet, striped bass, flounder, shrimp, blue crab, Alaska crab, and clams. Pollution, salinity changes, turbidity, and loss of food have reduced the fishery resources. What remains is often contaminated with toxic bacteria and concentrations of toxic elements such as mercury, cadmium, and lead. These contaminants, as well as oil that taints the flesh of fish and shellfish, make them unfit for consumption.

SALT MARSHES

On the alluvial plains about the estuary and in the shelter of spits and offshore bars and islands exists a unique community, the salt marsh. Although salt marshes appear as waving acres of grass, they are a complex of distinctive and clearly demarked plant associations. The reasons for this complex are tides and salinity. The tides play the most significant role in plant segregation, for two times a day the salt-marsh plants on the outermost fringes are submerged in salty water and then exposed to full sun. Their roots extend into poorly drained, poorly aerated soil in which the soil solution contains varying concentrations of salt. Only plant species with a wide range of salt tolerances can survive such conditions.

STRUCTURE

Plants

From the edge of the sea to the high land, zones of vegetation distinctive in form and color develop, reflecting a microtopography that lifts the plants to various heights within and above high tide (Figure 36.6). Most conspicuous on the seaward edge of the marsh and along tidal creeks are the deep green growths of *Spartina alterniflora*, which dominate the low marsh (Figure 36.7). Stiff, leafy, up to 3 meters tall, and submerged in salt water at each high tide, salt-marsh cordgrass forms a marginal strip between the open mud to the front and the high marsh behind. No litter accumulates beneath the stand. Strong tidal currents sweep the floor of the *Spartina* clean, leaving only thick, black mud.

Spartina alterniflora is well adapted to grow on the intertidal flats of which it has sole possession. It has a high tolerance for salt water and is able to live in a semisubmerged state. It can live in a saline environment by selectively concentrating sodium chloride in its cell at a level higher than the surrounding salt water, thus maintaining its osmotic integrity. To rid itself of excessive salts, salt-marsh cordgrass has special salt-secreting cells in the leaves. Water excreted with the salt evaporates, leaving behind sparkling crystals on the surface of leaves to be washed off by tidal water. To get air to its roots, buried in anaerobic mud, *Spartina alterniflora* has hollow tubes leading from the leaf to the root through which oxygen diffuses.

Above and behind the low marsh is the high marsh, standing at the level of mean high water. At this level the tall *Spartina* gives way rather sharply to a short form of *Spartina alterniflora*, yellowish, almost chloritic in appearance, contrasting with the tall, dark green form. This short form seemingly represents a phenotypic plastic response to environmental conditions of the high marsh. The high marsh has a higher salinity, a decreased input of nutrients, and an accumulation of toxic wastes, the result of lower tidal exchange rates. The shorter, more open canopy of the

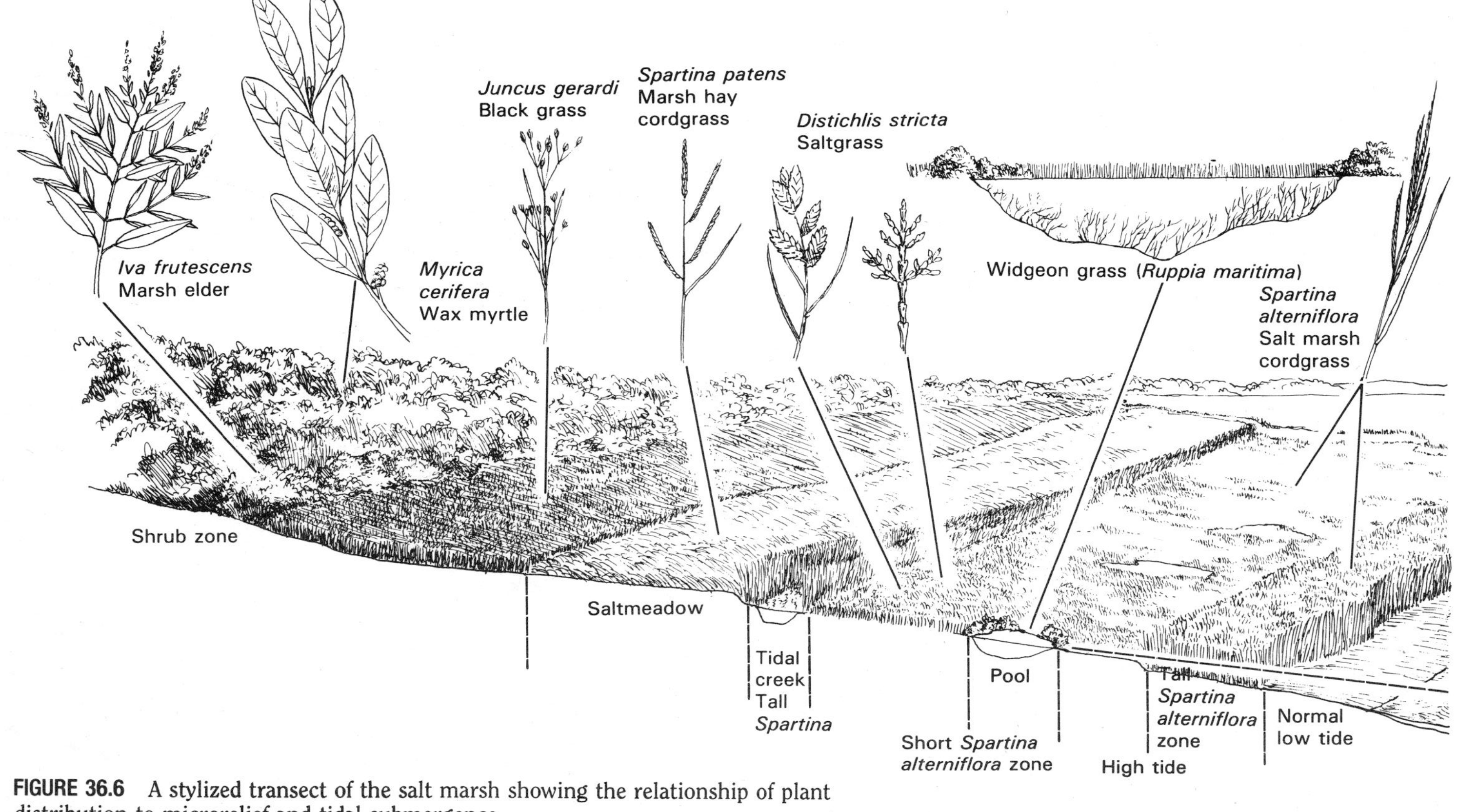

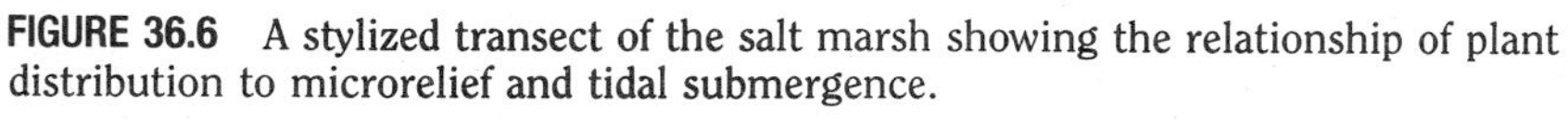

FIGURE 36.6 A stylized transect of the salt marsh showing the relationship of plant distribution to microrelief and tidal submergence.

FIGURE 36.7 Salt-marsh cordgrass (*Spartina alterniflora*) dominates the low marsh.

high marsh results in higher leaf and soil temperatures and a higher rate of evaporation than in the low marsh. These conditions, along with the decreased dominance of *Spartina*, allow other marsh plants to grow. Here are the fleshy, translucent glassworts (*Salicornia* spp.) (Figure 36.8) that turn bright red in fall, sea lavender (*Limonium carolinianum*), spearscale (*Atriplex patula*), and sea blite (*Suaeda maritima*).

Where the microelevation is about 5 cm above mean high water, short *Spartina alterniflora* and its associates are replaced by salt-meadow cordgrass, *Spartina patens*, and an associate spikegrass or saltgrass, *Distichlis spicata*. Salt-meadow cordgrass is a fine grass that grows so densely and forms such a tight mat that few other plants can grow with it (Figure 36.9).

As the microelevation rises several more centimeters above mean high tide and if there is some intrusion of fresh water, *Spartina* and *Distichlis* may be replaced by two species of black needlerush or black grass (*Juncus roemerianus* and *Juncus gerardi*), so called because their dark green color becomes almost black in the fall. Beyond the black grass and often replacing it is a

FIGURE 36.8 Glasswort (*Salicornia*) invades a saline bare spot on the high marsh.

FIGURE 36.9 A heavy growth of salt-marsh hay cordgrass (*Spartina patens*) on the high marsh. Note the typical cowlick formations caused by wind and water.

FIGURE 36.10 A tidal creek at low tide on the high marsh. Note the growth of tall *Spartina* along the banks.

shrubby growth of marsh elder (*Iva frutescens*) and groundsel (*Baccharis halimifolia*). These shrubs tend to invade the high marsh where a slight rise in microelevation exists, but such invasions are often short-lived, as storm tides sweep in and kill the plants. On the upland fringe grow bayberry (*Myrica pensylvanica*) and the pink-flowering sea holly (*Hibiscus palustris*).

Two conspicuous features of a salt marsh are the meandering creeks and pannes, or salt pans. The creeks form an intricate system of drainage channels that carry tidal waters back out to sea (Figure 36.10). The salt pans appear as circular to elliptical depressions. At high tide they are flooded; at low tide the depressions remain filled with salt water. If shallow enough, the water may evaporate completely, leaving an accumulating concentration of salt on the mud.

Pannes support distinctive vegetation, which varies with the depth of the water and salt concentration. Pools with a firm bottom and sufficient depth to retain tidal water support dense growths of widgeon grass (*Ruppia maritima*), whose small black seeds are relished by waterfowl.

Shallow depressions in which water evaporates are covered with a heavy algal crust and crystalized salt. The edges of these salt flats may be invaded by *Salicornia*, *Distichlis*, or even short *Spartina alterniflora* (see Figure 36.8).

The exposed banks of tidal creeks that braid through the salt marshes support a dense population of mud algae, the diatoms and dinoflagellates, photosynthetically active all year. Photosynthesis is highest in summer during high tides and in winter during low tides, when the sun warms the mud. Some of the algae are washed out at ebb tide and become part of the estuarine plankton available to such filter feeders as oysters.

The salt marsh described is typical of the North American Atlantic coast, but many variations exist locally and latitudinally around the world. North America has several distinctive types. Arctic salt marshes have few species. East and Gulf coast marshes dominated by *Spartina* and *Juncus* reach their best development on the heavy silt deposits from North Carolina south. Pacific coast marshes, poorly developed, are dominated up to mean high-water level by *Spartina foliosa*, and above that by *Salicornia*. Salt marshes on the western coast of Europe and in Great Britain are dominated by *Salicornia* and salt-marsh grasses *Puccinellia*. Those of northwestern Europe are dominated by *Spartina angelica*, a hybrid polyploid species that spontaneously formed from the native *Spartina townsendii*, and *Spartina alterniflora*, introduced from North America. Because of its vigorous growth, *S. angelica* is widely planted to stabilize and bind coastal mudflats.

Consumers

Animal life of the salt marsh, not noted for its diversity, is outstanding for its interest. Some of the inhabitants are permanent residents in sand and mud; others are seasonal visitors; and most are transients coming to feed at high and low tide.

Three dominant animals of the low marsh are the ribbed mussel (*Modiolus demissus*), fiddler crab (*Uca pugilator* and *U. pugnax*), and marsh periwinkle (*Littorina* spp.). The marsh periwinkle, related to the periwinkles of the rocky shore, moves up and down the stems of *Spartina* and onto the mud to feed on algae and detritus. Buried halfway in the mud is the ribbed mussel. At low tide the mussel is closed; at high tide the mussel opens to filter particles from the water, accepting some and rejecting others in a mucous ribbon known as pseudofeces. Running across the marsh at low tide like a vast herd of tiny cattle are the fiddler crabs. Among the marsh animals they are the most adaptable. They have both lungs and gills. They can endure periods of high tides and cold winters without oxygen. They have a salt-water control system that enables them to move from diluted seawater to briny pools. Omnivorous feeders, they eat detrital material, algae, and small animals. Fiddler crabs live in burrows, marked by mounds of freshly dug, marble-sized pellets. The burrowing activity of crabs is similar to that of earthworms, because in overturning the mud they bring up nutrients to the surface.

Prominent about the base of *Spartina* and under debris are the sandhoppers (*Orchestia*). These detrital-feeding amphipods may be abundant and are important in the diet of some marsh birds.

Three conspicuous vertebrate residents of the low marsh of eastern North America are the diamond-backed terrapin (*Malaclemys terrapin*), the clapper rail (*Rallus longirostris*), and the seaside sparrow (*Ammospiza maritima*). The terrapin feeds on fiddler crabs, small mollusks, marine worms, and dead fish. The rail eats fiddler crabs and sandhoppers. The sparrow eats sandhoppers and other small invertebrates.

On the high marsh, animal life changes as suddenly as the vegetation. The small, coffee-bean-colored pulmonate snail *Melampus*, found by the thousands under the low grass, replaces the marsh periwinkle. Meadow mice (*Microtus*) have a maze of runways within the matted growth. Replacing the clapper rail and seaside sparrow on the high marsh are the willet (*Catoptrophorus semipalmatus*) and the seaside sharp-tailed sparrow (*Ammospiza caudacuta*).

Among the shrubby fringes of the marsh, dense growth of marsh elder, groundsel, and wax myrtle gives nesting cover for blackbirds and provides sites for heron rookeries. Remote stands of these shrubs support the nests of smaller herons and egrets, whereas the tall dead pines and human-made structures support the nests of the fish-eating osprey.

Low tide brings a host of predaceous animals onto the marsh to feed. Herons, egrets, gulls, terns, willets, ibis, raccoons, and others spread over the exposed marsh floor and muddy banks of tidal creeks to feed. At high tide the food web changes as the tide waters flood the marsh. Such fish as silversides (*Menidia menidia*), killifish (*Fundulus heteroclitus*), and four-spined stickleback (*Apeltes quadracus*), restricted to channel waters at low tide, spread over the marsh at high tide, as does the blue crab.

FUNCTION

The salt marsh is a detrital system; grazing herbivores play a minuscule role. In Georgia salt marshes, the salt-marsh grasshopper (*Orchelimum*), which eats plant tissues, and the plant hopper (*Prokelisis*), which sucks plant juices, use only about 3 to 4 percent of net production. Most of the detrital material from *Spartina* of the low marsh is washed out by the tide; that of the high marsh is decomposed in place by bacteria. Feeding on the detritus and the bacteria it supports are fiddler crabs and mussels. On a New England marsh, which has few fiddler crabs and mussels, most of the *Spartina* is broken up by winter ice instead of fiddler crabs.

Much of the detrital material is carried out to the bay, where it is fed on by a major detritivore, the grass shrimp (*Palaemonetes pugio*). It breaks the coarse detritus into smaller pieces that are colonized by bacteria and diatoms. By its action the grass shrimp makes nutrients and biomass available for higher trophic levels.

The salt marsh, one of the most productive ecosystems, has become that way in part because of a tidal subsidy. Tidal flushing brings in new nutrients and sweeps out accumulated salts, metabolites, sulfides, and toxic wastes, and replaces anoxic interstitial water with oxygenated water. Added to this advantage is a tight internal nutrient cycling. Algal and bacterial populations turn over rapidly and detrital material is fragmented and decomposed. Up to 47 percent of net primary production is respired by microbes; part is grazed by nematodes and microscopic benthic organisms, both of which are consumed by deposit feeders. Depending on the nature of the salt marsh, between 11 and 66 percent of decomposed marsh grass is converted to microbial biomass. Belowground production enters an anaerobic food web of fermentation, dominated by sulfur-oxidizing bacteria that reduce sulfates and produce methane.

What happens to excess carbon production in a salt marsh is not well understood. Each salt marsh apparently differs in the way carbon is transformed in the food web and its route and amount of export. Some salt marshes are dependent on tidal exchanges and import more than they export, whereas others export more than they import. Some of the excess production goes into sediments; some may be transformed microbially in the water in the marsh and tidal creeks. A portion may be exported to the estuary physically as detritus, as bacteria, or as fish, crabs, and intertidal organisms in the food web.

The importance of salt marshes as a nutrient source and sink for the estuary, especially nitrogen, has been a question of long standing. In the salt marsh, denitrification often exceeds nitrogen fixation, and the marsh depends on inputs into the system. The amount of nitrogen cycled is determined by tidal input, physical and chemical exchanges with air and water, and biological fluxes. In some salt marshes, exemplified by Great Sippewissett marsh in Massachusetts, groundwater inflow brings in nitrates, some of which percolate through the peat and are exported to the estuary as organic nitrogen, ammonium, and nitrates. Of the total influx, one-third is exported by denitrification and two-thirds by tidal export. The much larger salt marshes of Sapelo Island, Georgia, depend on inputs of nitrogen from the associated river and tides. Only in summer does the system appear to export any nitrogen. Mostly the marsh is a net sink for nitrogen.

HUMAN IMPACT

Fifty-one percent of the population of the contiguous United States and 70 percent of all humanity live within 80 kilometers of the coastlines. With so much humanity clustered about the coast for many reasons, it is obvious why coastal ecosystems are threatened and are disappearing rapidly. In spite of some efforts at regulation and acquisition at state and federal levels to slow the loss, coastal wetlands in the United States are still disappearing at a rate of 8000 ha a year. Commonly regarded as economic wastelands, salt marshes have been and are being ditched, drained, and filled for real estate development (everyone likes to live at the water's edge), industrial development, and agriculture. Reclamation of marshes for agriculture is most extensive in Europe, where the high marsh is enclosed within a sea wall and drained. Most of the marshland and tideland in Holland has been reclaimed in this fashion. Many coastal cities such as Boston, Amsterdam, and much of London have been built on filled-in marshes. Salt marshes close to urban and industrial developments become polluted with spillages of oil, which become easily trapped within the vegetation.

The most rapid loss of wetlands in the United States is taking place on the coast of Texas, Mississippi, and Louisiana. Two-thirds of the loss has been to deep water, the remaining third to urban development. At the present rate of destruction, about 125 km^2/yr, most of the Louisiana coastal marshes will be gone in two decades. Canals laced through the marshes to accommodate oil-drilling rigs allow the intrusion of salt water and disrupt natural hydrology. Flood-control structures along the Mississippi interfere with the buildup of sediments in the delta areas necessary to maintain salt marshes along a sinking coastline. This sinking is further aggravated by extracting oil, gas, and groundwater, which allows the surface to subside.

Losses of coastal wetlands have a pronounced effect on the salt marsh and associated estuarine ecosystems. They are the nursery ground for commercial and recreational fisheries. There is, for example, a correlation between the expanse of coastal marsh and shrimp production in Gulf coastal waters. Oysters and blue crabs are marsh-dependent, and the decline of these important species is related to loss of salt marshes. Coastal marshes are major wintering grounds for waterfowl. One-half of the migratory waterfowl of the Mississippi flyway depend on Gulf Coast wetlands, and the bulk of the snow goose population winters on the coastal marshes from the Chesapeake Bay to North Carolina. These geese may remove by grazing or uprooting nearly 60 percent of the belowground production of marsh vegetation. Forced concentration of these wintering migratory birds into shrinking salt-marsh habitats could jeopardize marsh vegetation and the future of the birds.

MANGROVE FORESTS

Replacing salt marshes on tidal flats in tropical regions are *mangrove forests* or *mangals*, which cover 60 to 75 percent of the coastline of the tropical regions. Mangals develop where wave action is absent, sediments accumulate, and the muds are anoxic. They extend landward to the highest vertical tidal range, where they may be only periodically flooded. The dominant plants are mangroves, which include eight families and 12 genera dominated by *Rhizophora, Avicennia, Bruguiera,* and *Sonneratia*. Growing with them are other salt-tolerant plants, mostly shrubs. Mangals reach their finest development and have the most species in the Indo-Malaysian region.

In growth form, mangroves range from short, prostrate forms to timber-size trees 30 m high. All mangroves have shallow, widely spreading roots, many with prop roots coming from trunk and limbs (Figure 36.11). Many species have root extensions called *pneumatophores* that take in oxygen for the roots. Their fleshy leaves, although tough, are often succulent and may have salt glands. Many mangrove species have a unique method of reproduction, vivipary. Following fertilization, the embryos undergo uninterrupted development with no true resting seed. They grow into a seedling on the tree, drop to the water, and float upright until they reach water shallow enough for their roots to penetrate the mud.

STRUCTURE

The formation and physiognomy of mangrove forests are strongly influenced by the range and duration of tidal flooding. Mangals become established in areas where the lack of wave action allows fine sediments or mud to accumulate. The tangle of prop roots and pneumatophores further slows the movement of tidal waters, allowing more sediments to settle out. Land begins to march seaward, followed by colonizing mangroves.

A feature of this response is zonation, the changes in vegetation from seaward edge to true terrestrial environment. Although often used as an example, the mangals of the Americas, especially Florida, have the least pronounced zonation, largely because of the few species involved. The pioneering red mangrove (*Rhizophora mangle*) occupies the seaward edge and experiences the deepest tidal flooding. Red mangroves are backed by black mangroves (*Avicennia germinanas*), shallowly flooded by high tides. The landward edge is dominated by white mangroves (*Laguncularia racemosa*), along with buttonwood (*Conocarpus erectus*), a nonmangrove species that acts as a transition to terrestrial vegetation.

Mangals of the Indo-Malaysian region, containing up to 40 species, have a more pronounced, although variable, zonation. The seaward fringe is dominated by one or several species of *Avicennia* and perhaps trees of the genus *Sonneratia* that do not grow well in the shade of other species. Behind the *Avicennia* is a zone of *Rhizophora*, characteristic mangrove species with prop roots and pneumatophores that grow in areas covered by daily high tides up to a point covered only

(a)

(b)

FIGURE 36.11 (a) A red mangrove forest in Florida. (b) Interior of a red mangrove stand; note the prop roots.

by the highest spring tides. At and beyond the level of high spring tides is a broad zone of tall *Bruguiera*. The final and often indefinite mangrove zone is an association of small shrubs, mainly *Ceriops*.

Mangals are faunally rich, with a unique mix of terrestrial and marine life. Living and nesting in the upper branches are numerous species of birds, particularly herons and bitterns. As in the salt marsh, *Littorina* snails live on the prop roots and trunks of mangrove trees. Attached to the stems and prop roots are barnacles, and on the mud at the base of the roots are detrital-feeding snails. Fiddler crabs and tropical land crabs burrow into the mud during low tide and live on prop roots and high ground during high tide. In the Indo-Malaysian mangrove forests live mud skippers, fish of the genus *Periophthalmus*, with modified eyes set high on the head. They live in burrows in the mud and crawl about on the top of it. In many ways they act more like amphibians than fish. The sheltered waters about the roots provide a nursery and haven for the larvae and young of crabs, shrimp, and fish.

FUNCTION

Net productivity of mangrove forests is variable, ranging in Florida mangals from about 450 g C/m^2/yr to 2700 g C/m^2/yr. Productivity is influenced by tidal inflow and flushing, water chemistry, salinity, and soil nutrients, much as in the salt marsh. Highest rates of productivity occur in those mangrove forests under the influence of daily tides. For example, gross primary productivity of red mangroves, flooded daily by high tides, decreases with salinity, whereas gross primary productivity of white and black mangroves increases with increasing salinity. In areas of intermediate salinity, white mangroves have twice the productivity of red mangroves. Zonation of mangrove forests appears to reflect the optimal productivity niches of the species involved rather than physical or successional conditions.

Mangrove forests export a considerable portion of their production to the surrounding waters, largely as leaf fall and other detrital material. This material becomes the base of a detrital food web that supports an array of important commercial finfish and shellfish. In Florida, for example, these species include striped mullet, spotted seatrout, red drum, blue crab, and shrimp.

HUMAN IMPACT

Mangrove forests have been exploited to some degree, mostly as firewood, since antiquity. Such exploitation probably destroyed the mangrove forests of the Red Sea and Persian Gulf. Large mangroves have been cut for fence posts, poles, and timber. In the Indo-Malaysian region, where mangroves are an important timber resource, silvicultural management systems have been developed for them. Pulp of certain species is used in the manufacture of rayon, lacquers, and cellulose acetate, and tannin is extracted from the bark. In the past, at least, bark and fruit have been used as a source of medicinals for treatment of rheumatism, boils, and eye infections. In addition, mangroves stabilize and protect the coast against erosion and shelter and support important commercial fisheries.

In spite of their value, mangrove forests have been and are being destroyed by filling and dredging for commercial development, marinas, and condominiums. They even are used as solid-waste dumps. In parts of the Pacific, especially the Philippines, mangals have been cleared for rice lands and for mariculture—the raising of fish, crabs, and prawns in brackish pools. The most massive destruction of mangals took place in Vietnam, where an estimated 100,000 ha were destroyed by herbicidal spraying during the Vietnam War. The mangals have never recovered.

Summary

In estuaries, where fresh water meets the sea, and in their associated salt marshes, the nature and distribution of life are determined by salinity. As salinity declines from the estuary up through the river, so do estuarine fauna, chiefly marine species. The estuary serves as a nursery for many marine organisms, particularly a number of commercially important finfish and shellfish, for here the young can develop protected from predators and competing species unable to tolerate lower salinity. The estuary is both a detrital-based and a plankton-based ecosystem, in which much of the cycling of nutrients is internal between organisms and bottom sediments. Because of their location, estuaries have become highly developed commercially. Pollution from this development and that deposited by inflowing rivers, together with dredging and filling, has seriously impaired the natural structure and functioning of estuarine ecosystems.

Associated salt marshes are dominated by salt-tolerant grasses and flooded by daily tides. The interaction of salinity and tidal height and flow produces a distinctive zonation of vegetation from the low marsh to the high marsh. Because tidal flushing resupplies nutrients and carries away wastes, salt marshes are highly productive. Most of this primary production goes unharvested by herbivores and is converted to microbial biomass. Some production goes into the sediments and another portion is exported to the estuary. Because of their accessible coastal location, salt marshes are disappearing rapidly under the pressure of commercial and residential development. Because of their extreme importance to estuarine fisheries and wildlife, the loss of coastal wetlands can have serious ecological and economic implications.

In tropical regions mangrove forests or mangals replace salt marshes and cover up to 70 percent of coastlines. Uniquely adapted to a tidal environment, many mangrove species have supporting prop roots, pneumatophores that carry oxygen to the roots, and seeds that grow into seedlings on the tree and drop into the water to take root in the mud. Mangroves are economically important as sources of wood, tannin, and food, and as protection against coastal erosion. They support many species of wildlife and marine organisms. Like salt marshes, mangrove forests are being destroyed by development and draining for agriculture.

Review and Study Questions

1. What is an estuary?
2. What is the relationship of inflowing rivers and tides to salinity in the estuary?
3. What is tidal overmixing, and what influences govern it?
4. How is the estuary a nutrient trap?

5. Why is the estuary so vulnerable to pollution?
6. What influences the major structural features of a salt marsh?
7. What accounts for the high productivity of the salt marsh?
8. What are the typical fauna of the salt marsh?
9. What are the unique characteristics and adaptations of mangroves?
10. Describe the zonation of mangals.
11. What is the ecological and economic importance of mangals?
12. What is the fate of many mangrove forests?

*13. Finally there is growing concern over the condition of the estuaries. What has prompted states to act on the problem? What are some of the solutions? Are these solutions universally accepted? Why?

*14. A battle is forming between those who would preserve coastal wetlands and those who would destroy them. Build up a set of arguments for both sides. What side in the long term has the strongest case?

*15. If you live or vacation along the coast, investigate the conditions of and attitudes toward salt marshes in that area. Of what importance are those salt marshes to wintering waterfowl? To the local fishing industry?

*16. Investigate the importance of mangrove forests to selected areas of the tropical world: West Africa, Australia, Indo-Malaysia, the Philippines.

Selected References

Bildstein, K. L., G. T. Bancroft, P. J. Dugan, et al. 1991. Approaches to the conservation of coastal wetlands in the Western Hemisphere. *Wilson Bull.* 103:218–254. Excellent overview of problems.

Chapman, V. J. 1976. *Mangrove vegetation*. Leutershausen, Germany: J. Cramer. An authoritative reference on mangroves and mangrove forests.

Chapman, V. J., ed. 1977. *Wet coastal ecosystems*. Amsterdam: Elsevier. Covers salt marshes and mangals of the world. A basic reference.

Clark, J. 1974. *Coastal ecosystems: Ecological considerations for the management of the coastal zone*. Washington, DC: Conservation Foundation.

Haines, B. L., and E. L. Dunn. 1985. Coastal marshes. In B. F. Chabot and H. A. Mooney, eds., *Physiological ecology of North American plant communities*. New York: Chapman and Hall. Pp. 323–347.

Hopkinson, C. S., and J. P. Schubauer. 1984. Static and dynamic aspects of nitrogen cycling in the salt marsh graminoid *Spartina alterniflora. Ecology* 65:961–969.

Howarth, R. W., and J. Teal. 1979. Sulfate reduction in a New England salt marsh. *Limno. Oceanogr*. 24:999–1013.

Jefferies, R. L., and A. J. Davy, eds. 1979. *Ecological processes in coastal environments*. Oxford: Blackwell.

Josselyn, M. 1983. The ecology of San Francisco tidal marshes: A community profile. U.S. Fish and Wildlife Service Office of Biological Services. FWS/OBS–82/83.

Ketchum, B. H., ed. 1983. *Estuaries and enclosed seas*. Ecosystems of the World 26. Amsterdam: Elsevier. Major reference on world estuarine structure and function.

Long, S. P., and C. F. Mason. 1983. *Saltmarsh ecology*. New York: Chapman and Hall. Excellent introduction.

Lugo, A. E., and S. C. Snedaker. 1974. The ecology of mangroves. *Ann. Rev. Ecol. Syst.* 5:39–64.

McLusky, D. S. 1971. *Ecology of estuaries*. London: Heinemann Educational.

McLusky, D. S. 1981. *The estuarine ecosystem.* New York: Halsted Press, Wiley. This book and the above are excellent, accessible introductions to the estuary.

Nixon, S. W., and C. A. Oviatt. 1973. Ecology of a New England salt marsh. *Ecol. Monogr.* 43:463–498.

Nybakken, J. W. 1988. *Marine biology: An ecological approach,* 2nd ed. New York: Harper & Row. Chapters on estuaries, salt marshes, and mangrove forests.

Odum, W. E., C. C. McIvor, and T. J. Smith III. 1982. The ecology of the mangroves of South Florida: A community profile. U.S. Fish and Wildlife Service Office of Biological Services. FWS/OBS 81/24.

Perkins, E. J. 1974. *The biology of estuaries and coastal waters.* New York: Academic Press.

Pomeroy, L. R., and R. G. Wiegert, eds. 1981. *The ecology of a salt marsh.* New York: Springer-Verlag. Synthesis of a 20-year study of all aspects of a southern United States coastal marsh.

Stout, J. P. 1984. The ecology of irregularly flooded salt marshes of northeastern Gulf of Mexico: A community profile. U.S. Fish and Wildlife Service Biol. Rep. 85(7.1).

Teal, J. 1962. Energy flow in a salt marsh ecosystem of Georgia. *Ecology* 43:614–624.

Teal, J., and M. Teal. 1969. *Life and death of the salt marsh.* Boston: Little, Brown. A classic.

Valiela, I., and J. M. Teal. 1979. The nitrogen budget of a salt marsh ecosystem. *Nature* 47:337–371.

Valiela, I., J. M. Teal, and W. G. Denser. 1978. The nature of the growth forms in salt marsh grass *Spartina alterniflora. Am. Nat.* 112:461–470.

Wiley, M. 1976. *Estuarine processes.* New York: Academic Press.

Zedler, J., T. Winfield, and D. Mauriello. 1982. The ecology of southern California coastal marshes: A community profile. U.S. Fish and Wildlife Service Office of Biological Services. FWS/OBS 81/54.

GLOSSARY

abiotic Nonliving component of the environment including soil, water, air, light, nutrients, and the like.

abyssal Relating to the bottom waters of oceans, usually below 1000 m.

acclimatization Changes or differences in a physiological state that appear after exposure to different natural environments.

acid deposition Wet and dry atmospheric fallout with an extremely low pH. It is brought about by a combination of water vapor in the atmosphere with hydrogen sulfide and nitrous oxide vapors released to the atmosphere from the burning of fossil fuels. The results are sulfuric and nitric acid in rain, fog, and snow, as well as gases and particulate matter.

active transport Movement of ions and molecules across a cell membrane against a concentration gradient involving an expenditure of energy. The movement of the ion or molecule is in a direction opposite to that it would take under simple diffusion.

adaptation Genetically determined characteristic (behavioral, morphological, physiological) that improves an organism's ability to survive and successfully reproduce under prevailing environmental conditions.

adaptive radiation Evolution from a common ancestor of divergent forms adapted to distinct ways of life.

adiabatic cooling A decrease in air temperature that results when a rising parcel of warm air cools by expansion (which uses energy) rather than losing heat to the outside surrounding air. The rate of cooling is approximately 1° C/100 m for dry air and 0.6° C/100 m for moist air.

adiabatic lapse rate Rate at which a parcel of air loses temperature with elevation if no heat is gained from or lost to an external source.

adiabatic process One in which heat is neither lost to nor gained from the outside.

aerenchyma Plant tissue with large air-filled intercellular spaces, usually found in roots and stems of aquatic and marsh plants.

aerobic Living or occurring only in the presence of free uncombined molecular oxygen either as a gas in the atmosphere or dissolved in water.

aestivation Dormancy in animals through a drought or dry season.

aggregative response One in which consumers spend most of the time in food patches with the greatest density of prey.

aggressive mimicry Resemblance of a predator or parasite to a harmless species in order to deceive potential prey.

A horizon Surface stratum of mineral soil characterized by maximum accumulation of organic matter, maximum biological activity, and loss of such materials as iron, aluminum oxides, and clays.

alfisol Soil characterized by an accumulation of iron and aluminum in lower or B horizon.

allele One of two or more alternative forms of a gene that occupies the same relative position or locus on homologous chromosomes.

allelopathy Effect of metabolic products of plants (excluding microorganisms) on the growth and development of other nearby plants.

allopatric Having different areas of geographical distribution; possessing nonoverlapping ranges.

alluvial soil Soil developing from recent alluvium (material deposited by running water); exhibits no horizon development; typical of floodplains.

alpha diversity The variety of organisms occupying a given place or habitat.

altricial Condition among birds and mammals of being hatched or born usually blind and too weak to support their own weight.

altruism A form of behavior in which an individual increases the welfare of another at the expense of its own welfare.

ambient Refers to surrounding, external, or unconfined conditions.

amensalism Relationship between two species in which one is inhibited or harmed by the presence of another.

ammonification Breakdown of proteins and amino acids, especially by fungi and bacteria, with ammonia as the excretory by-product.

anadromous Refers to fish that typically inhabit seas or lakes but ascend streams to spawn, for example, salmon.

anaerobic Adapted to environmental conditions devoid of oxygen.

antibiotic Substance produced by a living organism that is toxic to organisms of different species.

apparent plants Large, easy to locate plants possessing quantitative defenses not easily mobilized at the point of attack, for example, tannins.

aquifer A porous and permeable underground stratum of rock or sand capable of holding water.

aridosol Desert soils characterized by little organic matter and high base content.

asexual reproduction Any form of reproduction, such as budding, that does not involve the fusion of gametes.

assimilation Transformation or incorporation of a substance by organisms; absorption and conversions of energy and nutrient uptake into constituents of an organism.

association A natural unit of vegetation characterized by a relatively uniform species composition and often dominated by a particular species.

aufwuchs Community of plants and animals attached to or moving about on submerged surfaces; also called *periphyton,* but that term more specifically applies to organisms attached to submerged plant stems and leaves.

autotrophy Ability of an organism to produce organic material from inorganic chemicals and some source of energy.

basal metabolic rate The minimal amount of energy expenditure needed by an animal to maintain vital processes.

bathyal Pertaining to anything, but especially organisms, in the deep sea, below the photic or lighted zone, and above 4000 m.

benthos Animals and plants living on the bottom of a lake or sea from high water mark to the deepest depths.

beta diversity Variety of organisms occupying a number of different habitats over a region; regional diversity compared to very local or alpha diversity.

B horizon Soil stratum beneath the A horizon characterized by accumulation of silica, clay, iron and aluminum oxides, and possessing blocky or prismatic structure.

biennial Plant that requires two years to complete a life cycle, with vegetative growth the first year and reproductive growth (flowers and seeds) the second.

biochemical oxygen demand (BOD) A measure of the oxygen needed in a specified volume of water to decompose organic materials. The greater the amount of organic matter in water, the higher the BOD.

biogeochemical cycle Movement of elements or compounds through living organisms and nonliving environment.

biological clock The internal mechanism of an organism that controls circadian rhythms without external time cues.

biological magnification Process by which pesticides and other substances become more concentrated in each link of the food chain.

biological species A group of potentially interbreeding populations reproductively isolated from all other populations.

bioluminescence Production of light by living organisms.

biomass Weight of living material, usually expressed as dry weight per unit area.

biome Major regional ecological community of plants and animals; usually corresponds to plant ecologists' and European ecologists' classification of plant formations and of life zones.

biophage Organism that feeds on living material.

biosphere Thin layer about Earth in which all living organisms exist.

biotic community Any assemblage of populations living in a prescribed area or physical habitat.

blanket mire Large areas of upland dominated by sphagnum moss and dependent on precipitation for a water supply; a moor.

bog Wetland ecosystem characterized by an accumulation of peat, acid conditions, and dominance of sphagnum moss.

bottleneck An evolutionary term for any stressful situation that greatly reduces a population.

browse Part of current leaf and twig growth of shrubs, woody vines, and trees available for animal consumption.

bryophyte Member of the division in the plant kingdom of nonflowering plants comprising mosses (Musci), liverworts (Hepaticae), and hornworts (Anthocerotae).

buffer A chemical solution that resists or dampens change in pH on addition of acids or bases.

C_3 plant Any plant that produces as its first step in photosynthesis the three-carbon compound phosphoglyceric acid.

C_4 plant Any plant that produces as its first step in photosynthesis a four-carbon compound malic or aspartic acid.

calcification Process of soil formation characterized by accumulation of calcium in lower horizons.

caliche An alkaline, often rocklike salt deposit on the surface of soil in arid regions; it forms at the level where leached Ca salts from the upper soil horizons are precipitated.

calorie Amount of heat needed to raise 1 gram of water 1° C, usually from 15° C to 16° C.

cannibalism Killing and consumption of one's own kind; intraspecific predation.

capillary water That portion of water in the soil held by capillary forces between soil particles.

carnivore Organism that feeds on animal tissue; taxonomically, a member of the order Carnivora (Mammalia).

carrying capacity (K) Number of individual organisms the resources of a given area can support, usually through the most unfavorable period of the year. The term has acquired so many meanings it is almost useless.

catadromous fish Fish that feed and grow in fresh water, but return to the sea to spawn.

catastrophic extinction A major episode of extinction involving many taxa occurring fairly suddenly in the fossil record.

catena A group of related soils.

cation Part of a dissociated molecule carrying a + electrical charge.

cation exchange capacity Ability of a soil

particle to absorb + charged ions.

chamaephyte Perennial shoots or buds on the surface of the ground to about 25 cm above the surface.

chilling tolerance Ability of a plant to carry on photosynthesis within a range of +5° to +10° C.

C horizon Soil stratum beneath the solum (A and B horizons) relatively little affected by biological activity and soil-forming process.

circadian rhythm Endogenous rhythm of physiological or behavioral activity of approximately 24 hours duration.

climax Stable end community of succession that is capable of self-perpetuation under prevailing environmental conditions.

climograph A diagram describing a locality based on the annual cycle of temperature and precipitation.

cline Gradual change in population characteristics over a geographical area, usually associated with changes in environmental conditions.

clone A population of genetically identical individuals resulting from asexual reproduction.

coevolution Joint evolution of two or more noninterbreeding species that have a close ecological relationship; through reciprocal selective pressures, the evolution of one species in the relationship is partially dependent on the evolution of the other.

coexistence Two or more species living together in the same habitat, usually with a form of competitive interaction.

cohort A group of individuals of the same age.

cold resistance Ability of a plant to resist low-temperature stress without injury.

colluvium Mixed deposits of soil material and rock fragments accumulated near the base of steep slopes through soil creep, landslides, and local surface runoff.

commensalism Relationship between species that is beneficial to one, but neutral or of no benefit to the other.

community A group of interacting plants and animals inhabiting a given area.

compensation intensity Light intensity at which photosynthesis and respiration balance each other so that net production is 0; in aquatic systems, usually the depth of light penetration at which oxygen utilized in respiration equals oxygen produced by photosynthesis.

competition Any interaction that is mutually detrimental to both participants; occurs between species that share limited resources.

competitive exclusion Hypothesis which states that when two or more species coexist using the same resource, one must displace or exclude the other.

conduction Direct transfer of heat from one substance to another.

consumer Any organism that lives on other organisms dead or alive.

contest competition Competition in which a limited resource is shared only by dominant individuals; this type of competition results in a relatively constant number of survivors, regardless of initial density.

continuum A gradient of environmental characteristics or changes in community composition.

convection Transfer of heat by the circulation of fluids, liquid, or gas.

convergent evolution Development of similar characteristics in different species living in different areas but under similar environmental conditions.

coprophagy Feeding on feces.

Coriolis effect Physical consequences of the law of conservation of angular momentum. As a result of Earth's rotation, a moving object veers to the right in the Northern Hemisphere and to the left in the Southern Hemisphere.

countercurrent circulation An anatomical and physiological arrangement by which heat exchange takes place between outgoing warm arterial blood and cool venous blood returning to the body core. It is important in maintaining temperature homeostasis in many vertebrates.

critical daylength The period of daylight, specific for any given species, that triggers a long-day or a short-day response in organisms.

cryptic coloration Coloration of organisms that makes them resemble or blend into their habitat or background.

cryptophyte Buds buried in the ground on a bulb or rhizome.

cyclic replacement Type of succession in which the sequence of seral stages is repeated by imposition of some disturbance so that the sere never arrives at a climax or stable sere.

day-neutral plant A plant that does not require any particular photoperiod to flower.

death rate Number of individuals in a population dying in a given time interval by the number alive at the midpoint of the time interval.

deciduous (of leaves) Shed during a certain season (winter in temperate regions, dry seasons in the tropics); (of trees) having deciduous parts.

decomposer Organism that obtains energy from the breakdown of dead organic matter to more simple substances; most precisely refers to bacteria and fungi.

decomposition Breakdown of complex organic substances into simpler ones.

definitive host Host in which a parasite reaches maturity and lives as an adult.

deme Local population or interbreeding group within a larger population.

denitrification Reduction of nitrates and nitrites to nitrogen by microorganisms.

density dependent Varying in relation to population density.

density independent Unaffected by population density.

detritivore Organism that feeds on dead

organic matter; usually applies to detritus-feeding organisms other than bacteria and fungi.

detritus Fresh to partly decomposed plant and animal matter.

dewpoint Temperature at which condensation of water in the atmosphere begins.

diapause A period of dormancy—usually seasonal—in the life cycle of an insect in which growth and development cease and metabolism is greatly decreased.

diffuse coevolution Coevolution involving the interactions of many organisms in contrast to pairwise interactions.

diffuse competition Type of competition in which a species experiences interference from numerous other species that deplete the same resources.

dimorphism Existing in two structural forms, two color forms, two sexes, and the like.

dioecious Plants in which male and female reproductive organs are borne on separate individuals.

diploid Having chromosomes in homologous pairs or twice the haploid numbers of chromosomes.

directional selection Selection favoring individuals at one extreme of the phenotype in the population.

disease Any deviation from normal state of health.

dispersal Leaving an area of birth or activity for another area.

dispersion Distribution of organisms within a population over an area.

disruptive selection Selection in which two extreme phenotypes in the population leave more offspring than the intermediate phenotype that has lower fitness.

diversity Abundance in number of species in a given location.

dominance (ecological) Control within a community over environmental conditions influencing associated species by one or several species, plant or animal, enforced by number, density, or growth form; (social) behavioral, hierarchial order in a population that gives high-ranking individuals priority of access to essential requirements; (genetic) ability of an allele to mask the expression of an alternative form of the same gene in a heterozygous condition.

dominant Population possessing ecological dominance in a given community and thereby governing type and abundance of other species in the community.

dormant State of cessation of growth and suspended biological activity during which life is maintained.

drought avoidance Ability of a plant to escape dry periods by becoming dormant or surviving the period as a seed.

drought resistance Sum of drought tolerance and drought avoidance.

drought tolerance Ability of plants to maintain physiological activity in spite of the lack of water or to survive the drying of tissues.

dystrophic Term applied to a body of water with a high content of humic organic matter, often with high littoral productivity and low plankton productivity.

ecocline A geographical gradient of communities or ecosystems produced by responses of vegetation to environmental gradients of rainfall, temperature, nutrient concentrations, and other factors.

ecological efficiency Percentage of biomass produced by one trophic level that is incorporated into biomass of the next highest trophic level.

ecological release Expansion of habitat or increase in food availability resulting from release of a species from interspecific competition.

ecosystem The biotic community and its abiotic environment functioning as a system.

ecotone Transition zone between two structurally different communities; see also *edge.*

ecotype Subspecies or race adapted to a particular set of environmental conditions.

ectothermy Determination of body temperature primarily by external thermal conditions.

edaphic Relating to soil.

edge Place where two or more vegetation types meet.

edge effect Response of organisms, animals in particular, to environmental conditions created by the edge.

effective population size The size of an ideal population that would undergo the same amount of random genetic drift as the actual population; sometimes used to measure the amount of inbreeding in a finite, randomly mating population.

egestion Elimination of undigested food material.

elaiosome Shiny, oil-containing ant-attracting tissue on the seed coat of many plants.

emigration Movement of part of a population permanently out of an area.

endemic Restricted to a given region.

endogenous Any process that arises within an organism.

endothermy Regulation of body temperature by internal heat production; allows maintenance of appreciable difference between body temperature and external temperature.

entrainment Synchronization of an organism's activity cycle with environmental cycles.

entropy Transformation of matter and energy to a more random, more disorganized state.

environment Total surroundings of an organism including other plants and animals embracing those of its own kind.

epidemic Rapid spread of a bacterial or viral disease in a human population. Compare with *epizootic.*

epifauna Benthic organisms that live on or move across the surface of a substrate.

epilimnion Warm, oxygen-rich, upper layer of water in a lake or other body of water, usually seasonal.

epiphyte Organism that lives wholly on the surface of plants, deriving support but not nutrients from the plants.

epizootic Rapid spread of a bacterial or viral disease in a dense population of animals.

equilibrium turnover rate Change in species composition per unit time when immigration equals extinction.

estivation Dormancy in animals during a period of drought or a dry season.

estuary A partially enclosed embayment where fresh water and sea water meet and mix.

euphotic zone Surface layer of water to the depth of light penetration where photosynthetic production equals respiration.

eutrophic Term applied to a body of water with high nutrient content and high productivity.

eutrophication Nutrient enrichment of a body of water; called *cultural eutrophication* when accelerated by introduction of massive amounts of nutrients by human activity.

evapotranspiration Sum of the loss of moisture by evaporation from land and water surfaces and by transpiration from plants.

evolution Change in gene frequency through time resulting from natural selection and producing cumulative changes in characteristics of a population.

exothermic A chemical reaction that releases heat to the environment.

exploitative competition Competition that results in a reduction of a resource level brought about by a group or groups of organisms adversely affecting other organisms.

exponential growth Instantaneous rate of population growth expressed as proportional increase per unit of time.

F_1 generation The first generation of offspring from a cross between individuals homozygous for contrasting alleles. The F_1 is necessarily heterozygous.

F_2 generation Offspring produced by selfing or by allowing the F_1 generation to breed among themselves.

facilitation model A model of succession in which a previous community prepares or "facilitates" the way for a succeeding community.

fecundity Potential ability of an organism to produce eggs or young; rate of production of young by a female.

fen Wetlands dominated by sedges in which peat accumulates.

fermentation Breakdown of carbohydrates and other organic matter under anaerobic conditions.

field capacity Amount of water held by soil against the force of gravity.

fitness Genetic contribution by an individual's descendants to future generations.

fixation Process in soil by which certain chemical elements essential for plant growth are converted from a soluble or exchangeable form to a less-soluble or nonexchangeable form.

floating reserve Individuals in a population of a territorial species that do not hold territories and remain unmated, but are available to refill territories vacated by death of an owner.

flux Flow of energy from a source to a sink or receiver.

food chain Movement of energy and nutrients from one feeding group of organisms to another in a series that begins with plants and ends with carnivores, detrital feeders, and decomposers.

food web Interlocking pattern formed by a series of interconnecting food chains.

foraging strategy Manner in which animals seek food and allocate their time and effort to obtaining it.

formation Classification of vegetation based on dominant life forms.

founder effect Population started by a small number of colonists, which contain only a small and often biased sample of genetic variation of the parent population. It may result in a markedly different new population.

fragmentation Reduction of a large habitat area into small, scattered remnants; reduction of leaves and other organic matter into smaller particles.

free-running cycle Length of a circadian rhythm in the absence of an external time cue.

frost pocket Depression in the landscape into which cold air drains, lowering the temperature relative to the surrounding area. Such pockets often support their own characteristic group of cold-tolerant plants.

frugivore Organism that feeds on fruit.

functional response Change in rate of exploitation of a prey species by a predator in relation to changing prey density.

fundamental niche Total range of environmental conditions under which a species can survive.

gamma diversity Diversity differences among similar habitats in widely separated geographic regions.

gap Opening made in a forest canopy by some small disturbance, such as windfall, death of an individual, or group of trees, that influences the development of vegetation beneath.

gap phase replacement Successional development in small disturbed areas within a stable plant community; filling in of a space left by a disturbance, but not necessarily by the species eliminated by the disturbance.

gene Unit material of inheritance; more specifically, a small unit of DNA molecule coded for a specific protein to produce one of the many attributes of a species.

gene flow Exchange of genetic material between populations.

gene frequency Actually, allele frequency; relative abundance of different alleles carried by an individual or a population.

gene pool The sum of all the genes of all individuals in a population.

genet A genetic individual that arises from a single fertilized egg.

genetic drift Random fluctuation in allele frequency over time, due to chance occurrence alone without any influence by natural selection. Important in small populations.

genetic feedback Evolutionary response of a population to adaptations of predators, parasites, or competitors.

genotype Genetic constitution of an organism.

geometric rate of increase Factor by which size of a population increases over a period of time.

gley soil Soil developed under conditions of poor drainage, resulting in reduction of iron and other elements and in gray colors and mottles.

global stability Ability of a community to withstand large disturbances and return to its original state.

gouger General group of stream invertebrates that lives and feeds on woody debris.

gradualism Hypothesis that evolution is a slow process of continuous change.

granivore Organism that feeds on seeds.

greenhouse effect Selective energy absorption by carbon dioxide in the atmosphere that allows short wavelength energy to pass through but absorbs longer wavelengths and reflects heat back to Earth.

gross production Energy fixed per unit area by photosynthetic activity of plants before respiration. Total energy flow at the secondary level is not gross production, but rather assimilation, because consumers use material already produced with respiratory losses.

group selection Elimination of one group of individuals by another group of individuals possessing superior genetic traits; not a widely accepted hypothesis.

growth form Morphological category of plants, such as tree, shrub, and vine.

guild A group of populations that utilizes a gradient of resources in a similar way.

gyre Circular motion of water in major ocean basins.

habitat Place where a plant or animal lives.

haploid Having a single set of unpaired chromosomes in each cell nucleus.

Hardy-Weinberg law The proposition that genotypic ratios resulting from random mating remained unchanged from one generation to another, provided natural selection, genetic drift, and mutation are absent.

hemicryptophyte Perennial shoots or buds close to the surface of the ground; often covered with litter.

herbivore Organism that feeds on plant tissue.

hermaphrodite Organism possessing the reproductive organs of both sexes.

heterogeneity State of being mixed in composition; can refer to genetic or environmental conditions.

heterotherm Organism that during part of its life history becomes either endothermic or ectothermic. Hibernating endotherms become ectothermic, and foraging insects such as bees become endothermic during periods of activity; they are characterized by rapid, drastic, repeated changes in body temperature.

heterotrophic Requiring a supply of organic matter or food from the environment.

heterozygous Containing two different alleles of a gene, one from each parent, at the corresponding loci of a pair of chromosomes.

hibernation Winter dormancy in animals, characterized by a great decrease in metabolism.

hierarchy A sequence of sets made up of smaller subsets.

histosol Soil characterized by high organic matter content.

homeostasis Maintenance of nearly constant conditions in functions of an organism or in interactions among individuals in a population.

homeotherm Animal with a fairly constant body temperature; also spelled *homoiotherm* and *homotherm.*

homeothermy Regulation of body temperature by physiological means.

home range Area over which an animal ranges throughout the year.

homologous chromosomes Corresponding chromosomes from male and female parents that pair during meiosis.

homozygous Containing two identical alleles of a gene at the corresponding loci of a pair of chromosomes.

horizon Major zones or layers of soil, each with its own particular structure and characteristics.

host Organism that provides food or other benefit to another organism of a different species; usually refers to an organism exploited by a parasite.

humus Organic material derived from partial decay of plant and animal matter.

hybrid Plant or animal resulting from a cross between genetically different parents.

hydrosphere Body of water on or near Earth's surface.

hypolimnion Cold, oxygen-poor zone of a lake that lies below the thermocline.

hypha Filament of a fungus thalli or vegetative body.

immigration Arrival of new individuals into a habitat or population.

immobilization Conversion of an element from inorganic to organic form in microbial or plant tissue, rendering the nutrient relatively unavailable to other organisms.

inbreeding Mating among close relatives.

inbreeding depression Detrimental effects of inbreeding.

inclusive fitness Sum of the total fitness of an individual and the fitness of its relatives, weighted according to the degree of relationship.

infaunal Organisms living within a substrate.

infralittoral Region below the littoral region of the sea.

inhibition model Model of succession proposing that the dominant vegetation occupying a site prevents colonization of that site by other plants of the next successional community.

instar Form of insect or other arthropod between successive molts.

interdemic selection Group selection of populations within a species.

interference competition Competition in which access to a resource is limited by the presence of a competitor.

intermediate host Hosts that harbor developmental phases of parasites; the infective stage or stages can develop only when the parasite is independent of its definitive host; see also *definitive host.*

intrinsic rate of increase (r) Intrinsic growth rate of a population under ideal conditions without inhibition from competition.

introgression Incorporation of genes of one species into the gene pool of another.

inversion (genetic) Reversal of part of a chromosome so that genes within that part lie in reverse order; (meteorological) increase rather than decrease in air temperature with height caused by radiational cooling of Earth (radiational inversion) or by compression and consequent heating of subsiding air masses from high pressure areas (subsidence inversion).

island biogeography Study of distribution of organisms and community structure on islands.

isolating mechanism Any structural, behavioral, or physiological mechanism that blocks or inhibits gene exchange between two populations.

isotherm Lines drawn on a map connecting points with the same temperature at a certain period of time.

iteroparous Multiple-brooded over a lifetime.

keystone species A species whose activities have a significant role in determining community structure.

kin selection Differential reproduction among groups of closely related individuals.

krummholz Stunted form of trees characteristic of transition zone between alpine tundra and subalpine coniferous forest.

***K*-selection** Selection under carrying capacity conditions and high level of competition.

landscape ecology Study of the structure, function, and change in a heterogeneous landscape composed of interacting ecosystems.

latent heat of fusion Amount of heat given up when a unit mass of a substance converts from a liquid to a solid state, or the amount of heat absorbed when a substance converts from the solid to the liquid state.

laterization Soil-forming process in hot, humid climates, characterized by intense oxidation resulting in loss of bases and in a deeply weathered soil composed of silica, sesquioxides of iron and aluminum, clays, and residual quartz.

leach Dissolving and removal of nutrients by water out of the soil, litter, and organic matter.

leaf area index Ratio of area of canopy foliage to ground area.

lentic Pertaining to standing water, as lakes and ponds.

life table Tabulation of mortality and survivorship schedule of a population.

life zone Major area of plant and animal life equivalent to a biome; transcontinental region or belt characterized by particular plants and animals and distinguished by temperature differences; applies best to mountainous regions where temperature differences accompany changes in altitude.

lignotuber Specialized structure on the roots of certain fire-adapted trees, particularly *Eucalyptus,* from which new growth sprouts following a fire.

limnetic Pertaining to or living in the open water of a pond or lake.

limnetic zone Shallow-water zone of lake or sea in which light penetrates to the bottom.

lithosol Soil showing little or no evidence of soil development and consisting mainly of partly weathered rock fragments or nearly barren rock.

lithosphere Rocky material of Earth's outer crust.

littoral Shallow water of lake in which light penetrates to the bottom, permitting submerged, floating, and emergent vegetative growth; also shore zone of tidal water between high-water and low-water marks.

local stability Ability of a system to return to its initial conditions following a small disturbance.

locus Site on a chromosome occupied by a specific gene.

loess Soil developed from wind-deposited material.

logistic curve S-shaped curve of population growth that slows at first, steepens, and then flattens out at asymptote, determined by carrying capacity.

logistic equation Mathematical expression for the population growth curve in which rate of increase decreases linearly as population size increases.

long-day organism Plant or animal that requires long days—days with more than a certain minimum of daylight—to flower or come into reproductive condition.

lotic Pertaining to flowing water.

macronutrients Essential nutrients needed in relatively large amounts by plants and animals.

macroparasite Parasitic worms, lice, fungi, and the like; have comparatively long generation time; spread by direct or indirect transmission; and may involve intermediate hosts or vectors.

marsh Wetland dominated by grassy vegetation such as cattails and sedges.

mediterranean-type climate Semiarid climate characterized by a hot, dry summer and a wet, mild winter.

meiofauna Benthic organisms within the size range of 1 to 0.1 mm; interstitial fauna.

meiosis Two successive divisions by gametic cells, with only one duplication of chromosomes so that the number of chromosomes in daughter cells is one-half the diploid number.

mesic Moderately moist habitat.

metalimnion Transition zone in lake between hypolimnion and epilimnion; region of rapid temperature decline.

micella Soil particle of clay and humus carrying + electrical charges at the surface.

microbivore Organism that feeds on microbes, especially in the soil and litter.

microclimate Climate on a very local scale that differs from the general climate of the area; influences the presence and distribution of organisms.

microflora Bacteria and certain fungi inhabiting the soil.

microhabitat That part of the general habitat utilized by an organism.

micronutrients Essential nutrients needed in very small quantities by plants and animals.

microparasite Viruses, bacteria, and protozoans characterized by small size, short generation time, and rapid multiplication.

migration Intentional, directional, usually seasonal movement of animals between two regions or habitats; involves departure and return of the same individual; a round-trip movement.

mimicry Resemblance of one organism to another or to an object in the environment evolved to deceive predators.

mineralization Microbial breakdown of humus and other organic matter in soil to inorganic substances.

minimum viable population Size of a population that with a given probability will ensure the existence of the population for a stated period of time.

mire Wetland characterized by an accumulation of peat.

mitosis Cell division involving chromosome duplication resulting in two daughter cells with a full complement of chromosomes, genetically the same as parent cells.

model In theoretical and systems ecology, an abstraction or simplification of a natural phenomenon developed to predict a new phenomenon or to provide insights into existing ones; in mimetic association, the organism mimicked by a different organism.

moder Type of forest humus layer in which plant fragments and mineral particles form a loose netlike structure held together by a chain of small arthropod droppings.

mollisol Soil formed by calcification, characterized by accumulation of calcium carbonate in lower horizons and high organic content in upper horizons.

monoecious In plants, occurrence of reproductive organs of both sexes on the same individual, either as different flowers (hermaphroditic) or in the same flower (diocious).

monogamy Mating of an animal and maintenance of a pair bond with only one member of the opposite sex at a time.

moor A blanket bog or peatland.

mor Type of forest humus layer of unincorporated organic matter, usually matted or compacted or both, and distinct from mineral soil; low in bases and acid in reaction.

morphology Study of the form of organisms.

mull Humus that contains appreciable amounts of mineral bases and forms a humus-rich layer of forested soil consisting of mixed organic and mineral matter; blends into the upper mineral layer without abrupt changes in soil characteristics.

mutation Transmissible changes in structure of gene or chromosome.

mutualism Relationship between two species in which both benefit.

mycelium Mass of hyphae that makes up the vegetative portion of a fungus.

mycorrhizae Association of fungus with roots of higher plants that improves the plants' uptake of nutrients from the soil.

natural selection Differential reproduction and survival of individuals that result in elimination of maladaptive traits from a population.

negative feedback Homeostatic control in which an increase in some substance or activity ultimately inhibits or reverses the direction of the processes leading to the increase.

nekton Aquatic animals that are able to move at will through the water.

neritic Marine environment embracing the regions where land masses extend outward as a continental shelf.

net production Accumulation of total biomass over a given period of time after respiration is deducted from gross production in plants and from assimilated energy in consumer organisms.

net reproductive rate Number of females produced per female per generation.

niche Functional role of a species in the community, including activities and relationships.

niche breadth Range of a single niche dimension occupied by a population.

nitrification Breakdown of nitrogen-containing organic compounds into nitrates and nitrites.

nitrogen fixation Conversion of atmospheric nitrogen to forms usable by organisms.

null hypothesis A statement of no difference between sets of values formulated for statistical testing.

numerical response Change in size of a population of predators in response to change in density of its prey.

nutrient Substance required by organisms for normal growth and activity.

nutrient cycle Pathway of an element or nutrient through the ecosystems from assimilation by organisms to release by decomposition.

old-growth forest Forest that has not been cut for decades or disturbed by humans for hundreds of years.

oligotrophic Term applied to a body of water low in nutrients and in productivity.

omnivore An animal that feeds on both plant and animal matter.

opportunistic species Organisms able to exploit temporary habitats or conditions.

optimal foraging Tendency of animals to harvest food efficiently—to select food sizes or food patches that will result in maximum food intake for energy expended.

optimum yield Amount of material that can be removed from a population and that will result in production of maximum amount of biomass on a sustained yield basis.

oscillation Regular fluctuation in a fixed cycle above or below some set point.

osmosis Movement of water molecules across a differentially permeable membrane in response to a concentration or pressure gradient.

outbreeding Production of offspring by the fusion of distantly related gametes.

overturn Vertical mixing of layers in a body of water brought about by seasonal changes in temperature.

paleoecology Study of ecology of past communities by means of fossil record.

parapatric Having ranges coming into contact but not overlapping by much more than the dispersal range of an individual in its lifetime.

parasitism Relationship between two species in which one benefits while the other is harmed (although not usually killed directly).

parasitoid Insect larva that kills its host by consuming completely the host's soft tissues before pupation or metamorphosis into an adult.

peat Unconsolidated material consisting of undecomposed and only slightly decomposed organic matter under conditions of excessive moisture.

pelagic Referring to the open sea.

periphyton In freshwater ecosystems, organisms that are attached to submerged plant stems and leaves; see also *aufwuchs.*

permafrost Permanently frozen soil.

permanent wilting point Point at which water potential in the soil and conductivity assume such low values that the plant is unable to extract sufficient water to survive and wilts permanently.

phagioclimax Vegetation type maintained as climax over long period of time by continued human activity.

phanerophyte Perennial buds carried high up in the air; trees, shrubs, and vines.

phenology Study of the seasonal changes in plant and animal life and the relationship of these changes to weather and climate.

phenotype Physical expression of a characteristic of an organism as determined by genetic constitution and environment.

phenotypic plasticity Ability to change form under different environmental conditions.

pheromone Chemical substance released by an animal that influences behavior of others of the same species.

photic zone Lighted water column of a lake or ocean inhabited by plankton.

photoperiodism Response of plants and animals to changes in relative duration of light and dark.

photorespiration Respiration that occurs in light in C_3 plants and is not coupled to oxidative phosphorylation and does not generate ATP; a wasteful process decreasing photosynthetic efficiency.

phreatophyte Type of plant that habitually obtains its water supply from zone of groundwater.

physiognomy Outward appearance of the landscape.

physiological longevity Maximum life span of an individual in a population under given environmental conditions.

phytoplankton Small, floating plant life in aquatic ecosystems; planktonic plants.

pioneer species Plants that are initial invaders of disturbed sites or early seral stages of succession.

plankton Small, floating or weakly swimming plants and animals in freshwater and marine ecosystems.

pneumatophore An erect respiratory root that protrudes above waterlogged soils; typical of bald cypress and mangroves.

podzolization Soil-forming process resulting from acid leaching of the A horizon and accumulation of iron, aluminum, silica, and clays in lower horizon.

poikilothermy Variation of body temperature with external conditions.

polyandry Mating of one female with several males.

polygyny Mating of one male with several females.

polymorphism Occurrence of more than one distinct form of individuals in a population.

polyploid Having three or more times the haploid number of chromosomes.

positive feedback Control in a system that reinforces the process in the same direction.

potential evapotranspiration Amount of water that would be transpired under constantly optimal conditions of soil moisture and plant cover.

precocial Young birds hatched with down, eyes open, and able to move about; also young mammals born with eyes open and able to follow their mothers after birth (for example, fawn deer, calves).

predation One living organism serves as a food source for another.

preferred temperature Range of temperatures within which poikilotherms function most efficiently.

primary production Production by green plants.

primary succession Vegetational development starting from a new site never before colonized by life.

producer Green plants and certain chemosynthetic bacteria that convert light or chemical energy into organismal tissue.

production Amount of energy formed by an individual, population, or community per unit time.

productivity Rate of energy fixation or storage per unit time; not to be confused with production.

profundal Deep zone in aquatic ecosystems below the limnetic zone.

punctualism Hypothesis that evolution occurs in a series of sudden bursts with long periods of stability.

rain shadow Dry area on lee side of mountains.

raised bog A bog in which the accumulation of peat has raised its surface above both the surrounding landscape and the water table; it develops its own perched water table.

ramet Any individual belonging to a clone.

realized niche Portion of fundamental niche space occupied by a population in face of competition from populations of other species; environmental conditions under which a population survives and reproduces in nature.

recombination Exchange of genetic material resulting from independent assortment of chromosomes and their genes during gamete production, followed by a random mix of different sets of genes at fertilization.

recruitment Addition of reproduction of new individuals to a population.

regolith Mantle of unconsolidated material below the soil from which soil develops.

reproductive isolation Separation of one population from another by the inability to produce viable offspring when the two populations are mated.

reproductive value Potential reproductive output of an individual at a particular age (x) relative to that of a newborn individual at the same time.

resilience Ability of a system to absorb changes and return to its original condition.

resistance Ability of a system to resist changes from a disturbance.

resource Environmental component utilized by a living organism.

resource allocation Action of apportioning the supply of a resource to a specific use.

respiration Metabolic assimilation of oxygen accompanied by production of carbon dioxide and water, release of energy, and breaking down of organic compounds.

restoration ecology Study of the application of ecological theory to the ecological restoration of highly disturbed sites.

rete A large network or discrete vascular bundle of intermingling small blood vessels carrying arterial and venous blood that acts as a heat exchanger in mammals and certain fish and sharks.

rhizome A horizontally growing underground stem that through branching gives rise to vegetative structures.

rhizosphere Soil region immediately surrounding roots.

richness A component of species diversity; the number of species present in an area.

riparian Along banks of rivers and streams; riverbank forests are often called gallery forests.

***r*-selection** Selection under low population densities; favors high reproductive rates under conditions of low competition.

ruminant Ungulate with a three- or four-chamber stomach. The large first chamber is known as the rumen in which bacterial fermentation of plant matter consumed occurs.

saprophage Organism that feeds on dead plant and animal matter.

savanna Tropical grassland, usually with scattered trees or shrubs.

sclerophyll Woody plant with hard, leathery, evergreen leaves that prevent moisture loss.

scramble competition Intraspecific competition in which limited resources are shared to the point that no individual survives.

scraper Aquatic insects that feed by scraping algae from a substrate.

search image Mental image formed in predators enabling them to find more quickly and to concentrate on a common type of prey.

secondary production Production by consumer organisms.

secondary substances Organic compounds produced by plants that are utilized in chemical defense.

secondary succession Plant succession taking place on sites that have already supported life.

seiche Oscillation of a structure of water about a point or node.

semelparity Having only a single reproductive effort in a lifetime over one relatively short period of time.

semiarid Region of fairly dry climate with precipitation between 25 and 60 cm a year and with an evapotranspiration rate high enough so that potential loss of water to the environment exceeds inputs.

senescence Process of aging.

serotinony Retention of seeds by cones until heated, usually by fire.

serpentine soil Soils derived from ultrabasic rocks that are high in iron, magnesium, nickel, chromium, and cobalt and low in calcium, potassium, sodium, and aluminum; support distinctive communities.

sessile Not free to move about; permanently attached to a substrate.

shade tolerant Plants that are able to grow and reproduce under low light conditions.

short-day organisms Plants and animals that come into reproductive condition under conditions of short days—days with less than a certain maximum length.

sibling species Species with similar appearance but unable to interbreed.

sigmoid curve S-shaped curve of logistic growth.

site Combination of biotic, climatic, and soil conditions that determine an area's capacity to produce vegetation.

snag Dead or partially dead tree at least 10.2 cm dbh and 1.8 m tall; important for cavity-nesting birds and mammals.

social parasite Animal that uses other individuals or species to rear its young, for example, the cowbirds.

soil association A group of defined and named soil taxonomic units occurring together in an individual and characteristic pattern over a geographic region.

soil horizon Developmental layer in the soil with its own characteristics of thickness, color, texture, structure, acidity, nutrient concentration, and the like.

soil profile Distinctive layering of horizons in the soil.

soil series Basic unit of soil classification, consisting of soils that are essentially alike in all major profile characteristics except texture of the A horizon. Soil series are usually named for the locality where the typical soil was first recorded.

soil structure Arrangement of soil particles and aggregates.

soil texture Relative proportions of the three particle sizes—sand, silt, and clay—in the soil.

soil type Lowest unit in the natural system of soil classification, consisting of soils that are alike in all characteristics, including texture of the A horizon.

speciation Separation of a population into two or more reproductively isolated populations.

species diversity Measurement that relates density of organisms of each type present in a habitat to the number of species in a habitat.

specific heat Amount of energy that must be added or removed to raise or lower temperature of a substance by a specific amount.

spiraling Mechanism of retention of nutrients in flowing-water ecosystems involving the interdependent processes of nutrient recycling and downstream transport.

stability Ability of a system to resist change or to recover rapidly after a disturbance; absence of fluctuations in a population.

stabilizing selection Selection favoring the middle in the distribution of phenotypes.

stable age distribution Constant proportion of individuals of various age classes in a population through population changes.

stand Unit of vegetation that is essentially homogenous in all layers and differs from adjacent types qualitatively and quantitatively.

standing crop Amount of biomass per unit area at a given time.

stationary age distribution Special form of stable age distribution in which the population has reached a constant size in which birthrate equals death rate and age distribution remains fixed.

stratification Division of an aquatic or terrestrial community into distinguishable layers on the basis of temperature, moisture, light, vegetative structure, and other such factors creating zones for different plant and animal types.

sublittoral Lower division of sea from about 40 m to 60 m to below 200 m.

subsidence inversion Atmospheric inversion produced by sinking air movement from aloft.

subspecies Geographical unit of a species population distinguishable by certain morphological, behavioral, or physiological characteristics.

succession Replacement of one community by another; often progresses to a stable terminal community called the climax.

sun plant Plant able to grow and reproduce only under high light conditions.

supercooling In ectotherms, lowering of body temperature below freezing without freezing body tissue; involves the presence of certain solutes, particularly glycerol.

sustained yield Yield per unit time from an exploited population equal to production per unit time.

swamp Wooded wetland in which water is near or above ground level.

switching A predator changing its diet from a less abundant to a more abundant prey species.

symbiosis Living together of two or more species.

sympatric Living in the same area; usually refers to overlapping populations.

system Set or collection of interdependent parts or subsystems enclosed within a defined boundary; the outside environment provides inputs and receives outputs transmitted to it by the system.

taiga The northern circumpolar boreal forest.

territory Area defended by an animal; varies among animals according to social behavior, social organization, and resource requirements of different species.

therophyte Life form of plants that survives unfavorable conditions in the form of a seed; annual and ephemeral species.

thermal conductance Rate at which heat flows through a substance.

thermal tolerance Range of temperatures in which an aquatic poikilotherm is most at home.

thermocline Layer in a thermally stratified body of water in which temperature changes rapidly relative to the remainder of the body.

thermogenesis Increase in production of metabolic heat to counteract the loss of heat to a colder environment.

threshold of security Point in local population density at which the predator turns its attention to other prey (see also *switching*) because of harvesting efficiency; the segment of prey population below the threshold is relatively secure from predation.

time lag Delay in a response to change.

toposequence A pattern of local soils whose development was controlled by topography of the landscape.

torpidity Temporary condition of an animal involving a great reduction in respiration; results in loss of power of motion and feeling; usually occurs in response to some unfavorable environmental condition, such as heat or cold, to reduce energy expenditure.

trace element Element occurring and needed in small quantities; see also *micronutrient.*

translocation Transport of material within a plant; absorption of minerals from soil into roots and their movement throughout the plant.

transpiration Loss of water vapor by land plants.

trophic Related to feeding.

trophic level Functional classification of organisms in an ecosystem according to feeding relationships, from first-level autotrophs through succeeding levels of herbivores and carnivores.

trophic structure Organization of a community based on the number of feeding or energy-transfer levels.

tundra Areas in arctic and alpine (high mountains) regions characterized by bare ground, absence of trees, and growth of mosses, lichens, sedges, forbs, and low shrubs.

turnover rate Rate of replacement of a substance or a species when losses to a system are replaced by additions.

upwelling Areas in oceans where currents force water from deep within the ocean into the euphotic zone.

vacuole Fluid-filled cavity within the cytoplasm.

vapor pressure The amount of pressure water vapor exerts independent of dry air.

vector Organism that transmits a pathogen from one organism to another.

vegetative reproduction Asexual reproduction in which plants propagate themselves by means of specialized multicellular organs such as bulbs, corms, rhizomes, stems, and the like.

viscosity Property of a fluid that resists the force within the fluid that causes it to flow.

Wallace's line Biogeographic line between the islands of Borneo and the Celebes that marks the eastward boundary of many landlocked Eurasian organisms and the boundary of the Asian region.

water potential Measure of energy in an aqueous solution needed to move water molecules across a semipermeable membrane; water tends to move from areas of high or less negative to areas of low or more negative potential.

watershed Entire region drained by a waterway that drains into a lake or reservoir; total area above a given point on a stream that contributes water to the flow at that point; the topographic dividing line from which surface streams flow in two different directions.

wilting point Moisture content of soil on an oven-dry basis at which plants wilt and fail to recover their turgidity when placed in a dark, humid atmosphere.

xeric Dry conditions, especially relating to soil.

xerophyte Plants adapted to life in a dry or physiologically dry (saline) habitat.

Zeitgeber The time setter, usually light, that entrains a circadian rhythm to environmental rhythms.

zoogeography Study of the distribution of animals.

zooplankton Floating or weakly swimming animals in freshwater and marine ecosystems; planktonic animals.

CREDITS AND SOURCES

I wish to thank the following for permission to adapt, reprint, or redraw the following figures and tables from their publications.

Illustrations

19 Figure 2.1 H. S. Dybas and M. Lloyd. 1962. Isolation by habitat in two synchronized species of periodical cicadas (Homoptera, Cicadidae, *Magicada*). *Ecology* 43: 444–459. Copyright © 1962 Ecological Society of America. Used with permission. *27* Figure 2.8 P. T. Boag and P. R. Grant. 1981. Intense natural selection in a population of Darwin's finches. *Science* 214: 83. Copyright © 1981 American Association for the Advancement of Science. Used with permission. *30* Figure 2.9 L. E. Mettler, T. G. Gregg, and Schaffer. 1969. *Population genetics and evolution,* 2nd ed., p. 83. Copyright © 1988. Reprinted by permission of Prentice-Hall, Englewood Cliffs, NJ. *31* Figure 2.10 O. H. Frankel and M. E. Soulé. 1981. *Conservation and evolution,* p. 32. Cambridge, England: Cambridge University Press. Used with permission. *32* Figure 2.11 T. J. Foose. 1983. The relevance of captive populations to the conservation of biotic diversity. In C. M. Schoenwald-Cox, S. M. Chambers, B. Macbryde, and W. L. Thomas, eds., *Genetics and conservation,* p. 376. Menlo Park, CA: Benjamin/Cummings. Used with permission. *41* Figure 3.2 Data from L. Clack. 1975. Subspecific intergradation and zoogeography of the painted turtle *(Chrysemys picta)* in northern West Virginia. MS thesis, West Virginia University. *49* Figure 3.6 Based on data from D. Amadon. 1947. Ecology and evolution of some Hawaiian birds. *Evolution* 1: 63–68; and 1950. The Hawaiian honeycreepers (Aves, Drepanidae), *Bull. Amer. Mus. Nat. Hist.* 100: 397–451. Also see R. J. Raikow. 1976. The origin and evolution of the Hawaiian honeycreepers (Drepanidae), *Living Bird* 15: 95–117. *59* Figure 4.1 Adapted from S. H. Schneider. 1987. Climate modeling. *Scientific American* 256: 5, 72–80. Copyright © 1987 by Scientific American. All rights reserved. *64* Figure 4.7 Adapted from M. S. Schroeder and C. C. Buck. 1970. *Fire weather.* Agricultural Handbook no. 360, p. 29. Washington, DC: U.S. Department of Agriculture. *65* Figure 4.8 Adapted from M. S. Schroeder and C. C. Buck. 1970. *Fire weather.* Agricultural Handbook no. 360, p. 62. Washington, DC: U.S. Department of Agriculture. *70* Figure 4.14 Adapted from R. E. Coker. 1947. *This great and wide sea.* Chapel Hill: University of North Carolina Press. *71* Figure 4.15 D. Gates. 1962. *Energy exchange and the biosphere,* p. 13. New York: Harper & Row. Copyright © 1962 David M. Gates. Modified with permission. *72* Figure 4.16 Adapted from J. N. Wolfe, R. T. Wareham, and H. T. Scofield. 1949. Microclimates and macroclimates of Neotoma, a small valley in central Ohio. *Ohio Biol. Surv. Bull.* no. 41. *73* Figure 4.17 Data from Dr. W. A. van Eck, West Virginia University. *74* Figure 4.18a, b Data from R. L. Smith, West Virginia University; *80* Figure 5.1b After A. Cernusca, 1976. Energy exchange within individual layers of a meadow. *Oecologica* 23: 148. *83* Figure 5.3 B. McNab. 1978. The evolution of endothermy in the phylogeny of mammals. *American Naturalist* 112: 9. Copyright © 1978 University of Chicago Press. Used with permission. *84* Figure 5.5 Adapted from H. B. Lillywhite. 1970. Behavioral temperature regulation in the bullfrog *Rana catesbeiana. Copeia* 1970: 158–168. *86* Figure 5.6b B. Heinrich. 1976. Heat exchange in relation to blood flow between thorax and abdomen in bumblebees. *Journal of Experimental Biology* 64: 564. Copyright © 1976 The Company of Biologists, Cambridge, England. Used with permission. *86* Figure 5.6c C. R. Taylor. 1972. The desert gazelle: A parody resolved. In G. M. O. Malory, ed., *Comparative physiology of desert animals,* Symposium of Zoological Society of London no. 31. New York: Academic Press. Used with permission. *86* Figure 5.6d K. Schmidt-Nielsen. 1970. *Animal physiology,* 3rd ed., p. 56. Englewood Cliffs, NJ: Prentice-Hall. Copyright © 1970 Prentice-Hall. Used with permission. *86* Figure 5.6e K. Schmidt-Nielsen. 1979. *Animal physiology: Adaptation and environment,* p. 272. New York: Cambridge University Press. Copyright © 1979 Cambridge University Press. Used with permission. *90* Figure 5.7 Data from H. Walter. 1973. *Vegetation of the Earth.* New York: Springer-Verlag. *91* Figure 5.8 W. Larcher. 1980. *Physiological plant ecology,* 2nd ed., pp. 46, 49. New York: Springer-Verlag. Copyright © 1980 Springer-Verlag. Used with permission. *110* Figure 7.1 W. E. Reifsnyder and H. W. Lull. 1965. *Radiant energy in relation to forests,* p. 21. U.S. Department of Agriculture Tech. Bull. no. 1344. *111* Figures 7.2 and 7.3 Adapted from W. Larcher. 1975. *Physiological plant ecology,* pp. 14, 15. New York: Springer-Verlag. Copyright © 1975 Springer-Verlag. Used with permission. *114* Figure 7.4 B. A. Hutchinson and D. R. Matt. 1977. The distribution of solar radiation within a deciduous forest. *Ecological Monographs* 47: 205. Copyright © 1977 Ecological Society of America. Used with permission. *116* Figures 7.6, 7.7, 7.9, and 7.10 M. G. Barbour, J. H. Burk, and W. D. Pitts. 1980. *Terrestrial plant ecology,* p. 310. Menlo Park, CA: Benjamin/Cummings. Copyright © 1980 Benjamin/Cummings. Used with permission. *118* Figure 7.11 M. Kluge. 1972. Crassulacean acid metabolism (CAM): CO_2 and water economy. In O. L. Lange, L. Kappen, and E. D. Schulze, eds., *Water and plant life,* p. 317, Ecological Studies no. 19. New York: Springer-Verlag. Copyright © 1972 Springer-Verlag. Used with permission. *119, 120* Figures 7.12 and 7.13 After O. Bjorkman. 1973. Comparative studies on photosynthesis in higher plants. In A. C. Geise, ed., *Photophysiology,* pp. 53, 56. Copyright © 1973 Academic Press. Used with permission. *125* Figure 8.1 F. A. Brown. 1959. Living clocks. *Science* 130: 1535–1544. *126* Figures 8.2 and 8.3 P. J. DeCoursey. 1960. Phase control of activity in a rodent. *Cold Spring Harbor Symposia on Quantitative Biology* 25: 51, 52, Biological Clocks. Adapted by permission. *127* Figure 8.4 E. Bunning. 1960. Circadian rhythms and the time measurement in photoperiodism. *Cold Spring Harbor Symposia on Quantitative Biology* 25: 253, Biological

Clocks. Used with permission. *128* Figure 8.5 Adapted from C. H. Johnson and J. W. Hastings. 1986. The elusive mechanism of the circadian clock. *American Scientist* 74: 29–36. Copyright © 1986 Sigma Xi. Adapted with permission. *129* Figure 8.7 J. D. Palmer. 1990. The rhythmic lives of crabs. *Bioscience* 40: 353, Figure 2. Copyright © American Institute of Biological Sciences. Used with permission. *143* Figure 9.3 P. A. Furley and W. W. Newey. 1983. *Geography of the biosphere,* p. 68. Copyright © 1983 Butterworth & Co., Ltd. Used with permission. *146* Figure 9.6 N. C. Brady. 1974. *The nature and properties of soil,* 8th ed., Figures 7.20, 7.21, pp. 190, 193. New York: Macmillan. Copyright © Macmillan. Used with permission. *148* Figure 9.7 Soil Survey Staff. USDA Soil Conservation Service. 1975. *Soil taxonomy.* Agricultural Handbook no. 436. *160* Figure 10.2 J. Wiens. 1973. Pattern and process in grassland bird communities. *Ecological Monographs* 43: 240, Figure 2. Copyright © 1973 Ecological Society of America. Used with permission. *162* Figure 10.4 F. H. Wagner, C. D. Besadny, and C. Kabat. 1965. Population ecology and management of Wisconsin pheasants. *Wisconsin Cons. Dept. Tech. Bull.* 34, p. 14. *163* Figure 10.5 D. B. Houston. 1982. *The northern Yellowstone elk: Ecology and management,* p. 54. New York: Macmillan. Copyright © 1982 Macmillan Publishing Co. Used with permission. *166* Figure 10.10a J. Hett and O. L. Loucks. 1976. Age structure models of balsam fir and eastern hemlock. *Journal of Ecology,* 64: 1035, Figure 1a. Copyright © 1976 Blackwell Scientific Publications. Used with permission. *173, 175* Figures 11.1a and 11.3a V. P. W. Lowe. 1969. Population dynamics of red deer (*Cervus elaphus* L) on Rhum. *Journal of Animal Ecology,* 38: 436, 437, Figure 3, Figure 4 (stags and hinds only). Copyright © 1969, Blackwell Scientific Publications. Used with permission. *173, 175* Figures 11.1b and 11.3b Data from M. C. Baker, L. R. Mewaldt, and R. M. Stewart. 1981. Demography of white-crowned sparrows *(Zonotricia leucophrys nuttali). Ecology* 62: 636–644. *173, 176* Figures 11.1c and 11.4a R. R. Scharitz and J. R. McCormick. 1973. Population dynamics of two competing plant species. *Ecology* 54: 729, 730, Figure 5. Table 2. Copyright © 1973 Ecological Society of America. Used with permission. *175* Figure 11.3c O. E. Sette. 1943. Biology of the Atlantic mackerel *Scomber scombus. U.S. Fish and Wildl. Serv. Fishery Bull.* 38. *176* Figure 11.4c M. Yarranton and G. A. Yarranton. 1975. Demography of a jack pine stand. Reproduced with permission of the National Research Board of Canada, from the *Canadian Journal of Botany,* Vol. 53, p. 311, 1975. *176* Figure 11.4b J. Sarukhan and J. L. Harper. 1973. Studies on plant demography: *Ranuculus repens* L., *R. bulbosa* L., and *R. acris* 1.—1, Population flux and survivorship. *Journal of Ecology* 61: 676–716. Copyright © 1973 Blackwell Scientific Publications. Used with permission. *177* Figure 11.5 P. Williamson. 1976. Aboveground primary production of chalk grassland allowing for leaf death. *Journal of Ecology* 64: 1063, in part. Copyright © 1973 Blackwell Scientific Publications. Used with permission. *182* Figure 12.2 Data from U.S. Census Bureau. Graph developed by R. L. Smith. *185* Figure 12.4 Data from U.S. Census Bureau. Graphs developed by R. L. Smith. *186* Figure 12.5 F. A. Pitelka. 1957. Some characteristics of microtine cycles in the Arctic. *Proceedings 18th Biology Colloquium,* pp. 79, 80, Oregon State College. Copyright © 1957 Oregon University Press. *192* Figure 13.2a M. C. Dash and A. R. Hota. 1980. Density effects on survival, growth rate, and metamorphosis of *Rana tigrina* tadpoles. *Ecology* 61: 1027, Figure 2. Copyright © 1980 Ecological Society of America. Used with permission. *192* Figure 13.2b After T. Backiel and L. D. LeCren. 1967. Some density relationships for fish population parameters. In S. D. Gerking, ed., *The biological basis of freshwater fish production,* pp. 261–293. New York: Wiley. Copyright © 1967 John Wiley. Used with permission. *193* Figure 13.3 A. R. E. Sinclair. 1977. *The African buffalo,* p. 140. Copyright © 1977 University of Chicago Press. Reprinted with permission. *194* Figure 13.4 Developed from data in A. R. E. Sinclair. 1977. *The African buffalo.* University of Chicago Press. *194* Figure 13.5a D. R. McCullough. 1981. Population dynamics of the Yellowstone grizzly. In C. W. Fowler and T. D. Smith, eds., *Dynamics of large mammal populations,* p. 177. New York: Wiley. Copyright © 1981 John Wiley. Used with permission. *194* Figure 13.5b C. W. Fowler. 1981. Density dependence as related to life history strategy. *Ecology* 62: 607, Figure 4. Copyright © 1981 Ecological Society of America. Used with permission. *198* Figure 13.6 M. W. Fox. 1980. *Soul of the wolf,* p. 61. Boston: Little, Brown. Copyright © 1980 M. W. Fox. Used with permission. *199* Figure 13.7 R. L. Smith. 1963. Some ecological notes on the grasshopper sparrow. *Wilson Bull.* 75: 159–165. *200* Figure 13.8 W. T. Jones. 1989. Dispersal distance and the range of nightly movements in Merriam's kangaroo rats. *Journal of Mammalogy* 20: 31. Copyright © 1989 American Society of Mammalogists. Used with permission. *201* Figure 13.9 A. S. Harestad and F. L. Bunnell. 1979. Home range and body weight—a reevaluation. *Ecology* 60: 390. Copyright © Ecological Society of America. Adapted with permission. *206* Figure 14.1 Based on data in V. P. W. Lowe. 1969. Population dynamics of red deer *(Cervus elaphus)* on Rhum. *J. Anim. Ecol.* 38: 425–457. *208* Figure 14.2 M. L. Cody. 1966. A general theory of clutch size. *Evolution* 20: 179. Copyright © 1966 Society of the Study of Evolution. Used with permission. *213* Figure 14.3 J. R. Krebs and N. B. Davies. 1981. *An introduction to behavioural ecology,* p. 117. Copyright © 1981 Blackwell Scientific Publications. Used with permission. *222* Figure 15.2 D. D. Tilman, M. Mattson, and S. Langer. 1981. Competition and nutrient kinetics along a temperature gradient: An experimental test of a mechanistic approach to niche theory. *Limno. Oceanogr.* 26: 1025, 1027. Copyright © 1981 Society of Limnology and Oceanography. With permission. *223* Figure 15.3 Adapted from H. C. Heller and D. Gates. 1971. Altitudinal zonation of chipmunks *(Eutamias):* Energy budgets. *Ecology* 52: 424, Figure 1. Copyright © 1971 Ecological Society of America. Used with permission. *224* Figure 15.4 N. A. Moran and T. G. Whigham. 1990. Interspecific competition between root-feeding and leaf-galling aphids mediated by host-plant resistance. *Ecology* 71: 1056, Figure 5. Copyright © 1990 Ecological Society of America. Used with permission. *227* Figure 15.6 N. K. Wieland and F. A. Bazzaz. 1975. Physiological ecology of three codominant successional annuals. *Ecology* 56: 686, Figure 6. Copyright © 1975 Ecological Society of America. Used with permission. *228* Figure 15.7 Adapted from M. A. Bowers and J. M. Brown. 1982. Body size and coexistence in desert rodents: Chance or community structure? *Ecology* 63: 396, Figure 1. Copyright © 1982 Ecological Society of America. Used with permission. *230* Figure 15.10 P. D. Putwain and J. L. Harper. 1970. Studies of dynamics of plant populations. 3. The influence of associated species on populations of *Rumex acetosa* and *R. acetosella* in grasslands. *Journal of Ecology* 58: 262, Figure 7. Copyright © 1970 Blackwell Scientific Publications. Used with permission. *230* Figure 15.11 E. R. Pianka. 1983. *Evolutionary ecology,* 3rd ed., Figure 7.4, p. 256. New York: Harper & Row. Copyright © 1983 Eric R. Pianka. Reprinted with permission. *231* Figure 15.13 P. Williamson. 1971. Feeding ecology of the red-eyed vireo *(Vireo olivaceous)* and associated foliage-gleaning birds. *Ecological Monographs* 41: 136, Figure 6. Copyright © 1971 Ecological Society of America. Used with permission. *232* Figure 15.14 B. J. Fox. 1981. Niche parameters and species richness. *Ecology* 62:

1418, Figure 1. Copyright © 1981 Ecological Society of America. Used with permission. *238* Figure 16.2 After C. C. Holling. 1959. The components of predation as revealed by a study of small mammal predation of the European pine sawfly. *Canadian Entomologist* 91: 293–320. *239* Figure 16.3 M. P. Hassell and R. M. May. 1974. Aggregation in predators and insect parasites and its effects on stability. *Journal of Animal Ecology* 43: 576, Figure 9. Copyright © 1974 Blackwell Scientific Publications. Used with permission. *241* Figure 16.6 M. P. Hassell, J. H. Lawton, and J. R. Beddington. 1976. The components of arthropod predation. *Journal of Animal Ecology* 45: 139, Figure 1a. Copyright © 1976 Blackwell Scientific Publications. Used with permission. *241* Figure 16.7 D. H. Rusch, E. C. Meslow, P. D. Doerr, and L. B. Keith. 1972. Response of great horned owl populations to changing prey densities. *Journal of Wildlife Management* 36: 291. Copyright © 1972 The Wildlife Society. Used with permission. *242* Figure 16.8 C. Holling. 1966. The functional response of inverterbrate predators to prey density. *Memoirs Entomological Society of Canada* 48: 19. *244* Figure 16.9 E. E. Werner and J. D. Hall. 1974. Optimal foraging and size selection of prey by the bluegill sunfish. *Ecology* 55: 1048, Figure 4. Copyright © 1974 Ecological Society of America. Used with permission. *245* Figures 16.10 and 16.11 N. B. Davies. 1977. Prey selection and social behavior in wagtails (Aves, Monticillidae). *Journal of Animal Ecology* 46: 48, Figure 8. Copyright © 1977 Blackwell Scientific Publications. Used with permission. *246* Figure 16.12 J. R. Krebs. 1978. Optimal foraging decision rules for predators. In J. R. Krebs and N. B. Davies, eds., *Behavioral ecology: An evolutionary approach,* 1st ed. Copyright © 1978 Blackwell Scientific Publications. Used with permission. *251* Figure 16.15 Adapted from L. B. Keith. 1983. Role of food in hare population cycles. *Oikos* 40: 385–395. *259* Figure 17.2 Adapted from R. C. Anderson and A. K. Prestwood. 1981. Lungworms. In W. R. Davidson, ed., *Diseases and parasites of white-tailed deer,* p. 277. Southeastern Cooperative Wildlife Disease Study, Mscl. Pub. No. 7. Tallahassee, FL: Tall Timbers Research Station. *260* Figure 17.3 S. E. Randolph. 1975. Patterns of the distribution of the tick *Ioxdes trianguliceps* Birula on its host. *Journal of Animal Ecology* 44: 454. Copyright © Blackwell Scientific Publications. Used with permission. *261* Figure 17.4 C. A. Lancinani. 1975. Parasite-induced alterations in host reproduction and survival. *Ecology* 56: 691. Copyright © 1975 Ecological Society of America. Used with permission. *275* Figure 18.2 P. DeBach. 1974. *Biological control by natural enemies,* p. 4. Copyright © 1974 Cambridge University Press, London. *278* Figure 18.3 Adapted from R. C. Thatcher, J. L. Searcy, J. E. Coster, and G. D. Hertel, eds. (n.d.), *The southern pine beetle.* U.S.D.A. Forest Service Science and Educational Tech. Bull. 1631. *280* Figure 18.4 G. Caughley. 1976. Wildlife management and the dynamics of ungulate populations. *Applied Biology* 1: 226. Copyright © 1976 Academic Press. *281* Figure 18.5a G. I. Murphy. 1967. Vital statistics of the Pacific sardine and the population consequences. *Ecology* 48: 734. Copyright © 1967 Ecological Society of America. Used with permission. *281* Figure 18.5b G. I. Murphy. 1966. Population biology of the Pacific sardine. *Proceedings California Academy Science* 4th Series 34: 1–84. *282* Figure 18.6 Data from G. L. Small. 1971. *The blue whale.* New York: Columbia University Press. *283* Figure 18.7 C. J. Walters. 1986. *Adaptive management of renewable resources,* p. 37. New York: Macmillan. Copyright © 1986 International Institute of Applied Systems Analysis. Used with permission. *284* Figure 18.8 H. A. Regier and W. L. Hartman. 1973. Lake Erie's fish community: 150 years of cultural stress. *Science* 180: 1248–1255. Copyright © 1973 American Association for the Advancement of Science. Used with permission. *307* Figure 19.5 After A. R. Kiester. 1971. Species density of North American amphibians and reptiles. *Syst. Zool.* 20: 131–132. *311* Figure 19.9 R. F. Whitcomb et al. 1981. Effects of forest fragmentation on avifauna of the eastern deciduous forest. In R. L. Burgess and D. M. Sharpe, eds., *Forest island dynamics in man-dominated landscapes,* p. 183, Figure 8.5c. Copyright © 1981 Springer-Verlag. Used with permission. *312* Figure 19.11 Based on R. H. MacArthur and E. O. Wilson. 1967. *The theory of island biogeography.* Princeton, NJ: Princeton University Press. *314, 315* Figures 19.12 and 19.13 Adapted from M. Williamson. 1981. *Island populations,* pp. 98, 101. Oxford, England: Oxford University Press. Copyright © 1981 Oxford University Press. Used with permission. *317* Figure 19.14 J. B. Levenson. 1981. Woodlots as biogeographic islands in southeastern Wisconsin. In R. L. Burgess and D. M. Sharpe, eds., *Forest island dynamics in man-dominated landscapes,* p. 34, Figure 3.9. Copyright © 1981 Springer-Verlag. Used with permission. *329* Figure 20.6 Adapted from R. Fortney. 1975. The vegetation of Canaan Valley: A taxonomic and ecological study. PhD. dissertation, West Virginia University. *331* Figure 20.7 A. G. Van der Valk and C. B. Davis. 1978. The role of seed banks in vegetation dynamics of prairie glacial marshes. *Ecology* 59: 333. Copyright © 1978 Ecological Society of America. Used with permission. *332* Figure 20.8 D. G. Sprugel. 1976. Dynamic structure of wave-generated *Abies balsamea* forests in northeastern United States. *Journal of Ecology* 64: 891. Copyright © 1976 Blackwell Scientific Publications. Used with permission. *334* Figure 20.9 Adapted from M. Davis. 1981. Quaternary history and the stability of forest communities. In D. C. West, H. H. Shugart, and D. Botkin, eds., *Forest succession: Concepts and applications,* p. 147. New York: Springer-Verlag. Copyright © 1981 Springer-Verlag. Used with permission. *340* Figure 21.2 P. L. Marks. 1974. The role of pin cherry (*Prunus pensylvanica* L) in the maintenance of stability in the northern hardwoods system. *Ecological Monographs* 44: 75, Figure 1. Copyright © 1974 Ecological Society of America. Used with permission. *346* Figure 21.6 L. F. DeBano, P. H. Dunn, and C. E. Conrad. 1977. Fire's effect on physical and chemical properties of chaparral soils. In H. A. Mooney and C. E. Conrad, eds., *Proceedings symposium environmental consequences of fire and fuel management in mediterranean ecosystems,* pp. 65–74. U.S.D.A. For. Serv. Gen. Tech. Rept. WO. 3. *359* Figure VI.1 Based in part on R. V. O'Neill. 1976. Ecosystem persistence and heterotrophic regulation. *Ecology* 57: 1244–1253. *365* Figures 22.1 and 22.2 D. F. Westlake. 1980. Primary production. In E. D. LeCren and R. H. Lowe-McConnell, eds., *The functioning of freshwater ecosystems,* Figure 5.3a, p. 166; Figure 5.9, p. 179. International Biological Programme no. 22. Copyright © 1980 Cambridge University Press. Reprinted with permission. *366* Figure 22.3 M. Monsi. 1968. Mathematical models of plant communities. In F. E. Eckardt, ed., *Functioning of terrestrial ecosystems at the primary production level,* pp. 131–149. Proceedings of the Copenhagen Symposium, Natural Resources Research V. Paris: UNESCO. *367* Figure 22.4 J. R. Etherington. 1982. *Environment and plant ecology,* 2nd ed., p. 355. New York: Wiley. Copyright © 1982 John Wiley and Sons. Used with permission. *368* Figure 22.5 Based on F. B. Golley and H. Leith. 1972. Basis of organic production in the tropics. In P. M. Golley and F. B. Golley, eds., *Tropical ecology with emphasis on organic production,* pp. 1–26. Athens: University of Georgia Press. *370* Figure 22.7 R. H. Whittaker, F. H. Bormann, G. E. Likens, and T. G. Siccama. 1974. The Hubbard Brook ecosystem study: Forest biomass and production. *Ecological Monographs* 44: 239, Figure 1. Copyright © 1974 Ecological Society of America. Used with permission. *370* Figure 22.8 G. D. Cooke.

1967. The pattern of autotrophic succession in laboratory microcosms. *Bioscience* 17: 719. Copyright © 1967 American Institute of Biological Sciences. Used with permission. *370* Figure 22.9 R. L. Cowan. 1962. Physiology of nutrition as related to deer. In *Proceedings 1st National White-tailed Deer Symposium,* pp. 1–8. Athens: Center for Continuing Education, University of Georgia. *371* Figure 22.10 Data from B. C. Moulder, D. E. Reichle, and S. L. Auerbach. 1970. Significance of spider predation in the energy dynamics of forest floor arthropod communities. *Oak Ridge National Laboratory Rept. ORNL 4452.* *376* Figure 23.1 Based on R. D. Bird. 1930. Biotic communities of the aspen parkland of central Canada. *Ecology* 11; 356–442. *378* Figure 23.2 Based on data from T. C. Scott. 1955. An evaluation of the red fox. *Ill. Nat. Hist. Surv. Biol. Notes* 35: 1–16. *379* Figure 23.3 R. G. Wiegert and D. F. Owen. 1971. Trophic structure, available resources, and population density in terrestrial vs. aquatic ecosystems. *J. Theoret. Biol.* 30: 69–81. Used with permission. *380* Figure 23.4 O. H. Paris. 1969. The function of soil fauna in grassland ecosystems. In R. L. Dix and R. L. Beidleman, eds., *The grassland ecosystem: A preliminary synthesis.* Range Sci. Dept. Sci. Ser. no. 2. Fort Collins: Colorado State University. *381* Figure 23.5 C. S. Gist and D. A. Crossley, Jr. 1975. A model of mineral element cycling for an invertebrate food web in a southeastern hardwood forest litter community. In F. G. Howell, J. B. Gentry, and M. H. Smith, eds., *Mineral cycling in southeastern ecosystem,* pp. 84–106. ERDA Symposium Series. Springfield, VA: National Technical Information Service, U.S. Department of Commerce. *383* Figure 23.6 J. D. Stout, K. R. Tate, and L. F. Molloy. 1976. Decomposition processes in New Zealand soils with particular respect to rates and pathways of plant degradation. In J. M. Anderson and A. Macfadyen, eds., *The role of terrestrial and aquatic organisms in decomposition processes,* Figure 5.1, p. 98. Copyright © 1976 Blackwell Scientific Publications. Used with permission. *385* Figure 23.7 R. J. Barsdate, T. Fenchel, and R. T. Prentki. 1974. Phosphorus cycles of model ecosystems; significance for decomposer food chains and effect of bacterial grazers. *Oikos* 25: 239–251. Used with permission. *387* Figure 23.8 Based on F. B. Golley. 1960. Energy dynamics of a food chain of an old field community. *Ecological Monographs* 30: 187–206. *390* Figure 23.9a Based on O. W. C. Park, V. E. Allee, and W. Shelford. 1939. *A laboratory introduction to animal ecology and taxonomy.* Chicago: University of Chicago Press. *390* Figure 23.9b N. R. French. 1979. Principal subsystem interactions in grasslands. In N. R. French, ed., *Perspectives in grassland ecology,* p. 185. Copyright © 1979 Springer-Verlag. Used with permission. *310* Figure 23.9c M. La Motte. 1979. The structure and function of a tropical savanna ecosystem. In F. B. Golley and E. Medina, eds., *Tropical ecological systems: Trends in terrestrial and aquatic research,* Figure 15.28a, p. 216. Copyright © 1975 Springer-Verlag. Used with permission. *406* Figure 24.5 W. M. Post, T. Peng, W. R. Emanuel, A. W. King, V. H. Dale, and D. L. DeAngelis. 1990. The global carbon cycle. *American Scientist,* 78: 312, 313, Figure 3a, 3b. Copyright © 1990 Sigma Xi. Use with permission. *407* Figures 24.6 and 24.7 Data from W. M. Post, T. H. Peng, W. R. Emanuel et al. 1990. The global carbon cycle. *American Scientist* 78: 310–326. *408* Figure 24.8 W. M. Post., T. Peng, W. R. Emanual et al. 1990. The global carbon cycle. *American Scientist* 78: 325, Figure 14. Copyright © 1990 Sigma Xi. Used with permission. *413* Figure 24.12 W. W. Kellogg, R. D. Cadle, E. R. Allen, A. L. Lazrus, and E. A. Martell. 1972. The sulfur cycle. *Science* 175: 594, Figure 2. Copyright © 1972 by the American Association for the Advancement of Science. Used with permission. *416* Figure 24.14 After J. P. Witherspoon, S. I. Auerbach, and J. S. Olson. 1962. Cycling of cesium-137 in white oak trees on sites of contrasting soil type and moisture. *Oak Ridge Nat. Lab. Rept.* 3328: 1–143. *447* Figure VII.5 R. H. Whittaker. 1970. *Communities and ecosystems,* Figure 4.10, p. 167. New York: Macmillan. Copyright © 1970 R. H. Whittaker. Used with permission. *448* Figure VII.6 L. R. Holdridge, W. C. Grenke, W. H. Hatheway, T. Liang, and J. A. Tosi, Jr. 1971. *Forest environments in tropical life zones.* New York: Pergamon. Used with permission. *457* Figure 26.5 W. K. Lavenroth. 1979. Grassland primary production: North American grasslands in perspective. In N. R. French, ed., *Perspectives in grassland ecology,* p. 10. New York: Springer-Verlag. Copyright © 1979 Springer-Verlag, New York. Used with permission. *458* Figure 26.6 Data from P. L. Sims and J. S. Singh. 1971. Herbage dynamics and net primary production in certain grazed and ungrazed grasslands in North America. In N. R. French, ed., *Preliminary analysis of structure and function in grasslands,* pp. 59–124. Range Sci. Dept. Sci. Ser. no. 10. Fort Collins: Colorado State University. *459* Figure 26.7 N. R. French. 1979. Principal subsystem interactions in grasslands. In N. R. French, ed., *Perspectives in grassland ecology,* p. 185. New York: Springer-Verlag. Copyright © 1979 Springer-Verlag. Used with permission. *462* Figure 26.10 R. W. Johnson and J. C. Tothill. 1985. Definition and broad geographic outline of savanna lands. In *Ecology and management of the world's savannas,* pp. 1–13. Copyright © 1985 by the Canberra Commonwealth Agricultural Bureau, Australian Academy of Science. Used with permission. *484* Figure 28.1 Adapted from P. L. Johnson and W. O. Billings. 1962. The alpine vegetation of the Beartooth Plateau and its relation to cyropedogenic processes. *Ecological Monographs* 32: 105–135. *505* Figure 29.4 D. W. Johnson, D. W. Cole, C. S. Bledso et al. 1982. Nutrient cycling in the forests of the Pacific Northwest. In R. L. Edwards, ed., *Coniferous forest ecosystems in the western United States,* p. 193. Stroudsburg, PA: Hutchinson Ross. Copyright © 1982 Hutchinson Ross. Used with permission. *505* Figure 29.5 J. W. Thomas, ed. 1979. *Wildlife habitats in managed forests,* p. 64. U.S.D.A. Agricultural Handbook no. 553. *512* Figure 29.11 Based on P. and S. Denaeyer-Desmet. 1967. Biomass, productivity, and mineral cycling in deciduous forests in Belgium. In *Symposium on primary productivity and mineral cycling in natural ecosystems,* pp. 167–183. Orono: University of Maine Press. *521* Figure 30.7 P. M. Golley and F. M. Golley, eds. 1972. *Tropical ecology with an emphasis on organic production.* Athens: University of Georgia Press. Used with permission. *531* Figure 31.3 G. E. Likens, ed. 1985. *An ecosystem approach to aquatic ecology,* Figure IV.B-11, p. 98; Figure IV.C-5, p. 116; Figure VI.D-1, p. 339. Copyright © 1985 Springer-Verlag. Used with permission. *534* Figure 31.5 L. Wells. 1960. Seasonal abundance and vertical movements of planktonic crustaceans in Lake Michigan. *U.S.D.I. Fish. Bull.* 60(172): 343–369. *535* Figure 31.6 G. E. Likens and F. H. Bormann. 1974. Linkages between terrestrial and aquatic ecosystems. *Bioscience* 24: 448, Figure 1. Copyright © 1974 American Institute of Biological Sciences. Used with permission. *537* Figure 31.7 H. L. Golterman and F. A. Kouwe. 1980. Chemical budgets and nutrient patterns. In E. D. LeCren and R. H. Lowe-McConnell, eds., *The functioning of freshwater ecosystems,* Figure 4.12, p. 138. International Biological Programme no. 22. Copyright © 1980 Cambridge University Press. Reprinted with permission. *544* Figure 32.1 R. W. Tiner. 1991. The concept of a hydrophyte for wetland identification. *Bioscience* 41: 328. Copyright © 1991 American Institute for Biological Sciences. Used with permission. *548* Figure 32.2 J. R. Gosselink and R. E. Turner. 1978. Role of hydrology in freshwater wetland ecosystems. In R. Good, D. F. Whigham, and R. L. Simpson, eds., *Freshwater wetlands,* Figure 6, p. 73. Copyright © 1978 Academic Press. Used with permission. *551* Figure 32.7 Based on P. Dansereau and F. Segadas-Vianna. 1952. Ecological study of peat bogs of eastern North America. *Canadian Journal of Botany* 30: 490–520. *552* Figure 32.8 D. F. Whigham, J. McCormick, R. E. Good, and R. L. Simpson. 1978. Biomass and primary production in freshwater tidal wetlands of the middle Atlantic Coast. In R. Good, D. F. Whigham, and R. L. Simpson, eds., *Freshwater wetlands,* Figure 1 (*Typha* and *Lythrum* only), p. 12. Copyright © 1978 Academic Press. Used with permission. *552* Figure 32.9 R. T. Prentki, T. D. Gufason, and M. S. Adams. 1978. Nutrient movement in lakeside marshes. In R. Good, D. F. Whigham, and R. L. Simpson, eds., *Freshwater wetlands,* Figure 3, p. 176. Copyright © 1978 Academic press. Used with permission. *553* Figure 32.10 W. Odum and M. Haywood. 1978. Decomposition of intertidal freshwater marsh plants. In R. Good, D. F. Whigham, and R. L. Simpson, eds., *Freshwater wetlands,* Figure 1, p. 92. Copyright © 1978 Academic Press. Used with permission. *555* Figure 32.12 T. E. Dahl. 1990. *Wetland losses in the United States: 1780's to 1980's.* Washington, DC: U.S. Department of the Interior, Fish and Wildlife Service. *564* Figure 33.5 Data from S. G. Fisher and G. E. Likens. 1973. Energy flow in Bear Brook, New Hampshire: An integrative approach to stream ecosystem metabolism. *Ecological Monographs* 43: 421–439. Copyright © 1973 Ecological Society of America. Used with permission. *565* Figure 33.6 K. W. Cummins. 1974. Structure and function of stream ecosystems. *Bioscience* 24: 633, Figure 3 (redrawn). Copyright © 1974 American Institute of Biological Sciences. Used with permission. *567* Figure 33.7 J. W. Newbold, R. V. O'Neill, J. W. Elwood, W. Van Winkle. 1982. Nutrient spiralling in streams: Implications for nutrients and invertebrate activity. *American Naturalist* 120: 630, Figure 1. Copyright © 1982 University of Chicago Press. Reprinted with permission. *583* Figure 34.3 Adapted from R. M. Hayman and K. C. McDonald. 1985. The geology of deep sea hot springs. *American Scientist* 73: 441–449, and other sources. *585* Figure 34.4 Adapted from J. E. G. Raymont. 1963. *Plankton and productivity in oceans,* p. 547. New York: Pergamon Press. Copyright © 1963 Pergamon Press. Used with permission. *586* Figure 34.5 J. H. Steele. 1974. *The structure of marine ecosystems.* Cambridge, MA: Harvard University Press. *586* Figure 34.6 T. Fenchel. 1988. Marine plankton food chains. Reproduced with permission from the *Annual review of ecology and systematics,* Vol. 19, Figure 1, p. 24. Copyright © 1988 by Annual Reviews. *592* Figure 35.2 Adapted from T. A. Stephenson and A. Stephenson. 1949. The universal features of zonation between tidemarks on rocky coasts. *Journal of Ecology* 37: 289–305. *597* Figure 35.7 S. K. Eltringham. 1971. *Life in mud and sand,* p. 203. New York: Crane, Russak. Copyright © 1971 Crane, Russak. Used with permission. *606* Figure 36.3 D. L. Correll. 1978, Estuarine productivity. *Bioscience* 28: 648. Copyright © 1978 American Institute of Biological Sciences. Used with permission. *606* Figure 36.4 Developed from P. C. Head. 1976. Organic processes in estuaries. In D. J. Burton and P. S. Liss, eds., *Estuarine chemistry,* pp. 54–91. London: Academic Press. *607* Figure 36.5 E. L. Cronin and A. J. Mansueti. 1971. The biology of the estuary. In P. A. Stroud and R. H. Stroud, eds., *A symposium on the biological significance of the estuary,* pp. 14–19. Washington, DC: Sport Fishing Institute.

Tables

76 Table 4.1 Adapted from H. E. Landsberg. 1970. Man-made climatic changes. *Science* 170: 1265–1274. *151* Table 9.2 D. L. Schertz, W. C. Moldenhauer, S. J. Livingston, G. A. Weesies, E. A. Hintz. 1989. Effect of past soil erosion on crop productivity in Indiana. *Journal of Soil and Water Conservation,* Vol. 44. *170* Tables 11.1 and 11.2 V. P. W. Lowe. 1969. Population dynamics of red deer (*Cervus elaphus* L) on Rhum. *Journal of Animal Ecology* 38: 435, Table 8, Table 9. Copyright © 1969 Blackwell Scientific Publications. Used with permission. *171* Table 11.3 R. W. Campbell. 1969. Studies on gypsy moth population dynamics. In W. E. Walters, ed., *Forest insect population dynamics: Proceedings of a workshop,* pp. 29–51. USDA Forest Service Res. Paper NW-125. *172* Table 11.4 R. R. Scharitz and J. R. McCormick. 1973. Population dynamics of two competing plant species. *Ecology* 54: 730. Copyright © 1973 Ecological Society of America. Used with permission. *177* Table 11.6 M. C. Baker, L. R. Mewaldt, and R. M. Stewart. 1981. Demography of white-crowned sparrows *(Zonotrichia leucophrys nuttali). Ecology* 62: 642, Table 6-1. Copyright © 1981 Ecological Society of America. Used with permission. *400* Table 24.2 G. R. Kalin and V. D. Bykov. 1969. The world's water resources, present and future. *Impact of Science on Society* 19: 135–150. Used with permission. *503* Table 29.1 Data from D. W. Cole and M. Rapp. 1981. Elemental cycling in forest ecosystems. In D. E. Reichle, ed., *Dynamic properties of forest ecosystems,* pp. 341–409. New York: Cambridge University Press. *510* Table 29.2 Data from Cole and Rapp, 1981. *510* Table 29.3 Data from G. E. Likens, F. H. Bormann, N. M. Johnson, and R. S. Pierce. 1967. The calcium, magnesium, potassium, and sodium budgets for a small forested ecosystem. *Ecology* 38: 46–49. *546* Table 32.1 Adapted from S. P. Shaw and C. G. Fredine. 1956. Wetlands of the United States. *U.S. Fish and Wildlife Service Circular* 39. *576* Table 34.1 After K. Kalle. 1971. Salinity: General introduction. In O. Kinne, ed., *Marine ecology: Vol. 1, environmental factors, Part 2.* pp. 683–688. New York: Wiley.

Photos

Page 1: NASA. *Page 2:* West Virginia Agricultural Forestry Department & U.S. Forestry Service. *Page 11:* Patti Murray/ANIMALS ANIMALS. *Page 15:* Mark N. Boulton/Photo Researchers. *Page 35:* Tom McHugh/Photo Researchers. *Page 55:* Erwin W. Cole/Soil Conservation Service/U.S. Dept. of Agriculture. *Page 58:* Jesse Lunger/Photo Researchers. *Page 78:* Donald C. Schuhart/Soil Conservation Service/U.S. Dept. of Agriculture. *Page 81, top:* R. L. Smith. *Page 81, bottom left:* R. M. Cady, Pennsylvania Game Commission. *Page 81, bottom right:* Edgar Munch/Photo Researchers. *Page 96:* Jen & Des Bartlett/Photo Researchers. *Page 109:* Pierre Berger/Photo Researchers. *Page 123:* Leonard Lee Rue III/ANIMALS ANIMALS. *Page 129:* R. L. Smith. *Page 136:* Omikron/Photo Researchers. *Page 152:* R. B. Branstead/Soil Conservation Service/U.S. Dept. of Agriculture. *Page 155:* Charlie Ott/Photo Researchers. *Page 157:* Leonard Lee Rue III/Photo Researchers. *Page 168:* Michigan Conservation Department. *Page 180:* Leonard Lee Rue III/ANIMALS ANIMALS. *Page 189:* Leonard Rue III/ANIMALS ANIMALS. *Page 204:* Bill Dyer/Photo Researchers. *Page 216:* Stouffer Productions/ANIMALS ANIMALS. *Page 235:* Suen-O Linblad/Photo Researchers. *Page 247:* R. L. Smith. *Page 254:* Leonard Lee Rue III/ANIMALS ANIMALS. *Page 258:* R. L. Smith. *Page 266:* R. L. Smith. *Page 271:* C. C. Lockwood/Earth Scenes. *Page 288:* George H. Harrison/Pennsylvania Game Commission. *Page 290:* R. L. Smith. *Page 295:* Mark Boulton/Natl. Audubon Society/Photo Researchers. *Page 297:* Irene Vandermolen/Leonard Rue Prod./Photo Researchers. *Page 320:* J. H. Robinson/Earth Scenes. *Pages 322,*

323: R. L. Smith. *Page 337:* C. C. Maxwell/Photo Researchers. *Page 339, top:* W. H. Jackson/The State Historical Society of Colorado/Courtesy George E. Gruell. *Page 339, bottom:* Courtesy George E. Bruell. *Page 341:* David M. Dennis/Tom Stack & Associates. *Page 344:* R. L. Smith. *Page 348:* T. M. Smith. *Page 349:* Clarence Krezevich/Soil Conservation Service/U.S. Dept. of Agriculture. *Page 353:* R. L. Smith. *Page 357:* Leonard Lee Rue IV/Earth Scenes. *Page 360:* David C. Fritts/ANIMALS ANIMALS. *Page 374:* Terry G. Murphy/ANIMALS ANIMALS. *Page 395:* M. Kleck/Terra Photographics/Biological Photo Service. *Page 420:* Soil Conservation Service/U.S. Dept. of Agriculture. *Page 429:* Hugh Morton. *Page 439:* R. L. Smith. *Page 449:* T. M. Smith. *Page 451:* R. L. Smith. *Page 453:* R. L. Smith. *Page 455:* South Dakota Fish & Game Commission. *Page 459:* Soil Conservation Service/U.S. Dept. of Agriculture. *Page 461:* T. M. Smith. *Page 463:* T. M. Smith. *Page 466:* Bill Bachman/Photo Researchers. *Page 470:* T. M. Smith. *Page 475, top left and right:* R. L. Smith; *bottom:* Mickey Gibson/Earth Scenes. *Page 476:* Arizona Fish and Game, Department. *Page 477, left:* J. A. L. Cooke/ANIMALS ANIMALS. *Page 477, right:* L. L. T. Rhodes/ANIMALS ANIMALS. *Page 479:* Holt Studios/Earth Scenes. *Page 482:* Jen & Des Bartlett/Photo Researchers. *Page 485:* U.S. Fish & Wildlife. *Page 486:* R. L. Smith *Page 487:*Mary M. Thacher/Photo Researchers. *Page 490:* Fred Whitehead/Earth Scenes. *Page 491:* T. M. Smith. *Page 492:* T. M. Smith. *Page 493:* Breck P. Kent/Earth Scenes. *Page 498:* Tom Edwards/Earth Scenes. *Page 500, top left:* T. M. Smith. *Page 500, top right:* R. L. Smith. *Page 500, bottom left:* R. L. Smith. *Page 500, bottom right:* R. L. Smith. *Page 501, left:* R. L. Smith. *Page 501, right:* U.S. Forestry Service. *Page 505:* Soil Conservation Service/U.S. Dept. of Agriculture. *Page 507, top:* Soil Conservation Service/U.S. Dept. of Agriculture. *Page 507, bottom:* R. L. Smith. *Page 508:* R. L. Smith. *Page 515:* Renee Lynn/Photo Researchers. *Page 516:* R. L. Smith. *Page 517:* T. M. Smith. *Page 519:* R. L. Smith. *Page 520, top:* T. M. Smith. *Page 520, bottom:* R. L. Smith. *Page 523, top left:* T. M. Smith. *Page 523, top right:* Alexine Keuroghlian. *Page 523, bottom left and bottom right:* R. L. Smith. *Page 524:* R. L. Smith. *Page 527:* Pierre Berger/Photo Researchers. *Page 529:* R. L. Smith. *Page 543:* Jack Dermid/Natl. Audubon Society/Photo Researchers. *Page 549, top:* Brock May/Photo Researchers. *Page 549, bottom:* U.S. Forestry Service. *Page 550, top left and top right:* T. M. Smith. *Page 550, bottom:* R. L. Smith. *Page 556:* Soil Conservation Service/U.S. Dept. of Agriculture. *Page 559:* R. L. Smith. *Page 561:* R. L. Smith. *Page 571:* R. L. Smith. *Page 572:* R. L. Smith. *Page 575:* Van Bucher/Photo Researchers. *Page 580:* Larry Cameron/Photo Researchers. *Page 590:* R. L. Smith. *Page 591:* R. L. Smith. *Page 594:* R. L. Smith. *Page 598:* Mickey Gibson/ANIMALS ANIMALS. *Page 603:* West Virginia University Agricultural & Forestry Experiment Station. *Page 604:* The Nature Conservancy. *Page 610:* R. L. Smith. *Page 611:* R. L. Smith. *Page 615:* R. L. Smith.

INDEX

Note: Boldface page numbers refer to pages containing tables or illustrations.